Atmospheric Dynamics

Atmospheric dynamics is a core component of all atmospheric science curricula. It is concerned with how and why different classes of geophysical disturbances form, what dictates their structure and movement, how the Earth's uneven surface impacts with them, how they evolve to mature stage, how they interact with the background flow, how they decay and how they collectively constrain the general circulation of the atmosphere.

Mankin Mak's new textbook provides a self-contained course on atmospheric dynamics. The first half of the book is suitable for undergraduates, and develops the physical, dynamical and mathematical concepts at the fundamental level. The second half of the book is aimed at more advanced students who are already familiar with the basics. The contents have been developed from many years of the author's teaching at the University of Illinois. The discussions are supplemented with schematics, weather maps and statistical plots of the atmospheric general circulation. Students often find the connection between theoretical dynamics and atmospheric observation somewhat tenuous, and this book demonstrates a strong connection between the key dynamics and real observations in the atmosphere, with many illustrative analyses in the simplest possible model settings. Physical reasoning is shown to be even more crucial than mathematical skill in tackling dynamical problems.

This textbook is an invaluable asset for courses in atmospheric dynamics for undergraduates as well as for graduate students and researchers in atmospheric science, ocean science, weather forecasting, environmental science and applied mathematics. Some background in mathematics and physics is assumed.

Mankin Mak is a Professor Emeritus in the Department of Atmospheric Sciences at the University of Illinois at Urbana Champaign, where he has taught and researched on atmospheric dynamics for many years. He was born in Hong Kong, China, where he completed his high school education, before moving on to undergraduate study in Engineering Physics at the University of Toronto, Canada, and graduate study in Meteorology at the Massachusetts Institute of Technology, USA. He has published extensively in various international journals of atmospheric sciences and served as an editor of the *Journal of Atmospheric Sciences*.

Praise for this book

'Professor Mak's new text provides a comprehensive and self-contained introduction to atmospheric dynamics. The latter chapters provide invaluable material on a range of topics not found together in other texts: baroclinic lifecycles, the dynamics of stationary waves, and moist baroclinic instability. It presents a forceful case for the ability of dynamical analysis to continue to clarify the atmosphere's behavior.'
Dr Daniel Kirk-Davidoff, *Department of Meteorology, University of Maryland*

'This book is a thoughtful distillation of decades of teaching. It includes figures from illustrative computations that are custom-fitted to the mathematical development. Although it is brisk and condensed, with so much material to cover, the book has a clear and distinctive voice, and offers some unique treatments and insights.'
Dr Brian Mapes, *Rosenstiel School of Marine and Atmospheric Science, University of Miami*

'This text neatly and logically outlines atmospheric dynamics, from the fundamental concepts in Chapter 1 and the development of the primitive equations in Chapter 2 to wave dynamics in Chapter 5 and overturning circulations in Chapter 10. All the while, walking the reader through the derivations and theory, yet referencing the physical applications with maps and diagrams or model outputs. It would be a helpful reference for any graduate student or scientist. Mak's text would quickly educate someone new to atmospheric science, but ready to jump in to the fray at the graduate level. … It really is a nicely written text.'
Dr Teresa Bals-Elsholz, *Department of Geography and Meteorology, Valparaiso University*

Atmospheric Dynamics

MANKIN MAK
University of Illinois at Urbana-Champaign

Shaftesbury Road, Cambridge CB2 8EA, United Kingdom

One Liberty Plaza, 20th Floor, New York, NY 10006, USA

477 Williamstown Road, Port Melbourne, VIC 3207, Australia

314–321, 3rd Floor, Plot 3, Splendor Forum, Jasola District Centre, New Delhi – 110025, India

103 Penang Road, #05–06/07, Visioncrest Commercial, Singapore 238467

Cambridge University Press is part of Cambridge University Press & Assessment,
a department of the University of Cambridge.

We share the University's mission to contribute to society through the pursuit of
education, learning and research at the highest international levels of excellence.

www.cambridge.org
Information on this title: www.cambridge.org/9780521195737

First published 2011

A catalogue record for this publication is available from the British Library

Library of Congress Cataloging-in-Publication data
Mak, Mankin, 1939–
Atmospheric dynamics / Mankin Mak.
p. cm.
ISBN 978-0-521-19573-7 (Hardback)
1. Atmospheric physics–Textbooks. 2. Meteorology–Textbooks. I. Title.
QC861.3.M335 2011
551.51′5–dc22
2010029996

ISBN 978-0-521-19573-7 Hardback

Additional resources for this publication at www.cambridge.org/9780521195737

To the memory of my parents who made it possible for me to pursue my dreams

Contents

Preface

Atmospheric dynamics is the foundation for understanding the movement of air currents. As such, it is a core component of all atmospheric science curricula. Atmospheric dynamics is the discipline concerned with what different classes of geophysical disturbances are made up of, how and why they form, what factors dictate their structure and movement, how the Earth's uneven surface impacts them, how they evolve to their mature stage, how they interact with the background flow, how they eventually decay and, above all, how they collectively constrain the atmospheric general circulation as a whole. An analysis of atmospheric dynamics can be process-oriented trying to address one or more of the questions above. Its goal would be to establish quantitative understanding of the nature of those processes. An analysis of atmospheric dynamics can be also phenomenon-oriented with the objective of developing a feel for why a specific phenomenon is as observed. We strive to cover both aspects of atmospheric dynamics in this book.

This book is mainly intended to serve students of atmospheric sciences. The goal is to come up with a book distinctly different from but complementary to the existing pool of texts on atmospheric dynamics. This book aims at being a resource valuable to instructors and self-explanatory to students as much as possible. It would be a bonus if weather forecasters as well as some practitioners in atmospheric science, ocean science, environmental science and applied mathematics find this book to be a useful reference.

Progress in atmospheric dynamics would not have been possible without the introduction of various insightful concepts and approximations. They greatly simplify the quantitative investigations of atmospheric disturbances with analytical and numerical methods. Computer graphics help visualize the structures and processes. For this reason, all of them receive due emphasis in the discussions. We extensively use schematics to depict the conceptual issues. We use different means to strengthen the connection between theoretical dynamics and atmospheric observations. For instance, real-time weather maps are generated to illustrate the various dynamical characteristics of atmospheric disturbances. The spatial and temporal statistical properties of various classes of disturbances are examined as background information for related theoretical discussions of their dynamical roles. Analyses with idealized models are used to highlight the essence of the phenomena under consideration.

The first half of the book discusses the physical concepts of atmospheric dynamics as well as the mathematical methodology at the fundamental level. The emphasis is to elucidate the physical nature of dynamical processes. The materials are written with undergraduates in mind who might have only cursory exposure to atmospheric sciences. With the exception of several sections on the more advanced topics, the first five chapters can be used in an introductory course on atmospheric dynamics. It would be helpful and/or necessary for an instructor to elaborate on the more technical aspects of the materials for the benefit of the uninitiated students. The more advanced topics are best reserved for a follow-up course of atmospheric dynamics. A second undergraduate course may begin with a review of the basic materials at a faster pace and proceed to cover a number of additional topics selected from the first eight chapters.

The second half of the book is suitable for graduate students who are already familiar with the basics. It is a collection upward of twenty-five illustrative model analyses of a wide variety of dynamical issues manifested in different interesting phenomena. The materials may be used as a graduate course sometimes called "Special Topics of Atmospheric Dynamics." The overriding objective is to show how one might tackle problems concerning the dynamics of different types of disturbances with the use of simple models. I show the merit of different approaches and mathematical methods dependent on the nature of the problem under consideration. Through the sample analyses, I hope to convey the idea that physical reasoning is even more crucial than mathematical skill in tackling dynamical problems. The discussions of several topics extensively make use of published research articles. The other illustrative model analyses and computations are specifically performed for this book. The physical meanings of the model results are discussed with generous use of figures. I strive to present each model analysis in a self-contained manner, so that they need not be read in sequence. The format of presentation in the second half of the book is therefore quite different from that used in the first half. These analyses are considerably more involved than typical homework assignments. They are more akin to term projects. I would encourage students, either individually or in groups, to reproduce some of the results as challenging exercises.

The book has twelve chapters starting with a synopsis in each:

- Chapter 1 introduces the basic concepts such as the continuum hypothesis of a fluid, Lagrangian versus Eulerian descriptions, non-inertial reference frame, stratification and baroclinicity. It reviews all the laws of physics needed for investigating the dry dynamics of atmospheric flows.
- Chapter 2 discusses the nature of different types of dynamical balance in atmospheric flows. The governing equations in different coordinate systems for atmospheric analyses are derived. We also elaborate on the general kinematic properties of atmospheric flows.
- Chapter 3 is a lengthy chapter focusing on the rotational properties of a geophysical flow. We discuss the concepts of vorticity, circulation and potential vorticity in conjunction with the related governing equations and mathematical theorems. The roles of different physical factors are delineated. This chapter also contains a detailed discussion of some advanced topics such as the impermeability theorem for potential vorticity and the notion of generalized potential vorticity for succinctly representing the impact of boundary from the perspective of vorticity dynamics.
- Chapter 4 discusses the impacts of small turbulent eddies on a background flow as a frictional force. The effect of such turbulence is discussed in the context of a simple parameterization. We introduce the notion of boundary layer. We analyze the structures of several different types of boundary layers and their implications, especially the atmospheric and oceanic Ekman layers.

- Chapter 5 discusses the fundamentals of wave dynamics concerned with internal gravity waves and Rossby waves. The characteristics of those free wave modes about their structure, propagation, dispersion and energetics are quantitatively examined. We also delineate the dynamical nature of the forced wave modes with two illustrative analyses. We examine some observed properties of Rossby waves in the format of general circulation statistics. We finally discuss the dynamics of edge waves.
- Chapter 6 presents the quasi-geostrophic theory for large-scale atmospheric flows, which heralded the major advances of dynamic meteorology. After showing some real-time weather maps of a representative event, we discuss the formulation of this theory. The particular form of this theory in a two-layer model setting is presented next. We illustrate all the dynamical concepts with an analysis of a baroclinic jet streak. The influences of the Earth's sphericity in the QG theory are finally discussed.
- Chapter 7 discusses how and why the velocity and pressure fields of a large-scale atmospheric flow would rapidly adjust towards a new balanced state whenever their existing balance is upset by unspecified causes. We analyze the adjustments from two canonical forms of initial imbalance.
- Chapter 8 is a lengthy chapter divided into four parts for clarity. It covers a potpourri of instability theories for disturbances of widely different sizes arising from different classes of shear flows. The first part deals with the instability of small-scale and meso-scale disturbances. The second and third parts discuss the transient and modal dynamics of cyclogenesis in the extratropics for basic flows of increasing structural complexity. The fourth part discusses the large-scale moist instability arising from self-induced condensational heating.
- Chapter 9 discusses the observed characteristics and the dynamics of stationary planetary waves. We examine the propagation of such waves through a weak horizontal or vertical background shear flow. The similarities and differences in the responses to large-scale thermal and/or topographic forcing are then delineated. An analysis of the summer mean Asian monsoon as a forced circulation serves to illustrate the dynamical properties of this important class of wave disturbances in a recognizable setting.
- Chapter 10 addresses the dynamics of interaction between a zonal mean flow and large-scale waves in the context of two intriguing phenomena. We begin by examining the mean meridional overturning circulation in both Eulerian and Lagrangian senses deduced from a global dataset. We perform a linear model analysis of them with the use of empirical forcing. Wave-mean flow interaction is then discussed in the context of stratospheric sudden warming.
- Chapter 11 discusses the equilibration dynamics of nonlinear baroclinic waves. We begin with a discussion of the rudiments of geostrophic turbulence in a two-layer model setting. It serves as background information for three specific analyses. The first analysis delineates the dynamics of life cycle of baroclinic waves. The second analysis brings to light a symbiotic relation between synoptic-scale waves and planetary-scale waves of a forced dissipative system. The third analysis looks into the dynamical nature of the relative intensity of the two major winter storm tracks.
- Chapter 12 discusses the nature of nongeostrophic dynamics in the context of three very different phenomena: surface frontogenesis, Hadley circulation and non-supercell

tornadogenesis. We present three illustrative model analyses of them using very different mathematical methods.

- An appendix summarizes the mathematical tools used in the book as a quick reference for some readers who might not have used them for a while.
- Problems and exercises are posted online in the website of the Cambridge University Press with the intention of updating them periodically in future. See www.cambridge.org/9780521195737.
- Only the research articles and books that readily came to mind during the writing of this book are included. I have made no special effort to comprise a comprehensive bibliography. There is no doubt that the resulting list of references is quite incomplete.

The framework of this book has been gradually taking shape over the course of many years of teaching and research at the University of Illinois, Urbana-Champaign, Illinois, USA. Zhenhua Li, Xian Lu, Sara T. Strey, Lantao Sun and Andy Vanloocke provided me with feedback on a number of chapters in the first half of the book from the students' perspective. Professors Ming Cai, Yi Deng and Lin Wang at different universities kindly gave me feedback on three chapters from the instructors' perspective. A number of anonymous reviewers solicited by the publisher provided some valuable suggestions. All this feedback has prompted significant revisions of the draft and for that, I am most appreciative. Joseph Grim and David Wojtowicz helped produce the high-resolution real-time weather maps. Lusheng Liang made the diagnosis of a five-year global dataset and wrote the code of model calculation that produces the figures in Chapter 10. Their technical contributions are indispensable and gratefully acknowledged. Finally, I would like to thank the editorial staff of Cambridge University Press – Matt Lloyd, Chris Hudson, Laura Clark and Sabine Koch – for their inputs and suggestions that helped make writing this book a wonderful experience of going through the maze of publication protocol.

1 Fundamental concepts and physical laws

We begin by discussing the fundamental concepts and the laws of physics that underlie the formulation of any dynamical analysis of atmospheric problems. We elaborate on how to conceptualize a fluid medium and how to mathematically quantify the state of a fluid in Section 1.1. We proceed to review a set of physical laws that govern the behavior of all disturbances in a dry atmosphere setting, detailing their physical meaning and mathematical representations in Sections 1.2 through 1.6. The observed distributions of two important mean characteristics of the atmosphere, namely its stratification and baroclinicity, are finally examined in Section 1.7.

1.1 Basic notions

1.1.1 Continuum hypothesis

Although the atmosphere may be visualized as a collection of randomly moving gas molecules in constant collision with one another, we conceptualize it as a fluid when we make dynamical atmospheric analyses by invoking a *continuum hypothesis*. We only seek to quantify the *macroscopic properties* of the atmosphere. One such property is the wind defined to be the average velocity of the molecules in a very small volume about each location. We call such a loosely defined amount of air molecules an atmospheric fluid parcel. We therefore think of the atmosphere as a *fluid medium* consisting of innumerable *parcels*. Furthermore, we assume that each property of the atmosphere varies continuously in time and space, whereby we may evaluate their derivatives with respect to the space variables and time in an analysis.

1.1.2 Lagrangian vs. Eulerian descriptions of a fluid

The fluid in a system can be described in principle in terms of the position of every one of its innumerably large number of fluid elements. The position may be measured by its Cartesian coordinates $(x(t), y(t), z(t))$ and its velocity components by $u(t) = dx/dt, v(t) = dy/dt, w(t) = dz/dt$ at all times. This is known as a *Lagrangian description* of a fluid with time being the only independent variable. A fluid element would deform indefinitely when it is subject to non-uniform stresses. Its shape and size would change continually making it impossible for us to track its identity for a long time. Hence, it is not feasible to perform a Lagrangian analysis of a fluid except for a short duration. A much more convenient means of describing a fluid is to quantify all of its fundamental properties at a fixed array of points in the domain at all times. We do not concern ourselves with the identity of the individual fluid parcels that happen to be at various points and time. This is known as an *Eulerian description* of a fluid. One fundamental property is the velocity field, $\vec{V} \equiv (u(x,y,z,t), v(x,y,z,t), w(x,y,z,t))$, where the independent variables are x, y, z and t. We will talk about the other fundamental properties shortly.

In a Lagrangian description the rate of change of any property of a fluid parcel, $\xi(t)$, in time as it moves about is called a *total derivative*, $D\xi/Dt$. In an Eulerian description, the rate of change of any property $\xi(x,y,z,t)$ at a fixed point is called a *local derivative*, $\partial\xi/\partial t$. The relationship between these two derivatives is established as follows. Suppose a generic property of an air parcel in a Lagrangian sense, $\xi(x(t), y(t), z(t), t)$, changes by $d\xi$ when its position has changed by (dx, dy, dz) in a time interval dt. They are interrelated by the chain rule of calculus

$$d\xi = \frac{\partial \xi}{\partial x}dx + \frac{\partial \xi}{\partial y}dy + \frac{\partial \xi}{\partial z}dz + \frac{\partial \xi}{\partial t}dt. \tag{1.1}$$

The partial derivatives in (1.1) are however in the Eulerian sense. Each of them is to be determined with the other three of the four independent variables (x, y, z, t) fixed. Upon dividing (1.1) through by dt, we get

$$\frac{D\xi}{Dt} \equiv \frac{d\xi}{dt} = \frac{\partial \xi}{\partial x}\frac{dx}{dt} + \frac{\partial \xi}{\partial y}\frac{dy}{dt} + \frac{\partial \xi}{\partial z}\frac{dz}{dt} + \frac{\partial \xi}{\partial t}. \tag{1.2}$$

This equation relates a Lagrangian description to an Eulerian description of a fluid. A more compact notation for (1.2) is

$$\frac{D\xi}{Dt} = \frac{\partial \xi}{\partial t} + \vec{V}\cdot\nabla\xi, \tag{1.3}$$

where $\vec{V} = (u, v, w) \equiv \left(\frac{dx}{dt}, \frac{dy}{dt}, \frac{dz}{dt}\right)$, $\nabla \equiv \left(\frac{\partial}{\partial x}, \frac{\partial}{\partial y}, \frac{\partial}{\partial z}\right)$ is the grad operator in Cartesian coordinates. Equation (1.3) can be of course applied in any coordinate system as long as $\vec{V}$ and ∇ are consistently defined. It is instructive to rewrite (1.3) as

$$\frac{\partial \xi}{\partial t} = \frac{D\xi}{Dt} - \vec{V}\cdot\nabla\xi. \tag{1.4}$$

Equation (1.4) says that:

Local rate of change of ξ at a point

= Rate of change of ξ in a parcel caused by external processes $\left(\frac{D\xi}{Dt}\right)$

+ Rate of change of ξ due to the advective process at a point, $(-\vec{V}\cdot\nabla\xi)$.

The advective rate of change of ξ at a point is positive when the velocity at that point is directed from a region of higher values of ξ towards a region of lower values of ξ. One part of it $(-u\xi_x - v\xi_y)$ is called horizontal advection of ξ and the other part $(-w\xi_z)$ is called vertical advection. The horizontal velocity components of air parcels in a large atmospheric disturbance are much larger than the vertical component. Those air parcels move quasi-horizontally. What we can reliably measure with a network of weather stations is only this component of airflow. It is what we usually call the *wind*,

$\vec{V}_2 \equiv (u, v)$. The advective rate of change in that case is mostly associated with the horizontal advection. If ξ stands for zonal velocity, u, then $-\vec{V} \cdot \nabla\xi \approx -\vec{V}_2 \cdot \nabla u$ would be advection of zonal momentum. If ξ stands for temperature, T, then $-\vec{V} \cdot \nabla\xi \approx -\vec{V}_2 \cdot \nabla T$ would be thermal advection.

1.1.3 Physical laws

Whatever happens in the atmosphere must be compatible with the laws of physics. The pertinent laws are listed below for quick reference:

- *Conservation of Mass* asserts that the mass of a fluid element remains unchanged no matter how it moves and how all its properties change.
- *Conservation of Momentum* is the underlying principle of Isaac Newton's three laws of mechanics. The latter are concerned with how the momentum of an object changes under the influence of forces.
- *Law of Gravitation* is a statement of the force acting on an object due to the presence of another object.
- The forces associated with the momentum exchange between a fluid parcel in contact with another fluid parcel are described by the notions of pressure and viscosity.
- *Equation of State* is the identity signature of the fluid medium under consideration.
- *Conservation of Energy* is the basis of the *First Law of Thermodynamics*. It quantifies how one might evaluate the rate of change of temperature of a fluid element in different circumstances.
- The laws of radiative transfer and phase transition of water in the atmosphere are approximately incorporated in an indirect manner when we analyze the basic dynamics of an atmospheric flow if necessary.

The mathematical expressions for these laws of physics take on the form of a complete set of partial differential equations. The solutions of those equations are very general for they could describe all classes of disturbances. The different concepts and physical laws are elaborated in the following sections of this chapter.

1.2 Laws of mechanics

The location and movement of an object are measured in the context of a reference frame. A special reference frame is an *inertial reference frame*, which is defined as one that moves at a constant speed in a certain direction. It follows that there is no unique inertial reference frame. A special one is absolutely at rest in space at all times. Such a reference frame presupposes that the notions of absolute space and absolute time are meaningful.

Newton formulated three laws of mechanics in conjunction with the use of an inertial reference frame. They are:

(1st Law) An object in uniform motion continues to be in the same uniform motion when it is not under the influence of a net force.

(2nd Law) When there is a force, $\vec{F}$, acting upon an object, the latter accelerates at a rate proportional to the force and inversely proportional to its mass, m,

$$\frac{d\vec{v}}{dt} = \frac{\vec{F}}{m} \tag{1.5}$$

where $\vec{v}$ is velocity in an inertial reference frame.

(3rd Law) If two bodies exert forces upon each other, these forces are equal in magnitude and opposite in direction. This law is applicable only if the force exerted by one object on another object is directed along the line connecting the two objects. Two known forces are of this type: gravitational force between two objects and electrostatic force between two electric charges. This law is not applicable for a case in which a force is velocity dependent.

Newton's three laws are in essence three statements concerned with the absolute momentum of an object defined as

$$\vec{P} = m\vec{v} \tag{1.6}$$

under three different circumstances. The 1st Law effectively says that the absolute momentum of an object conserves in the absence of external force. The 2nd Law effectively says that absolute momentum of an object under the influence of a force $\vec{F}$ changes in time at a rate equal to

$$\frac{d\vec{P}}{dt} = \vec{F}. \tag{1.7}$$

Suppose two objects with absolute momentum $\vec{P}_j$, $j = 1, 2$ interact with one another. According to Newton's 3rd Law, we have $\vec{F}_1 = -\vec{F}_2$ where the force acting on body-1 due to body-2 is $\vec{F}_1$ and vice versa is $\vec{F}_2$. Then by using (1.7) we have

$$\frac{d(\vec{P}_1 + \vec{P}_2)}{dt} = 0 \tag{1.8}$$

which simply means that the total momentum of a composite system would not change as a consequence of interactions among its members.

In passing, it should be emphasized that the concepts of absolute space and absolute time are ad hoc notions. Their logical implications are incompatible with the observation that the velocity of light is the same regardless of the movement of the light source. Einstein's special theory of relativity adopts that as a postulate and supersedes Newton's laws of mechanics since it accurately governs the movement of objects even in a speed close to the velocity of light. That theory establishes that the notions of absolute space and absolute time are superfluous and that space and time are intrinsically related depending on the motion of the observer. Newton's theory is nevertheless sufficiently accurate for atmospheric studies since wind speed is negligibly slow compared to the velocity of light.

1.2.1 Inertial vs. non-inertial reference frames

Wind is the movement of air parcels measured relative to a weather station fixed on Earth. The reference frame for measuring wind is therefore a non-inertial reference frame because

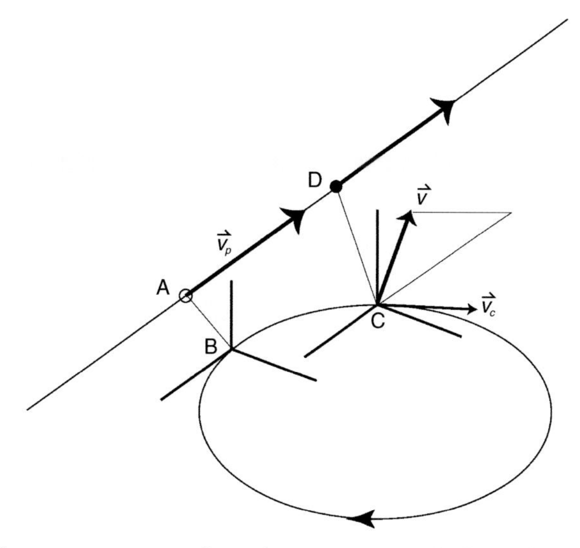

Fig. 1.1 Depiction of a particle moving in constant velocity along AD in an inertial reference frame and in a non-inertial reference frame (another observer moving in an elliptical orbit). The particle's velocity is zero to the second observer at B and is indicated by an arrow to this observer at C.

the Earth rotates about its own axis and also moves around the Sun. The total movement of each air parcel as viewed from space is quite complicated indeed. Let us compare the description of the movement of an object in the absence of any force $(\vec{F} = 0)$ in an inertial reference frame with the description in a non-inertial reference frame. The situation is depicted in Fig. 1.1. Suppose this object moves at a constant velocity $\vec{v}_p$ along a straight line with respect to an observer in an inertial reference frame. Its movement to an observer in an inertial reference frame is clearly very different to another observer in a reference frame that moves in an elliptical orbit with a constant speed $|\vec{v}_c| = |\vec{v}_p|$. Suppose the observer is at point B when the object is at point A and would have moved to C when the object moves to D. The object is momentarily stationary to observer at B as far as he can tell, $(\vec{v} = \vec{v}_p - \vec{v}_c = 0)$. Later at location C, the object appears to be moving in the direction indicated by the arrow, $(\vec{v} = \vec{v}_p - \vec{v}_c \neq 0)$. Hence, this observer concludes that the object continually accelerates as it moves. It follows that the equation of motion to him is NOT $\vec{F} = 0 = \frac{d}{dt}(m\vec{v})$ where $\vec{v}$ is the velocity with respect to him as the observer. He would conclude that there must be a force acting on the object if he wishes to use the form of Newton's formula (1.5) for analyzing its motion.

1.3 Equations of motion in a rotating reference frame

Since the Earth rotates about its own axis and moves around the Sun, it would be most convenient to use a non-inertial reference frame fixed on Earth when we analyze atmospheric disturbances. This reference frame rotates once a day. The movement of air parcels

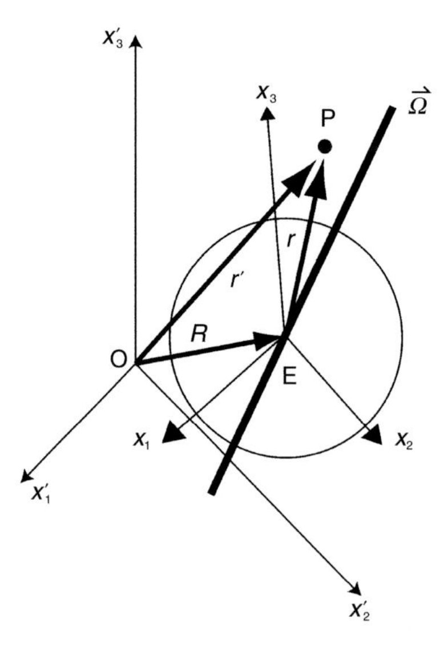

Fig. 1.2 The x_i' are coordinates of an inertial reference frame; x_i are coordinates of a rotating reference frame. E is center of such reference frame; $\vec{\Omega}$ is the vector representation of the rotation of the reference frame; $\vec{R}$ indicates the position of the origin of the non-inertial frame; $\vec{r}$ indicates the position of an object P in the non-inertial frame; $\vec{r}'$ indicates the position of the object in the inertial frame.

is measured relative to the surface of the Earth. We will derive in the following subsection an equation that adequately governs the motion of an air parcel in this framework. It is written in a vector form applicable to any choice of coordinate system. It takes on the form of $(d\vec{v}/dt)_{rotating} = \vec{F}/m - \vec{\Omega} \times \left(\vec{\Omega} \times \vec{r}\right) - 2\vec{\Omega} \times \vec{v}$ where $\vec{\Omega}$ is the vector representing the Earth's rotation about its own axis, $\vec{v}$ is the velocity of the air parcel in this rotating frame, $\vec{r}$ is the position vector of the air parcel from the center of the Earth, m the mass of the air parcel and $\vec{F}$ the net force acting on the air parcel.

1.3.1 General derivation

By definition, the real forces acting on an air parcel are the same regardless of the particular reference coordinates adopted in an analysis. So, we only need to transform the expression of acceleration from an inertial reference frame to a non-inertial reference frame. Without loss of generality, we use Cartesian coordinates to depict an object in both an inertial frame and a rotating reference frame as indicated in Fig. 1.2. Let us focus on an air parcel with mass m at point P. The x_i' $(i = 1,2,3)$ are the coordinates of an inertial reference frame according to the *right-hand convention*. The origin of this inertial reference frame is arbitrary. The x_i are the coordinates of a non-inertial reference frame whose origin is at the center of the Earth. $\vec{R}$ measures the position of the center of the Earth relative to the origin of the inertial frame. This non-inertial reference frame is completely general in that it may move and accelerate, $\dot{\vec{R}} \equiv \frac{d\vec{R}}{dt} \neq 0$, $\ddot{\vec{R}} \neq 0$. It also may rotate about an arbitrary axis in a varying rate, $\dot{\vec{\Omega}} \neq 0$.

The following notations are introduced for the derivation:
$\vec{R}$ = position vector for the origin of the rotating system in an inertial (fixed) frame
$\vec{e}_i$ = unit vectors of the rotating reference frame
$\vec{r} = \sum_{i=1}^{3} x_i\vec{e}_i$ = position vector of P in the rotating system
$\vec{r}\,' = \vec{R} + \vec{r}$ = position vector of P in a fixed system
$\vec{\Omega} = \sum_i \omega_i\vec{e}_i$ = rotation vector of the rotating system.
Then we have

$$\left(\frac{d\vec{r}\,'}{dt}\right) = \left(\frac{d\vec{R}}{dt}\right) + \frac{d}{dt}\sum_i (x_i\vec{e}_i) = \left(\frac{d\vec{R}}{dt}\right) + \sum_i \dot{x}_i\,\vec{e}_i + \sum_i x_i\,\dot{\vec{e}}_i. \tag{1.9}$$

The relative velocity is $(d\vec{r}/dt)_{rotating} = \sum_i \dot{x}_i\vec{e}_i$. We make use of two simple properties of vectors to determine $\dot{\vec{e}}_i$,

$$\vec{D} = \left(\vec{E} + \vec{F}\right) \times \vec{G} = \vec{E} \times \vec{G} + \vec{F} \times \vec{G},$$

$$\vec{C} = \vec{A} \times \vec{B} = (A_2B_3 - A_3B_2)\vec{e}_1 + (A_3B_1 - A_1B_3)\vec{e}_2 + (A_1B_2 - A_2B_1)\vec{e}_3.$$

Vector $\vec{C}$ is perpendicular to both $\vec{A}$ and $\vec{B}$ as evidenced by a special case $\vec{e}_1 \times \vec{e}_2 = \vec{e}_3$. It follows that the change of $\vec{e}_1$ in time arises from two components of $\vec{\Omega}$: ω_2 and ω_3. The change of $\vec{e}_1$ in time δt arising from ω_3 about the x_3 axis (pointing out of the paper) is $\delta\vec{e}_1 = \omega_3\delta t\vec{e}_2$. Likewise, the change of $\vec{e}_1$ in time δt arising from ω_2 about the x_2 axis (pointing into the paper) is $\delta\vec{e}_1 = -\omega_2\delta t\vec{e}_3$. Thus the total rate of change of $\vec{e}_1$ is

$$\dot{\vec{e}}_1 \equiv \frac{d\vec{e}_1}{dt} = \omega_3\vec{e}_2 - \omega_2\vec{e}_3 = \vec{\Omega} \times \vec{e}_1 = \begin{vmatrix} \vec{e}_1 & \vec{e}_2 & \vec{e}_3 \\ \omega_1 & \omega_2 & \omega_3 \\ 1 & 0 & 0 \end{vmatrix}. \tag{1.10}$$

Similarly $\dfrac{d\vec{e}_2}{dt} = -\omega_3\vec{e}_1 + \omega_1\vec{e}_3 \; \dfrac{d\vec{e}_3}{dt} = \omega_2\vec{e}_1 - \omega_1\vec{e}_2.$

Thus we have in general vector notation

$$\dot{\vec{e}}_i = \vec{\Omega} \times \vec{e}_i, \qquad i = 1, 2, 3.$$

By (1.9), (1.10)

$$\left(\frac{d\vec{r}\,'}{dt}\right)_{fixed} = \left(\frac{d\vec{R}}{dt}\right)_{fixed} + \left(\frac{d\vec{r}}{dt}\right)_{rotating} + \vec{\Omega} \times \vec{r}. \tag{1.11}$$

Equation (1.11) written in a more concise notation is $\vec{v}\,' = \dot{\vec{R}} + \vec{v} + \vec{\Omega} \times \vec{r}$. Taking the time derivative of (1.11) in the context of the inertial system, we get

$$\begin{aligned} \left(\frac{d\vec{v}\,'}{dt}\right)_{fixed} &= \ddot{\vec{R}} + \left(\frac{d\vec{v}}{dt}\right)_{fixed} + \dot{\vec{\Omega}} \times \vec{r} + \vec{\Omega} \times \left(\frac{d\vec{r}}{dt}\right)_{fixed} \\ &= \ddot{\vec{R}} + \left(\frac{d\vec{v}}{dt}\right)_{rotating} + \vec{\Omega} \times \vec{v} + \dot{\vec{\Omega}} \times \vec{r} + \vec{\Omega} \times \vec{v} + \vec{\Omega} \times \left(\vec{\Omega} \times \vec{r}\right). \end{aligned} \tag{1.12}$$

According to Newton's law, the equation of motion for P in an inertial reference frame is simply $\left(\frac{d\vec{v}'}{dt}\right)_{fixed} = \frac{\vec{F}}{m}$ where $\vec{F}$ is the net real force. With the use of (1.12), we can then write Newton's equation of motion in a general rotating reference frame as

$$\underset{(i)}{\frac{\vec{F}}{m}} - \underset{(ii)}{\ddot{\vec{R}}} - \underset{(iii)}{\dot{\vec{\Omega}} \times \vec{r}} - \underset{(iv)}{\vec{\Omega} \times \left(\vec{\Omega} \times \vec{r}\right)} - \underset{(v)}{2\vec{\Omega} \times \vec{v}} \equiv \frac{\vec{F}_{eff}}{m} = \left(\frac{d\vec{v}}{dt}\right)_{rotating}. \tag{1.13}$$

Equation (1.13) is written in the same format as that for an inertial reference frame. It should be emphasized that although Cartesian coordinates are used in the derivation above, we are now writing the final equation in vector notation. It would need to be written explicitly for a specific coordinate system in a particular application.

The five terms on the LHS of (1.13) are to be interpreted as "forces" acting on an air parcel. Since only term (i) represents the real forces by definition, the others are *apparent forces*, which stem solely from the use of the non-inertial reference frame under consideration. Their sum constitutes an effective net force that would give rise to a relative acceleration.

Term (ii) is associated with the sum of the Earth's movement around the Sun, the Sun's movement around the Milky Way, etc. A part of (i) must be responsible for (ii). That is essentially the net gravitational force from all celestial bodies exerted on the Earth. It must necessarily cancel (ii). If we neglect (ii) we must not include the gravitational force due to all celestial bodies except the Earth itself in (i) for consistency.

Term (iii) is proportional to $\dot{\Omega}$ associated with the precession of the axis of rotation of the Earth. The orientation of Earth's rotational axis changes in a cycle of approximately 26 000 years, like a wobbling top. This term is therefore very small compared to (i). There are additional details concerning the precession of the Earth's axis. We will neglect the precession of the rotation of the Earth altogether in atmospheric applications and thereby assume $\dot{\Omega} = 0$.

Term (iv) is the *centrifugal force* on the air parcel associated with the rotation of the non-inertial reference frame. It is in the direction outward normal to the axis of rotation.

Term (v) is called *Coriolis force* associated with the movement of the air parcel at P as measured in the rotating reference frame. Its direction is perpendicular to both the rotation vector and the relative velocity. Its magnitude is proportional to the speed of the object $|\vec{v}|$ and to twice the rotation rate of the Earth, 2Ω.

Summing up, we may simplify (1.13) to

$$\frac{\vec{F}}{m} - \vec{\Omega} \times \left(\vec{\Omega} \times \vec{r}\right) - 2\vec{\Omega} \times \vec{v} = \left(\frac{d\vec{v}}{dt}\right)_{rotating} \tag{1.14}$$

as an adequately accurate equation of motion for any air parcel in atmospheric studies.

1.3.2 Physical nature of the Coriolis force

Although the mathematical derivation of the expression for the Coriolis force in (1.14) is rigorous, it does not give us a feel for this force from a physical point of view. To develop a better feel, we re-derive the form of Coriolis force with more elemental but physical

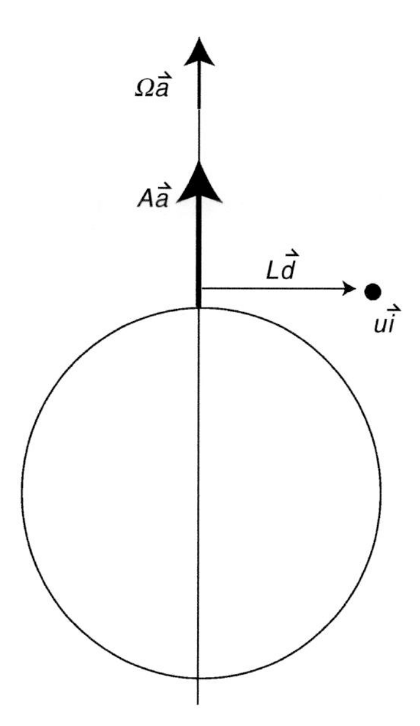

Fig. 1.3 Schematic of an air parcel on a meridional plane in the atmosphere. The Earth's rotation vector, position vector, relative zonal velocity and absolute angular momentum of the parcel are denoted by $\Omega\vec{a}$, $L\vec{d}$, $u\vec{i}$ and $A\vec{a}$ respectively; $\vec{i}$ is a unit vector pointing into the plane.

reasoning in two scenarios. In the first scenario, we consider an air parcel of unit mass initially at rest at a latitude φ. Its location is at a distance L from the Earth's axis of rotation. If its absolute motion synchronizes with the Earth's rotation, there would be no relative motion and there would be a radially inward acceleration (known as centripetal acceleration). A change of its sign can be reinterpreted as a force in a radially outward direction acting on the air parcel (known as *centrifugal force*, $\vec{F}_{cen}$). This is the force required to sustain it at this location in a rotating coordinate. For convenience, we will make use of two different coordinate systems in the following discussion. We will consider two scenarios in which the air parcel is set to simple motion. We first describe what happens in a *polar-cylindrical coordinate system* $(\vec{i}, \vec{d}, \vec{a})$ where $\vec{i}$ is a unit vector pointing eastward, $\vec{d}$ a unit vector pointing away from the Earth's axis of rotation, and $\vec{a}$ a unit vector directed along the rotation axis. The rotation vector of the Earth is then $\vec{\Omega} = \Omega\vec{a}$. The initial position vector of the air parcel can be designated by $\vec{L} = L\vec{d}$. Its absolute velocity is $\vec{V} = \vec{\Omega} \times \vec{L} = \Omega L\vec{i}$ in the azimuthal direction and its centripetal acceleration is $\vec{\Omega} \times \vec{V} = -\Omega^2 L\vec{d}$ pointing towards the axis of rotation. The centrifugal force is then $\vec{F}_{cen} = \Omega^2 L\vec{d}$. The mathematical expression for the *angular momentum*, $\vec{A}$, of this air parcel is

$$\vec{A} = \vec{L} \times \vec{V} = \Omega L^2 \vec{a}, \tag{1.15}$$

where $\vec{A}$ is a vector in the direction of the rotation axis with a magnitude equal to the product of the tangential velocity about the axis of rotation and the distance from the axis (moment of momentum). According to Newton's law, the angular momentum $\vec{A}$ of the air parcel would not change in time if it is not acted upon by an external torque. The configuration is depicted in Fig. 1.3.

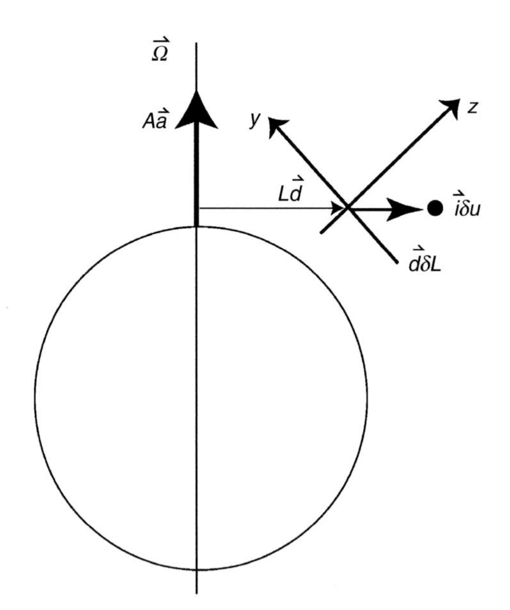

Fig. 1.4 Schematic of a displaced air parcel on a meridional plane in the atmosphere.

Suppose the air parcel is somehow displaced infinitesimally outward to a distance $L + \delta L$ in a short time interval δt. We want to deduce what might be the consequence of this displacement. Since no torque is required for this displacement, its absolute angular momentum would not change. This implies a change in its azimuthal (zonal) velocity, δu, since its distance from the axis has changed:

$$
\begin{aligned}
(A\vec{a})_{initial} &= (A\vec{a})_{later} \\
\Omega L^2 \vec{a} &= (L + \delta L)\vec{d} \times (\Omega(L + \delta L) + \delta u)\vec{i} \\
&= \left(\Omega L^2 + 2\Omega L\, \delta L + L\, \delta u + \delta L\, \delta u\right)\vec{a}.
\end{aligned}
\tag{1.16}
$$

For an infinitesimal displacement, the product term $\delta L \delta u$ would be negligibly smaller than other terms. Then it follows

$$\delta u = -2\Omega\, \delta L. \tag{1.17}$$

Equation (1.17) says that if an air parcel is displaced outward from the axis of rotation, it would necessarily move to the west, $\delta u < 0$. This movement would be observable as a zonal velocity in a rotating reference frame. This relation is analogous to an ice skater who could slow down his spinning motion by stretching his arms outward.

Now let us represent this radial displacement in a local Cartesian coordinate system with its origin collocated at the air parcel. $\left(\vec{i}, \vec{j}, \vec{k}\right)$ are the unit vectors. The result is $\delta L\vec{d} = \delta y\vec{j} + \delta z\vec{k}$ (Fig. 1.4; δy would have a negative value in this case). The contribution to (1.17) from δL comes from two parts: $\delta L_1 = -\delta y \sin\varphi$ and $\delta L_2 = \delta z \cos\varphi$ where φ is the latitude. The contributions to δu induced by the two parts of δL are additive. Thus, (1.17) may be written as

$$\delta u = 2\Omega \sin\varphi\, \delta y - 2\Omega \cos\varphi\, \delta z. \tag{1.18}$$

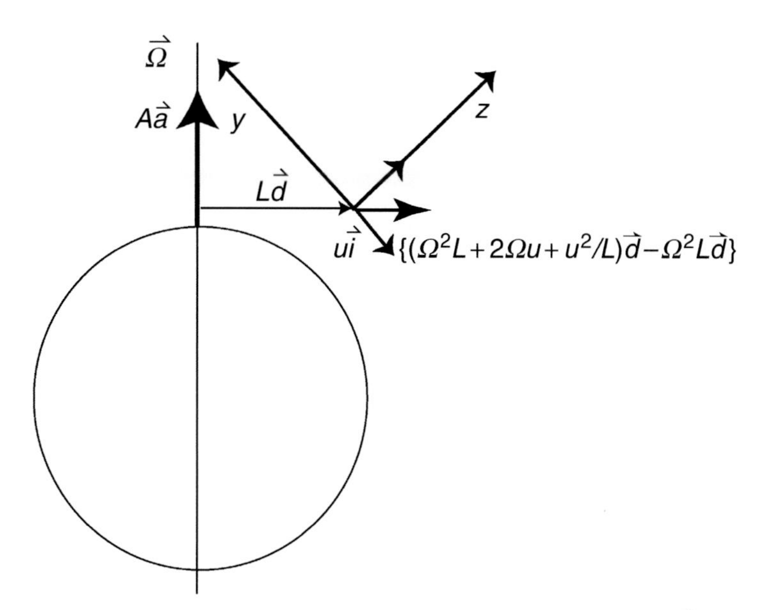

Fig. 1.5 Schematic of the change of force on an air parcel impulsively given a relative zonal velocity, $u\vec{i}$ in the atmosphere. The force indicated by the arrow in the direction of $\vec{d}$ is equivalent to the vectorial sum of the arrows in the *y*- and *z*-directions.

Defining $v = dy/dt$ and $w = dz/dt$ as the relative meridional and vertical velocity components, we finally obtain from (1.18)

$$\frac{du}{dt} = 2\Omega v \sin\varphi - 2\Omega w \cos\varphi. \tag{1.19}$$

Equation (1.19) tells us that an air parcel in the northern hemisphere $(\varphi > 0)$ must be accelerated eastward in a rotating reference frame whenever it moves northward $(v > 0)$ or/and downward $(w < 0)$. The RHS of (1.19) plays the role of a force. This is then the zonal component of a Coriolis force associated with such relative motions.

In the second scenario, a small *velocity* in the zonal direction $u\vec{i}$ is impulsively imparted to the air parcel by unspecified means (Fig. 1.5). This velocity would be a relative velocity in a reference frame rotating with the Earth. Its absolute rotational speed about the axis then becomes $(\Omega + u/L)$. The new centrifugal force in *an inertial reference frame* is

$$\left(\vec{F}_{cen}\right)_{new} = \left(\Omega + \frac{u}{L}\right)^2 L\vec{d} = \left(\Omega^2 L + 2\Omega u + \frac{u^2}{L}\right)\vec{d}. \tag{1.20}$$

The difference between the new and initial centrifugal forces would necessarily accelerate the air parcel in the direction of $\vec{d}$. Equation (1.20) says that the new centrifugal force is made up of three parts. The first part $(\Omega^2 L\vec{d})$ is equal to the initial centrifugal force. The third part, $\frac{u^2}{L}\vec{d}$, is a centrifugal force associated with the relative zonal velocity in a direction outward normal to the axis of rotation. By making use of the relation $\vec{d} = -\vec{j}\sin\varphi + \vec{k}\cos\varphi$ where $\vec{j}$ and $\vec{k}$ have been defined earlier, we can rewrite the second term in (1.20) as

$$2\Omega\, u\vec{d} = (-2\Omega\, u \sin\varphi)\vec{j} + (2\Omega\, u\cos\varphi)\vec{k}. \tag{1.21}$$

These are two components of an additional force associated with a zonal relative velocity in a local Cartesian coordinate system. This force is the Coriolis force. The first component of this force would accelerate the air parcel of unit mass in the y-direction in a rotating reference frame. The second component of this force would accelerate the air parcel in the z-direction. Their relations are then

$$\frac{dv}{dt} = -2\Omega\, u \sin\varphi, \qquad \frac{dw}{dt} = 2\Omega\, u \cos\varphi. \tag{1.22}$$

These two components of Coriolis force manifest in a rotating reference frame, apart from the centrifugal force, $\frac{u^2}{L}\vec{d}$.

Finally, let us summarize all the results deduced above in the context of a *spherical coordinate system*(λ, φ, r) for longitude, latitude and radial distance from the center of the Earth respectively. The corresponding unit vectors are $\left(\vec{\lambda}, \vec{\varphi}, \vec{r}\right)$ which coincide with $\left(\vec{i}, \vec{j}, \vec{k}\right)$ of a local Cartesian system. The rotation vector of the Earth is $\vec{\Omega} = (0, \Omega\cos\varphi, \Omega\sin\varphi)$. The position vector of a fluid parcel is $r\vec{r}$ and its velocity is defined as $\vec{v} \equiv (u, v, w) = \left(r\cos\varphi\frac{d\lambda}{dt}, r\frac{d\varphi}{dt}, \frac{dr}{dt}\right)$. The Coriolis force deduced above is therefore

$$\vec{F}_{co} = (-2\Omega w\cos\varphi + 2\Omega v\sin\varphi)\vec{\lambda} - (2\Omega u\sin\varphi)\vec{\varphi} + (2\Omega u\cos\varphi)\vec{r} = -2\vec{\Omega}\times\vec{v}. \tag{1.23}$$

This expression of Coriolis force is identical to the one obtained by formal coordinate transformation in Section 1.3.1. We conclude by saying that the Coriolis force in a rotating reference frame must take on this particular form as a logical consequence of the fundamental principle of conservation of absolute angular momentum.

1.3.3 Equation of motion in spherical coordinates

The spherical coordinate system would be a natural choice of coordinate system for analyzing a flow of global extent. Equation (1.14) simply takes on the following form

$$\frac{D\left(u\vec{\lambda}\right)}{Dt} + \frac{D(v\vec{\varphi})}{Dt} + \frac{D(w\vec{r})}{Dt} = -2\vec{\Omega}\times\vec{v} - \vec{\Omega}\times\left(\vec{\Omega}\times r\vec{r}\right) + real\,forces, \tag{1.24}$$

where $\frac{D}{Dt} = \frac{\partial}{\partial t} + \vec{v}\cdot\nabla = \frac{\partial}{\partial t} + \frac{u}{r\cos\varphi}\frac{\partial}{\partial\lambda} + \frac{v}{r}\frac{\partial}{\partial\varphi} + w\frac{\partial}{\partial r}$. To determine $D\vec{\lambda}/Dt$, we need to work out $\partial\vec{\lambda}/\partial\lambda$. Consider two points $\delta\lambda$ longitude apart on latitude φ, A and B (Fig. 1.6).

Using notation $\vec{\Delta}$ for a unit vector in the direction of $\overrightarrow{\delta\lambda}$, we get

$$\frac{\partial\vec{\lambda}}{\partial\lambda} = \lim\frac{\overrightarrow{\delta\lambda}}{\delta\lambda} = \frac{\delta\lambda\vec{\Delta}}{\delta\lambda} = (\sin\varphi\vec{\varphi} - \cos\varphi\,\vec{r}).$$

Hence we get

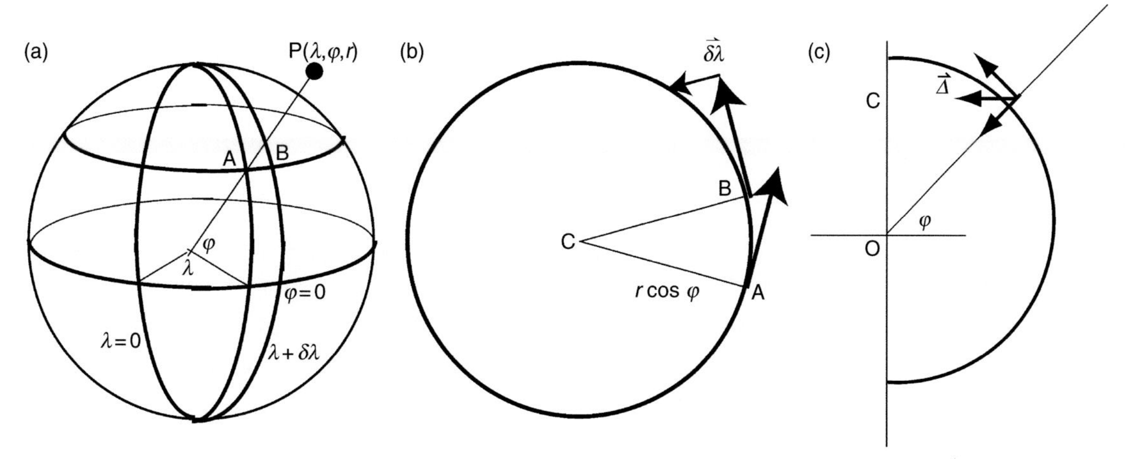

Fig. 1.6 (a) Coordinates (λ, φ, r), (b) unit vectors $\vec{\lambda}$ at A and B (exaggerated length), their difference $\overrightarrow{\delta\lambda}$ and (c) side view of unit vector $\vec{\Lambda} = (\sin\varphi\vec{\varphi} - \cos\varphi\vec{r})$ where $\vec{\Lambda}$ is in the direction of $\overrightarrow{\delta\lambda}$.

$$u\frac{D\vec{\lambda}}{Dt} = \frac{u^2}{r\cos\varphi}\frac{\partial\vec{\lambda}}{\partial\lambda} = \frac{u^2}{r\cos\varphi}(\sin\varphi\vec{\varphi} - \cos\varphi\vec{r}).$$

Similarly $v\frac{D\vec{\varphi}}{Dt} = v\left(\frac{u}{r\cos\varphi}\frac{\partial\vec{\varphi}}{\partial\lambda} + \frac{v}{r}\frac{\partial\vec{\varphi}}{\partial\varphi}\right) = -\frac{uv\tan\varphi}{r}\vec{\lambda} - \frac{v^2}{r}\vec{r},$

$$w\frac{D\vec{r}}{Dt} = w\left(\frac{u}{r\cos\varphi}\frac{\partial\vec{r}}{\partial\lambda} + \frac{v}{r}\frac{\partial\vec{r}}{\partial\varphi}\right) = \frac{uw}{r}\vec{\lambda} + \frac{vw}{r}\vec{\varphi}.$$

Then the explicit form of relative acceleration is

$$\frac{D\vec{v}}{Dt} = \left(\frac{Du}{Dt}\vec{\lambda} + \frac{Dv}{Dt}\vec{\varphi} + \frac{Dw}{Dt}\vec{r}\right) + \left(\frac{-uv\tan\varphi}{r} + \frac{uw}{r}\right)\vec{\lambda} + \left(\frac{u^2\tan\varphi}{r} + \frac{vw}{r}\right)\vec{\varphi} - \frac{u^2+v^2}{r}\vec{r}. \quad (1.25)$$

We have previously shown that the Coriolis force in spherical coordinates is

$$-2\vec{\Omega}\times\vec{v} = 2\Omega(v\sin\varphi - w\cos\varphi)\vec{\lambda} - (2\Omega u\sin\varphi)\vec{\varphi} + (2\Omega u\cos\varphi)\vec{r}. \quad (1.26)$$

Thus, the three components of the equation of motion in spherical coordinates are

$$\frac{Du}{Dt} + \left(\frac{-uv\tan\varphi}{r} + \frac{uw}{r}\right) = 2\Omega(v\sin\varphi - w\cos\varphi) + \vec{\lambda}\cdot\vec{F},$$

$$\frac{Dv}{Dt} + \left(\frac{u^2\tan\varphi}{r} + \frac{vw}{r}\right) = -(2\Omega u\sin\varphi) + \vec{\varphi}\cdot\vec{F}, \quad (1.27)$$

$$\frac{Dw}{Dt} - \left(\frac{u^2+v^2}{r}\right) = -(2\Omega u\sin\varphi) + \vec{r}\cdot\vec{F},$$

where $\vec{F}$ = (sum of real forces $-\vec{\Omega}\times(\vec{\Omega}\times\vec{r})$). The bracketed terms on the LHS are called metric terms associated with the curvature of the Earth.

1.4 Forces

In this section, we will elaborate on the nature and representation of the forces that need to be considered in a dynamical analysis of the atmosphere.

1.4.1 Gravitational force and gravity

According to *Newton's law of gravitation*, any two objects mutually pull toward one another by a force. Specifically, the Earth's *gravitational force* pulling an air parcel of unit mass towards the center of mass of the Earth is

$$\vec{g}_o = G\frac{M_E}{r^2}\vec{e}_r, \tag{1.28}$$

where the position of the air parcel relative to the center of the Earth is $\vec{r} = -r\vec{e}_r$ with $\vec{e}_r$ being a unit vector directed from the air parcel towards the Earth's center of mass, $M_E = 5.97 \times 10^{24}$ kg the mass of the Earth and $G = 6.67 \times 10^{-11}\ \mathrm{m^3\,kg^{-1}\,s^{-2}}$ the gravitational constant. Since the solid Earth is so much more massive than all other substance in the atmosphere combined, (1.28) is effectively the net gravitational force experienced by an air parcel.

The gravitational pull on an air parcel with mass m as indicated by a plumb bob is however not in the same direction as that of the gravitational force itself because the air parcel is also subject to the centrifugal force. Such gravitational pull is referred to as "gravity," which is directed perpendicular to the ground below. We represent it as

$$m\vec{g} = m\vec{g}_o - m\vec{\Omega} \times \left(\vec{\Omega} \times \vec{r}\right). \tag{1.29}$$

The term $-m\vec{\Omega} \times \left(\vec{\Omega} \times \vec{r}\right)$ is the centrifugal force experienced by an air parcel located at $\vec{r}$. It is directed outward normal to the rotation axis. Hence, the Earth could not be a perfect sphere, but is instead an ellipsoid bulging slightly at the equator by the time the planet solidified so that the net force on the Earth's surface is locally perpendicular to it. This feature has been known through observation for centuries. The gravitational pull on the surface of the Earth would not be the same everywhere. It follows that one's weight would vary slightly depending on where one stands on the Earth's surface. None other than Isaac Newton himself deduced this characteristic.

An exaggerated picture of the Earth with the three vectors in (1.29) at an air parcel is shown in Fig. 1.7. Under the influence of the centrifugal force, the Earth progressively deformed after formation until its crust eventually solidified. The current equatorial radius is 21.4 km longer than its polar radius. One consequence of this is that gravity at the surface $|\boldsymbol{g}|$ is greater at the poles than at the equator by about $0.052\ \mathrm{m\,s^{-2}}$; $|\boldsymbol{g}|$ is also smaller at greater elevations from the surface because of its dependence on the distance from the center of mass of the Earth. But for the air in the troposphere ($z \leq 12$ km $\ll$

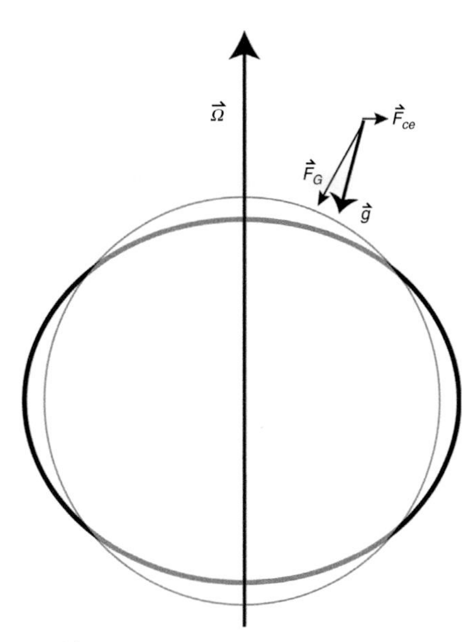

Fig. 1.7 Schematic of the gravitational force $\vec{F}_G$ (perpendicular to the spheroid), the centrifugal force associated with the Earth's rotation alone, $\left(\vec{F}_{ce}\right) = \Omega^2 r \cos\varphi$ and gravity $\vec{g}$ (perpendicular to the ellipsoid).

radius of the Earth), the variations of $|\boldsymbol{g}|$ with latitude/height is negligibly small. Using a constant value $|\boldsymbol{g}| = 9.8\,\mathrm{m\,s^{-2}}$ is sufficient for most analyses of wind. The Earth may be treated as a sphere.

The following equation of motion would be appropriate for making atmospheric analyses using a coordinate system that rotates with the Earth,

$$m\frac{D\vec{\mathbf{v}}}{Dt} = \vec{\hat{F}} + m\vec{g} - m2\vec{\Omega} \times \vec{v}, \tag{1.30}$$

where $\vec{\hat{F}}$ contains all real forces other than gravity.

1.4.2 Pressure gradient force

Fluid parcels press against one another. When a neighboring parcel on one side presses harder against it than its neighbor on the opposite side, it would experience a net force. We call the force at an interface a "stress" acting on one parcel by the other parcel. The component of this stress in the direction normal to the surface is called "pressure" which has the dimension of force per unit area. Let us consider such stress on all surfaces of a fluid parcel in the shape of a cube (Fig. 1.8). We simply sum over all such stresses to get the net force on this fluid parcel.

Mass of this fluid parcel $= \rho\,\delta x\,\delta y\,\delta z$.

Pressure on surface A $p_A = p_B + \dfrac{\partial p}{\partial x}\delta x$.

Net force on surfaces A and B in the x-direction $= (p_B - p_A)\delta y\,\delta z = -\dfrac{\partial p}{\partial x}\delta x\,\delta y\,\delta z$.

Similarly, such force in the y- and z-direction are $-\dfrac{\partial p}{\partial y}\delta x\,\delta y\,\delta z$ and $-\dfrac{\partial p}{\partial z}\delta x\,\delta y\,\delta z$.

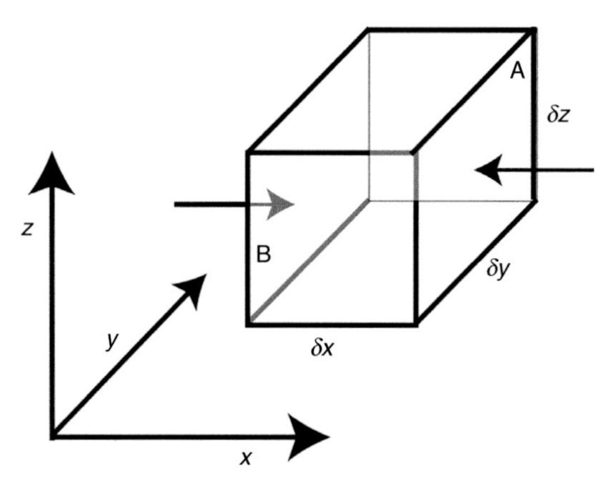

Fig. 1.8 Schematic of the pressure on surfaces A and B of a fluid parcel normal to the x-axis.

The whole net force due to pressure on fluid parcel $= -\nabla p\,\delta x\,\delta y\,\delta z$.

$$\text{Net force due to pressure per unit mass of fluid parcel} = -\frac{1}{\rho}\nabla p. \tag{1.31}$$

Since this force is proportional to the gradient of pressure, we call it *pressure gradient force*. The negative sign signifies that it is directed from a region of high pressure to a region of low pressure.

1.4.3 Molecular viscous force

There can be an exchange of momentum at the molecular level between two fluid parcels in contact. We say the fluid parcels rub against one another. There would be a macroscopic stress acting on the contact surface of a fluid parcel if it moves relative to its neighboring fluid parcels. When a fluid parcel moves faster than an adjacent parcel in a direction, the molecules in the former on the average move faster in that direction than the molecules in the latter. An exchange of molecules between these two parcels would result in a flux of momentum of that direction across the interface. Consequently the faster moving parcel would slow down and the slower one would speed up. This process is represented by a molecular frictional force. This force intrinsically tends to reduce the spatial variation of the velocity field in a fluid.

We now derive the mathematical expression for the molecular frictional force. Consider three adjacent blocks of infinitesimal fluid parcels A, B, C in Fig. 1.9. Their velocities in the x-direction are u_A, u_B, u_C.

First, consider the contact surface between fluid parcel B and fluid parcel A normal to the z-direction. Experiments reveal that there is a stress in the x-direction on such a surface when the x-velocity of B is not equal to that of C. The notation for such stress (force per unit area) in the x-direction is τ_{zx}. Experiments further show that this stress is proportional to such shear with a proportionality constant μ called dynamic viscosity coefficient in kg $\mathrm{m^{-1}\,s^{-1}}$. The contribution to such force exerted on B by A is then

$$F^{(top)} = \mu\frac{u_A - u_B}{\delta z}\delta x\,\delta y = \mu\frac{\partial u}{\partial z}\delta x\,\delta y = \tau_{zx}\big|_{top}\,\delta x\,\delta y.$$

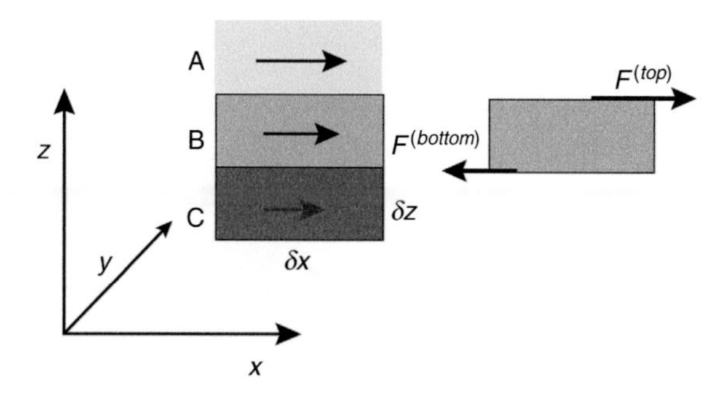

Fig. 1.9 Schematic of surface stresses acting on contact surfaces of three fluid parcels in differential motion.

By the same token, there is a similar force exerted on B by fluid parcel C, namely

$$F^{(bottom)} = \mu \frac{u_C - u_B}{\delta z} \delta x \, \delta y = -\mu \frac{\partial u}{\partial z} \delta x \, \delta y = -\tau_{zx}\big|_{bottom} \delta x \, \delta y.$$

Total force on B in the x-direction on such a surface is $\left(\tau_{zx}\big|_{top} - \tau_{zx}\big|_{bottom}\right)\delta x \delta y = \frac{\partial \tau_{zx}}{\partial z} \delta x \delta y \delta z$.

There are additional contributions to the viscous force on the two other pairs of surfaces of B. By inference, we conclude that the total viscous force on the fluid parcel B in the x-direction by all neighboring fluid parcels $= \left(\frac{\partial \tau_{xx}}{\partial x} + \frac{\partial \tau_{yx}}{\partial y} + \frac{\partial \tau_{zx}}{\partial z}\right) \delta x \, \delta y \, \delta z$.

The viscous force in the x-direction per unit mass is then

$$F^{(x)} = \frac{1}{\rho}\left(\frac{\partial}{\partial x}\left(\mu \frac{\partial u}{\partial x}\right) + \frac{\partial}{\partial y}\left(\mu \frac{\partial u}{\partial y}\right) + \frac{\partial}{\partial z}\left(\mu \frac{\partial u}{\partial z}\right)\right),$$

where μ is almost a constant for air. We often use a *kinematic viscosity coefficient* instead defined by $\frac{\mu}{\rho} \equiv \kappa = 10^{-5}\,\mathrm{m}^2\,\mathrm{s}^{-1}$ for air. The three components of viscous force can be written as $F^{(x)} = \kappa \nabla^2 u$, $F^{(y)} = \kappa \nabla^2 v$, $F^{(z)} = \kappa \nabla^2 w$. The molecular frictional force can be compactly written as

$$\vec{F} = \kappa \nabla^2 \vec{v}. \tag{1.32}$$

The small value of κ suggests that it is typically very weak except where the velocity varies greatly over a short distance, such as next to a solid surface. The fluid may be assumed to be at rest at a solid surface. This is called the no-slip boundary condition.

1.5 Conservation of mass

According to classical physics, the mass of an object can neither be created nor destroyed as it moves around. The mathematical expression of this conservation law for a fluid can be derived as follows. Consider a fluid parcel as a small cube of lengths δx, δy and δz without loss of generality (Fig. 1.10). Denoting its density by ρ, we have $\rho \delta x \, \delta y \, \delta z$ for its mass. Then conservation of mass of this fluid parcel means

$$\frac{D}{Dt}(\rho \, \delta x \, \delta y \, \delta z) = 0. \tag{1.33}$$

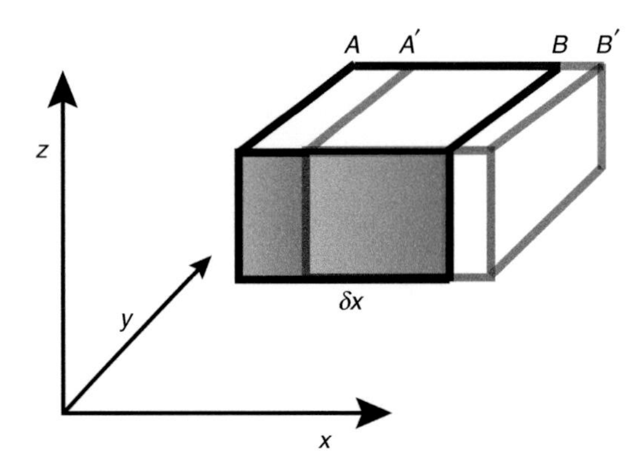

Fig. 1.10 Depiction of the change in a fluid parcel: Surface A moves to A' and surface B moves to B'.

The explicit form of (1.33) is

$$\delta x\,\delta y\,\delta z\frac{D}{Dt}(\rho)+\rho\delta y\,\delta z\frac{D}{Dt}(\delta x)+\rho\delta x\,\delta z\frac{D}{Dt}(\delta y)+\rho\delta x\,\delta y\frac{D}{Dt}(\delta z)=0. \tag{1.34}$$

The shape and size of a fluid parcel in a flow continually change in time.

The velocity is denoted by $\vec{v}=(u,v,w)$. When its surfaces A and B move at different speeds, u_A and u_B, its length in the x-direction would change.

The rate of change of the length of the fluid parcel in the x-direction would be

$$\frac{D}{Dt}(\delta x)=u_B-u_A=u_A+\left(\frac{\partial u}{\partial x}\right)_A\delta x-u_A$$

$$\frac{1}{\delta x}\frac{D}{Dt}(\delta x)=\frac{\partial u}{\partial x}. \tag{1.35}$$

$$\text{Similarly,}\quad \frac{1}{\delta y}\frac{D}{Dt}(\delta y)=\frac{\partial v}{\partial y}\quad\text{and}\quad\frac{1}{\delta z}\frac{D}{Dt}(\delta z)=\frac{\partial w}{\partial z}.$$

Dividing (1.34) by $\rho\delta x\,\delta y\,\delta z$ and making use of the results in (1.35), we get

$$\frac{D\rho}{Dt}=-\rho\nabla\cdot\vec{v}. \tag{1.36}$$

The term $\nabla\cdot\vec{v}=u_x+v_y+w_z$ is called the divergence of the 3-D velocity field in Cartesian coordinates. Equation (1.36) is often called the *continuity equation*, which gives us a Lagrangian representation of mass conservation. Equation (1.36) says that the density of a fluid parcel would increase in time if its volume shrinks and would decrease if its volume expands. Another form of the continuity equation gives us an Eulerian representation of mass conservation, namely

$$\frac{\partial\rho}{\partial t}=-\nabla\cdot(\vec{v}\rho), \tag{1.37}$$

where $(\vec{v}\rho)$ stands for mass flux. Equation (1.37) says that the density of the fluid at a point would increase in time if there is a convergence of mass flux towards that point and vice versa.

Special case: incompressible fluid

When a fluid is very weakly compressible, it may be idealized as an incompressible fluid. By that we mean its density does not change in time at all,

$$\frac{D\rho}{Dt} = 0 \Rightarrow \nabla \cdot \vec{v} = 0. \tag{1.38}$$

Water is much less compressible than air. Therefore, (1.38) may be used in analyses of currents in lakes and oceans. Oddly as it might sound, we even use $\nabla \cdot \vec{v} = 0$, or less restrictively $\nabla \cdot (\rho\vec{v}) = 0$, as an approximation of (1.37) in some analyses of atmospheric flow even though air is decidedly compressible. This is called the *Boussinesq approximation* to be further elaborated later in Section 5.3.

1.6 Thermodynamics and equation of state

Thermodynamics is the branch of physics concerned with the relationship between the bulk properties of a *system* and its *surroundings* involving heat exchange and mechanical work done on or by it, with or without chemical reactions among its constituents. In the context of a dry atmosphere, we think of an air parcel as a system consisting of a mixture of chemically non-interacting gases. Its surroundings would be the remaining part of the atmosphere. The bulk properties of this system are pressure, temperature and density. They are referred to as *thermodynamic variables*. The kinetic theory of gases provides a molecular interpretation of pressure and temperature. Pressure is the force on a surface arising from the collisions of the gas molecules on it. Temperature is a measure of the average kinetic energy of the gas molecules rapidly moving about in a fluid parcel.

One important notion is *thermodynamic equilibrium*. When a system reaches thermodynamic equilibrium, its thermodynamic variables would cease to change in time and its internal energy would be a unique function of the thermodynamic variables. The thermodynamic variables would be interrelated by the so-called Maxwell's relations. A full discussion of Maxwell's relations can be found in standard physics textbooks on thermodynamics. In analyzing the dynamics of atmospheric disturbances, we assume the air parcels to be in *local thermodynamic equilibrium*. By that we mean the thermodynamic properties of any air parcel change continuously in time from one state of thermodynamic equilibrium to another state of thermodynamic equilibrium. Other concepts of thermodynamics that we need to invoke in atmospheric analyses will be introduced in conjunction with the laws of thermodynamics in the following sections. Those laws provide the basis for determining how the thermodynamic variables of each air parcel might change in time.

1.6.1 Equation of state

Each type of fluid is characterized by a unique relationship among its thermodynamic variables. We may think of air in the troposphere as a mixture of "*ideal gases*" with a certain mean molecular weight. The molecules within such an air parcel individually move at great speeds and collide frequently with one another. When an ideal gas in a container is in thermodynamic equilibrium, it is experimentally known to have a unique relation among its thermodynamic variables: pressure p, temperature T and density ρ, viz.

$$p = R\rho T, \tag{1.39}$$

where $R = \dfrac{R^*}{M} = 287\,\mathrm{m}^2\,\mathrm{s}^{-2}\,\mathrm{K}^{-1}$ is the gas constant for dry air; R^* is called the universal gas constant and M is the mean molecular weight for the mixture of gases in air. Equation (1.39) is the *equation of state for dry air*. This relation still holds when the gas, such as the air in the atmosphere, is only in local thermodynamic equilibrium discussed earlier. We often use the same value of gas constant for both dry air and moist air in an analysis of atmospheric dynamics because the amount of water vapor is proportionally very small in an air parcel. It should be noted in passing that an incompressible fluid has no thermodynamic variables and we have no equation of state to deal with.

1.6.2 First Law of Thermodynamics

A fundamental principle of nature is that the total energy of a system and its surroundings is conserved: *conservation of energy*. The First Law of Thermodynamics is a mathematical statement of that principle. We make use of it explicitly in every dynamical analysis if we need to determine the change of temperature of the system under consideration. A key notion is that heat and mechanical work done by an air parcel onto its surroundings or vice versa are two inter-changeable forms of energy. The energy content in an air parcel would increase when it absorbs some heat from its environment, and would decrease when it loses some heat to the environment. The energy content of an air parcel would also decrease when it does mechanical work on its environment or increase when the environment does mechanical work upon it. The change of its temperature is determined by the sum of these two effects, i.e.

Increase in total energy of a system
= *Net heat added to the system* + *Net work done on the system*.

By definition
(*Work done by an air parcel on its environment*)
= – (*Work done on an air parcel by the environment*).

To mathematically formulate the First Law of Thermodynamics is a complex task when the energy of a parcel is defined to include all conceivable forms of energy. One form of energy of an air parcel is called *internal energy* which is proportional to the temperature because the temperature of an air parcel in thermodynamic equilibrium is a measure of the mean kinetic energy of its molecules.

The following notations are used:

$I = c_v T =$ internal energy of a unit mass of air parcel (for ideal gas).

$c_v =$ specific heat of air at constant volume, $716\,\mathrm{m}^2\,\mathrm{s}^{-2}\,\mathrm{K}^{-1}$.

$T =$ temperature of air in K.

$Q =$ net rate of heat gained by unit mass of air parcel in $\mathrm{J\,s}^{-1}(\mathrm{kg})^{-1}$. It is called the diabatic heating rate. It is a function of space and time in general.

$\alpha = \dfrac{1}{\rho} = \dfrac{RT}{p} =$ volume of unit mass of air parcel (specific volume).

For an electrically neutral moving air parcel, its total energy consists of internal energy, kinetic energy and gravitational potential energy since it is under the influence of gravity. An air parcel with $T = 300\,\mathrm{K}$ has internal energy equal to $I = 2.15 \times 10^5\,\mathrm{m^2\,s^{-2}}$. Let us ask the question: "What would be the velocity of an air parcel if its kinetic energy is equal to its internal energy?" The answer is $|\vec{v}| = \sqrt{2c_v T} = 655\,\mathrm{m\,s^{-1}}$. Such velocity is much faster than any wind speed in the part of atmosphere we are dealing with. In other words, the internal energy of an air parcel is typically much larger than its kinetic energy. In considering the gravitational potential energy of an air parcel, we may use the surface of the Earth, $z = 0$, as a reference level. Let us also ask the question, "At what height, h, would this air parcel be for its potential energy to be equal to its internal energy?" The answer is $h = \dfrac{c_v T}{g} = 2.15 \times 10^4\,\mathrm{m} = 21.5\,\mathrm{km}$. Weather disturbances typically extend up to about 12 km. Thus, the change in the potential energy of an air parcel in a weather disturbance may be as much as 50 percent of its internal energy. Therefore, the variation of potential energy of an air parcel in a disturbance should be taken into consideration in general.

Suppose we leave out the kinetic energy and the gravitational energy of an air parcel as parts of its total energy for the moment. Then the internal energy would be its total energy. The derivation of the First Law of Thermodynamics in this case is very simple.

Work done on unit mass of air parcel
= (pressure) × (rate of decrease in the volume of unit mass of air)
$= p\left(-\dfrac{D\alpha}{Dt}\right)$.

The mathematical statement of the principle of conservation of energy is then

$$\frac{D(c_v T)}{Dt} = Q - p\frac{D\alpha}{Dt}. \tag{1.40}$$

This is the *First Law of Thermodynamics*.

If we include kinetic energy and potential energy in the definition of its total energy, the expression for the work done on an air parcel should for consistency include additional terms that represent the work done by gravity as well as the work done by the pressure gradient force. The explicit derivation of the thermodynamic equation for a moving air parcel would have more technical details. But the additional terms related to the processes mentioned above necessarily cancel themselves out. So in the end, we would obtain exactly the same form as before. Equation (1.40) is therefore perfectly general. In using (1.40) in an analysis of a disturbance does not mean that we disregard kinetic energy and potential energy of air parcels altogether. We should examine them in conjunction with the equation of motion.

The thermodynamic equation can be written in several different forms. One of them is particularly instructive. Since

$$\left[\frac{D\alpha}{Dt} = \frac{D}{Dt}\left(\frac{RT}{p}\right) = \frac{R}{p}\frac{DT}{Dt} - \left(\frac{RT}{p^2}\right)\frac{Dp}{Dt}\right],$$

we may write (1.40) as

$$\left[\frac{(c_v + R)}{T}\frac{DT}{Dt} - \frac{R}{p}\frac{Dp}{Dt} = \frac{Q}{T}\right].$$

For diatomic gases such as oxygen and nitrogen, which account for more than 98 percent of the atmosphere by mass, we have $c_p = c_v + R$, specific heat at constant pressure. It follows that

$$\frac{D}{Dt}\left(\ln\left(\frac{T}{p^{R/c_p}}\right)\right) = \frac{Q}{c_p T}. \tag{1.41}$$

It is convenient to introduce a new dependent variable

$$\theta = T\left(\frac{p_{oo}}{p}\right)^{R/c_p}, \tag{1.42}$$

where p_{oo} can be any constant, normally taken to be 1000 mb. Then, the First Law of Thermodynamics can be written as

$$\frac{D\theta}{Dt} = \frac{\theta Q}{c_p T}. \tag{1.43}$$

A moving fluid which does not gain or lose heat ($Q = 0$) is said to be in *adiabatic motion*. Equation (1.43) says that the value of θ in an air parcel in adiabatic motion is invariant no matter how complicated the flow might be. No matter what temperature and pressure an air parcel starts out with, its temperature would be equal to its value θ when it adiabatically moves to a level of pressure equal to p_{oo}. That is why θ is called *potential temperature*.

The physical reason for the invariance of potential temperature of an air parcel as it moves around adiabatically is easy to appreciate. When an air parcel adiabatically goes up in the atmosphere, it naturally expands because its surrounding air has a lower pressure. This amounts to saying that the air parcel does some mechanical work on the environment. Consequently, its internal energy decreases as manifested by a drop in its temperature. Invariance of potential temperature merely reflects the fact that the decrease in its internal energy is exactly equal to the work that it has done. The reasoning applies equally well in reverse sense for a descending air parcel. Its potential temperature also remains unchanged in that case because the increase of its internal energy is exactly equal to the work that the environment does on it. The change in its temperature is completely reversible.

There are occasions when it would be meaningful to pose a dynamical problem for a system where the total energy only consists of kinetic energy and gravitational potential energy. One such example would be the flow in a layer of homogeneous fluid. In that case, the First Law of Thermodynamics would be irrelevant.

1.6.3 Second Law of Thermodynamics

Another law of thermodynamics is known as the *Second Law of Thermodynamics*. On the basis of observations, it asserts that the irreversible physical processes in nature are an indication of a unique direction in the change of time, the so-called *thermodynamic arrow of time*. It can be expressed in a number of different ways. The most general statement about this law is that every system in nature would evolve towards a more and more disordered state if there is no exchange of matter and energy across its boundary (i.e. an *isolated system*). A familiar manifestation of the Second Law of Thermodynamics is that when someone smokes a cigarette anywhere in a closed room, the smoke would irreversibly spread to the whole room. Another familiar manifestation is that ice cubes floating in a cup of warm water would eventually melt resulting from an inexorable flux of heat from the warm water to the colder ice. The degree of disorder in each case is measured with a concept called *entropy*, S. The entropy of a thermodynamic system may change due to two distinctly different categories of mechanisms: one external and one internal. The external mechanism is associated with an exchange of heat with the surrounding of the system. The internal mechanism stems from irreversible adiabatic process(es), viz.

$$\begin{aligned} \Delta S &= (\Delta S)_{ext} + (\Delta S)_{int}, \\ (\Delta S)_{ext} &= \frac{\Delta Q}{T}, \end{aligned} \tag{1.44}$$

where T is the temperature of the system when the exchange of heat, ΔQ, takes place. For example, if 1 kg of water is evaporated at 100 °C, the entropy of the water would increase by $(\Delta S)_{ext} = \frac{mL}{T} = \frac{2.52 \times 10^6}{373} = 0.67 \times 10^4\,\mathrm{J\,K^{-1}}$. An example of irreversible adiabatic change of entropy is the following. Suppose one mole of air is originally confined by a partition to one-half of a thermally insulated container. After the partition is removed, the air will eventually spread to the rest of the container. The entropy of the air will increase by $(\Delta S)_{int} = R \ln 2 = 199\,\mathrm{J\,K^{-1}\,kg^{-1}}$.

Observation reveals that there is a zonal-time average net radiative heating in the atmosphere–Earth system in the region between about 40° S and 40° N due to absorption of solar radiation and emission of terrestrial radiation to space. There is a compensating zonal-time average net radiative cooling in the regions poleward of 40 degree latitude of the two hemispheres. Since the temperature at high latitudes is much colder, the radiative exchange of energy with space tends to reduce the entropy of the atmosphere–Earth system. On the other hand, the global circulation constantly mixes colder air with warmer air, thereby tending to increase the entropy of the atmosphere–Earth system. The balance between the two tendencies would give rise to no net change of entropy in a particular climate. In other words, there exists a certain level of entropy in the atmosphere–Earth system in each climatic state. This is possible because the Earth is an *open system*, rather than an isolated system.

We note in passing that there is a *Third Law of Thermodynamics*. It is about the nature of matter with temperature approaching absolute zero degrees Kelvin. Since we never encounter such a low temperature in atmospheric studies, it is not of our concern.

1.7 Stratification and baroclinicity

In this section, we briefly touch upon two important aspects of the spatial distribution of potential temperature in the atmosphere: stratification and baroclinicity. From the definition $\theta = \frac{(p_{oo})^{\kappa} p^{1-\kappa}}{R} \frac{1}{\rho}$, we see that a larger value of θ corresponds to a smaller value of ρ at a

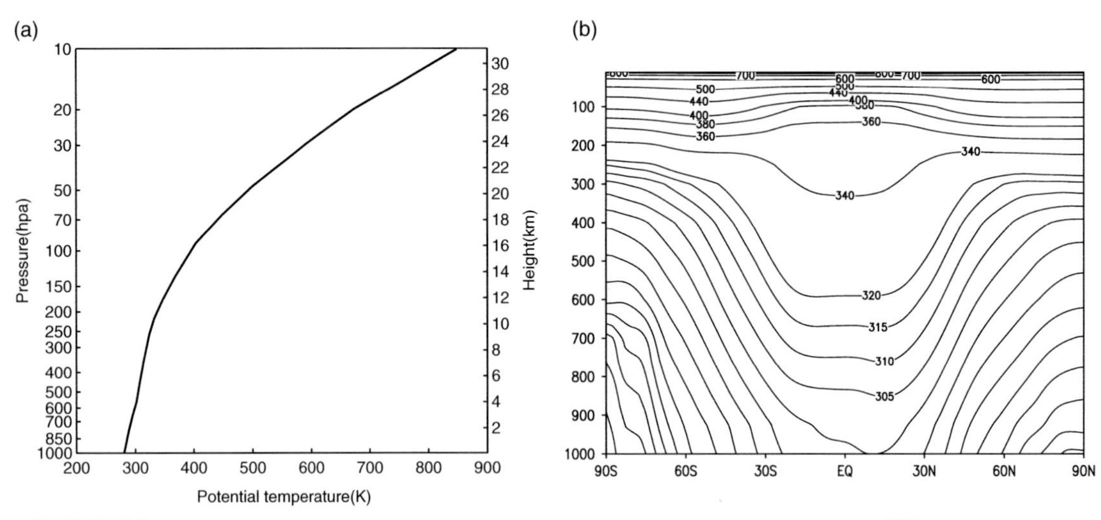

Fig. 1.11 (a) Vertical profile of the climatological annual-horizontal average potential temperature, $\overline{\langle\theta\rangle}$, and (b) distribution of the climatological annual-zonal average potential temperature, $\overline{[\theta]}$, on a latitude–pressure cross-section in K.

lower pressure level. Since pressure decreases exponentially with increasing elevation, density must also decrease exponentially with increasing elevation if θ increases with height. It follows that $\theta_z > 0$ means that the heavier air parcels are located at lower elevations. That would be a stable configuration of mass. The local and instantaneous value of θ_z may however be positive or negative, such as next to an asphalt road on a hot summer day or within a thick cloud in stormy weather. The quantity θ_z, or equivalently $(-\theta_p)$, is therefore a measure of *stratification* of air in the atmosphere. A layer of air with $\theta_z > 0$ is said to have a *stable stratification*, and a layer with $\theta_z < 0$ an *unstable stratification.* A mathematical analysis of this type of instability will be discussed in Part A of Chapter 8 for small disturbances. On the other hand, the potential temperature averaged over a large area is found to be higher at greater elevation. Thus the large-scale environment has a stable stratification, which is a stabilizing factor for a large-scale disturbance. This property has important ramifications for the dynamics of large-scale disturbances to be elaborated in other chapters.

We *define* a background state of the atmosphere with respect to a disturbance. We may think of a short-term background state or a long-term background state. The observed time and global average distribution of potential temperature in the atmosphere is denoted by $\overline{\langle\theta\rangle}$, with a bar referring to time averaging and an angular bracket referring to the horizontal averaging over the whole globe at each elevation. Figure 1.11a shows the distribution of the climatological $\overline{\langle\theta\rangle}$ as a function of pressure or height. (The dynamical reason behind the use of pressure as a vertical coordinate will be discussed in Chapter 2.) This result is obtained with the use of five years (2001 to 2005) of data taken from a dataset known as NCEP/NCAR (National Center for Environmental Prediction/National

Center for Atmospheric Research) Reanalysis Data. We see that $\overline{\langle\theta\rangle}$ increases with height at all levels in the troposphere. In other words, the global-time mean background state of the atmosphere does have a stable stratification at all levels. This characteristic arises from the fact that air is a compressible fluid under the strong gravitational influence of a massive planet. It is estimated from Fig. 1.11a that the average value of $\frac{d\overline{\langle\theta\rangle}}{dp}$ is about $8\,\mathrm{K}/100\,\mathrm{mb}$ in the troposphere corresponding to $\frac{d\overline{\langle\theta\rangle}}{dz} \sim 6\times10^{-3}\,\mathrm{m}^{-1}\,\mathrm{K}$. We will later see that a convenient measure of stratification is $N=\sqrt{\frac{g}{\theta_{oo}}\frac{d\overline{\langle\theta\rangle}}{dz}}$ called *Brunt–Vaisala frequency*, where g is gravity and θ_{oo} is a reference value. Its corresponding value is $N\sim10^{-2}\,\mathrm{s}^{-1}$. It is also seen in Fig. 1.11a that $\overline{\langle\theta\rangle}$ begins to increase sharply with height above about 12 km. That level is known as *tropopause* separating the *troposphere* below from the *stratosphere* above. In other words, the stratosphere has a much greater stratification than the troposphere.

The distribution of the time-zonal mean potential temperature, $[\bar{\theta}]$, is shown in Fig. 1.11b. It is found that $[\bar{\theta}]$ varies very little with latitude at each pressure level in the tropical region (between 30° S and 30° N), significantly decreases with increasing latitude between 30° and 55° latitude and varies by a smaller amount with latitude in the polar region. The variation of temperature with latitude is called *baroclinicity.* In other words, only the mid-latitudinal belt is a strong baroclinic zone in the troposphere. We estimate from Fig. 1.11b that $\left(-\frac{\partial[\bar{\theta}]}{\partial\varphi}\right)\sim 1\,\mathrm{K}$ per degree latitude in the mid-latitudes, corresponding to $\left(-\frac{\partial[\bar{\theta}]}{\partial y}\right)\sim10^{-5}\,\mathrm{K\,m}^{-1}$. Figure 1.11b also reveals that there is rather small variation in the stratification $\frac{\partial[\bar{\theta}]}{\partial z}$ with latitudes.

The dynamical significance of baroclinicity will be elaborated in Chapter 2 in the form of the dynamical relationship between the latitudinal variation of temperature and the vertical variation of wind. We will also elaborate in Chapter 8 the significance of baroclinicity as an important factor in dynamic instability of atmospheric flows.

1.8 Summary of the equations for a dry atmospheric model

For easy reference, we summarize here all the equations for a dry atmosphere. The fundamental characteristics of dry air are: gas constant, $R=287\,\mathrm{m^2\,s^{-2}\,K^{-1}}$, specific heats, $c_p=c_v+R=1003\,\mathrm{m^2\,s^{-2}\,K^{-1}}$, kinematic viscous coefficient, $\kappa=10^{-5}\,\mathrm{m^2\,s^{-1}}$. There are six dependent variables (u,v,w,T,p,ρ) in a dry atmosphere model. They are governed by the following equations in spherical coordinates:

$$\begin{aligned}
&\frac{Du}{Dt} - \frac{uv\tan\varphi}{r} + \frac{uw}{r} = -\frac{1}{\rho r\cos\varphi}\frac{\partial p}{\partial\lambda} + 2\Omega(v\sin\varphi - w\cos\varphi) + \kappa\nabla^2 u,\\
&\frac{Dv}{Dt} + \frac{u^2\tan\varphi}{r} + \frac{vw}{r} = -\frac{1}{\rho r}\frac{\partial p}{\partial\varphi} - 2\Omega u\sin\varphi + \kappa\nabla^2 v,\\
&\frac{Dw}{Dt} - \frac{u^2+v^2}{r} = -\frac{1}{\rho}\frac{\partial p}{\partial r} + 2\Omega u\cos\varphi - g + \kappa\nabla^2 w,\\
&\frac{D\rho}{Dt} + \rho\nabla\cdot\vec{v} = 0,\\
&c_v\frac{DT}{Dt} + p\frac{D}{Dt}\left(\frac{1}{\rho}\right) = Q,\\
&p = R\rho T,
\end{aligned} \tag{1.45}$$

where $\vec{v} \equiv (u, v, w) = \left(r\cos\frac{d\lambda}{dt}, r\frac{d\varphi}{dt}, \frac{dr}{dt}\right)$

$$\frac{D}{Dt} = \frac{\partial}{\partial t} + \vec{v}\cdot\nabla = \frac{\partial}{\partial t} + \frac{u}{r\cos\varphi}\frac{\partial}{\partial\lambda} + \frac{v}{r}\frac{\partial}{\partial\varphi} + w\frac{\partial}{\partial r},$$

$$\nabla^2 = \frac{1}{r^2}\frac{\partial}{\partial r}\left(r^2\frac{\partial}{\partial r}\right) + \frac{1}{r^2\cos\varphi}\frac{\partial}{\partial\varphi}\left(\cos\varphi\frac{\partial}{\partial\varphi}\right) + \frac{1}{r^2\cos^2\varphi}\frac{\partial^2}{\partial\lambda^2},$$

$$\nabla\cdot\vec{v} = \frac{1}{r\cos\varphi}\left(\frac{\partial u}{\partial\lambda} + \frac{\partial}{\partial\varphi}(v\cos\varphi)\right) + \frac{1}{r}\frac{\partial}{\partial r}(wr),$$

The model domain is $0 \leq \lambda \leq 2\pi,\ -\pi/2 \leq \varphi \leq \pi/2,\ r_o \leq r < \infty$ where $r_o(\lambda, \varphi)$ is the radial distance from the center of the Earth to its surface; $r_o(\lambda, \varphi)$ contains the information of the topography. To complete the formulation, we need:

i. To specify the distribution of the six dependent variables at a particular time, $t = 0$.
ii. To impose physically meaningful conditions at the boundaries (surface and "top" of atmosphere).
iii. To prescribe the diabatic heating, Q.

This set of equations subject to these auxiliary conditions can be numerically solved for a future state of the model atmosphere. Solutions of this model are capable of describing different classes of disturbances. It is however a daunting task to interpret such numerical solutions from a physical point of view. The discussions in the rest of the book would help make such a task more feasible.

2 Basic approximations and elementary flows

It would not have been possible to make progress in atmospheric sciences if researchers had not been innovative in introducing various types of approximations in their analyses. Approximations enabled them to investigate the dynamical nature of qualitatively different types of disturbances separately. Approximations are therefore both necessary and valuable. They must rest upon observational information. In this chapter, we will discuss some rudimentary approximations and their implications. We will also show how to transform the governing equations from using height as a vertical coordinate to using either pressure or potential temperature as a vertical coordinate. The dynamics of some simple balanced flows as well as the kinematic properties of a flow in general will be discussed.

2.1 Sphericity of the Earth and thin-atmosphere approximation

The most elementary approximations are related to the physical characteristics of the planet Earth. The Earth is a planet rapidly rotating about an axis through the north and south poles once every 24 hours. It also slowly rotates around the Sun once every 365 days. Its axis tilts at an angle of $\sim 23°$ relative to the plane of rotation about the Sun. The duration of solar radiation reaching each location on Earth therefore varies throughout a year. This gives rise to four seasons. The Earth is an ellipsoid. Its equatorial radius is only 7.1 km longer than its mean radius (6371 km) and its polar radius is about 14.2 km shorter. Most weather disturbances occur in a layer of atmosphere (known as troposphere) of which the thickness varies from about 9 km in high latitudes to about 12 km in low latitudes. About 90 percent of its mass exists in that layer. In other words, the Earth is almost a sphere and the atmosphere is a very thin layer of air. For the purpose of analyzing the dynamics of weather related disturbances, two approximations are clearly justifiable:-

(1) Sphericity approximation:
   Earth is treated as a sphere with a radius $a = 6370\,\text{km}$;
   gravity is treated as a constant, $g \approx 9.8\,\text{m}\,\text{s}^{-2}$.

(2) Thin atmosphere approximation:

Radial distance of an air parcel from the center of Earth is $r = a + z \approx a$ and $\partial/\partial r = \partial/\partial z$, where z is the elevation from the Earth's surface. It follows that we may use $\vec{V} = (u, v, w) = \left(a\cos\varphi\frac{d\lambda}{dt}, a\frac{d\varphi}{dt}, \frac{dz}{dt}\right)$, $\nabla\cdot\vec{V} = \frac{1}{a\cos\varphi}\left(\frac{\partial u}{\partial\lambda} + \frac{\partial(v\cos\varphi)}{\partial\varphi}\right) + \frac{\partial w}{\partial z}$, $\nabla^2 = \frac{1}{a^2\cos\varphi}\frac{\partial}{\partial\varphi}\left(\cos\varphi\frac{\partial}{\partial\varphi}\right) + \frac{1}{a^2\cos^2\varphi}\frac{\partial^2}{\partial\lambda^2} + \frac{\partial^2}{\partial z^2}$.

The momentum equations in a spherical-height coordinate system can be simplified to

$$\begin{aligned}
\frac{Du}{Dt}-\frac{uv\tan\varphi}{a}+\frac{uw}{a}&=-\frac{1}{\rho a\cos\varphi}\frac{\partial p}{\partial\lambda}+2\Omega(v\sin\varphi-w\cos\varphi)+\kappa\nabla^2u,\\
\frac{Dv}{Dt}+\frac{u^2\tan\varphi}{a}+\frac{vw}{a}&=-\frac{1}{\rho a}\frac{\partial p}{\partial\varphi}-2\Omega u\sin\varphi+\kappa\nabla^2v,\\
\frac{Dw}{Dt}-\frac{u^2+v^2}{a}&=-\frac{1}{\rho}\frac{\partial p}{\partial z}-g+2\Omega u\cos\varphi+\kappa\nabla^2w,\\
\text{where }\frac{D}{Dt}&=\frac{\partial}{\partial t}+\frac{u}{a\cos\varphi}\frac{\partial}{\partial\lambda}+\frac{v}{a}\frac{\partial}{\partial\varphi}+w\frac{\partial}{\partial z}.
\end{aligned}\tag{2.1}$$

2.2 Hydrostatic balance, implications and applications

There are days when there is virtually no wind in a region. The pressure and temperature are pretty much horizontally uniform, i.e. $(u,v,w,p,T,\rho)=(0,0,0,p_o(z),\ T_o(z),\rho_o(z))$. For such an atmospheric state to persist even for just a short time, there must be a balance between the vertical pressure gradient force and gravity at each air parcel,

$$0=-\frac{1}{\rho}\frac{\partial p}{\partial z}-g.\tag{2.2}$$

This is called *hydrostatic balance*.

We may always define a background state of the atmosphere characterized by hydrostatic balance with a density decreasing exponentially upward, $\rho_o=\rho_{oo}\exp(-z/H)$, where H is a constant, known as *scale height* and ρ_{oo} is the surface density of this background state. Denoting the corresponding pressure and temperature distributions by p_o and T_o, we get $p_o=p_{oo}+gH\rho_{oo}\left(e^{-z/H}-1\right)$ and $T_o=gH/R+(T_{oo}-gH/R)e^{z/H}$ by requiring them to satisfy both the equation of state and hydrostatic balance. Here p_{oo} and T_{oo} refer to the surface pressure and temperature of this state. Furthermore, we must require p_o to vanish as $z\rightarrow\infty$ and T_o to be finite so that they are physically relevant. It follows that we have $p_{oo}=gH\rho_{oo}$ and $T_{oo}=gH/R$. Then such a background state is necessarily isothermal $T_o=gH/R$ with $\rho_o=\rho_{oo}e^{-z/H}$ and $p_o=gH\rho_{oo}e^{-z/H}$.

Now consider the presence of a large-scale disturbance embedded in such a background state. The component of density associated with the disturbance is much smaller than that of the background state at the same level, $(\rho=\rho_o+\rho',\ |\rho'|\ll\rho_o)$. Likewise, we may decompose pressure as $p=p_o+p'$ with $|p'|\ll p_o$. By making use of the facts that $\frac{|\rho'|}{\rho_o}\ll 1$ and $0=-\frac{1}{\rho_o}\frac{\partial p_o}{\partial z}-g$, we can approximate the vertical momentum equation as

$$\cdots\approx-\frac{1}{\rho_o}\frac{\partial p'}{\partial z}-\frac{g\rho'}{\rho_o}+\frac{\rho'}{\rho_o^2}\frac{\partial p'}{\partial z}+\cdots\tag{2.3}$$

where $\cdots$ stands for the terms other than the pressure gradient force and gravity. The third term on the RHS is negligible compared to the other two terms since it is the product of two vanishingly small quantities. Furthermore, if the first two terms are more than one order of magnitude larger than all the remaining terms, then (2.3) may be approximated to

$$0 \approx -\frac{1}{\rho_o}\frac{\partial p'}{\partial z} - \frac{g\rho'}{\rho_o}. \tag{2.4}$$

Equation (2.4) says that hydrostatic balance would be also applicable to this class of disturbances per se. The balance is between the disturbance vertical pressure gradient force $\left(-\frac{1}{\rho_o}\frac{\partial p'}{\partial z}\right)$ and a *buoyancy force* $\left(-\frac{g\rho'}{\rho_o}\right)$. This is a very good approximation for large-scale disturbances like mid-latitude cyclones and anticyclones. It is even an acceptable approximation for considerably smaller disturbances such as hurricanes, but it is definitely not justifiable for smaller disturbances such as thunderstorms or individual clouds.

2.2.1 Implications and applications

Several deductions can be made from hydrostatic balance:

(i) Vertical acceleration

The vertical acceleration is negligibly small, but the vertical velocity needs not be zero and may even change in time. That means that the vertical velocity field instantaneously adjusts from one hydrostatically balanced state to another such state.

(ii) Surface pressure

By integrating (2.2) from the surface to the top of atmosphere, we get

$$p_s = g\int_0^\infty \rho \, dz. \tag{2.5}$$

Equation (2.5) says that surface pressure at a point is equal to the weight of the air molecules in the whole atmospheric column of unit cross-section over that point. Indeed, the pressure at any elevation is equal to the weight of air molecules in a column above that level. It follows that pressure is higher at a lower elevation. This is the underlying reason as to why you would feel a pressure change in an elevator when it rapidly moves in a very tall building.

(iii) Surface pressure tendency equation

The time derivative of (2.5) tells us how the change in surface pressure would come about,

$$\begin{aligned}\frac{\partial p_s}{\partial t} &= g\int_0^\infty \frac{\partial \rho}{\partial t}dz = -g\int_0^\infty \nabla\cdot\left(\rho\vec{V}\right)dz = -g\int_0^\infty \frac{\partial(\rho w)}{\partial z}dz - g\int_0^\infty \nabla_2\cdot\left(\rho\vec{V}_2\right)dz \\ &= g\rho_s w_s - g\int_0^\infty \nabla_2\cdot\left(\rho\vec{V}_2\right)dz \approx -g\int_0^\infty \nabla_2\cdot\left(\rho\vec{V}_2\right)dz,\end{aligned} \tag{2.6}$$

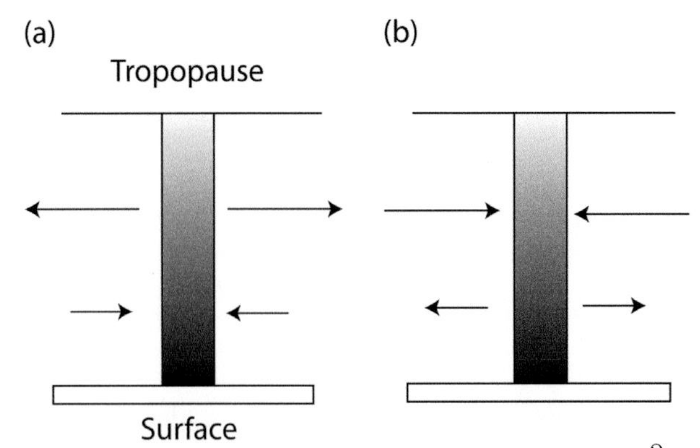

Fig. 2.1 (a) Divergence in an upper layer larger than convergence in a lower layer, resulting in $\frac{\partial p_s}{\partial t} < 0$, (b) vice versa.

where $\vec{V}_2$ refers to the horizontal velocity. We have imposed zero vertical velocity at the surface, $w_s = 0$, and $\rho \to 0$ as $z \to \infty$. It says that the surface pressure at a location increases at a rate equal to the total horizontal mass flux convergence in the atmospheric column above it. Note that we typically observe $p_s \sim 10^6$ Pa m^{-2} and $\nabla_2 \cdot \vec{V}_2 \sim 10^{-6}\,\mathrm{s}^{-1}$. If the horizontal divergence in a large-scale disturbance had the same sign at all levels, we would get

$$-g\int_0^{\infty} \nabla_2 \cdot (\rho \vec{V}_2) dz \approx \int_o^{ps} \nabla_2 \cdot \vec{V}_2\, dp$$
$$= p_s \nabla_2 \cdot \vec{V}_2 \sim 1\,\mathrm{Pa\,m^{-2}\,s^{-1}} = 10^5\,\mathrm{Pa\,m^{-2}\,day^{-1}} = 100\,\mathrm{mb\,day^{-1}}.$$

The observed surface pressure tendency is never so big. It means that the sign of $\nabla_2 \cdot \vec{V}_2$ could not be the same at all levels in the troposphere. In large-scale disturbances, a layer of convergence of $|\nabla \cdot \vec{V}| \sim 10^{-6}\,\mathrm{s}^{-1}$ typically overlies a layer of divergence of comparable magnitude and vice versa. The column average divergence is consequently an order of magnitude smaller than the convergence or divergence itself at an individual level. This is schematically illustrated in Fig. 2.1.

(iv) Thickness of an atmospheric layer in hydrostatic balance is proportional to its average temperature.

Integrating (2.2) across a layer of atmosphere of thickness Δz between $(z_1 - \Delta z)$ and z_1 where the pressures are $p_1 + \Delta p$ and p_1 respectively, we get

$$\int_{p1}^{p1+\Delta p} \frac{dp}{p} = -\int_{z1}^{z1-\Delta z} g\frac{1}{RT}\, dz$$

$$\left(\ln\left(\frac{p_1 + \Delta p}{p_1}\right)\right) = \frac{g}{R}\overline{\left(\frac{1}{T}\right)}\Delta z \approx \frac{g}{R\bar{T}}\Delta z \tag{2.7}$$

$$\bar{T} = \frac{g}{R\ln\left(\frac{p_1+\Delta p}{p_1}\right)}\Delta z.$$

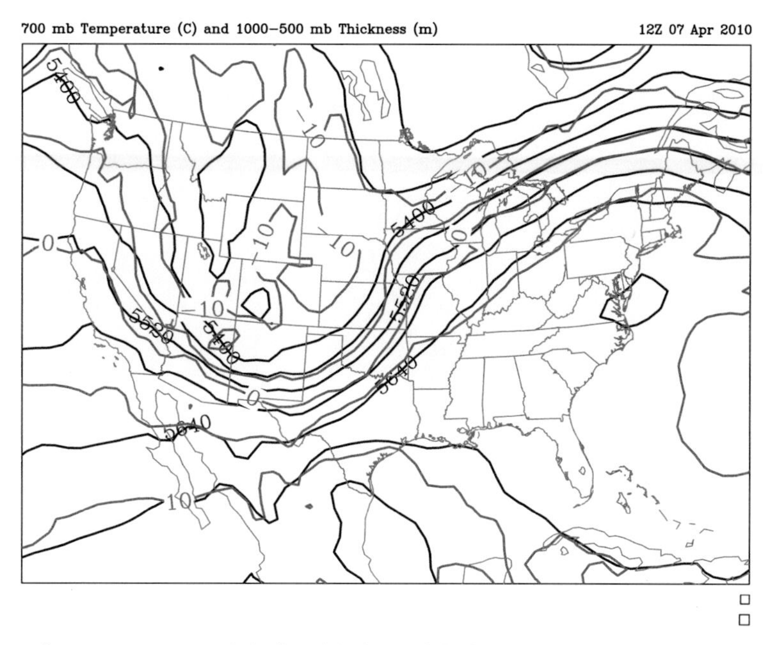

Fig. 2.2 Distributions of temperature at 700 mb (red) and thickness of the layer between 500 mb and 1000 mb (black) on April 7, 2010. See color plates section.

The bar denotes the average value of a quantity in a layer. This relationship is useful for weather forecasting, especially for the thickness of the layer between the 500-mb and 1000-mb surfaces, $\Delta Z = Z_{500-1000}$. A map of ΔZ is called a *thickness map*. The spacing between two adjacent contours of a thickness map is a visual indication of the spatial gradient of the mean temperature in the lower half of the troposphere. The smaller such spacing is, the stronger the thermal gradient would be. Figure 2.2 shows that the contours of the 700 mb mean temperature and those of the 500–1000 mb thickness on April 7, 2010 are indeed virtually parallel to one another.

(v) Sea level pressure differential over continent versus over oceans

It is well known observationally that the surface pressure over a large continent such as N. America or Asia is generally high in winter and low in summer. The reverse is true over the adjacent N. Pacific and N. Atlantic oceans. Before we can meaningfully compare the pressure distribution over continents with that over oceans, we first need to take into account the dependency of surface pressure on the elevation of the surface. We do so by deducing a corresponding sea level pressure (SLP) from a surface pressure at each location. We add to the surface pressure the weight of a hypothetical air column underneath each location that extends to a reference sea level. A global SLP map can be thereby constructed.

Figure 2.3a shows that the winter (DJF) average SLP is as high as 1035 mb over central Asia and 1020 mb over central N. America, whereas that over the central part of Pacific and Atlantic is below 1005 mb. In contrast, Fig. 2.3b shows that the summer (JJA) average SLP

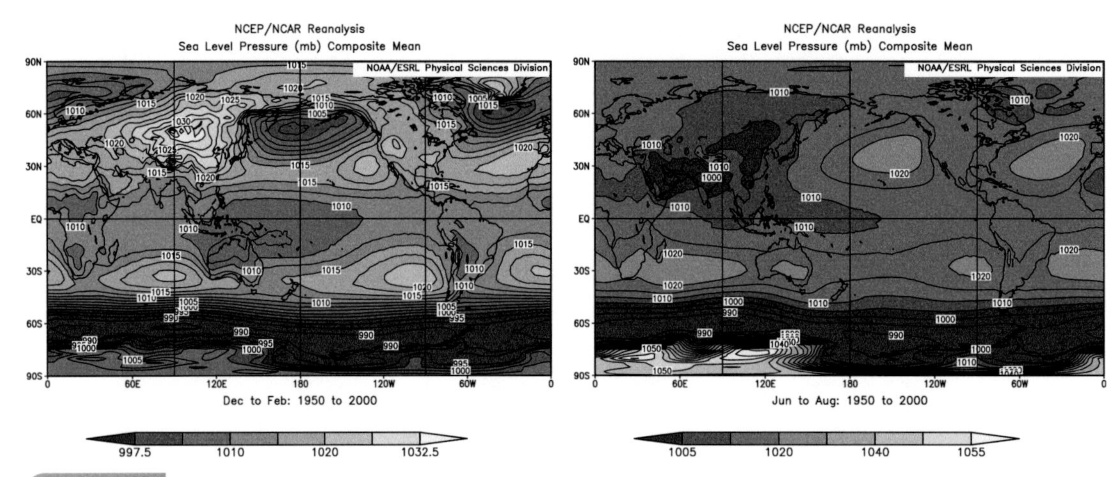

Fig. 2.3 Climatological mean sea level pressure over the globe in (a) winter (DJF) and (b) summer (JJA) in mb deduced from 50 years of NCEP/NCAR Reanalysis data.

is as low as 1003 mb over Asia and about 1012 mb over N. America, with the corresponding values over the N. Pacific and Atlantic oceans reaching 1025 mb in the subtropical latitudes. In other words, there is a pronounced differential in the seasonal average SLP over continent versus ocean. This is a rudimentary feature of the global monsoon phenomenon. A similar pressure differential over land versus ocean is observed in the southern hemisphere (SH), albeit in much smaller magnitude because the continental masses in the southern hemisphere are considerably smaller. For example, the SLP over the Indian Ocean and South Pacific in DJF (the summer season of SH) is noticeably lower than that over Africa, Australia and South America. In JJA, the area of maximum SLP moves over to the land masses. An elementary account for this broad feature can be given on the basis of hydrostatic balance.

Monsoon is a multifaceted complex phenomenon associated with the continent–ocean contrast ultimately due to the seasonal variation of the solar heating. The SLP distribution is naturally an integral feature of monsoon. Without going into the details of the monsoonal circulation, we can gain some appreciation of its rudimentary feature with the notion of hydrostatic balance and qualitative consideration of the net diabatic heating/cooling. Let us visualize what happens in the atmosphere over a continent and its adjacent oceans as a season begins to change. By inference, there is practically no SLP differential between a continent and ocean in a transition season, say the autumn (SON). As the winter season begins in a hemisphere, the solar radiation reaching the Earth's surface per unit area substantially diminishes day by day. The Earth nevertheless continues to emit longwave radiation to space nearly at the same pace. This would give rise to a net deficit in the surface radiation budget. The air just above the surface would therefore begin to drop in response. But since an ocean is an enormous heat reservoir, its surface temperature would not decrease nearly as much as the land surface temperature. Consequently, the air temperature over the land surface would drop faster and become greater than that over the ocean surface. Since an air parcel shrinks when its temperature drops, we would expect a greater reduction of the thickness of a shallow surface layer over land than over ocean. As this layer shrinks, less mass of air would remain in the column above a particular level. Recall that under hydrostatic balance, the pressure at a level is equal to the weight of all the air parcels above that level. Thus, the pressure at higher levels in winter would be lower over a continent than over an ocean. The corresponding horizontal pressure differential would induce a flow of air from ocean to land. The convergence of air towards an atmospheric column over land would in turn increase the SLP over land. There should be at the same time a divergent outflow from the land to the ocean near the surface because of a surface pressure differential in the opposite direction. However, under the influence of surface friction, the surface divergent outflow would be weaker than the convergent flow aloft. The net result would be a net build-up of mass of air in an atmospheric column and hence a net increase in the surface SLP. This process would continue throughout the first half of the winter season leading to a progressive build-up of the surface SLP over a continent. This is the fundamental reason why we observe distinctly higher mean SLP over a continent than over the adjacent oceans in winter. The larger a continent is, the greater would be such surface pressure differential as observed over Asia and N. America.

In summer, all processes mentioned above occur in a reverse sense. The qualitative account for the observed lower SLP over continents than over oceans in summer is the same. But, there is an additional complication in summer. The expected inflow towards a continent from its nearby ocean surface brings in a lot of water vapor. Such an influx of water vapor eventually precipitates somewhere over the continent. There is consequently an important diabatic heating as an integral part of the summer monsoon. The copious monsoonal precipitation greatly complicates the dynamics of the circulation. Nevertheless, the surface pressure differential between continent and ocean may be expected even without explicitly taking into account the moist dynamics.

2.2.2 Isobaric coordinates and governing equations

Since pressure monotonically decreases upward in the atmosphere as long as hydrostatic balance is applicable, there is one-to-one correspondence between the pressure at a point and the elevation of that point. Pressure therefore can be used as a vertical coordinate in an analysis. It is known as the *isobaric coordinate*. This coordinate is used in all operational weather forecast models. The advantage in using it is that meteorological instruments are designed to make measurements of weather data at a number of preset pressure values, e.g. 950, 900, 850, 700, 600, 500, . . . , etc. (mb). Such data can be directly used in a forecast model using an isobaric coordinate. This avoids the errors one would inevitably introduce if data at pressure levels are interpolated to height levels in a model using a height coordinate.

In trying to use pressure as a vertical coordinate in a model analysis, we need to transform all governing equations from height coordinates to pressure coordinates. Hydrostatic balance is invoked at the outset. Let us use local Cartesian coordinates as the horizontal spatial variables. We denote the height coordinates by $(\tilde{x}, \tilde{y}, \tilde{z}, \tilde{t})$ and the pressure coordinates by (x, y, p, t). The transformation relations are: $x = \tilde{x}$, $y = \tilde{y}$, $p = \tilde{p}(\tilde{x}, \tilde{y}, \tilde{z}, \tilde{t})$, $t = \tilde{t}$. The height of pressure surfaces, $z(x, y, p, t)$, is a dependent variable in an isobaric coordinate system. It is the counterpart of pressure at a point, $\tilde{p}(\tilde{x},\tilde{y},\tilde{z},\tilde{t})$, as a dependent variable in a height coordinate system. The velocity components in the $(\tilde{x},\tilde{y},\tilde{z},\tilde{t})$ coordinates are denoted by $(\tilde{u},\tilde{v},\tilde{w})$. The consistent notation for the horizontal velocity components in (x,y,p,t) coordinates is (u,v). The transformation relationships for the derivatives are

$$\frac{\partial}{\partial\tilde{x}}=\frac{\partial}{\partial x}+\frac{\partial\tilde{p}}{\partial\tilde{x}}\frac{\partial}{\partial p},\ \frac{\partial}{\partial\tilde{y}}=\frac{\partial}{\partial y}+\frac{\partial\tilde{p}}{\partial\tilde{y}}\frac{\partial}{\partial p},\ \frac{\partial}{\partial\tilde{z}}=\frac{\partial\tilde{p}}{\partial\tilde{z}}\frac{\partial}{\partial p},$$
$$\frac{\partial}{\partial\tilde{t}}=\frac{\partial}{\partial t}+\frac{\partial\tilde{p}}{\partial\tilde{t}}\frac{\partial}{\partial p}.$$

Recall that the total derivative in height coordinates is $\frac{D}{Dt}=\frac{\partial}{\partial\tilde{t}}+\tilde{u}\frac{\partial}{\partial\tilde{x}}+\tilde{v}\frac{\partial}{\partial\tilde{y}}+\tilde{w}\frac{\partial}{\partial\tilde{z}}$. It follows that the total derivative is transformed as

$$\frac{D}{Dt}=\frac{\partial}{\partial\tilde{t}}+\tilde{u}\frac{\partial}{\partial\tilde{x}}+\tilde{v}\frac{\partial}{\partial\tilde{y}}+\tilde{w}\frac{\partial}{\partial\tilde{z}}=\frac{\partial}{\partial t}+u\frac{\partial}{\partial x}+v\frac{\partial}{\partial y}+\left(\frac{\partial\tilde{p}}{\partial\tilde{t}}+\tilde{u}\frac{\partial\tilde{p}}{\partial\tilde{x}}+\tilde{v}\frac{\partial\tilde{p}}{\partial\tilde{y}}+\tilde{w}\frac{\partial\tilde{p}}{\partial\tilde{z}}\right)\frac{\partial}{\partial p}.$$

We now introduce the notation $\omega\equiv Dp/Dt=D\tilde{p}/D\tilde{t}$ in (x,y,p) coordinates for the "p-velocity" of a fluid parcel in the direction of increasing pressure. The p-velocity is in the vertical direction and is in downward motion when it has a positive value. Therefore, according to the relation obtained above, we may write the total derivative in (x,y,p) coordinates as $\frac{D}{Dt}=\frac{\partial}{\partial t}+u\frac{\partial}{\partial x}+v\frac{\partial}{\partial y}+\omega\frac{\partial}{\partial p}$.

By using $\alpha=\frac{RT}{p}$, we can readily rewrite the thermodynamic equation as

$$c_v\frac{DT}{Dt}+p\frac{D\alpha}{Dt}=Q\ \rightarrow\ c_p\frac{DT}{Dt}-\alpha\omega=Q\rightarrow$$
$$\frac{\partial T}{\partial t}+u\frac{\partial T}{\partial x}+v\frac{\partial T}{\partial y}-\frac{pS}{R}\omega=\frac{Q}{c_p}, \tag{2.8}$$

where $S=\frac{R}{p}\left(-\frac{\partial T}{\partial p}+\frac{\alpha}{c_p}\right)=\frac{R}{gp\rho}\left(\frac{g}{c_p}+\frac{\partial T}{\partial z}\right)\equiv\frac{R(\Gamma_d-\Gamma)}{gp\rho}$. Here Γ_d is known as the dry adiabatic lapse rate with a value $\Gamma_d=\frac{g}{c_p}\approx 10\,\mathrm{K\,km^{-1}}$; $\Gamma=-\partial T/\partial z$ is called the lapse rate typically $\approx 6\,\mathrm{K\,km^{-1}}$; S is a measure of the stratification with a value of $\sim 0.5\times10^{-5}\mathrm{m^2\,s^{-2}\,Pa^{-2}}$ in mid-troposphere, and typically increases with height.

To transform the pressure gradient force (PGF) from $(\tilde{x},\tilde{y},\tilde{z},\tilde{t})$ coordinates to (x,y,p,t) coordinates, we make use of the inverse transformation relationships for the derivatives, namely $\frac{\partial}{\partial x}=\frac{\partial}{\partial\tilde{x}}+\frac{\partial\tilde{z}}{\partial x}\frac{\partial}{\partial\tilde{z}}$ and $\frac{\partial}{\partial y}=\frac{\partial}{\partial\tilde{y}}+\frac{\partial\tilde{z}}{\partial y}\frac{\partial}{\partial\tilde{z}}$. Taking the derivative of p with respect to x, we get $\frac{\partial p}{\partial x}=\frac{\partial\tilde{p}}{\partial\tilde{x}}+\frac{\partial z}{\partial x}\frac{\partial\tilde{p}}{\partial\tilde{z}}$. By definition, we have $\frac{\partial p}{\partial x}=0$. Since we invoke hydrostatic balance as a fundamental assumption in using the isobaric coordinates, we have $\frac{\partial\tilde{p}}{\partial\tilde{z}}=-g\rho$, and therefore obtain

$$-\frac{1}{\rho}\frac{\partial\tilde{p}}{\partial\tilde{x}}=-g\frac{\partial z}{\partial x}. \tag{2.9}$$

The notation $\Phi = gz$ is often used in conjunction with the isobaric coordinates. It is called geopotential height. Thus, the horizontal PGF in p-coordinates is $-\nabla_{(2)}\Phi \equiv -\vec{i}\Phi_x - \vec{j}\Phi_y$ where $(\vec{i}, \vec{j}\,)$ are the unit vectors. The operator $\nabla_{(2)}$ is a 2-D grad operator performing on an isobaric surface. Strictly speaking, we should define $\Phi = g_o\xi$ where $\xi = \frac{1}{g_o}\int_0^z g\,dz$ is geopotential height. Since we use $g \approx g_o = 9.8\ \mathrm{m\,s^{-2}}$ as an elementary approximation, we assume $\xi = z$ in meteorological applications. Furthermore, we may use $\omega = \frac{dp}{dt} = \frac{\partial p}{\partial t} + u\frac{\partial p}{\partial x} + v\frac{\partial p}{\partial y} + w\frac{\partial p}{\partial z} \approx w\frac{\partial p}{\partial z} \approx -g\rho w$ for a large-scale flow. Hence, the horizontal momentum equations can be then written as

$$\frac{D\vec{V}_2}{Dt} = -\nabla_{(2)}\Phi - 2\vec{\Omega} \times \vec{V} + \kappa\nabla^2\vec{V}_2, \tag{2.10a,b}$$

where $\vec{V}_2 = (u, v)$, $\vec{V} = \left(u, v, -\frac{\omega}{g\rho}\right)$. The vertical momentum equation is of course just the hydrostatic balance

$$\frac{\partial\Phi}{\partial p} = -\frac{RT}{p}. \tag{2.11}$$

The equation expressing mass conservation can be written as

$$\frac{D}{Dt}(\rho\,\delta x\,\delta y\,\delta z) = \frac{D}{Dt}\left(-\frac{\delta x\,\delta y\,\delta p}{g}\right) = 0,$$

$$\frac{1}{\delta x}\frac{D}{Dt}(\delta x) + \frac{1}{\delta y}\frac{D}{Dt}(\delta y) + \frac{1}{\delta p}\frac{D}{Dt}(\delta p) = 0,$$

$$\frac{\delta u}{\delta x} + \frac{\delta v}{\delta y} + \frac{\delta\omega}{\delta p} = 0.$$

Hence, the continuity equation is

$$u_x + v_y + \omega_p = 0. \tag{2.12}$$

The form of the equation of state is unchanged, viz.

$$p = R\rho T. \tag{2.13}$$

In summary, the six dependent variables in the (x, y, p) system are $(u, v, \omega, \Phi, T, \rho)$. They are governed by Eqs. (2.8), (2.10a,b), (2.11), (2.12) and (2.13).

We may integrate (2.12) and get (using $\omega = 0$ at $p = 0$ as a boundary condition)

$$\int_0^\omega d\omega = -\int_0^p (u_x + v_y)dp, \text{ and hence}$$

$$\omega = -\int_0^p (u_x + v_y)dp. \tag{2.14}$$

This equation says that the value of ω at a pressure level is equal to the integrated horizontal convergence in the atmospheric column above that pressure level. However, we do not normally estimate ω by applying this equation because the typical 10 percent error in the wind data would accumulatively give rise to a 100 percent error in its horizontal divergence. We will discuss a more reliable method of determining ω in Chapter 6.

2.2.3 Primitive equations

When one replaces the vertical momentum equation by the hydrostatic balance, the uw/a and $2\Omega w\cos\varphi$ terms in the u-momentum equation and the vw/a term in the v-momentum equation with the use of spherical-height coordinates should be also dropped. It suffices to say that this is done for self-consistency from a consideration of the energetics of a flow. The resulting set of governing equations is known as *primitive equations*. Those equations written in spherical-isobaric coordinates are

$$
\begin{aligned}
\frac{Du}{Dt} - \frac{uv\tan\varphi}{a} &= -\frac{1}{a\cos\varphi}\frac{\partial\Phi}{\partial\lambda} + fv + F, \\
\frac{Dv}{Dt} + \frac{u^2\tan\varphi}{a} &= -\frac{1}{a}\frac{\partial\Phi}{\partial\varphi} - fu + G, \\
0 &= -\frac{\partial\Phi}{\partial p} - R_{(p)}\theta, \\
\frac{1}{a\cos\varphi}\frac{\partial u}{\partial\lambda} + \frac{1}{a\cos\varphi}\frac{\partial(v\cos\varphi)}{\partial\varphi} + \frac{\partial\omega}{\partial p} &= 0, \\
\frac{D\theta}{Dt} = \frac{\theta\hat{Q}}{c_pT} &\equiv Q,
\end{aligned}
\tag{2.15}
$$

where $\frac{D}{Dt} = \frac{\partial}{\partial t} + \frac{u}{a\cos\varphi}\frac{\partial}{\partial\lambda} + \frac{v}{a}\frac{\partial}{\partial\varphi} + \omega\frac{\partial}{\partial p}$, a is the mean radius of the Earth, $\Phi = gz$, z being the height of pressure surfaces, $\theta = T\left(\frac{p_{oo}}{p}\right)^{R/c_p}$, $R_{(p)} = \frac{R}{p_{oo}}\left(\frac{p_{oo}}{p}\right)^{c_v/c_p}$, $f = 2\Omega\sin\varphi$, F and G the components of the horizontal frictional force, $\hat{Q}$ diabatic heating rate in $\mathrm{J\,kg^{-1}\,s^{-1}}$, and Q diabatic heating rate in $\mathrm{K\,s^{-1}}$.

2.2.4 Isentropic coordinates and governing equations

We have mentioned in Section 1.8 that potential temperature θ increases monotonically with height in a stably stratified fluid as in a large-scale atmospheric layer. That also makes θ another acceptable vertical coordinate. It can be convenient for making deductions in theoretical analyses. The explicit form of the total derivative is now $\frac{D}{Dt} = \frac{\partial}{\partial t} + u\frac{\partial}{\partial x} + v\frac{\partial}{\partial y} + \dot{\theta}\frac{\partial}{\partial\theta}$. $\dot{\theta} = \frac{d\theta}{dt}$ is the "θ-velocity" which is a measure of vertical motion in a (x, y, θ) coordinate system. It is proportional to the net diabatic heating rate at a point according to the thermodynamic equation. The height of isentropic surfaces $z(x, y, \theta, t)$ is a dependent variable.

To transform the pressure gradient force (PGF) from $(\tilde{x},\tilde{y},\tilde{z},\tilde{t})$ coordinates to (x,y,θ,t) coordinates, we make use of the inverse transformation relationships for the derivatives, namely $\frac{\partial}{\partial x}=\frac{\partial}{\partial\tilde{x}}+\frac{\partial\tilde{z}}{\partial x}\frac{\partial}{\partial\tilde{z}}$ and $\frac{\partial}{\partial y}=\frac{\partial}{\partial\tilde{y}}+\frac{\partial\tilde{z}}{\partial y}\frac{\partial}{\partial\tilde{z}}$. Taking the derivative of p with respect to x, we get $\frac{\partial p}{\partial x}=\frac{\partial\tilde{p}}{\partial\tilde{x}}+\frac{\partial\tilde{z}}{\partial x}\frac{\partial\tilde{p}}{\partial\tilde{z}}$. By hydrostatic balance, we have $\frac{\partial\tilde{p}}{\partial\tilde{z}}=-g\rho$, and therefore obtain $\frac{\partial\tilde{p}}{\partial\tilde{x}}|_z=\frac{\partial p}{\partial x}|_\theta+g\rho\frac{\partial z}{\partial x}|_\theta$. Hence, the PGF in the x-direction is

$$-\frac{1}{\rho}\left(\frac{\partial\tilde{p}}{\partial\tilde{x}}\right)_z=-g\left(\frac{\partial z}{\partial x}\right)_\theta-\frac{1}{\rho}\left(\frac{\partial p}{\partial x}\right)_\theta. \tag{2.16}$$

From the definition $\theta=T\left(\frac{p_{oo}}{p}\right)^\kappa$, we have $0=\frac{1}{T}\left(\frac{\partial T}{\partial x}\right)_\theta-\frac{\kappa}{p}\left(\frac{\partial p}{\partial x}\right)_\theta$. Thus,

$$\begin{aligned}-\frac{1}{\rho}\left(\frac{\partial\tilde{p}}{\partial\tilde{x}}\right)_z&=-g\left(\frac{\partial z}{\partial x}\right)_\theta-c_p\left(\frac{\partial T}{\partial x}\right)_\theta\equiv-\left(\frac{\partial M}{\partial x}\right)_\theta,\\ M&=c_pT+gz.\end{aligned} \tag{2.17}$$

This is called the *Montgomery streamfunction,* which is the counterpart of the geopotential height Φ in an isobaric coordinate. Similarly, we have $-\frac{1}{\rho}\left(\frac{\partial p}{\partial y}\right)_z=-\left(\frac{\partial M}{\partial y}\right)_\theta$. The horizontal PGF is then $-\frac{1}{\rho}\nabla_{(2)}p=-\nabla_{(2)\theta}M$ where $\nabla_{(2)}$ is the horizontal gradient operator at constant height and $\nabla_{(2)\theta}$ is the horizontal gradient operator at fixed θ. The horizontal momentum equations in isentropic coordinates then take on the form of

$$\begin{aligned}\frac{Du}{Dt}&=-\frac{\partial M}{\partial x}+fv+\kappa\nabla^2u,\\ \frac{Dv}{Dt}&=-\frac{\partial M}{\partial y}-fu+\kappa\nabla^2v.\end{aligned} \tag{2.18a, b}$$

Hydrostatic balance can be written in terms of $\partial M/\partial\theta$ as well. We have $\frac{\partial M}{\partial\theta}=g\frac{\partial z}{\partial\theta}+c_p\frac{\partial T}{\partial\theta}=g\frac{\partial z}{\partial p}\frac{\partial p}{\partial\theta}+c_p\frac{\partial T}{\partial\theta}=-\frac{1}{\rho}\frac{\partial p}{\partial\theta}+c_p\frac{\partial T}{\partial\theta}$. From the definition $\theta=T\left(\frac{p_{oo}}{p}\right)^\kappa$, we get $1=\frac{\partial T}{\partial\theta}\left(\frac{p_{oo}}{p}\right)^\kappa+T(-\kappa)\left(\frac{p_{oo}}{p}\right)^\kappa\left(\frac{1}{p}\right)\frac{\partial p}{\partial\theta}\rightarrow c_p\left(\frac{p}{p_{oo}}\right)^\kappa=c_p\frac{\partial T}{\partial\theta}-\left(\frac{1}{\rho}\right)\frac{\partial p}{\partial\theta}$. Hence, we get

$$\frac{\partial M}{\partial\theta}=c_p\left(\frac{p}{p_*}\right)^\kappa. \tag{2.19}$$

The continuity equation in isentropic coordinates can be deduced as follows. Suppose an elemental volume of air has dimensions $(\delta x,\delta y,\delta z)$ in height coordinates, $(\delta x,\delta y,\delta p)$ in isobaric coordinates and $(\delta x,\delta y,\delta\theta)$ in isentropic coordinates. They are all interrelated subject to hydrostatic balance,

$$\delta m = \rho\, \delta x\, \delta y\, \delta z = -\frac{1}{g}\delta x\, \delta y\, \delta p = \sigma\, \delta x\, \delta y\, \delta\theta.$$
$$\text{Hence } \sigma = -\frac{1}{g}\frac{\partial p}{\partial \theta}, \tag{2.20}$$

where σ therefore stands for mass per unit "volume" in isentropic coordinates. It is the equivalence of "density" although its dimension is $\mathrm{kg\,K^{-1}\,m^{-2}}$ rather than $\mathrm{kg\,m^{-3}}$. It follows that conservation of mass means

$$\frac{D}{Dt}(\sigma\, \delta x\, \delta y\, \delta\theta) = \frac{1}{\sigma}\frac{D}{Dt}\sigma + \frac{1}{\delta x}\frac{D}{Dt}\delta x + \frac{1}{\delta y}\frac{D}{Dt}\delta y + \frac{1}{\delta\theta}\frac{D}{Dt}\delta\theta = 0,$$
$$\frac{D\sigma}{Dt} + \sigma\left(\left(\frac{\partial u}{\partial x}\right) + \left(\frac{\partial v}{\partial y}\right) + \left(\frac{\partial \dot{\theta}}{\partial \theta}\right)\right) = 0.$$

An alternative form of the continuity equation in isentropic coordinates is

$$\frac{\partial \sigma}{\partial t} + \left(\frac{\partial (u\sigma)}{\partial x}\right) + \left(\frac{\partial (v\sigma)}{\partial y}\right) + \left(\frac{\partial (\dot{\theta}\sigma)}{\partial \theta}\right) = 0. \tag{2.21}$$

The thermodynamic equation is simply

$$\dot{\theta} = \frac{1}{c_p}\left(\frac{p_{oo}}{p}\right)^{\kappa} Q, \tag{2.22}$$

where Q/c_p is the net diabatic heating rate in $\mathrm{K\,s^{-1}}$. In summary, the six dependent variables in isentropic coordinates are $(u, v, \dot{\theta}, M, p, \sigma)$. They are governed by (2.18a,b), (2.19), (2.20), (2.21) and (2.22). The equation of state is

$$p = R\rho\theta\left(\frac{p}{p_{oo}}\right)^{\kappa}. \tag{2.23}$$

We would need to use it only if we want to calculate ρ.

2.3 Geostrophic balance

The wind vectors on a typical weather map at a pressure level are approximately parallel to the height contours on that pressure level, $z(x, y, p, t)$. This is illustrated by the map for 300 mb on April 7, 2010 (Fig. 2.4). This feature is a manifestation of a particularly simple balance of horizontal forces on air parcels. This is most valid in regions where the wind varies little in its direction. The values of $z(x, y, p, t)$ are smaller on the left side than on the right side relative to the stream-wise direction in the northern hemisphere (NH). The pressure gradient force therefore is directed toward the left of the wind. The Coriolis force associated with this wind is directed to the right. These two forces are approximately in balance because they are much larger than all other terms in the horizontal momentum equations. This kind of balance is called

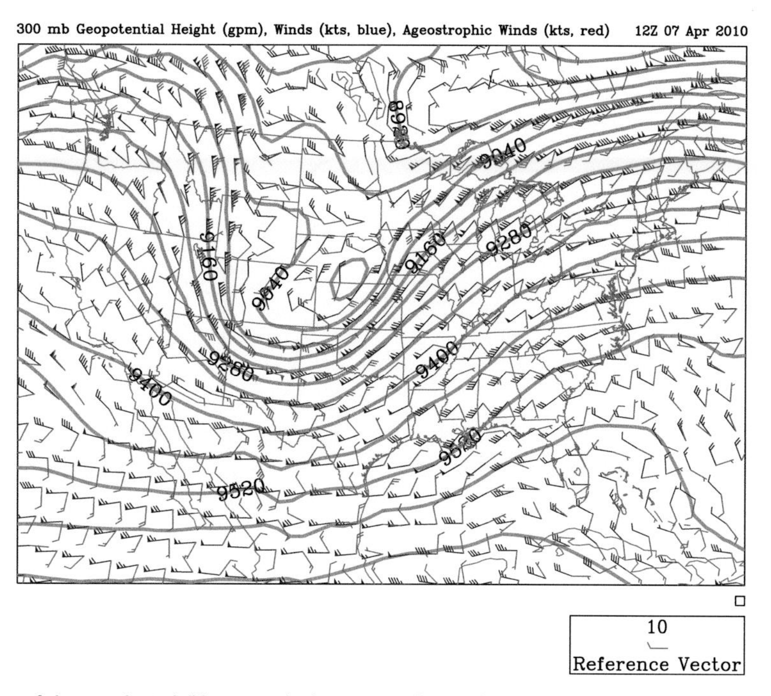

Fig. 2.4 Distributions of the actual wind (blue arrows), the ageostrophic wind (red arrows) and the contours of the geopotential height at 300 mb on April 7, 2010. See color plates section.

geostrophic balance. The wind characterized by such balance is called *geostrophic wind*, $\vec{V}_g$. The difference between the actual wind and the geostrophic wind is called *ageostrophic wind*, $\vec{V}_a = \vec{V} - \vec{V}_g$. Where a velocity vector on a map is significantly not parallel to a geopotential contour, the ageostrophic wind $\left(\vec{V}_a = \vec{V} - \vec{V}_g\right)$ is large. It is also more pronounced at lower latitudes where the Coriolis parameter is smaller. We see that the geostrophic balance is not good at those locations where the direction of wind varies significantly over a short distance. The momentum advection, $-\vec{V}\cdot\nabla\vec{V}$, would be large where there is strong curvature in the wind field. Nevertheless, a large-scale disturbance as a whole evolves slowly because the temporal acceleration of the wind is generally small compared to the rotation rate of the Earth.

The quantitative relationship to express geostrophic balance in p-coordinates is

$$0 \approx -\nabla_2 \Phi - f\vec{k} \times \vec{V}_2 \tag{2.24}$$

where $\vec{V}_2 = (u, v)$, $\Phi = gz$, $f = 2\Omega \sin\varphi$, $\vec{k}$ is the unit vector pointing *upward*. The two components of geostrophic balance in Cartesian coordinates take on the form

$$0 = -\frac{\partial \Phi}{\partial x} + fv_g; \qquad 0 = -\frac{\partial \Phi}{\partial y} - fu_g. \tag{2.25}$$

When we use $\vec{V} \approx \vec{V}_g$, we introduce an error of ~ 10 percent. The wind near a trough, say, on 500 mb is sub-geostrophic (weaker than the geostrophic wind), whereas the wind near a ridge is super-geostrophic (stronger than the geostrophic wind) (Fig. 2.5).

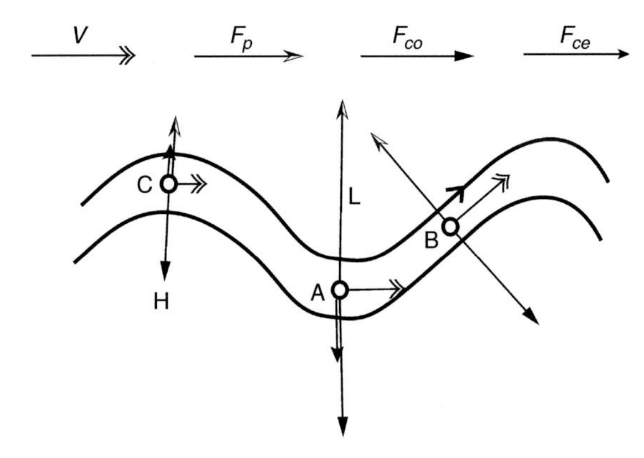

Fig. 2.5 Balance of forces in different parts of a large-scale flow in NH. L low pressure (smallest value of z); H high pressure (largest value of z); F_{co} Coriolis force; F_p pressure gradient force; F_{ce} centrifugal force. Around a trough, $F_p > F_{co}$, the flow is sub-geostrophic. Around a ridge, $F_p < F_{co}$, the flow is super-geostrophic. At B flow is geostrophic.

2.4 Thermal wind relation

When there are both geostrophic balance and hydrostatic balance, there necessarily exists a quantitative relationship between the vertical gradient of the wind and the horizontal gradient of the temperature field. Using the two balances as a starting point,

$$0 = -\frac{\partial \Phi}{\partial p} - \frac{RT}{p}; \qquad 0 = -\frac{\partial \Phi}{\partial x} + fv; \qquad 0 = -\frac{\partial \Phi}{\partial y} - fu, \tag{2.26}$$

we readily obtain

$$v_p = -\frac{R}{fp} T_x; \qquad u_p = \frac{R}{fp} T_y. \tag{2.27}$$

Thus, u would increase upward if T decreases northward in the northern hemisphere and v would increase upward if T increases eastward. We could get a better feel for the thermal wind by examining it in vector form. Equation (2.27) can be written as $\frac{\partial \vec{V}}{\partial p} = -\frac{R}{fp}(-\vec{i}T_y + \vec{j}T_x) = -\frac{R}{fp}\vec{k} \times \nabla_2 T$, where $\vec{V} = (u, v)$ is the horizontal velocity vector. The finite difference form of this equation is $\frac{\vec{V}_{lower} - \vec{V}_{upper}}{p_{lower} - p_{upper}} \equiv -\frac{\overrightarrow{\delta V}}{\delta p} = -\left(\frac{R}{fp}\vec{k} \times \nabla_2 T\right)_{layer}$. The difference between upper level wind and lower level wind, $(\vec{V}_{upper} - \vec{V}_{lower}) \equiv \overrightarrow{\delta V}$, is referred to as *thermal wind.* Since $\delta p = (p_{lower} - p_{upper}) > 0$, $\overrightarrow{\delta V}$ is directed 90° to the left of $(\nabla_2 T)_{layer}$ implying that *the thermal wind vector is parallel to the isotherms with the colder air in the layer under consideration being on the left side.* Another way of describing the relation is that, if the wind has cold advection at a level, the wind vector would turn anticlockwise to become one at a higher level. Both descriptions are equivalent as illustrated in Fig. 2.6. Such wind is said to be *backing* with increasing height. The reverse is called *veering* with height.

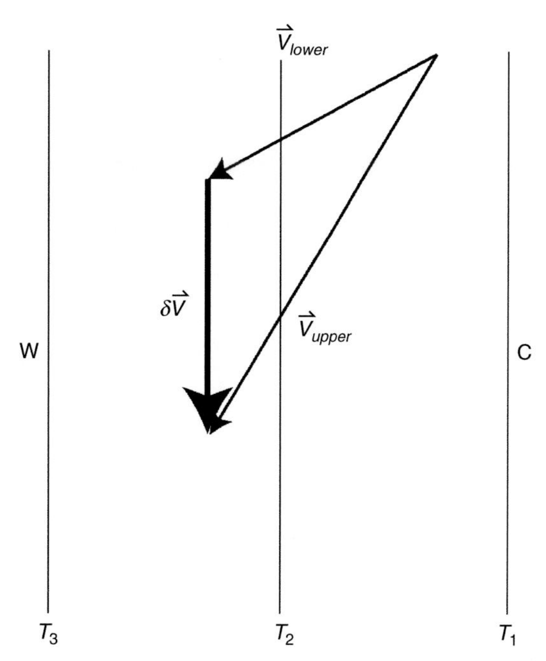

Fig. 2.6 Schematic of velocity at a low level, velocity at an upper level, the thermal wind $\delta\vec{V} = \vec{V}_{upper} - \vec{V}_{lower}$; W, warm and C, cold; $T_3 > T_2 > T_1$.

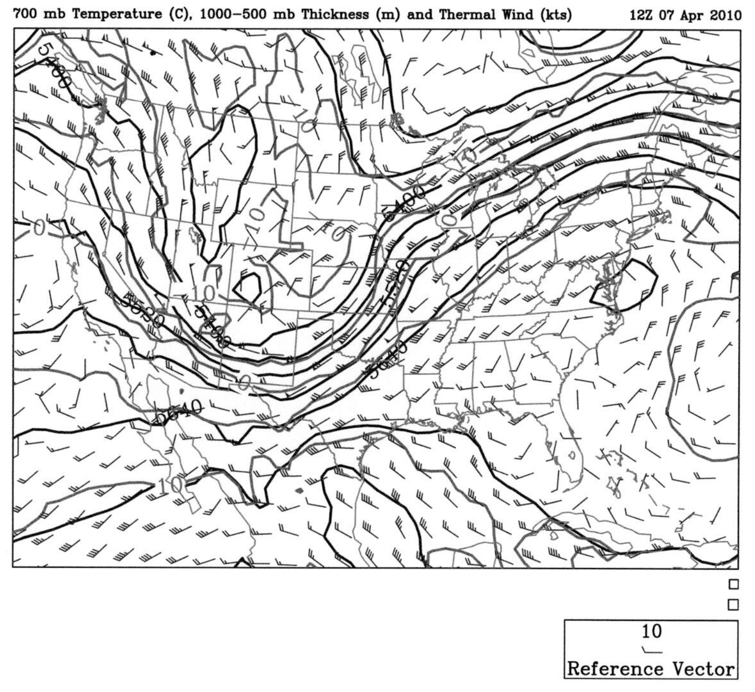

Fig. 2.7 Distributions of 700 mb temperature (red contour), thickness of the layer between 500 mb and 1000 mb (black), thermal wind vector field of this layer (arrows) on April 7, 2010. See color plates section.

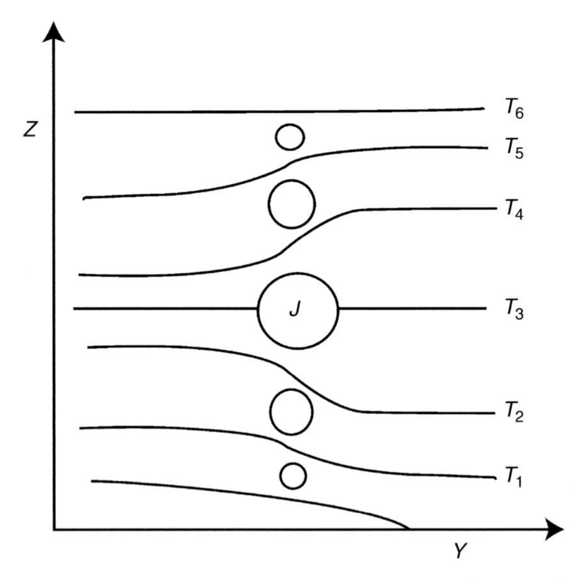

Fig. 2.8 Schematic of thermal wind in a latitude–height cross-section of temperature $(T_j > T_{j+1})$ and zonal wind.

Thermal wind relation is routinely observed on weather maps simply because large-scale flow is indeed approximately in geostrophic and hydrostatic balance (Fig. 2.7). For example, the red contours (temperature at 700 mb) are roughly parallel to the black contours (thickness of 500–1000 mb layer). The arrows (thermal wind) are indeed approximately tangential to the contours.

It is common to find a jet stream in mid-latitude in winter. That is a manifestation of the existence of thermal wind relation in the atmosphere. The relationship between the temperature and wind of a jet stream in a latitudinal zone is schematically shown in Fig. 2.8. The core of the jet is labeled by J. Such a zone is called a *baroclinic zone* and such a jet is called a *baroclinic jet.*

2.5 Balanced flows

If the local acceleration $\partial\vec{V}/\partial t$ of the horizontal velocity is much smaller than the inertial acceleration associated with its curvature $(\vec{V}\cdot\nabla\vec{V})$, the flow is said to be a *balanced flow.* In that case, there would be a balance between the inertial acceleration and the forces. The simplest type of balanced flow is of course a unidirectional flow, which has no inertial acceleration at all. It is a geostrophic flow. When there is a pronounced curvature in the wind field, the momentum advection is important such as the wind in a mature hurricane or tornado. The dynamical character of a balanced flow can be most succinctly analyzed with the use of a *natural coordinate.*

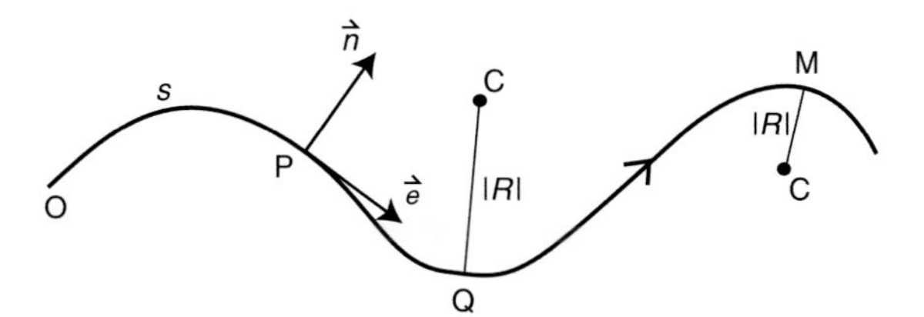

Fig. 2.9 Schematic of natural coordinates: see the text for the nomenclature of all symbols shown.

2.5.1 Natural coordinates

The natural coordinate system is useful for depicting and analyzing a general 2-D flow. The trajectory of an air parcel on a plane may be adopted as one coordinate, known as the s-axis. A unit vector tangential to the s-axis at each point is denoted by $\vec{e}$. The other coordinate at each point along a trajectory is directed 90 degrees to the left of the s-axis (*right-hand-convention*). It is referred to as the n-axis with a corresponding unit vector $\vec{n}$. We may examine the balance of forces on an air parcel in this coordinate system depicted in Fig. 2.9.

The location of an air parcel at a particular time is measured from a chosen reference point, O, on the s-axis. Two pieces of information (s, R) are associated with each point on a trajectory, P: s is the distance along the trajectory from the reference point to P; R is called the *radius of curvature* with $|R|$ being the distance from a point under consideration such as Q to a point referred to as the center of curvature C for that point. A trajectory on a (x, y) plane can be described by an algebraic expression, $y = G(x)$. The radius of curvature at any point of the trajectory can be evaluated with the formula $R = (1 + G'^2)^{3/2}/G''$ where $G' = dG/dx$ and $G'' = d^2G/dx^2$. The radius R can have a positive or negative value since the sign of G'' changes across an inflexion point. Geometrically speaking, R is positive in a segment of convex curvature such as around Q in Fig. 2.9, and negative in a segment of concave curvature such as around M. Note that $\vec{n}$ at Q is directed towards a center of curvature C, whereas $\vec{n}$ at M points away from its C. Therefore, we may adopt a rule of thumb that R has a positive value where $\vec{n}$ is directed towards C, and a negative value where $\vec{n}$ points away from C. Then we can write down the following fundamental relations:

$$
\begin{aligned}
&\frac{d\vec{e}}{ds} = \lim \frac{\Delta\vec{e}}{\Delta s} = \frac{\vec{n}}{R} \text{ at each point along the } s\text{-axis} \\
&\text{Velocity: } \vec{V} = V\vec{e} \\
&\text{Speed: } V = \frac{ds}{dt} > 0 \text{ by definition, } t \text{ being time} \\
&\text{Acceleration: } \frac{D\vec{V}}{Dt} = \frac{dV}{dt}\vec{e} + V\frac{d\vec{e}}{dt} = \frac{dV}{dt}\vec{e} + V\frac{d\vec{e}}{ds}\frac{ds}{dt} \\
&\qquad = \frac{dV}{dt}\vec{e} + \frac{V^2}{R}\vec{n}.
\end{aligned}
\tag{2.28}
$$

The last formula expresses the well-known concept that acceleration stems from a change of speed (first term) and/or a change of direction (second term).

2.5.2 Gradient wind balance and special flows

The equation of motion in isobaric-natural coordinates in the absence of viscosity is

$$\begin{aligned} \frac{dV}{dt}\vec{e}+\frac{V^2}{R}\vec{n} &= -\frac{\partial\Phi}{\partial s}\vec{e}-\frac{\partial\Phi}{\partial n}\vec{n}-fV\vec{n} \\ \rightarrow \frac{dV}{dt} &= -\frac{\partial\Phi}{\partial s}, \\ \frac{V^2}{R} &= -\frac{\partial\Phi}{\partial n}-fV. \end{aligned} \tag{2.29}$$

The third equation above may be thought of as a quadratic equation of V for given values of R and $\partial\Phi/\partial n$. There are then two roots namely

$$V_{\pm} = -\frac{fR}{2}\pm\sqrt{\left(\frac{fR}{2}\right)^2-R\frac{\partial\Phi}{\partial n}}. \tag{2.30}$$

Physically meaningful solution(s) of (2.30) are to be compatible with the definition that V has a positive value. We need to make a judgment as to whether a particular solution is physically meaningful or not on a case-by-case basis.

A flow would be a balanced flow when there is no acceleration along a trajectory, $\frac{DV}{Dt}=0$, which requires that the s-axis is parallel to a contour of the geopotential height. If the s-axis is almost parallel to a contour of geopotential height, the flow is a *quasi-balanced* flow. Furthermore, if $\frac{\partial\Phi}{\partial n}=$ constant on an s-axis, R would have the same value at all points in a balanced flow. In such a case, the trajectory would be a circle and the flow would be a vortex.

Three special types of flow

(i) Geostrophic wind

A unidirectional flow corresponds to the case of an infinitely large radius of curvature. Then the balance of forces would be reduced to that between Coriolis force and pressure gradient force. Such balance is the familiar *geostrophic balance* and the corresponding wind is geostrophic wind,

$$V = -\frac{1}{f}\frac{\partial\Phi}{\partial n}. \tag{2.31}$$

It would require $\partial\Phi/\partial n < 0$ in the northern hemisphere. In other words, lower pressure must be located to the left of the wind vector.

(ii) Cyclostrophic wind

The centrifugal force associated with a vortex of small radius of curvature is much larger than the Coriolis force. Then the balance of forces would be reduced to that between the centrifugal force and pressure gradient force. Such balance is called *cyclostrophic balance* and such wind is called cyclostrophic wind,

$$V = \sqrt{-R\frac{\partial \Phi}{\partial n}}. \tag{2.32}$$

Note that the wind can be either cyclonic $\left(\frac{\partial \Phi}{\partial n} > 0, R < 0\right)$ or anticyclonic $\left(\frac{\partial \Phi}{\partial n} < 0, R > 0\right)$ around a small low-pressure center. By the same token, there can be no cyclostrophic balance around a small high-pressure center since the two forces would be both directed radially outward.

Cyclostrophic wind is observed in a tornado, which has very low pressure at the center. The vast majority of tornadoes have cyclonic wind, but rare exceptions have been reported. What might be the source of the observed directional bias? The answer lies at the source of its angular momentum. Observations clearly suggest that most tornadoes spawn from a larger background disturbance that has significant rotation, called a supercell. In the northern hemisphere, supercells themselves rotate cyclonically.

Another type of disturbance in cyclostrophic balance are called "dust devils." They arise from intense surface heating in summer often in the absence of a significant background wind. The surface heating induces a strong convergence in the flow near the surface. This would lead to rising motion. Unless a particular configuration of the surface heating and roughness introduces a bias to one direction of rotation in the converging inflow, there is generally no systematic bias in the direction of rotation.

To a lesser extent, the wind in a mature hurricane is also in cyclostrophic balance because the pressure gradient force is the smallest (but not negligible) of the three forces. The source of angular momentum of a hurricane is the rotation of the Earth. A hurricane is energetically fueled by latent heat released from precipitation. To have precipitation, there must be convergent wind at the low levels. The Coriolis force acting on such convergent wind would give rise to low-level cyclonic wind in the northern hemisphere. Hence, hurricanes in NH are always in cyclonic rotation.

(iii) Inertial oscillation

A steady flow at a level where the pressure is uniform has to be circular. The balance of forces in the $\vec{n}$-direction only involves the Coriolis force and the centrifugal force

$$0 = -\frac{V^2}{R} - fV \Rightarrow V = -fR. \tag{2.33}$$

Therefore, R has to be negative in NH where $f > 0$. That means that this wind is in *anti-cyclonic (clockwise) motion in the northern hemisphere*. The opposite should occur in SH. This motion is called *inertial oscillation*, even though the flow itself is steady. The balance of forces in this case is shown in Fig. 2.10.

The period of oscillation of an individual air parcel is $2\pi R/V = |2\pi/f|$, half of a pendulum day (one day at latitude 30 degrees). How large the radius of inertial circle of oscillation at a location would be depends on the speed of fluid parcel, $|R| = V/|f|$. This

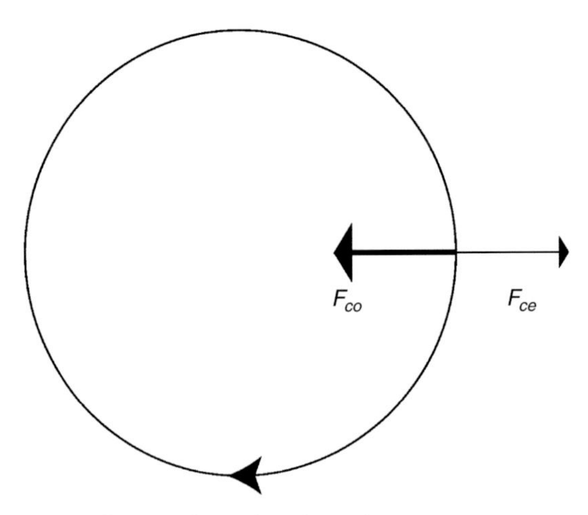

Fig. 2.10 Forces associated with inertial oscillation in the northern hemisphere.

type of motion is seldom observed in the atmosphere because the atmospheric pressure field is seldom approximately uniform. In contrast, inertial oscillation is more commonly detected in oceanographic data because the spatial variation of pressure in the ocean is usually much weaker.

2.5.3 Dynamical nature of four conceivable configurations of circular flows

Flows with circular symmetry are common in the atmosphere. The pressure at the center of a vortex may be higher or lower than the surroundings. There are a number of possible circular flows.

(i) Cyclonic flow around a low-pressure center in the northern hemisphere (NH)

The radius of curvature R of a cyclonic flow has a positive value. The directions of the three forces are indicated in Fig. 2.11 suggesting that this is a possible configuration. For a known value of $R > 0, f > 0$ and $\partial\Phi/\partial n < 0$, there is only one possible solution of V because the other root has a negative value. It is $V = -\frac{fR}{2} + \sqrt{\left(\frac{fR}{2}\right)^2 - R\frac{\partial\Phi}{\partial n}} > 0$. It would have a relatively small value. It corresponds to the wind that we normally observe around a low-pressure center.

Numerical example:

Parameter values representative of an extratropical cyclone are $f = 10^{-4}\,\mathrm{s}^{-1}$, $|R| = 10^6\,\mathrm{m}$, $|\partial\Phi/\partial n| = 10^{-3}\,\mathrm{m\,s}^{-2}$.

The two terms of the solution are equal to $-50\,\mathrm{m\,s}^{-1}$ and $59\,\mathrm{m\,s}^{-1}$ leading to a gradient wind of $V = 9\,\mathrm{m\,s}^{-1}$. To appreciate its physical nature, it is noteworthy that $\left(-\frac{fR}{2}\right)^2 = 2.5 \times 10^5\,\mathrm{m}^2\,\mathrm{s}^{-2}$, $-R\frac{\partial\Phi}{\partial n} = 10^3\,\mathrm{m}^2\,\mathrm{s}^{-2}$, hence $\left(-\frac{fR}{2}\right)^2 \gg \left(-R\frac{\partial\Phi}{\partial n}\right)$.

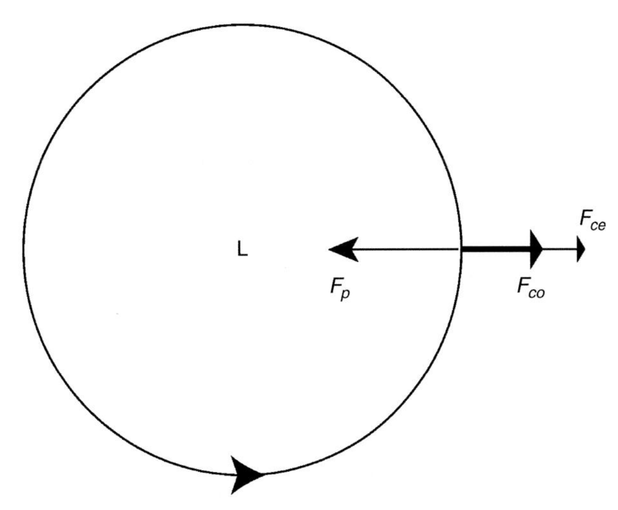

Fig. 2.11 Forces on a cyclonic flow around a low-pressure center.

Since $(1+x)^n = 1 + nx + \frac{n(n-1)}{2}x^2 + \cdots$, we have

$$V = \frac{-fR}{2} + \frac{fR}{2}\left(1 - \left(\frac{R\partial\Phi/\partial n}{(fR/2)^2}\right)^2\right)^{1/2}$$

$$\approx \frac{-fR}{2} + \frac{fR}{2}\left(1 + \frac{-R\partial\Phi/\partial n}{2(fR/2)^2} - \frac{1}{8}\left(\frac{-R\partial\Phi/\partial n}{(fR/2)^2}\right)^2\right) = \frac{\partial\Phi/\partial n}{f} - \frac{(\partial\Phi/\partial n)^2}{f^3 R}.$$

Noting that $\frac{\partial\Phi/\partial n}{f} \sim 10, \frac{(\partial\Phi/\partial n)^2}{f^3R} \sim 1$, we see that this particular gradient wind solution V is close to a geostrophic wind with a correction arising from the curvature. When this correction is small, (i.e. when the curvature is not too large), it is called a *quasi-geostrophic wind.*

(ii) Anticyclonic flow around a low-pressure center in NH

For such a flow, R would have a negative value. The directions of the three forces are indicated in Fig. 2.12 suggesting that this is also a possible configuration. For a known value of $R < 0, f > 0$ and $\partial\Phi/\partial n > 0$, there is also only one possible solution of V. It is $V = -\frac{fR}{2} + \sqrt{\left(\frac{fR}{2}\right)^2 - R\frac{\partial\Phi}{\partial n}} > 0$. Here V must have a fairly large value in order for F_{ce}, which depends on V quadratically, to be able to balance the sum of the other two forces. This is virtually never observed around a large-scale low-pressure center and is therefore called "anomalous flow."

The reason why we do not observe anticyclonic wind around a low-pressure center is that the rotation of the Earth is so strong that the flow seldom departs too greatly from a geostrophically balanced state. In the example that we have considered, the two terms are equal to 50 m s^{-1} and 59 m s^{-1} leading to $V = 109$ m s^{-1}.

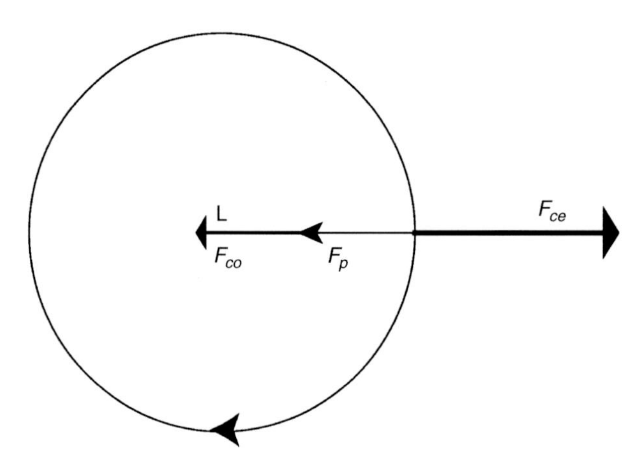

Fig. 2.12 Forces on an anticyclonic flow around a low-pressure center.

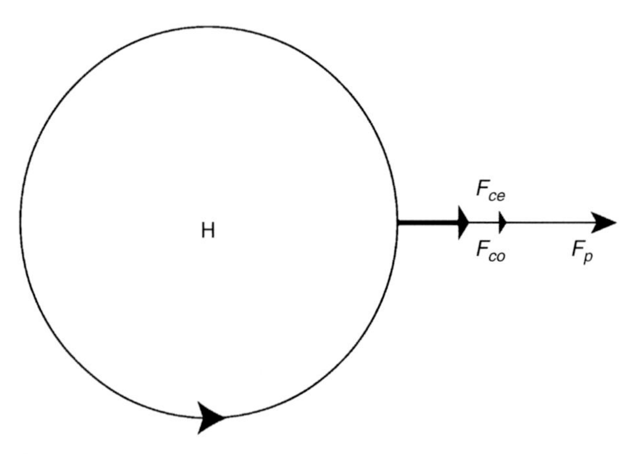

Fig. 2.13 Forces on a cyclonic flow around a high-pressure center. A physically unrealizable configuration.

(iii) A cyclonic flow around a high-pressure center in NH

Such a flow is dynamically impossible because the directions of all three forces are in the same direction as indicated in Fig. 2.13. The two solutions are therefore not physically meaningful. Indeed, we never observe a cyclonic flow around a high-pressure center.

(iv) An anticyclonic flow around a high-pressure center in NH

For such a flow, R would have a negative value. The directions of the three forces are indicated in Fig. 2.14, suggesting that this is a possible configuration. In this case, for a known value of $R < 0$, $f > 0$ and $\partial\Phi/\partial n < 0$, there is one physically meaningful solution of V. It is $V = -\dfrac{fR}{2} - \sqrt{\left(\dfrac{fR}{2}\right)^2 - R\dfrac{\partial\Phi}{\partial n}} > 0$. Here V can have a small value. A weak flow is indeed normally observed around a large-scale high-pressure center.

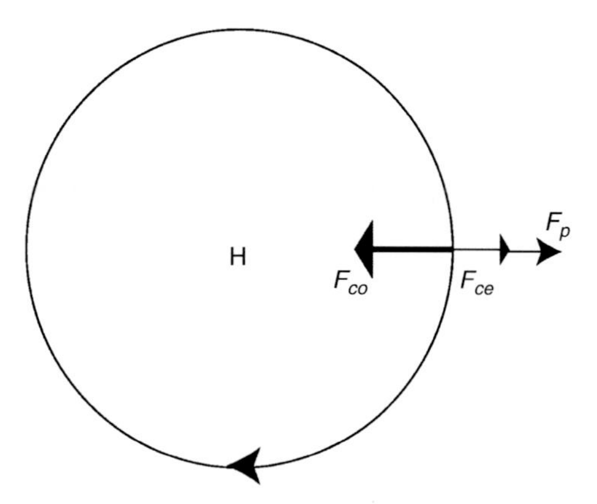

Fig. 2.14 Forces on a normal anticyclonic flow around a high-pressure center.

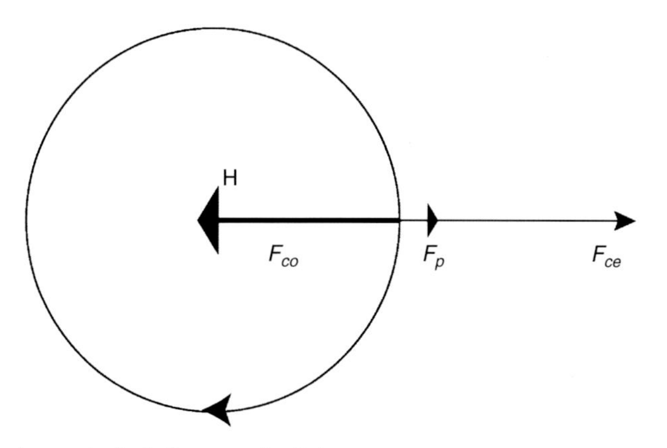

Fig. 2.15 Forces on an anomalous anticyclonic flow around a high-pressure center.

There is also the possibility of a relatively strong wind in this case, corresponding to the balance qualitatively depicted in Fig. 2.15.

2.6 Kinematic properties of wind

We next diagnose the characteristics of a general wind field in Cartesian coordinates, $\vec{V} = (u, v)$. The structure of wind in a small neighborhood around a point of interest (x_o, y_o) can be adequately described by a truncated Taylor series, viz.

$$\begin{aligned} u(x,y) &\approx u(x_o,y_o) + (x - x_o)u_x + (y - y_o)u_y, \\ v(x,y) &\approx v(x_o,y_o) + (x - x_o)v_x + (y - y_o)v_y, \end{aligned} \tag{2.34}$$

where the partial derivatives above are to be evaluated at (x_o, y_o). Thus, the local properties of wind in a small neighborhood around a point (x_o, y_o) stem from six pieces of information of the wind field: $u|_o$, $v|_o$, $u_x|_o$, $u_y|_o$, $v_x|_o$ and $v_y|_o$.

The $u(x_o, y_o)$ and $v(x_o, y_o)$ in (2.34) are in essence the components of the average wind in that neighborhood. Wind at one location carries all properties of air parcels to a downstream location, thereby giving rise to the *advective effect*. Other information about the wind field is contained in different combinations of $u(x_o, y_o)$ and $v(x_o, y_o)$. For example, the quantity $\frac{\rho}{2}(u^2 + v^2)$ is the kinetic energy of air in unit volume around (x_o, y_o), where ρ is the density of air. The quantity $uv\rho$ can be interpreted as a transport of u-momentum in the y-direction through a vertical unit cross-section centering at the point (x_o, y_o). It may be alternatively interpreted as a flux of v-momentum in the x-direction. The average of $uv\rho$ over an appropriately defined distance in the x-direction at y_o, $\frac{1}{L}\int_0^L uv\rho\, dx$, would represent the average flux of u-momentum in the y-direction by a disturbance through a vertical strip of unit length. This flux gives rise to a redistribution of u-momentum in the region under consideration and would have significant dynamical consequences.

It is instructive to rewrite (2.34) as

$$u(x,y) \approx u_o + \frac{1}{2}(x - x_o)\delta_o + \frac{1}{2}(x - x_o)D_o^{(1)} + \frac{1}{2}(y - y_o)D_o^{(2)} - \frac{1}{2}(y - y_o)\zeta_o,$$

$$v(x,y) \approx v_o + \frac{1}{2}(x - x_o)D_o^{(2)} + \frac{1}{2}(x - x_o)\zeta_o + \frac{1}{2}(y - y_o)\delta_o - \frac{1}{2}(y - y_o)D_o^{(1)}, \qquad (2.35)$$

where a quantity with a subscript "o" means that it is to be evaluated at the point (x_o, y_o). The four quantities ($\delta = u_x + v_y = \nabla \cdot \vec{V}$, $\zeta = v_x - u_y = \vec{k} \cdot \nabla \times \vec{V}$, $D^{(1)} = u_x - v_y$, $D^{(2)} = v_x + u_y$) are called *horizontal divergence*, *vorticity*, *stretching deformation* and *shearing deformation* respectively. These are four independent combinations of the derivatives in the sense that none of them can be constructed as a combination of the other three. The four panels of Fig. 2.16 illustrate the flow configurations associated with these properties in a small neighborhood of an air parcel and the influences they would have on an air parcel centered at (x_o, y_o). The air parcel is depicted as a greatly magnified square for clarity. The solid outline in each panel indicates the shape of an air parcel centered at (x_o, y_o) in each case. The light outline in each panel indicates the shape it would tend to change to resulting from the influence of the quantity. The arrows indicate the velocity of fluid elements in the immediate neighborhood associated with the local property under consideration.

We see that horizontal divergence, δ, is a measure of the rate of horizontal spreading of air parcels. A negative value of δ is convergence. Therefore, δ indicates the possible change in the horizontal length of a fluid element in the directions normal to its sides. Vorticity, ζ, is a measure of the rotational property of this horizontal flow about the direction normal to the horizontal plane. A positive (negative) value of vorticity ζ in the northern hemisphere is called cyclonic (anticyclonic) vorticity. It should be noted in passing that vorticity in this case only refers to the vertical component of a general vorticity vector. Stretching deformation, $D^{(1)}$, is a measure of the deformation of a fluid element due to compression in one direction and simultaneous dilatation in a direction orthogonal (90 degrees) to it. It depends on the spatial variation of a velocity component in its own direction, e.g. u_x. Shearing deformation,

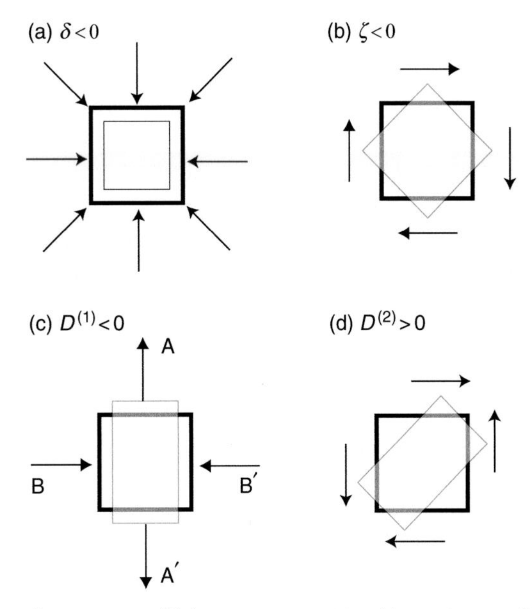

Fig. 2.16 A flow that is (a) horizontally convergent, (b) has negative vorticity, (c) negative stretching deformation and (d) positive shearing deformation. All but the one property indicated above are zero in each panel. AA′ in panel (c) is the axis of dilatation; BB′ axis of contraction.

$D^{(2)}$, is a measure of the deformation of a fluid element arising from the shear of a flow. It depends on the spatial variation of a velocity component in the direction normal to itself, e.g. u_y. Imagine the possible consequence to a temperature field under the influence of a wind field with strong deformation. It is not hard to visualize the temperature contours would soon come together as a consequence of deformation, leading to an atmospheric front.

Panels (c) and (d) are intrinsically similar. The line parallel to the long sides of the light rectangle in (d) can be also visualized as an axis of dilatation and the one along the short sides as an axis of contraction. In other words, the values of $D^{(1)}$ and $D^{(2)}$ depend on the coordinate system used for depicting the flow. As such, the individual parts have no unique significance. But the total deformation as measured by $((D^{(1)})^2 + (D^{(2)})^2)$ does not depend on the coordinate system. That is why the deformation process does have physical significance. As an example, let us consider a simple 2-D flow given by its streamfunction as $\psi = \sin x \cos y$. The corresponding stretching deformation is $D^{(1)} = u_x - v_y = -2\psi_{xy} = 2\cos x \sin y$ and the corresponding shearing deformation is $D^{(2)} = u_y + v_x = \psi_{xx} - \psi_{yy} = 0$. Suppose we examine this flow in another coordinate system which is rotated by an angle ϕ relative to the x-axis defined by $x' = x\cos\phi - y\sin\phi$ and $y' = x\sin\phi + y\cos\phi$. $\psi(x,y)$ would be transformed to $\psi'(x',y')$. The stretching deformation in the (x',y') coordinate system would be $D'^{(1)} = u'_{x'} - v'_{y'} = 2(\sin^2\phi - \cos^2\phi)\psi_{xy} = (\cos^2\phi - \sin^2\phi)D^{(1)}$. Thus, there would be no stretching deformation for $\phi = 45°$. On the other hand, the shearing deformation in the (x',y') coordinate system would be $D'^{(2)} = u'_{y'} + v'_{x'} = (\cos^2\phi - \sin^2\phi)(\psi_{xx} - \psi_{yy}) -4\psi_{xy}\sin\phi\cos\phi$. For $\phi = 45°$, we would have $D'^{(2)} = -2\psi_{xy} = D^{(1)}$. In other words, the deformation would appear entirely in the form of shearing deformation in this new coordinate system.

2.6.1 Structural properties of a relevant idealized flow

A hurricane is a highly axisymmetric disturbance. At low levels, the wind is virtually zero near the center and increases from the center to a maximum value at about $r_o = 50$ to 100 km. The wind decreases monotonically further outward at $r > r_o$. Such distribution of wind can be approximately described by an idealized flow known as *Rankin vortex.* Denote the polar coordinates by (φ, r) and the corresponding velocity components by (v, u). A Rankin vortex has $u = 0$ and the following azimuthal velocity

$$\begin{aligned} v &= \frac{Vr}{r_o} \quad \text{in } r \leq r_o, \\ v &= \frac{Vr_o}{r} \quad \text{in } r > r_o, \end{aligned} \tag{2.36}$$

where V is the maximum velocity at radial distance $r = r_o$. In polar coordinates, we calculate vorticity by $\frac{1}{r}\frac{d(vr)}{dr} - \frac{1}{r}\frac{\partial u}{\partial \varphi}$, horizontal divergence by $\frac{1}{r}\frac{\partial(ur)}{\partial r} + \frac{1}{r}\frac{\partial v}{\partial \varphi}$, stretching deformation by $\frac{1}{r}\frac{\partial(ur)}{\partial r} - \frac{1}{r}\frac{\partial v}{\partial \varphi}$ and shearing deformation by $\frac{1}{r}\frac{d(vr)}{dr} + \frac{1}{r}\frac{\partial u}{\partial \varphi}$.

It follows that the vorticity of a Rankin vortex is equal to $2V/r_o$ in the inner core $r \leq r_o$, and is zero at $r > r_o$. In using $V = 30\,\mathrm{m\,s^{-1}}$ for a hurricane, the vorticity in the inner core is equal to $2V/r_o \sim 10^{-3}\,\mathrm{s^{-1}}$ which is more than one order of magnitude larger than the local value of Coriolis parameter. Horizontal divergence is zero everywhere, so is stretching deformation. There is uniform shearing deformation equal to $2V/r_o$ in the inner core $r \leq r_o$ but shearing deformation is zero at $r > r_o$.

The forces acting on the fluid elements are in gradient wind balance

$$\frac{\partial \Phi}{\partial r} = \frac{v^2}{r} + fv. \tag{2.37}$$

By (2.36) and (2.37), we obtain the following distribution of geopotential height of a pressure level subject to boundary conditions $\Phi = \Phi_{oo}$ at $r = 0$, and continuous Φ at $r = r_o$,

$$\begin{aligned} \Phi &= \Phi_{oo} + \frac{V^2}{2}\left(1 + \frac{fr_o}{V}\right)\left(\frac{r}{r_o}\right)^2 \quad \text{in } r \leq r_o, \\ \Phi &= \Phi_{oo} + \frac{V^2}{2}\left(2 + \frac{fr_o}{V}\right) + fVr_o \ln\left(\frac{r}{r_o}\right) - \frac{V^2}{2}\left(\frac{r_o}{r}\right)^2 \quad \text{in } r > r_o. \end{aligned} \tag{2.38}$$

It is convenient to measure r in units of r_o, v in units of V and Φ in units of Φ_{oo}. Two non-dimensional parameters are particularly relevant, namely the Rossby number $R = V/fr_o$ and $\lambda = fr_o/\sqrt{\Phi_{oo}}$ which is the ratio of r_o to $\sqrt{\Phi_{oo}}/f$ known as a form of radius of deformation. Then the non-dimensional properties of a Rankin vortex are

$$\begin{aligned} v &= r, \ \Phi = 1 + \frac{\lambda^2 R^2}{2}\left(1 + \frac{1}{R}\right) r^2 \quad \text{in } r \leq 1, \\ v &= \frac{1}{r}, \ \Phi = 1 + \frac{\lambda^2 R^2}{2}\left(2 - \left(\frac{1}{r}\right)^2\right) + R\lambda^2\left(\frac{1}{2} + \ln(r)\right) \quad \text{in } r \geq 1. \end{aligned} \tag{2.39}$$

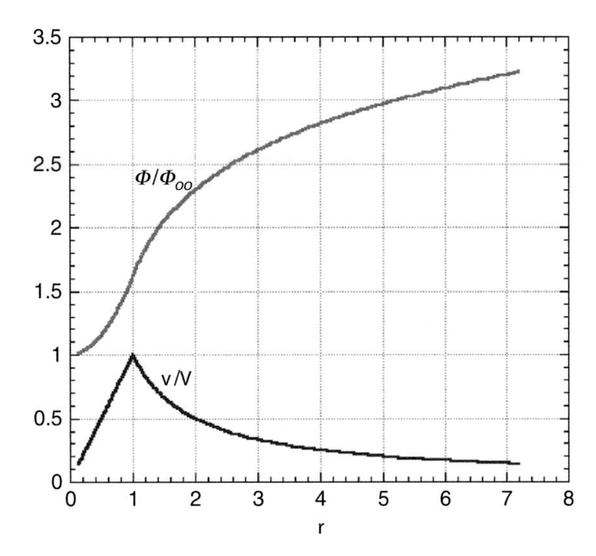

Fig. 2.17 Radial distribution of non-dimensional velocity and pressure of a Rankin vortex as a function of (r/r_o).

The non-dimensional velocity and total pressure Φ of a Rankin vortex for $R = 0.83$ and $\lambda = 0.9$ are shown in Fig. 2.17. Rankin vortex is often used in theoretical studies of vortices.

2.7 Divergent wind and vertical motion

As a 2-D vector field, a general wind field can be always partitioned into two parts. One is a rotational part and the other a divergent part, $\vec{V} \equiv (u, v) = \vec{V}_{rot} + \vec{V}_{div}$. They can be defined in terms of two scalar functions, ψ the streamfunction and ϕ the velocity potential, by $\vec{V}_{rot} = \vec{k} \times \nabla_{(2)}\psi = (-\psi_y, \psi_x)$ and $\vec{V}_{div} = \nabla_{(2)}\phi = (\phi_x, \phi_y)$ where $\vec{k}$ is the vertical unit vector and $\nabla_{(2)} = \left(\frac{\partial}{\partial x}, \frac{\partial}{\partial y}\right)$. It follows that $\nabla_{(2)} \cdot \vec{V}_{rot} = 0$ and $\vec{k} \cdot \nabla_{(2)} \times \vec{V}_{div} = 0$. Geostrophic wind is an example of the rotational part of the wind (not considering the variation of the Coriolis parameter). Therefore, the divergent part of a large-scale wind field is much weaker than its rotational part, typically by one order of magnitude. The vertical motion of air in turn is associated with the divergent wind as suggested by the continuity equation in isobaric coordinates, $u_x + v_y + \omega_p = 0$. The vertical motion associated with the divergence/convergence in a large-scale atmospheric flow is therefore very weak ($\sim$cm s^{-1}) compared to the horizontal motion. An example of an "observed" ω field will be shown in Fig. 6.3. Note that this "observed" ω field has a lot of small-scale features, implying that it is a noisy field. The error in it must be substantial. Weak as it is, a large-scale vertical velocity field can lead to formation of stratiform clouds as well as precipitation when the atmospheric layer under consideration is sufficiently moist.

It is instructive to estimate the vertical motion induced by a diabatic heating rate due to a rainfall rate of 1 inch/day. Suppose that the released latent heat warms up half of an atmospheric column, i.e. an atmospheric layer of 500 mb thick. The heating rate would be $Q = \dfrac{\rho_w PL}{\Delta p/g} = 1.2 \times 10^4\,\mathrm{J\,(kg\,day)^{-1}}$. The induced ascending motion would be $\omega = -\dfrac{Q}{Sc_p} = -1.2\times10^4\,\mathrm{Pa\,day^{-1}} = -1.4\times10^{-3}\,\mathrm{mb\,s^{-1}}$; $w \approx -\frac{\omega}{g\rho} = 2.8\times10^{-2}\,\mathrm{m\,s^{-1}}$.

If 1 inch of rain falls in one hour, then the estimate would be $w = 0.67\ \mathrm{m\,s^{-1}}$.

The vertical velocity of a large-scale flow is so weak that it cannot be measured directly with the meteorological instruments used for collecting daily weather data. For synoptic-scale disturbances, the following approximation is justifiable,

$$\omega \equiv \frac{Dp}{Dt} = \frac{\partial p}{\partial t} + u\frac{\partial p}{\partial x} + v\frac{\partial p}{\partial y} + w\frac{\partial p}{\partial z} \approx w\frac{\partial p}{\partial z} \approx w(-g\rho). \tag{2.40}$$

Then at 500 mb, $\omega \sim 10^{-2}(10)(0.5)\ \mathrm{Pa\,s^{-1}} = 5 \times 10^{-4}\ \mathrm{mb\,s^{-1}} \Rightarrow S\omega \approx 5\ \mathrm{K\,day^{-1}}$ which contributes significantly to the change in the temperature field. It is then not surprising to find, as we will elaborate in Chapter 6, that the divergent wind also plays an important dynamical role in the evolution of the rotational wind itself.

2.8 Summary: *z*-, *p*- and *θ*-coordinates and equations of balance

	p-coordinate	z-coordinate	θ-coordinate
Independent variables	(x, y, p) p decreasing upward	(x, y, z) z increasing upward	(x, y, θ) θ increasing upward
Unit vectors	$(\vec{i}, \vec{j}, \vec{k})$ $\vec{k}$ in the direction of decreasing p	$(\vec{i}, \vec{j}, \vec{k})$ $\vec{k}$ in the direction of increasing z	$(\vec{i}, \vec{j}, \vec{k})$ $\vec{k}$ in the direction of increasing θ
Velocity	(u, v, ω) $\omega = \dfrac{dp}{dt}$ $\omega < 0 \Rightarrow$ upward motion	(u, v, w) $w = \dfrac{dz}{dt}$ $w > 0 \Rightarrow$ upward motion	$(u, v, \dot{\theta})$ $\dot{\theta} > 0 \Rightarrow$ upward motion, also a measure of heating rate
Total derivative, $\dfrac{D}{Dt}$	$\dfrac{\partial}{\partial t} + u\dfrac{\partial}{\partial x} + v\dfrac{\partial}{\partial y} + \omega\dfrac{\partial}{\partial p}$	$\dfrac{\partial}{\partial t} + u\dfrac{\partial}{\partial x} + v\dfrac{\partial}{\partial y} + w\dfrac{\partial}{\partial z}$	$\dfrac{\partial}{\partial t} + u\dfrac{\partial}{\partial x} + v\dfrac{\partial}{\partial y} + \dot{\theta}\dfrac{\partial}{\partial \theta}$
Thermal variables	(T, ρ, Φ) $\Phi = gz$; $z(x, y, t)$ is height of isobaric surface	(T, ρ, p)	(M, σ, p) $M = c_pT + gz$ $z(x, y, t)$ is height of isentropic surface $\sigma = -g\dfrac{\partial p}{\partial \theta}$, "density"
Hydrostatic balance	$\dfrac{\partial \Phi}{\partial p} = -\dfrac{RT}{p}$	$\dfrac{\partial p}{\partial z} = -g\rho$	$\dfrac{\partial M}{\partial \theta} = c_p\left(\dfrac{p}{p_*}\right)^{\kappa}$
Geostrophic balance	$fv = \Phi_x$ $fu = -\Phi_y$	$fv = \rho^{-1}p_x$ $fu = -\rho^{-1}p_y$	$fv = M_x$ $fu = -M_y$
Thermal wind balance	$v_p = -\dfrac{R}{fp}T_x$ $u_p = \dfrac{R}{fp}T_y$	$fv_z = \dfrac{fvT_z}{T} + \dfrac{g}{T}T_x$ $fu_z = \dfrac{fuT_z}{T} - \dfrac{g}{T}T_y$	$v_\theta = \dfrac{c_p}{f\theta}T_x$ $u_\theta = -\dfrac{c_p}{f\theta}T_y$

Definition $\theta = T\left(\dfrac{p_*}{p}\right)^{\kappa}$, $\kappa = R/c_p$, $p_* = 1000\,\mathrm{mb}$.

3 Vorticity and potential vorticity dynamics

This chapter discusses the fundamental concepts (vorticity, circulation and potential vorticity) used for quantifying the rotational property of a flow and analyzing its evolution. The mathematical and physical characteristics of vorticity and circulation are elaborated in Sections 3.1 through 3.6. Those of potential vorticity are elaborated in Sections 3.8.2, 3.8.4 and 3.8.6. We elaborate on how the various mechanisms could change these different measures of the rotational property. We illustrate circulation with sea-breeze in Section 3.4 and potential vorticity with an eddy-driven jet in Section 3.8.5 and hurricane in Section 3.8.7. The emphasis of most discussions is on large-scale flows. To get a feel for the large-scale potential vorticity in the atmosphere, we show several instantaneous as well as statistical distributions of it in Section 3.8.3. Section 3.9 shows how we can succinctly represent the effects of diabatic heating and friction at the boundaries on the potential vorticity dynamics. It is done with the use of additional notions of generalized potential vorticity and generalized potential vorticity flux.

3.1 Vorticity and circulation of a three-dimensional flow

There are two complementary concepts that quantify the rotational property of a flow: *vorticity* and *circulation*. Vorticity is a local measure, whereas circulation is a bulk (integral) measure. Each can be defined either in the Lagrangian sense or in the Eulerian sense.

Vorticity

Vorticity is defined as the curl of the velocity, $\vec{\zeta} = \nabla \times \vec{V}$. Using the notation $\vec{V} = (u, v, w)$ in Cartesian coordinates (x, y, z) with unit vectors $\left(\vec{i}, \vec{j}, \vec{k}\right)$, we get an explicit expression for vorticity as

$$\vec{\zeta} = \nabla \times \vec{V} = \begin{vmatrix} \vec{i} & \vec{j} & \vec{k} \\ \dfrac{\partial}{\partial x} & \dfrac{\partial}{\partial y} & \dfrac{\partial}{\partial z} \\ u & v & w \end{vmatrix} = \left(w_y - v_z\right)\vec{i} + \left(u_z - w_x\right)\vec{j} + \left(v_x - u_y\right)\vec{k}. \tag{3.1}$$

We may always describe the velocity vector $\vec{V}$ in another local Cartesian frame (x', y', z') with the corresponding unit vectors $\left(\vec{i}', \vec{j}', \vec{k}'\right)$. If we choose $\vec{i}'$ to be parallel to $\vec{V}$ itself, the velocity would be $\vec{V} = V\vec{i}'$ and the vorticity vector is $\vec{\zeta} = V_{z'}\vec{j}' - V_{y'}\vec{k}'$. It follows that we have $\vec{V}\cdot\vec{\zeta}=0$. It means that any velocity vector is intrinsically orthogonal to the corresponding vorticity vector.

In spherical coordinates (λ,φ,r), the velocity vector is $\vec{V}=u\vec{\lambda}+v\vec{\varphi}+w\vec{r}$. The corresponding expression for the vorticity vector would be

$$\vec{\zeta}=\nabla\times\vec{V}=\frac{1}{r^2\cos\varphi}\begin{vmatrix} r\cos\varphi\vec{\lambda} & r\vec{\varphi} & \vec{r} \\ \dfrac{\partial}{\partial\lambda} & \dfrac{\partial}{\partial\varphi} & \dfrac{\partial}{\partial r} \\ ur\cos\varphi & vr & w \end{vmatrix}$$

$$=\left(\frac{1}{r}\frac{\partial w}{\partial\varphi}-\frac{1}{r}\frac{\partial(rv)}{\partial r}\right)\vec{\lambda}+\left(\frac{1}{r}\frac{\partial(ru)}{\partial r}-\frac{1}{r\cos\varphi}\frac{\partial w}{\partial\lambda}\right)\vec{\varphi}+\left(\frac{1}{r\cos\varphi}\frac{\partial v}{\partial\lambda}-\frac{1}{r\cos\varphi}\frac{\partial(u\cos\varphi)}{\partial\varphi}\right)\vec{r}. \quad (3.2)$$

The part of velocity and vorticity associated with the Earth's rotation are

$$\vec{V}_e=\vec{\Omega}\times r\vec{r}=\Omega r\cos\varphi\vec{\lambda}, \quad (3.3)$$

$$\vec{\zeta}_e=\nabla\times\vec{V}_e=\frac{1}{r^2\cos\varphi}\begin{vmatrix} r\cos\varphi\vec{\lambda} & r\vec{\varphi} & \vec{r} \\ \dfrac{\partial}{\partial\lambda} & \dfrac{\partial}{\partial\varphi} & \dfrac{\partial}{\partial r} \\ \Omega r^2\cos^2\varphi & 0 & 0 \end{vmatrix}=2\Omega\cos\varphi\vec{\varphi}+2\Omega\sin\varphi\vec{r}=2\vec{\Omega}. \quad (3.4)$$

The term $\vec{\zeta}_e$ is called planetary vorticity, which is simply a vector equal to twice the Earth's rotation vector. Corresponding to absolute velocity $\vec{V}_{absolute}=\vec{V}_e+\vec{V}$ is absolute vorticity $\vec{\zeta}_{absolute}=\nabla\times\vec{V}_{absolute}=2\vec{\Omega}+\vec{\zeta}$. The components of vorticity could significantly change when its magnitude and/or its direction changes appreciably due to various processes.

Circulation

Circulation is a scalar that measures the rotational property of a flow stemming from the notions of vortex lines and vortex tubes. A *vortex line* is defined to be a line to which the vorticity vector is tangential at every point of it as depicted in Fig. 3.1. It is analogous to a *streamline,* which is a line to which the velocity vector is tangential at every point of it. While the velocity field may have nonzero 3-D divergence, $\nabla\cdot\vec{V}\neq 0$, vortex lines are always nondivergent, since $\nabla\cdot\vec{\zeta}=\nabla\cdot\nabla\times\vec{V}=0$ whatever $\vec{V}$ might be. It follows that a vortex line cannot begin or terminate at any point in the interior of a fluid, but it may form a closed loop. If a vortex line does not form a closed loop, its two ends must terminate at the boundaries of a domain. Each vortex line meanders in a fluid during the evolution of a time dependent flow.

Let us focus on an arbitrary small closed curve Γ in a fluid. There is a certain vortex line passing through each point of this curve. The totality of all vortex lines passing through Γ would make up the surface of a tube, which is called a *vortex tube* (Fig. 3.1). The fluid inside such a tube is called a *vortex-filament.* We now define a line integral of the velocity component tangential to this chosen closed curve. The sign convention is that this integration is performed in an anticlockwise direction:

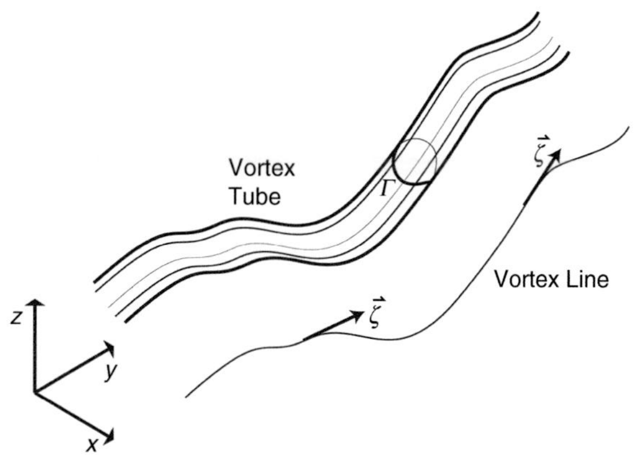

Fig. 3.1 Schematic of vortex lines and a vortex tube.

$$C = \oint_{\Gamma} \vec{V} \cdot d\vec{\ell}. \tag{3.5}$$

This is what we call *circulation*. Its dimension is $\mathrm{m^2\,s^{-1}}$. If $\vec{V}$ is $\vec{V}_{absolute}$, C would be absolute circulation, $C_{absolute}$. If $\vec{V}$ is $\vec{V}_{relative}$, C would be relative circulation, $C_{relative}$. If the curve Γ in (3.5) lies entirely on the surface of a vortex tube without enclosing it, the circulation would be zero. If Γ encircles a vortex tube, C is a measure of the strength of the vortex-filament. Any closed curve in 3-D space surrounding the same vortex tube encloses the same bundle of vortex lines. Thus, the circulation of all such closed curves would have the same value. Circulation is therefore a bulk measure of the rotational property of the fluid in a vortex-filament.

3.2 Relationship between vorticity and circulation

There must be a unique relationship between vorticity and circulation, since they are two alternative measures of the same rotational property of a flow. A simple form of such relation can be deduced in the context of a 2-D steady flow with the use of natural coordinates. Let us focus on the fluid within a closed curve ABCD straddling two adjacent trajectories in a segment where the curvature is concave (Fig. 3.2).

The circulation for a fluid parcel ABCD is

$$\begin{aligned}
\delta C &= \oint_{\Gamma} \vec{V} \cdot \overrightarrow{d\ell} = V_{\mathrm{AB}} \ell_{\mathrm{AB}} + V_{\mathrm{CD}} \ell_{\mathrm{CD}} \\
&= V|R|\delta\theta - \left(V + \frac{\partial V}{\partial n} \delta n \right) (|R| - \delta n) \delta\theta \\
&\approx V \delta n \, \delta\theta - \frac{\partial V}{\partial n} |R| \delta n \, \delta\theta.
\end{aligned} \tag{3.6}$$

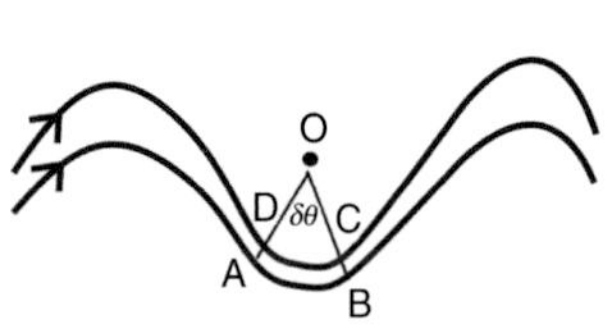

Fig. 3.2 Circulation and vorticity of a fluid element ABCD between two streamlines in a trough region.

Hence, we have

$$\frac{\delta C}{|R|\delta n\,\delta\theta} = \frac{V}{|R|} - \frac{\partial V}{\partial n}, \tag{3.7}$$

where the area of ABCD is $|R|\delta n\,\delta\theta$. Similar consideration of a fluid parcel in a region where the curvature of a flow is convex would yield $\frac{\delta C}{|R|\delta n\,\delta\theta} = -\frac{V}{|R|} - \frac{\partial V}{\partial n}$. Recall the sign convention adopted for natural coordinates, namely R is positive if $\vec{n}$ points toward the center of curvature (i.e. concave curvature; trough region), and vice versa. It follows that

$$\frac{\delta C}{|R|\delta n\,\delta\theta} = \frac{V}{R} - \frac{\partial V}{\partial n} \equiv \zeta \tag{3.8}$$

can be used at all locations. The two terms on the RHS of (3.8) constitute two parts of vorticity. One part is associated with the curvature and the other with the shear across the direction of a flow. In short, *circulation per unit area about a point is equal to the vorticity at that point in this flow.* Such vorticity may be interpreted as the *areal average vorticity around that point.*

The curvature term and shear term in (3.8) for a curved flow may be equal in magnitude but opposite in sign. In that case, the two parts would cancel and the vorticity of such a flow would be zero everywhere. It is called a *potential flow* (also referred to as *irrotational flow*). An example of a potential flow in a 2-D fluid has the following velocity components $u = (\cos x)(\cos y)$; $v = -(\sin x)(\sin y)$; $\zeta = v_x - u_y = 0$.

In general, a 2-D flow can be decomposed into two parts: one part associated with its rotational property and another part associated with its divergent property. The former is expressible in terms of a scalar function called *streamfunction* $\psi(x, y)$. The latter is expressible in terms of another scalar function called *velocity potential* $\chi(x, y)$. The mathematical relationships are $u = -\psi_y + \chi_x$, $v = \psi_x + \chi_y$. The field ψ in a potential flow is identically zero. The velocity potential of the sample flow mentioned above is $\chi = (\sin x)(\cos y)$. A large-scale atmospheric flow is far from being an irrotational flow.

There is a mathematical theorem, known as *Stokes' theorem*, in vector calculus. It says

$$\oint_{\Gamma} \vec{V} \cdot d\vec{\ell} = \iint_{A} (\nabla \times \vec{V}) \cdot \vec{n}\, da, \tag{3.9}$$

valid for *any* 3-D vector field, $\vec{V}$, and *any* closed curve Γ which encloses an area A; $\vec{n}$ is the unit vector at each point locally normal to the surface A. When $\vec{V}$ is the velocity field of a

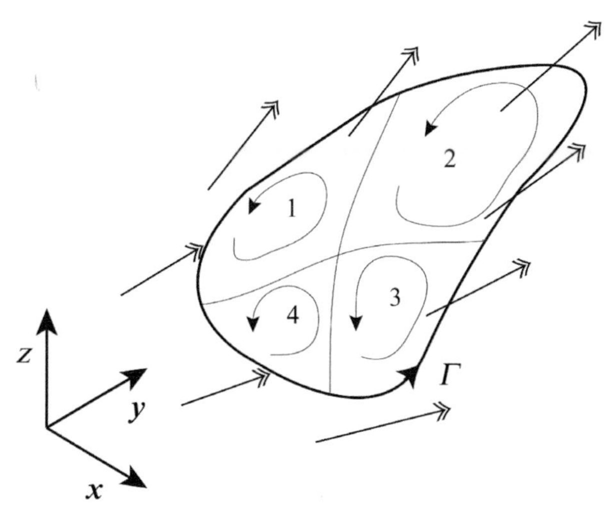

Fig. 3.3 Illustration of Stokes' theorem for the circulation of a velocity field defined for a closed curve Γ in space. The two thin lines divide area A enclosed by Γ into four subareas.

fluid, the integrand on the RHS of (3.9) at each point is the component of vorticity in the direction outward normal to an elemental area da centered at that point. If a fixed closed curve in space is defined for Γ, the LHS of (3.9) would be circulation in the Eulerian sense. If on the other hand, a closed material curve is chosen for Γ, the quantity is circulation in the Lagrangian sense. Hence, the average vorticity over the area A enclosed by a closed curve Γ is in general the circulation defined for that curve per unit area.

A heuristic illustration of Stokes' theorem for a general velocity field is given in Fig. 3.3. Imagine that we divide the enclosed area into four pieces as shown. The circulation of the whole area is equal to the sum of the circulation of the four subareas because the contributions at the interior boundaries would cancel one another. If we repeat this procedure of dividing up each area indefinitely, each piece of the surface would be infinitesimally small. Each piece may be regarded as a tiny circle centered about a point with a radius, R_i. The rotational property of the fluid enclosed by each C_i is effectively associated with certain rotating motion. Let us denote such rotating rate by ω_i. It follows that

$$C_i = \oint_{\Gamma_i} \vec{V} \cdot d\vec{\ell} = \omega_i R_i 2\pi R_i. \tag{3.10}$$

Hence, C is equal to the sum of the rotational property of all fluid elements enclosed by Γ,

$$C = \sum_i C_i = \sum_i 2\pi R_i^2 \omega_i = \sum_i \zeta_i \pi R_i^2, \tag{3.11}$$

where $\zeta_i = 2\omega_i$. In other words, circulation is indeed equal to the sum of the product of vorticity and the area of all constituent areas. The circulation per unit area is then the average vorticity in the enclosed area. This is the essence of Stokes' theorem.

3.3 Kelvin circulation theorem

A problem of general interest is to determine how the circulation of a flow associated with a material curve that encloses a vortex tube, $C(t)$, would change in time:

$$\frac{dC}{dt} \equiv \frac{DC}{Dt} = \frac{D}{Dt}\oint_{\Gamma} \vec{V} \cdot d\vec{\ell} = \oint_{\Gamma} \frac{D\vec{V}}{Dt} \cdot d\vec{\ell} + \oint_{\Gamma} \vec{V} \cdot \frac{D}{Dt}\left(d\vec{\ell}\right). \tag{3.12}$$

The second term vanishes because $\oint_{\Gamma} \vec{V} \cdot \frac{D}{Dt}\left(d\vec{\ell}\right) = \oint_{\Gamma} \vec{V} \cdot d\vec{V} = 0$. We then have for an inviscid flow in an inertial reference frame

$$\begin{aligned}\frac{dC_a}{dt} &= \oint_{\Gamma} \frac{D\vec{V}_a}{Dt} \cdot d\vec{\ell} = \oint \left(-\frac{1}{\rho}\nabla p + \vec{g}\right) \cdot d\vec{\ell} \\ &= -\oint_{\Gamma} \frac{dp}{\rho}. \end{aligned} \tag{3.13}$$

Equation (3.13) is known as *Kelvin circulation theorem*. Gravitational force does not contribute to $\frac{dC_a}{dt}$ because it is a conservative force. As such, it is a function of position only, meaning that it may be written as the gradient of a scalar potential function, ϕ, namely $\vec{g} = \nabla\phi = \phi_\ell \vec{\ell} + \phi_n \vec{n}$. It follows that $\oint \vec{g} \cdot d\vec{\ell} = \oint \nabla\phi \cdot d\vec{\ell} = \oint \phi_\ell \, d\ell = 0$.

When the RHS of (3.13) for a closed contour in a fluid is not zero, such a fluid is called a *baroclinic fluid* and the absolute circulation of a flow in it would change in time. To develop a feel for the influence of baroclinicity, let us consider the distributions of atmospheric temperature and pressure at a level near the surface shown in Fig. 3.4. Suppose the temperature decreases eastward and the pressure increases northward as we often observe in summer over the USA.

The baroclinic effect on the rotational property of the air enclosed by a closed contour ABCD in Fig. 3.4 is

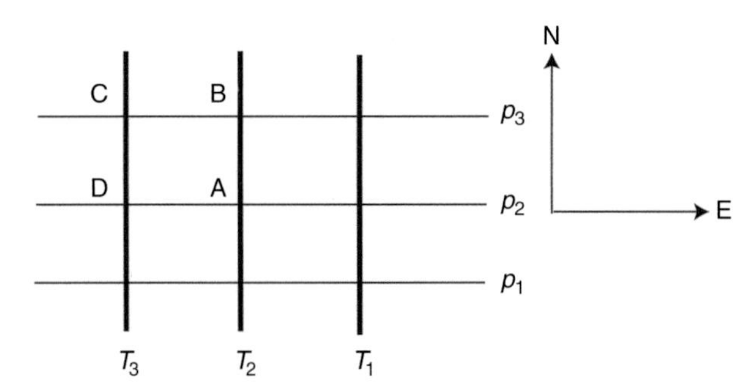

Fig. 3.4 Configuration of temperature and pressure fields on a horizontal plane where $p_j = p_o + j\Delta p$ and $T_j = T_o + j\delta T$; $T_{j+1} > T_j$ and $p_{j+1} > p_j$.

$$\frac{dC}{dt} = -\oint \frac{dp}{\rho} = -\oint \frac{RT}{p} dp = -RT_{\mathrm{AB}}(\xi_{\mathrm{B}} - \xi_{\mathrm{A}}) - RT_{\mathrm{CD}}(\xi_{\mathrm{D}} - \xi_{\mathrm{C}}),$$

where $\xi = \ln(p)$. Since $\Delta\xi \equiv (\xi_{\mathrm{B}} - \xi_{\mathrm{A}}) = -(\xi_{\mathrm{D}} - \xi_{\mathrm{C}}) > 0$, $T_{\mathrm{AB}} = T_2$ and $T_{\mathrm{CD}} = T_3$, we have $dC/dt = -R\Delta\xi(T_2 - T_3) > 0$. Thus, we conclude that the baroclinic effect would increase the circulation of the air on the closed curve ABCD. It follows from the discussion in Section 3.2 that the average vorticity of the air enclosed by ABCD would tend to become more cyclonic due to such configuration of temperature and pressure fields. Another way of looking at this is that pressure gradient force tends to accelerate air on CD as well as AB southward. The average density of air on AB is however larger than that on CD according to the equation of state because the air on AB is colder. It follows that the air on CD would be accelerated more compared to the air on AB. Such differential acceleration would give rise to a cyclonic shear. Hence, the circulation would increase confirming that this configuration of temperature and pressure is favorable for generating cyclonic vorticity in the air.

If the pressure in a fluid is solely a function of density, $p(\rho)$, such a fluid would be called a *barotropic fluid*. There exists an inverse relationship in this fluid, $\rho(p)$, implying the existence of a function $H(p)$ such that $dH/dp = 1/\rho$. It follows that we have $\oint dp/\rho = \oint dH = 0$. The absolute circulation for any closed curve in a barotropic fluid would not change in time and (3.13) would be reduced to

$$\frac{dC_a}{dt} = 0. \tag{3.14}$$

The simplest barotropic fluid model is a 2-D fluid. Since circulation for an infinitesimally small closed curve is vorticity per unit area, conservation of absolute circulation in turn implies conservation of absolute vorticity of each fluid element in a 2-D fluid.

$$\text{Special case, a 2-D fluid model:} \ \frac{D\zeta_a}{Dt} = 0. \tag{3.15}$$

This result can be, of course, derived directly from the momentum equations for a 2-D flow.

3.4 Dynamics of sea-breeze from the circulation perspective

The concepts we have just learned could help us get some insight into a local wind system in coastal areas during the daytime. It is known as *sea-breeze circulation*. It blows from the sea towards land at low levels. It typically penetrates about 10 km inland, L. The center of a related return flow aloft is at a height of about 1 km, h. A closed circulation is formed by rising motion over land and sinking motion over water. Sea-breeze is strongest in summer, especially in low latitude locations such as Hawaii or Florida where the solar heating is strong and the background wind is weak. It takes several hours after sunrise to develop to its maximum intensity. At night, there is a circulation in the reverse direction, known as *land-breeze circulation*.

The physics of sea-breeze is very simple. The sun warms up the coastal land surface in the morning much faster than the water next to it because land has a smaller specific heat.

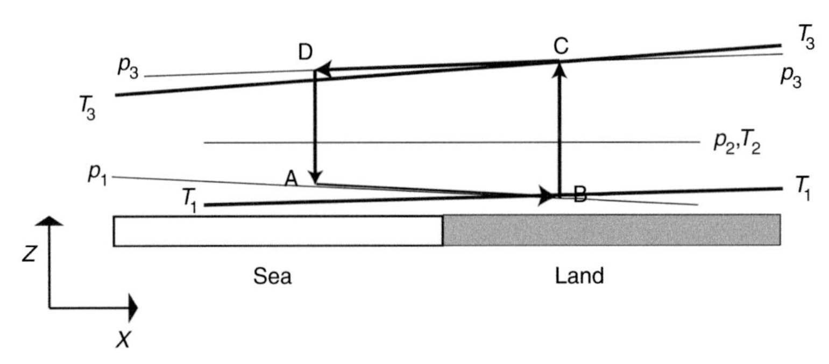

Fig. 3.5 Temperature and pressure surfaces over land and sea with $p_1 > p_2 > p_3$ and $T_1 > T_2 > T_3$. Circulation is defined for a closed curve, ABCD. AB is along $p = p_1$ and CD is along $p = p_3$.

Some heat is soon diffused upward by small turbulent eddies warming up a shallow layer of atmosphere over land more so than over water. That layer expands vertically causing pressure surfaces higher up to slightly slope down toward the sea. Air up there would begin to flow outward to sea creating a lower surface pressure over land. The resulting pressure gradient force at the surface in turn pushes the wind toward land. This process intensifies throughout the morning. More surface air rises as the wind blows inland because surface friction progressively inhibits the penetration of surface flow inland. Such rising air would further support the wind out toward the sea at higher levels. The vertical extent of this circulation is limited partly by the presence of a typical background stable stratification and partly by the diminishing heat diffusion with height. The level of strongest return flow outward to the sea is the top of the sea-breeze circulation. The Earth's rotation only has a minor effect partly because the time scale is only a few hours.

The dynamical nature of sea-breeze can be more rigorously examined in terms of the baroclinic effect on a flow. Since solar heating differentially raises the surface temperature and lowers the surface pressure on land, a plausible configuration of isotherms and isobars is sketched in Fig. 3.5. We can deduce a quantitative relationship between the intensity of a sea-breeze and the pertinent parameters by applying Kelvin's theorem to the circulation on a closed curve such as ABCD.

Making use of the fact that the pressure along two segments (AB and CD) of the closed curve is uniform, we can write

$$\begin{aligned}\frac{dC_a}{dt} &= -\oint \frac{dp}{\rho} = -\oint \frac{RT}{p}\,dp \\ &= R\bar{T}_{\mathrm{BC}} \ln \frac{p_1}{p_3} + R\bar{T}_{\mathrm{AD}} \ln \frac{p_3}{p_1} \\ &= R(\bar{T}_{\mathrm{BC}} - \bar{T}_{\mathrm{AD}}) \ln\left(\frac{p_1}{p_3}\right),\end{aligned} \tag{3.16}$$

$$\frac{d\langle u\rangle}{dt} = \frac{R(\bar{T}_{\mathrm{BC}} - \bar{T}_{\mathrm{AD}})}{2(h+L)} \ln \frac{p_1}{p_3}, \tag{3.17}$$

where $\langle u \rangle$ is the average speed along ABCD. For pertinent parameter values $p_1 = 100\,\text{kPa}$, $p_3 = 90\,\text{kPa}, \bar{T}_{\text{BC}} - \bar{T}_{\text{AD}} = 2\,\text{K}, L = 20\,\text{km}, h = 1\,\text{km}$, we get $d\langle u \rangle / dt = 1.3 \times 10^{-3}\,\text{m}\,\text{s}^{-2}$; $\langle uht \rangle$ would then increase by $4.5\,\text{m}\,\text{s}^{-1}$ in 1 hr. This value is unrealistically large, but we have not taken friction into consideration. The effect of friction may be evaluated as $\oint_{\Gamma} \vec{F}_{friction} \cdot d\vec{\ell}$ by using an appropriate expression for $\vec{F}_{friction}$ as a function of velocity. This is a limiting factor for the intensity of sea-breeze.

3.5 Tendency of relative circulation

Absolute velocity, absolute vorticity and absolute circulation of an atmospheric flow each consists of two parts. We use the following explicit notations in the discussion here: $\vec{V}_a = \vec{V}_e + \vec{V}_r$, $\vec{\zeta}_a = \vec{\zeta}_e + \vec{\zeta}_r$, $C_a = C_e + C_r$. Subscript "a" refers to value of a quantity in an inertial reference frame. Subscript "e" refers to the part associated with the rotation of the Earth. Subscript "r" refers to the part measured in a rotating coordinate. The planetary vorticity is simply equal to twice the Earth's rotation vector, $\vec{\zeta}_e = 2\vec{\Omega}$ according to (3.3). It is pertinent to ask: What is the circulation associated with the Earth's rotation defined for a contour Γ that encloses an area A as indicated by the sketch in Fig. 3.6?

We work out the expression for C_e as follows:

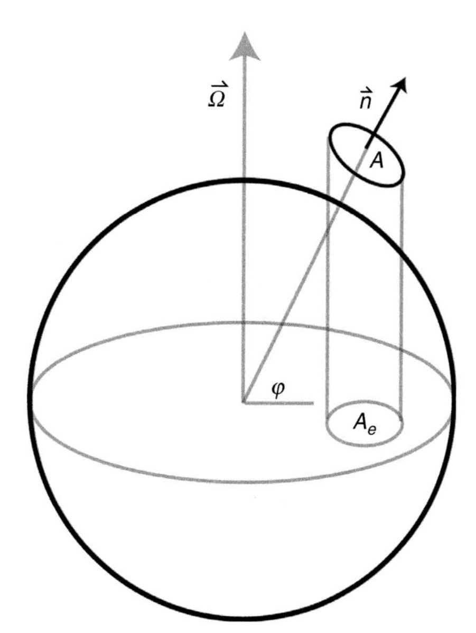

Fig. 3.6 Circulation associated with Earth's rotation for a closed curve enclosing area A in the atmosphere at latitude φ. The projection A_e is on the equatorial plane.

$$\begin{aligned} C_e &= \oint_\Gamma \vec{V}_e \cdot d\vec{\ell} = \iint \nabla \times \vec{V}_e \cdot \vec{n}\, da \\ &= \int_{\varphi l}^{\varphi l+\Delta\varphi} \int_{\lambda l}^{\lambda l+\Delta\lambda} = 2\Omega \sin\varphi (a\, d\varphi)(a \cos\varphi\, d\lambda) \\ &= 2\Omega a^2 \sin\varphi_1 \cos\varphi_1 \Delta\varphi\Delta\lambda \\ &= 2\Omega A_e. \end{aligned} \tag{3.18}$$

In other words,

(circulation for a closed contour Γ associated with the rotation of the Earth)
$= 2 \times$ (area A) $\times$ (local vertical component of Earth's rotation)
$= 2 \times$ (projected area of A on equatorial plane) $\times$ (Earth's rotation rate).

Hence, we have the following equation

$$\frac{dC_e}{dt} = 2\Omega \frac{dA_e}{dt}. \tag{3.19}$$

It says that the rate of change of C_e is proportional to the rate of change of A_e. It may arise from a change of the mean latitude of the contour or/and a change of the size of the area within Γ itself.

Kelvin circulation theorem can then be written with (3.19) as

$$\frac{dC_r}{dt} = -2\Omega \frac{dA_e}{dt} - \oint_\Gamma \frac{dp}{\rho}. \tag{3.20}$$

This equation says that the relative circulation for a contour could change by baroclinic effect and/or by a change in the projected area of the contour on a plane perpendicular to the rotation axis (i.e. "equatorial plane"). If such area decreases in time, C_r would tend to increase as in the case if the closed contour moves equatorward.

3.6 General vorticity equation

The equation that governs the rate of change of the vorticity vector at each point in the atmosphere is to be derived from the momentum equation,

$$\frac{\partial \vec{v}}{\partial t} = -\vec{v} \cdot \nabla \vec{v} - 2\vec{\Omega} \times \vec{v} - \frac{\nabla p}{\rho} + \nabla\phi + \vec{F}, \tag{3.21}$$

where $\nabla\phi$ is gravity with ϕ being the gravitational potential; $\vec{F}$ is frictional force per unit mass. Using $(\vec{v} \cdot \nabla)\vec{v} = \left((\nabla \times \vec{v}) \times \vec{v} + \nabla\left(\frac{1}{2}\vec{v} \cdot \vec{v}\right)\right)$ as a vector identity, taking the curl of (3.21) and making use of another vector identity $\nabla \times (\vec{U} \times \vec{V}) = (\vec{V} \cdot \nabla)\vec{U} - (\vec{U} \cdot \nabla)\vec{V} + \vec{U}(\nabla \cdot \vec{V}) - \vec{V}(\nabla \cdot \vec{U})$, we obtain the vorticity equation as

$$\frac{\partial \vec{\zeta}}{\partial t} = -(\vec{v} \cdot \nabla)\vec{\zeta}_a + \left(\vec{\zeta}_a \cdot \nabla\right)\vec{v} - \vec{\zeta}_a \nabla \cdot \vec{v} + \frac{\nabla \rho \times \nabla p}{\rho^2} + \nabla \times \vec{F}, \tag{3.22}$$

where $\vec{\zeta} = \nabla \times \vec{v}$ and $\vec{\zeta}_a = \vec{\zeta} + 2\vec{\Omega}$. The vorticity vector field of a flow therefore would change in time under the combined influences of five distinct processes represented by the five terms on the RHS of (3.22). The physical meanings of these terms may be briefly stated as follows:

(i) advection of absolute vorticity at the point under consideration by the flow field itself, $-(\vec{v} \cdot \nabla)\vec{\zeta}_a$,

(ii) stretching or compression of the vortex tube passing through that point (arising from the spatial variation of the velocity vector in the direction of the vortex tube), $(\vec{\zeta}_a \cdot \nabla)\vec{v}$,

(iii) 3-D convergence/divergence at that point, $-\vec{\zeta}_a(\nabla \cdot \vec{v})$, reflecting what would happen when a flow redistributes the mass of fluid under the inherent constraint of conservation of angular momentum,

(iv) baroclinic effect at that point, $\frac{\nabla \rho \times \nabla p}{\rho^2}$. A barotropic fluid is characterized by $\frac{\nabla \rho \times \nabla p}{\rho^2} = 0$ since density surfaces are parallel to isobaric surfaces.

The sea-breeze problem nicely illustrates this effect. Note that

$$\frac{1}{\rho^2}\nabla \rho \times \nabla p = \frac{1}{R\rho^2}\nabla\left(\frac{p}{T}\right) \times \nabla p = \frac{1}{R\rho^2}\left(\frac{1}{T}\nabla p - \frac{p}{T^2}\nabla T\right) \times \nabla p = \frac{R}{p}\nabla p \times \nabla T. \tag{3.23}$$

The y-axis in Fig. 3.5 is parallel to the coastline with unit vector $\vec{j}$. It follows that $\frac{R}{p}(\nabla p \times \nabla T) \cdot \vec{j} < 0$. Thus, the baroclinic effect tends to generate vorticity in the $(-y)$-direction. The corresponding circulation for a closed curve ABCD would be in a counterclockwise direction just like a sea-breeze.

(v) Frictional effect on the vorticity at a point is the curl of the frictional force, $\nabla \times \vec{F}$.

3.7 Vorticity dynamics of a large-scale flow

The typical values of the three components of relative vorticity in a large-scale flow of the extratropical atmosphere are $\left(-v_z + w_y\right) \approx -v_z \sim 10^{-3}\,\text{s}^{-1}$, $(u_z - w_x) \approx u_z \sim 10^{-3}\,\text{s}^{-1}$, $\left(v_x - u_y\right) \approx 10^{-5}\,\text{s}^{-1}$. The horizontal components of vorticity are therefore typically 100 times larger than the vertical component in such flows. The vorticity vectors then generally lie close to a horizontal plane. According to the thermal wind relation, the horizontal components of vorticity in a large-scale flow are also proportional to the horizontal variation of the temperature field: $u_z - w_x \approx u_z \propto -T_y$ and $-v_z + w_y \approx -v_z \propto -T_x$.

Although the vertical component of vorticity is the smallest of the three, it has special significance for a large-scale flow because such a flow is quasi-horizontal and its

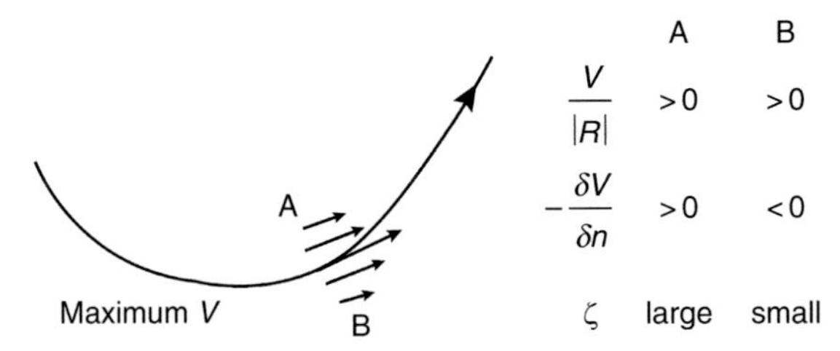

Fig. 3.7 Schematic of a flow and vorticity near a trough. A contour indicates the loci of maximum speed.

fluctuation would give a clear indication of how the wind field is changing. That is why close examination is traditionally given to this component alone. Unless specified otherwise, operational meteorologists use the word "vorticity" and notation ζ as synonym with the vertical component of the relative vorticity vector. The term "absolute vorticity" of a large-scale disturbance similarly refers to its vertical component alone, $\zeta_{abs} = 2\Omega \sin\varphi + v_x - u_y \equiv f + \zeta$ where f is likewise called planetary vorticity (twice the local vertical component of the Earth's rotation). The values of ζ_{abs} are generally larger at higher latitudes due to an increase of the planetary vorticity with latitude. The value of $f = 2\Omega \sin\varphi$ at $30, 40, 50, 60°$ N is 0.73, 0.94, 1.1, 1.26 in units of $10^{-4}\,\mathrm{s}^{-1}$ respectively. So the value of f in mid-latitude is $\sim 1 \times 10^{-4}\,\mathrm{s}^{-1}$ which is usually an order of magnitude larger than the vertical component of the relative vorticity vector, ζ, in large-scale disturbances. The absolute vorticity in an intense trough region however may exceed $2 \times 10^{-4}\,\mathrm{s}^{-1}$; ζ can be therefore quite large locally. We hardly ever observe negative values of ζ_{abs}. The observed distribution of absolute vorticity of a disturbance over N. America on April 7 is shown in Fig. 6.4 in the context of some related dynamical discussion.

The relative vorticity in a trough (region of low geopotential height) of an upper tropospheric flow is characteristically large and positive. The center of curvature is located to the left (poleward) side of a trough. The radius of curvature at points near a trough has small positive values. Thus, the curvature contribution to vorticity is large and positive around a trough. Furthermore, the spatial variation of wind speed is strongest in the area adjacent to a trough. The local shear, $\partial V/\partial n$, would then have large negative values on the left side of the wind maximum. Hence, the shear contribution to the vorticity is also large and positive. It follows that the total value of vorticity in a trough region is large. So, we expect to find much larger vorticity on the poleward side of a jet streak than on its equatorward side. These characteristics are summarized in Fig. 3.7.

Conversely, the ridges of a large-scale atmospheric flow are generally much broader than the troughs. We have $R < 0$ around a ridge. The curvature contribution is then negative and small. The wind around a ridge is also quite weak because the height contours are typically far apart. It follows that $\partial V/\partial n$ would be positive but small in the inside region of a ridge leading to a negative shear contribution. The two parts of vorticity around a ridge are therefore negative and small. The distribution of absolute vorticity of a representative large-scale flow field is given in Fig. 6.4.

We next elaborate on the intrinsic aspects of the vorticity dynamics of a large-scale disturbance with the equation that governs the variations in the vertical component of vorticity. Let us use a local Cartesian coordinate system (x, y, z) with its origin at (λ_o, φ_o) to depict such a flow. The unit vectors are $(\vec{i}, \vec{j}, \vec{k})$. The corresponding three components of a vector are systematically denoted by numeric indices 1, 2 and 3. For example, the vorticity is $\vec{\zeta} \equiv (\zeta_1, \zeta_2, \zeta_3) = (w_y - v_z)\vec{i} + (u_z - w_x)\vec{j} + (v_x - u_y)\vec{k}$ and the frictional force is $\vec{F} = (F_1, F_2, F_3)$. Taking the dot product of $\vec{k}$ and (3.23), we get

$$\frac{\partial \zeta_3}{\partial t} = -\vec{v} \cdot \nabla(\zeta_3 + f) + (\zeta_1 w_x + (\zeta_2 + h) w_y + (\zeta_3 + f) w_z) - (\zeta_3 + f)(u_x + v_y + w_z) + \frac{1}{\rho^2}(\rho_x p_y - \rho_y p_x) + ((F_2)_x - (F_1)_y), \tag{3.24}$$

where $f = 2\Omega \sin \varphi$ is the vertical component of Earth's rotation and $h = 2\Omega \cos \varphi$ is the poleward component of Earth's rotation. Hence, we can rewrite it as

$$\frac{\partial \zeta_3}{\partial t} = -\vec{v} \cdot \nabla(\zeta_3 + f) - (\zeta_3 + f)(u_x + v_y) + (\zeta_1 w_x + (\zeta_2 + h) w_y) + \frac{1}{\rho^2}(\rho_x p_y - \rho_y p_x) + ((F_2)_x - (F_1)_y). \tag{3.25}$$

The five terms on the RHS of (3.25) represent five distinct physical processes that could change the vertical component of vorticity. They are:

(i) Advection of absolute vorticity, $-(\vec{v} \cdot \nabla)(\zeta_3 + f)$

The latitudinal range of a large-scale flow is large enough that the variation of f should be taken into consideration. A northerly flow in northern hemisphere would have a positive contribution since f is larger at higher latitudes. As a first approximation, we may use

$$f = f_o + \beta y \tag{3.26}$$

where $f_o = 2\Omega \sin \varphi_o$, $\beta = \dfrac{2\Omega \cos \varphi_o}{a}$ and φ_o is the latitude of the center of the reference frame, $y = 0$. The advection of planetary vorticity with the approximate representation of the Coriolis parameter (3.26) is referred to as the *beta-effect*. We will see that this effect plays an important dynamical role in various types of disturbances.

(ii) Vortex stretching, $-(\zeta_3 + f)(u_x + v_y)$

Although this term is called "vortex stretching" in meteorology by tradition, it is actually a part of the 3-D divergence effect as noted in (3.24). It says that a vortex tube would be stretched or compressed in the vertical direction when the flow has horizontal convergence or divergence. Then ζ_3 would be increased or decreased correspondingly. An analogy of this process is that a skater can increase his spinning rate by bringing his arms towards his body by virtue of conservation of angular momentum about the vertical axis.

(iii) Tilting of the horizontal component of vorticity vector, $(\zeta_1 w_x + (\zeta_2 + h) w_y)$

This term is a part of the stretching/compression of a vortex tube through the point under consideration as noted earlier. However, meteorologists prefer to call this a *tilting term* for

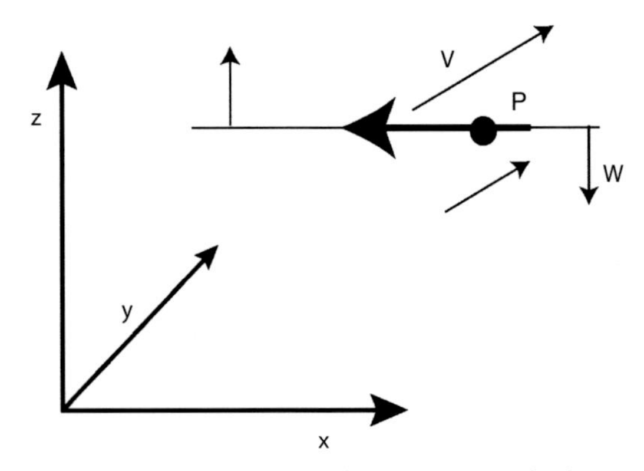

Fig. 3.8 Schematic of variation of meridional velocity and vertical velocity near point P for the case that the tilting term has a positive value.

short. The vertical component of vorticity at a point would be changed when horizontal variation in the vertical velocity tends to tilt the horizontal components of the absolute vorticity vector to the vertical direction. This interpretation is illustrated in Fig. 3.8 in the context of a velocity field that is characterized by $u = 0, v \neq 0, w \neq 0$.

Here the velocity field in a neighborhood of P is characterized by $v_z > 0$ and $w_y = 0$. There is then a vorticity component in the negative x-direction as represented by a bold arrow, $\zeta_1 = w_y - v_z < 0$. The vertical velocity decreases in the x-direction, $w_x < 0$. It follows that $\zeta_1 w_x > 0$ in this case. This picture suggests that the bold arrow would be "tilted" upward under this circumstance and ζ_3 would be increased. In general, $\zeta_1 w_x$ together with the term $\zeta_2 w_y$ jointly represent a process of tilting the horizontal part of the vorticity vector. It is instructive to make use of the fact that the term $\left(\zeta_1 w_x + (\zeta_2 + h) w_y\right)$ is negligible for large-scale disturbances because their w is much smaller than u and v. For this reason, we may focus on ζ_3 and its dependence on $f = 2\Omega \sin\varphi$.

(iv) Baroclinic effect

The term, $\vec{k} \cdot \left(\nabla p \times \nabla\left(\frac{1}{\rho}\right)\right) = \frac{1}{\rho^2}\left(\rho_x p_y - \rho_y p_x\right)$, stands for baroclinic effect on the vertical component of vorticity. It is not zero if the density surfaces do not coincide with the pressure surfaces. The pressure contours would intersect the density contours. For this reason, it is known as "solenoid term" in fluid dynamics. Its physical nature can be readily illustrated in Fig. 3.9.

In this configuration, we have

$$p_y = 0, \qquad p_x > 0, \qquad \rho_y < 0 \Rightarrow \rho_x p_y - \rho_y p_x = -\rho_y p_x > 0.$$

It follows that this term gives a positive contribution to $\partial\zeta_3/\partial t$. To develop a feel for this term, let us consider the pressure gradient force acting on two parcels of equal mass, A and B in Fig. 3.9. The pressure gradient is the same at A and B, but the density is smaller at A than at B. The pressure gradient force is therefore larger at A than at B. Mass A would be

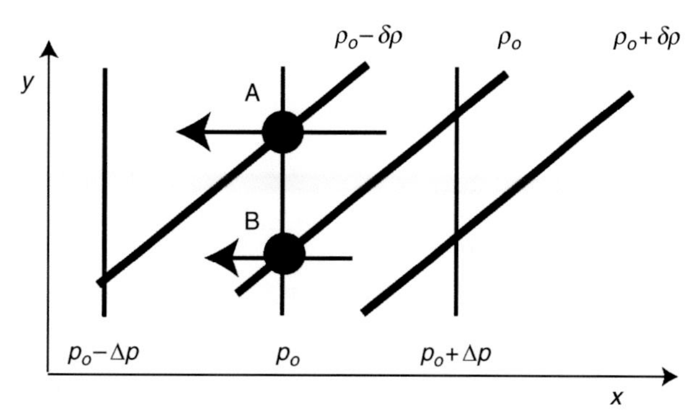

Fig. 3.9 Horizontal distribution of pressure and density fields for the case of a positive baroclinic effect. The z-coordinate points upward from the page.

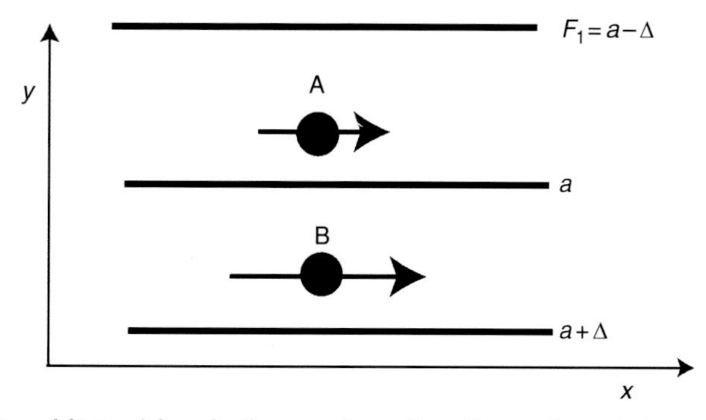

Fig. 3.10 Horizontal distribution of frictional force for the case of a positive effect on the vertical component of vorticity.

therefore accelerated more to the negative x-direction than mass B. That would give rise to cyclonic shear and hence positive vorticity. We may then conclude that $\zeta = v_x - u_y$ would be increased, as confirmed by the fact $\left(\rho_x p_y - \rho_y p_x\right) > 0$ in this case. In other words, positive value of the solenoid term at a point means that the pressure gradient force exerts a torque on a fluid parcel at that point and thereby increases its spin.

(v) Frictional effect

The term, $((F_2)_x - (F_1)_y)$, is illustrated by the following consideration. Suppose the frictional force at a level is directed in the x-direction ($F_2 = 0, F_1 \neq 0$) and F_1 is smaller at larger y as shown in Fig. 3.10. Now consider two parcels A and B of equal mass. Since the frictional force is stronger at B than at A, parcel B would be accelerated by a greater extent. Hence, we conclude that the vertical component of vorticity would be increased in this situation consistent with the condition of $\left(\frac{\partial F_2}{\partial x} - \frac{\partial F_1}{\partial y}\right) > 0$. In other words, the frictional force exerts a torque about the z-axis on a fluid parcel and thereby increases its spin.

The vorticity vector for a large-scale flow in isobaric coordinates (with the implicit application of the hydrostatic balance) may be approximately written as

$$\vec{\zeta} \equiv (\zeta_1, \zeta_2, \zeta_3) = (v_p)\vec{i} + (-u_p)\vec{j} + (v_x - u_y)\vec{k}. \tag{3.27}$$

Then the vorticity equation (the equation for the vertical component of the vorticity vector) in isobaric coordinates is

$$\begin{aligned}\frac{\partial \zeta}{\partial t} = &-(\zeta + f)(u_x + v_y) - u(\zeta)_x - v(\zeta)_y - \omega(\zeta)_p - v\beta \\ &- v_p\omega_x + u_p\omega_y + \left((F_2)_x - (F_1)_y\right),\end{aligned} \tag{3.28}$$

where $\zeta = v_x - u_y$. Same physical interpretations can be given to the terms in (3.28). Note that the baroclinic term does not appear explicitly in (3.28).

3.8 Potential vorticity dynamics

In 1942, H. Ertel identified a remarkable property of a flow in a rotating compressible fluid. It is called *potential vorticity* (PV). This is a generalization of simpler expressions of this property derived by C. G. Rossby several years earlier for fluid systems subject to hydrostatic balance. We will show in Section 3.8.2 that Ertel's PV for air is $q \equiv \frac{\vec{\zeta}_a \cdot \nabla\theta}{\rho}$, where $\vec{\zeta}_a = 2\vec{\Omega} + \nabla \times \vec{V}$ is absolute vorticity, $\vec{\Omega}$ the Earth's rotation vector, $\vec{V}$ velocity, θ potential temperature and ρ density. In spite of its name, its dimension is not that of vorticity but is rather $m^2\ s^{-1}\ K\ kg^{-1}$. It is convenient to use $10^{-6}\ m^2\ s^{-1}\ K\ kg^{-1}$ as one unit of PV (PVU). The PV of air near the tropopause turns out to be approximately equal to 2 PVU, increasing greatly with height. The air of tropospheric origin can be therefore readily distinguished from the air of stratospheric origin either in a map of PV distribution on a 320 K isentropic surface or in a map of potential temperature field on a 2-PVU surface. Such a pattern also broadly portrays the structure of the circulation because of the unique property of PV.

3.8.1 Preliminary remarks

Potential vorticity is a function of all dependent variables of the fluid ($\vec{V}$, θ and ρ). This means that the full set of governing equations for the fluid under consideration must be used in order to derive the equation that governs its temporal rate of change. As such, it is not a self-contained equation in general. We will go over the derivation in Section 3.8.6.

One might wonder at the outset: Why would anyone want to deal with a complicated looking quantity, such as $\frac{\vec{\zeta}_a \cdot \nabla\theta}{\rho}$? There are several good reasons for it. One reason is that under the condition of negligible dissipation and negligible net diabatic heating, this property of an air parcel would not change in time. We will elaborate on this characteristic of *material invariance* of a fluid in Section 3.8.2. It is a general principle in physics that each type of conservation is associated with a particular form of symmetry. The conservation of potential vorticity in a fluid is no exception. It can be shown to be intrinsically associated

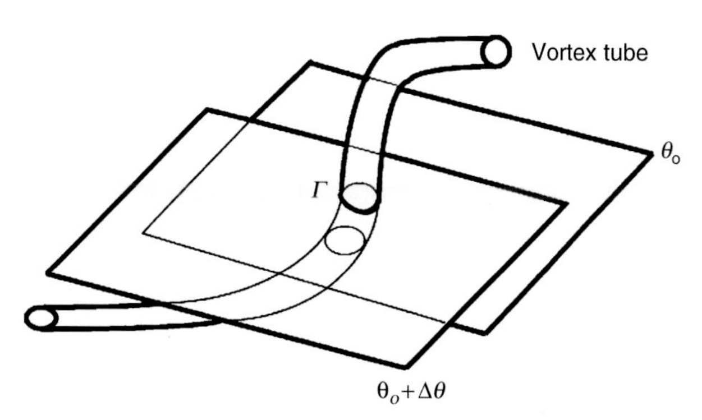

Fig. 3.11 Schematic of a vortex tube intersecting two isentropic surfaces; Γ is the interface curve of the vortex tube on the $\theta_o + \Delta\theta$ surface that encloses an area A; ℓ is the distance between the two isentropic surfaces.

with *particle-relabeling symmetry property* of a fluid (see Salmon (1998; Section 4.3) for details). Another reason is that one can deduce all other properties of a flow $(p, \rho, T, \vec{v})$ from its PV field together with the distribution of potential temperature on the lower and upper boundaries of a domain if it is a balanced flow. This is typically the case for a large-scale atmospheric disturbance. This result is so fundamental that it has come to be known as the *invertibility principle* (Hoskins *et al.*, 1985). Furthermore, the effects of diabatic heating and friction at the boundaries on a flow can be formally incorporated into the framework of a conservative property by introducing the notions of *generalized potential vorticity* and *generalized potential vorticity flux*. We will elaborate on such treatment in Section 3.9.

3.8.2 Physical nature of PV of a compressible fluid

We first prove that the material invariance property of $q \equiv \frac{\vec{\zeta}_a \cdot \nabla\theta}{\rho}$ in air parcels is a simple logical consequence of the Kelvin circulation theorem and mass conservation under adiabatic and inviscid conditions. Let us consider an air parcel on a certain isentropic surface, θ_o. A small vortex tube passes through this surface. The closed curve of intersection is labeled Γ which encloses an area, A. This vortex tube intersects another adjacent isentropic surface, $\theta_o + \Delta\theta$, at a distance ℓ apart from θ_o (Fig. 3.11).

By Kelvin circulation theorem *under inviscid condition*, we have

$$\frac{dC_a}{dt} = -\oint_{\Gamma} \frac{dp}{\rho}, \tag{3.29}$$

where $C_a = \oint_{\Gamma} \vec{V}_a \cdot d\vec{s}$ is absolute circulation. Since this Γ lies on an isentropic surface, the density along it would be only a function of pressure, viz.

$$\rho = \frac{p}{\theta_o R}\left(\frac{p_{oo}}{p}\right)^{R/c_p}. \tag{3.30}$$

It follows that the integral on the RHS of (3.29) would be zero even in a baroclinic fluid. Moreover, *under adiabatic condition*, θ of each fluid element conserves according to the First Law of Thermodynamics. Hence, Γ would be a material curve as well. Then we have

$$\frac{dC_a}{dt} = 0, \tag{3.31}$$

in a Lagrangian sense. The strength of such a vortex tube is therefore invariant,

$$C_a = \text{constant}. \tag{3.32}$$

Since $C_a = \oint_\Gamma \vec{V}_a \cdot d\vec{s} = \iint_A (\nabla \times \vec{V}_a) \cdot \vec{n}\, da$, the absolute vorticity normal to the area enclosed by Γ is equal to the circulation per unit area. (3.32) amounts to

$$\vec{\zeta}_a \cdot \vec{n} A = \text{constant} \tag{3.33}$$

for the elemental area A. Note that $\vec{n} = \frac{\nabla\theta}{|\nabla\theta|} = \frac{\nabla\theta}{(\Delta\theta)/\ell}$ is a unit vector normal to the enclosed area. *Under adiabatic condition, $\Delta\theta$ of the fluid column does not change,* (3.33) implies $\left(\vec{\zeta}_a \cdot \nabla\theta\right) A\ell =$ constant. The amount of mass in the infinitesimal fluid column would be also unchanged,

$$\rho A \ell = \text{constant}. \tag{3.34}$$

Then (3.33) and (3.34) jointly imply

$$\frac{\vec{\zeta}_a \cdot \nabla\theta}{\rho} = \text{constant}. \tag{3.35}$$

We therefore conclude that Ertel's potential vorticity, $q \equiv \frac{\vec{\zeta}_a \cdot \nabla\theta}{\rho}$, *of a compressible fluid parcel of unit mass conserves under adiabatic and inviscid conditions*. It warrants emphasizing that no additional approximation has been incorporated in the derivation above.

3.8.3 Observed characteristics of potential vorticity in the atmosphere

The distribution of potential vorticity field associated with the large-scale disturbance over North America on April 7, 2010 is shown in Fig. 3.12. There is a pronounced PV anomaly on the eastern side of the Rocky Mountain associated with a pronounced trough. The maximum value exceeds $10 \times 10^{-4}\,\mathrm{K\,m\,s^{-3}mb^{-1}} = 10 \times 10^{-6}\,\mathrm{K\,m^2\,s^{-1}/kg^{-1}} \equiv 10$ PV units. The map suggests significant exchange of stratospheric air and the tropospheric air in the 300–100 mb layer during this episode. The northerly flow on the west side of the trough brings large PV air to the south. It is noteworthy that the PV field has positive value effectively everywhere. We will see the implication of such a feature in the context of instability to be discussed in Chapter 8A. The map of PV for the 600–400 mb layer (not shown) reveals that this anomaly is still quite pronounced in mid-troposphere. Its intensity is significant with a maximum value reaching 2 PV units.

Time mean distributions of PV $q|_{x,y,\theta}$ tell us other information about atmospheric flow. Much of the day-to-day features in PV field are absent in the January mean map (Fig. 3.13). The contour of $\bar{q}|_{x,y,\theta} = 2$ is taken as an indication of the location of the mean tropopause. We see that the mean PV gradient is particularly strong off the east coast of Asia (western Pacific) and is much weaker over the eastern half of the Pacific. The gradient

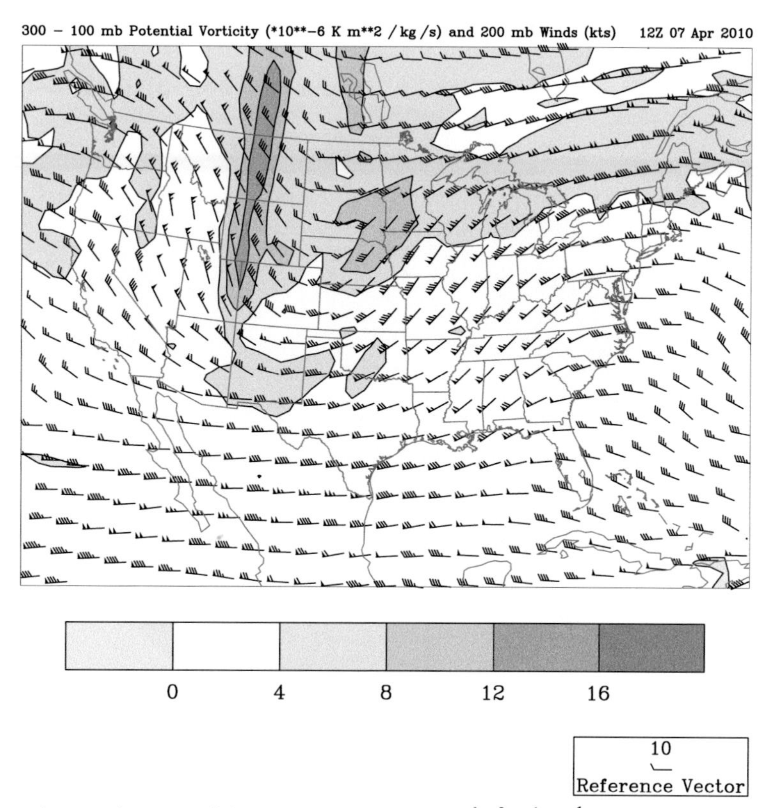

Fig. 3.12 Distributions of potential vorticity of the 300–100 mb layer in $10^{-6}\,m^2\,s^{-1}\,kg^{-1}\,K$ and the 200 mb wind field in knots on April 7, 2010. See color plates section.

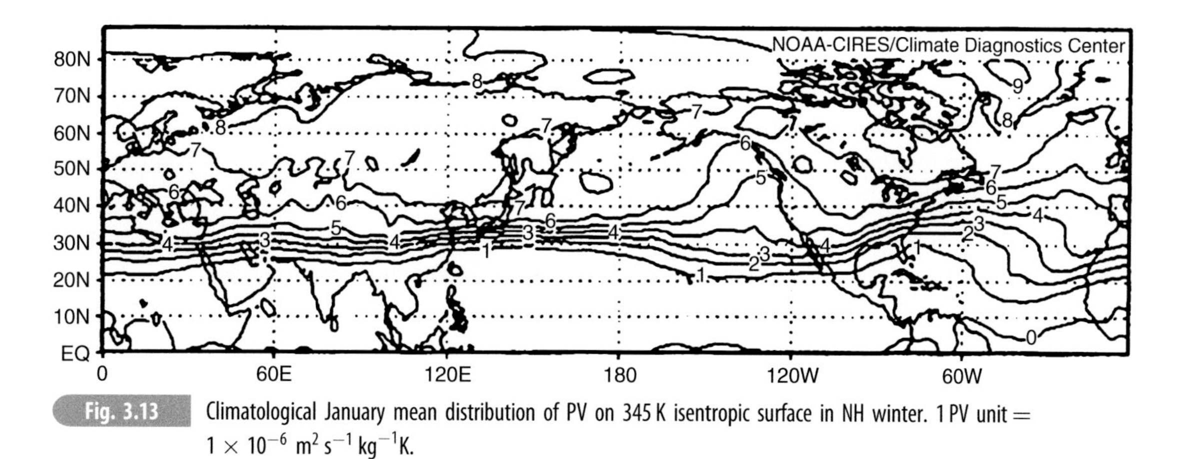

Fig. 3.13 Climatological January mean distribution of PV on 345 K isentropic surface in NH winter. 1 PV unit = $1 \times 10^{-6}\,m^2\,s^{-1}\,kg^{-1}K$.

is substantially weaker over the Atlantic. Furthermore, the PV gradient over the western Pacific is distinctly stronger in mid-winter (January) than in early winter (November) or late-winter (March). We will discuss some of the dynamical implications of such PV distributions in Chapter 11.

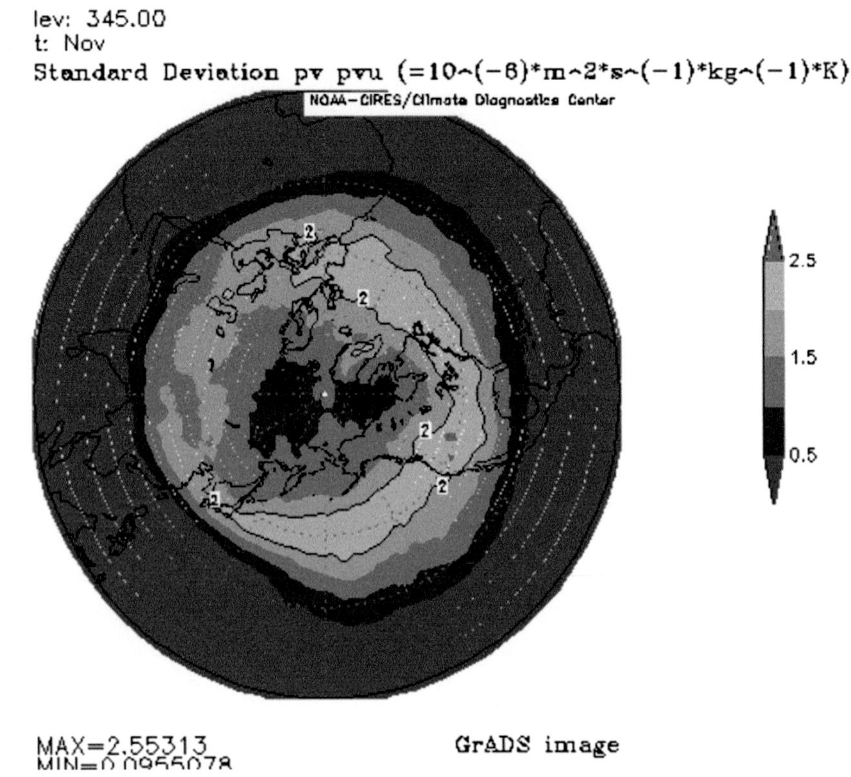

Fig. 3.14 Distribution of the standard deviation of PV on 345 K surface in January. See color plates section.

A statistical measure of the fluctuations in the PV field can be quantified by its root mean square value at each point with respect to the January mean, $\left[\overline{(q-\bar{q})^2}\right]^{0.5}$ (Fig. 3.14). It reveals that the variability of the isentropic PV is largest at the general location of the tropopause. It stems primarily from the movement of the tropopause during each winter month.

3.8.4 Physical nature of PV in a shallow-water model

In 1936, C. G. Rossby derived a mathematical expression of potential vorticity for a *shallow layer of rotating homogeneous fluid*. No thermodynamic process is at work in this fluid, which is therefore a *barotropic fluid*. Nevertheless, such a model is a simple setting suitable for investigating certain large-scale dynamical problems. As such, it is a valuable learning tool. The thickness of a shallow layer of water is much smaller than the horizontal scale of a disturbance under consideration by definition. It follows that we may introduce *hydrostatic balance a priori*. This in turn implies that the horizontal velocity of a disturbance does not vary with height in the layer. Each fluid column would move as an entity. The vorticity vector would be pointing upward in the vertical direction.

We now seek to similarly derive Rossby's result as we have derived the expression of PV for a compressible fluid. The closed curve around a fluid column under consideration is Γ that encloses an area A. We neglect friction for the moment. By Kelvin circulation theorem, we have

$$\frac{DC_a}{Dt} = 0 \Rightarrow C_a = \text{constant}. \tag{3.36}$$

Since vorticity is circulation per unit area enclosed by the closed curve, it means

$$\zeta_a A = \text{constant}, \tag{3.37}$$

where $\zeta_a = \zeta + f = v_x - u_y + 2\Omega \sin \varphi$. Denoting the length of the column by h, we get from mass conservation of this column of fluid

$$hA = \text{constant}. \tag{3.38}$$

Hence, we conclude that

$$\frac{\zeta + f}{h} = \text{constant},$$

$$\rightarrow \left(\frac{\partial}{\partial t} + u\frac{\partial}{\partial x} + v\frac{\partial}{\partial y}\right)\left(\frac{\zeta + f}{h}\right) = 0. \tag{3.39}$$

The quantity $q \equiv \dfrac{\zeta + f}{h}$ is the *potential vorticity of a fluid column in a shallow-water model* with a free surface. This invariance means that as a fluid column moves around in this layer, the change of its absolute vorticity would be subject to this strong constraint. If we take into account the variation of f with latitude, its relative vorticity and/or its length would change in such a way that its potential vorticity as expressed by (3.39) remains unchanged. We will apply this model in Section 5.8 to examine the dynamics of orographically forced disturbances after the basic Rossby wave dynamics has been discussed in Sections 5.6 and 5.7.

3.8.5 An intriguing deduction: eddy-driven jet

We next show the powerful constraint arising from the conservation of potential vorticity in the context of a forced flow. Let us consider for simplicity a shallow-water channel model on a beta-plane ($f = f_o + \beta y$) with a constant uniform depth (h=constant). It is effectively a 2-D fluid model; PV is reduced to absolute vorticity in this system. Equation (3.39) would be reduced to (3.15), namely $\dfrac{D}{Dt}(\zeta + f) = 0$, with $\zeta = \nabla^2 \psi$ where ∇^2 is the 2-D Laplacian. The domain is bounded by two parallel rigid lateral boundaries, say $y = \pm Y$. The flow is initially at rest. The boundary conditions are $\psi(-X, y) = \psi(X, y)$ and $\psi(x, \pm Y) = 0$ at all time. The domain average zonal momentum in the fluid under consideration is therefore zero. The zonal average zonal velocity at the lateral boundaries would remain unchanged, ($\int u dx$ = constant at $y = \pm Y$).

The following thought experiment is of some interest. Suppose a row of externally controlled paddles steadily stir the fluid in a narrow latitude belt. The stirring is random in the sense that it does not add or remove any net zonal momentum. All of the moving fluid

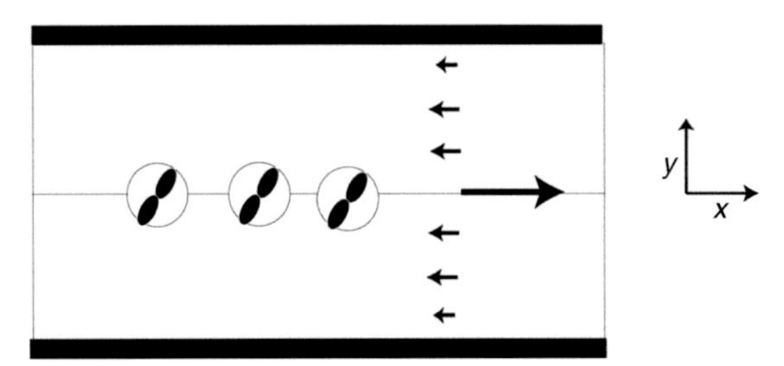

Fig. 3.15 Eddy-driven westerly jet on a reentrant beta-plane 2-D fluid model.

parcels not directly affected by the localized stirring would conserve their absolute vorticity. There would be an exchange of some fluid parcels across a *particular latitude to the north away from the stirring, say* $y = y_o$. Some of the fluid parcels in the region to the north of it must have originated from latitudes further south where f is smaller. In order for such a fluid parcel to conserve its original absolute vorticity, it must now have a negative value of relative vorticity. It follows that the circulation $\oint_{\Gamma} \vec{V} \cdot d\vec{\ell}$ defined for a closed path Γ consisting of four segments: $(x,y) = (x,y_o)$ with $-X \leq x \leq X$, $(x,y) = (X,y)$ with $y_o \leq y \leq Y$, $(x,y) = (x,Y)$ with $-X \leq x \leq X$ and $(x,y) = (-X,y)$ with $y_o \leq y \leq Y$, must have a negative value because the average vorticity in the enclosed area is circulation per unit area. Since the zonal mean zonal velocity at the boundary $y = Y$ remains zero by assumption, a negative circulation means that *there must be a zonal mean westward flow at* $y = y_o$. Similarly, consider the fluid parcels that have crossed any latitude to the south away from the stirring, say $y = -y_o$. The circulation in such a region has to be positive for the same reason. It follows that *there must also be a mean westward flow at* $y = -y_o$ so that the circulation for the fluid in the region $-Y \leq y \leq -y_o$ is positive. Since the stirring introduces no net zonal momentum into the fluid by assumption, there must be a compensating *zonal mean eastward flow in the narrow latitude belt of stirring*. Hence, we may conclude that random stirring would induce *a westerly jet* at the center of the domain in this model setting as schematically shown in Fig. 3.15. Notice that although there is large vorticity in the zonal mean flow near the two edges of the westerly jet as well as near the lateral boundaries, the domain integrated vorticity is zero.

The overall conclusion for this thought experiment would not change when we take into account the influence of weak viscous damping in a real fluid. Since molecular friction necessarily tends to smooth out a relative velocity field, the turbulent eddies must transport zonal momentum in an opposite direction so that a zonal jet could be sustained. In other words, the turbulent eddies transport momentum in an up-gradient direction without violating any laws of physics. This is a counter-intuitive relation between turbulent eddies and a mean flow. It should also be emphasized that these eddies owe their energetic existence to the external forcing. This is an example of what Starr (1968) tantalizingly referred to as the *phenomena of negative viscosity* observed in a number of natural settings. We now know that it stems from the fundamental constraint of conservation of potential vorticity.

The validity of the deduction above is quite general because it does not hinge upon any detailed properties of the eddies that do the stirring. Nor is it restricted to a barotropic fluid because the principle of conservation of potential vorticity applies to a baroclinic fluid as well. So if we associate such stirring with the action of the prevalent synoptic-scale eddies in the extratropics, we would come to the intriguing conclusion that a westerly jet in the extratropical atmosphere is in a local sense eddy-driven. The position of the observed mean jet stream also partly depends on, among other factors, the broad latitudinal distribution of the solar heating. Indeed, eddy-driven jets have been numerically simulated in simple baroclinic models that internally generate baroclinic eddies via baroclinic instability as long as a sufficiently strong broad north–south thermal contrast is imposed as an external forcing. We will discuss in Chapter 8 the topic of baroclinic instability in full. Furthermore, applying this argument to an equatorial region on a rotating planet, one may anticipate that if there are sufficiently strong random equatorial waves, they might even induce a mean westerly flow over the equator. Such a flow would appear to have *superrotation* in the planetary atmosphere.

3.8.6 General potential vorticity equation

We next derive the equation that governs the rate of change of the potential vorticity field in a general fluid system including the influences of diabatic heating and friction. It is to be derived from the full set of governing equations. The functional form of PV, namely $q = \dfrac{\vec{\zeta}_a \cdot \nabla\theta}{\rho}$, suggests that we might follow the following procedure. A starting point is the vorticity equation (3.17) repeated below for easy reference

$$\frac{\partial\vec{\zeta}}{\partial t} = -(\vec{v}\cdot\nabla)\vec{\zeta}_a + \left(\vec{\zeta}_a\cdot\nabla\right)\vec{v} - \vec{\zeta}_a\nabla\cdot\vec{v} + \frac{\nabla\rho\times\nabla p}{\rho^2} + \nabla\times\vec{F}, \tag{3.40}$$

where $\vec{F}$ is frictional force in m s^{-2}, $\vec{\zeta}_a = \vec{\zeta} + 2\vec{\Omega}$ with $\partial\vec{\Omega}/\partial t = 0$. Another governing equation is mass conservation, $D\rho/Dt = \rho\nabla\cdot\vec{v}$, with which we can rewrite (3.40) as

$$\frac{D}{Dt}\left(\frac{\vec{\zeta}_a}{\rho}\right) = \left(\frac{\vec{\zeta}_a}{\rho}\cdot\nabla\right)\vec{v} + \frac{\nabla\rho\times\nabla p}{\rho^3} + \frac{\nabla\times\vec{F}}{\rho}. \tag{3.41}$$

We next make use of the thermodynamic equation,

$$\frac{D\theta}{Dt} = Q, \tag{3.42}$$

where Q stands for diabatic heating rate in unit of K s^{-1}. Performing the operation $\left(\dfrac{\vec{\zeta}_a}{\rho}\cdot\nabla\right)$ on (3.42) and noting that

$$\left(\frac{\vec{\zeta}_a}{\rho}\cdot\nabla\right)\frac{D\theta}{Dt} = \frac{\vec{\zeta}_a}{\rho}\cdot\frac{D}{Dt}\nabla\theta + \left(\left(\frac{\vec{\zeta}_a}{\rho}\cdot\nabla\right)\vec{v}\right)\cdot\nabla\theta, \tag{3.43}$$

we get

$$\frac{\vec{\zeta}_a}{\rho}\cdot\frac{D}{Dt}\nabla\theta = \left(\frac{\vec{\zeta}_a}{\rho}\cdot\nabla\right)Q - \left(\left(\frac{\vec{\zeta}_a}{\rho}\cdot\nabla\right)\vec{v}\right)\cdot\nabla\theta. \tag{3.44}$$

Now take the dot product of (3.41) with $\nabla\theta$

$$\nabla\theta\cdot\frac{D}{Dt}\left(\frac{\vec{\zeta}_a}{\rho}\right)=\nabla\theta\cdot\left(\frac{\vec{\zeta}_a}{\rho}\cdot\nabla\right)\vec{v}+\nabla\theta\cdot\frac{\nabla\rho\times\nabla p}{\rho^3}+\nabla\theta\cdot\frac{\nabla\times\vec{F}}{\rho}. \tag{3.45}$$

The second term on the RHS is zero in light of the functional dependence $\theta(\rho,p)$. Adding (3.44) to (3.45) finally gives us the PV equation

$$\frac{D}{Dt}\left(\frac{\vec{\zeta}_a\cdot\nabla\theta}{\rho}\right)=\frac{\vec{\zeta}_a\cdot\nabla Q}{\rho}+\frac{\nabla\theta\cdot(\nabla\times\vec{F})}{\rho}. \tag{3.46}$$

It says that the PV of a fluid parcel would increase if (a) the gradient of heating has a component in the direction of the absolute vorticity vector and/or (b) the curl of the frictional force has a component in the direction of the gradient of potential temperature. Note that θ may be replaced by any scalar function of ρ and p, $\chi(p,\rho)$, as long as Q stands for the rate of gain/loss of χ.

3.8.7 Dynamics of a hurricane from the PV perspective

We illustrate one aspect of PV dynamics with (3.46) in the context of a thermally excited disturbance, such as a hurricane. Let us consider the wind and thermal structure of a hurricane from the PV perspective. There is strong diabatic heating in a hurricane associated with the copious precipitation. Its vertical variation is particularly large, implying $\nabla Q\approx\vec{k}Q_z$. As condensational heating in a hurricane is strongest at mid-tropospheric levels, we have $Q_z<0$ in the upper troposphere and $Q_z>0$ in the lower troposphere. The condensational heating would then tend to increase PV in its lower layer where $\vec{\zeta}_a\cdot\nabla Q\approx(v_x-u_u+f)$ $Q_z\approx fQ_z>0$ and to decrease PV in its upper layer where $(v_x-u_u+f)\,Q_z\approx fQ_z<0$ in the northern hemisphere. Cyclonic (anticyclonic) vorticity would be then progressively generated at the lower (upper) levels. As a hurricane intensifies, the frictional force associated with the surface wind also becomes progressively stronger. A simple valid representation of the frictional force is a Newtonian drag linearly proportional to the velocity, $\vec{F}=-\kappa\vec{V}$ with κ being a coefficient. Then the frictional effect on the PV of a fluid parcel, $\frac{1}{\rho}\nabla\theta\cdot\left(\nabla\times\vec{F}\right)\approx-\frac{\kappa}{\rho_o}\theta_z\zeta$, would be increasingly negative (positive) in the lower (upper) half of a mature hurricane. The heating effect and the friction effect on a developing hurricane are opposite. A hurricane would eventually reach a mature stage when these opposite effects on its PV field become comparable in magnitude.

The structure of pressure, horizontal rotational wind and vertical motion in a hurricane in the northern hemisphere are therefore interrelated through the condensational heating as summarized in Fig. 3.16. We will show a quantitative analysis of this problem in Section 3.9.2 using a two-layer model to verify the broad aspects of the deduction above, albeit in a larger scale setting.

Condensational heating can also be a significant factor in the growth of large-scale disturbances in the southern hemisphere, including hurricane-like disturbances. There must

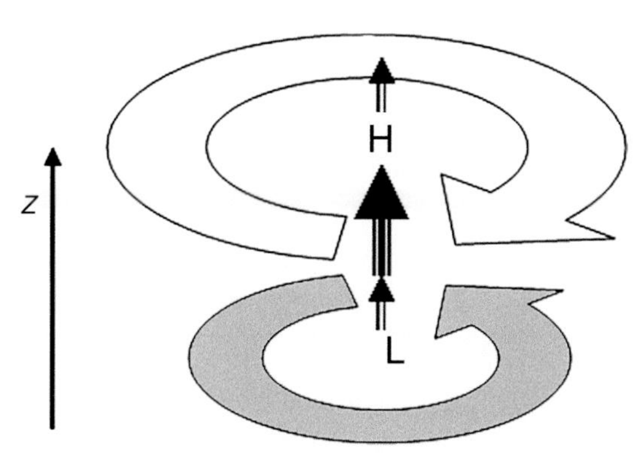

Fig. 3.16 Schematic of the distributions of pressure, horizontal wind and vertical motion in a hurricane in northern hemisphere.

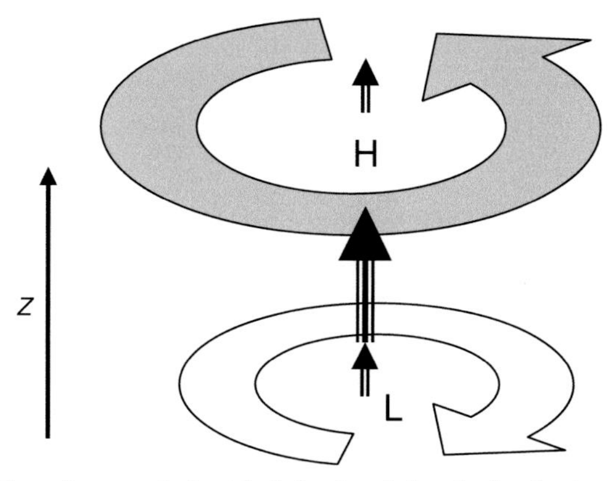

Fig. 3.17 Schematic of the distributions of pressure, horizontal wind and vertical motion in a hurricane in southern hemisphere.

be ascending motion to support the convection near the center of these disturbances. The heating would give rise to a low-pressure center near the surface. We would still find $Q_z > 0$ in the lower troposphere and $Q_z < 0$ in the upper troposphere. Since the Coriolis parameter is negative in the southern hemisphere, the heating would generate large negative values of low-level PV ($fQ_z < 0$ at low level). The low- (high-) level wind in such a disturbance must be in a clockwise direction (anticlockwise) in contrast to a similar disturbance in the northern hemisphere. The wind structure would be therefore opposite of that shown in Fig. 3.16, while the pressure distributions are similar. A schematic is shown in Fig. 3.17. Such low-level wind distribution is plainly evident in the satellite image of a low-pressure weather disturbance off Australia's southern coast.

3.9 Impermeability theorem and generalized potential vorticity

Useful as the principle of conservation of potential vorticity is, the Earth surface, the diabatic and the dissipative processes however do change the potential vorticity in an atmospheric flow. We now elucidate the dynamical nature of these influences from the PV perspective. Some intrinsic characteristics of PV with or without the influence of non-conservative processes are first discussed. A representation of the impacts of a domain's boundary, the diabatic and frictional processes on the PV dynamics will be next discussed by invoking the additional notions of generalized PV and generalized PV flux.

3.9.1 PV substance and impermeability theorem

From the definition of potential vorticity, we have $\rho q = \vec{\zeta}_a \cdot \nabla\theta = \nabla \cdot \left(\vec{\zeta}_a \theta\right)$ because $\nabla \cdot \vec{\zeta}_a = \nabla \cdot (\nabla \times \vec{v}_a) = 0$ for any absolute velocity. Haynes and McIntyres (1990; HM) called $s \equiv \rho q$ *PV concentration*. It is instructive to consider the integral of s over a *volume enclosed by an isentropic surface.* Let this volume be referred to by A and the enveloping surface area by B. By repeated applications of the divergence theorem, we would get

$$I = \iiint_A s\, dV = \iiint_A \nabla \cdot \left(\theta \vec{\zeta}_a\right) dV = \iint_B \theta \vec{\zeta}_a \cdot \vec{n}\, db = \theta \iint_B \vec{\zeta}_a \cdot \vec{n}\, db = \theta \iiint_A \nabla \cdot \vec{\zeta}_a\, dV = 0, \tag{3.47}$$

where $\vec{n}$ is unit vector outward normal to the elemental surface db. This is so because, unlike the concentration of a chemical tracer, s is not a positive definite property of a fluid. The quantity $(s\delta V)$ is simply PV of the fluid in an elemental volume δV. Haynes and McIntyres call it *PV-substance* (PVS). The total PV-substance in such a volume is always zero implying that there must be an equal amount of positive PV-substance and negative PV-substance in such a volume. This result is based entirely on the definition of potential vorticity and no physical argument has been invoked in the course of establishing (3.47). That means that (3.47) is valid under ANY conditions regardless of whether or not there is diabatic heating/cooling and friction.

Since the large-scale atmosphere is stably stratified, the potential temperature at each location increases monotonically with height. Consider a layer of the atmosphere between two different isentropic surfaces with values θ_1 and θ_2. Each surface envelops the whole planet Earth and the lower surface θ_2 does not intersect the ground. The two surfaces under consideration are therefore topologically spherical. Let us refer to the volume by A, the upper and lower isentropic surfaces by B_1 and B_2. The entire volume enclosed by B_1 is referred to as A_1 and that by B_2 as A_2, so that $A = A_1 - A_2$. Then we have

$$\begin{aligned} I = \iiint_{A_1 - A_2} s\, dV &= \iiint_{A_1} \nabla \cdot \left(\theta \vec{\zeta}_a\right) dV - \iiint_{A_2} \nabla \cdot \left(\theta \vec{\zeta}_a\right) dV \\ &= \iint_{B_1} \theta \vec{\zeta}_a \cdot \vec{n}\, db - \iint_{B_2} \theta \vec{\zeta}_a \cdot \vec{n}\, db \\ &= \theta_1 \iiint_{A_1} \nabla \cdot \vec{\zeta}_a\, dV - \theta_2 \iiint_{A_2} \nabla \cdot \vec{\zeta}_a\, dV \\ &= 0. \end{aligned} \tag{3.48}$$

This layer is equivalent to a volume enclosed by a single isentropic surface since its total PVS is exactly zero. The PV of the air over the northern and southern hemispheres largely has opposite signs stemming from the rapid rotation of the Earth about a fixed axis. That is the underlying physical basis of (3.48). It warrants emphasizing again that we have not imposed any physical restriction for establishing (3.48). It is therefore valid under ANY conditions.

Why does a volume of air enclosed by an isentropic surface always have a zero value of PVS under all circumstances? To address this question, we need to consider it in conjunction with the governing PV equation (3.46). The diabatic heating in units of $\mathrm{K\,s^{-1}}$ is denoted as $\dot{\theta}$. The PV equation can be rewritten as

$$\begin{aligned} &\rho\frac{Dq}{Dt}+\nabla\cdot\vec{N}=0, \\ &\vec{N}=-\dot{\theta}\vec{\zeta}_a-\vec{F}\times\nabla\theta, \end{aligned} \tag{3.49}$$

because $\vec{\zeta}_a\cdot\nabla\dot{\theta}=\nabla\cdot\left(\vec{\zeta}_a\dot{\theta}\right)-\dot{\theta}\nabla\cdot\vec{\zeta}_a=\nabla\cdot\left(\vec{\zeta}_a\dot{\theta}\right)$ and $\nabla\theta\cdot\left(\nabla\times\vec{F}\right)=\vec{F}\cdot(\nabla\times\nabla\theta)-\nabla\cdot\left(\nabla\theta\times\vec{F}\right)=\nabla\cdot\left(\vec{F}\times\nabla\theta\right)$. The influences of diabatic heating, $\dot{\theta}$, and frictional force, $\vec{F}$, on the PV of a fluid are now expressed in the form of divergence of a vector $\vec{N}$. This vector has the dimension of PV flux per unit volume, (velocity)∗(density)∗ (PV per unit mass) = $\mathrm{K\,s^{-2}}$. By making use of the continuity equation, we can further rewrite (3.49) as

$$\begin{aligned} &\frac{\partial(\rho q)}{\partial t}+\nabla\cdot\vec{J}=0, \\ &\vec{J}=\vec{v}\rho q+\vec{N}=\vec{v}\rho q-\dot{\theta}\vec{\zeta}_a-\vec{F}\times\nabla\theta. \end{aligned} \tag{3.50}$$

So one may interpret $\vec{J}$ as a total flux of s consisting of an *advective flux* $\vec{v}\rho q$ and a *non-advective flux* $\vec{N}$. Noting that $\dot{\theta}=\theta_t+\vec{v}\cdot\nabla\theta$, we may introduce the following identity

$$\begin{aligned} \vec{v}\rho q-\dot{\theta}\vec{\zeta}_a &=\vec{v}\rho q-\dot{\theta}\vec{\zeta}_a+\rho q\frac{\nabla\theta}{|\nabla\theta|^2}\left(\dot{\theta}-\frac{\partial\theta}{\partial t}-\vec{v}\cdot\nabla\theta\right) \\ &=-\rho q\left(\frac{\partial\theta}{\partial t}\Big/|\nabla\theta|\right)\frac{\nabla\theta}{|\nabla\theta|}+\rho q\left(\vec{v}-\frac{(\vec{v}\cdot\nabla\theta)}{|\nabla\theta|}\frac{\nabla\theta}{|\nabla\theta|}\right)-\dot{\theta}\left(\vec{\zeta}_a-\frac{\vec{\zeta}_a\cdot\nabla\theta}{|\nabla\theta|}\frac{\nabla\theta}{|\nabla\theta|}\right) \\ &\equiv\rho q\vec{v}_\perp+\rho q\vec{v}_{||}-\dot{\theta}\vec{\zeta}_{||}. \end{aligned} \tag{3.51}$$

A vector which is parallel to a local isentropic surface is indicated by the symbol || in the last expression. Likewise, a vector that is perpendicular to a local isentropic surface is indicated by the symbol $\perp$; $\vec{v}_\perp$ is a velocity that measures the local movement of an isentropic surface. Now integrating (3.50) over a volume A enclosed by an isentropic surface B and making use of (3.51), we get

$$\iiint\limits_A\frac{\partial s}{\partial t}dV=\iint\limits_B\left(\rho q\vec{v}_\perp+\rho q\vec{v}_{||}-\dot{\theta}\vec{\zeta}_{||}-\vec{F}\times\nabla\theta\right)\cdot\vec{n}\,db. \tag{3.52}$$

The integrals of the second, third and fourth terms on the RHS are zero since those three vectors are parallel to the isentropic surface B. The first term on the RHS represents a flux of s due to the velocity component normal to the surface B. The LHS is equal to $\frac{d}{dt}\iiint_A s\,dV + \iint_B s\vec{v}_\perp \cdot \vec{n}\,db$. It follows that

$$\frac{d}{dt}\iiint_A s\,dV = 0. \tag{3.53}$$

Thus, the PVS of volume A defined by a particular isentropic surface does not change at all, meaning that there is no source/sink of PVS within such a volume. This conclusion is referred to by HM as the *impermeability theorem for PV substance*. They suggest thinking of PVS as a "notional substance" that is trapped on isentropic surfaces. A few words of elaboration are in order. The enclosing isentropic surface under consideration is made up of different fluid parcels before and after the heating. Impermeability for PV-substance is not synonymous with impermeability for PV. The mathematical deduction leading up to (3.53) clearly tells us that the movement of the enclosing isentropic surface is equal to the transport of PVS in the direction normal to the isentropic surface itself. Therefore, the observable transport of PVS can only be along an isentropic surface. Also, since no isentropic surface encloses a volume within a single hemisphere, it would be meaningless to speak of impermeability of PVS for a portion of a domain. Impermeability for PV-substance is meaningful only in a global context.

Now consider the impacts of local diabatic and frictional processes near the ground over a certain area. The isentropic surfaces would be intersecting the surface of the Earth. We refer to that surface including this limited portion of the ground by $B_{surface}$. The potential temperature on $B_{surface}$ then is non-uniform yielding

$$\iint_{B_{surface}} \theta\vec{\zeta}_a \cdot \vec{n}\,db \neq 0. \tag{3.54}$$

The value of this integral can change in time. We may describe this situation by saying that the lower boundary of the atmosphere is a source/sink of PVS. This interpretation highlights the important role of the physical processes at the surface of the Earth from the perspective of PV dynamics.

3.9.2 Influence of boundaries and generalized potential vorticity

Since the surface of the Earth is not an isentropic surface, (3.47) would not hold for a volume enclosed by a surface including a part of the ground. The boundary of a domain per se therefore plays an important role in PV dynamics. Physically, the typical variations of the potential temperature at a boundary arise from the diabatic and frictional processes. The impacts of those processes are expressible in terms of equivalent fluxes of PV according to (3.50). We make use of such formalism to establish a succinct representation of the influences of non-conservative processes at the boundary. Schneider *et al.* (2003) show that this is feasible by virtue of a mathematical analogy between the definition of PV together with the PV governing equation and two of Maxwell's equations in the theory of electromagnetism.

We begin by noting that PV concentration is equal to the divergence of a vector, $\rho q = \nabla \cdot \left(\vec{\zeta}_a \theta\right)$. However, there is no unique vector for a specific distribution of ρq, since an additional non-divergent vector may be added to it without affecting this relation. We refer to the unrestricted vector field by $\vec{D} = \vec{\zeta}_a \theta + \vec{B}$, where $\vec{B}$ can be *an arbitrary non-divergent vector field*. A more general form of the relation is then

$$\rho q = \nabla \cdot \vec{D}. \tag{3.55}$$

For example, if we choose $\vec{B} = -\nabla \times (\theta \vec{v}_a)$, the corresponding $\vec{D}$ would be $\vec{D} = \vec{v}_a \times \nabla \theta$. With (3.55) we can also rewrite the PV equation (3.46) in the form of divergence of a vector field,

$$\nabla \cdot \left(\vec{D}_t + \vec{J}\right) = 0, \tag{3.56}$$

where $\vec{J} = \vec{v}\rho q - \dot{\theta}\vec{\zeta}_a - \vec{F} \times \nabla \theta$. Since the divergence of this combined vector $\left(\vec{D}_t + \vec{J}\right)$ is zero, we can always represent it as the curl of another vector field $\vec{H}$. A more general form of the PV equation is then

$$\nabla \times \vec{H} = \vec{D}_t + \vec{J}, \tag{3.57}$$

where $\vec{H}$ is the counterpart of the vector potential in electromagnetism. Furthermore, the gradient of an arbitrary scalar function ψ may be added to $\vec{H}$ for given $\vec{D}_t$ and $\vec{J}$ without affecting the validity of (3.57). The transformation $\vec{H} \to \vec{H} + \nabla \psi$ is called a *gauge transformation*. Such transformations on $\vec{H}$ are possible because the sum of $\vec{D}_t$ and $\vec{J}$ specifies only the curl of $\vec{H}$.

We now examine the PV dynamics from the perspective of the two field equations, (3.55) and (3.57). Equation (3.55) has the form of Coulomb's law in electrostatics and (3.57) has the form of Ampere's law generalized by Maxwell for a time-dependent electric field. Therefore, they are identical in form to two of the four Maxwell equations; ρq is the analog of electric charge density, $\vec{J}$ of the current density, $\vec{D}$ of the electric field and $\vec{H}$ of the magnetic field. The term $\partial \vec{D}/\partial t$ in (3.57) corresponds to the *displacement current* in the theory of electromagnetism; ρq and $\vec{J}$ may be conceptually thought of as the sources of the vector fields $\vec{D}$ and $\vec{H}$. The analogy between the equations for PV dynamics and the Maxwell equations is obviously not complete. After all, there is no counterpart of Faraday's law in PV dynamics; $\vec{H}$ is also not constrained to be non-divergent as the case of a magnetic field. Although the relationship between $(\rho q)_t$ and $\nabla \cdot \vec{J}$ is unique according to (3.50), the relationship between $\vec{D}_t$ and $\vec{J}$ according to (3.55) and (3.57) is not so because we are free to specify the arbitrary vectors $\vec{B}$ and $\nabla \psi$.

A general gauge transformation can be introduced as

$$\begin{aligned} \vec{D} &\to \vec{D}' = \vec{D} + \nabla \times \vec{B}, \\ \vec{H} &\to \vec{H}' = \vec{H} + \vec{B}_t + \nabla \psi. \end{aligned} \tag{3.58a,b}$$

Here $\vec{B}$ can be any vector field since $\nabla \times \vec{B}$ is non-divergent and ψ can be any scalar field. If we apply this gauge transformation to (3.55) and (3.57), we get

$$\nabla \cdot \vec{D}' = \rho q$$
$$\nabla \times \vec{H}' = \vec{D}'_t + \vec{J} \tag{3.59a,b}$$

for any generating function ψ. In other words, the form of the equations for $\vec{D}'$ and $\vec{H}'$ is identical to that of those for $\vec{D}$ and $\vec{H}$. Corresponding to a solution of $\vec{D}$ and $\vec{H}$, there can be an infinite number of possible solutions of $\vec{D}'$ and $\vec{H}'$ depending on the choices of $\vec{B}$ and ψ. The choice of $\vec{B}$ and ψ is to be made for convenience as in an analysis of the Maxwell equations.

For a particularly simple gauge $\vec{B} = 0$ and $\psi = 0$, we have

$$\vec{D} = \vec{\zeta}_a \theta,$$
$$\nabla \times \vec{H} = \vec{\zeta}_a \frac{\partial \theta}{\partial t} + \theta \frac{\partial \vec{\zeta}_a}{\partial t} + \vec{v}\rho q - \dot{\theta}\vec{\zeta}_a - \vec{F} \times \nabla\theta. \tag{3.60a,b}$$

The immediate task is to establish the functional relationship between $\vec{H}$ and the state variables by rewriting the RHS of (3.60b) as the curl of a vector. First, note that

$$-\dot{\theta}\vec{\zeta}_a + \vec{\zeta}_a \frac{\partial \theta}{\partial t} = -\vec{\zeta}_a(\vec{v} \cdot \nabla)\theta,$$
$$\vec{v}\rho q - \vec{F} \times \nabla\theta = \vec{v}\left(\vec{\zeta}_a \cdot \nabla\theta\right) + \nabla\theta \times \vec{F}$$
$$= \vec{v}\left(\nabla \cdot \left(\theta\vec{\zeta}_a\right)\right) + \nabla \times \left(\theta\vec{F}\right) - \theta\left(\nabla \times \vec{F}\right).$$

By the vorticity equation (3.17), we also have

$$\theta \frac{\partial \vec{\zeta}_a}{\partial t} = -\theta(\vec{v} \cdot \nabla)\vec{\zeta}_a + \theta\left(\vec{\zeta}_a \cdot \nabla\right)\vec{v} - \theta\vec{\zeta}_a \nabla \cdot \vec{v} + \theta \frac{\nabla\rho \times \nabla p}{\rho^2} + \theta\nabla \times \vec{F}$$
$$= -(\vec{v} \cdot \nabla)\left(\theta\vec{\zeta}_a\right) + \vec{\zeta}_a(\vec{v} \cdot \nabla)\theta + \theta\left(\vec{\zeta}_a \cdot \nabla\right)\vec{v} - \theta\vec{\zeta}_a \nabla \cdot \vec{v} + \frac{\theta}{\rho^2}\nabla\rho \times \nabla p + \theta\nabla \times \vec{F}.$$

Furthermore, using the equation of state for ideal gas and the definition of θ, we get

$$\frac{\theta}{\rho^2}\nabla\rho \times \nabla p = -\frac{R\theta}{p}\nabla T \times \nabla p = -\frac{R\theta}{p}\nabla T \times \left(\frac{c_p p}{RT}\nabla T - \frac{c_p p}{R\theta}\nabla\theta\right)$$
$$= c_p \nabla T \times \nabla\theta$$
$$= \nabla \times (T\nabla\theta).$$

Upon substituting the four expressions above into (3.60b), cancelling some terms and using the vector identity, $\nabla \times \left(\vec{U} \times \vec{V}\right) = \left(\vec{V} \cdot \nabla\right)\vec{U} - \left(\vec{U} \cdot \nabla\right)\vec{V} + \vec{U}\left(\nabla \cdot \vec{V}\right) - \vec{V}\left(\nabla \cdot \vec{U}\right)$, we can rewrite (3.59b) as $\nabla \times \vec{H} = \nabla \times \left(\vec{v} \times \vec{\zeta}_a\theta + c_p T\nabla\theta + \vec{F}\theta\right)$. It follows that

$$\vec{H} = \left(\vec{v} \times \vec{\zeta}_a\theta + c_p T\nabla\theta + \vec{F}\theta\right). \tag{3.61}$$

Equations (3.60a) and (3.61) relate $\vec{D}$ and $\vec{H}$ to the state variables for this particular gauge. These are merely relations and not solutions since the state variables themselves are still unknown. The vector fields $\vec{D}$ and $\vec{H}$ remain to be determined by solving (3.55) and (3.57).

We next apply Eqs. (3.60a) and (3.61) to formulate a representation of the boundary effects on PV dynamics. First note that the condition at a surface for the field $\vec{D} = \vec{\zeta}_a\theta$ is typically inhomogeneous since potential temperature varies and fluctuates at the surface. It would be much simpler to perform an analysis for the field subject to a homogeneous boundary condition. The *normal component of an electric field* (the counterpart of $\vec{D}$) is known to be discontinuous on a surface that has electric charge (the counterpart of ρq). It suggests that we could seek to pose a problem such that $\vec{D}$ may be regarded as zero inside a surface, but nonzero barely above the surface. Let us denote the $\vec{D}$ inside the surface by $\vec{D}_b$, the $\vec{D}$ immediately above the surface by $\vec{D}_s$, and the unit vector normal to the Earth surface by $\vec{n}$. So, the problem amounts to determining the surface PV required for forcing the normal component of the field $\vec{D}$ from $\vec{n}\cdot\vec{D} = \vec{n}\cdot\vec{D}_b = 0$ inside the surface to certain $\vec{n}\cdot\vec{D} = \vec{n}\cdot\vec{D}_s \neq 0$ barely above the surface. This can be done by integrating (3.55) over a small volume, A, enclosed by a surface area B with an infinitesimal thickness Δ across the surface $z = z_s$. We then get

$$\iiint_A \rho q\, dV = \iiint_A \nabla\cdot\vec{D}\, dV = \iint_B \vec{D}\cdot\vec{n}\, dS = \vec{n}\cdot\left(\vec{D}_s - \vec{D}_b\right)B. \tag{3.62}$$

The LHS would be nonzero for an infinitestimal volume A only if q has a singularity (extremely large) in that small volume. In other words, an inhomogeneous boundary condition for $\vec{D}$ immediately above the surface can be replaced by a homogeneous boundary condition inside the surface if we introduce an additional infinitely thin sheet of PV, q^*, at the surface such that $\rho q^* B\Delta = \vec{n}\cdot\vec{D}_s B$. That leads to

$$q^* = \lim_{\Delta \to 0} \frac{\vec{n}\cdot\vec{D}_s}{\rho\Delta} = \frac{\vec{n}\cdot\vec{D}_s}{\rho}\delta(z - z_s), \tag{3.63}$$

where $\delta(z - z_s)$ is a Dirac delta function at $z = z_s$. By (3.60a), we have $\vec{D} = \vec{\zeta}_a\theta$ and therefore $q^* = \dfrac{\vec{\zeta}_a\cdot\vec{n}}{\rho}\theta\delta(z - z_s)$. Thus, it warrants introducing an abstract notion of *generalized potential vorticity* equal to

$$q_G = q + \frac{\vec{n}\cdot\vec{\zeta}_a}{\rho}\theta\delta(z - z_s) \tag{3.64}$$

for the gauge $\vec{B} = 0$. It is the sum of the conventional PV and an additional infinitely thin sheet of PV immediately above the surface. The latter is proportional to the normal component of absolute vorticity and the potential temperature at the surface. The inhomogeneous boundary condition for a flow is equivalent to a flow component associated with this additional surface potential vorticity element. The use of (3.64) enables us to use a homogeneous boundary condition at the surface when we solve (3.55) for the flow field associated with a given PV distribution.

To analyze the evolution of a flow, we need to determine the other vector field $\vec{H} = \left(\vec{v}\times\vec{\zeta}_a\theta + c_pT\nabla\theta + \vec{F}\theta\right)$. Its boundary condition is obviously also inhomogeneous. Again, it would be desirable to determine $\vec{H}$ subject to an equivalent homogeneous boundary condition. We now make use of another known fact in electromagnetism that the *tangential component of a magnetic field* (the counterpart of $\vec{H}$) is discontinuous on a boundary surface where there is a current density (the counterpart of $\vec{J}$). Let us denote $\vec{H}$ inside the surface by $\vec{H}_b$ and immediately above the surface by $\vec{H}_s$. It suggests that we could seek to pose a problem such that the tangential component of $\vec{H}$ would be zero inside a surface ($\vec{n} \times \vec{H} = \vec{n} \times \vec{H}_b = 0$), but nonzero barely above the surface ($\vec{n} \times \vec{H} = \vec{n} \times \vec{H}_s \neq 0$). The problem then amounts to determining the surface PV flux compatible with such a discontinuity. Note that $\vec{H}$ lies on a plane normal to the vector $\nabla \times \vec{H}$. This plane intersects the boundary surface. The required surface PV flux can be obtained by integrating (3.57) over a small rectangular area A enclosed by a closed curve Γ on this plane straddling across the surface. The long side of the rectangle is ℓ and the other side straddling across the boundary is Δ which is infinitesimally short. By Stokes' theorem, we then get

$$\iint_A \vec{J}\, dS = -\iint_A \vec{D}_t\, dS + \iint_A (\nabla \times \vec{H}) dS = \oint_\Gamma \vec{H} \cdot d\vec{r} = \vec{n} \times (\vec{H}_s - \vec{H}_b)\ell = \vec{n} \times \vec{H}_s \ell. \tag{3.65}$$

The LHS would be nonzero for an infinitesimally small area A only if $\vec{J}$ has a singularity. In other words, the inhomogeneous boundary condition for $\vec{H}$ barely above the surface can be replaced by a homogeneous boundary condition inside the surface if we introduce an additional infinitesimally thin sheet of surface PV density flux, $\vec{J}^*$, so that $\vec{J}^* \ell \Delta = \vec{n} \times \vec{H}_s \ell$. That leads to

$$\vec{J}^* = \lim_{\Delta \to 0} \frac{\vec{n} \times \vec{H}_s}{\Delta} = \vec{n} \times \vec{H}_s \delta(z - z_s). \tag{3.66}$$

The corresponding *generalized PV density flux* would be

$$\vec{J}_G = \vec{J} + \vec{J}^*. \tag{3.67}$$

For the gauge $\vec{B} = 0$ and $\psi = 0$, the singular surface PV density flux according to (3.64) and (3.67) is

$$\vec{J}^* = \vec{v}\rho q^* + \vec{J}^*_{bc} + \vec{J}^*_F, \tag{3.68}$$

where $q^* = \dfrac{\vec{\zeta}_a \cdot \vec{n}}{\rho} \theta \delta(z - z_s)$, $\vec{J}^*_{bc} = c_p T (\vec{n} \times \nabla\theta) \delta(z - z_s)$ and $\vec{J}^*_F = \theta(\vec{n} \times \vec{F}) \delta(z - z_s)$. The singular potential vorticity density flux at the surface has three components: (i) $\vec{v}\rho q^*$ an advective flux of the generalized potential vorticity, (ii) $\vec{J}^*_{bc}$ a baroclinic component of the flux directed along the lines of intersection between isentropes and the surface, and (iii) $\vec{J}^*_F$ a frictional flux component. If we impose a no-slip condition for the flow at the surface, then $\vec{J}^*$ would only consist of the baroclinic and frictional components.

In the special case of a quasi-geostrophic flow under adiabatic and inviscid conditions in a Boussinesq fluid over a flat surface, $z_s = 0$, the fluctuating part of PV is simplified to $q' = \left(\dfrac{\vec{\zeta}_a \cdot \nabla\theta}{\rho_o}\right)' \approx \dfrac{f_o}{\rho_o} \dfrac{\partial \theta'}{\partial z} + \dfrac{(v_x - u_y)}{\rho_o} \dfrac{d\theta_o}{dz}$ where $\theta' = \theta - \theta_o$, $\rho \approx \rho_o$ a reference density, u and v are velocity components. Hydrostatic and geostrophic approximations provide a basis to relate θ' to u and v in the form of thermal wind equations. The boundary contribution to the potential vorticity is $q^* = \frac{\vec{\zeta}_a \cdot \vec{n}}{\rho}\theta'\delta(z) \approx \frac{f_o}{\rho_o}\theta'\delta(z)$. The boundary contribution to the surface PV density flux is $\vec{J}^* = f_o\vec{v}\theta'\delta(z)$. The quasi-geostrophic generalized potential vorticity is then $q_G = \frac{f_o}{\rho_o}\frac{\partial\theta'}{\partial z} + \frac{(v_x - u_y)}{\rho_o}\frac{d\theta_o}{dz} + \frac{f_o\theta'}{\rho_o}\delta(z)$.

The PV dynamics is then equivalently described by

$$\frac{\partial(\rho q_G)}{\partial t} + \nabla \cdot (\vec{J}_G) = 0, \tag{3.69}$$

subject to a homogeneous boundary condition. Upon integrating (3.69) over the whole atmosphere, we would get

$$\frac{d}{dt} \oiiint \rho q_G \, db = 0. \tag{3.70}$$

This simply says that the total generalized PV in the atmosphere conserves. The conservation property of generalized PV holds true in a general flow.

The notion of generalized potential vorticity in a quasi-geostrophic setting was first demonstrated to be useful by Bretherton (1966) for interpreting the dynamical nature of the instability of a baroclinic shear flow between the rigid top and bottom of a domain. Such an instability is called baroclinic instability and will be discussed in detail in Chapter Part 8B. We will interpret baroclinic instability as a process of mutual positive reinforcement of phase-locked counter-propagating constituent Rossby waves associated with such generalized PV anomalies at the two boundaries. This process is called "wave resonance mechanism" for short in Section 8B.5.6. This interpretation is applicable to not only the instability of a baroclinic flow but also the instability of a barotropic (horizontal) shear flow. We will elaborate on this matter in Section 8B.2.4.

Summing up, the generalized potential vorticity is a sum of the interior PV and a singular surface PV. Its use allows one to replace inhomogeneous conditions by simpler homogeneous boundary conditions. This is applicable to balanced as well as non-balanced flows. The concept of generalized potential vorticity helps us understand better the nature of the dynamics of even a non-geostrophic flow strongly influenced by boundary effects. An illustrative demonstration is given in Schneider *et al.* (2003) in the context of a topographic meso-scale flow.

4 Friction and boundary layers

The impact of subgrid-scale motions on a background flow, including the molecular motion and small turbulent eddies, is referred to as the frictional effect. The atmosphere is a weakly viscous fluid, but friction is important wherever the velocity gradient is very large. The frictional effect is therefore strong at least in a layer next to the surface of the Earth. Such a layer is called the boundary layer elaborated in Section 4.2. The representation of the subgrid-scale influence of small turbulent eddies on a background flow is elaborated in Section 4.3. A simple representation of small turbulence eddies is given in Section 4.5. We define and examine the dynamics of several types of boundary layers in Sections 4.6, 4.9 and 4.10. The dynamics of the atmospheric Ekman layer and the oceanic Ekman layer is particularly important for large-scale geophysical flows and is elaborated in Sections 4.7 and 4.8. Although the frictional effect is generally unimportant in the short-term evolution of a geophysical flow, it would exert a strong influence on the long-term evolution through its coupling with the boundary layer.

4.1 Scale and estimate of the frictional force

The frictional force in large-scale disturbances at a level one-kilometer or more above the surface is weak compared to other forces. That is evident in the fact that wind at those levels is approximately in geostrophic balance. Nevertheless, the frictional force becomes progressively stronger at lower and lower levels. The significance of the frictional force at a point may be assessed in terms of a ratio between acceleration and viscous force. This dimensionless number is known as *Reynolds number*, $Re = \dfrac{acceleration}{viscous\ force}$. The viscous force would be negligible for $Re \geq 10$. The numerator and denominator are to be estimated using the characteristics of the disturbance under consideration. The characteristics include the representative values of its size, velocity in each direction, its rate of temporal change, temperature, pressure etc. These values are called *scales*.

An estimate of a scale of a quantity may be made as follows. For example, suppose we want to estimate $\partial u/\partial x$ for the flow associated with winter cyclones and anticyclones in mid-latitude. The zonal velocity u has magnitude in the range of 20 to $0\,\mathrm{m\,s^{-1}}$. Therefore its mean magnitude is $10\,\mathrm{m\,s^{-1}}$ with fluctuations of $\pm 10\,\mathrm{m\,s^{-1}}$. The distance in the x-direction between a trough and a ridge is $\sim 2000\,\mathrm{km}$. We denote the scale of u of those disturbances by V and their horizontal length scale by L. The numerical values of the scales are not meant to be precise, but they should be estimates of the correct order of magnitude. So we may set $V = 10\,\text{m s}^{-1}$ and $L = 10^6$ m. Then the scale of $\partial u/\partial x$ is $\partial u/\partial x \sim V/L$. The scale of the vertical component of vorticity $\partial v/\partial x - \partial u/\partial y$ is empirically known to be V/L because the two parts $\partial v/\partial x$ and $(-\partial u/\partial y)$ have the same sign. In contrast, the scale of horizontal divergence, $(\partial u/\partial x + \partial v/\partial y)$, is not V/L, because the two parts turn out to have opposite signs. Observation indeed suggests that its scale should be $(\partial u/\partial x + \partial v/\partial y) \sim (10^{-1})(V/L)$.

If the time scale is equal to L/V, it is called *advective time scale* because it is the time taken for a fluid parcel to travel a distance of L in a speed V. It follows the scale of the acceleration would be $\partial u/\partial t \sim V/LV^{-1} \sim V^2/L$. By the same token, the molecular frictional force is $viscous\,force = \kappa\nabla^2\vec{V} \sim \kappa V/\ell^2$ where $\kappa = 10^{-5}\,\text{m}^2\,\text{s}^{-1}$ = kinematic molecular viscous coefficient for air and ℓ is the shortest length scale of the flow in a certain direction. Hence, we would have

$$Re = \frac{acceleration}{viscous\,force} = \frac{V^2/L}{\kappa V/\ell^2} = \frac{V\ell^2}{\kappa L}.$$

In considering the effect of molecular friction on the wind in an extratropical cyclone, we set $\ell = H = 10^4$ m as its vertical length scale. Then the expression for Reynolds number becomes

$$Re = \frac{VH^2}{\kappa L} = 10^8 \gg 1.$$

This estimate strongly suggests that molecular frictional force is utterly negligible compared to the acceleration at levels high above the surface.

Wind inevitably decreases towards zero at the Earth's surface. In other words, the vertical length scale of the wind ℓ near a surface is much smaller than the shortest length scale of a large-scale disturbance. Then the Reynolds number Re at a distance sufficiently close to a surface would become order unity.

4.2 Concept of boundary layer

An atmospheric layer in which frictional force is important is called a *boundary layer* in which we have $Re \sim 1$; Re can be in turn used to estimate the thickness of a boundary layer. Associated with molecular friction, the latter would be $\ell = (\kappa L/V)^{1/2}$ with $\kappa = 10^{-5}\,\text{m}^2\,\text{s}^{-1}$. Then we would get $\ell \sim 1$ m for the boundary layer associated with an extratropical cyclone which have scales $V = 10\,\text{m s}^{-1}$ and $L = 10^6$ m. In other words, molecular friction in such a flow should be negligible above 1 meter or so.

There are lots of small turbulent eddies embedded in atmospheric flow. Such eddies are particularly prevalent and intense near the surface of the Earth. They may stem from stirring by the mean wind blowing over irregularities on land or from non-uniform surface heating. Some eddies may originate from the clouds above. Such small eddies can transport momentum of different directions. The flux of momentum of each direction generally is non-uniform. That would have the effect of a force acting on the background flow. We call it a frictional force, but it is distinctly different from molecular friction. In the next three sections, we will discuss how one might statistically represent the effect of eddies. Those results will enable us to estimate the thickness of the boundary layer associated with small turbulent eddies in Section 4.6.

4.3 Reynolds averaging

The weather stations of existing operational networks are too far apart to adequately quantify the spatial distribution of the total velocity of air. The data collected at those stations only allow us to deduce the properties of disturbances that have a length scale that is long compared to the typical distance between two weather stations and fluctuate more slowly than the time interval between two measurements. Those disturbances constitute the resolvable component of an atmospheric flow. The small and rapidly fluctuating turbulent eddies constitute the irresolvable component of a flow. We can only attempt to assess the irresolvable component in a statistical manner.

The irresolvable component can have an important dynamical impact on the resolvable component. Reynolds averaging is a method for quantifying such an impact. *It is essentially a method of running-average.* If we want to determine the Reynolds-averaged velocity at a particular instant t_o, we take an average of the velocity data at each location over an interval, τ, centering about t_o. The interval τ should be long compared to the time scale of the turbulent eddies, but short compared to the time scale of the weather disturbance. Similar consideration applies to Reynolds average with respect to space.

We denote the resolvable part of wind by a bar, and the irresolvable component (turbulent eddies) by a prime. For example, the total zonal wind is decomposed into two parts, viz.

$$\begin{aligned} u_{total} &= \bar{u} + u' \\ \bar{u} &= \frac{1}{\tau} \int_{t-\tau/2}^{t+\tau/2} u dt; \quad \bar{u'} = 0. \end{aligned} \tag{4.1}$$

It is more instructive to decompose each of the thermodynamical variables, such as T, p, ρ and θ, into three parts, viz.

$$\rho = \rho_o + \bar{\rho} + \rho', \tag{4.2}$$

where $\rho_o(z)$ is a background component present even in the absence of any disturbance, $\bar{\rho}$ is the Reynolds average density associated with a large-scale disturbance and ρ' is associated with the turbulent eddies. We will make use of decompositions of (4.1) and (4.2) shortly when we apply Reynolds averaging to the governing equations themselves.

4.4 Boussinesq approximation

Air may be regarded as effectively incompressible for some disturbances. If so, we would have $D\rho/Dt = 0$ implying according to the continuity equation

$$u_x + v_y + w_z = 0. \tag{4.3}$$

In addition, the density ρ may be approximated by its background value ρ_o wherever it appears in the governing equations with one important exception. We must retain the fluctuation of density in a term where it is multiplied by gravity. Such a product can be significant because gravity is strong. Furthermore, ρ_o may be even simplified by a constant, ρ_{oo}, in some problems. This package of approximations is collectively referred to as *Boussinesq approximation*.

In light of the discussion in Section 4.2, we may dismiss molecular friction as negligibly weak at the outset. Let us see how we can represent the impacts of small eddies on the resolvable component of a flow. By invoking Boussinesq approximation, the u-momentum equation is written with the use of (4.3) as

$$u_t + (uu)_x + (vu)_y + (wu)_z = -\frac{1}{\rho_o} p_x + fv. \tag{4.4}$$

First expand each variable according to (4.1). We get from (4.4)

$$\bar{u}_t + u'_t + ((\bar{u}+u')(\bar{u}+u'))_x + ((\bar{v}+v')(\bar{u}+u'))_y + ((\bar{w}+w')(\bar{u}+u'))_z$$
$$= -\frac{1}{\rho_o}(\bar{p}+p')_x + f(\bar{v}+v'). \tag{4.5}$$

Note that $\overline{(\bar{u}+u')(\bar{u}+u')_x} = (\bar{u}\bar{u})_x + \overline{(u'u')_x}$ because $\overline{(\bar{u}u')_x} = 0$. Thus Reynolds averaging (4.5) term by term would yield

$$\bar{u}_t + (\bar{u}\bar{u})_x + \left(\overline{u'u'}\right)_x + (\bar{v}\bar{u})_y + \left(\overline{u'v'}\right)_y + (\bar{w}\bar{u})_z + \left(\overline{u'w'}\right)_z = -\frac{1}{\rho_o}\bar{p}_x + f\bar{v}. \tag{4.6}$$

The form of (4.6) is similar to (4.4) except for two aspects: (i) the total dependent variable in each term of (4.4) is replaced by its Reynolds-averaged counterpart, and (ii) there are three extra terms containing products of the turbulent velocity components. We can rewrite (4.6) as

$$\bar{u}_t + (\bar{u}\bar{u})_x + (\bar{v}\bar{u})_y + (\bar{w}\bar{u})_z = -\frac{1}{\rho_o}\bar{p}_x + f\bar{v} - \nabla\cdot\left(\overline{\mathbf{v}'u'}\right). \tag{4.7}$$

The last term on the RHS of (4.7) represents a convergence of u-momentum fluxes per unit mass due to the turbulent eddies. It is a frictional force per unit mass in the x-direction, arising from the turbulent eddies. The quantities: $-\rho_o\overline{u'u'}, -\rho_o\overline{u'v'}$ and $-\rho_o\overline{u'w'}$ have the dimension of force per unit area $\mathrm{m\,s^{-2}\,kg\,m^2}$ and are called *Reynolds stresses*.

When we similarly perform Reynolds averaging of the v-momentum and w-momentum equations, we would get three extra terms in each resulting equation. The corresponding

Reynolds stresses are defined to be: $-\rho_o\overline{v'u'}$, $-\rho_o\overline{v'v'}$, $-\rho_o\overline{v'w'}$ and $-\rho_o\overline{u'w'}$, $-\rho_o\overline{w'v'}$, $-\rho_o\overline{w'w'}$. All Reynolds stresses are collectively referred to by a 3-by-3 stress tensor:

$$-\rho_o\begin{pmatrix}\overline{u'u'} & \overline{u'v'} & \overline{u'w'}\\ \overline{u'v'} & \overline{v'v'} & \overline{v'w'}\\ \overline{u'w'} & \overline{v'w'} & \overline{w'w'}\end{pmatrix}.$$

When we perform Reynolds averaging of the thermodynamic equation, we similarly get extra terms that represent turbulent fluxes of heat in the x-, y- and z-directions, viz. $-c_p\overline{u'T'}$, $-c_p\overline{v'T'}$, $-c_p\overline{w'T'}$. Note that all eddy terms are themselves unknowns. Therefore, the Reynolds-averaged governing equations as exemplified by (4.7) do not form a closed system of equations. It would be necessary to represent the eddy fluxes in terms of the Reynolds-averaged dependent variables.

4.5 Flux-gradient theory of turbulence

Turbulence is not well understood and the mathematical theories of turbulence have limited success. We only discuss a simple and very crude theory known as *flux-gradient theory* (*K-theory* for short). It postulates that turbulent eddies mix the properties of their environment. The momentum fluxes of turbulent eddies are assumed to be proportional to the vertical gradient of the corresponding velocity component in the down-gradient direction. The mathematical relations are then hypothesized to be

$$\begin{aligned}(\overline{u'u'}, \overline{u'v'}, \overline{u'w'}) &= -K_{(x)}\left(\frac{\partial\bar{u}}{\partial x}, \frac{\partial\bar{u}}{\partial y}, \frac{\partial\bar{u}}{\partial z}\right),\\ (\overline{u'v}, \overline{v'v'}, \overline{v'w'}) &= -K_{(y)}\left(\frac{\partial\bar{v}}{\partial x}, \frac{\partial\bar{v}}{\partial y}, \frac{\partial\bar{v}}{\partial z}\right),\\ (\overline{u'w'}, \overline{v'w'}, \overline{w'w'}) &= -K_{(z)}\left(\frac{\partial\bar{w}}{\partial x}, \frac{\partial\bar{w}}{\partial y}, \frac{\partial\bar{w}}{\partial z}\right),\end{aligned} \tag{4.8}$$

where $K_{(x)}$, $K_{(y)}$ and $K_{(z)}$ are eddy viscosity coefficients for the three directions. They may vary in space. The turbulent heat fluxes are similarly represented by

$$(\overline{u'\theta'}, \overline{v'\theta'}, \overline{w'\theta'}) = -K_{(\theta)}\left(\frac{\partial\bar{\theta}}{\partial x}, \frac{\partial\bar{\theta}}{\partial y}, \frac{\partial\bar{\theta}}{\partial z}\right), \tag{4.9}$$

where $K_{(\theta)}$ is the eddy thermal coefficient. All such eddy coefficients are to be established empirically.

Underlying this crude representation of turbulence is the idea that turbulent eddies are analogous to molecules in a volume of gas. They collide over an average distance and exchange properties via such collisions. The average distance between two collisions is referred to as a *mixing length* in analogy to the *mean-free path in the kinetic theory of gas*. The turbulent coefficients are in principle functionally dependent on the mixing length.

The eddy coefficients are often treated as a single constant. We would then have

$$-\nabla\cdot\left(\overline{\mathbf{v}'u'}\right)=K\nabla^2\bar{u},\qquad -\nabla\cdot\left(\overline{\mathbf{v}'v'}\right)=K\nabla^2\bar{v},\qquad -\nabla\cdot\left(\overline{\mathbf{v}'w'}\right)=K\nabla^2\bar{w},\tag{4.10}$$

where $\mathbf{v}'=(u',v',w')$. The turbulent frictional force then takes on the same mathematical form as that of molecular friction except for a much larger value of coefficient, K, which must be determined by observation. A pertinent value is $K\sim 5\,\mathrm{m^2\,s^{-1}}$.

Furthermore, the vertical variations of wind and temperature near the surface are much larger than the counterpart horizontal variations. We may introduce additional approximation as

$$\begin{aligned}
-\nabla\cdot\left(\overline{\mathbf{v}'u'}\right)&\approx-\frac{\partial\overline{u'w'}}{\partial z}=K\frac{\partial^2\bar{u}}{\partial z^2},\\
-\nabla\cdot\left(\overline{\mathbf{v}'v'}\right)&\approx-\frac{\partial\overline{v'w'}}{\partial z}=K\frac{\partial^2\bar{v}}{\partial z^2},\\
-\nabla\cdot\left(\overline{\mathbf{v}'w'}\right)&\approx\frac{\partial\overline{w'w'}}{\partial z}=K\frac{\partial^2\bar{w}}{\partial z^2}.
\end{aligned}\tag{4.11}$$

Hence, we finally can write the momentum equations as

$$\begin{aligned}
\bar{u}_t+(\bar{u}\bar{u})_x+(\bar{v}\bar{u})_y+(\bar{w}\bar{u})_z&=-\frac{1}{\rho_o}\bar{p}_x+f\bar{v}+K(\bar{u})_{zz},\\
\bar{v}_t+(\bar{u}\bar{v})_x+(\bar{v}\bar{v})_y+(\bar{w}\bar{v})_z&=-\frac{1}{\rho_o}\bar{p}_y-f\bar{u}+K(\bar{v})_{zz},\\
\bar{w}_t+(\bar{u}\bar{w})_x+(\bar{v}\bar{w})_y+(\bar{w}\bar{w})_z&=-\frac{1}{\rho_o}\bar{p}_z-g\frac{\bar{\rho}}{\rho_o}+K(\bar{w})_{zz}.
\end{aligned}\tag{4.12}$$

It should be noted that if the turbulence is very vigorous in a region, the effect of its mixing could completely smooth out the spatial structure of the background variable under consideration. Then the linear relation between the turbulent flux and the background gradient would no longer hold. In other words, the K-theory is adequate only when the turbulence is not too vigorous.

4.6 Types of boundary layer

It is meaningful to define distinctly different boundary layers in the atmosphere for different circumstances. One type of boundary layer is called a *mixed layer* where the turbulence is intense underneath an inversion. All properties of the atmosphere within such a layer are thoroughly mixed in the vertical direction. An example is the relatively thin layer of low-level trade wind in the tropics. The turbulent eddies are capped within a trade wind layer by the air above that is strongly stratified, i.e. large values of $(\theta_o)_z$.

The turbulence in the boundary layer of a large-scale disturbance in mid-latitude is moderately intense. As such, it can be represented by the K-theory. Such a boundary layer is called a *planetary boundary layer* (PBL). The Coriolis force plays an important role.

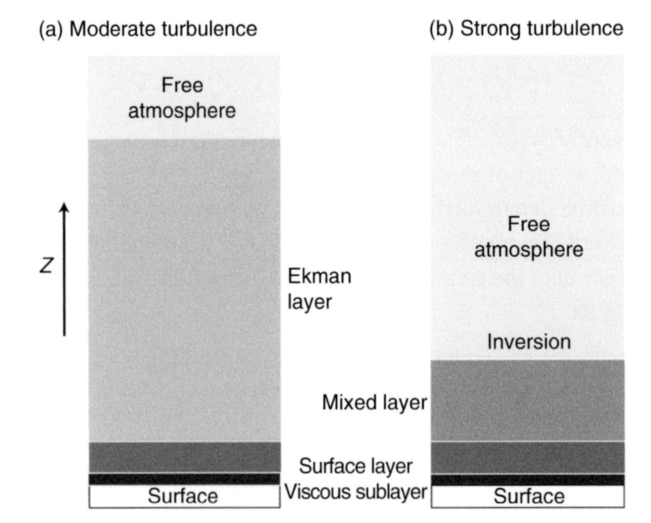

Fig. 4.1. Schematic depiction of different boundary layers.

A rough estimate of its thickness is $(KL/V)^{1/2} \sim 700$ m for $K = 5\,\mathrm{m^2\,s^{-1}}$, $L = 10^6\,\mathrm{m\,s^{-1}}$ and $V = 10\,\mathrm{m\,s^{-1}}$. The eddy coefficient may vary with the stratification. We will however only discuss the effect of turbulence in the context of a constant eddy coefficient. This simplest version of PBL was first analytically investigated by V. W. Ekman and is referred to as the Ekman layer.

Below either a mixed layer or a PBL, there is usually a distinctly different boundary layer about one order of magnitude thinner. It is referred to as a *surface layer*. It is associated with the roughness of the surface and is at most several tens of meters thick. The unique aspect of a surface layer is that the eddy coefficient K cannot be assumed to be constant because eddies closer to the surface are smaller since they are kinematically constrained by the presence of the surface. One would have to invoke a different assumption about the character of the turbulent eddies in a surface layer.

Finally, we may visualize an extremely thin *viscous sublayer* right next to a surface associated with the molecular friction. The velocity scale and the horizontal length scale are much smaller than those of a weather disturbance aloft. For $\kappa = 10^{-5}\,\mathrm{m^2\,s^{-1}}$, $L = 1$ m and $V = 0.1\,\mathrm{m\,s^{-1}}$ the thickness would be $(\kappa L/V)^{1/2} \sim 10^{-2}$ m. The various possible types of boundary layers mentioned above are schematically summarized in Fig. 4.1.

4.7 Atmospheric Ekman layer

The most important planetary boundary layer is the one driven by the large-scale wind aloft. Such a PBL is dynamically similar to the boundary layer in the wind-driven surface ocean current first analyzed by Ekman (1905). We refer to both as Ekman layers (EL). The

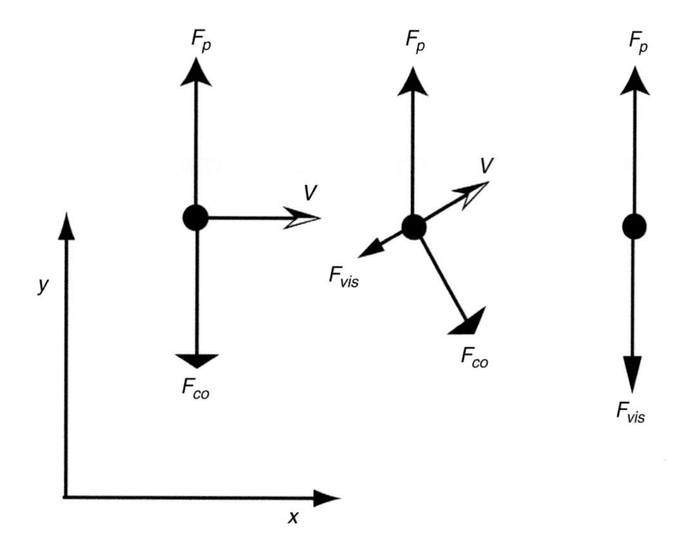

Fig. 4.2 Schematic showing the balance of forces in an Ekman layer in the northern hemisphere at three levels from left to right: near the top, at mid-level and near the surface (F_p pressure gradient force, F_{co} Coriolis force, F_{vis} frictional force, V velocity).

velocity in an Ekman layer is the Reynolds average velocity. Let us consider a quasi-steady wind aloft which is in geostrophic balance between a pressure gradient force and the Coriolis force. The acceleration is therefore weak and would also be so within an EL. But the frictional force becomes increasingly important at lower and lower levels. The Coriolis force is expected to become weaker and weaker towards the surface. So an EL is characterized by a balance among these three forces. The balance of forces at different levels of an EL is shown in Fig. 4.2.

The turbulent frictional force is assumed to be adequately represented by the K-theory. We also consider a neutral stratification so that the pressure gradient force does not vary with height within the EL. The pressure gradient force can be written in terms of the geostrophic velocity (u_g,v_g) aloft with Boussinesq approximation. The governing equations for the flow in an EL are then

$$\begin{aligned} 0 &= -fv_g + fv + Ku_{zz}, \\ 0 &= fu_g - fu + Kv_{zz}. \end{aligned} \tag{4.13}$$

The boundary conditions (b.c.) are: (i) at the surface, $z = 0$, $u = v = 0$ and (ii) at the top of the EL, formally at $z \to \infty$, $u = u_g$, $v = v_g$. The solution of (4.13) is dictated by a smooth transition of the balance of three forces from the surface to the top. These are two coupled second-order ordinary differential equations (ODEs) governing two unknowns u and v. The horizontal structures of (u,v) are left unspecified. Only their z dependence is explicitly governed according to (4.13).

It is convenient to introduce a complex dependent variable $\xi = u + iv$, $i = \sqrt{-1}$. By combining (4.13) accordingly, we may get a second-order ODE for ξ as a single unknown, viz.

$$\begin{aligned}
&\xi_{zz} - b^2\xi = -b^2\xi_g,\\
&\text{b.c. at } z = 0,\ \xi = 0\\
&\quad\text{at } z \to \infty,\ \xi = \xi_g,\\
&\text{where } b^2 = \frac{if}{K} = \frac{f}{K}e^{i\pi/2}.
\end{aligned} \tag{4.14}$$

The general solution is

$$\begin{aligned}
&\xi = Ae^{bz} + Be^{-bz} + \xi_g,\\
&b = \gamma(1+i),\ \gamma = \sqrt{\frac{f}{2K}}.
\end{aligned} \tag{4.15}$$

To satisfy the boundary conditions, we must choose $A = 0$, $B = -\xi_g$. Hence, we get

$$\begin{aligned}
&\xi = \xi_g\left(1 - e^{-bz}\right)\\
\Rightarrow\ &u = \mathrm{Re}\{\left(u_g + iv_g\right)(1 - e^{-\gamma z}(\cos(\gamma z) - i\ \sin(\gamma z)))\},\\
&v = \mathrm{Im}\{\left(u_g + iv_g\right)(1 - e^{-\gamma z}(\cos(\gamma z) - i\ \sin(\gamma z)))\}.
\end{aligned} \tag{4.16}$$

4.7.1 Hodograph of an Ekman layer

The explicit form of the EL solution is

$$\begin{aligned}
u &= u_g(1 - \exp(-\gamma z)\cos(\gamma z)) - v_g \exp(-\gamma z)\sin(\gamma z)\\
v &= v_g(1 - \exp(-\gamma z)\cos(\gamma z)) + u_g \exp(-\gamma z)\sin(\gamma z).
\end{aligned} \tag{4.17}$$

A curve plotting $\vec{V} = (u, v)$ at different vertical levels projected on a (x,y) plane is called a *hodograph*. Because of its spiral shape, it is called an Ekman spiral. An example of the structure of wind in an Ekman layer driven by a non-dimensional geostrophic flow aloft equal to $(u_g, v_g) = (\cos\phi, \sin\phi)$ with $\phi = 30°$ is shown in Fig. 4.3. The key parameters are $f = 10^{-4}\,\mathrm{s}^{-1}$ and $K = 5\,\mathrm{m}^2\,\mathrm{s}^{-1}$. The points refer to the values of the wind at heights of about 6 m apart. We see that while the wind diminishes towards the surface, it is directed 45° to the left relative to the geostrophic wind aloft. Recall that the pressure is lower to the left of a geostrophic wind in the northern hemisphere. Therefore, the wind at all levels of an EL has a component in the direction from high towards low pressure. This feature of cross-isobar flow has two important implications, which will be elaborated shortly.

The top of an EL may be defined as the lowest level $z_* \equiv D_e$ where $\vec{V}$ is parallel to $\vec{V}_g$. That level is defined by the condition $z_*\gamma = \pi$. Hence the thickness of an EL is

$$D_e = \frac{\pi}{\gamma} = \pi\left(\frac{2K}{f}\right)^{1/2}. \tag{4.18}$$

For $f = 10^{-4}\,\mathrm{s}^{-1}$ and $K = 5\,\mathrm{m}^2\,\mathrm{s}^{-1}$, we have $D_e = 10^3\,\mathrm{m}$. Note that this measure of the thickness of an Ekman layer is close to a rough estimate of it on the basis of the velocity and length scales of a synoptic disturbance ($(KL/V)^{1/2} \sim 700$ m). Furthermore, if we choose

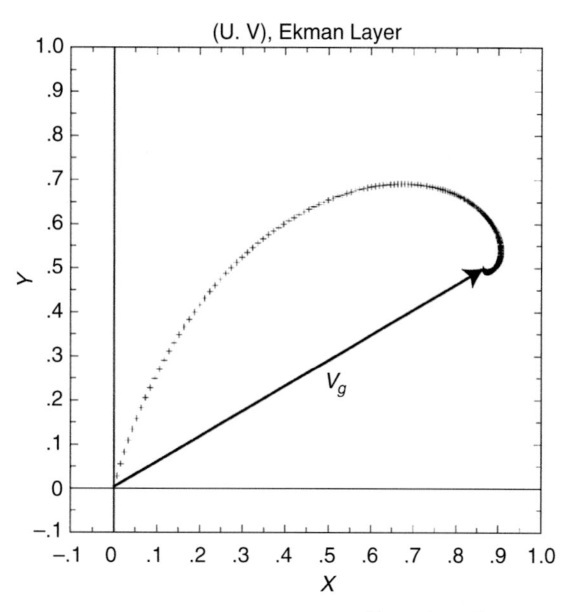

Fig. 4.3 Hodograph of an Ekman layer driven by a geostrophic wind aloft, $\vec{V}_g = (u_g, v_g)$.

the x-axis to be in the direction of $\vec{V}_g$, then $v_g = 0$ and $u_g = |\vec{V}_g|$. It follows that the solution takes on a simpler form,

$$\begin{aligned} u &= u_g(1 - \exp(-\gamma z)\cos(\gamma z)), \\ v &= u_g \exp(-\gamma z)\sin(\gamma z). \end{aligned} \tag{4.19}$$

The wind just above the surface, $z \to 0$, would be oriented to the left of the geostrophic wind by an angle ϑ satisfying $\lim_{z\to 0} \frac{v}{u} = \tan\vartheta$. The solution above yields $\vartheta = 45°$ as seen in Fig. 4.3.

4.7.2 Energetics of an Ekman layer

Frictional force continually destroys kinetic energy of a flow in an Ekman layer. An obvious question then is how can an Ekman layer remain steady? The answer is that the frictional loss of kinetic energy is continually compensated by a production of energy by the pressure gradient force. For that to happen, there must be a cross-isobar flow component from high pressure toward low pressure. That is precisely what we find in the analytic solution of an Ekman layer. This balance of energy in an EL can be analytically demonstrated as follows. The local rate of generation of kinetic energy by the pressure gradient force at any level $z*$ inside an Ekman layer is

$$\begin{aligned} \left(\frac{d(KE)}{dt}\right)_{PGF} &= (\text{pressure gradient force at } z^*)\cdot(\text{velocity at } z^*) \\ &= -\frac{1}{\rho}p_x u - \frac{1}{\rho}p_y v \\ &= -fv_g u + fu_g v. \end{aligned} \tag{4.20}$$

On the other hand, the local rate of change of KE due to frictional dissipation at a level z^* within the Ekman layer can be computed as

$$\left(\frac{d(KE)}{dt}\right)_{friction} = (\text{frictional force at } z^*)\cdot(\text{velocity at } z^*) \\ = (Ku_{zz})(u) + (Kv_{zz})(v). \tag{4.21}$$

The sum of (4.20) and (4.21) is therefore zero at every level

$$\left(\frac{d(KE)}{dt}\right)_{PGF} + \left(\frac{d(KE)}{dt}\right)_{friction} = \left(-fv_g + Ku_{zz}\right)u + \left(fu_g + Kv_{zz}\right)v \\ = -fuv + fuv = 0. \tag{4.22}$$

It follows that the flow in an Ekman layer can be steady. Ultimately, the PGF itself has to be maintained by physical processes associated with the dynamics of the large-scale disturbance.

4.7.3 Ekman suction/pumping

Another important implication of the cross-isobar flow in an EL is that it gives rise to a mass flux from a high pressure region towards a low pressure region. The wind in an Ekman layer has the same horizontal structure as the geostrophic wind aloft. To get a feel for the overall structure of the wind in an Ekman layer, let us examine the wind at a sample level within an EL, say $z = 300$ m, driven by a representative geostrophic flow aloft such as at 900 mb (Fig. 4.4). The analytic expression for the geopotential height of this surface is chosen to be $Z(x,y) = A - By/S + C\sin(x/S + \phi)\exp(-(y/S)^2)$ where A,B,C,S,ϕ are constants.

The conspicuous cross-isobar component everywhere towards lower pressure gives rise to a mass convergence in the region of a low-pressure center. It must be compensated by a corresponding vertical mass flux coming out of the top of an EL in such a region. This upward mass flux is referred to as *Ekman pumping*. By the same token, we conclude that there must be also a mass flux downward into an Ekman layer through its top over a high-pressure center (*Ekman suction*). Air is therefore recycled between an Ekman layer and the atmosphere above. This is the mechanism by which an inviscid synoptic disturbance in the free atmosphere "feels" the influence of the turbulent eddies in an Ekman layer.

Now let us quantitatively analyze this mass flux and the related vertical motion. Integrating the continuity equation through the EL, we get

$$\int_0^\infty w_z dz = -\int_0^\infty \left(u_x + v_y\right) dz \\ w(\infty) - w(0) = -\frac{\partial}{\partial x}\int_0^\infty u\,dz - \frac{\partial}{\partial y}\int_0^\infty v\,dz. \tag{4.23}$$

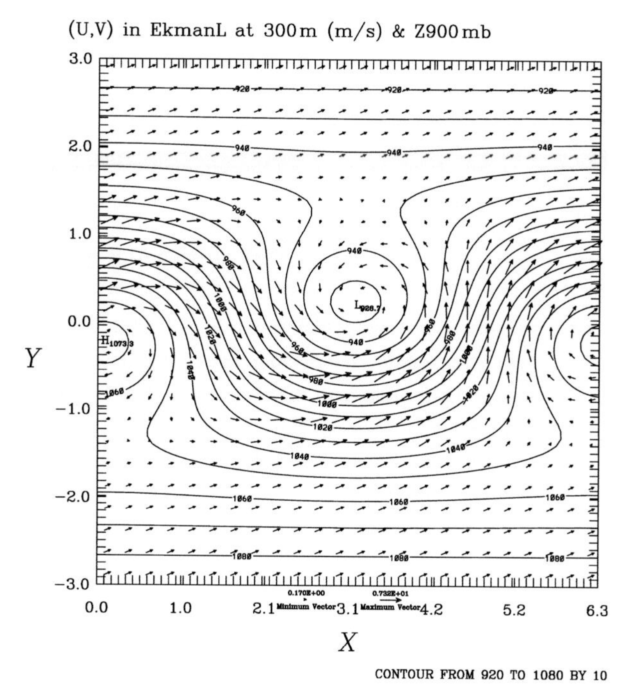

Fig. 4.4 The velocity field at $z = 300$ m in the Ekman layer below a geostrophic flow indicated by the geopotential height field at 900 mb. Distance is in units of 1000 km.

We may use $w\,(0) = 0$ as a lower boundary condition for a flat horizontal surface; $w\,(\infty)$ is the vertical velocity through the top of an EL. As noted earlier, we are free to choose the x-axis along the geostrophic wind aloft for convenience. Then we have $v_g = 0$ and (4.19) for the velocity field in an EL without loss of generality. In this case, $\partial u_g/\partial x$ stands for the horizontal divergence of the geostrophic wind aloft and $(-\partial u_g/\partial y = \zeta_g)$ for its vorticity. Horizontal divergence is known to be much smaller than vorticity in a synoptic-scale disturbance. Hence, making use of (4.19) and (4.23), we get

$$w(\infty) \approx -\frac{\partial}{\partial y}\int_0^\infty v\,dz = \zeta_g \int_0^\infty e^{-\gamma z}\sin(\gamma z)dz = \frac{\zeta_g}{2\gamma} = \frac{\zeta_g D_e}{2\pi}, \qquad (4.24)$$

where $D_e = \pi/\gamma$ is the definition of the thickness of an EL. The key finding is that the vertical velocity at the top of an EL is proportional to the vorticity of the geostrophic wind aloft and the thickness of the EL. A scaling argument can be used to estimate w_∞ in comparison to the vertical velocity of a large-scale disturbance:

$$\frac{w_\infty}{w_I} \sim \frac{D_e \zeta_g}{2\pi W} \sim \frac{D_e V}{W 2\pi L}$$

since $\dfrac{W}{H} \sim \dfrac{V}{L} Ro$, $\dfrac{D_e}{H} \sim Ro$, Ro being the Rossby number

$$\frac{w_\infty}{w_\mathrm{I}} \sim \frac{D_e}{HRo2\pi} \sim \frac{1}{2\pi}. \qquad (4.25)$$

The vertical velocity at the top of an EL turns out to be smaller than the vertical velocity of the synoptic disturbance by a factor of 2π. Since the latter is only of order cm s^{-1}, the vertical velocity at the top of an EL is really quite small. Nevertheless, it plays a dynamically very important role in influencing atmospheric flow. The accumulative effect of friction is significant and limits the intensity of the circulation of the atmosphere.

4.7.4 Observational evidence

An examination of the wind and height fields at 950 mb on April 7, 2010 reveals that the theory of the Ekman layer is indeed relevant to the atmospheric flow near the surface (Fig. 4.5). We see that the Z_{950} field has a long stretch of area oriented from SSW to NNE in the Midwest of the USA with relatively smaller values. The wind at 950 mb level indeed converges towards this region. We will examine among other things, the vertical motion field on that day in Section 6.1. We will see a close connection between the vertical motion associated with the Ekman layer and the vertical motion field of the disturbance as a whole. The Z_{950} field on April 7, 2010 also has a maximum center over the Atlantic Ocean and another one over the west coast of the USA. The 950 mb wind diverges from those two regions consistent with the Ekman theory.

How realistic is the Ekman solution?

A spiral-looking hodograph is often observed in the wind of an atmospheric PBL. But the angle between the surface wind and geostrophic wind aloft is usually much smaller than 45°. This discrepancy stems from the surface boundary condition used in the analysis. We may consider a less restrictive surface boundary condition, namely the surface wind vector is parallel to the stress vector

$$\begin{aligned} &\text{at } z = 0, \quad u = au_z, \; v = av_z \\ &\text{hence} \qquad \xi = a\xi_z, \end{aligned} \tag{4.26}$$

where a is a constant coefficient. The corresponding solution satisfying (4.26) is (with $v_g = 0$)

$$\begin{aligned} u &= u_g\left[1 - \exp(-\gamma z)\frac{(1+\gamma a)\cos(\gamma z) - \gamma a \sin(\gamma z)}{(1+\gamma a)^2 + (\gamma a)^2}\right], \\ v &= u_g\left[\exp(-\gamma z)\frac{(1+\gamma a)\sin(\gamma z) + \gamma a \cos(\gamma z)}{(1+\gamma a)^2 + (\gamma a)^2}\right]. \end{aligned} \tag{4.27}$$

The wind profile still has a spiral structure. The surface wind and surface angle with respect to the geostrophic wind in this case are

$$\begin{aligned} u_o = u_g\left[1 - \frac{1+\gamma a}{(1+\gamma a)^2 + (\gamma a)^2}\right], \qquad v_o = u_g\left[\frac{\gamma a}{(1+\gamma a)^2 + (\gamma a)^2}\right], \\ \varphi = \tan^{-1}\left(\frac{v_o}{u_o}\right) = \tan^{-1}\left(\frac{1}{1+2\gamma a}\right). \end{aligned} \tag{4.28}$$

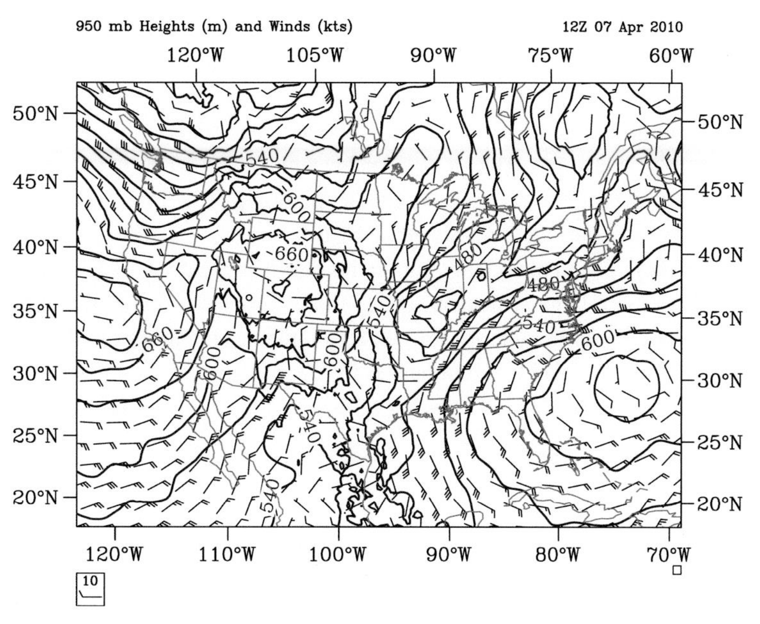

Fig. 4.5 Distributions of the 950 mb height and wind fields on April 7, 2010. See color plates section.

In order to have $\phi = 15°$ with $f = 10^{-4}\,\mathrm{s}^{-1}$ and $K = 5\,\mathrm{m}^2\,\mathrm{s}^{-1}$, we need to use $a = 450\,\mathrm{m}$ according to (4.28). This is a large value leading to a strong surface wind, $u_o = 0.76\,u_g$, $v_o = 0.087\,u_g$. This is an unrealistic value in itself, showing the limitation of this modified Ekman layer solution.

The thickness of this EL, z^*, is again defined as the level where $v_o = 0$ for the first time. Using (4.27), we get

$$\begin{aligned} \tan(\gamma z^*) &= -\frac{\gamma a}{1+\gamma a} \\ \tan(\pi - \gamma z^*) &= -\tan(\gamma z^*) = \frac{\gamma a}{1+\gamma a} \\ z^* &= \frac{\pi - 0.59}{\gamma} = 826\,\mathrm{m}. \end{aligned} \tag{4.29}$$

In other words, the less restrictive lower boundary condition also leads to a somewhat shallower EL.

The vertical velocity at the top of EL for the case of using $z = 0$, $\vec{V} = a\vec{V}_z$ as the surface boundary condition is

$$w|_D = \frac{\zeta_g}{(1+\gamma a)^2 + (\gamma a)^2}\left[a\left(1 - e^{-\gamma D}\cos\gamma D\right) + \frac{1}{2\gamma}(\cos\gamma D + \sin\gamma D)e^{-\gamma D}\right]. \tag{4.30}$$

It is found that $w|_D = 0.8 \times 10^2\,\zeta_g$ for the numerical values of parameters considered before.

4.7.5 Spin-down time

Air with relatively strong wind in the interior atmosphere enters into the EL in a high-pressure region via Ekman suction. The friction in the EL removes some momentum from such wind. The air that has been slowed down returns to the interior atmosphere over a region of low pressure via Ekman pumping. The time it takes for the interior flow to slow down due to this process alone by a factor of e ~2.73 is called spin-down time (e-folding time). The removal of momentum from the air in an EL by friction could eventually destroy a large-scale disturbance in the absence of external forcing.

Consider the interior flow of a synoptic-scale barotropic disturbance between the top of the EL, D_e, and the tropopause, H. We may use the condition $w = 0$ at $z = H$. The vorticity equation for a weak disturbance in the absence of forcing can be approximated by

$$\frac{\partial \zeta}{\partial t} = f w_z. \tag{4.31}$$

Since the horizontal flow and hence vorticity are independent of height, we get by integrating (4.31) with the boundary condition

$$(H - D_e)\frac{\partial \zeta}{\partial t} = -f w|_{De}$$

$$\frac{\partial \zeta}{\partial t} \approx -f\frac{w|_{De}}{H} = -\left(\frac{fK}{2H^2}\right)^{1/2}\zeta \quad \text{since } H \gg D_e \tag{4.32}$$

$$\therefore \zeta = \zeta_o \exp\left(-\frac{t}{\tau}\right), \qquad \tau = \left(\frac{2H^2}{fK}\right)^{1/2}.$$

The disturbance in the interior atmosphere therefore decays exponentially as a consequence of the damping within the Ekman layer associated with it.

The spin-down time is estimated to be $\tau = H(2/fK)^{1/2} \sim 6 \times 10^5$ s ~ 7 days. It is instructive to compare this with the diffusion time in the troposphere. Even if we apply the eddy coefficient to the whole troposphere, the diffusion time would be very long, $\tau_{diffusion} = H^2/K \sim 200$ days. So, we conclude that the recycling of air through the Ekman layer is a much more effective means of slowing down a disturbance aloft.

Influence of stable stratification on the spin-down process

A typical large-scale atmospheric disturbance in the lower troposphere has a cyclonic flow around a low-pressure center in an environment that has a stable stratification. The rising air out of such an EL would encounter resistance. The layer of frictionally induced vertical motion in the interior in this case would be thinner than that in a neutrally stratified atmosphere. The spin-down time of a thinner layer should then be shorter. Furthermore, the upward motion would necessarily give rise to an adiabatic cooling in the low-pressure center. Such a thermal effect would tend to increase the stratification and weaken a large-scale disturbance itself. Both effects contribute to an enhancement of the damping of the disturbance in the interior of the atmosphere. We may then infer that the spin-down time in this more general case would be shorter than in the case of a barotropic interior flow.

4.8 Oceanic Ekman layer

The surface wind of a large-scale atmospheric disturbance over an oceanic region exerts a large-scale stress on the water. It also generates turbulent eddies in a relatively shallow layer of water. Such surface stress in conjunction with the turbulent eddies would give rise to an oceanic Ekman layer (OEL). Similar to an atmospheric EL, there is mass flux of water within the OEL. We will see in a detailed analysis that the wind induced mass flux is directed to the right of the surface stress in the northern hemisphere. Dependent upon the distribution of the surface stress, such mass flux would induce either up-welling or down-welling at the bottom of the OEL. The surface water temperature in a region could become significantly colder or warmer. This could in turn reduce or enhance the evaporation rate at the ocean surface. In other words, there is in general a significant interaction between an atmospheric disturbance and the ocean underneath it. For example, the evolution of a hurricane is strongly affected by such sea–air interaction. Furthermore, the mass exchange between an OEL and the interior of the oceans would give rise to a global system of ocean currents. It is therefore important to learn about the essential properties of an OEL.

OEL model formulation

An OEL is also characterized by a balance of three forces: pressure gradient force, Coriolis force and frictional force. As demonstrated in Section 4.7, we may introduce $\xi = u + iv$ as a single dependent variable for such a layer. It is a good approximation to adopt hydrostatic balance and to treat the density of water, ρ_w, as a constant. We in effect consider a barotropic model of OEL. The pressure gradient at any depth would then depend only on the known slope of the ocean surface, $z_o(x, y)$. There is a corresponding geostrophic velocity, which is used to define $\xi_{(g)} = u_{(g)} + iv_{(g)}$. Hence, $\xi_{(g)}$ would not vary with depth. The governing equations are then the same as (4.14) in Section 4.7 and can be combined to a single equation cited below for easy reference:

$$\xi_{zz} - b^2 \xi = -b^2 \xi_{(g)}, \tag{4.33}$$

where $b^2 = if/K$, $b = \gamma(1 + i)$, $\gamma = \sqrt{f/2K}$ and K is the eddy coefficient of water. The domain is $z_o \geq z \geq -H$. The thickness of the OEL is supposed to be a fraction of the depth, $(-H)$. So, we may formally treat the domain as $0 \geq z \geq -\infty$ without loss of generality whenever it is convenient to do so. We choose the x-direction as the direction of a surface stress τ which is proportional to the shear at the surface. We may also use $u_{(g)} \neq 0$ and $v_{(g)} = 0$. The boundary conditions are then: at $z = 0$, $\rho_w K u_z = \tau$ and $v_z = 0$; at $z \to -\infty$, $u = u_{(g)}$ and $v = 0$. It is straightforward to obtain the following solution of (4.33)

$$u = \frac{\tau}{2\rho_w K\gamma} e^{\gamma z}(\cos(\gamma z) + \sin(\gamma z)) + u_{(g)},$$
$$v = \frac{\tau}{2\rho_w K\gamma} e^{\gamma z}(-\cos(\gamma z) + \sin(\gamma z)). \tag{4.34}$$

This hodograph has a spiral structure. The current at the surface is $u_o = \frac{\tau}{2\rho_w K\gamma} + u_{(g)} = \frac{\tau}{\rho_w\sqrt{2fK}} + u_{(g)}$ and $v_o = -\frac{\tau}{2\rho_w K\gamma}$. This analysis reveals that a surface wind stress would contribute to a component of the surface flow in the x-direction, but also drive the surface water of the ocean toward the right of it. The surface current would be faster for a stronger stress, at lower latitude, or a smaller eddy coefficient. The current veers with depth.

It is easy to see that the hodograph for the case of $\tau^{(y)} = 0$, $\tau^{(x)} = \tau$ and $u_{(g)} = 0$ has a spiral structure. The surface current directs 45° to the right of the surface wind in the northern hemisphere. The current veers further to the right at greater depths. Let us define the thickness of this OEL, D, as the first level at which the current is in the opposite direction of the surface current. Since $\left(\frac{v}{u}\right)_{z=-D} = -1$, we therefore have

$$\left(\frac{v}{u}\right)_{z=-D} = -1$$
$$\frac{-\cos\gamma D - \sin\gamma D}{\cos\gamma D - \sin\gamma D} = -1 \tag{4.35}$$
$$\sin\gamma D = 0 \Rightarrow D = \frac{\pi}{\gamma} = \pi\sqrt{\frac{2K}{f}}.$$

Then the thickness of an oceanic Ekman layer is $D \approx 140$ m for $K = 10^{-1}\,\text{m}^2\,\text{s}^{-1}$ and $f = 10^{-4}\,\text{s}^{-1}$.

4.8.1 Mass transport in oceans

It is valuable to know the mass transport in the global oceans. The vertical average current in the model under consideration is

$$\langle u\rangle = \frac{1}{H}\int_{-H}^{z_o} u\,dz + u_{(g)} \approx \frac{1}{H}\int_{-\infty}^{0} u\,dz + u_{(g)} = \frac{\tau}{2\rho_w K\gamma H}\int_{-\infty}^{0} e^{\gamma z}(\cos\gamma z + \sin\gamma z)dz + u_{(g)} = u_{(g)},$$
$$\langle v\rangle = \frac{1}{H}\int_{-H}^{z_o} v\,dz \approx \frac{1}{H}\int_{-\infty}^{0} v\,dz = \frac{\tau}{2\rho_w K\gamma H}\int_{-\infty}^{0} e^{\gamma z}(-\cos\gamma z + \sin\gamma z)dz = \frac{-\tau}{\rho_w f H}, \tag{4.36}$$

since $\int_{-\infty}^{0} e^{\gamma z}\cos\gamma z\,dz = -\int_{-\infty}^{0} e^{\gamma z}\sin\gamma z\,dz = 1/2\gamma$ and $\gamma^2 = f/2K$. The total mass transport consists of two parts. One is a geostrophic transport and the other is a wind-driven (Ekman) transport in the direction 90° to the right of the surface stress. The mass transport (flow of the mass of water in a column of unit horizontal area per unit time) in the x-direction in this case is entirely geostrophic transport, $M_{(x)} = M_{(x)}^{(G)} = \rho_w H\langle u\rangle = \rho_w H u_{(g)}$, in units of kg m s^{-1} m^{-2}. The mass transport in the y-direction is entirely wind-driven (Ekman) transport, $M_{(y)} = M_{(y)}^{(E)} = \rho_w H\langle v \rangle = -\tau/f$. It is proportional to the magnitude of the surface stress and inversely proportional to the Coriolis parameter.

In general, a surface wind and the corresponding wind stress can have any direction denoted by $\vec{V}_{(g)} = \left(u_{(g)}, v_{(g)}\right)$ and $\vec{\tau} = \left(\tau_{(x)}, \tau_{(y)}\right)$. The boundary conditions in a general case are then: at $z = 0$, $\rho_w K u_z = \tau_{(x)}$ and $\rho_w K v_z = \tau_{(y)}$; at $z \to -\infty$, $u = u_{(g)}$ and $v = v_{(g)}$. The corresponding solution of (4.33) can be similarly shown to be

$$\begin{aligned} u &= \frac{1}{2\rho_w K\gamma} e^{\gamma z}\left[\left(\tau^{(x)} + \tau^{(y)}\right)\cos\gamma z + \left(\tau^{(x)} - \tau^{(y)}\right)\sin\gamma z\right] + u_{(g)}, \\ v &= \frac{1}{2\rho_w K\gamma} e^{\gamma z}\left[\left(\tau^{(x)} + \tau^{(y)}\right)\sin\gamma z - \left(\tau^{(x)} - \tau^{(y)}\right)\cos\gamma z\right] + v_{(g)}. \end{aligned} \tag{4.37}$$

Using (4.37), we get for the mass transport

$$\begin{aligned} M_{(x)} &= \frac{\tau_{(y)}}{f} + \rho_w H u_{(g)}, \\ M_{(y)} &= -\frac{\tau_{(x)}}{f} + \rho_w H v_{(g)}. \end{aligned} \tag{4.38a,b}$$

It follows that

$$\frac{\partial M_{(x)}}{\partial x} + \frac{\partial M_{(y)}}{\partial y} = \frac{1}{f}\left(-\frac{\partial \tau_{(x)}}{\partial y} + \frac{\partial \tau_{(y)}}{\partial x}\right) + \frac{\beta}{f^2}\tau_{(x)} + \rho_w H\left(\frac{\partial u_{(g)}}{\partial x} + \frac{\partial v_{(g)}}{\partial y}\right). \tag{4.39}$$

Since the vertically integrated mass flux of a stationary flow has to be non-divergent because there is no mass source or sink, the LHS of (4.39) must be zero. Although geostrophic wind is quasi-non-divergent, we should take into account its divergence associated with the variation of the Coriolis parameter with latitude in a time mean surface wind of planetary scale such as that over an ocean basin. Since $\rho f v_{(g)} = p_x$, $\rho f u_{(g)} = -p_y$, we get

$$\frac{\partial u_{(g)}}{\partial x} + \frac{\partial v_{(g)}}{\partial y} = -\frac{\beta}{f} v_{(g)}. \tag{4.40}$$

Substituting (4.40) into (4.39), we finally get

$$\begin{aligned} 0 &= \frac{\vec{k}\cdot\nabla\times\vec{\tau}}{f} - \frac{\beta}{f}\left(\frac{-\tau_{(x)}}{f} + \rho_w H v_{(g)}\right), \\ \therefore M_{(y)} &= \frac{\vec{k}\cdot\nabla\times\vec{\tau}}{\beta}. \end{aligned} \tag{4.41a,b}$$

Equation (4.41b) is known as the *Sverdrup relation*, which says that the total meridional mass flux at a location in an ocean is proportional to the dot product of the vertical unit vector and the curl of the surface stress and also inversely proportional to $\beta = df/dy$. This formula is actually applicable for a baroclinic ocean model as well, although it is derived above in the context of a barotropic Ekman layer model. Note that Eqs. (4.38b) and (4.41b) are two different formulas for the same quantity, $M_{(y)}$. According to (4.38b), one needs data of $\tau_{(x)}$ and $v_{(g)}$ (hence the pressure data in the ocean) at a point to estimate $M_{(y)}$. According to (4.41b), we would not need the data of the pressure field in the ocean at all, and only need the data of the curl of wind stress. While one can readily get a reasonable estimate of the wind stress by parameterizing it in terms of the surface wind, the pressure data are less accurate. That is why (4.41b) has been a very useful formula in oceanography. It should also be noted that estimating the mass transport in the zonal direction, $M_{(x)}$, does require data of wind stress as well as the pressure field in the ocean.

The northeasterly trade wind over the North Atlantic exerts a surface stress such as is characterized by $\vec{k}\cdot\nabla\times\vec{\tau}<0$. Hence, we would expect an equatorward vertically integrated mass flux in the North Atlantic Ocean. Furthermore, since the beta parameter is larger at lower latitude, the southward mass flux would tend to be weaker at lower latitude. That means $\partial M_{(y)}/\partial y<0$ at the low tropical latitudes. There should be a corresponding $\partial M_{(x)}/\partial x>0$.

The Sverdrup relation (4.41b) is a vertically integrated property of a wind-driven circulation, depending only upon the externally imposed surface stress. As such, the details concerning the Ekman layer dynamics should be immaterial even though we have made use of the explicit solution of the Ekman layer flow in the derivation above. It is indeed the case as verified below. Recall that the frictional force associated with the small turbulent eddies can be generically written as

$$\left(-\frac{\partial\left(\overline{u'w'}\right)}{\partial z},-\frac{\partial\left(\overline{v'w'}\right)}{\partial z}\right)\equiv\frac{1}{\rho_w}\left(\frac{\partial\tau^{(x)}}{\partial z},\frac{\partial\tau^{(y)}}{\partial z}\right).$$

If we cross-differentiate the two horizontal momentum equations with the use of this form of frictional force, combine them and integrate the resulting equation over the depth of a model ocean, H, we would obtain

$$0=Hf_o\langle u_x+v_y\rangle+\beta H\langle v\rangle-\frac{1}{\rho_w}\left(\left(\vec{k}\cdot\nabla\times\vec{\tau}\right)_{surface}-\left(\vec{k}\cdot\nabla\times\vec{\tau}\right)_{bottom}\right).$$

Since the vertically integrated horizontal divergence is zero and the frictional stress at the bottom of the ocean is negligibly weak, we would indeed get the mass transport in the meridional direction as $\rho_w H\langle v\rangle=\dfrac{\vec{k}\cdot\nabla\times\vec{\tau}_{surface}}{\beta}$.

4.8.2 Oceanic Ekman pumping/suction

We can deduce the vertical mass flux through the *bottom* of an oceanic Ekman layer by integrating the continuity equation from $z=-D$ to $z=0$. The *mass transport within the oceanic Ekman layer* is then

$$\begin{aligned}M_{(x)}^{(E)}&=\int_{-D}^{0}\rho_w u\,dz=\frac{\tau_{(y)}}{f}\left(1+e^{-\pi}\right)+\rho_w Du_{(g)},\\M_{(y)}^{(E)}&=\int_{-D}^{0}\rho_w v\,dz=-\frac{\tau_{(x)}}{f}\left(1+e^{-\pi}\right)+\rho_w Dv_{(g)}.\end{aligned}\tag{4.42}$$

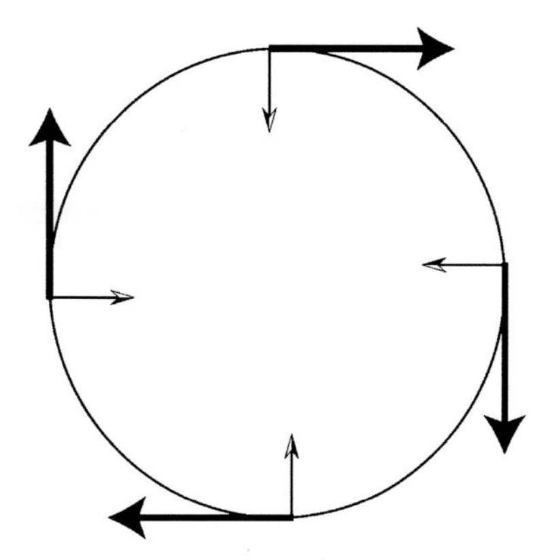

Fig. 4.6 Thick arrows stand for the stress vectors at the ocean surface where there is an anticyclonic surface wind system. Thin short arrows stand for Ekman layer-average mass flux.

By mass continuity, writing the density of water simply as ρ_* instead of ρ_w, we have

$$-\int_{-D}^{0} w_z dz = \int_{-D}^{0} (u_x + v_y) dz,$$

$$\begin{aligned} w|_{-D} &= \frac{1}{\rho_*}\left(\frac{\partial M_{(x)}^{(E)}}{\partial x} + \frac{\partial M_{(y)}^{(E)}}{\partial y}\right) \\ &\approx \frac{1}{\rho_*}\left(\frac{\partial}{\partial x}\left(\frac{\tau_{(y)}}{f} + \rho D u_{(g)}\right) + \frac{\partial}{\partial y}\left(\frac{-\tau_{(x)}}{f} + \rho D v_{(g)}\right)\right) \approx \frac{1}{\rho_* f_o}\left(\frac{\partial \tau_{(y)}}{\partial x} - \frac{\partial \tau_{(x)}}{\partial y}\right), \\ w|_{-D} &= \frac{1}{\rho_* f_o} \vec{k} \cdot \nabla \times \vec{\tau}. \end{aligned} \tag{4.43}$$

The climatological surface wind over the N. Atlantic is anticyclonic around a high-pressure center. It follows that $\vec{k} \cdot \nabla \times \vec{\tau} < 0$ implying a downward velocity at the bottom of the Ekman layer, $w|_{-D} < 0$. It is associated with a mass convergence in the OEL towards the center of the Sargasso Sea (Fig. 4.6). This downward mass flow into the interior of the Atlantic Ocean in turn feeds an interior flow, part of which goes equatorward. This would support up-welling in the equatorial latitudes. The latter in turn compensates the Ekman transport from the south towards the center of the Sargasso Sea.

In passing, it is noteworthy that the curl of the surface stress of a hurricane is cyclonic. According to (4.43), there would be up-welling at the bottom of the OEL (Ekman suction) to compensate for the radial outward flux of water.

4.9 Surface layer

An atmospheric Ekman layer does not actually extend to the surface of the Earth because the turbulent eddies near the surface have different characteristics. The turbulence near the surface is increasingly constrained by its presence, thereby giving rise to a boundary layer of different character. It is referred to as a *surface layer.* In particular, the mixing length of turbulence ℓ becomes shorter at elevations closer to a surface. A first approximation would be that ℓ is a linear function of distance from the ground in a surface layer,

$$\ell = kz. \tag{4.44}$$

A relevant empirical observation is that the Reynolds stress is virtually constant in a surface layer, viz.

$$\overline{u'w'} = u_*^2, \tag{4.45}$$

where u_* is called the *friction velocity.* It is assumed that the vertical shear in the surface layer is linearly proportional to the friction velocity, and that the proportionality constant is the reciprocal of the mixing length. Thus we use

$$\frac{\partial u}{\partial z} = \frac{u_*}{\ell} = \frac{u_*}{kz}. \tag{4.46}$$

We impose a boundary condition that u is zero at a certain height z_o which is called *surface roughness*. Then, integrating (4.46) from z_o to any height z in the surface layer yields a logarithmic function for the wind profile,

$$u = \frac{u_*}{k}\ln\left(\frac{z}{z_o}\right), \tag{4.47}$$

where k is a constant called the von Karman constant. It is noteworthy that the frictional force in a surface layer is zero because the momentum flux by the turbulent eddies is non-divergent. The value of z_o depends on the type of surface. It is equal to a few cm for grassland and a larger value for a wooded area. The value of $\overline{u'w'}$ is often observed to be about $0.1\ \mathrm{m^2\,s^{-2}}$ leading to $u_* \approx 0.3\ \mathrm{m\,s^{-1}}$. The slope of the mean wind in a plot using logarithmic scale is generally confirmed to be almost a constant. It is also found that $k \approx 0.4$ on the basis of such slope and u_*.

4.10 Mixed layer

The turbulence over land on a hot summer afternoon is intense and convectively driven. It is typically capped by distinctly stable air above. The turbulence gives rise to a different boundary layer in which the wind and potential temperature are virtually uniform. Such a layer is called a *mixed layer*, schematically depicted in Fig. 4.7. It is of the order of 100 m thick. Since this vertical mixing is local, it has little impact on the structure of the horizontal pressure gradient.

There is no frictional stress at the top of the mixed layer. The Reynolds stresses at the bottom of a mixed layer are parameterized with a bulk-aerodynamic formula as follows,

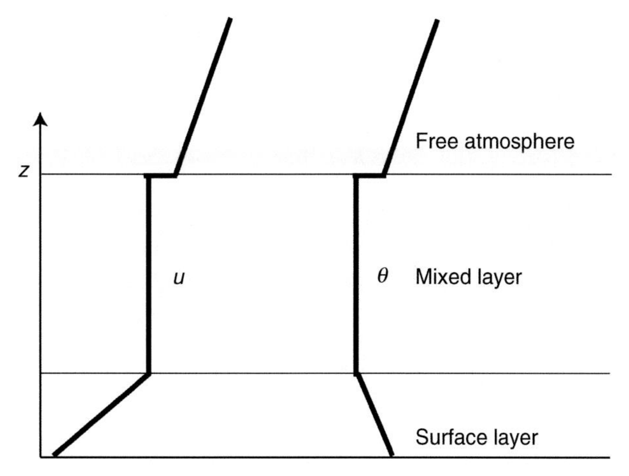

Fig. 4.7 Schematic of the structure of the velocity and potential temperature in a mixed layer.

$$\begin{aligned}\tau_B^{(x)} &= -C_D\left|\vec{V}\right|u,\\ \tau_B^{(y)} &= -C_D\left|\vec{V}\right|v,\end{aligned} \tag{4.48}$$

where C_D is the drag coefficient, 1.5×10^{-3} for an ocean surface, 5.0×10^{-3} for a land surface and $\left|\vec{V}\right|$ is the wind speed. Let h be the thickness of a mixed layer. There is a balance of three forces in a mixed layer: PGF, Coriolis force and frictional force. The PGF is written in terms of a geostrophic wind aloft. The velocity of the air in this layer is independent of height. Integrating the momentum equations from the bottom to the top of a mixed layer, we get

$$\begin{aligned}0 &= hf\left(v - v_g\right) + \tau_B^{(x)},\\ 0 &= -hf\left(u - u_g\right) + \tau_B^{(y)}.\end{aligned} \tag{4.49}$$

Substituting (4.48) into (4.49), we get

$$\begin{aligned}0 &= \left(v - v_g\right) - K_S\left|\vec{V}\right|u,\\ 0 &= -\left(u - u_g\right) - K_S\left|\vec{V}\right|v,\end{aligned} \tag{4.50}$$

where $K_S = C_D/hf \sim 0.05\ \mathrm{m\,s^{-1}}$ for an ocean surface with $h = 300$ m. We set the x-axis in the direction of the geostrophic wind aloft, so that $v_g = 0$. From (4.50) we get

$$\begin{aligned}v &= K_S\left|\vec{V}\right|u,\\ u &= u_g - K_S\left|\vec{V}\right|v,\\ \left|\vec{V}\right| &= \left(u^2 + v^2\right)^{1/2} = \left(uu_g\right)^{1/2},\\ v &= K_S u_g^{1/2} u^{3/2},\\ u &= u_g - K_S^2 u_g u^2,\\ u &= \frac{1}{2K_S^2 u_g}\left(-1 + \sqrt{1 + 4K_S^2 u_g^2}\right).\end{aligned} \tag{4.51}$$

Using $u_g = 10\,\mathrm{m\,s^{-1}}$, we then have $u = 8.3\,\mathrm{m\,s^{-1}}$, $v = 3.9\,\mathrm{m\,s^{-1}}$, $|\vec{V}| = 9.1\,\mathrm{m\,s^{-1}}$. The result indicates that turbulence in a mixed layer slightly reduces the wind speed from u_g and generates a cross-isobar flow component towards the lower pressure. If the turbulence is very strong, (i.e. large K_S), then $u \approx \frac{1}{K_S}$, $v \approx \left(\frac{u_g}{K_S}\right)^{1/2}$, $\frac{v}{u} \approx \sqrt{K_S u_g}$. Thus, the velocity would be weaker and the cross-isobar angle would be larger.

A similar mixed layer exists at the top of the ocean associated with intense convection. The turbulence can be wind-driven or convection-driven. Such a layer is typically tens of meters thick.

5 Fundamentals of wave dynamics

This chapter is concerned with the dynamics of atmospheric motions that characteristically have cyclical structure in space and periodic fluctuations in time. Such type of motion is what we call *waves*. As far as weather disturbances are concerned, the two most important types of waves in the atmosphere are known as *internal gravity waves* and *Rossby waves*. Following some brief preliminary remarks in Section 5.1, we discuss the physical nature and the restoring mechanisms responsible for the existence of these waves in Sections 5.2, 5.5 and 5.6. We formulate the simplest possible models for them and deduce the intrinsic properties of the free wave modes in Sections 5.3, 5.4 and 5.7. We also delineate their structures arising from a given external forcing in Section 5.8. A brief discussion of the observed statistical properties of Rossby waves in the atmosphere is given in Section 5.9. The chapter ends in Section 5.10 with a discussion of the fundamental dynamics of a unique subclass of wave motions known as *edge waves* that turns out to have wide ramifications.

5.1 Preliminary remarks

Wave is a generic term for vibrations in a physical medium, whether it is a string, a membrane, a three-dimensional structure, water in an ocean or air in the atmosphere. A wave has cyclic structure in space and/or periodic fluctuations in time. A localized disturbance in the atmosphere may be thought of as an ensemble of constituent waves. There are distinctly different types of waves in the atmosphere and oceans. Familiar examples are the sound that we make and hear (acoustic waves); the undulating patterns visible in photographic images of cloud from space and the ground (internal gravity waves, e.g. Fig. 5.1); the tsunamis in oceans and the ripples on the surface of a lake (external gravity waves); the meandering temperature and pressure contours on a weather map (Rossby waves). Each type of wave is associated with a unique restoring force. It follows that each has its own unique intrinsic properties. Waves in the atmosphere are important in that they can transport large amounts of energy and momentum to great distances from their regions of origin. The background flow could then be altered accordingly.

This chapter focuses on two types of waves most important for weather forecasting: internal gravity waves (IGW) and Rossby waves. We touch upon their basic dynamics. We show how to establish the intrinsic properties of prototype IGW with or without the

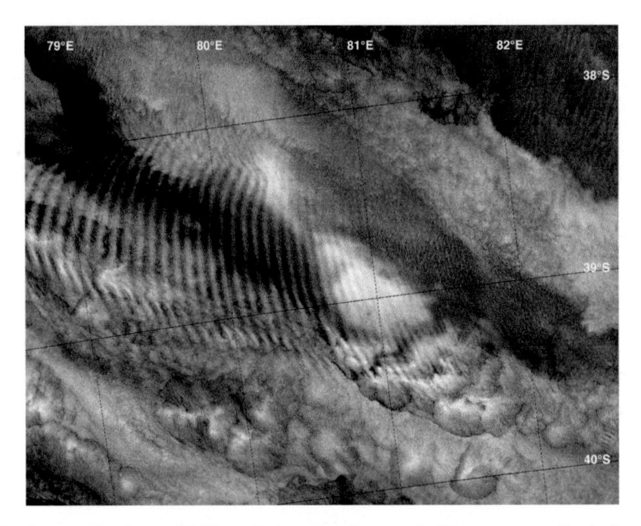

Fig. 5.1 An image taken by the satellite-borne Multi-angle Imaging Spectro-Radiometer over the Indian Ocean showing atmospheric internal gravity waves (courtesy of Larry De Girolamo and Guangyu Zhao).

Earth's rotation in a simplest possible model setting. The orographically forced IGW is discussed as an example. Some more complex aspects of IGW are discussed in Part A of Chapter 8 and Section 12.3. The intrinsic properties of prototype Rossby waves in a barotropic model setting either with or without a uniform basic zonal flow are next discussed. We also examine the dispersion of a particular wave-packet. The intrinsic characteristic of Rossby waves is highlighted by an analysis of the forced response to either a westerly or an easterly basic flow going over a large-scale orography in a simplest possible model.

The fundamental difference in the dynamical nature of internal gravity waves and Rossby waves makes it possible to design a model that would expunge the gravity waves a priori. The construction of such a model is historically significant for two reasons. On the one hand, it enabled scientists to make numerical weather forecasts in the early days when computers were slow and small in capacity. On the other hand, such a model greatly facilitates the researches on the fundamental dynamics of large-scale flows. We elaborate on how to construct such a model in Chapter 6. Many more subtle aspects of Rossby waves are elaborated in the context of such a model as presented in Section 6.6, Parts B and C of Chapter 8, Sections 9.3, 9.4, 9.5 and Chapter 11.

There is one exception to the rule that gravity waves and Rossby waves are inherently separable. In a model that includes the equator there exists a subclass of waves with characteristics of both gravity waves and Rossby waves. While the eastward-propagating branch of this set of waves resembles gravity waves, the westward-propagating branch of them resembles Rossby waves. They are known as *mixed Rossby–gravity waves* (Matsuno, 1966). We do not go over them in this book in the interest of brevity. A detailed discussion of the properties of this set of special wave motions can be found in several texts (e.g. Gill, 1982).

5.2 Physical nature of internal gravity waves

The restoring force that gives rise to internal gravity waves (IGW) is gravity acting on elements of a stably stratified fluid. A stably stratified atmospheric layer has positive values of $\partial\theta/\partial z$, where θ is potential temperature. When an air parcel is adiabatically displaced upward from its equilibrium position, its potential temperature would remain unchanged and therefore would be lower than that of its new surroundings. The displaced air parcel would then have a negative θ anomaly. This is equivalent to having a positive density anomaly since the corresponding pressure change is relatively small. Gravity acting on such a density anomaly would be a downward pointing *buoyancy force*, which would pull the air parcel back toward its original position. The descending air parcel naturally overshoots past its original level due to its inertia. Then, the buoyancy force would change sign and oppose continual descent of the air parcel until it stops and moves upward subsequently. This sequence of motion of the air parcel would repeat, resulting in oscillations back and forth about its original level. There is actually an accompanying change in the pressure. The pressure differential would set the neighboring air parcels to motion as well. Oscillating motion will then quickly spread to adjacent horizontal locations and to other levels as well, generating a wave field. This overall movement is referred to as internal gravity waves (IGW).

Vertical displacement of air parcels can be initiated by a number of different processes. For example, when wind blows over an undulating surface, the air next to the ground would be forced to follow the terrain. This influence is transmitted to the fluid at the next level above. The resulting disturbance is an orographic IGW. An IGW may also be excited by dynamic instability of a flow in a stably stratified fluid with a sufficiently strong vertical shear.

A number of well-known weather phenomena are closely related to IGW. For example, strong down-slope wind in mountainous areas (e.g. the Santa Anna wind in southern California) is a manifestation of orographic IGW. A severe storm generates strong IGW by virtue of the heat associated with precipitation. Those IGWs could penetrate deeply into the stratosphere and beyond. New severe convective storms in the troposphere may even be initiated by nearby IGW generated at an earlier time.

5.3 Generic model of IGW

The satellite image in Fig. 5.1 depicts an extensive area of narrowly spaced, long cloud bands over the Indian Ocean. Such cyclical structure suggests the presence of waves in this atmospheric region. The separation between two bands in the east–west direction is about 11 km. It is progressively wider in the western part of the domain. In contrast, each cloud band in the approximately north–south direction is over 120 km long. The height of the cloud bands is estimated to be 2.2 km. These cloud bands suggest the presence of essentially two-dimensional IGW. But we cannot infer from such an image the dynamical and complete structural properties of basic IGW. A simple model however may be used to establish such information. A generic model of an IGW is *a layer of 2-D non-rotating stably stratified atmosphere on a vertical (x,z) plane in an unbounded domain with no heating/cooling, no viscosity and no background wind.* It is a simplest possible but adequate model setting for delineating the intrinsic dynamics of this class of wave motion. It would be appropriate to refer to the IGW in such a model as *prototype IGW.* The governing equations for this atmospheric model written in standard notations are

$$\begin{aligned}
u_t + uu_x + wu_z &= -\frac{1}{\rho}p_x,\\
w_t + uw_x + ww_z &= -\frac{1}{\rho}p_z - g,\\
\rho_t + u\rho_x + w\rho_z &= -\rho(u_x + w_z),\\
\theta_t + u\theta_x + w\theta_z &= 0,\\
\theta = \frac{p}{R\rho}\left(\frac{p_{oo}}{p}\right)^{\kappa}, &\quad \kappa = R/c_p.
\end{aligned} \tag{5.1}$$

The subscripts *x,z,t* stand for partial derivatives with respect to those independent variables. The model atmosphere has a stably stratified background thermodynamic state: $\theta_o(z), p_o(z), \rho_o(z)$. It is necessarily in hydrostatic balance, $dp_o/dz = -g\rho_o$ with $d\theta_o/dz > 0$. We may decompose each thermodynamic variable as a sum of a background component and a departure

$$\rho = \rho_o(z) + \tilde{\rho}(x,z,t) \approx \rho_{oo} + \tilde{\rho},\ p = p_o(z) + \tilde{p},\ \theta = \theta_o(z) + \tilde{\theta}, \tag{5.2}$$

where $\tilde{\rho}, \tilde{p}, \tilde{T}$ stand for the thermodynamic properties of a disturbance. They are assumed to be small compared to the corresponding background values: $\rho_{oo} \gg \tilde{\rho}$, $p_o \gg \tilde{p}$, $\theta_o \gg \tilde{\theta}$.

We first simplify (5.1) by invoking *Boussinesq approximation* which says that any term containing the density disturbance can be neglected except when it is multiplied by gravity. Hence, the RHS of the second equation in (5.1) can be simplified as follows

$$\begin{aligned}
-\frac{1}{\rho}\frac{\partial p}{\partial z} - g &= -\frac{1}{\rho_o(1+\tilde{\rho}/\rho_o)}\left(\frac{\partial p_o}{\partial z} + \frac{\partial \tilde{p}}{\partial z}\right) - g\\
&\approx -\frac{1}{\rho_o}\left(1 - \frac{\tilde{\rho}}{\rho_o}\right)\left(\frac{\partial p_o}{\partial z} + \frac{\partial \tilde{p}}{\partial z}\right) - g\\
&\approx -\frac{1}{\rho_o}\frac{\partial \tilde{p}}{\partial z} + \frac{\tilde{\rho}}{\rho_o^2}\frac{\partial p_o}{\partial z}\\
&\approx -\frac{1}{\rho_o}\frac{\partial \tilde{p}}{\partial z} - \frac{g\tilde{\rho}}{\rho_o}\\
&= -\frac{\partial}{\partial z}\left(\frac{\tilde{p}}{\rho_o}\right) - \frac{\tilde{p}}{\rho_o^2}\frac{d\rho_o}{d_z} - \frac{g\tilde{\rho}}{\rho_o}.
\end{aligned}$$

We next rewrite the last two terms on the RHS above in terms of the departure in potential temperature, $\tilde{\theta}$. In light of the definition $\theta = \frac{p_{oo}^k\, p^{1-k}}{R\quad \rho}$, we have $d\theta = \frac{\partial\theta}{\partial\rho}d\rho + \frac{\partial\theta}{\partial p}dp$, where $\frac{\partial\theta}{\partial\rho} = -\frac{\theta}{\rho}$ and $\frac{\partial\theta}{\partial p} = \frac{c_v\,\theta}{c_p\, p}$, leading to $\frac{\tilde{\theta}}{\theta_o} = -\frac{\tilde{\rho}}{\rho_o} + \frac{c_v\,\tilde{p}}{c_p\, p_o}$. While the pertinent $p_o(z)$ and $\rho_o(z)$ vary exponentially with height, the corresponding θ_o varies much less so in the troposphere. Thus, it would be appropriate to use $\theta_{oo} \equiv \theta_o = \frac{p_{oo}^k\, p_o^{1-k}}{R\quad \rho_o}$ to be a constant in this specific consideration so that we have $0 = -\frac{p_o}{\rho_o^2}\frac{d\rho_o}{dz} - \frac{gc_v}{C_p}$ and therefore

$-\frac{\tilde{p}}{\rho_o^2}\frac{d\rho_o}{dz}-\frac{g\tilde{\rho}}{\rho_o}=\frac{gc_v\tilde{p}}{c_p p_o}-\frac{g\tilde{\rho}}{\rho_o}=\frac{g\tilde{\theta}}{\theta_{oo}}$. Although $\theta_o \approx \theta_{oo}$ is introduced in obtaining this expression, we must not neglect the variation of θ_o with z altogether. It is imperative to retain $\frac{d\theta_o}{dz}\neq 0$ in the thermodynamic equation. We may additionally use $\rho_o \approx \rho_{oo} =$ constant when we consider a disturbance in a thin layer of atmosphere. This would enable us to simplify the continuity equation as well as the pressure gradient force terms to the greatest possible extent. Thus, we finally may approximate the RHS of the second equation in (5.1) as

$$-\frac{1}{\rho}\frac{\partial p}{\partial z}-g\approx -\frac{1}{\rho_{oo}}\frac{\partial \tilde{p}}{\partial z}+\frac{g\tilde{\theta}}{\theta_{oo}}. \tag{5.3}$$

The set of equations (5.1) can be then simplified to

$$\begin{aligned} u_t+uu_x+wu_z &= -\frac{1}{\rho_{oo}}\tilde{p}_x, \\ w_t+uw_x+ww_z &= -\frac{1}{\rho_{oo}}\tilde{p}_z+g\frac{\tilde{\theta}}{\theta_{oo}}, \\ u_x+w_z &= 0, \\ \tilde{\theta}_t+u\tilde{\theta}_x+w\left(\theta_{oz}+\tilde{\theta}_z\right) &= 0. \end{aligned} \tag{5.4a,b,c,d}$$

The quantity $g\tilde{\theta}/\theta_{oo}$ is called *buoyancy force* indispensable for the existence of IGW. It should be emphasized that the use of ρ_{oo} instead of $\rho_o(z)$ in the first two equations in (5.4) is not a crucial assumption. If an IGW propagates through a deep layer of atmosphere, the exponential decrease of ρ_o with z should not be neglected. The consequence is that the wave amplitude would increase exponentially with height in compensation for the decrease in density. Those waves would break up into turbulence where its amplitude becomes sufficiently large.

5.3.1 Method of linearization

Let us use a prime superscript to denote the properties of a weak disturbance. In the absence of a background flow, the nonlinear advective terms in (5.4) would be a product of two prime quantities and are therefore negligibly small compared to the linear terms. Equation (5.4) may be linearized by dropping the nonlinear advective terms. The resulting equations are called *perturbation equations*

$$\begin{aligned} u'_t &= -\frac{1}{\rho_{oo}}p'_x, \\ w'_t &= -\frac{1}{\rho_{oo}}p'_z+g\frac{\theta'}{\theta_{oo}}, \\ u'_x+w'_z &= 0, \\ \theta'_t+w'\theta_{oz} &= 0. \end{aligned} \tag{5.5a,b,c,d}$$

By eliminating u', p', θ' among (5.5), we get one single equation for one unknown,

$$\frac{\partial^2}{\partial t^2}\left(\frac{\partial^2}{\partial x^2}+\frac{\partial^2}{\partial z^2}\right)w'+N^2\frac{\partial^2}{\partial x^2}w'=0. \tag{5.6}$$

Note that there is only one parameter, $N^2 = \frac{g}{\theta_{oo}} \frac{d\theta_o}{dz}$, in (5.6), suggesting that this is the sole controlling factor for the dynamics of the disturbance in this model. N has the dimension of s^{-1} and is called the *Brunt–Vaisala frequency*. It embodies the joint influence of gravity and stratification. We do not need to take into consideration any boundary conditions when we wish to learn about the intrinsic characteristics of possible motions in this system. When the length scales of disturbances are much shorter than the size of a domain under consideration, the domain is effectively unbounded. We will return later to examine IGW under the influence of an undulating surface in the presence of a basic flow.

5.3.2 Intrinsic properties of prototype IGW

Dispersion relation of IGW

Since (5.6) is a linear partial differential equation with a constant coefficient, it can be solved with the method of separation of variables. An elementary solution for the vertical velocity is a sinusoidal function of x, z and t written in the form of complex variables as

$$w' = A\exp(i(kx + mz - \sigma t + \xi)) \equiv Ae^{i\phi} \tag{5.7}$$

with $i = \sqrt{-1}$. It contains five constants of real values: A, k, m, σ and ξ. It is understood that the vertical velocity in physical space is to be identified with the real part of (5.7), $A\cos\phi$. It is a wave solution since it has a cyclical spatial structure and fluctuates periodically in time. This is the simplest model solution for an internal gravity wave, IGW.

The quantity $\phi \equiv kx + mz - \sigma t + \xi$ is called the *phase function* of a 2-D plane wave. The amplitude of a wave perturbation A can have any value. Any property of the wave has the same value along a line on the x–z plane at a particular time where ϕ has a specific value. If the pressure has a minimum (maximum) value along such a line, we call it a trough (ridge). Any of those lines is called a *wave-front*; $k = \partial\phi/\partial x$ is called the *horizontal wavenumber in the x-direction*, $m = \partial\phi/\partial z$ the *vertical wavenumber in the z-direction* and $\sigma = \partial\phi/\partial t$ the *frequency* of the wave. We may use only positive values of k without loss of generality. If we choose to do that, m and σ must be allowed to have either sign so that the function $\phi =$ constant could depict a wave-front with any orientation moving in any direction. For example, a wave-front for $k > 0$ and $m < 0$ is oriented in the "west-up" direction on a x–z plane; a wave-front for $k > 0$ and $m > 0$ is oriented in the "west-down" direction. The *wavelength in the x-direction* is $L_{(x)} = 2\pi/k$, its *wavelength in the z-direction* is $L_{(z)} = 2\pi/|m|$ and its period is $T = 2\pi/|\sigma|$.

The parameter ξ is called the *phase angle*. It contains the information about where each wave-front is located at a particular time. For example, the vertical velocity at $z = 0$ and $t = 0$ varies along the x-axis as $w' = A\cos(kx + \xi)$. It would have a maximum value equal to A at the locations of x where $kx + \xi = 2n\pi$, where $n = 0, 1, 2\ldots$ The quantity $B \equiv Ae^{i\xi}$ is called the *complex amplitude* of the wave.

If (5.7) is a non-trivial solution of (5.6), meaning $A \neq 0$, then the wave parameters k,m,σ must be related to the external parameter N by the following formula

$$\sigma = \pm\frac{Nk}{(k^2 + m^2)^{1/2}}. \tag{5.8}$$

This is called the *dispersion relation of IGW.* These are two possible values of σ for each set of parameters: k,m,N. Equation (5.8) suggests $|\sigma| \leq N$ meaning that the Brunt–Vaisala frequency is the upper bound of the frequency of a prototype IGW. In particular, we expect $|\sigma| \approx N$ when the horizontal wavelength of the wave under consideration is much shorter than its vertical wavelength, $(k \gg m)$.

Phase velocity of IGW

One thing we would like to know about a wave is where its wave-fronts are heading and how fast they move. It is common to find in textbooks of dynamics the use of definitions $\sigma/k \equiv c_{(x)}$ and $\sigma/m \equiv c_{(z)}$, for so-called "*phase speeds* of a wave in the x- and z-directions," as measures of the movement of a 2-D wave. However, such definitions do not serve the stated purpose very well for two reasons. First, while the notion of "speed" is a positive definite quantity according to common usage, $c_{(x)}$ and/or $c_{(z)}$ may have negative values because σ and m can be of either sign. Second, neither $c_{(x)}$ nor $c_{(z)}$ individually or in combination directly tell us about the direction and rate of the movement of a wave-front. In other words, $c_{(x)}$ and $c_{(z)}$ are not components of a vector that measures the movement of a wave-front.

Each wave-front of a plane wave in this generic model setting moves in a direction normal to itself. It moves either in the direction of the *wavenumber vector*, $\vec{j} = (k, m)$, or opposite to the direction of the wavenumber vector, $-\vec{j} = (-k, -m)$. The distance between two wave-fronts differing in their phase function by 2π is $2\pi/j$. Hence, the propagation of a 2-D plane wave is most logically measured by the following phase velocity

$$\vec{c} = \frac{\sigma\vec{j}}{jj} = \left(\frac{\sigma k}{k^2+m^2}, \frac{\sigma m}{k^2+m^2}\right) \equiv (c_x, c_z). \tag{5.9}$$

The magnitude of this vector would be truly a phase speed, $c \equiv |\vec{c}| = |\sigma|/j$. The $\vec{c}$ of a wave with a positive value of σ is in the direction of the wavenumber vector. This $\vec{c}$ would have an eastward component. Its vertical component would be upward if its vertical wavenumber m has a positive value. Then, we have

$$\vec{c} = \left(\frac{Nk^2}{(k^2+m^2)^{3/2}}, \frac{Nkm}{(k^2+m^2)^{3/2}}\right), \quad \text{for an } \textit{eastward-propagating IGW}. \tag{5.10}$$

By the same token, the $\vec{c}$ of a wave with a negative value of σ is in the opposite direction of the wavenumber vector. Such $\vec{c}$ would have a westward component. Its vertical component would be downward if its vertical wavenumber m has a positive value. Then, we have

$$\vec{c} = \left(\frac{-Nk^2}{(k^2+m^2)^{3/2}}, \frac{-Nkm}{(k^2+m^2)^{3/2}}\right), \quad \text{for a } \textit{westward-propagating IGW}. \tag{5.11}$$

Note that the phase velocity of IGW is a function of its wavenumbers. In an ensemble, such waves with different wavelengths would propagate at different speeds and in different directions. This ensemble of waves would therefore disperse in time.

It should be added that model analyses of linear wave dynamics often take into consideration a constant basic flow $\bar{u}$. The acceleration term in the governing equations would be $(\partial/\partial t + \bar{u}\partial/\partial x)$. Applying to a wave solution, $\sim e^{i(kx - t\sigma)}$, the acceleration term would be proportional to $ik(\bar{u} - \sigma/k) \equiv ik\left(\bar{u} - c_{(x)}\right)$. The behavior of the wave solution would be sensitive to the value of $\left(\bar{u} - c_{(x)}\right)$, which is referred to as "Doppler-shifted zonal phase speed." For this reason, an emphasis is often put on the phase speed in the x-direction $c_{(x)}$ rather than on the x-component of the phase velocity, c_x. This is understandable, especially when an analyst does not intend to put emphasis on the movement of wave-fronts in the direction perpendicular to it.

Group velocity of IGW

Another property we would like to know about a wave is where its energy is heading and how fast it moves. A vector called *group velocity* is introduced to quantify such information. It is defined for a 2-D wave by

$$\vec{c}_g = \left(\frac{\partial \sigma}{\partial k}, \frac{\partial \sigma}{\partial m}\right). \tag{5.12}$$

Hence, we get

$$\begin{aligned} \vec{c}_g &= \left(\frac{Nm^2}{(k^2+m^2)^{3/2}}, \frac{-Nmk}{(k^2+m^2)^{3/2}}\right), \quad \text{for eastward-propagating IGW,} \\ \vec{c}_g &= \left(\frac{-Nm^2}{(k^2+m^2)^{3/2}}, \frac{Nmk}{(k^2+m^2)^{3/2}}\right), \quad \text{for westward-propagating IGW.} \end{aligned} \tag{5.13}$$

The group velocity of a prototype IGW may be faster or slower than its phase velocity. It follows from (5.10), (5.11) and (5.13) that

$$\vec{c} \cdot \vec{c}_g = 0. \tag{5.14}$$

The group velocity and phase velocity of any propagating IGW are therefore orthogonal to one another. In other words, the group velocity of IGW is parallel to a wave-front. This is a unique signature of a prototype IGW.

Group velocity is typically illustrated in textbooks using a wave field that consists of two constituent waves of equal amplitude with two slightly different wavenumbers and frequencies. For example, suppose one considers two 1-D waves in the x-direction: one characterized by k and σ, another by $k + \Delta k$ and $\sigma + \Delta\sigma$ with $k \gg \Delta k$ and $\sigma >> \Delta\sigma$. The mathematical representation of this total wave field is $\{\exp(i(kx - \sigma t))\}$ $\{1 + \exp(i(x\Delta k - t\Delta\sigma))\}$. A plot of this wave field gives the appearance of a short wave with wavenumber k and frequency σ rapidly moving through a wavy outline that has a much smaller wavenumber Δk and smaller frequency $\Delta\sigma$. The former is called a *carrier wave* and the latter an *envelope wave*. The envelope wave moves in a speed equal to $\lim \Delta\sigma/\Delta k \to \partial\sigma/\partial k$ for infinitesimally small Δk and $\Delta\sigma$. This velocity corresponds to the *group velocity* defined in (5.12). It is often *asserted* without proof that such a quantity measures the propagation of energy of the wave field under consideration. This is merely a kinematic illustration of group velocity. It is noteworthy that the concept of group velocity is applicable even to a monochromatic wave because a single wave still transmits energy across space. But there is no "wave group" to speak of in a single wave. Hence, to convincingly prove that the concept of group velocity truly measures the direction and speed of energy propagation, we need to diagnose the energetics of each type of wave under consideration. We will do so in Section 5.4.2 after we have discussed the structure of IGW.

The stratification of the troposphere is typically characterized by $N = 10^{-2}\,\mathrm{s}^{-1}$. There can be a wide range of values for the phase velocity and group velocity of IGW dependent upon the combination of its wavenumbers.

For $L_{(x)} = 100\,\mathrm{km}, L_{(z)} = 1\,\mathrm{km}$, we get $|\vec{c}| \approx Nk/m^2 \sim 0.016\,\mathrm{m\,s^{-1}}, |\vec{c}_g| \sim \frac{N}{m} \sim 1.6\,\mathrm{m\,s^{-1}}$,
for $L_{(x)} = 1\,\mathrm{km}$, $L_{(z)} = 1\,\mathrm{km}$, we get $|\vec{c}| = |\vec{c}_g| = \frac{N}{2k} \sim 0.8\,\mathrm{m\,s^{-1}}$,
for $L_{(x)} = L_{(z)} = 10\,\mathrm{km}$, we get $|\vec{c}| = |\vec{c}_g| = \frac{N}{2k} \sim 8\,\mathrm{m\,s^{-1}}$.

An atmosphere with an inversion in its stratification may have a much larger value of Brunt–Vaisala frequency.

5.4 Properties of prototype IGW

5.4.1 Structure of prototype IGW

To see what a prototype IGW actually looks like, we construct its u', w', p', θ' fields on a x–z plane at a particular time instant. The *relative configurations* of these four properties are unique. Thus, we may use a particular structure of one of them, say w', as reference and determine the structure of the corresponding u', p', θ' without loss of generality. Denoting the solutions of all dependent variables by

$$(u', w', p', \theta') = \left(\hat{u}, A, \hat{p}, \hat{\theta}\right) e^{i\phi}, \tag{5.15}$$

we obtain the following relations among the amplitude functions from the perturbation equations, (5.5),

$$\hat{u} = -\frac{m}{k}A, \quad \hat{\theta} = -\frac{i}{\sigma}\frac{d\theta_o}{dz}A, \quad \hat{p} = -\frac{\rho_o m\sigma}{k^2}A. \tag{5.16}$$

These relations are called *polarization relations*. Notice that the ratio $\hat{u}/\hat{w} = -m/k$ is a positive real number for a negative value of m and is a negative real number for a positive value of m. The direction of the velocity vector is therefore constant. The velocity vectors lie on wave-fronts. Such a polarization state of this wave is called *linear polarization*. This characteristic of the IGW under consideration stems from (5.5c), the mass conservation in terms of the non-divergence condition.

The analytic form for the structure of the IGW under consideration is

$$\begin{aligned} &w' = A\cos\phi, \; u' = -\frac{m}{k}A\cos\phi, \; p' = -\frac{m\rho_o\sigma}{k^2}A\cos\phi, \\ &\theta' = \frac{1}{\sigma}\frac{d\theta_o}{dz}A\cos\left(\phi - \frac{\pi}{2}\right). \end{aligned} \tag{5.17}$$

We may deduce from (5.17) the relative phase relationships among all these properties for different cases:

u' and w' are in phase for IGW that have $m < 0$;

p' and w' are in phase for IGW if m and σ have opposite signs; and θ' leads w' by a quarter of a wavelength for IGW if $\sigma > 0$.

These relative phase relationships in turn have important implications on the direction of the momentum, energy and heat fluxes by IGWs. An example of the structure of IGW is shown in Section 5.4.3.

A property of a wave represented by the product of two of its dependent variables is equal to the average of that property *over its wavelength in either the x-direction or z-direction, or over its period*. All such averages are equivalent and amount to an average over its phase function for 2π. We denote this average by a square bracket, $[\xi] = \frac{1}{2\pi}\int_0^{2\pi} \xi \, d\phi$. For example, the vertical flux of zonal momentum and heat would be measured by $[u'w']$ and $[\theta'w']$ respectively. The directions of the various fluxes of the prototype IGWs are summarized in Table 5.1. Readers are encouraged to verify them using the solution of the wave for each case in terms of its dependent variables, (5.17).

Case I in Table 5.1 shows that $[u'w']$ and $[p'w']$ are in the same direction for eastward-propagating IGW. Specifically, they are downward for $m > 0$ meaning that the wave-fronts are in a "west-up" vs. "east-down" direction. Before we graphically depict all properties of an IGW, let us first perform a diagnosis of its energetics whereby we will clearly establish the physical meaning of the group velocity $\vec{c}_g$ and the interpretation of $[p'w']$.

5.4.2 Energy flux and group velocity of prototype IGW

We readily obtain from (5.5) the following equations for the energetics of a perturbation

$$\frac{1}{2}\left(u'^2 + w'^2\right)_t + \frac{1}{\rho_{oo}}\left((u'p')_x + (w'p')_z\right) - \frac{g}{\theta_{oo}} w'\theta' = 0,$$
$$\frac{1}{2}\left(\frac{g}{\theta_{oo}\frac{d\theta_o}{dz}}\theta'^2\right)_t + \frac{g}{\theta_{oo}} w'\theta' = 0. \qquad (5.18a,b)$$

We introduce two notations to rewrite (5.18) in a concise form: local total energy per unit volume, E, and local energy flux vector, $\vec{S}$, as

$$E = \frac{\rho_{oo}}{2}\left(u'^2 + w'^2\right) + \frac{\rho_{oo}}{2}\left(\frac{g}{\theta_{oo}\frac{d\theta_o}{dz}}\theta'^2\right),$$
$$\vec{S} = \vec{v}'p', \qquad (5.19)$$

where $\vec{v}' = (u', w')$ is perturbation velocity. The total energy is the sum of kinetic energy $\frac{\rho_{oo}}{2}\left(u'^2 + w'^2\right)$ and potential energy $\frac{\rho_{oo} g}{2\frac{1}{\theta_{oo}}\frac{d\theta_o}{dz}}\theta'^2$. Equation (5.18a) suggests that the quantity $\frac{g}{\theta_{oo}} w'\theta'$ represents the local rate of conversion from potential energy to kinetic energy.

Upon summing (5.18a) and (5.18b), we get

$$E_t = -\nabla \cdot \vec{S}. \qquad (5.20)$$

Table 5.1 Summary of vertical fluxes of momentum, energy and heat by IGWs with different characteristics of σ and m

Case	σ	c_x	m	$[u'w']$	$[p'w']$	$[\theta'w']$
I	>0	eastward	>0	<0	<0	0
II	>0	eastward	<0	>0	>0	0
III	<0	westward	<0	>0	<0	0
IV	<0	westward	>0	<0	>0	0

It says that the rate of increase in the local energy of a perturbation, E, is equal to the convergence of energy flux, S, through each location.

Using the solution of the prototype IGW above, (5.16), and the dispersion relation of IGW, (5.8), we obtain for the *energy of an IGW*

$$
\begin{aligned}
[E] &= \frac{\rho_{oo}}{2}\left([u'^2]+[w'^2]\right)+\frac{\rho_{oo}g}{2\frac{1}{\theta_{oo}}\frac{d\theta_o}{dz}}[\theta'^2] \\
&= \frac{\rho_{oo}}{4}\left(1+\frac{m^2}{k^2}\right)A^2+\frac{\rho_{oo}g}{4\sigma^2}\frac{1}{\theta_{oo}}\frac{d\theta_o}{dz}A^2 \\
&= \frac{\rho_{oo}}{2}\left(\frac{k^2+m^2}{k^2}\right)A^2.
\end{aligned}
\tag{5.21}
$$

Equation (5.21) reveals that the energy of a prototype IGW is partitioned equally between its kinetic energy and potential energy. When we specifically apply this to an eastward-propagating IGW, we obtain for its energy flux components

$$
\begin{aligned}
[u'p'] &= \frac{\rho_{oo}Nm^2}{2k^2(k^2+m^2)^{0.5}}A^2 = \frac{Nm^2}{(k^2+m^2)^{3/2}}\frac{\rho_{oo}(k^2+m^2)A^2}{2k^2}, \\
[w'p'] &= \frac{-\rho_{oo}Nmk}{2k^2(k^2+m^2)^{0.5}}A^2 = \frac{-Nmk}{(k^2+m^2)^{3/2}}\frac{\rho_{oo}(k^2+m^2)A^2}{2k^2}.
\end{aligned}
\tag{5.22}
$$

Recall that the group velocity of an eastward-propagating IGW is, $\vec{c}_g = \left(\frac{Nm^2}{(k^2+m^2)^{3/2}}, \frac{-Nkm}{(k^2+m^2)^{3/2}}\right)$. It follows from (5.21) and (5.22) that

$$
\left[\vec{S}\right] = [\vec{v}'p'] = \vec{c}_g[E]. \tag{5.23}
$$

Equation (5.23) establishes that the group velocity of an IGW is indeed the propagation velocity of its energy. Note that $[E]$ is simply a function of its wavenumbers and amplitude; so is $[S]$ aside from its dependence on the background stratification. Hence, the energy of a prototype IGW as an entity does not change in time $(\partial/\partial t[E]=0)$ because there is no convergence of its energy flux $(\nabla\cdot[S]=0)$ as implied by (5.20).

5.4.3 Graphical depiction of all properties of prototype IGW

We are now in a position to concisely depict all properties of a prototype IGW. Let us consider the IGW referred to as case I in Table 5.1. Its vertical velocity is $w' = \mathrm{Re}\{Ae^{i\phi}\}$

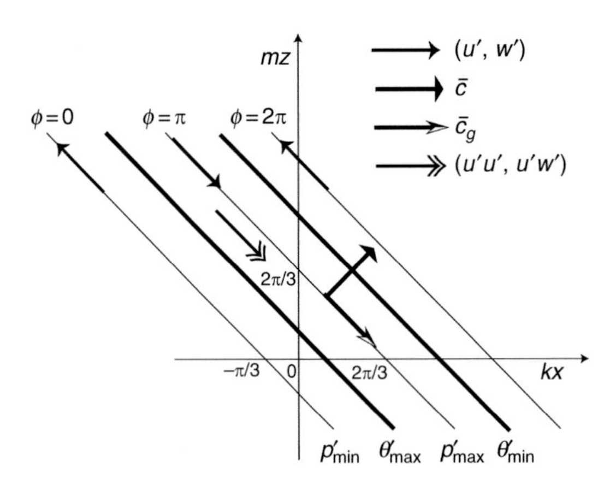

Fig. 5.2 Graphical depiction of a generic IGW characterized by $m = k > 0$, $\sigma > 0$, $\xi = \pi/3$ and $A = 1$.

where $\phi = kx + mz - \sigma t + \xi$ characterized by $m = k > 0$, $\sigma > 0$, $\xi = \pi/3$ and $A = 1$. At $t = 0$, the wave-front designated as $\phi = 0$ intersects the x-axis at $x = -\pi/3$. The relations among the properties of this IGW discussed in Sections 5.4.1 and 5.4.2 enable us to depict all of them at this time instant in Fig. 5.2. The wave-fronts have an "east-down west-up" orientation. The wavenumber vector (k, m) points to the "east-up" direction and is parallel to the phase velocity. Its group velocity $\vec{c}_g = \left(\dfrac{Nm^2}{(k^2+m^2)^{3/2}}, \dfrac{-Nmk}{(k^2+m^2)^{3/2}} \right)$ points to the "east-down" direction. The u-momentum flux and energy flux of this IGW are in the same direction. The maximum potential temperature perturbation is located a quarter-wavelength east of the nearest minimum pressure perturbation.

After half of a period of the wave, the wave-front designated as $\phi = -\pi$ will have moved to the position of the current wave-front designated as $\phi = 0$. The whole pattern would be displaced by that amount in the direction of $\vec{c}$. In other words, every property at a point in space fluctuates sinusoidally in time reflecting the passage of a wave perturbation through that point. This prototype IGW has the following transport properties: $[u'w'] < 0$, $[p'w'] < 0$ and $[w'\theta'] = 0$ as noted earlier in Table 5.1.

It would be instructive to verify that this structure of IGW makes physical sense. When a θ'_{max} wave-front moves in the direction indicated by the arrow $\vec{c}$, the buoyancy at a point just to the east of a θ'_{max} line would increase. It is then consistent to find the vertical velocity at that point changing from descent to ascent. At the same time, when the p'_{max} contour moves in the direction indicated by $\vec{c}$, the pressure gradient in the x-direction at that point would become smaller. Then it would be consistent to find the air just to the east of the θ'_{max} line changing from $u' > 0$ to $u' < 0$. Furthermore, the negative correlation between p' and w' and the positive correlation between p' and u' is consistent with the direction of $\vec{c}_g$. In other words, the structure of all properties of a wave are consistent with its phase propagation. Such internal consistency provides a means to conceptually check the correctness of the solution and sketch.

The key mathematical results of the prototype IGW are $[w'\theta'] = 0$ and $[p'w'] = \frac{\sigma}{k}\rho_{oo}[u'w']$. The results of any prototype IGW can be concisely summarized as follows:

Prototype IGW transports energy, but not heat per se. Downward energy propagating IGW with $\sigma > 0$ ($\sigma < 0$) carries westerly (easterly) momentum in the downward direction; the same applies to upward propagating IGW.

5.4.4 Application: orographic IGW

We may develop a feel for IGW with a quantitative analysis in an idealized but physically realizable model setting. The setting consists of a steady uniform basic flow, U, going over an undulating terrain in a vertically unbounded domain. The stratification in the model measured by Brunt–Vaisala frequency, N, has a certain value below $z = H$ and a larger value above. They are denoted by N_1^2 and N_2^2 with $N_1^2 < N_2^2$ (subscripts 1 and 2 refer to the lower and upper layers respectively). Our task is to deduce the forced response.

5.4.4.1 Sinusoidal orography

Let us first consider the simpler case of a terrain of sinusoidal shape, $h = h_o e^{ikx}$, with a maximum height h_o and a wavelength $L = 2\pi/k$. Suppose that there is a sufficiently weak steady basic flow U so that the forced disturbance may be expected to be a steady IGW with the same wavenumber k. The wave would be refracted and/or reflected at $z = H$ across which there is a change in the stratification. The governing equations for perturbations in the two layers are then

$$\begin{aligned}
&\left(U\frac{\partial}{\partial x}\right)^2\left(\frac{\partial^2}{\partial x^2}+\frac{\partial^2}{\partial z^2}\right)w_1 + N_1^2\frac{\partial^2}{\partial x^2}w_1 = 0 \quad \text{for} \quad 0 \le z \le H, \\
&\left(U\frac{\partial}{\partial x}\right)^2\left(\frac{\partial^2}{\partial x^2}+\frac{\partial^2}{\partial z^2}\right)w_2 + N_2^2\frac{\partial^2}{\partial x^2}w_2 = 0 \quad \text{for} \quad H \le z.
\end{aligned} \tag{5.24}$$

The equations in (5.24) are similar to (5.6) except that $U\partial/\partial x$ takes the place of $\partial/\partial t$. The solution of (5.24) has the form

$$w_j = W_j(z)\exp(ikx), \qquad j = 1, 2. \tag{5.25}$$

The amplitude functions satisfy

$$\begin{aligned}
&\frac{d^2 W_j}{dz^2} + m_j^2 W_j = 0, \\
&m_j^2 = \left(\frac{N_j^2}{U^2k^2} - 1\right)k^2, \qquad j = 1 \text{ for } h \le z \le H \text{ and } j = 2 \text{ for } H \le z < \infty.
\end{aligned} \tag{5.26}$$

An elementary solution of (5.26) would be $W_j \propto \exp(\pm im_j z)$ if m_j^2 are both positive, or else it would be $W_j \propto \exp(\pm n_j z)$ if $m_j^2 = -n_j^2$ are negative. In other words, the disturbance would have an undulating (wave-like) structure in the z-direction above $z = H$ only if the parameters are such that $m_2^2 > 0$ (i.e. for a sufficiently strong stratification in the upper layer, a sufficiently weak basic flow or/and sufficiently long wavelength). Physical consideration suggests that there should be neither mass accumulation nor pressure jump at the level $z = H$. We therefore impose two matching conditions at $z = H$, viz.

$$\begin{aligned} &\text{at } z = H, \qquad w_1 = w_2 \rightarrow W_1 = W_2, \\ &\text{at } z = H, \qquad \frac{dw_1}{dz} = \frac{dw_2}{dz} \rightarrow \frac{dW_1}{dz} = \frac{dW_2}{dz}. \end{aligned} \tag{5.27a,b}$$

In the context of linear dynamics, the lower boundary condition at the terrain may be incorporated effectively at $z = 0$,

$$\text{at } z = 0, \quad w_1 = U\frac{\partial h}{\partial x} \rightarrow W_1 = ikh_oU. \tag{5.27c}$$

Equation (5.27c) means that the velocity component normal to the undulating terrain is zero. The remaining boundary condition at $z \rightarrow \infty$ depends on whether or not there is total reflection of the wave at the interface. We will elaborate on this issue as we present the solutions for two different situations below.

Case (1)

The parameter condition of this case is $1 > N_2^2/U^2k^2$ and $1 > N_1^2/U^2k^2$. This would be the case if the undulating terrain has a sufficiently short wavelength, or the basic flow is sufficiently strong or the stratification in both layers is sufficiently weak. It follows that $m_2^2 \equiv -n_2^2 < 0$ and $m_1^2 \equiv -n_1^2 < 0$. The general solution of (5.26) is then

$$\begin{aligned} W_1 &= A\exp(n_1 z) + B\exp(-n_1 z), \\ W_2 &= C\exp(n_2 z) + D\exp(-n_2 z). \end{aligned} \tag{5.28}$$

At $z \rightarrow \infty$, the solution must be finite for it to be physically meaningful. Thus, we set $C = 0$. By applying (5.27a,b,c), it is straightforward to determine A, B and D on the basis of the matching conditions (5.27a,b). The solution is then

$$\begin{aligned} W_1 &= F\{(n_1 - n_2)\exp(n_1(z-H)) + (n_1 + n_2)\exp(-n_1(z-H))\}, \\ W_2 &= F2n_1\exp(-n_2(z-H)), \\ \text{with } F &= \frac{ikh_oU}{n_2(e^{n_1H} - e^{-n_1H}) + n_1(e^{n_1H} + e^{-n_1H})}. \end{aligned} \tag{5.29}$$

This forced disturbance is *a vertically trapped IGW.*

In light of the continuity equation, we may also define a streamfunction of the disturbance $\psi(x,z) = \varphi(z)e^{ikx}$ such that $w = \psi_x$ and $u = -\psi_z$. It follows that $\varphi_j = W_j/ik$. The amplitude functions of pressure and temperature fields of the wave can be readily shown to be

$$\hat{p} = U\rho_{oo}\frac{d\varphi}{dz}, \quad \hat{\theta} = -\frac{\theta_{oo}N^2}{gU}\varphi. \tag{5.30}$$

The streamfunction associated with the basic flow is $\Psi = -Uz$.

The system has six parameters: k, h_o, H, U, N_1, N_2. Thus, we may define four independent non-dimensional parameters, such as $\tilde{h}_o = h_o/H, \tilde{k} = kH, \tilde{N}_1 = N_1/Uk, \tilde{N}_2 = N_2/Uk$. We use them to introduce the following non-dimensionalization: $\tilde{x} = kx$, $(\tilde{h}_o, \tilde{z}) = \frac{1}{H}(h_o, z)$, $(\tilde{\Psi}, \tilde{\psi}_j) = \frac{1}{UH}(\Psi, \psi_j)$. Then the non-dimensional total streamfunctions in the two layers are

$$\begin{aligned}
\tilde{\Psi} + \tilde{\psi}_1 &= -\tilde{z} + \tilde{F}\{(\tilde{n}_1 - \tilde{n}_2)\exp(\tilde{n}_1(\tilde{z}-1)) + (\tilde{n}_1 + \tilde{n}_2)\exp(-\tilde{n}_1(\tilde{z}-1))\}\cos(\tilde{x}),\\
\tilde{\Psi} + \tilde{\psi}_2 &= -\tilde{z} + \tilde{F}2\tilde{n}_1\exp(-\tilde{n}_2(\tilde{z}-1))\cos(\tilde{x}),\\
\tilde{F} &= \frac{\tilde{h}_o}{\tilde{n}_2(\exp(\tilde{n}_1) - \exp(-\tilde{n}_1)) + \tilde{n}_1(\exp(\tilde{n}_1) + \exp(-\tilde{n}_1))},\\
\tilde{n}_j &= \tilde{k}\sqrt{1 - \tilde{N}_j^2}.
\end{aligned} \tag{5.31}$$

A meaningful set of parameter values are: $k = 2\pi/L$, with $L = 10^4$ m, $H = 10^3$ m, $h_o = 300$ m, $U = 10\,\mathrm{m\,s^{-1}}$, $N_1 = 0.2 \times 10^{-2}\,\mathrm{s^{-1}}$, $N_2 = 0.4 \times 10^{-2}\,\mathrm{s^{-1}}$. The values of the non-dimensional parameters are $\tilde{h}_2 = 0.3, \tilde{k} = 0.628, \tilde{N}_1^2 = 0.1, \tilde{N}_2^2 = 0.4$. The total streamfunction field is shown in Fig. 5.3. It depicts a forced wave trapped close to the surface. The disturbance at $\tilde{z} = 3.0$ is already greatly attenuated. The loci of maximum and minimum pressure are indicated by a solid line and a thin line respectively. It is noteworthy that the distribution of the surface pressure is identical to the orography implying that there is no exchange of zonal momentum across the orography. It follows that there is no continual input of energy into the fluid in this steady state consistent with the condition that there is no dissipation.

Case (2)

The parameter condition of this case is $N_1^2 < U^2k^2 < N_2^2$. This would be the case if the stratification of the upper layer is sufficiently large, while the stratification of the lower layer is sufficiently small. It follows that $m_2^2 > 0$ and $m_1^2 = -n_1^2 < 0$. The disturbance has a wavy structure in the upper layer. The general solution is

$$\begin{aligned}
W_1 &= A\exp(n_1 z) + B\exp(-n_1 z),\\
W_2 &= C\exp(im_2 z) + D\exp(-im_2 z).
\end{aligned} \tag{5.32}$$

The boundary condition at $z \to \infty$ in this case would be that the forced wave would only transmit energy upward because the wave source is at the surface. This is known as the *radiation condition*.

The dispersion relation for an elementary solution of the form $e^{i(kx+mz)}$ in this problem is $\sigma = 0 = Uk - Nk/(k^2+m^2)^{1/2}$. The group velocity of this forced wave is $(\partial\sigma/\partial k, \partial\sigma/\partial m) = \frac{Nk}{(k^2+m^2)^{3/2}}(k, m)$. Therefore, a wave that transmits energy upward must have a positive value of vertical wavenumber. The correct solution is then

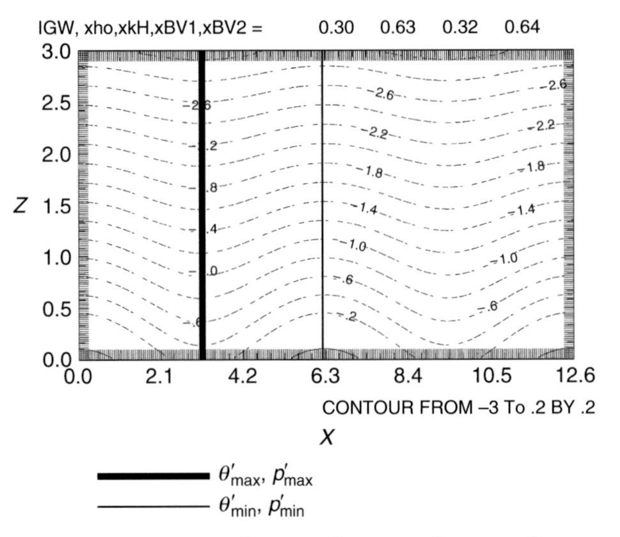

Fig. 5.3 Total streamfunction for orographic forcing with $\tilde{h}_2 = 0.3$, $\tilde{k} = 0.628$, $\tilde{N}_1^2 = 0.1$, $\tilde{N}_2^2 = 0.4$.

proportional to $\exp(im_2 z)$ instead of $\exp(-im_2 z)$. Hence, we set $D = 0$; A,B,C can be determined on the basis of (5.27a,b,c). It is straightforward to get the solution as

$$W_1 = G\Big(-(im_2 + n_1)e^{n_1(z-H)} + (im_2 - n_1)e^{-n_1(z-H)}\Big),$$
$$W_2 = -G2n_1 e^{im_2(z-H)}, \tag{5.33}$$
$$G = \frac{ikh_o U}{(im_2 - n_1)e^{n_1 H} - (im_2 + n_1)e^{-n_1 H}}.$$

The total non-dimensional streamfunction is then

$$\tilde{\Psi} + \tilde{\psi}_1 = -\tilde{z} + \mathrm{Re}\big\{\big[\tilde{G}(-(i\tilde{m}_2 + \tilde{n}_1))\exp(\tilde{n}_1(\tilde{z}-1)) + (i\tilde{m}_2 - \tilde{n}_1)\exp(-\tilde{n}_1(\tilde{z}-1))\big]\exp(i\tilde{x})\big\},$$
$$\tilde{\Psi} + \tilde{\psi}_2 = -\tilde{z} + \mathrm{Re}\big\{\big[-\tilde{G}2\tilde{n}_1\exp(i\tilde{m}_2(\tilde{z}-1))\big]\exp(i\tilde{x})\big\}, \tag{5.34}$$
$$\tilde{G} = \frac{\tilde{h}_o}{(i\tilde{m}_2 - \tilde{n}_1)\exp(\tilde{n}_1) - (i\tilde{m}_2 + \tilde{n}_1)\exp(-\tilde{n}_1)}.$$

We use $N_2 = 0.7 \times 10^{-2}\ \mathrm{s}^{-1}$ but the other parameters have the same values as in case (1). The corresponding non-dimensional parameters are $\tilde{h}_o = 0.3$, $\tilde{k} = 0.628$, $\tilde{N}_1^2 = 0.1$, $\tilde{N}_2^2 = 1.24$. The non-dimensional total streamfunction is shown in Fig. 5.4. The amplitude of the wave does not attenuate in the layer $\tilde{z} \geq 1$. It should be emphasized that the amplitude $\tilde{G}$ has a complex value giving rise to two important features in the solution. First, the surface pressure to the west of a topographic ridge is higher than to the east. It implies a zonal momentum transfer into the surface. Such loss of momentum in the absence of viscosity is referred to as a *form drag*. Second, the forced wave has a distinct westward vertical tilt in the layer above whereby energy continually propagates upward. The imposed basic zonal flow ultimately compensates for the loss of momentum at the surface and the loss of energy to afar above.

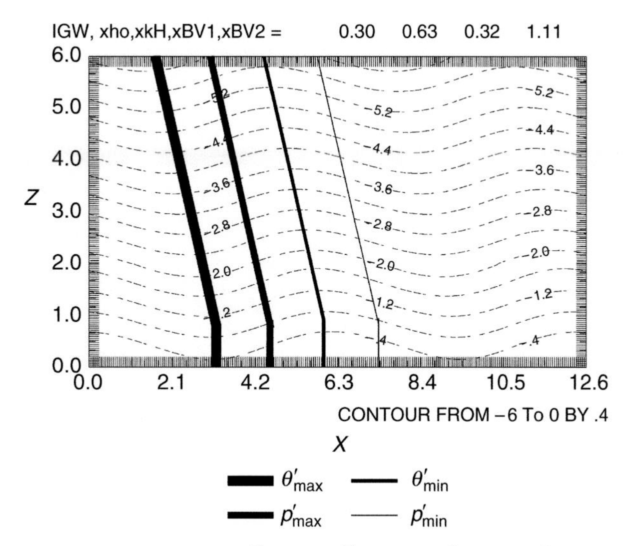

Fig. 5.4 Total streamfunction for orographic forcing with $\tilde{h}_o = 0.3$, $\tilde{k} = 0.628$, $\tilde{N}_1^2 = 0.1$, $\tilde{N}_2^2 = 1.24$.

5.4.4.2 Localized orography

A more recognizable setting would be a localized terrain. Let us consider an idealized form of it in this analysis, $h(x) \neq 0$ only for $|x| \leq L$. It may be thought of as a combination of constituent wave components according to the Fourier theory of representation of a function. The response to each Fourier orographic component may be separately determined by the analysis above. The sum of all the spectral responses would then give us a total disturbance field. In light of the results in Section 5.4.4.1, we may expect a forced disturbance in the form of steady wave-packets if the parameter condition for the dominant spectral components are favorable. Those wave-packets would be able to transmit energy upward.

For illustration, we consider a terrain of Gaussian shape, $h = h_o \exp\left(-(x/L)^2\right)$ with a half-width L. We introduce the following non-dimensional variables: $\tilde{x} = x/L$, $\left(\tilde{h}_o, \tilde{z}\right) = \frac{1}{H}(h_o, z)$, $\left(\tilde{\Psi}, \tilde{\psi}_j\right) = \frac{1}{UH}\left(\Psi, \psi_j\right)$. Then the non-dimensional orography is $\tilde{h} = \tilde{h}_o \exp(-\tilde{x}^2)$. If we use I grid points to depict $-X \leq \tilde{x} \leq X$, the spectrum of wave-number would be $k = nk_{\min}$, $n = 0, 1, 2, \ldots (I-1)/2$, $k_{\min} = 2\pi/2X$. We use a domain much wider than the terrain, $X = 15$. The other parameters are $h_o = 300\,\text{m}$, $H = 1000\,\text{m}$, $L = 1000\,\text{m}$, $U = 10\,\text{m}\,\text{s}^{-1}$, $N_1 = 1 \times 10^{-2}\,\text{s}^{-1}$ and $N_2 = 2 \times 10^{-2}\,\text{s}^{-1}$. When we compute each spectral response, it is necessary to check whether N_j^2/U^2k^2 is greater than or less than unity for each spectral component so that the appropriate form of the solution for each layer would be used. There is no zonal mean component in the response of a linear dynamical model.

The non-dimensional streamfunction of the forced disturbance is shown in Fig. 5.5. The analysis in the last section has prepared us to interpret this more complex solution. It depicts a pronounced wave-packet in the immediate downstream region of the terrain. It has a westward vertical tilt signifying that it indeed transmits energy upward. We see that there can be strong updraft and downdraft in this wave-packet. They have a strong bearing

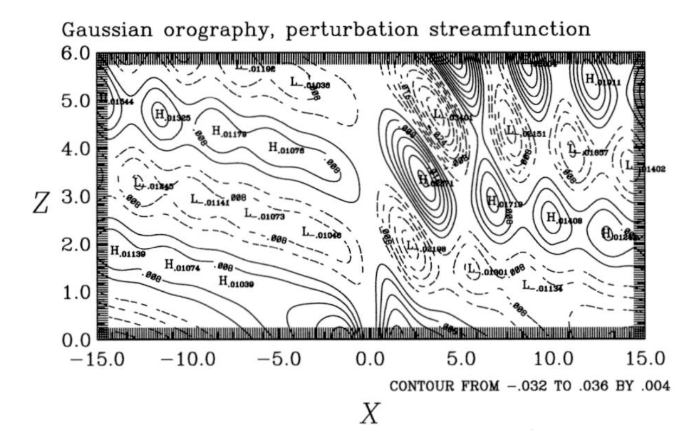

Fig. 5.5 Streamfunction of forced disturbance for a Gaussian orography, $\tilde{h} = \tilde{h}_o \exp(-\tilde{x}^2)$ with $\tilde{h}_o = 0.3$, $\frac{L^2N_1^2}{4\pi^2U^2} = 2.5$, $\frac{L^2N_2^2}{4\pi^2U^2} = 5.0$.

on the weather immediately downstream of a mountain ridge, such as eastern Colorado in the USA. A second much weaker wave-packet is also present upstream of the terrain. Because of the use of a reentrant channel domain, the upstream perturbation appears to be a continuation of the downstream perturbation at higher levels. As such, this feature is a model artifact.

5.4.5 Inertio-internal gravity wave

Some IGW in the atmosphere have a horizontal wavelength several hundred times longer than its vertical wavelength. The corresponding frequency would be small enough that the effect of the Earth's rotation is no longer negligible. Those IGW are called *inertio-internal gravity waves*. The Coriolis force must be included in the momentum equations. For comparison with the results above, we again consider perturbations that do not vary in the y-direction, even though the velocity field is necessarily three-dimensional in this case. A straightforward extension of the previous analysis would yield the following single governing equation for a perturbation in this case

$$\left(\frac{\partial^2}{\partial t^2}\left(\frac{\partial^2}{\partial x^2}+\frac{\partial^2}{\partial z^2}\right)+f^2\frac{\partial^2}{\partial z^2}+N^2\frac{\partial^2}{\partial x^2}\right)w' = 0. \tag{5.35}$$

Equation (5.35) has an extra term $f^2\partial^2/\partial z^2$ compared to (5.6). The normal mode solution of (5.35) is $w' = Be^{i(kx+mz-\sigma t)}$. All properties of inertio-IGW can be similarly deduced as in Sections 5.3.3 and 5.4.1. They are summarized in Table 5.2 for easy reference. It would be a good exercise for the readers to verify these expressions for themselves. The solutions show that the Earth's rotation increases the frequency of IGW, increases its phase velocity and reduces its group velocity. The phase velocity and group velocity are still orthogonal to one another. The influences of the Earth's rotation are also felt in its structural properties. The amplitudes of the temperature and pressure fields of the wave are weaker. The meridional velocity arises from the rotation.

Table 5.2

Dispersion relation:	$\sigma^2 = \frac{m^2 f^2 + N^2 k^2}{k^2 + m^2}$
Phase velocity:	$(c_x, c_z) = \frac{\sigma}{k^2 + m^2}(k, m)$
Group velocity:	$(c_{gx}, c_{gz}) = \frac{km(N^2 - f^2)}{\sigma(k^2 + m^2)^2}(m, -k)$
	$\vec{c} \cdot \vec{c}_g = 0$ (5.36)
Polarization relations:	$\left(u', v', w', \frac{\theta'}{\theta_{oo}}, \frac{p'}{\rho_o}\right) = \left(\hat{u}, \hat{v}, \hat{w}, \hat{\theta}, \hat{p}\right) e^{i(kx+mz-\sigma t)}$
	$\hat{v} = \frac{ifm}{\sigma k}\hat{w}, \hat{u} = -\frac{m}{k}\hat{w}, \hat{\theta} = \frac{-iN^2}{\sigma g}\hat{w}, \hat{p} = \frac{-m(N^2 - f^2)}{\sigma(k^2 + m^2)}\hat{w}$

The structures of all properties of a sample inertio-IGW, $w' = A\exp(i(kx + mz - \sigma t + \xi))$, characterized by $m = k > 0$, $\sigma > 0$, $\xi = \pi/3$ and $A = 1$ are the same as the counterparts of a non-rotating IGW (Fig. 5.3) with one exception. There is an additional meridional velocity component in an inertio-IGW. The location of maximum northward flow (v'_{max}) coincides with the minimum temperature θ'_{min} and extreme southward flow (v'_{min}) coincides with the warmest temperature θ'_{max}.

5.5 Rudimentary characteristics of wave motions in large-scale flows

A distinctly different class of wave motions is evident in any daily weather map such as that of the geopotential height field at a particular pressure level in the troposphere. Such northern hemispheric and southern hemispheric 300 mb maps on January 1, 1996 are shown in Fig. 5.6a and b for illustration. They are constructed with the website of the NOAA/ESRL Physical Sciences Division. This is a day of winter (summer) in NH (SH). The contour interval is 175 m in Fig. 5.6. The range of values is from 9700 m to 8100 m in NH and from 9700 m to 8400 m in SH. The airflow associated with these wavy disturbances is quasi-horizontal. On this particular day, the flow has a characteristically broad wavy pattern in the extratropical region of both hemispheres, suggesting the presence of synoptic-scale and planetary-scale waves embedded in a strong zonal mean flow. For example, there is a pronounced trough spanning much of North America and another trough to the east of the International Dateline, with a ridge in between over the eastern N. Pacific. There is another trough of moderate intensity over the N. Atlantic as well as one over the western N. Pacific. There are minor wavy features over

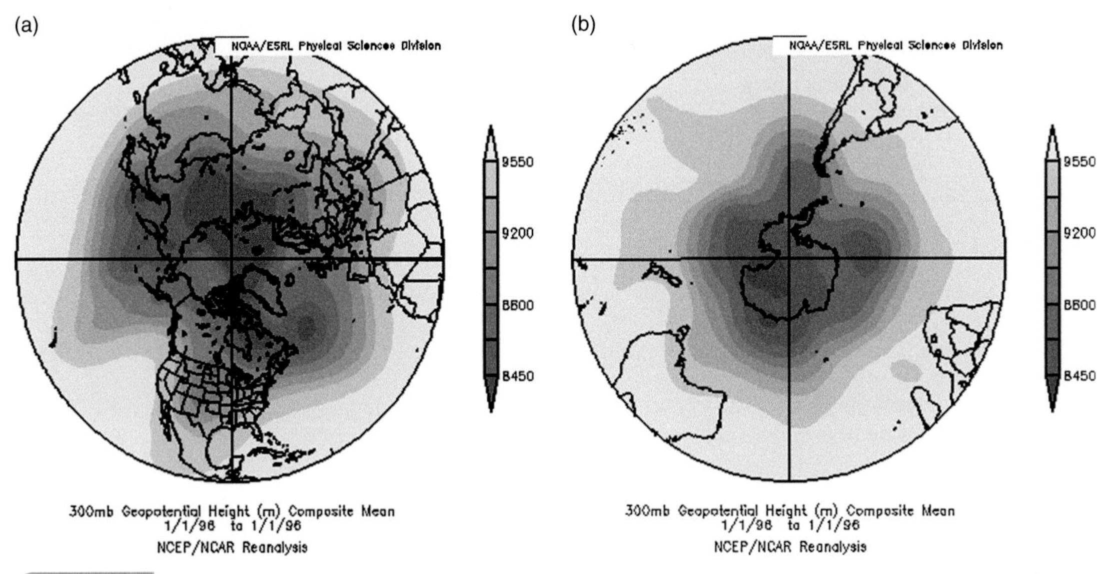

Fig. 5.6 The 300 mb geopotential height field on 1/1/1996 over (a) northern hemisphere and (b) southern hemisphere.

Asia and the Middle East as well. There is a smoother large-scale wavy pattern in the southern hemispheric flow on that day (Fig. 5.6b), again mostly concentrated in the extratropical region. The waves in NH as a whole are stronger as they extend from the polar region to lower latitudes than those in SH. This family of waves is known as Rossby waves named after Carl-Gustaf Rossby. The figures suggest that the most essential dynamical factor responsible for Rossby waves is present in both hemispheres. We will soon elaborate that the rotation of the Earth is such a factor for the existence of these waves.

Maps like these over several consecutive days reveal that these waves typically propagate eastward. An examination of such maps over several days tells us that there is also a quasi-stationary component in both the wave field and the zonal mean part of the flow. The differences in the gross features of the wave fields in the two hemispheres suggest that some waves additionally depend on the seasonal difference in the solar heating and the size and configuration of the land masses.

It is instructive to compare a weather map of the upper troposphere (300 mb map) with a counterpart of the lower troposphere (700 mb map) (Fig. 5.7). Each major feature in one field at the upper level has a counterpart feature at the lower level in each hemisphere. In other words, Figs. 5.6 and 5.7 together suggest that Rossby waves have a broad vertical structure typically spanning the whole troposphere.

The fundamental dynamics of Rossby waves will be discussed in the next three sections. We will later review some observed properties of Rossby waves from a statistical point of view in Section 5.9. The dynamics of Rossby wave generation will be separately discussed in Chapter Parts 8B and 8C. Additional aspects of Rossby wave dynamics will be discussed in Chapters 9, 10 and 11.

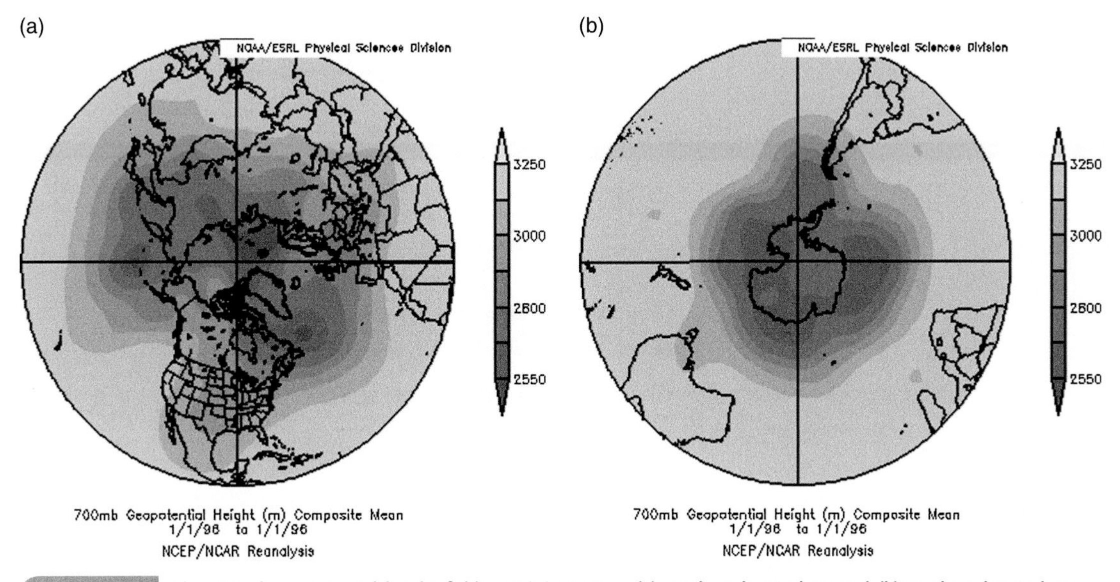

Fig. 5.7 The 700 mb geopotential height field on 1/1/1996 over (a) northern hemisphere and (b) southern hemisphere.

5.6 Physical nature of Rossby waves in a simplest possible model

Rossby (1939) obtained a solution for the intrinsic wave motions in a simplest possible model setting, namely a 2-D rotating fluid layer without dissipation and forcing. This class of wave motions exists in a fluid system, which has a non-uniform distribution of potential vorticity (PV). Their existence stems from conservation of PV in each fluid element. The PV in Rossby's generic 2-D fluid model is simply the absolute vorticity. Its spatial variation is solely attributable to the latitudinal variation of the planetary vorticity (Coriolis parameter). Rossby introduced an approximate representation of the latter in a channel domain as a linear function of the meridional distance from a reference latitude φ_o,

$$f = 2\Omega \sin\varphi \approx f_o + \beta y, \tag{5.37}$$

where $f_o = 2\Omega \sin\varphi_o$, $\beta = \frac{2\Omega}{a}\cos\varphi_o$, $y = a(\varphi - \varphi_o)$, $a =$ radius of Earth. This is known as the "*beta-plane*" approximation.

A qualitative application of the notion of conservation of PV in this generic model, namely $D/Dt(\zeta + f) = 0$, would illustrate the physical nature of Rossby waves. It suffices to consider a cyclical 1-D disturbance that sinusoidally varies in the x-direction in the northern hemisphere at a certain time instant referred to as $t = 0$, $\psi(x, 0)$, (Fig. 5.8).

Let us consider what would happen to a southward displacement of fluid parcel A. When it moves to a more southern location where the planetary vorticity is smaller, its relative vorticity would increase in order to conserve its absolute vorticity. This change of vorticity would be effectively the same as a westward displacement of the vorticity

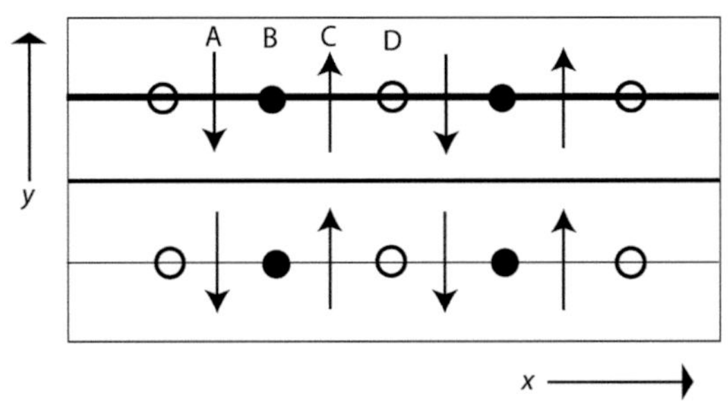

Fig. 5.8 Schematic of a 1-D Rossby wave indicated by a continuous sequence of relative vorticity anomalies of alternate signs. The contours are $(f - f_0)$ with solid lines increasing northward.

anomaly south of B. By the same token, when fluid parcel C moves to a northern location, its relative vorticity would decrease. This change of vorticity would be also equivalent to a westward displacement of vorticity anomaly north of D. Furthermore, when the position of a vorticity anomaly is displaced, so would be the velocity field because of the inherent relationship between vorticity and velocity, $\psi = \nabla^{-2}\zeta$. The evolution of the flow field therefore amounts to a *westward movement of the whole sinusoidal pattern*. Such movement is the signature of a Rossby wave. The directional bias of this movement towards the west stems from the fact that the Coriolis parameter increases northward. Therefore, Rossby waves in both southern and northern hemispheres intrinsically propagate westward.

5.7 Properties of prototype Rossby waves

Recall that the governing equation for a 2-D rotating fluid model is conservation of absolute vorticity. With the beta-plane approximation it is

$$\left(\frac{\partial}{\partial t} + u\frac{\partial}{\partial x} + v\frac{\partial}{\partial y}\right)\zeta + v\beta = 0, \tag{5.38}$$

where $\zeta = v_x - u_y$. A flow in a 2-D fluid is necessarily non-divergent, $u_x + v_y = 0$. This constraint allows us to introduce a streamfunction ψ such that $u = -\psi_y$, $v = \psi_x$. It follows that $\zeta = v_x - u_y = \nabla^2\psi$ and (5.38) becomes

$$\left(\frac{\partial}{\partial t} - \psi_y\frac{\partial}{\partial x} + \psi_x\frac{\partial}{\partial y}\right)\nabla^2\psi + \beta\psi_x = 0. \tag{5.39}$$

By applying the method of linearization discussed in Section 5.3.1 to Equation (5.39), we get the following governing equation for a weak disturbance $\psi'(x, y, t)$.

$$\nabla^2\psi'_t + \beta\psi'_x = 0. \tag{5.40}$$

It suffices to consider an unbounded domain. The elementary solution of (5.40) is a plane wave with a frequency σ and a zonal wavenumber k and a meridional wavenumber ℓ,

$$\psi = \mathrm{Re}\left\{Ae^{i(kx+\ell y-\sigma t)}\right\}, \tag{5.41}$$

where A is a complex amplitude. Equation (5.41) is a non-trivial solution, $A \neq 0$, only if the following condition is satisfied,

$$\sigma = -\frac{\beta k}{k^2 + \ell^2}. \tag{5.42}$$

Equation (5.42) is the *dispersion relation of prototype Rossby waves*, with β the only external parameter. We use only $k > 0$ and allow positive or negative values for ℓ so that the wave-fronts with any orientation can be represented by (5.41). It should be emphasized that σ can only have a negative value. We sometimes interchangeably use in this book the phrases "large-scale waves" and "large-scale eddies," although "eddies" generally carry the connotation of turbulence and "waves" imply motions with a high degree of orderliness. It would be helpful to think of a wave as an eddy that satisfies a dispersion relation.

It is instructive to examine the dependence of the frequency of a Rossby wave upon its zonal wavenumber. Those Rossby waves with $\ell \gg k$ have frequency $\sigma \approx -\beta k/\ell^2$ which is a linear function of k. In other words, in the range of $\ell \gg k$, a longer Rossby wave would have a lower frequency. In contrast, in the range of $\ell \ll k$, the frequency is inversely proportional to k, $\sigma \approx -\beta/k$. This means that a shorter Rossby wave would also have a smaller frequency in this range. By inference, the magnitude of the frequency of Rossby waves is largest at a certain intermediate value of the zonal wavenumber k. That turns out to be $k = |\ell|$. It is convenient to show the dependence of the frequency of Rossby waves on the zonal wavenumber in a non-dimensional form. Equation (5.42) can be rewritten as $\dfrac{\sigma}{\beta/|\ell|} = \dfrac{-k/|\ell|}{1 + (k/\ell)^2}$. The functional dependence of $\dfrac{\sigma}{\beta/|\ell|}$ on $k/|\ell|$ is depicted in Fig. 5.9, confirming that $|\dfrac{\sigma\ell}{\beta}|$ has a maximum value at $k/|\ell| = 1$.

Relevant parameter values in mid-latitude: $|\ell| = 2\pi/2 \times 10^6 = \pi \times 10^{-6}\,\mathrm{m}^{-1}$, $\beta = 1.6 \times 10^{-11}\,\mathrm{m}^{-1}\,\mathrm{s}^{-1}$, $\beta/|\ell| = 0.5 \times 10^{-5}\,\mathrm{s}^{-1}$. A wave with $2k = |\ell|$ has a zonal wavelength of 4×10^6 m and $\dfrac{|\sigma|}{\beta/|\ell|} \approx 0.4 \to |\sigma|_{\mathrm{max}} \sim 0.2 \times 10^{-5}\,\mathrm{s}^{-1} \to$ minimum period is $2\pi/|\sigma| \sim 30 \times 10^5\,\mathrm{s} \sim 30$ days. Thus, the frequency of Rossby waves is typically much smaller than the Coriolis parameter, $|\sigma| \ll f$.

5.7.1 Phase velocity of prototype Rossby waves

The generic formula for the phase velocity of a 2-D wave is $\vec{c} \equiv (c_x, c_y) = \dfrac{\sigma\vec{j}}{j\,j}$, where $\vec{j} = (k, \ell), j^2 = k^2 + \ell^2$. The two components of the phase velocity of a prototype Rossby wave are therefore

$$c_x = -\frac{\beta k^2}{(k^2 + \ell^2)^2}, \quad c_y = -\frac{\beta k\ell}{(k^2 + \ell^2)^2}. \tag{5.43}$$

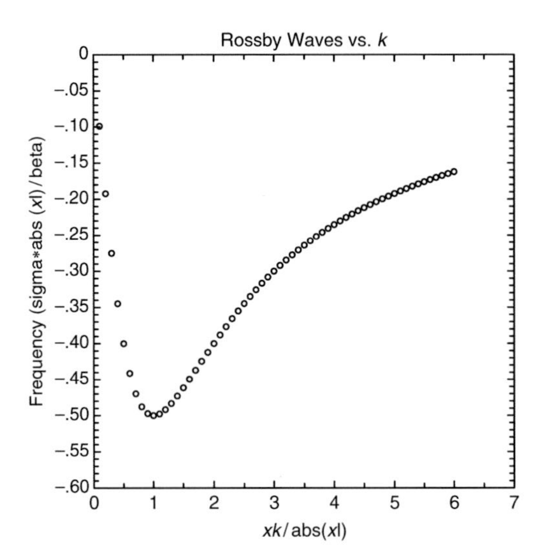

Fig. 5.9 Variation of the frequency of Rossby wave $\frac{\sigma}{\beta/|\ell|}$ with $k/|\ell|$.

The zonal component is necessarily westward, $c_x < 0$. This phase velocity has a northward component if $\ell < 0$ or a southward component if $\ell > 0$. In other words, a Rossby wave with a wave-front oriented in the NW–SE direction $(k > 0, \ell > 0)$ would propagate towards the southwest. A Rossby wave with a wave-front oriented in the SW–NE direction $(k > 0, \ell < 0)$ would propagate towards the northwest. What dynamical processes would dictate a particular orientation of a Rossby wave in the atmosphere is a separate issue.

Some relevant numerical values for synoptic-scale Rossby waves at 45° N are $k = \ell \sim \dfrac{2\pi}{4 \times 10^6} \sim 1.6 \times 10^{-6}\,\mathrm{m}^{-1}$, $\beta = 1.6 \times 10^{-11}\,\mathrm{m}^{-1}\,\mathrm{s}^{-1}$. Then we have $c_x = c_y = -\beta k^2/\left(k^2 + \ell^2\right)^2 \sim -2\,\mathrm{m\,s}^{-1}$ indicating that the intrinsic *westward* movement of the wave-front of a typical Rossby wave is quite slow.

The expression for the group velocity of the prototype Rossby waves is

$$\vec{c}_g = \left(c_{gx}, c_{gy}\right) = \left(\frac{\partial \sigma}{\partial k}, \frac{\partial \sigma}{\partial \ell}\right) = \left(\frac{\beta(k^2 - \ell^2)}{(k^2 + \ell^2)^2}, \frac{2\beta k \ell}{(k^2 + \ell^2)^2}\right). \tag{5.44}$$

Unlike the prototype IGW, the phase velocity and group velocity of any prototype Rossby wave are not orthogonal because

$$\vec{c} \cdot \vec{c}_g = \frac{-\beta^2 k^2}{(k^2 + \ell^2)^3} < 0. \tag{5.45}$$

The angle between $\vec{c}$ and $\vec{c}_g$ of any prototype Rossby wave is therefore greater than 90 degrees. Denoting the angle between $\vec{c}$ and $\vec{c}_g$ by γ, we have for a wave with $|\ell| = k$, $\cos\gamma = \frac{\vec{c} \cdot \vec{c}_g}{|\vec{c}_g||\vec{c}|} = -1/\sqrt{2} \therefore \gamma = 135°$.

5.7.2 Momentum flux, energy flux and group velocity of prototype Rossby waves

We next examine the energetics of Rossby waves. Multiplying the governing equation (5.40) by ψ, (dropping the prime notation) we get

$$\nabla\cdot(\psi\nabla\psi_t) - \frac{1}{2}(\nabla\psi\cdot\nabla\psi)_t + \frac{\beta}{2}(\psi^2)_x = 0. \tag{5.46}$$

We rewrite (5.46) as an equation which tells us about the rate of change of the local energy per unit volume, $E = \frac{\rho_o}{2}\nabla\psi\cdot\nabla\psi = \frac{\rho_o}{2}(u'^2 + v'^2)$:

$$\begin{aligned}\frac{\partial E}{\partial t} &= \rho_o\nabla\cdot\left(\psi\nabla\psi_t + \frac{\beta}{2}\psi^2\vec{i}\right)\\ &= -\nabla\cdot\vec{S},\end{aligned} \tag{5.47}$$

where $\vec{S} = (p'u', p'v')$. It follows that $d/dt[E] = 0$ and $\nabla\cdot\left[\vec{S}\right] = 0$.

The last equality in (5.47) stems from the following deduction,

$$\begin{aligned}\nabla\cdot(\psi\nabla\psi_t) &= \nabla\cdot\left(\psi\left(v_t'\vec{i} - u_t'\vec{j}\right)\right)\\ &= \nabla\cdot\left(\psi\left(-\frac{1}{\rho_o}p_y' + f\psi_y\right)\vec{i} - \psi\left(-\frac{1}{\rho_o}p_x' + f\psi_x\right)\vec{j}\right)\\ &= \left(-\frac{1}{\rho_o}(\psi p')_{yx} + \frac{1}{\rho_o}\left(p'\psi_y\right)_x + \frac{f}{2}(\psi^2)_{yx}\right) + \left(\frac{1}{\rho_o}(\psi p')_{xy} - \frac{1}{\rho_o}(p'\psi_x)_y - \frac{f}{2}(\psi^2)_{xy} - \frac{\beta}{2}(\psi^2)_x\right)\\ &= -\frac{1}{\rho_o}(p'u')_x - \frac{1}{\rho_o}(p'v')_y - \frac{\beta}{2}(\psi^2)_x.\end{aligned}$$

Thus we get $\rho_o\nabla\cdot\left(\psi\nabla\psi_t + \frac{\beta}{2}\psi^2\vec{i}\right) = -\nabla\cdot(p'u', p'v') \equiv -\nabla\cdot\vec{S}$. Similar to (5.20) for a prototype IGW, (5.47) then says that for a prototype Rossby wave the local rate of change in its energy is equal to the convergence of its energy flux through that location. For a prototype Rossby wave, $\psi = A\cos(kx + \ell y - t\sigma) \equiv A\cos\phi$ with $\sigma = -\frac{\beta k}{k^2 + \ell^2}$, the average of E over a wavelength is

$$\begin{aligned}[E] &= \frac{\rho_o}{2}[\nabla\psi\cdot\nabla\psi]\\ &= \frac{\rho_o}{2}(k^2 + \ell^2)\left[\sin^2\phi\right] = \frac{\rho_o}{4}(k^2 + \ell^2)A^2.\end{aligned} \tag{5.48}$$

The energy of a prototype Rossby wave as an entity is a function of its wavenumbers and amplitude; so is its *energy flux* $\left[\vec{S}\right]$ apart from the dependence on the beta parameter. Likewise, we get

$$\begin{aligned}\left[\vec{S}\right] &= [\vec{v}'p'] = -\left[\rho_o\psi\nabla\psi_t + \rho_o\frac{\beta}{2}\psi^2\vec{i}\right]\\ &= -\vec{i}\rho_0\left(k\sigma + \frac{\beta}{2}\right)\frac{A^2}{2} - \vec{j}\rho_o\ell\sigma\frac{A^2}{2}\\ &= \left(\vec{i}\left(\frac{\beta(k^2 - \ell^2)}{(k^2 + \ell^2)^2}\right) + \vec{j}\left(\frac{2\beta k\ell}{(k^2 + \ell^2)^2}\right)\right)\frac{\rho_0 A^2}{4}(k^2 + \ell^2)\\ &= \vec{c}_g[E].\end{aligned} \tag{5.49}$$

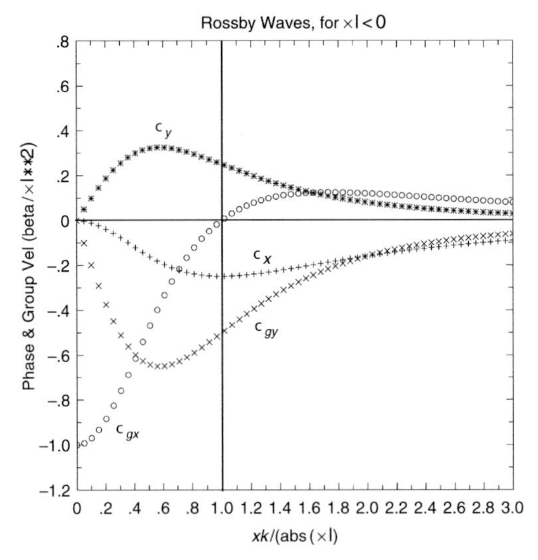

Fig. 5.10 Variation of the components of the non-dimenional phase velocity $\frac{\vec{c}}{\beta/\ell^2}$ and group velocity $\frac{\vec{c}_g}{\beta/\ell^2}$ of 2-D Rossby waves with the non-dimensional zonal wavenumber $\frac{k}{|\ell|}$ for $\ell < 0$.

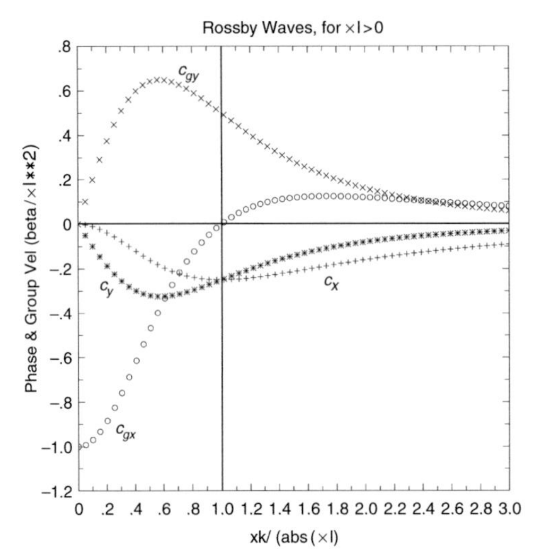

Fig. 5.11 Variation of the components of the non-dimenional phase velocity $\frac{\vec{c}}{\beta/\ell^2}$ and group velocity $\frac{\vec{c}_g}{\beta/\ell^2}$ of 2-D Rossby waves with the non-dimensional zonal wavenumber $\frac{k}{|\ell|}$ for $\ell > 0$.

Equation (5.49) confirms that the group velocity $\vec{c}_g$ indeed is a proper measure of the velocity of energy propagation associated with a Rossby wave. Since $|\vec{c}| = \frac{\beta k}{(k^2+\ell^2)^{3/2}}$, $|\vec{c}_g| = \frac{\beta}{(k^2+\ell^2)}$, we have $\frac{|\vec{c}_g|}{|\vec{c}|} = \sqrt{1+\frac{\ell^2}{k^2}} > 1$. So the energy of any Rossby wave propagates faster than its wave-front since $|\vec{c}_g| > |\vec{c}|$.

It is convenient to examine the phase velocity and group velocity in units of β/ℓ^2 as a function of non-dimensional k in units of $|\ell|$. The variations of such non-dimensional phase velocity and group velocity components with the zonal wavenumber for positive and negative ℓ are shown separately in Figs. 5.10 and 5.11.

For $\ell < 0$, c_y is positive and c_{gy} is negative for all values of k. The orientation of the wave-front of this subset of waves is generally in the SW to NE direction. For the other subset of waves with $\ell > 0$, the orientation of the wave-front is generally in the SE to NW direction with negative c_y and positive c_{gy}. On the other hand, it would be more instructive to think of the subset of relatively short waves (zonal wavelength versus the meridional wavelength) and the other subset of relatively long waves separately. Here c_x and c_{gx} are both negative (both westward) for the long waves ($k < |\ell|$). The short Rossby waves ($k > |\ell|$) have a positive value of c_{gx} (eastward) and negative value of c_x (westward). In other words, the energy of short Rossby waves ($k > |\ell|$) propagates eastward; and that of long Rossby waves ($k < |\ell|$) propagates westward. The energy of Rossby waves with $\ell > 0$ also propagates northward; and that of $\ell < 0$ propagates southward.

Intrinsic relation between meridional fluxes of zonal momentum flux and energy by Rossby waves

It is noteworthy that

$$\begin{aligned} [u'v'] &= -\left[\psi_y \psi_x\right] = -\frac{k\ell}{2}A^2, \\ [p'v'] &= -\rho_o\left[\psi\psi_{ty}\right] = -\frac{\rho_o \ell \sigma}{2}A^2 = \frac{\rho_o \beta k \ell}{2(k^2+\ell^2)}A^2. \end{aligned} \tag{5.50a,b}$$

The important conclusion from (5.50a,b) is that $[u'v']$ and $[p'v']$ have opposite signs. It is the same as saying that $[u'v']$ and c_{gy} have opposite signs according to (5.50a) and (5.44). *In other words, the meridional fluxes of zonal momentum and energy by a Rossby wave are intrinsically opposite in direction.* For example, a Rossby wave that propagates energy equatorward would transport zonal momentum poleward or vice versa. Knowing one would enable us to deduce the other. This is a signature characteristic of Rossby waves in the atmosphere.

We will discuss the excitation mechanisms of Rossby waves in Chapters 8 and 9. For the time being, it suffices to say that the Rossby waves generated in a mid-latitude zone must have a structure such that they would propagate energy away from the zone. In other words, we would expect $[p'v'] < 0$ to the south and $[p'v'] > 0$ to the north of the zone. According to (5.50a,b), we would then expect $[u'v'] > 0$ to the south and $[u'v'] < 0$ to the north. It follows that there would be a convergence of zonal momentum flux, leading to the formation of a westerly zonal jet in this zone. Since the excitation of these Rossby waves does not generate or remove zonal momentum, there must be a compensating easterly zonal jet on either side of the zone of wave excitation. This deduction is fully consistent with the discussion in Section 3.8.5 of an eddy-driven jet from the perspective of conservation of potential vorticity.

5.7.3 Graphical depiction of all properties of a prototype Rossby wave

We graphically depict all properties of a prototype Rossby wave with $\ell = k$ and an unspecified phase angle in Fig. 5.12. Such a Rossby wave is described by $\psi = A\cos\phi$, $(u', v') = \left(-\psi_y, \psi_x\right)$ and $\zeta' = -A(k^2 + \ell^2)\cos\phi$ with $\phi = kx + \ell y - \sigma t$.

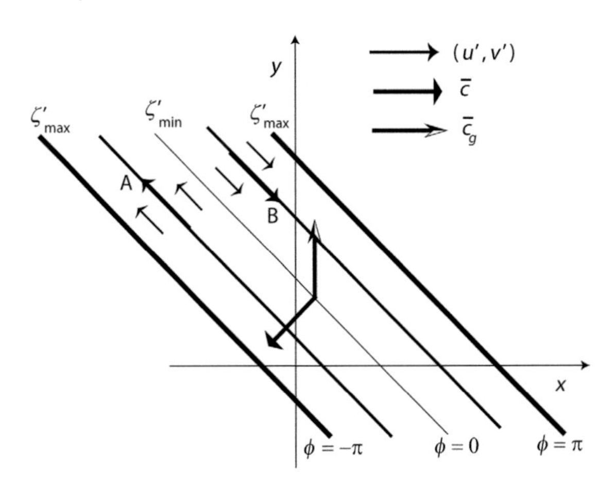

Fig. 5.12 Graphical summary of all properties of a prototype Rossby wave with $\ell = k > 0$.

The wave-fronts of this Rossby wave propagate toward the southwest, whereas its energy propagates northward. It confirms that advection of planetary vorticity would reduce the vorticity at the point A and would increase the vorticity at the point B. This change in vorticity amounts to a systematic westward displacement of the whole wave pattern. This Rossby wave is also a linearly polarized transverse wave.

5.7.4 Dispersion of Rossby wave-packet

A localized disturbance consists of constituent Rossby waves that intrinsically tend to disperse because the phase velocity of each Rossby wave is a function of its wavenumber. We illustrate the characteristics of Rossby waves dispersing from a simple 1-D weak localized disturbance, $\psi(x,t)$ in a beta-plane model without a basic flow. Recall that the perturbation governing equation in this case is

$$\frac{\partial}{\partial t}\frac{\partial^2\psi}{\partial x^2} + \beta\frac{\partial\psi}{\partial x} = 0. \tag{5.51}$$

Suppose an initial disturbance has a Gaussian structure centered at $x = x_0$ with a half-width, L,

$$\psi'(x,0) = \exp\left(-((x-x_o)/L)^2\right). \tag{5.52}$$

The disturbance field would remain independent of the y-coordinate at all times. The Fourier transform of $\psi(x,0)$ is

$$\hat{\psi}(k,0) = \int_{-\infty}^{\infty} \psi'(x,0)e^{-ikx}dx. \tag{5.53}$$

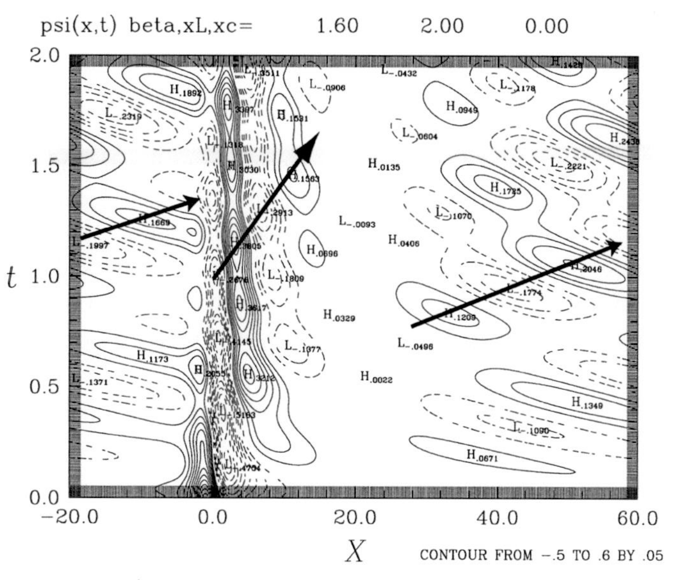

Fig. 5.13 Evolution of non-dimensional $\psi'(x,t)$ in the case of $x_o = 0$ and $\beta = 1.6$.

The Fourier transform of $\psi(x,t)$ is denoted by $\hat{\psi}_k(k,t)$ which must satisfy $-k^2\frac{d\hat{\psi}_k}{dt} + ik\beta\hat{\psi}_k = 0$ according to (5.51). The solution of $\hat{\psi}_k$ can be solved for each value of k separately. The solution is

$$\hat{\psi}(k,t) = \hat{\psi}(k,0)\exp\left(\frac{i\beta}{k}t\right). \tag{5.54}$$

The solution of the disturbance field in physical space is simply the inverse Fourier transform $\psi'(x,t) = \frac{1}{2\pi}\int_{-\infty}^{\infty}\hat{\psi}(k,t)e^{ikx}dk$.

The evolution of such an initial disturbance is illustrated with a sample calculation. The result is presented in a non-dimensional form for convenience. Distance, velocity and time are measured in units of $L/2$, V and $2/(VL)$ respectively. We use $L/2 = 1 \times 10^6$ m and $V = 10\,\mathrm{m\,s^{-1}}$, a unit of time would be $\sim$ 1 day. Let us consider a very long channel domain, $-20 \leq x \leq 60$, so that the disturbance field barely recycles across the zonal boundaries in the time interval under consideration. We use the following non-dimensional parameters in the calculation: $x_o = 0.0$, $L = 2.0$, $\beta = 1.6$. The evolution of the wave field is shown in Fig. 5.13.

The meridional wavenumber of all constituent waves under consideration is zero in this case. It follows that $c_x = -\beta/k^2$, $c_{gx} = \beta/k^2$ and $c_y = c_{gy} = 0$. All constituent waves of the wave-packet therefore have an eastward group velocity, whereas the phase velocities of all constituent waves are westward. The constituent waves propagate at different speeds, leading to continual change in the shape of the wave-packet as a whole in time. The plot confirms that the group velocity (energy propagation) of all constituent waves is indeed eastward, whereas the phase velocity of all constituent waves is westward. The dispersion leads to the emergence of a well-defined wave-packet far to the east of the initial localized disturbance at $t \approx 0.2$. It moves fairly slowly eastward and eventually recycles around the channel domain. A similar wave-packet repeatedly emerges subsequently.

We will defer until Chapter 6 discussion of the dynamics of three-dimensional Rossby waves in a more general model known as the quasi-geostrophic model. An analysis of the fundamental wave modes in that model will show that a complete set of fundamental wave modes consists of one barotropic Rossby wave mode and a set of baroclinic Rossby wave modes. The prototype Rossby wave mode discussed in this section corresponds to the barotropic Rossby wave mode as a special case.

5.7.5 Rossby wave propagation through a background uniform zonal flow

If there is a uniform basic flow in the zonal direction, $\bar{u}$, the perturbation governing equation in a 2-D beta-plane model becomes

$$\nabla^2\psi'_t + \bar{u}\nabla^2\psi'_x + \beta\psi'_x = 0. \tag{5.55}$$

An elementary solution of (5.55) is again $\psi' = \Phi \exp(i(kx + \ell y - \sigma t))$ with σ, k, ℓ and Φ all being constants. The dispersion relation is

$$\sigma = \bar{u}k - \frac{\beta k}{k^2 + \ell^2}. \tag{5.56}$$

The basic flow therefore Doppler shifts the frequency of a Rossby wave. The phase velocity and group velocity of a Rossby wave in this basic zonal flow are

$$\begin{aligned} \vec{c} \equiv (c_x, c_y) &= \left(\frac{\sigma k}{k^2 + \ell^2}, \frac{\sigma \ell}{k^2 + \ell^2}\right) = \left(\frac{\left(\bar{u} - \frac{\beta}{k^2+\ell^2}\right)k^2}{k^2 + \ell^2}, \frac{\left(\bar{u} - \frac{\beta}{k^2+\ell^2}\right)k\ell}{k^2 + \ell^2}\right), \\ \vec{c}_g \equiv (c_{gx}, c_{gy}) &= \left(\bar{u} + \frac{\beta(k^2 - \ell^2)}{(k^2 + \ell^2)^2}, \frac{\beta 2k\ell}{(k^2 + \ell^2)^2}\right). \end{aligned} \tag{5.57a,b}$$

Equation (5.57a) shows that the basic flow affects both components of the phase velocity equally, and therefore does not alter the direction of propagation of a wave-front. On the other hand, (5.57b) shows that the basic flow only affects the zonal component of the group velocity. From the point of view of energy propagation, this basic flow simply advects a wave in its direction. For a typical basic wind of $\bar{u} = 10\,\mathrm{m\,s^{-1}}$, the zonal component of Rossby waves in the phase velocity is eastward. It is also useful to consider *the so-called phase speed in the x-direction* of this wave, as defined by $c_{(x)} = \sigma/k = \bar{u} - \dfrac{\beta}{k^2 + \ell^2}$. We will examine the spectrum of Rossby waves as a function of $c_{(x)}$ at each latitude obtained from a spectral diagnosis of the departure of wind data from the zonal mean wind.

A Rossby wave for a given zonal wavenumber in the presence of a basic flow can be stationary ($\sigma = 0$) when its meridional wavenumber is equal to

$$\ell = \pm\sqrt{\frac{\beta}{\bar{u}} - k^2}. \tag{5.58}$$

For a representative background wind in mid-latitude, $\bar{u} = 10\,\mathrm{m\,s^{-1}}$, the wavelength of a stationary Rossby wave with $k \sim \ell$ has a *wavelength* $\sim 8.4 \times 10^6$ m. A planetary scale Rossby wave may even retrograde to the west in the presence of a typical westerly basic flow. It should be emphasized that although the wave-fronts of a stationary Rossby wave do not move by definition, it can still transmit energy from one location to another in the atmosphere because its group velocity is not zero.

$$\begin{aligned}
c_{gx} &= \frac{\partial \sigma}{\partial k} = \bar{u} + \frac{\beta(k^2 - \ell^2)}{(k^2 + \ell^2)^2} = \frac{2\beta k^2}{(k^2 + \ell^2)^2}, \\
c_{gy} &= \frac{\partial \sigma}{\partial \ell} = \frac{2\beta k \ell}{(k^2 + \ell^2)^2}, \\
\frac{c_{gx}}{c_{gy}} &= \frac{k}{\ell}.
\end{aligned} \tag{5.59}$$

Equation (5.59) says that a stationary Rossby wave transmits energy in the direction of its wavenumber vector with an *eastward* component. For example, a stationary Rossby wave with a wave-front oriented in the SE–NW direction would transmit energy in the NE direction. We will return to analyze and discuss the dynamics of propagation of stationary waves in Section 2 of Chapter 9.

5.8 Forced orographic Rossby waves in a shallow-water model

We next consider a simple example of forced Rossby waves which are excited by a steady uniform zonal flow U flowing over a localized orography $z_B(x, y)$ in a shallow-water model. The domain is a zonally unbounded channel. As an analogy of the Rocky Mountain in N. America, we prescribe a model orography that has a meridional scale much longer than its zonal scale. But we only consider a modest height so that the forced disturbance can be adequately examined in the context of linear dynamics. Such a response would be a steady disturbance. The task is to determine its structure. We will see that the orographic disturbance excited by a basic westerly zonal flow is qualitatively different from that excited by a basic easterly zonal flow. We will interpret such difference in terms of the intrinsic property of Rossby waves.

The appropriate governing equation for this model can be formulated as follows. We have shown in Section 3.8.4 that the potential vorticity of a fluid column in a shallow-water model is $\frac{\zeta + f}{h}$ where ζ is relative vorticity, f the Coriolis parameter and $h = (z_{surface} - z_B)$ its length. This potential vorticity of each fluid column would conserve when there is no damping. In considering a relatively low orography of large scale, the flow may be assumed to be in geostrophic balance. In that case, we may write the height of the free surface as $z_{surface} = H + \eta$ with $\eta/H \ll 1$ where H is the mean thickness of the layer. The relative vorticity is then $\zeta = \nabla^2(g\eta/f_o)$ where g is gravity and f_o the mean Coriolis parameter. Then by invoking the conditions $\eta/H \ll 1$ and $z_B/H \ll 1$, we may approximate PV for such a flow as

$$q = \frac{\nabla^2\left(\frac{g\eta}{f_o}\right) + f}{H\left(1 + \frac{\eta}{H} - \frac{z_B}{H}\right)} \approx \frac{1}{H}\left(\nabla^2\left(\frac{g\eta}{f_o}\right) + f - f_o\left(\frac{\eta}{H} - \frac{z_B}{H}\right)\right).$$

The constant factor $1/H$ is inconsequential and may be dropped from the expression. With the use of beta-plane representation of the Coriolis parameter, we can then compactly write the quasi-geostrophic PV as $q = f_o + \beta y + \nabla^2\psi - \lambda^2(\psi - \psi_B)$, where $\psi = g\eta/f_o$ is a geostrophic streamfunction, $\psi_B = gz_B/f_o$ a counterpart quantity for the orography and $\lambda^2 = f_o^2/gH$ with λ^{-1} referred to as *a radius of deformation*. Since the orography under consideration has relatively low height and long meridional scale, the meridional derivative of the basic potential vorticity would be primarily associated with the beta effect, $\partial q_{basic}/\partial y = \beta + \lambda^2 \partial\psi_B/\partial y \approx \beta$. It follows that free Rossby waves in this model would still intrinsically propagate westward.

The appropriate governing PV equation with the frictional effect of an Ekman layer is then

$$\left(\frac{\partial}{\partial t} - \psi_y \frac{\partial}{\partial x} + \psi_x \frac{\partial}{\partial y}\right) q = \varepsilon \nabla^2 \psi,$$
$$q = f_o + \beta y + \nabla^2\psi - \lambda^2(\psi - \psi_B), \tag{5.60}$$

where $u = -\psi_y$, $v = \psi_x$; $\varepsilon = \frac{f_o}{H}\left(\frac{\kappa}{2f_o}\right)^{1/2}$ is the frictional coefficient. The domain has two rigid boundaries at $y = \pm Y$.

We decompose every quantity into a basic part and a departure from it: $\psi = \bar{\psi} + \psi' = -Uy + \psi'$; $q = \bar{q} + q'$; $\bar{q} = f_o + \beta y - \lambda^2(-Uy - \psi_B)$, $q' = \nabla^2\psi' - \lambda^2\psi'$. A steady perturbation is governed by the linearized form of (5.60) as

$$Uq'_x + U\bar{q}_x - \psi'_y\bar{q}_x + \bar{q}_y\psi'_x = \varepsilon\nabla^2\psi',$$
$$U\left(\nabla^2\psi' - \lambda^2\psi'\right)_x + U\lambda^2\psi_{Bx} + \left(\beta + \lambda^2\left(U + \psi_{By}\right)\right)\psi'_x = \varepsilon\nabla^2\psi'. \tag{5.61a,b}$$

The term $-\psi'_y\bar{q}_x$ has been neglected consistent with the linearization, since $U \gg u' = -\psi'_y$. If we further limit to the parameter condition $\frac{\lambda^2\psi_{By}}{\beta} \ll 1$, (5.61b) may be further simplified to

$$U\nabla\psi'_x + \beta\psi'_x = \varepsilon\nabla^2\psi' - U\lambda^2\psi_{Bx}. \tag{5.62}$$

Equation (5.62) says that the PV dynamics of this orographic disturbance is dictated by a balance among the orographic forcing $(-U\lambda^2\psi_{Bx})$, dissipation $(\varepsilon\nabla^2\psi')$, advection of perturbation PV by the basic flow $\left(-U\left(\nabla^2\psi' - \lambda^2\psi'\right)_x\right)$ and advection of basic PV by the perturbation $\left(-\left(\beta + \lambda^2 U\right)\psi'_x\right)$. Note that the zonal mean component of the topography is not a forcing under the influence of a zonal basic flow. Thus, there is no zonal mean component in the forced response.

The domain is specifically $-\infty < x < \infty$, $-Y \leq y \leq Y = \pi/2\ell$. Let us consider the following north–south oriented localized orography

$$z_B = z_* \exp\left(-\left(\frac{x}{L}\right)^2\right) \cos \ell y \equiv S(x) \cos \ell y, \tag{5.63}$$

with $2\pi/\ell \gg L$. We may seek a solution in the following form

$$\psi' = \xi(x) \cos \ell y. \tag{5.64}$$

Using Fourier representation for $S(x)$ and $\xi(x)$, we write

$$\begin{aligned} \psi_B &= \frac{g \cos \ell y}{f_o} \int_{-\infty}^{\infty} \hat{S}(k) e^{ikx} dx, \\ \hat{S}(k) &= \frac{z_*}{2\sqrt{\pi}} \exp\left(-\frac{k^2}{4}\right), \\ \xi &= \int_{-\infty}^{\infty} \hat{\xi}(k) e^{ikx} dx. \end{aligned} \tag{5.65a,b,c}$$

In other words, the forced disturbance may be thought of as an ensemble of wave components, which are Rossby waves. The Fourier transform of (5.62) establishes the relationship between the spectral amplitudes $\hat{\xi}$ and $\hat{S}$, viz.

$$\begin{aligned} &(-k^2 - \ell^2)\hat{\xi} + \frac{\beta}{U}\hat{\xi} - i\frac{\varepsilon(k^2 + \ell^2)}{kU}\hat{\xi} - \frac{f_o}{H}\hat{S} = 0, \\ &\hat{\xi}(k) = \frac{f_o\left(-k^2 - \ell^2 + \frac{\beta}{U} + i\frac{\varepsilon(k^2 + \ell^2)}{Uk}\right)}{\left(\left(k^2 + \ell^2 - \frac{\beta}{U}\right)^2 + \left(\frac{\varepsilon(k^2 + \ell^2)}{Uk}\right)^2\right)} \frac{\hat{S}(k)}{H}. \end{aligned} \tag{5.66a,b}$$

Each spectral component of the response depends only on the corresponding spectral component of the orography. Using (5.66b) in conjunction with (5.64) and (5.65c), we can construct the forced disturbance. One may add that if we retain the term $\lambda^2 \psi_{Bx} \psi'_x$ in (5.61), the analysis would become technically more involved because all spectral components would be dynamically coupled and the whole ensemble of waves would have to be evaluated simultaneously.

Equation (5.66b) reveals that one Rossby wave with wavenumber $k = k_r$ such that $\frac{\beta}{U} - k_r^2 - \ell^2 = 0$, would be strongest. Indeed it is only because of damping that it is kept to finite magnitude. The location of the forced response as a whole depends on the relative phases of the constituent wave components, which are functions of all parameters. From a physical point of view, the influence of damping is significant only where the response is particularly strong. It should be stressed that there is no resonant response even without damping in the event that the basic flow is an easterly $(U < 0)$ since $\beta/U - k^2 - \ell^2 \neq 0$

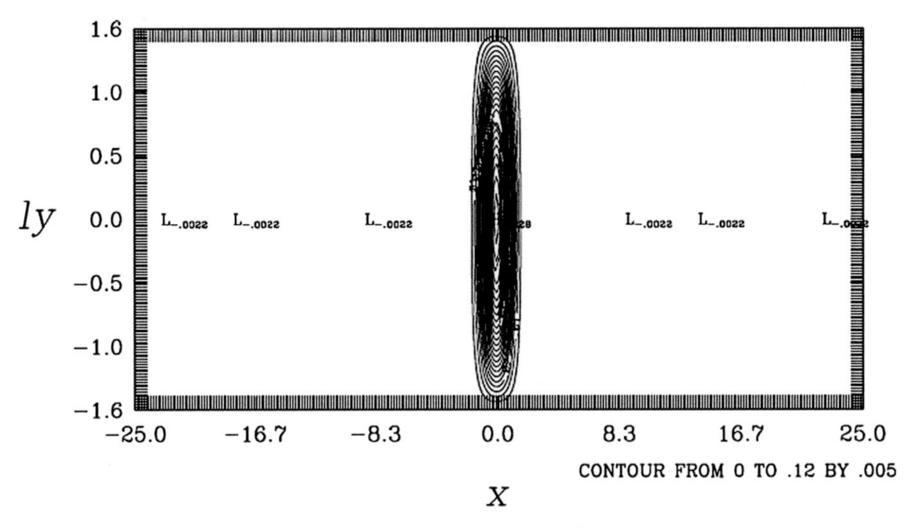

Fig. 5.14 Topography defined by (5.63) as a function of non-dimensional x and ℓy.

for any spectral component. Summing up, the orographically forced response is qualitatively different dependent upon the direction of the basic flow and damping plays an important role near the orography.

It is convenient to present the result in non-dimensional form. Hence, we measure $x, y, \psi', k, \ell, \beta, \varepsilon, \hat{S}, U$ in units of $L, L, f_o L^2, L^{-1}, L^{-1}, |U|L^{-2}, |U|L^{-1}, H, |U|$ respectively and $U = 1$ ($U = -1$) is used as an indication of a westerly (easterly) zonal flow. We illustrate the response with two computations for a relevant setting. Suppose the topography has half-width equal to 10° longitude at 43° N ($f_o = 1 \times 10^{-4}\,\mathrm{s}^{-1}$), corresponding to $L = 8.0 \times 10^5$ m. The other parameters are: $\varepsilon_{\mathrm{dim}} = 0.23 \times 10^{-5}\,\mathrm{s}^{-1}$ (5-day damping time), $|U| = 15\,\mathrm{m\,s}^{-1}$, $\beta_{\mathrm{dim}} = 1.6 \times 10^{-11}\,\mathrm{m}^{-1}\,\mathrm{s}^{-1}$, $H = 8 \times 10^3$ m, $h_* = 10^3$ m. Thus, the non-dimensional parameters are $\beta = 0.68$, $\varepsilon = 0.12$ and $\frown{h} = 0.125$. Computations are made for a particular long channel ($-25 \le \tilde{x} \le 25$). Its width is $-5 \le \tilde{y} \le 5$ with $-\pi/2 \le \ell y \le \pi/2$, corresponding to $\ell = 0.25$. We would have $\lambda^2 \psi_{By}/\beta \sim 0.4$ for this set of parameter conditions. The topography under consideration is plotted in Fig. 5.14.

5.8.1 Forced disturbance excited by a westerly flow

For a westerly basic flow, $U = 1$, the response is a Rossby wave-packet in the downstream region (Fig. 5.15). The dominant wavelength in the response is equal to ~ 8, which corresponds to that of a quasi-resonant response, satisfying $k = \pm\sqrt{\frac{\beta}{U} - \ell^2}$. The corresponding dimensional value of this wavelength is $\sim 6 \times 10^6$ m. The wave train progressively weakens in the downstream direction obviously due to dissipation. The disturbance has cyclonic vorticity immediately downstream of the orography. This means that the surface pressure is higher on the west side of the orography than on the east side. There is a transfer of zonal momentum from the fluid to the surface apart from the momentum transfer due to friction. This is a form drag for this model setting.

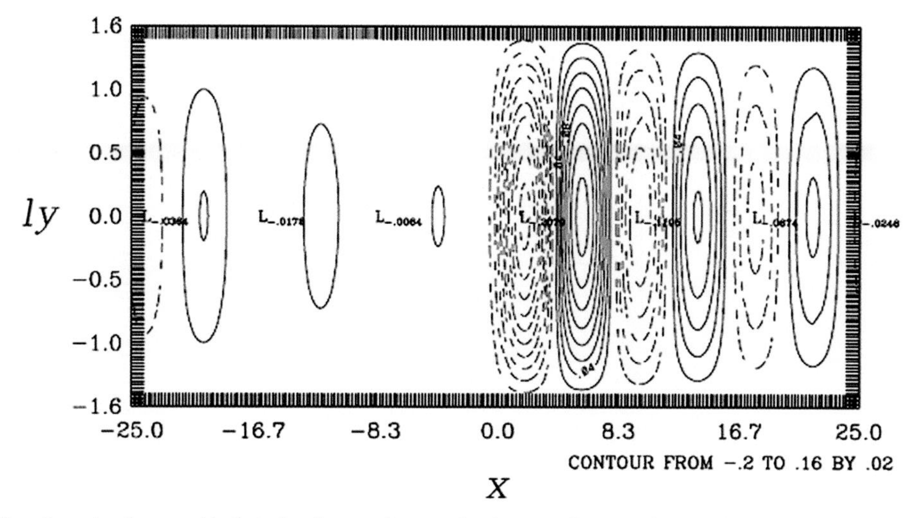

Fig. 5.15 Non-dimensional orographically induced streamfunction for the case of a westerly basic flow. See text for the parameter setting.

Fig. 5.16 shows the total streamfunction of the flow computed according to

$$\psi_{\text{total}} = -RUy + \psi' \tag{5.67}$$

where $R = |U_{\text{dim}}|/f_o L = 0.19$ for the parameter condition under consideration. Its key features are understandable from the perspective of PV dynamics. Under a condition of weak friction, the PV of each fluid column, $\left(\dfrac{f+\zeta}{h-z_B}\right)$, is quasi-conservative. As a fluid column is advected to the orography by the basic flow, its bottom rises with the topographic surface. Together with a lateral expansion of the fluid column, its length, $(h - z_B)$ would become shorter. Due to conservation of its PV, such a fluid column would move southward. Since the value of $df/dy = \beta$ is large in the analysis, a large decrease of $(\zeta + f)$ would induce cyclonic vorticity $(\zeta > 0)$. The southward displacement stops at some distance further east where its relative vorticity attains a maximum value. The meridional displacement would be reversed. A fluid column therefore oscillates about its initial latitude in a wavy manner. Its length simultaneously fluctuates. This wavy motion is what we call Rossby wave motion. The intrinsic westward propagation of Rossby waves can be opposed by the advective influence of the westerly basic flow making it possible to force a pattern of stationary waves downstream of the orography. The influence of damping is accumulative. Hence, the excited waves are progressively weaker further downstream as seen in Fig. 5.16.

5.8.2 Forced disturbance excited by an easterly flow

We repeat the calculation for the case of an easterly basic flow ($U = -1$). The perturbation streamfunction for the same set of parameters is shown in Fig. 5.17. We see that there is no wavy response at all. There is instead a very localized anticyclonic response over the orography itself. It is noteworthy that this forced response is wider than the orography

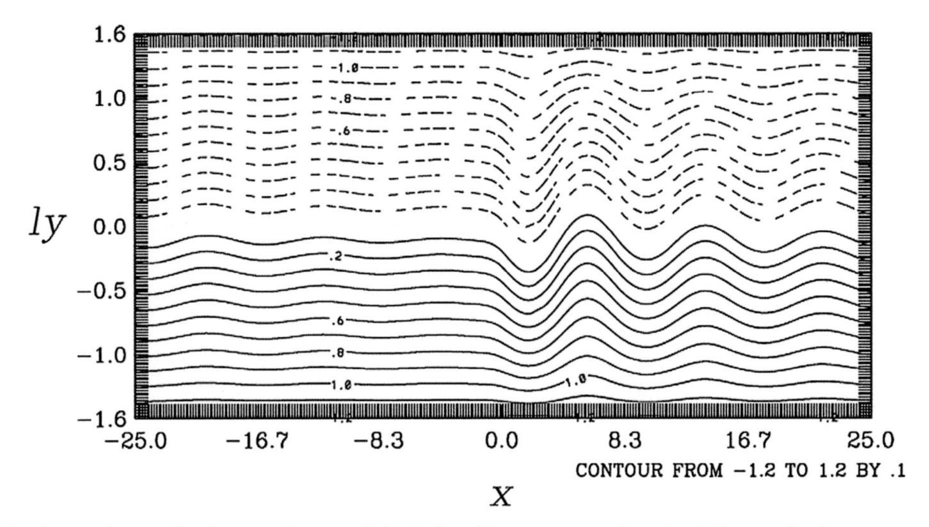

Fig. 5.16 Total streamfunction for the case of a westerly basic flow. The zero contour is omitted. See text for the parameter setting.

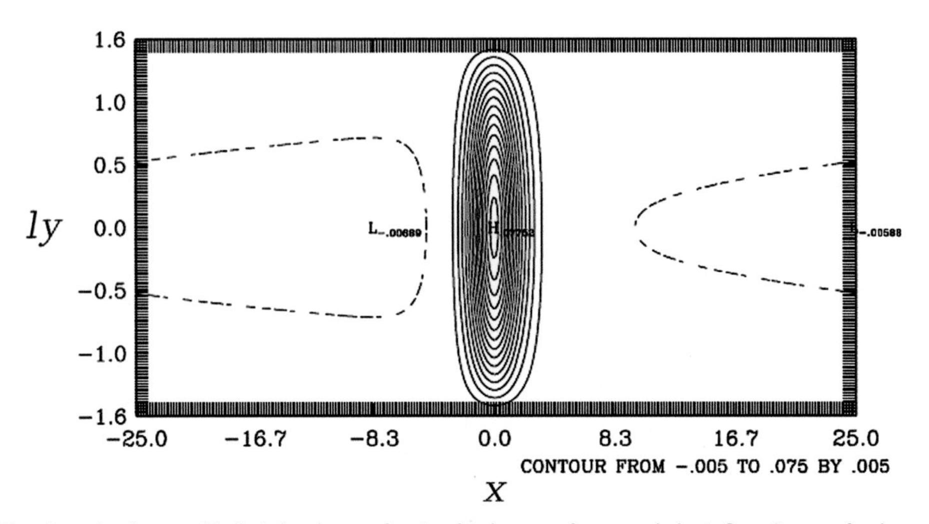

Fig. 5.17 Non-dimensional orographically induced streamfunction for the case of an easterly basic flow. See text for the parameter setting.

itself. This means that there is also an upstream impact on a column before it reaches the foothill. Quantitatively speaking, this response is only about one-third as strong as that in the case of the counterpart westerly basic flow (0.075 vs. 0.2).

The corresponding total streamfunction is shown in Fig. 5.18. The key features can be again succinctly interpreted from the perspective of PV dynamics. Figure 5.18 reveals that a fluid column first moves southward with a cyclonic curvature as it approaches the orography from the east. The cyclonic curvature and the southward displacement together

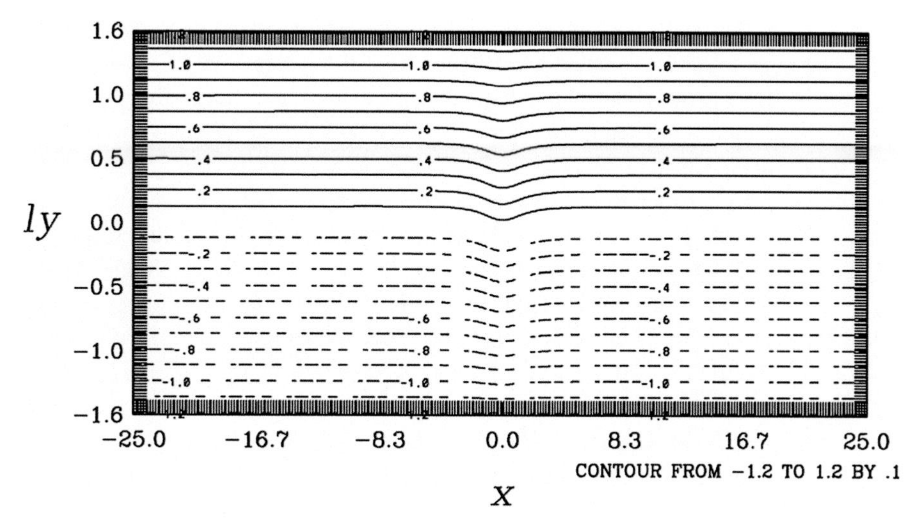

Fig. 5.18 Total streamfunction for the case of an easterly basic flow. The zero contour is suppressed. See text for the parameter setting.

with a small change in the column's length conserve its PV. The curvature soon changes to anticyclonic. The southward and anticyclonic movement accelerates as the fluid column rises up the orography. It reaches the extreme southern location where it is a peak. It begins to turn northward and eventually returns to its original latitudinal location. The entire trajectory is exactly symmetrical about the axis of the orography. The absence of a wavy character in the response is attributable to the fact that Rossby waves cannot remain stationary in the presence of an easterly background flow.

A number of illustrative analyses of orographically excited disturbances in a stratified fluid with different basic shear flows are presented in Section 9.4 with a more general model.

5.9 Some observed statistical properties of Rossby waves

We have so far only discussed the rudimentary properties of Rossby waves in the context of a generic barotropic model. The counterpart properties of Rossby waves in a generic baroclinic model will be discussed in Section 6.3.4. The actual Rossby waves in the atmosphere are much more complicated both structurally and dynamically speaking. They are continually excited and decay. They may be thought of as building blocks of the turbulent large-scale eddies in the atmospheric circulation. It is convenient to divide the eddies and hence the Rossby waves into two broad categories: stationary Rossby waves and transient Rossby waves. The former corresponding to the wavy structure of a time mean atmospheric flow. The latter is the departure of an instantaneous flow from this time mean structure. Before we examine the structural and dynamical properties of these categories of Rossby waves in more general model settings in subsequent chapters, we need to learn about some statistical properties of these waves deducible from operational weather data.

The statistical properties of the waves that we examine in this section are often referred to as general circulation statistics from the perspective of zonal and time averages. The time mean zonal momentum flux, heat flux and energy flux by transient waves across latitudes are denoted by $\left[\overline{u'v'}\right]$, $\left[\overline{v'\theta'}\right]$ and $\left[\overline{v'p'}\right]$ respectively. To the extent that WKBJ approximation is meaningful, (5.48) tells us that $\left[\overline{u'v'}\right]$ and $\left[\overline{v'p'}\right]$ are in opposite directions for Rossby waves of any wavenumbers k and ℓ. For the purpose of illustration, it suffices to examine $\left[\overline{u'v'}\right]$ and $\left[\overline{v'\theta'}\right]$ deduced with the use of five years of data (2001 to 2005) from a comprehensive global dataset known as NCEP/NCAR (National Center for Environmental Prediction/National Center for Atmospheric Research) Reanalysis Data. Fig. 5.19 shows that the eddy momentum flux is poleward in mid-latitudes (between $\sim 20°$ and $\sim 50°$) and is strongest at the upper tropospheric levels in both hemispheres. There is a high degree of symmetry between the two hemispheres. The maximum value in each hemisphere is located at about the 200 mb level over 30° latitude. Rossby waves are therefore most intense in the upper troposphere over the mid-latitudes. It is also found that $\left[\overline{u'v'}\right]$ has negative values north of about 60° N and positive values south of 60° S. In light of (5.48), we may infer that Rossby waves in both hemispheres tend to propagate energy equatorward ($\left[\overline{p'v'}\right] < 0$ in NH and $\left[\overline{p'v'}\right] > 0$ in SH) and poleward from the mid-latitude zone. By the same token, some Rossby waves propagate energy into the polar regions. This means that the transient Rossby waves in the atmosphere propagate energy away from the mid-latitude zone which is the primary source region of Rossby waves.

The distribution of time-zonal mean eddy flux of heat in terms of $\left[\overline{v'\theta'}\right]$ is shown in Fig. 5.20. This observed measure of eddy heat flux is towards the pole in each hemisphere virtually at all latitudes. It is strongest in mid-latitudes, particularly at the lower tropospheric levels, centering at about 50° latitude of both hemispheres. The values are even larger at the lower stratospheric levels. While both synoptic-scale and planetary-scale waves contribute significantly to $\left[\overline{v'\theta'}\right]$ in the troposphere, only the latter do so at the stratospheric levels. We will discuss the dynamical reason for it in Chapter 9.

It is also instructive to examine the distribution of eddy momentum flux at each latitude as a function of the zonal phase speed of the Rossby waves. This format is known as space-time cross-spectrum. The first part of such diagnosis is to decompose the longitudinal variation of the daily wind data (u and v) at each latitude into Fourier components. The second part is to perform a time series analysis of each of such Fourier coefficients. The cospectrum of u and v at each latitude as a function of frequency can be converted to a function of zonal phase speed for each zonal wave component. Adding up the results for all wave components at each latitude would yield the cospectrum. It was found by Randel and Held (1991) that virtually all Rossby waves in the atmosphere propagate eastward in both seasons (Fig. 5.21). The plot also shows the distribution of time-zonal mean background velocity, $[\bar{u}]$. It reaches a maximum value of $30\,\mathrm{m\,s^{-1}}$ at about 30°. On the other hand, waves have a zonal phase speed $c_{(x)}$ of at most $20\,\mathrm{m\,s^{-1}}$. It follows that the waves have an intrinsic westward propagation, $\left(c_{(x)} - [\bar{u}]\right) < 0$ and that the eastward propagation of the waves is a consequence of advection by the westerly zonal mean flow. This observational result, therefore, corroborates the validity of the theoretical results of Rossby waves discussed in Section 5.7.2.

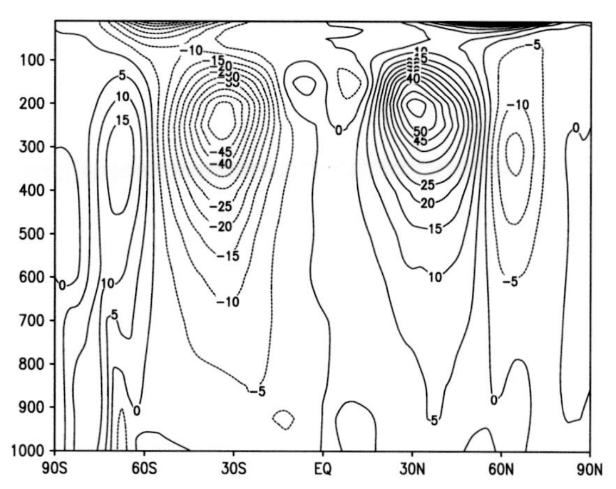

Fig. 5.19 Distribution of the climatological annual-zonal mean eddy zonal momentum flux $\overline{[u'v']}$ in m^2 s^{-2}.

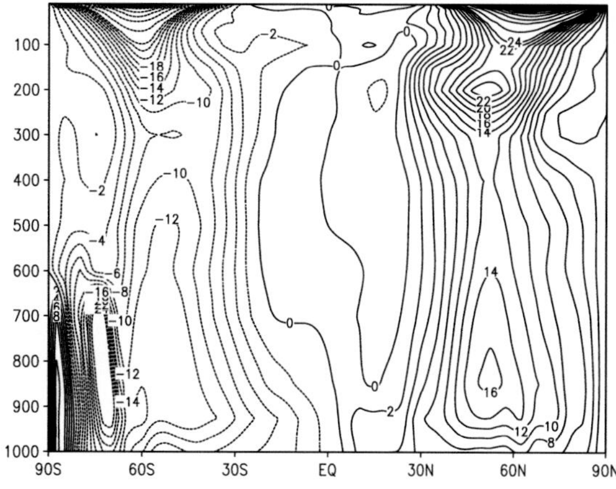

Fig. 5.20 Distribution of the climatological annual-zonal mean eddy heat flux, $\overline{[v'\theta']}$ in m s^{-1} K.

Figure 5.21 also shows that the observed Rossby waves mostly exist in the mid-latitudes between two latitudes where the phase speed of the waves is equal to the corresponding time-zonal mean flow. Those latitudes are known as critical latitudes. It is also clear that the atmospheric Rossby waves are more intense in the winter than in the summer of the northern hemisphere. The seasonal difference in the southern hemisphere is less pronounced.

5.10 Edge waves

The boundary of a domain exerts strong influence on wave motions by virtue of its kinematic constraint. Waves can be reflected and/or absorbed at a boundary. However, there are wave motions that intrinsically propagate along a boundary in a rotating fluid so

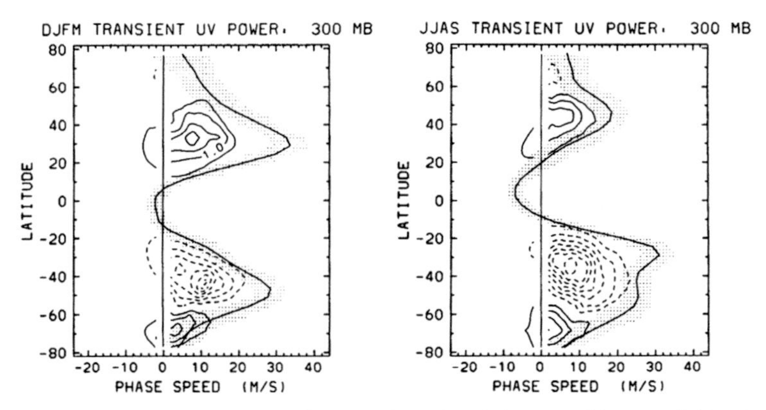

Fig. 5.21 Contours of 300 mb transient eddy momentum flux versus latitude and phase speed for DJFM (left) and JJAS (right). Contours interval is 0.50, with zero contours omitted. Heavy lines in each panel denote seasonal zonal mean zonal wind and shading denotes plus and minus one daily standard deviation. (Taken from Randel and Held, 1991.)

that they do not reflect at all. These waves are known as edge waves or boundary waves. They may be fundamentally of the gravity wave type or Rossby wave type in terms of physical nature. Kelvin (Thomson, 1879) first discovered an example of edge waves in a layer of incompressible fluid between two straight boundaries analogous to the British Channel. These waves are fundamentally gravity waves in nature, but the Coriolis force plays an essential role in conjunction with the boundary. Such waves are now named the Kelvin waves in his honor. Matsuno (1966) found wave solutions dynamically similar to the Kelvin waves in a totally different model setting. He analyzed the normal modes of a shallow-water model on an equatorial beta-plane with a domain including the equator. Even though there is no physical boundary in that model, the flow of a wave on both sides of the equator can be entirely zonal and in geostrophic balance. Those waves are now known as *equatorial Kelvin waves*. Less obvious are the edge waves of the Rossby wave type. For instance, such edge waves concentrating at the top and bottom boundaries of a vertical domain exist in a quasi-geostrophic model with a basic flow of uniform baroclinic shear (Bretherton, 1966). The dynamic instability of such a flow can be succinctly interpreted in terms of reinforcing interaction between such edge waves as elaborated in Chapter Part 8B. More generally, edge waves exist in the interior of a fluid along a line of discontinuity in the distribution of its basic potential vorticity as elaborated in Chapter Part 8C. In light of the broad relevance of edge waves, it would be valuable to learn about their fundamental dynamics.

It suffices to discuss the dynamics of edge waves in the context of a simplest possible model. Let us consider a layer of rotating incompressible fluid in a semi-infinite domain (say $-\infty < x < \infty$, $0 \leq y < \infty$) bounded by a straight boundary along $y = 0$. The Coriolis parameter is treated as a constant. The layer of fluid has a flat bottom and a constant depth H in its unperturbed state. If the horizontal scale of the disturbances under consideration is much longer than H, the flow would be in hydrostatic balance. This model is then a shallow-water model governed by the following equations

$$\begin{aligned} u_t + uu_x + vu_y &= -gh_x + fv, \\ v_t + uv_x + vv_y &= -gh_y - fu, \\ h_t + uh_x + vh_y + h(u_x + v_y) &= 0, \end{aligned} \tag{5.68a,b,c}$$

where h is the height of the free surface, g is gravity, u and v are the velocities in the x and y directions. For small amplitude perturbations, (5.68a,b,c) may be linearized to

$$\begin{aligned} u_t - fv &= -gh_x, \\ v_t &= -gh_y - fu, \\ h_t + H(u_x + v_y) &= 0, \end{aligned} \tag{5.69a,b,c}$$

where h now refers to the perturbation of the free surface from the mean depth. A solution compatible with the presence of a boundary along $y = 0$ would have identically zero meridional velocity, $v = 0$. Equation (5.69a,b,c) would be then further reduced to

$$\begin{aligned} u_t &= -gh_x, \\ 0 &= -gh_y - fu, \\ h_t + Hu_x &= 0. \end{aligned} \tag{5.70a,b,c}$$

Note that (5.70a) and (5.70c) form a closed set of equations for two unknowns, u and h. They can be combined to

$$h_{tt} - gHh_{xx} = 0. \tag{5.71}$$

Equation (5.71) is a wave equation governing one-dimensional external gravity waves. It is noteworthy that neither the Coriolis parameter nor the y-dependence of the disturbances under consideration appears in (5.71). We may seek normal mode solutions of (5.71) in the form of

$$h = A(y)\exp(i(kx - \sigma t)). \tag{5.72}$$

According to (5.71), the dispersion relation of these modes is

$$\begin{aligned} &\sigma^2 = gHk^2 \\ &\rightarrow \sigma = \pm k\sqrt{gH}. \end{aligned} \tag{5.73}$$

The amplitude of the wave $A(y)$ must be also compatible with (5.70b), which states that its zonal velocity is in geostrophic balance. Similarly writing the normal mode solution of u as $u = U(y)\exp(i(kx - \sigma t))$, we get from (5.70a) the relation $U = \frac{gk}{\sigma}A$. Using this relation, we get from (5.70b)

$$\frac{dA}{dy} = -\frac{fk}{\sigma}A. \tag{5.74}$$

The solution of (5.74) is $A = B\exp\left(-\frac{fk}{\sigma}y\right)$ where B is an integration constant of an arbitrary value. A physically meaningful normal mode must have finite values in the domain under consideration, $y > 0$. This condition leads us to using only the positive root of σ for the solution if the domain is in the northern hemisphere, $f > 0$. Then the solution is finally

$$h = B\exp\left(-\frac{fy}{\sqrt{gH}}\right)\exp(i(kx - \sigma t)),$$
$$u = \frac{B}{H}\sqrt{gH}\exp\left(-\frac{fy}{\sqrt{gH}}\right)\exp(i(kx - \sigma t)), \tag{5.75a,b}$$

with $\sigma = k\sqrt{gH}$ for the northern hemisphere. This wave propagates eastward with a speed $c = \frac{\sigma}{k} = \sqrt{gH}$. The amplitude of this wave mode is largest at the coastline, $y = 0$, decreasing exponentially away from the coastline at a rate of $\sqrt{gH}/f$. The wave hugs along the coastline as it propagates eastward. This is therefore a bona fide edge wave. Equation (5.75a,b) is the complete solution of a Kelvin wave. The maximum westerly (easterly) flow in such a wave is located where the surface elevation is maximum (minimum) whereby there is geostrophic balance everywhere. For a layer of depth $H = 2\,\text{km}$ in the extratropics $f = 10^{-4}\,\text{s}^{-1}$, we get $\sqrt{gH}/f \sim 40\,\text{km}$. The e-folding distance also naturally appears in the problem of geostrophic adjustment to be discussed extensively in Chapter 7. It is known as the "Rossby radius of deformation."

It should be added that the normal modes of (5.69a,b,c) are known as *Poincare waves*. They can be determined in the same way as we did earlier for analyzing the prototype IGW and Rossby waves. In the event that the domain is enclosed by rigid boundaries, we would need to impose the condition of zero normal velocity at the boundaries. That would lead to a discrete set of permissible wavenumbers for the modes. These waves are in essence inertio-external gravity waves. The Kelvin waves and the Poincare waves together form a complete set of modes in this simple model.

6 Quasi-geostrophic theory and two-layer model

This chapter discusses the single most important theory in dynamic meteorology for large-scale flows in the extratropics. It is known as the quasi-geostrophic theory. One example of such flow is briefly presented in Section 6.1 for the uninitiated readers. Section 6.2 discusses a method known as scale analysis for estimating the order of magnitude of each term of a governing equation. Such analysis is applied in Section 6.3 to derive a simplified version of the complete set of governing equations, known as the *quasi-geostrophic (QG) system of equations*. In Section 6.4, we combine them to get a diagnostic equation that highlights the symbiotic relationship between the rotational and divergent wind components. The QG system of equations is next reduced to one prognostic equation for a single unknown in Section 6.5. Not surprisingly, that is the QG version of the potential vorticity equation. We verify that only Rossby waves could exist in a QG model in Section 6.6. Section 6.7 shows the formulation of a two-layer version of the QG model, which will be used extensively throughout the book. Here we apply it to quantitatively illustrate all concepts encountered so far in the context of a developing baroclinic jet streak. Finally, Section 6.8 discusses how a QG model is cast in a global spherical domain.

6.1 Observed features of a synoptic disturbance

Disturbances with a horizontal dimension of the order of a thousand kilometers are ubiquitous in the extratropical atmosphere. They are referred to as synoptic-scale cyclones/anticyclones. Let us begin by getting a feel for the overall structural characteristics of one such disturbance over North America. We have already examined maps showing some dynamical properties of this disturbance in Sections 2.3 (actual and ageostrophic wind fields, Fig. 2.4) and 2.4 (thermal wind field, Fig. 2.7). We next examine (i) its mean sea level pressure field (Fig. 6.1), (ii) the relative configuration of its height fields at a lower tropospheric and an upper tropospheric pressure level (Fig. 6.2) and (iii) its omega field (Fig. 6.3). Figure 6.1 also indicates the movement of the main low-pressure center of this disturbance by several letters of "L." Its movement from April 5 to April 8, 2010 is indicated by arrows. The center moves in the general direction of WSW to ENE starting from the southwest corner of Oklahoma towards eastern Michigan. The lowest value of this pressure center is 996 mb on April 6. It changes little in the next 3 days.

Figure 6.2 depicts the geopotential height fields at 850 mb and 300 mb levels, Z_{850} and Z_{300}, on April 7. There is a trough on the 850 mb map with a minimum value of about 1350 m at the northwest corner of Missouri. There is a corresponding trough at the 300 mb

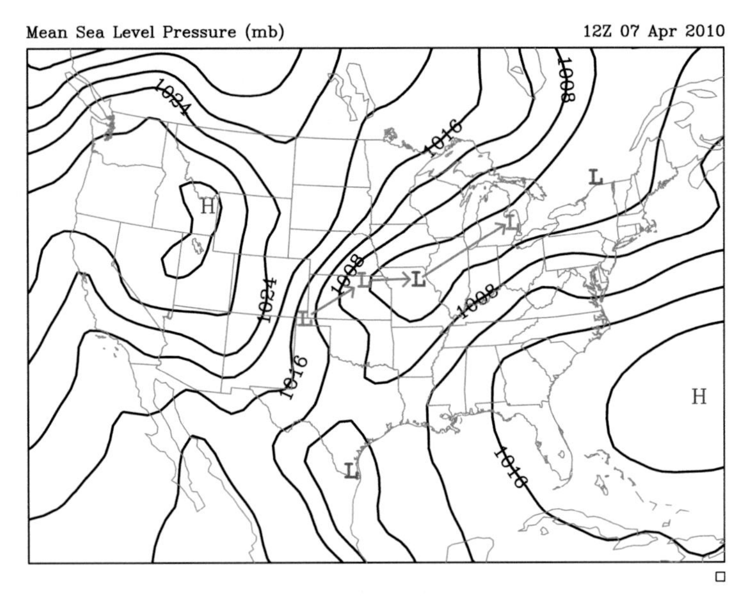

Fig. 6.1 Distribution of Mean Sea Level Pressure over N. America at 12Z on April 7, 2010 and the track of the major surface-low center from April 5 to April 8. See color plates section.

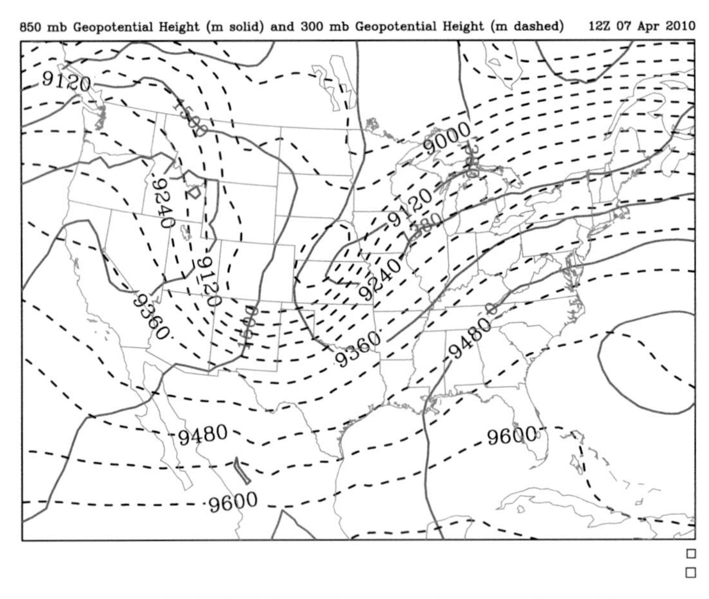

Fig. 6.2 Distributions of the geopotential height field of 850 mb and 300 mb surfaces, Z_{850} and Z_{300}, in meters on April 7, 2010. See color plates section.

level oriented in the SSW to NNE direction with a minimum value of about 8950 m at central Nebraska. An important feature is that the upper level trough (such as at 300 mb) is distinctly located to the west of the lower level trough (at 850 mb). We will see in the discussion of dynamic instability that this westward tilt of the trough is a tell tale feature about this type of

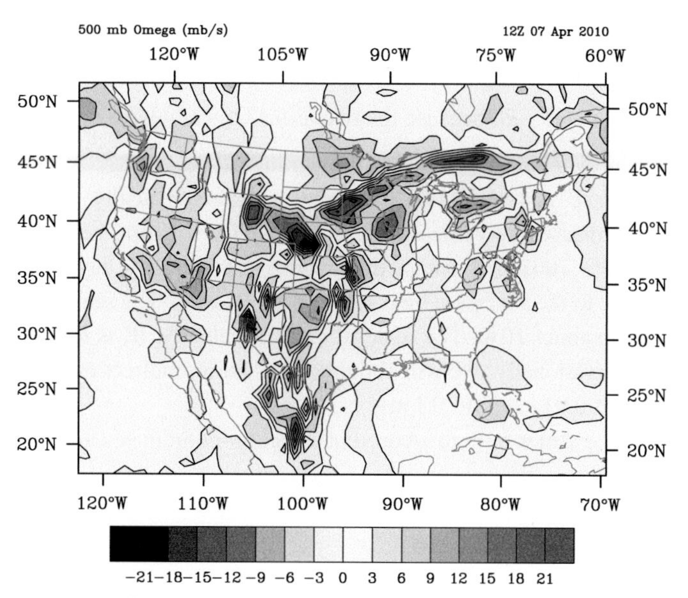

Fig. 6.3 The omega field, ω in mb s^{-1} at 500 mb on April 7, 2010. See color plates section.

instability. The trough on all levels propagates as a unit eastward together with the surface low-pressure center because they are all integral parts of the disturbance.

Figure 6.3 shows the vertical motion field in terms of ω at the 500 mb level. Unlike the geopotential field, the omega field is much more noisy. Nevertheless, a broad pattern of ω is recognizable with a region of distinct ascending motion to the east of the developing trough seen in Fig. 6.2.

We will shortly examine in Section 6.4.2 several maps that depict additional more subtle dynamical properties of the disturbance.

6.2 Scale analysis

Scale analysis is a method for estimating the order of magnitude of all terms of a governing equation in the context of a specific class of disturbances. The first step is to identify the scales of all properties of the class of disturbances under consideration. For example, the maps shown in Section 6.1 can be used to estimate the scales of synoptic disturbances in the extratropical atmosphere. One fourth of the distance between two adjacent troughs is the horizontal length scale, $L \sim 1000$ km. Half of the range of wind speed is the horizontal velocity scale, $U \sim 10\,\mathrm{m\,s^{-1}}$. The height field of the 500 mb surface varies by ~ 340 m from ridge to trough along a latitude. The amplitude of the variation δz is therefore ~ 170 m. The scale of the height of an isobaric surface is then ~ 100 m. It follows that the scale of $\Phi = gz$ is $\delta\Phi = 1000\,\mathrm{m^2\,s^{-1}}$.

Suppose we wish to estimate $u\partial u/\partial x$. The derivative can be estimated as of the order of $\partial u/\partial x \sim \delta U/L$ with $\delta U \sim U$ implying that $u\partial u/\partial x \sim U^2/L$. One characteristic of this class of disturbances is indicated by a non-dimensional number $Ro \equiv \frac{acceleration}{Coriolis\ force} \sim U/fL \ll 1$. It is called the Rossby number. The scale of relative vorticity is $\zeta = v_x - u_y \sim U/L$ because v_x and $(-u_y)$ typically have the same sign. In contrast, the scale of the horizontal divergence is $u_x + v_y \sim \frac{U}{L}Ro$ because the two terms typically have opposite signs and thereby are larger than their sum by an order of magnitude.

Since this class of disturbances characteristically extends from the surface to the tropopause, the vertical length scale is $H \sim 10^4$ m. The vertical scale in terms of pressure is $\Delta P = 700$ mb. Since the disturbance lasts for a few days, the time scale is ~1 day which is close to the advective time scale $\tau = L/U = 10^5$ s. The scale of temperature may be estimated to be about 10 K. The scale of vertical velocity, W, is too weak to be reliably established observationally. We will show how to get an indirect estimate of it, which turns out to be about 0.01 m s^{-1} to 0.1 m s^{-1}.

The scales in both z-coordinates and p-coordinates are summarized below.

Scale		z-coordinate	p-coordinate
horizontal length	L	10^6 m	10^6 m
vertical length	H	10^4 m	ΔP 700×10^2 Pa
time	τ	10^5 s	10^5 s
horizontal velocity	U	10 m s^{-1}	10 m s^{-1}
vertical velocity	W	0.01 m s^{-1} to 0.1 m s^{-1}	$\hat{\omega}$ 0.5 Pa s^{-1} to 0.05 Pa s^{-1}
pressure	δp	0.5×10^3 Pa	$\delta\Phi$ 10^3 m^2 s^{-2}
temperature	δT	10 K	10 K
density	$\frac{\delta\rho}{\rho_o}$	10^{-2}	10^{-2}

When we perform a scale analysis, we estimate the order of magnitude of every term of each governing equation. If one term is smaller than another by more than one order of magnitude, the former may be regarded negligibly small in a relative sense. All terms of the same order of magnitude should be either retained or neglected as a group. The simplest approximate form of that equation would contain only the terms of largest order of magnitude. One example is the hydrostatic balance between gravity and the vertical pressure gradient force in the vertical momentum equation. Another example is the geostrophic balance between the Coriolis force and the pressure gradient force in each horizontal momentum equation as discussed in Sections 2.2 and 2.3. These balance relations however contain no information about the evolution of the wind and temperature fields. We need to apply a scale analysis to the vorticity equation and thermodynamic equation in order to obtain prognostic equations. We will do so in the next section.

It should be added that a scale analysis may enable us to deduce the scale of a property which is too small to be reliably measured with operational weather data. Such a scale may be expressible in terms of the observable scales of other quantities which can be measured directly with acceptable accuracy if they are related through a quasi-balance condition. For example, if the vertical thermal advection balances the diabatic heating ($w\theta_z \sim Q$), we would have $W \sim Q/\theta_z$ as the scale of vertical velocity in terms of the scales of heating and stratification.

6.3 Quasi-geostrophic system of equations

To derive the simplified prognostic equations for synoptic disturbances, we apply a scale analysis to the vorticity equation and thermodynamic equation. The resulting equations shed light on the different and complementary roles of the strong rotational component and the weak divergent component of the wind field.

6.3.1 Scale analysis of the vorticity equation

Let us consider the governing equation for the vertical component of vorticity ($\zeta \equiv \zeta_3$) in Cartesian coordinates. It is Eq. (3.25) with the approximation that $\zeta_2 \gg h \equiv 2\Omega \cos \varphi$, which is repeated below for convenient reference.

$$\frac{\partial \zeta}{\partial t} = -(\zeta + f)(u_x + v_y) - \vec{V} \cdot \nabla(\zeta + f) + (\zeta_1 w_x + \zeta_2 w_y) + \frac{1}{\rho^2}(\rho_x p_y - \rho_y p_x) + \left((F_2)_x - (F_1)_y\right) \tag{6.1}$$

$$\vec{\zeta} \equiv (\zeta_1, \zeta_2, \zeta_3) = (w_y - v_z)\vec{i} + (u_z - w_x)\vec{j} + (v_x - u_y)\vec{k},$$

$$\vec{V} \cdot \nabla = u\frac{\partial}{\partial x} + v\frac{\partial}{\partial y} + w\frac{\partial}{\partial z},$$

$$\vec{F} \equiv (F_1, F_2, F_3), \text{ frictional force,}$$

$$2\Omega \sin \varphi \approx f = f_o + \beta y,\ f_o = 2\Omega \sin \varphi_o,\ \beta = \frac{2\Omega \cos \varphi_o}{a}.$$

We have noted earlier that $\zeta = (v_x - u_y) \sim \frac{U}{L}$, $(u_x + v_y) \sim \frac{U}{L} Ro$. It follows that $\zeta_t \sim u\zeta_x \sim v\zeta_y \sim \frac{U^2}{L^2} \sim 10^{-10}\,\text{s}^{-2}$, $w\zeta_z \sim \frac{UW}{LH} \sim 10^{-11}\,\text{s}^{-2}$, $vf_y \sim U\beta \sim 10^{-10}\,\text{s}^{-2}$, $f(u_x + v_y) \sim f_o \frac{U}{L} Ro \sim 10^{-10}\,\text{s}^{-2}$, $\zeta(u_x + v_y) \sim \frac{U}{L}\frac{URo}{L} \sim 10^{-11}\,\text{s}^{-2}$, $\zeta_1 w_x + \zeta_2 w_y \approx w_x v_z - w_y u_z \sim \frac{WU}{LH} \sim 10^{-11}\,\text{s}^{-2}$, $\frac{1}{\rho^2}(\rho_x p_y - \rho_y p_x) \sim \frac{\delta\rho\,\delta p}{\rho_o^2 L^2} \sim \frac{10^{-2}(0.5 \times 10^3)}{0.5 \times 10^{12}} \sim 10^{-11}\,\text{s}^{-2}$. The largest terms are therefore ζ_t, $u\zeta_x$, $v\zeta_y$, $f_o(u_x + v_y)$ and $v\beta$. They are of the same order of magnitude, $10^{-10}\,\text{s}^{-2}$. The terms smaller by one order of magnitude are $w\zeta_z$, $\zeta(u_x + v_y)$, $w_x v_z - w_y u_z$ and $\frac{1}{\rho^2}(\rho_x p_y - \rho_y p_x)$. The frictional term is much smaller above the planetary boundary layer. We explicitly identify the geostrophic and ageostrophic components of the horizontal velocity as $u = u_{(g)} + u_{(a)}$, $v = v_{(g)} + v_{(a)}$.

We now invoke additional approximations for the largest terms mentioned above:

(i) u and v in the advection terms of (6.1) are approximated by their geostrophic component, $u_{(g)}$ and $v_{(g)}$, which is non-divergent. The divergence term in (6.1) is associated with the ageostrophic wind, $u_{(a)}$ and $v_{(a)}$;

(ii) ζ are approximated by the part associated with $u_{(g)}$ and $v_{(g)}$;

(iii) f is approximated by f_o in the first term on the RHS of (6.1), but its variation with latitude should be retained in the second term representing advection of planetary vorticity. In this way (6.1) can be simplified to

$$\zeta_{(g)t} + u_{(g)}\zeta_{(g)x} + v_{(g)}\zeta_{(g)y} = -f_o\left(u_{(a)x} + v_{(a)y}\right) + v_{(g)}\beta. \tag{6.2}$$

This is called the *quasi-geostrophic vorticity equation.*

Equation (6.2) can be rewritten in a more compact form. Let us decompose pressure and density as

$$\begin{aligned} p &= p_o(z) + p_*(x,y,z,t), \\ \rho &= \rho_o(z) + \rho_*(x,y,z,t) \approx \rho_{oo} + \rho_*(x,y,z,t), \end{aligned} \tag{6.3}$$

where $p_o(z)$ and $\rho_o(z)$ are associated with a background thermodynamic state; $p_*(x,y,z,t)$ and $\rho_*(x,y,z,t)$ are associated with the large-scale disturbances. Since the velocity of a large-scale disturbance is approximately in geostrophic balance, we have

$$u_{(g)} = -\frac{1}{f\rho}\frac{\partial p}{\partial y} \approx -\frac{1}{f_o\rho_o}\frac{\partial p_*}{\partial y} \approx -\frac{1}{f_o\rho_{oo}}\frac{\partial p_*}{\partial y}, \qquad v_{(g)} \approx \frac{1}{f_o\rho_{oo}}\frac{\partial p_*}{\partial x}, \tag{6.4}$$

implying $u_{(g)x} + v_{(g)y} = 0$. This suggests that we may define a geostrophic streamfunction

$$\psi = \frac{p_*}{f_o\rho_{oo}}. \tag{6.5}$$

It follows that

$$u_{(g)} = -\psi_y, \qquad v_{(g)} = \psi_x, \qquad \zeta_{(g)} = \psi_{xx} + \psi_{yy}. \tag{6.6}$$

Then (6.2) can be rewritten as

$$\nabla^2\psi_t - \psi_y\nabla^2\psi_x + \psi_x\nabla^2\psi_y = -f_o\left(u_{(a)x} + v_{(a)y}\right) - \psi_x\beta \tag{6.7}$$

where $\nabla^2 = \frac{\partial^2}{\partial x^2} + \frac{\partial^2}{\partial y^2}$. Furthermore, horizontal divergence is associated with the vertical velocity as expected from the continuity equation, which may be simplified to $u_{(a)x} + v_{(a)y} = -\frac{1}{\rho_o}\frac{\partial}{\partial z}(\rho_o w)$, by invoking the Boussinesq approximation. Thus, there are effectively two unknowns in (6.7), ψ and w.

A scale analysis of the vorticity equation in isobaric coordinates yields a counterpart of (6.7). In that case, we would define the geostrophic streamfunction as $\psi = \frac{\Phi_*}{f_o} = \frac{gz_*}{f_o}$ where $\Phi = \Phi_o + \Phi_*$ is the geopotential of a pressure surface. Thus, the counterpart of (6.7) in isobaric coordinates is

$$\nabla^2\psi_t - \psi_y\nabla^2\psi_x + \psi_x\nabla^2\psi_y = f_o\omega_p - \psi_x\beta. \tag{6.8}$$

This form of equation naturally also contains two unknowns, ω and ψ.

6.3.2 Scale analysis of the thermodynamic equation

The thermodynamic equation in p-coordinates in the absence of diabatic heating is

$$T_t + uT_x + vT_y - \frac{pS}{R}\omega = 0, \tag{6.9}$$

where $S = \frac{R}{p}\left(\frac{RT}{pc_p} - \frac{\partial T}{\partial p}\right) = \frac{1}{gT\rho^2}\left(\frac{g}{c_p} + \frac{\partial T}{\partial z}\right) = \frac{1}{gT\rho^2}(\Gamma_d - \Gamma)$. We may similarly decompose T as

$$T = T_o(p) + T_*(x, y, p, t), \tag{6.10}$$

where T_o is the background component of the temperature field and T_* is the temperature associated with the disturbance. Observation suggests that $T_o \gg T_*$. Hence, it would be justifiable to evaluate S using the background thermal state T_o and ρ_o, $S \approx \frac{1}{gT_o\rho_o^2}\left(\frac{g}{c_p} + \frac{dT_o}{dz}\right) \sim 0.5 \times 10^{-5}\,\mathrm{m^2\,s^{-2}\,Pa^{-2}}$. For large-scale disturbances, the numerical values of the scales are $\delta T_* = 5\,\mathrm{K}$ to $10\,\mathrm{K}$, $\frac{pS}{R} \sim 10^{-3}\,\mathrm{K\,Pa^{-1}}$, $T_t \sim uT_x \sim vT_y \sim \frac{U\delta T}{L} \sim 5 \times 10^{-5}\,\mathrm{K\,s^{-1}}$, $\omega \sim 0.05\,\mathrm{Pa\,s^{-1}}$, $\frac{pS}{R}\omega \sim 5 \times 10^{-5}\,\mathrm{K\,s^{-1}}$. Hence, all terms are of the same order of magnitude and none can be neglected. However, we may still introduce the approximation $u \approx u_{(g)}$, $v \approx v_{(g)}$ and treat S as only a function of T_o. Temperature can be also expressed in terms of geopotential height by virtue of the hydrostatic balance, $0 = -\Phi_{o\ p} - RT_o/p$ and $0 = -\Phi_{*p} - RT_*/p$. The definition $\psi = \Phi_*/f_o$ leads to

$$-\psi_p = \frac{RT_*}{f_o p}. \tag{6.11}$$

Substituting (6.11) into the simplified thermodynamic equation yields

$$\left(\frac{\partial}{\partial t} - \frac{\partial\psi}{\partial y}\frac{\partial}{\partial x} + \frac{\partial\psi}{\partial x}\frac{\partial}{\partial y}\right)\frac{\partial\psi}{\partial p} + \frac{S}{f_o}\omega = 0. \tag{6.12}$$

Should we include diabatic heating, the RHS of (6.12) would have a term $\left(\frac{-RQ}{f_o p}\right)$ where Q is the heating rate in $\mathrm{K\,s^{-1}}$ and should be $\leq 10\,\mathrm{K\,s^{-1}}$. Equation (6.12) is called the *quasi-geostrophic thermodynamic equation*.

6.3.3 QG system of equations

The two equations (6.8) and (6.12) form a closed set of equations relating two unknowns, ω and ψ. They are known as the *quasi-geostrophic system of equations*. They would be energetically self-consistent only if S is a function of p. It should be added that scale analysis is a heuristic form of a more rigorous method known as singular perturbation method with the use of a small parameter, $\varepsilon \ll 1$, which may be identified with the Rossby number U/f_oL. According to this method, each dependent variable is represented as a power series of ε, such as $\psi = \psi_o + \varepsilon\psi_1 + \varepsilon^2\psi_2 + \cdots$. After substituting such an expansion of all dependent variables into each governing equation, we equate the terms of each order of ε. The ε^0 th order equations contain the terms of the geostrophic balance. The ε-order equations can be combined to get (6.8) and (6.12). The details of that approach are elaborated in the text by Pedlosky (1979).

A less restrictive set of approximate equations can be derived with the use of the *geostrophic momentum approximation*. According to that formulation, while the total derivative of the horizontal velocity, $D\vec{v}/Dt$, in the momentum equation is replaced by $D\vec{v}_g/Dt$, no approximation is invoked in the operator D/Dt itself. In other words, the advective acceleration is to be approximated by advection of the geostrophic velocity by the total velocity including the ageostrophic component of the velocity. The same treatment is also applied to the thermodynamic equation. The resulting equations are more suitable for investigating the dynamics of an atmospheric front. Such a model is referred to as a *semi-geostrophic model*. We will elaborate on that approximation and apply it in Section 12.2 in an illustrative analysis of frontogenesis.

6.4 Diagnostic function of the QG theory

While the large-scale wind field can be measured with about 90 percent accuracy, the vertical velocity field is too weak to be measured directly. The wind represented by the geostrophic streamfunction ψ is a rotational wind. Since it is the dominant part of the wind, it is the *primary circulation* of a large-scale flow. The divergent component of the wind is an order of magnitude weaker. Together with the vertical motion field, ω, it is called the *secondary circulation* of a large-scale flow. We now show that the rotational and divergent parts of the wind are intrinsically interrelated by a diagnostic equation with which we can deduce the divergent wind on the basis of the data concerning the rotational wind. This is a *symbiotic relation* in that one cannot exist without the other in a representative large-scale disturbance. Variation of ω in the vertical direction influences the primary circulation through the process of vortex stretching by virtue of the Earth's strong rotation. The vertical motion itself ω significantly affects the temperature via adiabatic heating/cooling associated with a stable stratification. When significant diabatic heating is also present, it would affect the horizontal variation of ω and temperature directly. In either case, the vertical structure of ψ would be changed by virtue of the thermal wind relation. The net effect is that the secondary circulation continually helps maintain the primary circulation in quasi-geostrophic balance and hydrostatic balance.

6.4.1 Omega equation (secondary circulation)

We can combine (6.8) and (6.12) by eliminating the time derivative term between them. That would yield an equation that functionally relates ω to ψ. It would be meaningful to assume S to be constant for simplicity. Treating it as a function of pressure would only introduce minor complications. Hence, the operation $\frac{f_o}{S}\left(\frac{\partial}{\partial p}(6.8) - \nabla^2(6.12)\right)$ leads to

$$\left(\nabla^2 + \frac{f_o^2}{S}\frac{\partial^2}{\partial p^2}\right)\omega = \frac{f_o}{S}\frac{\partial}{\partial p}\left(\vec{V}\cdot\nabla(\nabla^2\psi + f)\right) - \frac{f_o}{S}\nabla^2\left(\vec{V}\cdot\nabla\frac{\partial\psi}{\partial p}\right). \tag{6.13}$$

The geostrophic velocity is written here as $\vec{V} \equiv \left(-\psi_y, \psi_x\right)$ in the interest of clarity. This diagnostic equation is called the *omega equation*. It is a linear inhomogeneous partial differential equation (PDE) for ω. The operator on the LHS is effectively a three-dimensional Laplacian operator because the static stability S has positive values in a large-scale environment. The two terms on the RHS are viewed as *forcing functions* in the sense that the ω field is uniquely associated with them, but not in the sense of causality. Qualitatively speaking, we may expect negative values of ω in or near a region of positive values of net forcing since we may think of a large-scale flow as consisting of sinusoidal components.

Equation (6.13) can be solved numerically as a boundary value problem when the RHS is known. If we regard the tropopause and the Earth's surface as the vertical boundaries of a domain, $\omega = 0$ would be meaningful boundary conditions at those levels. If we want to include the effect of surface friction, we need to use a different lower boundary condition so that it would incorporate the frictional convergence in an Ekman layer. For a cyclical channel domain, $-X \leq x \leq X$, bounded at two latitudes, $y = \pm Y$, we may use $\omega(-X,y) = \omega(X,y)$ and $\omega = 0$ at $y = \pm Y$ as the horizontal boundary conditions. An example calculation will be presented in Section 6.7.1.

6.4.2 Qualitative deductions on the basis of the omega equation

It is instructive to delineate the physical meaning of the forcing functions. The two terms on the RHS of (6.13) are denoted by $F_j, j = 1,2$. The symbolic form of the omega equation for each part of $\omega = \sum_{j=1}^{2} \omega_j$ is

$$\left(\nabla^2 + \frac{f_o^2}{S}\frac{\partial^2}{\partial p^2}\right)\omega_j = F_j, \; j = 1, 2 \tag{6.14}$$

with $F_1 = \frac{f_o}{S}\frac{\partial}{\partial p}\left(\vec{V}\cdot\nabla(\nabla^2\psi + f)\right)$, $F_2 = -\frac{f_o}{S}\nabla^2\left(\vec{V}\cdot\nabla\frac{\partial\psi}{\partial p}\right)$. Each of F_j "forces" a component of ω, which may be evaluated separately; $F_1 \equiv \frac{f_o}{S}\frac{\partial}{\partial p}\left(\vec{V}\cdot\nabla(\nabla^2\psi + f)\right)$ stems from the process of "vertical differential vorticity advection" (dif-VAd) because $\frac{\partial}{\partial p}\left(\vec{V}\cdot\nabla(\nabla^2\psi + f)\right) = -\frac{\partial}{\partial p}\left(-\vec{V}\cdot\nabla(\nabla^2\psi + f)\right) = \frac{1}{g\rho}\frac{\partial}{\partial z}(\textit{absolute vorticity advection})$,

$F_2 \equiv -\frac{f_o}{S}\nabla^2\left(\vec{V}\cdot\nabla\frac{\partial\psi}{\partial p}\right)$ is proportional to the "negative-Laplacian of thermal advection" because $-\nabla^2\left(\vec{V}\cdot\nabla\frac{\partial\psi}{\partial p}\right) \propto \left(\vec{V}\cdot\nabla\frac{\partial\psi}{\partial p}\right) \propto (-\vec{V}\cdot\nabla T) = (\textit{thermal advection})$.

The signature of these quantities can be seen in their distributions in an observed disturbance at 00Z on April 7, 2010. Figure 6.4 shows the ζ_{abs} and $\vec{V}$ fields at 500 mb. The distribution of ζ_{abs} in this flow has the shape of a gigantic hook located at about the middle of the trough. There is a long strip of large values of ζ_{abs} along the west side of the trough and the values at the southern tip of the trough are particularly large.

The advection of absolute vorticity at 500 mb and 850 mb are shown in Figs. 6.5 and 6.6 respectively. We may use them to qualitatively deduce the distribution of vertical differential vorticity advection at 700 mb. There is significant positive absolute vorticity advection in a long strip on the western side of the trough at 500 mb. The important thing to note is that the absolute vorticity advection at 850 mb is much smaller. It follows that the *dif-VAd in that long strip* has negative values. Little can be said with much confidence about the dif-VAd on the eastern side of the strip because it appears to be rather noisy at both 850 and 500 mb levels.

The temperature field at 700 mb is shown in Fig. 6.7. Cold thermal air is advected from Canada to almost the Gulf of Mexico on the west side of the trough, but the associated thermal advection is relatively weak. There is significant warm advection on the eastern side of the trough.

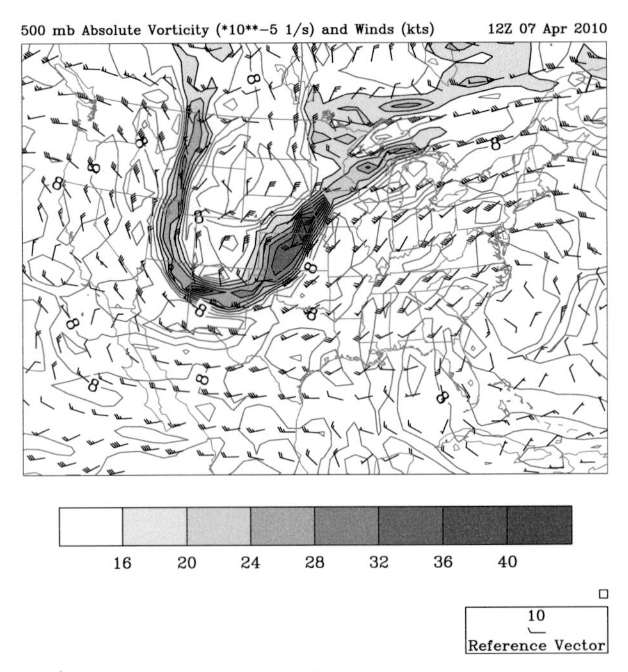

Fig. 6.4 Distributions of 500 mb $\vec{V}$ in knots and ζ_{abs} in $10^{-5}\ \mathrm{s}^{-1}$ on April 7, 2010. See color plates section.

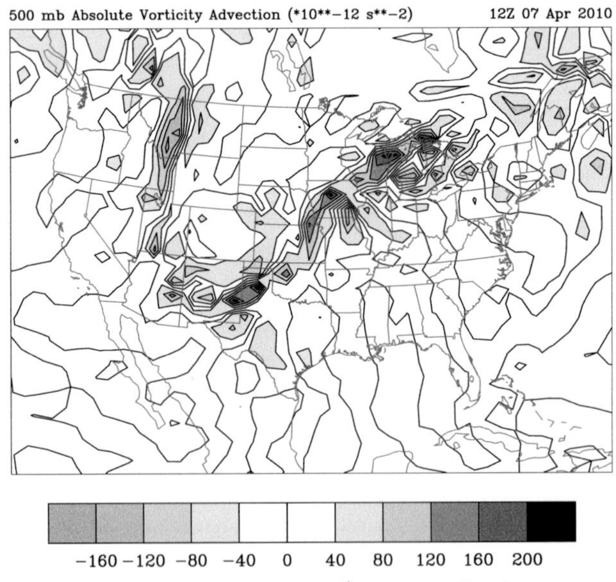

Fig. 6.5 Distribution of advection of absolute vorticity at 500 mb, $-\vec{V}\cdot\nabla\zeta_{abs}(10^{-12}\ \mathrm{s}^{-2})$ on April 7, 2010. See color plates section.

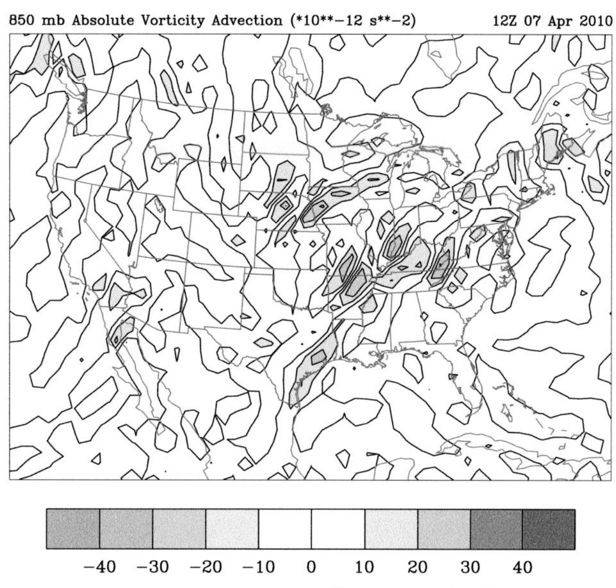

Fig. 6.6 Distribution of advection of absolute vorticity at 850 mb, $-\vec{V}\cdot\nabla\zeta_{abs}(10^{-12}\ \mathrm{s}^{-2})$ on April 7, 2010. See color plates section.

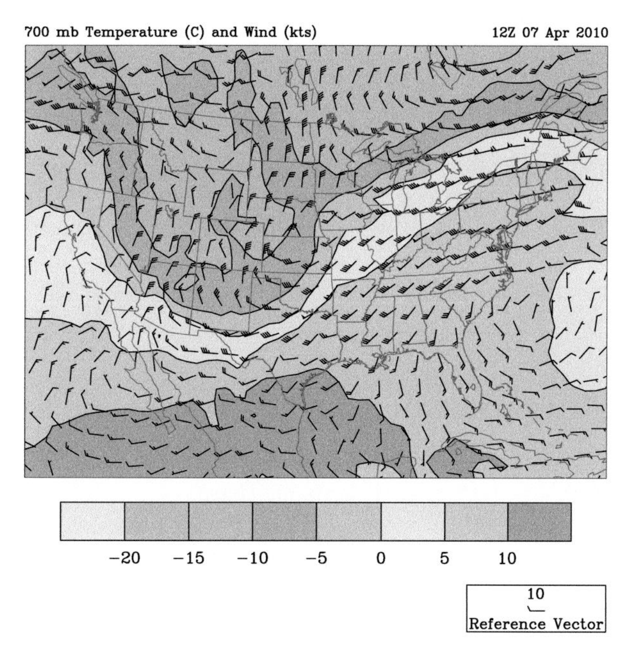

Fig. 6.7 Distributions of wind in knots and temperature in °C at 700 mb on April 7, 2010. See color plates section.

On the basis of these observed features, we may expect that positive values of ω at 700 mb are primarily associated with the differential vorticity advection on the western side of the trough, and negative values of ω at 700 mb are primarily associated with thermal advection on the eastern side. The ω field shown earlier in Fig. 6.3 indicates a long narrow region of positive ω at 700 mb (descent) on the west side and a broad region of negative ω at 700 mb (ascent) on the east side. In other words, the sign of ω is indeed opposite of the sign of the net forcing.

6.4.3 Q-vector

The processes of differential vorticity advection and thermal advection are clearly not independent of one another because they are both functions of the streamfunction. Therefore, it would be meaningless to give too much emphasis on the influence of each of the two processes in isolation. An instructive way of examining them jointly is to work with an alternative form of the omega equation. That is to be derived directly from the momentum equations and thermodynamic equation with quasi-geostrophic approximation as follows.

$$\left(\frac{\partial}{\partial t}+u\frac{\partial}{\partial x}+v\frac{\partial}{\partial y}\right)u=f_o v_a+\beta y v,$$
$$\left(\frac{\partial}{\partial t}+u\frac{\partial}{\partial x}+v\frac{\partial}{\partial y}\right)v=-f_o u_a-\beta y u, \tag{6.15a,b,c}$$
$$\left(\frac{\partial}{\partial t}+u\frac{\partial}{\partial x}+v\frac{\partial}{\partial y}\right)T-\frac{Sp}{R}\omega=0,$$

where u and v are understood to be the geostrophic velocity component; u_a and v_a are the ageostrophic velocity component. The thermal wind relations are

$$\frac{\partial u}{\partial p}=\frac{R}{pf_o}\frac{\partial T}{\partial y},\quad \frac{\partial v}{\partial p}=-\frac{R}{pf_o}\frac{\partial T}{\partial x}. \tag{6.16a,b}$$

Eliminating the $\partial/\partial t$ term between (6.15a) and (6.15c) with the operation $p\frac{\partial}{\partial p}(6.15\text{a})-\frac{R}{f_o}\frac{\partial}{\partial y}(6.15\text{c})$ and making use of (6.16a), we obtain

$$S\frac{\partial\omega}{\partial y}-f_o^2\frac{\partial v_a}{\partial p}=-2Q_2, \tag{6.17}$$

with

$$Q_2=-\frac{R}{p}\left(\frac{\partial\vec{V}}{\partial y}\cdot\nabla T-\frac{\beta y}{2}T_x\right), \tag{6.18}$$

where $\nabla=\left(\frac{\partial}{\partial x},\frac{\partial}{\partial y}\right)$, $\vec{V}=(u,v)$. Thus,

$$S\frac{\partial^2\omega}{\partial y^2}-f_o^2\frac{\partial}{\partial p}\left(\frac{\partial v_a}{\partial y}\right)=-2\frac{\partial Q_2}{\partial y}. \tag{6.19}$$

By making use of (6.15b), (6.15c) and (6.16b), we similarly get

$$S\frac{\partial^2\omega}{\partial x^2}-f_o^2\frac{\partial}{\partial p}\left(\frac{\partial u_a}{\partial x}\right)=-2\frac{\partial Q_1}{\partial x}, \tag{6.20}$$

with

$$Q_1=-\frac{R}{p}\left(\frac{\partial\vec{V}}{\partial x}\cdot\nabla T+\frac{\beta y}{2}T_y\right). \tag{6.21}$$

The sum of (6.18) and (6.19) is

$$S\nabla^2\omega + f_o^2\omega_{pp} = -2\nabla\cdot\vec{Q}, \tag{6.22}$$

where $\vec{Q} = (Q_1, Q_2)$ is the *Q-vector*. Equation (6.22) is, in essence, the same as the omega equation (6.14). The advantage of (6.22) is that the net forcing is proportional to a single quantity, namely the convergence of a vector field, the Q-vector. The sign of the forcing function at a location can be visually ascertained from a plot of the vector field. That would enable one to immediately deduce the sign of ω in that neighborhood by inspection. One part of the Q-vector is proportional to the dot product of a corresponding spatial derivative of the geostrophic velocity and the temperature gradient. The other part of the Q-vector depends on the beta parameter and the temperature derivative. The disadvantage is that the physical interpretation of the Q-vector is more obscure.

6.5 Prognostic function of the QG theory

We next show that the evolution of large-scale flows is governed by a single prognostic equation for one unknown in the context of the QG theory.

6.5.1 QG-streamfunction tendency equation for $\partial\psi/\partial t$

An equation for a single unknown that governs the evolution of the QG-streamfunction can be derived by eliminating ω between (6.8) and (6.12). The operation $\frac{\partial}{\partial p}\left(\frac{f_o^2}{S}(6.12)\right)$ yields

$$f_o^2\frac{\partial}{\partial p}\left(\frac{1}{S}\left(\psi_{pt} - \psi_y\psi_{px} + \psi_x\psi_{py}\right)\right) + f_o\omega_p = 0. \tag{6.23}$$

Since S is at most a function of p by assumption, we can write (6.21a) as

$$\frac{\partial}{\partial t}\frac{\partial}{\partial p}\left(\frac{f_o^2}{S}\psi_p\right) + \frac{\partial}{\partial p}\left(\vec{V}\cdot\nabla\left(\frac{f_o^2}{S}\psi_p\right)\right) = -f_o\omega_p. \tag{6.24}$$

Adding (6.8) to (6.21b) would yield

$$\frac{\partial}{\partial t}\left(\nabla^2\psi + \frac{\partial}{\partial p}\left(\frac{f_o^2}{S}\frac{\partial\psi}{\partial p}\right)\right) = -\vec{V}\cdot\nabla(\nabla^2\psi + f) - \frac{\partial}{\partial p}\left(\vec{V}\cdot\nabla\left(\frac{f_o^2}{S}\frac{\partial\psi}{\partial p}\right)\right). \tag{6.25}$$

Equation (6.25) is traditionally called the *QG-streamfunction tendency equation* because it enables forecasters to estimate the tendency $\chi \equiv \partial\psi/\partial t$ whenever the instantaneous wind field and hence $\psi\ (x,y,p,t_o)$ field is known. In using a constant value for the stratification S, (6.25) is reduced to

$$\underbrace{\left(\nabla^2 + \frac{\partial}{\partial p}\left(\frac{f_o^2}{S}\frac{\partial}{\partial p}\right)\right)\chi}_{A} = \underbrace{\vec{V}\cdot\nabla(\nabla^2\psi + \beta y)}_{B} - \underbrace{\frac{\partial}{\partial p}\left[\frac{f_o^2}{S}\vec{V}\cdot\nabla\left(-\frac{\partial\psi}{\partial p}\right)\right]}_{C}, \tag{6.26}$$

which is an inhomogeneous linear partial differential equation for χ with two forcing functions. Mathematically speaking, χ is "forced" by two identifiable physical processes which are functions of the ψ field alone.

6.5.2 Qualitative deductions on the basis of the tendency equation

The change of geopotential at a location, and hence the streamfunction tendency, might arise from the passage of a disturbance through that location and/or from a change of the intensity of the disturbance itself. The movement of a disturbance depends on its intrinsic dynamical characteristic as exemplified by the propagation of Rossby waves. A change in the intensity of a disturbance under consideration could result from interaction with its background flow. The representation of those processes is embodied in the terms B and C of (6.26).

A qualitative knowledge of the terms B and C would enable us to deduce the instantaneous streamfunction tendency by the same reasoning used in our discussion of the omega equation. According to the term B, $-f_o\vec{V}\cdot\nabla(\nabla^2\psi+\beta y)$, positive advection of absolute vorticity would tend to decrease $\chi\equiv\partial\psi/\partial t$. This is known as the *effect of absolute vorticity advection*. The term C

$$-\frac{\partial}{\partial p}\left[-\frac{f_o^2}{S}\vec{V}\cdot\nabla\left(-\frac{\partial\psi}{\partial p}\right)\right]=-\frac{\partial}{\partial p}\left[-\frac{Rf_o^2}{pS}\vec{V}\cdot\nabla T\right]\approx\frac{Rf_o^2}{g\rho_o pS}\frac{\partial}{\partial z}\left[-\vec{V}\cdot\nabla T\right]$$

is known as the *effect of differential thermal advection*. Specifically, it is proportional to the vertical differential thermal advection (dif-TAd) being positive when thermal advection is greater at higher elevation. Such a process would give rise to vertical stretching of an air column and hence an increase in vorticity. That would imply a decrease in ψ and equivalently a negative value of $\chi\equiv\partial\psi/\partial t$.

6.5.3 Potential vorticity perspective of the streamfunction tendency equation

The tendency equation (6.26) can be rewritten as

$$\frac{\partial}{\partial t}\left(\beta y+\nabla^2\psi+\frac{\partial}{\partial p}\left(\frac{f_o^2}{S}\frac{\partial\psi}{\partial p}\right)\right)+\vec{V}\cdot\nabla\left(\beta y+\nabla^2\psi+\frac{\partial}{\partial p}\left(\frac{f_o^2}{S}\frac{\partial\psi}{\partial p}\right)\right)=0 \tag{6.27}$$

because $\frac{\partial\vec{V}}{\partial p}\cdot\nabla\left(\frac{f_o^2}{S}\frac{\partial\psi}{\partial p}\right)=\frac{f_o^2}{S}\left(-\psi_{yp}\psi_{xp}+\psi_{xp}\psi_{yp}\right)=0$. So, we see that the QG tendency equation is nothing but a potential vorticity equation written as

$$\left(\frac{\partial}{\partial t}-\psi_y\frac{\partial}{\partial x}+\psi_x\frac{\partial}{\partial y}\right)q=0,$$
$$q=\nabla^2\psi+\beta y+\frac{\partial}{\partial p}\left(\frac{f_o^2}{S}\frac{\partial\psi}{\partial p}\right). \tag{6.28a,b}$$

Here q is known as *quasi-geostrophic potential vorticity* (QG-PV). The corresponding expression of QG-PV in height coordinates is

$$q=\nabla^2\psi+\beta y+\frac{f_o^2}{\rho_o}\frac{\partial}{\partial z}\left(\frac{\rho_o}{N^2}\frac{\partial\psi}{\partial z}\right). \tag{6.28c}$$

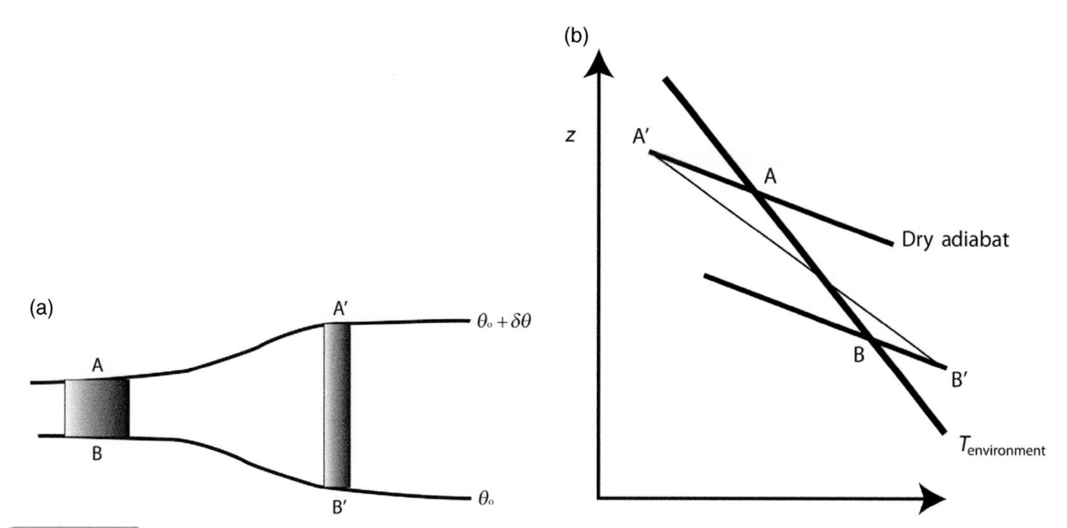

Fig. 6.8 Schematic to illustrate the physical nature of QG-potential vorticity of a fluid column AB between two isentropic surfaces moving adiabatically and frictionlessly to A′ B′; (a) cross-section of two isentropic surfaces and (b) a sounding with two dry adiabat lines.

The expression for S transformed to height coordinates is $N^2/g^2\rho_o^2$ where N is the Brunt–Vaisala frequency and ρ_o is the background density. The quantity NH/F is the radius of deformation for a continually stratified fluid. The fact that QG-PV is a function of the streamfunction alone, ψ, makes it possible for us to compute ψ by solving a boundary value problem whenever the q field and the temperature field at the boundaries are known. This is a simpler form of the *invertibility principle* discussed in Chapter 3. If we need to include diabatic heating, the RHS of (6.27) would have one more term $\left(-f_o R\frac{\partial}{\partial p}\left(\frac{Q}{Sp}\right)\right)$. If we further include friction, the RHS of (6.27) would have another additional term. We will elaborate on these issues in conjunction with illustrative analyses in later chapters.

Equation (6.28a) says that q of a fluid parcel conserves while it moves about geostrophically under adiabatic and inviscid conditions. We may visualize the invariance of QG-PV as follows. It consists of three parts. For it to conserve, there must be compensating changes in the other two parts when one of them changes. It is instructive to rewrite the third part of QG-PV in terms of the vertical derivative of the departure of temperature from the background state as follows

$$\frac{\partial}{\partial p}\left(\frac{f_o^2}{S}\frac{\partial\psi}{\partial p}\right) = -\frac{\partial}{\partial p}\left(\frac{Rf_o}{pS}T_*\right) \approx -\frac{Rf_o}{pS}\frac{\partial T_*}{\partial p}. \tag{6.29}$$

Observation suggests that the product (pS) roughly has the same value at all tropospheric levels. In other words, the third part is proportional to the lapse rate. Let us see how the third part would change when a fluid column between two isentropic surfaces adiabatically and frictionlessly moves from one location to another location on the same latitude as indicated in Fig. 6.8.

Since every fluid parcel conserves its potential temperature as well as its QG-PV, its temperature would change along its own adiabat line. Parcel A rises up to position A′ and parcel B sinks to B′. The changes of the temperature in between are indicated by the thin line in the sketch. In other words, the lapse rate in the column, $\partial T/\partial p$, increases from that of the sounding (solid line) to that of the thin line. As $\partial T/\partial p = \partial/\partial p(T_o + T_*)$ and $\partial T_o/\partial p$ do not change, we conclude that $\partial T_*/\partial p$ would increase when this fluid column moves to position (A′B′). The third part of QG-PV then would have a larger negative value. It follows that there must be a compensating increase in its geostrophic vorticity $\nabla^2\psi$. The process of vortex stretching is therefore embodied in the definition of QG-PV. This is one fundamental reason why q can remain invariant. Another fundamental reason is that a northward movement of a fluid parcel would be accompanied by a decrease of $\nabla^2\psi$ and vice versa. Northward displacement of a fluid parcel could also lead to a decrease in $\frac{\partial}{\partial p}\left(\frac{f_o^2}{S}\frac{\partial\psi}{\partial p}\right)$ without changing $\nabla^2\psi$.

6.6 Intrinsic wave modes in a QG model

In this section, we will discuss the intrinsic characteristics of the wave components in a QG model.

6.6.1 Wave modes in the absence of a basic flow

A perturbation is governed by the linearized form of the PV equation. In the absence of a basic flow in a quasi-geostrophic model the governing equation in height coordinates is

$$\begin{gathered} q_t^* + \beta\psi_x^* = 0 \\ q^* = \psi_{xx}^* + \psi_{yy}^* + \frac{f_o^2}{\rho_o}\left(\frac{\rho_o}{N^2}\psi_z^*\right)_z \end{gathered} \tag{6.30}$$

where $\rho_o = \rho_{oo}\exp(-z/H)$ is the density of the basic state. An elementary solution of (6.30) in a wave form is

$$\psi^* = \Phi e^{z/2H}\exp(i(kx + \ell y + mz - \sigma t)), \tag{6.31}$$

where Φ, k, ℓ, m, σ are constants. The $e^{z/2H}$ factor in (6.31) is introduced in anticipation of the dependence of a solution upon the height varying background density. For (6.31) to be a solution of (6.30), the wave parameters must satisfy the following dispersion relation

$$\sigma = -\frac{k\beta}{k^2 + \ell^2 + \frac{f_o^2}{N^2}\left(m^2 + \frac{1}{4H^2}\right)}. \tag{6.32}$$

The solution for $m = 0$ represents a special case of *barotropic Rossby wave modes* for different values of horizontal wavenumbers k and ℓ. The properties of these modes have been discussed in Section 5.7. The solutions with $m \neq 0$ represent *baroclinic Rossby wave modes* that have a non-trivial vertical structure. One important point warrants emphasis. Although the background state is stably stratified and gravity is present, internal gravity waves (IGW) do not exist in a QG system. Internal gravity waves have been expunged from the system as a consequence of the geostrophic and hydrostatic approximations which are invoked in the course of deriving the QG equations. This is the underlying reason for the great simplicity of a QG model.

6.6.2 Wave modes in the presence of a basic zonal flow

When there is basic zonal shear flow $U(y,z)$, the corresponding QG-PV varies in both y- and z-directions, $[q]\,(y,z)$. A perturbation would be governed by

$$\begin{aligned} q^*_t + Uq^*_x + [q]_y\psi^*_x &= 0, \\ [q]_y = \beta - U_{yy} - \frac{f^2}{\rho_o}\left(\frac{\rho_o}{N^2}U_z\right)_z&, \\ q^* = \psi^*_{xx} + \psi^*_{yy} + \frac{f^2}{\rho_o}\left(\frac{\rho_o}{N^2}\psi^*_z\right)_z&. \end{aligned} \tag{6.33}$$

If the shear of the basic flow were sufficiently strong, one would intuitively expect it to be dynamically unstable. It would require a separate careful analysis to delineate the instability characteristics. This will be done in Chapter 8. We have a more limited objective here, namely to identify what might happen to disturbances propagating through a basic state with long meridional and vertical scales. In other words, we want to consider a situation in which there is a distinct scale separation between a wave disturbance and the basic flow. It follows that the basic shear would be weak and the basic flow would be dynamically stable. Nevertheless, an elementary solution of (6.30) may be still regarded as a wave with a basic flow being effectively *uniform in a local sense*. This is known as *WKB approximation for a wave* solution. It is referred to as a wave-packet. The wave parameters must also satisfy a local dispersion relation stemming from (6.30), viz.

$$\begin{aligned} \psi^* &= \Phi e^{z/2H}\exp(i(kx + \ell y + mz - \omega t)), \\ \omega &= Uk - \frac{[q]_y k}{K_T^2}, \end{aligned} \tag{6.34a,b}$$

where $K_T^2 = k^2 + \ell^2 + \left(m^2 + \frac{1}{4H^2}\right)\frac{f_o^2}{N^2}$. The zonal wavenumber k is a constant since the basic flow is independent of x. The wave parameters (ℓ, m, ω and Φ) in (6.34a) for a given k are now smooth functions of y, z and t resulting from the impact of the basic shear as a wave-packet propagates through it.

The three components of the group velocity of a wave under consideration are then

$$\begin{aligned} c_{gx} &= \frac{\partial\omega}{\partial k} = U - \frac{[q]_y(K_T^2 - 2k^2)}{K_T^4}, \\ c_{gy} &= \frac{\partial\omega}{\partial \ell} = \frac{2[q]_y k\ell}{K_T^4}, \\ c_{gz} &= \frac{\partial\omega}{\partial m} = \frac{f_o^2}{N^2}\frac{2[q]_y km}{K_T^4}. \end{aligned} \tag{6.35}$$

The seasonal and annual mean values of $[q]_y$ in the atmosphere are observed to be positive. So the group velocity component c_{gy} would be northward if k and ℓ have the same sign. Likewise, c_{gz} would be upward if k and m have the same sign.

We highlight the variables of a disturbance under consideration with an asterisk to emphasize that it is a wave disturbance. We can readily obtain the expressions for the zonal average of any quadratic function of disturbance variables in terms of the wave parameters. For example, the meridional flux of zonal momentum by this Rossby wave-packet at a latitude and vertical level is

$$\rho_o[u^*v^*] = -\frac{\rho_{oo}}{2}|\Phi|^2 k\ell. \tag{6.36}$$

It would be northward if k and ℓ have opposite signs. The horizontal tilt of a local wave-front in such a case would be oriented in the SW–NE direction. Likewise, the corresponding heat flux is

$$\rho_o[v^*\theta^*] = \frac{\rho_{oo} f_o \theta_{oo}}{2g}|\Phi|^2 km. \tag{6.37}$$

The heat flux would be northward for a wave-packet that has same sign in k and m. Its local wave-front would tilt westward with height on a (x,z) plane.

In light of (6.28c), we can write the zonal mean PV flux of the wave as

$$[v^*q^*] = -[u^*v^*]_y + \frac{f_o g}{\rho_o \theta_{oo}}\left(\frac{\rho_o}{N^2}[v^*\theta^*]\right)_z. \tag{6.38}$$

In other words, the wave flux of PV $\overline{[v^*q^*]}$ is made up of a combination of meridional gradient of the time mean eddy momentum flux per unit mass $\overline{[u^*v^*]}$ and vertical gradient of the time mean eddy heat flux per unit mass $\overline{[v^*\theta^*]}$ weighted by stratification. The fact that the RHS of (6.38) has the form of divergence of a quantity on the (y,z) cross-section suggests that it would be useful to introduce a vector $\vec{E}$ so that (6.38) can be rewritten as

$$[v^*q^*] = \frac{1}{\rho_o}\nabla\cdot\vec{E}, \tag{6.39}$$

where $\vec{E} = \left(-\rho_o[u^*v^*], \frac{f_o g \rho_o}{\theta_{oo}N^2}[v^*\theta^*]\right)$ with $\nabla = \left(\frac{\partial}{\partial y}, \frac{\partial}{\partial z}\right)$; $\vec{E}$ is called the *Eliassen–Palm vector* (EP-vector). Note that the y-component of $\vec{E}$ is the Reynolds stress of large-scale eddies with a dimension of kg m s^{-2} m^{-2}. Its z-component is proportional to the Coriolis parameter and eddy heat flux and inversely proportional to stratification.

By (6.36), (6.37) and (6.39), the EP-vector for this wave is then equal to

$$\vec{E} = \left(-\rho_o[u^*v^*], \frac{f_o g}{\theta_{oo}}\left(\frac{\rho_o}{N^2}[v^*\theta^*]\right)\right) = \frac{|\Phi^2|}{2}\rho_{oo}k\left(\ell, \frac{f_o^2}{N^2}m\right)$$
$$= \frac{\rho_{oo}|\Phi^2|K_T^4}{4[q]_y}\left(c_{gy}, c_{gz}\right). \tag{6.40}$$

Hence, the EP-vector of a Rossby wave is parallel to its group velocity on the (y,z) plane. The $\vec{E}$ vectors of a Rossby wave field give us a general indication of the direction of its propagation on a meridional plane.

From the definition of PV, we also have $q^* = -K_T^2\psi^*$ for the wave represented by (6.34a). An important property of a wave disturbance is known as *enstrophy*. It is defined as

$$\rho_o\left[q^{*2}\right] = \frac{\rho_{oo}}{2}K_T^4|\Phi|^2. \tag{6.41}$$

Another property of a wave disturbance is known as *wave-activity density, A*. It is defined as

$$A = \frac{\rho_o\left[q^{*2}\right]}{2[q]_y} = \frac{\rho_{oo}K_T^4|\Phi|^2}{4[q]_y}. \tag{6.42}$$

In light of (6.40) and (6.42), we finally get the following important relation

$$\vec{E} = \vec{c}_g A \tag{6.43}$$

and $\vec{E}$ can be therefore interpreted as a flux of wave-activity density. The relationship $\rho_o[v^*q^*] = \nabla\cdot\vec{E} = \nabla\cdot\left(\vec{c}_g A\right)$ tells us that the PV flux per unit volume is equal to the divergence of wave-activity density flux.

The wave-activity density of a Rossby wave is also related to its energy density. The energy density of a Rossby wave ε is defined as the average of the sum of kinetic and potential energy over a wavelength of the wave. For a Rossby wave represented by (6.34a), it is

$$\varepsilon = \frac{1}{2}\left[\rho_o\left(\psi_x^2 + \psi_y^2 + \frac{f_o^2}{N^2}\psi_z^2\right)\right] = \frac{1}{4}\rho_{oo}K_T^2|\Phi|^2. \tag{6.44}$$

Then by (6.42) and (6.44) we obtain another expression for A,

$$A = \frac{\varepsilon}{U - k^{-1}\omega}. \tag{6.45}$$

Equation (6.45) says that the wave-activity density is equal to wave energy density divided by the velocity of a wave-front in the x-direction relative to the basic flow. It has the dimensions of *(energy per unit volume)*time/length*. Then by (6.43) and (6.45) the Eliassen–Palm vector can be alternatively written as

$$\vec{E} = \frac{\vec{c}_g\varepsilon}{U - k^{-1}\omega}. \tag{6.46}$$

It says that the Eliassen–Palm vector can also be thought of as a flux of energy divided by the velocity of a wave-packet relative to the basic flow.

Finally, we show that wave-activity density of a wave is expressible in terms of the displacement of the parcels associated with a wave. From a Lagrangian point of view, q^* is the PV anomaly of a fluid parcel. It is a function of the displacement in the y-direction, η^* and the meridional gradient of the zonal mean PV, viz.

$$\eta^* = \frac{q^*}{[q]_y}. \tag{6.47}$$

So the wave-activity density can be also written as

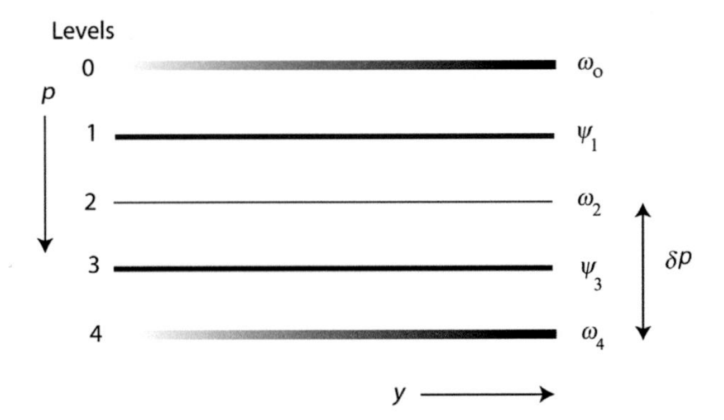

Fig. 6.9 Schematic of a two-layer quasi-geostrophic model (see text for the meanings of the symbols).

$$A = \frac{1}{2}\rho_o\left[\eta^{*2}\right][q]_y. \tag{6.48}$$

In summary, there are three different but equivalent expressions for A, the wave-activity density:

$$A = \frac{\rho_o[q^{*2}]}{2[q]_y} = \frac{\varepsilon}{U - k^{-1}\omega} = \frac{1}{2}\rho_o\left[\eta^{*2}\right][q]_y. \tag{6.49}$$

The EP-vector is just the flux of A by a wave-packet in a smoothly varying basic zonal flow. We will examine the characteristics of several representative observed $\vec{E}$ fields in the atmosphere as well as discuss their dynamical implications in the context of wave-mean flow interaction in Chapter 10.

6.7 Evolution of a baroclinic jet streak in a quasi-geostrophic two-layer model

Since large-scale disturbances span the whole vertical extent of the troposphere, they can be adequately depicted with only two degrees of freedom in the vertical direction. A two-layer version of the QG model constitutes the simplest possible model setting suitable for analyzing such flows. The words "layer" and "level" are used interchangeably in the following discussion, intentionally overlooking the subtle but minor difference. Figure 6.9 is a schematic depiction of a two-layer model in isobaric coordinates. The thickness of each layer in pressure may be set to δp = 400 mb with the mid-level pressure as p_2 = 600 mb. A mid-latitude zone is approximately represented by a reentrant channel in the zonal direction, $-X \le x \le X$, between two lateral rigid boundaries, $y = \pm Y$. The domain should be sufficiently wide so that the lateral boundaries would have little impact on the behavior of the flow in the model. The consistent boundary conditions are: $\omega_0 = \omega_4 = 0$ in the vertical, cyclical condition for all variables at $x = \pm X$, and $\psi_{13} = \psi_{3x} = 0$ at $y = \pm Y$ meaning no mass flux across the lateral boundaries.

The derivative of any property ξ with respect to p at level-j is represented by center-difference, $\left(\frac{\partial \xi}{\partial p}\right)_j = \frac{\xi_{j+1} - \xi_{j-1}}{\delta p}$. Applying (6.8) at level-1 and level-3 and making use of the vertical boundary conditions, we get

$$\frac{\partial}{\partial t}\nabla^2\psi_1 - \psi_{1y}\nabla^2\psi_{1x} + \psi_{1x}\nabla^2\psi_{1y} + \beta\psi_{1x} = f_o\frac{\omega_2}{\delta p}, \tag{6.50}$$

$$\frac{\partial}{\partial t}\nabla^2\psi_3 - \psi_{3y}\nabla^2\psi_{3x} + \psi_{3x}\nabla^2\psi_{3y} + \beta\psi_{3x} = -f_o\frac{\omega_2}{\delta p}. \tag{6.51}$$

Applying (6.12) at level-2 and using $\psi_2 = \frac{1}{2}(\psi_1 + \psi_3)$ by interpolation, we get

$$\frac{\partial}{\partial t}\frac{\psi_3 - \psi_1}{\delta p} - \frac{1}{2}(\psi_1 + \psi_3)_y\left(\frac{\psi_3 - \psi_1}{\delta p}\right)_x + \frac{1}{2}(\psi_1 + \psi_3)_x\left(\frac{\psi_3 - \psi_1}{\delta p}\right)_y = -\frac{S_2}{f_o}\omega_2. \tag{6.52}$$

The advection terms may be written more compactly in the form of a Jacobian. The Jacobian of any two functions $\xi(x,y)$ and $\eta(x,y)$ is $J(\xi,\eta) \equiv \xi_x\eta_y - \xi_y\eta_x$. The Jacobian involving any four functions A,B,C,D and constant r has the following properties: $J(A,rA) = 0$, $J(A,B) = -J(B,A)$, and $J(A+B,C+D) = J(A,C)+J(A,D)+J(B,C)+J(B,D)$. Equations (6.50), (6.51) and (6.52) can be then written as

$$\frac{\partial \nabla^2\psi_1}{\partial t} + J(\psi_1, \nabla^2\psi_1 + \beta y) = \frac{f_o\omega_2}{\delta p}, \tag{6.53}$$

$$\frac{\partial \nabla^2\psi_3}{\partial t} + J(\psi_3, \nabla^2\psi_3 + \beta y) = \frac{-f_o\omega_2}{\delta p}, \tag{6.54}$$

$$\frac{\partial}{\partial t}\left[\lambda^2(\psi_3 - \psi_1)\right] + J\left(\frac{1}{2}(\psi_1 + \psi_3), \lambda^2(\psi_3 - \psi_1)\right) = \frac{-f_o\omega_2}{\delta p}, \tag{6.55}$$

where $\lambda^2 = \frac{f_o^2}{S_2(\delta p)^2}$ which has a dimension of m^{-2}; λ^{-1} is called the *radius of deformation* which is an important intrinsic length scale of this system. A more compact form of (6.55) is

$$\frac{\partial}{\partial t}\left[\lambda^2(\psi_3 - \psi_1)\right] + J(\psi_1, \lambda^2\psi_3) = \frac{-f_o\omega_2}{\delta p}. \tag{6.56}$$

Equations (6.53), (6.54) and (6.56) can be readily combined to yield the quasi-geostrophic potential vorticity equations at level-1 and level-3 by eliminating ω_2 among them. The resulting equations are

$$\frac{\partial q_j}{\partial t} + J(\psi_j, q_j) = 0, \; j = 1, 3,$$
$$q_1 = \beta y + \nabla^2\psi_1 - \lambda^2(\psi_1 - \psi_3); \; q_3 = \beta y + \nabla^2\psi_3 + \lambda^2(\psi_1 - \psi_3). \tag{6.57a,b}$$

The quantities q_1 and q_3 are QG-PV. Equations (6.57a,b) are the two-layer version of (6.28a,b) the QG-PV equations. If diabatic heating is included, the RHS of (6.57a) would have one more term, ($-G \equiv -\lambda^2 RQ_2/f_o$), for the upper layer and G for the lower layer. If the frictional force at level-j is included and represented by $(-\varepsilon_j\vec{V}_j)$ with ε_j as a friction coefficient, the RHS of (6.57a,b) would have one more additional term $(-\varepsilon_j\nabla^2\psi_j)$. These additional terms would be consistent with the QG framework as long as the diabatic heating effect and frictional effect are at most the same order of magnitude as the advection terms.

The quasi-geostrophic omega equation for a two-layer model is obtained by eliminating the time derivative terms among (6.53), (6.54) and (6.56), viz.

$$\frac{f_o}{\lambda^2 \delta p}(\nabla^2 - 2\lambda^2)\omega_2 = J(\psi_3, \nabla^2\psi_3 + \beta y) - J(\psi_1, \nabla^2\psi_1 + \beta y) - \nabla^2 J(\psi_1, \psi_3). \quad (6.58)$$

The boundary conditions are $\omega_2(-X,y) = \omega_2(X,y)$ and $\partial\omega_2/\partial y = 0$ at $y = \pm Y$ and (6.58) enables us to determine $\omega_2(x,y,t)$ whenever $\psi_1(x,y,t)$ and $\psi_3(x,y,t)$ are known. The sum of the first two terms on the RHS of (6.58) represents a forcing associated with vertical differential absolute vorticity advection. The third term on the RHS stands for the negative Laplacian of thermal advection. Inclusion of diabatic heating and friction in the model would give rise to two additional terms on the RHS, $\left(-\frac{1}{\lambda^2}\nabla^2 G\right)$ and $(\varepsilon_3\nabla^2\psi_3 - \varepsilon_1\nabla^2\psi_1)$ respectively.

The alternative form of the omega equation in terms of the Q-vector for a two-layer QG model can be derived using the momentum equations as follows

$$\begin{aligned}
&u_{1t} + \vec{V}_1\cdot\nabla u_1 = f_o\hat{v}_1 + \beta y v_1, \qquad u_{3t} + \vec{V}_3\cdot\nabla u_3 = f_o\hat{v}_3 + \beta y v_3,\\
&v_{1t} + \vec{V}_1\cdot\nabla v_1 = -f_o\hat{u}_1 - \beta y u_1, \qquad v_{3t} + \vec{V}_3\cdot\nabla v_3 = -f_o\hat{u}_3 - \beta y u_3,\\
&T_{2t} + \vec{V}_2\cdot\nabla T_2 - \frac{Sp_2}{R}\omega_2 = 0,\\
&\frac{u_3 - u_1}{\delta p} = \frac{R}{f_o p_2}T_{2y}, \qquad \frac{v_3 - v_1}{\delta p} = -\frac{R}{f_o p_2}T_{2x},\\
&\hat{u}_{1x} - \hat{u}_{3x} + \hat{v}_{1y} - \hat{v}_{3y} = \frac{2\omega_2}{\delta p}.
\end{aligned} \quad (6.59)$$

The superscript $\wedge$ refers to the ageostrophic component of velocity. Upon eliminating the time-dependent terms among these equations, we obtain the resulting equations

$$\begin{aligned}
&(\nabla^2 - 2\lambda^2)\omega_2 = -2\nabla\cdot\vec{Q},\\
&\vec{Q} = (Q_{(x)}, Q_{(y)}),\\
&Q_{(x)} = -\frac{R}{2Sp_2}\left(\frac{\partial}{\partial x}(\vec{V}_2\cdot\nabla T_2) + \frac{f_o p_2}{R\delta p}(\vec{V}_3\cdot\nabla v_3 - \vec{V}_1\cdot\nabla v_1 + \beta y(u_3 - u_1))\right),\\
&Q_{(y)} = -\frac{R}{2Sp_2}\left(\frac{\partial}{\partial y}(\vec{V}_2\cdot\nabla T_2) - \frac{f_o p_2}{R\delta p}(\vec{V}_3\cdot\nabla u_3 - \vec{V}_1\cdot\nabla u_1 - \beta y(v_3 - v_1))\right).
\end{aligned} \quad (6.60)$$

The units of $\vec{Q}$ are m^{-1} s^{-1} Pa. With $\tilde{\vec{Q}} = S\delta p\vec{Q}/V^2\lambda f_o$ and the other previously defined non-dimensional quantities, we may write the non-dimensional form of (6.60) as (after dropping the tilda notation):

$$\begin{aligned}
&(\nabla^2 - 2)\omega_2 = -2\nabla\cdot\vec{Q},\\
&Q_{(x)} = -\frac{1}{2}\left[\left(\frac{\delta p}{p_2}\right)^2\frac{\partial}{\partial x}(\vec{V}_2\cdot\nabla T_2) + \vec{V}_3\cdot\nabla v_3 - \vec{V}_1\cdot\nabla v_1 + \beta y(u_3 - u_1)\right],\\
&Q_{(y)} = -\frac{1}{2}\left[\left(\frac{\delta p}{p_2}\right)^2\frac{\partial}{\partial y}(\vec{V}_2\cdot\nabla T_2) - \vec{V}_3\cdot\nabla u_3 + \vec{V}_1\cdot\nabla u_1 + \beta y(v_3 - v_1)\right],\\
&u_j = -\frac{\partial\psi_j}{\partial y}, \qquad v_j = \frac{\partial\psi_j}{\partial x}, \qquad T_2 = \psi_1 - \psi_3.
\end{aligned} \quad (6.61)$$

This two-layer QG model is a simplest possible but very powerful tool for learning about the fundamental large-scale dynamics. In Chapter 8, we will use it to closely examine the intricacy of dynamic instability. In Chapter 9, it will be used for analyzing the dynamics of stationary waves. In Chapter 11, we will use it to learn about the nonlinear dynamics of large-scale waves.

6.7.1 Illustrative analysis of a baroclinic jet streak

We illustrate the intrinsic aspects of the quasi-geostrophic dynamics with an analysis of the short-term evolution of an idealized localized jet. This type of jet usually propagates as an entity at a speed slower than the wind. A jet streak can be as much as a thousand kilometers long. Its vertical structure usually spans the whole troposphere, being relatively weak at the lower levels and considerably stronger at the upper levels. In light of these gross characteristics, a two-layer QG model should be capable of capturing the essence of the dynamics of a baroclinic jet streak.

Let us consider a zonally oriented baroclinic jet streak in a reentrant channel of 30 000 km long and 10 000 km wide. We measure distance, velocity and time in units of the radius of deformation $\lambda^{-1} = \frac{\delta p\sqrt{S}}{f_o}$, V and $V^{-1}\lambda^{-1}$ respectively. The non-dimensional quantities are defined as follows: $(\tilde{x}, \tilde{y}) = \lambda(x, y)$, $\tilde{t} = V\lambda t$, $\tilde{\psi}_j = \frac{\psi_j \lambda}{V}$ for $(j = 1,3)$, $\tilde{\omega}_2 = \frac{\omega_2 f_o}{V^2\lambda^2\delta p}$, $\tilde{T}_2 = \frac{T_2 R p_2}{f_o V\lambda^{-1}\delta p}$, $\tilde{\beta} = \frac{\beta}{V\lambda^2}$. In particular, for $V = 10\,\mathrm{m\,s^{-1}}$, $\beta = 1.6\times10^{-11}\,\mathrm{m^{-1}\,s^{-1}}$, $p_2 = 600\times10^1\,\mathrm{N\,m^{-2}}$, $\delta p = 400\times10^2\,\mathrm{N\,m^{-2}}$, $S = 6.25\times10^{-6}\,\mathrm{N^{-2}\,m^6\,s^{-2}}$, $f_o = 10^{-4}\,\mathrm{s^{-1}}$, we would have $\lambda^{-1} = 10^6$ m and $\tilde{\beta} = 1.6$. A unit of time is then $V^{-1}\lambda^{-1} = 10^5\,\mathrm{s} \sim 1$ day. The units of Ψ, ω and T are $10^7\,\mathrm{m^2\,s^{-1}}$, $4\times10^{-4}\,\mathrm{mb\,s^{-1}}$ and 2.3 K respectively. The corresponding units of $w = -\frac{\omega}{g\rho}$ are $1\times10^{-2}\,\mathrm{m\,s^{-1}}$. The tilda superscript is dropped from here on for convenience. The non-dimensional domain is $-X \le x \le X$, $-Y \le y \le Y$ with $X = 15.0$, $y = 5$. The flow is depicted with 150 grid points in the x-direction and 101 grid points in the y-direction.

The idealized jet streak is prescribed with the following relative vorticity field, ζ_j,

$$\begin{aligned}
&\zeta_1 = Ay\exp\left(-y^2\right)F(x),\\
&F(x) = (1 + \tanh(a(x - x_W))) \text{ for } -X \le x \le 0,\\
&F(x) = (1 - \tanh(a(x - x_E))) \text{ for } X \ge x \ge 0,\\
&\zeta_3 = \varepsilon\zeta_1,\ \varepsilon < 1,
\end{aligned} \tag{6.62}$$

where $a = 2$, $A = 6$, $x_W = -2$, $x_E = 2$, $\varepsilon = 0.2$. It is located at the center of the domain, being symmetrical in both zonal and meridional directions. The relative vorticity field of this jet streak in the upper layer, ζ_1, is shown in Fig. 6.10. There is a region of positive (negative) values to the north (south) of the axis of the jet streak. The ζ_3 field is identical to ζ_1 except that it is weaker by a factor of five. This jet streak is embedded in a broad zonal baroclinic shear flow, U_j with $U_1 = 1.0$, $U_3 = 0.3$ which has zero relative vorticity.

Structural properties of the initial flow

The first task is to diagnose the initial structural characteristics of the flow. The streamfunction associated with the jet streak alone is $\psi'_j = \nabla^{-2}\zeta_j$ in symbolic form subject to the

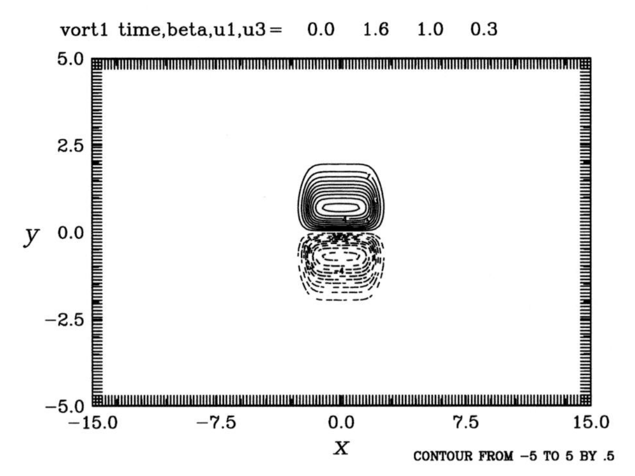

Fig. 6.10 Relative vorticity field in the upper layer of a zonally oriented baroclinic jet streak in units of $V\lambda = 10^{-5}\ \mathrm{s}^{-1}$.

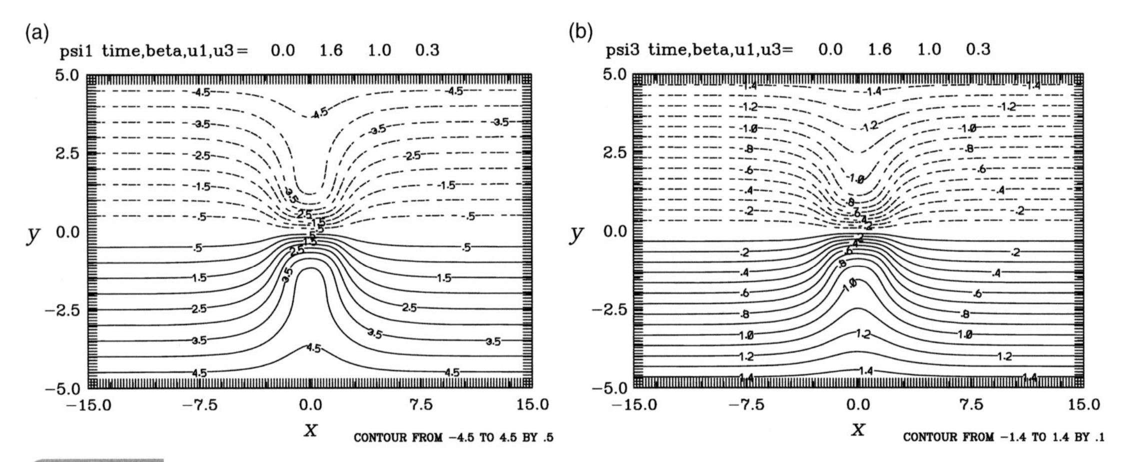

Fig. 6.11 Streamfunction field associated with a zonal jet streak in the (a) upper and (b) lower layer in units of $V\lambda^{-1} = 10^7\ \mathrm{m}^2\ \mathrm{s}^{-1}$. Zero contours not shown.

boundary conditions: $\psi_j'(-X, y) = \psi_j'(X, y)$ and $\psi_j'(x, \pm Y) = 0$. The total streamfunction field, ψ_j, is therefore

$$\psi_j = -U_j y + \nabla^{-2}\zeta_j, \quad j = 1, 3. \tag{6.63}$$

The ψ_j fields are shown in Fig. 6.11. The flow is confluent at the jet entrance region and diffluent at the jet exit region. The wind is strongest at the center of the jet streak with a maximum speed of $52\ \mathrm{m\,s}^{-1}$ in the upper layer and $11\ \mathrm{m\,s}^{-1}$ in the lower layer.

The non-dimensional temperature field is $(\psi_1 - \psi_3)$. Its structure (not shown for brevity) looks similar to Fig. 6.11(a) because ψ_1 is three times stronger than ψ_3.

Evolution of the jet streak

By numerically integrating the non-dimensional form of (6.57), we establish how this jet streak evolves over a few days. The algorithm consists of evaluating the streamfunction

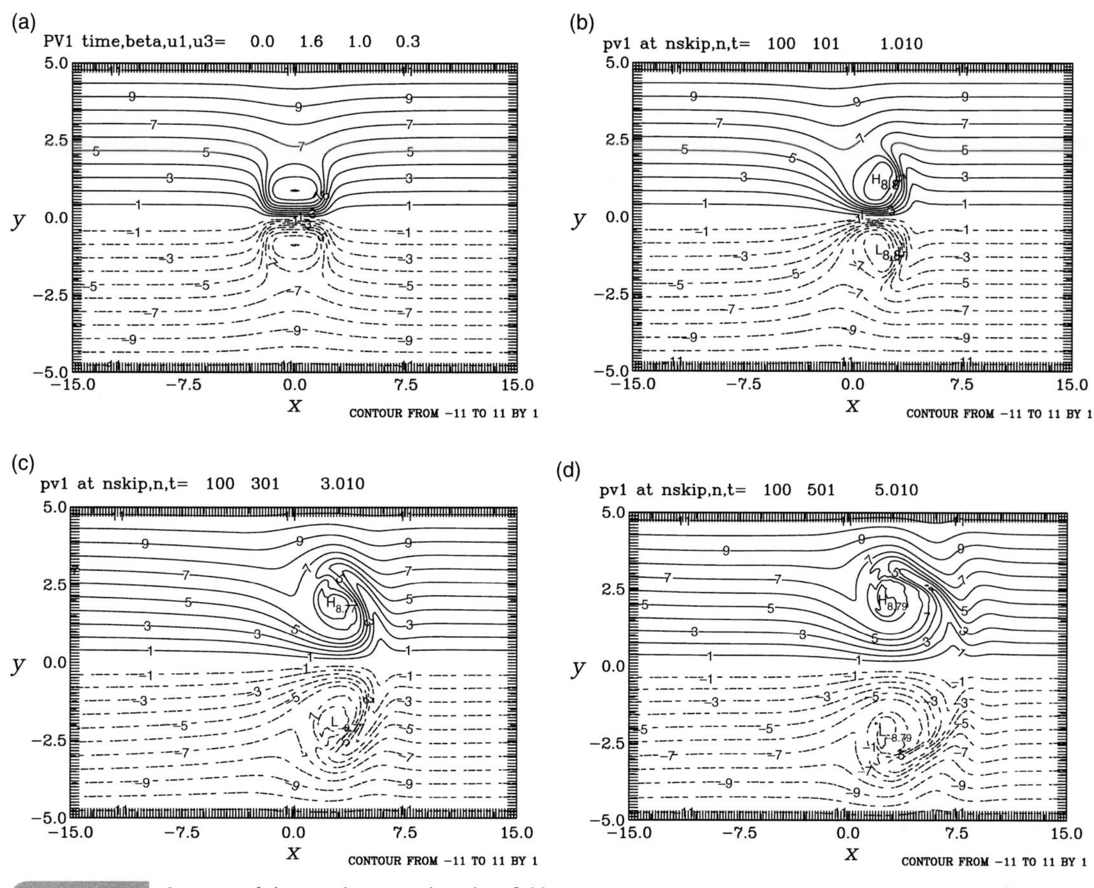

Fig. 6.12 Structure of the non-dimensional total q_1 field at $t = 0,1,3,5$.

tendency field, $\partial\psi_j(x,y)/\partial t$, at each time step on the basis of the ψ_j field and integrating a system of ordinary differential equations at each grid point in the form of

$$\frac{df}{dt} = g(f) \tag{6.64}$$

where $g(f)$ is a nonlinear function of f. Equation (6.64) is solved with a predictor-corrector scheme discussed in Section A.13 with a time step of $\Delta t = 0.01$. It corresponds to a 15-minute time step. All other properties of the flow at any time may be diagnosed if so desired.

The evolution is shown in terms of the changes in the structure of the non-dimensional PV in the upper layer over five days (Fig. 6.12). The local extreme values of q_1 near the jet streak are $\pm\, 8 \times 10^{-5}\ \text{s}^{-1}$ and those of q_3 are twenty times weaker, $\pm\, 0.4 \times 10^{-3}\ \text{s}^{-1}$. The value of the Coriolis parameter at a reference mid-latitude corresponding to $y = 0$ in the model, $f_o \sim 10$, has not been included in these plots. Including this constant value would make the total PV values positive everywhere.

These panels reveal that the jet streak progressively moves eastward. The changes of its structure (shape) are quite dramatic, from a zonally symmetrical localized structure to that of a "breaking wave" in five days. The ψ_1 field (not shown) has the appearance of a smoothed version of the q_1 field. The diffluent flow at the jet exit advects a narrow strip of relatively small values of PV air northwestward in the northern half of the domain and another counterpart strip southwestward in the southern half of the domain. At the same time, the confluent flow at the jet entrance region advects larger PV air toward the center of the jet streak in the north and smaller PV air in the south. The forward part of the jet streak appears to move faster than the hind part, leading to an enhancement of the diffluence in the front and a weakening of the confluence at the back. The strengthening diffluent flow progressively wraps the relatively low (high) PV air around a local pool of maximum (minimum) PV air. A new trough (ridge) develops to the south (north) in the front giving clear evidence of Rossby wave generation and propagation. The evolving PV field to the south of the jet axis is a mirror image of that to the north in each layer. By $t = 5$, the upper layer streamfunction field corresponding to a "breaking wave" in the PV field evolves to a cutoff low in the northern half and a cutoff high in the southern half of the domain.

It is noteworthy that the distribution of each property of this flow initially has either even or odd symmetry. The initial zonal symmetry with respect to the center of the jet does not persist under the influence of the beta effect. The intrinsic westward propagation of the constituent Rossby waves spontaneously breaks its zonal symmetry. But, there are no dynamical processes in the model that would alter its initial meridional symmetry. Consequently, we expect that while its initial zonal symmetry breaks as soon as it starts evolving, the flow remains symmetrical with respect to the axis of the jet at all times. This intrinsic aspect of symmetry is evident in Fig. 6.12.

The evolution of the secondary circulation in terms of the non-dimensional ω_2 is shown in Fig. 6.13.

The ω_2 field initially has the appearance of a pair of dipoles, one in the jet entrance region and the other in the jet exit region. It shows that the maximum ascent or descent is of the order of 1 cm s^{-1}. New dipoles emerge further downstream as the jet streak evolves in the form of a wave train. They propagate eastward slightly away from the axis of the jet. They are associated with baroclinic Rossby waves. While the constituent waves disperse during evolution, the flow as a whole is at the same time gradually attenuated partly due to the damping effect of the numerical scheme. From this figure one can deduce that an observer facing downstream of the flow would see a clockwise overturning circulation $(\hat{v}, -\omega)$, in the (y, z) cross-section over the jet exit region. There is ascending motion to the north, a secondary northerly flow component at the upper level and descending motion to the south. The Coriolis force acting on that northerly divergent flow component in the upper layer would decelerate the zonal velocity, thereby supporting the eastern portion of the jet streak. The Coriolis force acting on the southerly divergent flow component in the lower layer would accelerate the zonal flow there. These processes would decrease the vertical shear and make the exit part of the jet more barotropic. This same observer also would see an overturning circulation of anticlockwise direction in the (y,z) cross-section over the jet entrance region. The Coriolis force acting on such a southerly divergent flow component in the upper layer would accelerate the zonal velocity, thereby supporting the eastward movement of the jet streak. The Coriolis force acting on such a northerly divergent flow component in the lower layer would decelerate the zonal velocity there.

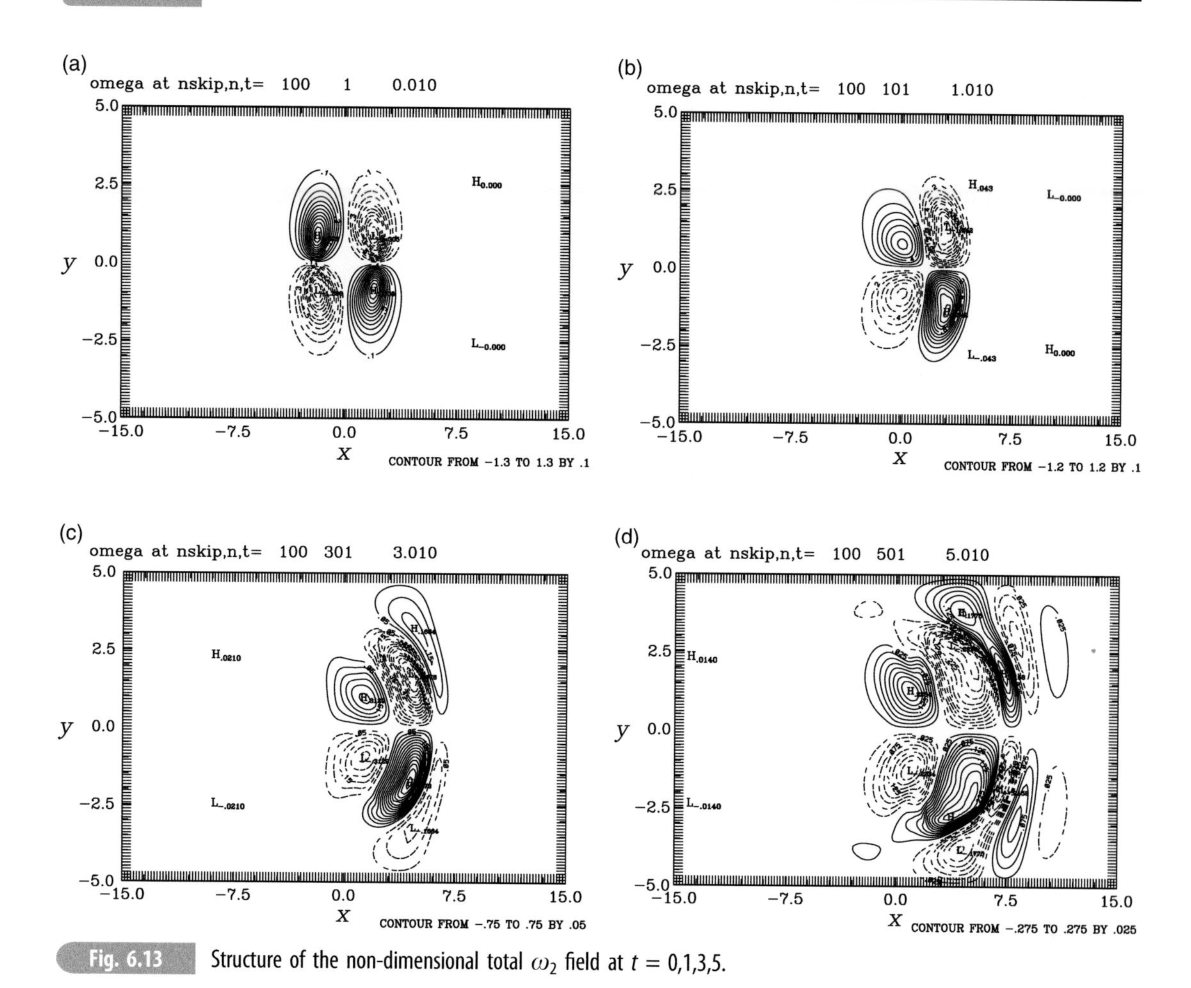

Fig. 6.13 Structure of the non-dimensional total ω_2 field at t = 0,1,3,5.

The area of secondary circulation at the jet entrance region becomes more zonally elongated, whereas that in the jet exit region becomes progressively more compressed. In other words, the secondary circulation plays a crucially important dynamical role in changing the shape of the evolving jet streak.

Initial advective properties and their implications

We get some insight into why the flow evolves in the way it does by examining all advective properties of the flow. It suffices to simply examine the advective properties of the initial flow. The distribution of the advection of PV, $(-J(\psi_j,q_j))$, are shown in Fig. 6.14. In the jet entrance region at the upper level, there is a relatively small region of positive (negative) PV advection to the southwest (northwest). The pattern of PV advection is symmetrically opposite in the jet exit region but with a more complex distribution. This result suggests that the jet streak would propagate downstream (eastward in Fig. 6.12a). The PV advection in the lower layer is about thirty times smaller, but with more complex

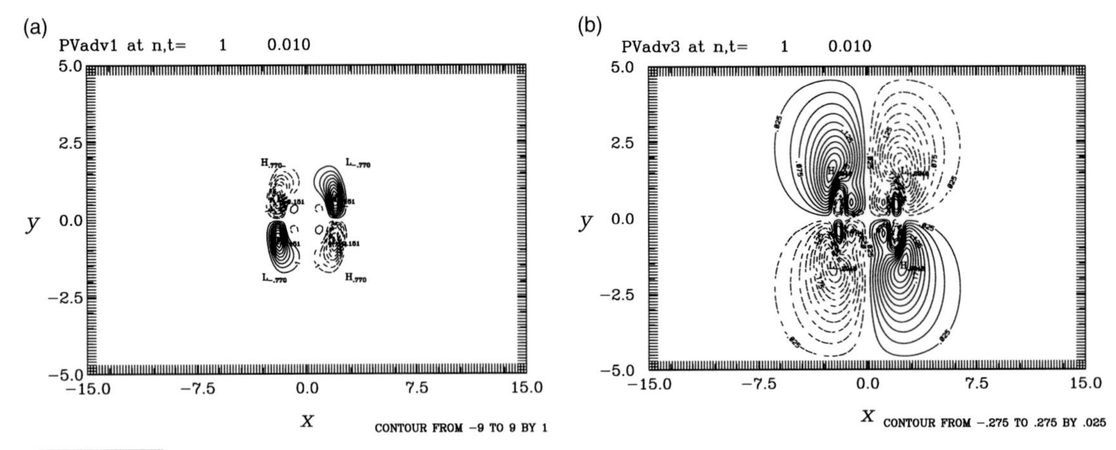

Fig. 6.14 Distribution of advection of potential vorticity, $(-J(\psi_j,q_j))$, in (a) upper layer and (b) lower layer in units of $V^2\lambda^2 = 10^{-10}\,s^{-2}$.

details. There are counterpart features similar to that in the upper layer next to the jet streak, but these features are embedded in a much broader pattern of PV advection of opposite sign in the far field of the jet streak. The result also suggests a downstream propagation of the jet streak at the lower level.

The implication of these PV advection fields is ascertained by computing the initial tendency of the ψ_1 and ψ_3 fields. The results would validate the initial evolution of q_1 as seen in Fig. 6.12. The QG-PV equations for the two layers can be rewritten in terms of the *tendency of a barotropic streamfunction*, $\eta \equiv \frac{\partial}{\partial t}(\psi_1 + \psi_3)$, and the *tendency of a baroclinic streamfunction*, $\xi \equiv \frac{\partial}{\partial t}(\psi_1 - \psi_3)$, as

$$\begin{aligned} \nabla^2\eta &= -J(\psi_1, q_1) - J(\psi_3, q_3), \\ (\nabla^2 - 2\lambda^2)\xi &= -J(\psi_1, q_1) + J(\psi_3, q_3). \end{aligned} \tag{6.65a,b}$$

Using the results of the PV advection in the two layers, we solve (6.65) for η and ξ subject to the same boundary conditions for the streamfunction itself. The tendencies of the streamfunction in the two layers are then

$$\frac{\partial\psi_1}{\partial t} = \frac{1}{2}(\eta + \xi), \quad \frac{\partial\psi_3}{\partial t} = \frac{1}{2}(\eta - \xi). \tag{6.66}$$

The tendencies in both layers are quite similar in the near neighborhood of the jet streak because the controlling factor is the large PV advection in the upper layer. The results shown in Fig. 6.15 suggest that the jet streak in each layer would move eastward in agreement with what we have inferred from the PV advection directly. The values in the upper layer are about five times larger.

The thermal advection $(-J(\psi_2, \psi_1 - \psi_3))$ at the mid-level is shown in Fig. 6.16. In the jet entrance region, there is an area of cold (warm) advection to the southwest (northwest). There is opposite distribution of thermal advection in the jet exit region. Such thermal advection would suggest an initial eastward displacement of the temperature field as well. It is then compatible with the eastward propagation of the jet streak itself.

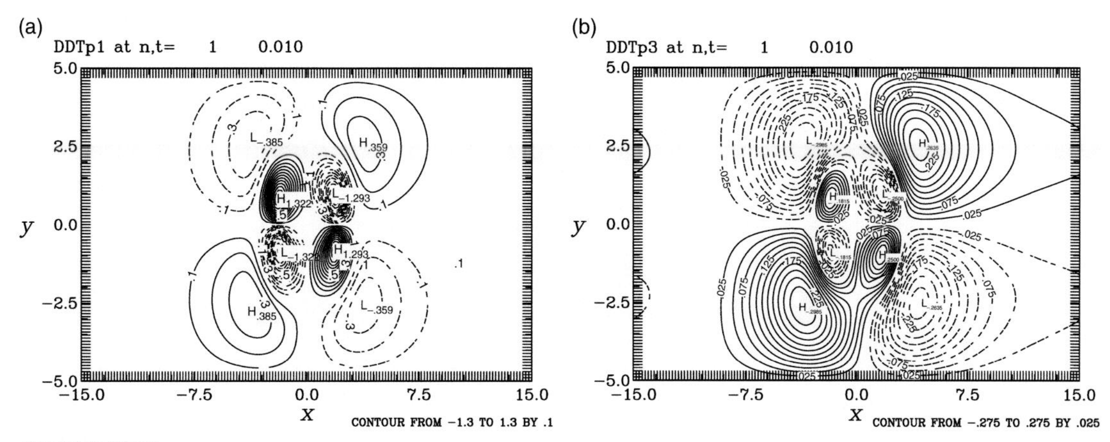

Fig. 6.15 Initial tendency of the non-dimensional streamfunctions of a zonal jet streak, (a) $\partial\psi_1/\partial t$ and (b) $\partial\psi_3/\partial t$ in units of $V^2 = 100\,\mathrm{m}^2\,\mathrm{s}^{-2}$.

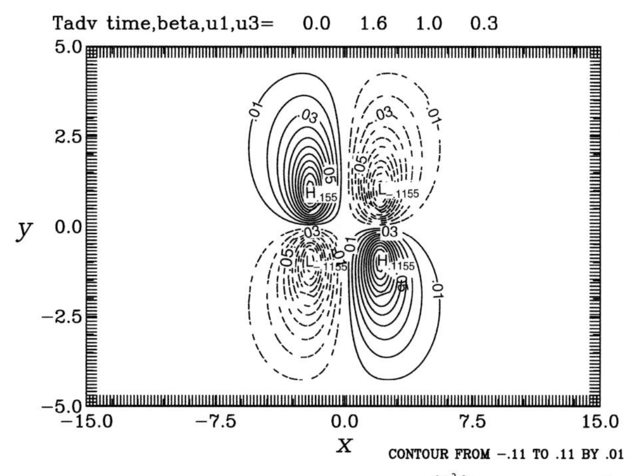

Fig. 6.16 Distribution of the thermal advection, $-J(\psi_2,\psi_1-\psi_3)$, in units of $\frac{f_0 V^2 \delta p}{R p_2} = 2.5\times10^{-5}\,\mathrm{K\,s}^{-1}$.

The absolute vorticity advection in each layer, $(-J(\psi_j,\beta y+\nabla^2\psi_j))$, differs from the advection of potential vorticity $(-J(\psi_j,q_j))$ by a term $\pm\, J(\psi_1,\psi_3)$ which is small in comparison. Those fields are not shown for brevity. The initial vertical velocity field arises from those fields of thermal advection and vorticity advection. It is found that the net forcing in the omega equation has a region of significant positive (negative) values in the southwest and northeast (northwest and southeast) sectors next to the jet streak. We would then expect negative (positive) values of ω_2 in those two sectors. Rising motion should be then expected in the southwest sector and descending motion in the northwest sector in the jet entrance region. For the same reason, we expect ascent in the northeast sector and descent in the southeast sector in the jet exit region. This reasoning confirms the initial omega field shown in Fig. 6.13(a).

We also compute the distribution of the Q-vector field of this zonal baroclinic jet streak. The Q-vectors shown in Fig. 6.17 are strong in the entrance and exit regions. They are

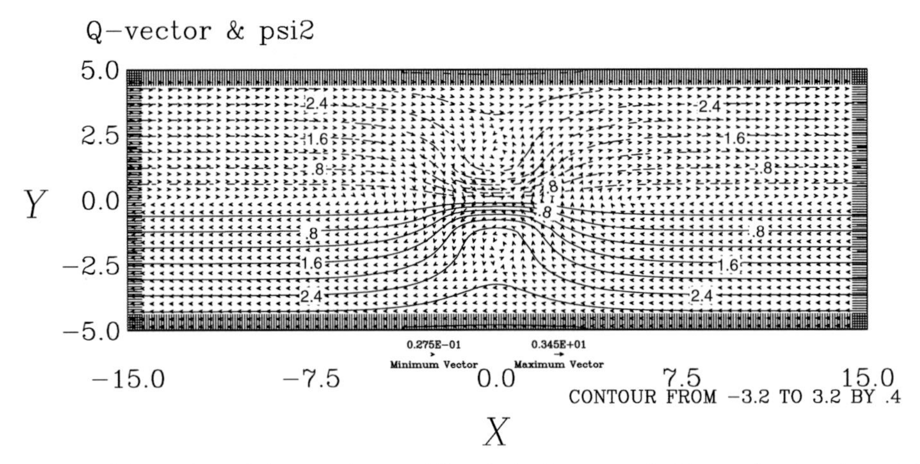

Fig. 6.17 The Q-vector field together with the streamfunction field at the level-2 associated with a zonally oriented baroclinic jet streak.

divergent in the northwest and southeast regions next to the jet streak. It follows according to (6.61) that these should be regions of descending motion. On the other hand, the Q-vectors are convergent in the southwest and northeast regions next to the jet streak. Those are then regions of ascent. These features of the ω_2 field inferred from the Q-vector field are indeed consistent with those seen in Fig. 6.13(a) corroborating the fact that (6.58) and (6.61) are equivalent. The advantage of (6.61) is that it is the most compact form of the omega equation and highlights the need of considering the influences of vorticity advection and thermal advection jointly.

6.8 Influences of the Earth's sphericity in the QG theory

A flow with a length scale longer than the Earth's radius is still quasi-geostrophic in nature. We would want to analyze such a flow with a quasi-geostrophic model in a global domain so that the full impact of the Earth's sphericity would be incorporated. Such a flow would be in *local geostrophic balance* at all latitudes. The geostrophic velocity component of the flow then is said to satisfy

$$0 = -\nabla Z - f\vec{k} \times \vec{V}_g \tag{6.67}$$

in isobaric coordinates. Here we use a variable Coriolis parameter, $f = 2\Omega \sin \varphi$ with φ being the latitude; ∇ refers to the *horizontal gradient operator*, $\vec{k}$ a unit vector in the local vertical direction, $Z = gz$ with z being the height of pressure surfaces, $\vec{V}_g = (u_g, v_g)$. It follows from (6.67) $\nabla \cdot (\vec{V}_g f) = \nabla \cdot (\vec{k} \times \nabla Z) = 0$ implying that the horizontal divergence of this geostrophic velocity is not zero

$$\nabla \cdot \vec{V}_g = -\frac{1}{f}(\vec{V}_g \cdot \nabla f) = -\frac{v_g \cot \varphi}{a}. \tag{6.68}$$

Although we have $\vec{V}_g \gg \vec{V}_a$, $\nabla \cdot \vec{V}_g$ and $\nabla \cdot \vec{V}_a$ have comparable magnitude since both are of the order $10^{-6}\ \mathrm{s}^{-1}$, particularly so at lower latitudes. From (6.67) we also get

$$\frac{\partial \vec{V}_g}{\partial p} \cdot \nabla \left(\frac{\partial Z}{\partial p} \right) = 0. \tag{6.69}$$

We will make use of (6.69) shortly. The continuity equation is

$$\nabla \cdot \left(\vec{V}_a + \vec{V}_g \right) + \frac{\partial \omega}{\partial p} = 0, \tag{6.70}$$

where $\vec{V}_a = (u_a, v_a)$ refers to the ageostrophic velocity component. Note that the divergence of the geostrophic velocity is retained in (6.70). On the basis of scaling considerations, the consistent form of quasi-geostrophic vorticity equation in this case is

$$\left(\frac{\partial}{\partial t} + \vec{V}_g \cdot \nabla \right) (\zeta + f) + (\nabla \cdot \vec{V}_g + \nabla \cdot \vec{V}_a) f = 0, \tag{6.71}$$

where $\zeta = \vec{k} \cdot \nabla \times \vec{V}_g$. We retain the effect of vortex stretching due to divergence of the geostrophic flow through the additional term $\left(\nabla \cdot \vec{V}_g \right) f$. The QG thermodynamic equation is

$$\left(\frac{\partial}{\partial t} + \vec{V}_g \cdot \nabla \right) \frac{\partial Z}{\partial p} + \omega S = 0, \tag{6.72}$$

where $S = -\frac{R}{p} \left(\frac{p_{oo}}{p} \right)^{cv/cp} \frac{d\Theta}{dp}$ represents the stratification with θ being a reference potential temperature profile unrelated to the disturbance. We still treat S as only a function of p. Equations (6.69), (6.70), (6.71), (6.72) are a closed set of four equations for four unknowns $\left(\vec{V}_g, \nabla \cdot \vec{V}_a, \omega, Z \right)$. The equations are written in vector notation so that they are applicable to any horizontal coordinates. From (6.72), we get

$$f \frac{\partial \omega}{\partial p} = -f \frac{\partial}{\partial p} \left(\frac{1}{S} \left(\frac{\partial}{\partial t} + \vec{V}_g \cdot \nabla \right) \frac{\partial Z}{\partial p} \right). \tag{6.73}$$

It follows from (6.70), (6.71) and (6.73)

$$\begin{aligned} f(\nabla \cdot \vec{V}_g + \nabla \cdot \vec{V}_a) &= f \frac{\partial}{\partial p} \left(\frac{1}{S} \left(\frac{\partial}{\partial t} + \vec{V}_g \cdot \nabla \right) \frac{\partial Z}{\partial p} \right) \\ &= \left(\frac{\partial}{\partial t} + \vec{V}_g \cdot \nabla \right) \left(\frac{\partial}{\partial p} \left(\frac{f}{S} \frac{\partial Z}{\partial p} \right) \right) - (\vec{V}_g \cdot \nabla f) \frac{\partial}{\partial p} \left(\frac{1}{S} \frac{\partial Z}{\partial p} \right). \end{aligned} \tag{6.74}$$

Upon substituting (6.74) into (6.71), we effectively eliminate ω between (6.71) and (6.72). The resulting equation is the *QG potential vorticity equation in a spherical domain* containing only one unknown Z, viz.

$$\left(\frac{\partial}{\partial t} + \vec{V}_g \cdot \nabla \right) \left(\zeta + f + \frac{\partial}{\partial p} \left(\frac{f}{S} \frac{\partial Z}{\partial p} \right) \right) + (\nabla \cdot \vec{V}_g) \frac{\partial}{\partial p} \left(\frac{f}{S} \frac{\partial Z}{\partial p} \right) = 0. \tag{6.75}$$

It would be reduced to the QG-PV equation for a beta-plane by setting $\nabla \cdot \vec{V}_g = 0$ and replacing f by f_o in the third component of the PV expression. In considering the full sphericity of the Earth, we retain the advection of planetary vorticity by the geostrophic flow component and the vortex stretching associated with the divergence of the geostrophic velocity. The two influences can be represented in one single term as $(\vec{V}_g \cdot \nabla f)\left(1-\frac{\partial}{\partial p}\left(\frac{1}{S}\frac{\partial Z}{\partial p}\right)\right)$. Therefore (6.75) differs from the QG-PV equation for beta-plane geometry in that the advection of planetary vorticity is weighted by a factor of $\left(1-\frac{\partial}{\partial p}\left(\frac{1}{S}\frac{\partial Z}{\partial p}\right)\right)$.

In spherical coordinates, the relation between the geostrophic vorticity and geopotential is

$$\begin{aligned}\zeta &= \frac{1}{a\cos\varphi}\frac{\partial v_g}{\partial \lambda} - \frac{1}{a\cos\varphi}\frac{\partial}{\partial\varphi}\left(u_g\cos\varphi\right)\\ &= \frac{1}{f}\left(\frac{1}{a^2\cos^2\varphi}\frac{\partial^2 Z}{\partial\lambda^2} + \frac{\tan\varphi}{a^2}\frac{\partial}{\partial\varphi}\left(\frac{\partial Z}{\partial\varphi}\cot\varphi\right)\right)\\ &\equiv \frac{1}{f}L\{Z\}.\end{aligned} \tag{6.76}$$

Note that because of the variation of the Coriolis parameter with latitude, we cannot express the velocity components and vorticity in terms of a geostrophic streamfunction as we do when we use beta-plane geometry. We must instead use the geopotential Z as the unknown dependent variable. The operator $L\{\}$ takes the place of the Laplacian operator. By eliminating the time derivative term between (6.71) and (6.72) we get an extended QG omega equation as

$$f^2\frac{\partial^2\omega}{\partial p^2} + SL\{\omega\} = f\frac{\partial}{\partial p}\left((\vec{V}_g\cdot\nabla)(\zeta+f)\right) - L\left\{(\vec{V}_g\cdot\nabla)\frac{\partial Z}{\partial p}\right\}. \tag{6.77}$$

It differs from the counterpart in beta-plane geometry in two ways. The coefficient of the $\frac{\partial^2\omega}{\partial p^2}$ term is f^2 instead of f_o^2.

6.8.1 Fundamental modes of quasi-geostrophic motion on a sphere

The linearized form of (6.75) about a basic state at rest governs a perturbation in the spherical QG system,

$$\frac{\partial}{\partial t}\left(L\{Z\} + f\frac{\partial}{\partial p}\left(\frac{f}{S}\frac{\partial Z}{\partial p}\right)\right) + \frac{fv_g}{a}\frac{df}{d\varphi} = 0. \tag{6.78}$$

Note that the influence of divergence involves the product of two perturbation quantities and is therefore a nonlinear term. As such, it has negligible effect on the fundamental modes and does not appear in (6.78).

It is convenient to use the following non-dimensionalization:

$$\tilde{t} = 2\Omega t,\ \tilde{Z} = \frac{Z}{(2\Omega\, a)^2},\ \tilde{p} = \frac{p}{p_{oo}},\ \tilde{S} = \frac{Sp_{oo}^2}{(2\Omega\, a)^2},\ 0\le\tilde{p}\le 1,\ 0\le\lambda\le 2\pi,\ -\frac{\pi}{2}\le\varphi\le\frac{\pi}{2}.$$

The non-dimensional form of (6.78) is

$$\begin{aligned}&\frac{\partial}{\partial \tilde{t}}\left(\tilde{L}\{\tilde{Z}\}+\sin^2\varphi\frac{\partial}{\partial \tilde{p}}\left(\frac{1}{\tilde{S}}\frac{\partial \tilde{Z}}{\partial \tilde{p}}\right)\right)+\frac{\partial \tilde{Z}}{\partial \lambda}=0,\\ &\tilde{L}\{\tilde{Z}\}=\left(\frac{1}{\cos^2\varphi}\frac{\partial^2}{\partial\lambda^2}+\frac{\partial^2}{\partial\varphi^2}-\frac{1}{\cos\varphi\sin\varphi}\frac{\partial}{\partial\varphi}\right)\{\tilde{Z}\}.\end{aligned} \tag{6.79}$$

Separation of variables suggests a solution of (6.79) in the form of

$$\tilde{Z}=\xi(\tilde{p})\Psi(\varphi)\exp(i(m\lambda-\sigma t)), \tag{6.80}$$

where m is an integer, the azimuthal wavenumber and σ the frequency. Substituting (6.80) into (6.79) leads to a vertical structure equation,

$$\frac{d}{d\tilde{p}}\left(\frac{1}{\tilde{S}}\frac{d\xi}{d\tilde{p}}\right)=\varepsilon\xi, \tag{6.81}$$

where ε is the separation constant. The boundary conditions would be

$$\begin{aligned}&\frac{d\xi}{d\tilde{p}}=0 \text{ at } \tilde{p}=1,\\ &\xi \text{ finite as } \tilde{p}\to 0.\end{aligned} \tag{6.82}$$

There is a set of eigenvalues ε for each specified profile of the static stability $\tilde{S}(\tilde{p})$. This eigenvalue is called *equivalent depth*.

The horizontal structure equation is

$$\begin{aligned}&\frac{d^2\Psi}{d\varphi^2}-\frac{1}{\cos\varphi\sin\varphi}\frac{d\psi}{d\varphi}-\left(\frac{m^2}{\cos^2\varphi}+\frac{m}{\sigma}-\varepsilon\sin^2\varphi\right)\Psi=0,\\ &\varphi=\pm\pi/2, \quad \Psi=0.\end{aligned} \tag{6.83}$$

With $\mu=\sin\phi$, (6.83) can be rewritten as

$$\begin{aligned}&(1-\mu^2)\frac{d^2\Psi}{d\mu^2}-\left(\mu+\frac{1}{\mu}\right)\frac{d\Psi}{d\mu}-\left(\frac{m^2}{1-\mu^2}+\frac{m}{\sigma}-\mu^2\varepsilon\right)\Psi=0,\\ &\Psi(\pm 1)=0.\end{aligned} \tag{6.84}$$

Equation (6.84) defines another eigenvalue–eigenfunction problem with σ being the eigenvalue for a given ε and m. The symmetry property of (6.84) with respect to μ might give one the impression that its solution could be either an even or an odd function of μ. However, only the odd solutions are compatible with the geostrophic relations. It amounts to requiring $\psi=0$ and $d\Psi/d\mu=0$ at $\mu=0$ in order to be compatible with a finite velocity at the equator. Since the even solutions are not zero at $\mu=0$ they are not admissible solutions. The solutions must then be constructed as a linear combination of the normalized associated Legendre polynomial, $P_n^m(\mu)$ with $n-m=$ odd integer,

$$\Psi=\sum_{\substack{n=|m|+1\\(2)}} B_n^m P_n^m. \tag{6.85}$$

Here n indicates the number of nodes of this polynomial between the two poles. The notation (2) below the summation sign refers to an increment of 2. We substitute (6.85) into (6.84) and make use of the following properties of $P_n^m(\mu)$:

$$(1-\mu^2)\frac{d^2P_n^m}{d\mu^2} = 2\mu\frac{dP_n^m}{d\mu} - \left[n(n+1) - \frac{m^2}{1-\mu^2}\right]P_n^m,$$

$$(1-\mu^2)\frac{dP_n^m}{d\mu} = (n+1)d_n^m P_{n-1}^m - m d_{n+1}^m P_{n+1}^m,$$

$$\mu P_n^m = d_{n+1}^m P_{n+1}^m + d_n^m P_{n-1}^m,$$

$$\text{with } d_n^m = \sqrt{\frac{n^2-m^2}{4n^2-1}}.$$

We would get a recurrence relation. The values of σ are such that they would make the determinant of the coefficient matrix of infinite rank vanish. In general, these eigenvalues can only be evaluated numerically.

Fortunately, analytic results for both eigenvalues and eigenfunctions can be obtained for the special case of $\varepsilon = 0$. It corresponds to the case of an infinite equivalent depth. The eigenvalues of (6.84) for this case are

$$\sigma = -\frac{m}{n^2}, \tag{6.86}$$

for $n = |m| + 2$, $|m| + 4$, ... The corresponding eigenfunctions are

$$\Psi = A_{n-1}^m P_{n-1}^m + A_{n+1}^m P_{n+1}^m. \tag{6.87}$$

With the ratio of the two coefficients being

$$\frac{A_{n-1}^m}{A_{n+1}^m} = -\left(\frac{d_{n+1}^m g_{n+2}^m}{d_n^m g_{n-1}^m}\right), \quad g_n^m = -n^2 - \frac{m}{\sigma}. \tag{6.88}$$

This analytic solution for the special case should be compared with the exact solution of the fundamental modes in a non-divergent barotropic fluid on a sphere. The latter is known as a Rossby–Haurwitz (RH) wave, which has a dispersion relation

$$\sigma = -\frac{m}{(n+1)n}, \quad \text{for integers } n \geq |m|. \tag{6.89}$$

A Rossby–Haurwitz wave has a structure

$$\Psi_{RH} = P_n^m. \tag{6.90}$$

Thus, the difference between a quasi-geostrophic mode and a corresponding RH mode is larger for smaller n (i.e. larger meridional scale). It is generally small enough that quasi-geostrophic theory can be meaningfully applied for studying large-scale motions even in a global domain.

7 Dynamic adjustment

Dynamic adjustment refers to the simultaneous changing of the pressure and velocity fields of a large-scale flow from an initially unbalanced state to a new balanced state. The conceptual issues concerning such adjustment are discussed in Section 7.1. In Sections 7.2 and 7.3, we solve the adjustment problem for two canonical forms of initial unbalanced disturbance in the context of a shallow-water model: a disturbance that has a velocity field without pressure variations or vice versa. We show that the new balanced state can be deduced without having to work out the transient evolution by virtue of conservation of potential vorticity in each fluid column. The analytic solutions reveal that whether the mass field or the velocity field would do most of the adjustment depends strongly on the scale of the initial disturbance. An extension of the geostrophic adjustment analysis to a gradient-wind adjustment analysis is finally discussed in Section 7.4.

7.1 Problem of rotational adjustment

From the perspective of large-scale geophysical flow, dynamic adjustment is synonymous with *rotational adjustment*, which is a generic term for the process of establishing a near-balance state from an imbalanced state in a rotating fluid. The observed large-scale wind vectors everywhere on an upper-level isobaric surface in mid-latitude are always approximately parallel to the height contours as evidenced in Fig. 2.5. In other words, a geostrophic balance between the wind field and the pressure field appears to be prevalent all the time. It is actually more precise to say that they are approximately in gradient-wind balance since the curvature effect is significant in the close neighborhood of a trough or ridge as we have discussed in Section 2.5.

There is a subtle distinction between the continual evolution of a quasi-balance flow in a closed system and the rapid adjustment of the pressure and wind fields in response to an impulsive local external forcing. The former can be adequately described in the context of *quasi-geostrophic theory* elaborated at length in Chapter 6. According to that theory, the rotational wind is much stronger and remains in quasi-balance all the time due to the effectively instantaneous feedback influence of the much weaker divergent wind. Internal gravity waves do not play a role at all in such evolution. The situation is qualitatively different when a localized external forcing of impulsive nature significantly disrupts the balance between the pressure field and wind field. For example, heavy precipitation in a locale would release a large amount of latent heat over a short time interval. It would momentarily change the local configurations of pressure and temperature but not the wind, and thereby the pre-existing balance. Internal gravity waves would be inevitably excited. They would quickly propagate to distant locations possibly leaving behind a new balanced flow in the original location of unbalanced disturbance.

What would the new balanced state be like for a given initial unbalanced state in a geophysical fluid? Rossby (1938) hypothesized that it is possible to deduce the adjusted state without having to work out the intermediate states. This is the classical problem of *geostrophic adjustment*. We can visualize two diametrically different types of unbalanced initial states. The initial disturbance of one type only consists of a velocity field but no pressure field. In contrast, the initial disturbance of the other type only consists of a pressure field but no velocity field. These are two canonical unbalanced initial states. Rossby tackled the first problem, although textbooks typically go over a linearized analysis of the second problem and call it the "Rossby problem." It is historically more accurate to refer to the adjustment problem for the first canonical initial unbalanced state as the "Rossby problem," and to refer to the adjustment problem for the second canonical initial unbalanced state as the "complementary Rossby problem." We examine both problems in this chapter.

A more general form of rotational adjustment is *gradient-wind adjustment*. It is a technically more difficult problem because the relationship between the centrifugal force and velocity is nonlinear. We will also discuss an analysis of such an adjustment in the context of an unbalanced axisymmetric vortex in this chapter.

The transient aspect of the adjustment problem entails inertial oscillation as well as gravity wave excitation and propagation. There may even be turbulence. An investigation of such detailed motions would require a mathematically more involved analysis. Evidence suggests that the transient stage of adjustment is relatively brief. As such, it ought to be a secondary aspect of the adjustment process. This does not mean we should belittle the transient part of the adjustment process because meso-scale disturbances with severe weather consequences are sometimes excited during geostrophic adjustment. This topic lies beyond the scope of this chapter.

7.2 Rossby problem of geostrophic adjustment

The pressure field stems from the distribution of mass of air in the atmosphere by virtue of hydrostatic balance. Under the influence of rapid rotation of the Earth, the large-scale mass field (pressure field) and velocity field cannot help but be in a quasi-balance state most of the time. Rossby (1938) posed an adjustment problem in the form of a localized ocean current in an unbounded shallow-water model (Section 3.8.4). The fluid in an infinite strip of width equal to $2a$ is initially moving at a constant speed in a particular direction designated here as the y-direction (northward; $v_o = V$) and is at rest elsewhere ($v_o = 0$). The initial thickness of the layer is assumed to be uniform everywhere, $h_o = H$. This initial state is clearly not in balance. The task is to deduce the flow configuration at large time.

7.2.1 An exact analysis

Friction may be assumed to have negligible influence in the subsequent response if the adjustment time is relatively short compared to the diffusion time scale. The upper panel of Fig. 7.1 is a schematic of the initial state under consideration. It is possible to deduce the adjusted state solely on the basis of the unbalanced initial state without having to work out the intermediate states. Such a possibility stems from the fact that the initial and final states are inseparably linked by a fundamental dynamical constraint.

We may qualitatively anticipate the following development. The initial northward flow in the central strip in the northern hemisphere would be deflected eastward by the Coriolis force immediately after $t = 0$. This would create mass convergence next to its eastern boundary and mass divergence next to its western boundary. The elevation of the fluid surface would then rise in the east and sink in the west. This change of pressure field is a reaction against the initial tendency. The resulting pressure gradient force would in turn set the fluid in motion in the outer regions of the domain. In other words, the initial imbalance would induce oscillatory motion of the fluid in the central strip and excite external gravity waves in the outer regions. The gravity waves would begin to propagate outward to far distance. The mass field and velocity field in the whole system would continually adjust until a new steady balanced state eventually emerges. The rectilinear geometry of the system suggests that the adjusted state must be in geostrophic balance in a near neighborhood of the central strip.

Since the initial state is symmetric in the x-direction with respect to the center of moving fluid, the Coriolis force and gravity do not alter the intrinsic symmetry of the system. That point of symmetry is designated as $x = 0$ in the adjusted state without loss of generality. The surface elevation at $x = 0$ should be H by symmetry argument. The fluid in a nearly balanced flow of a shallow-water model moves as a column. For the reason mentioned earlier, we may expect odd-symmetry in the surface elevation of the adjusted state with respect to the center point and even-symmetry in the corresponding velocity field. But, the

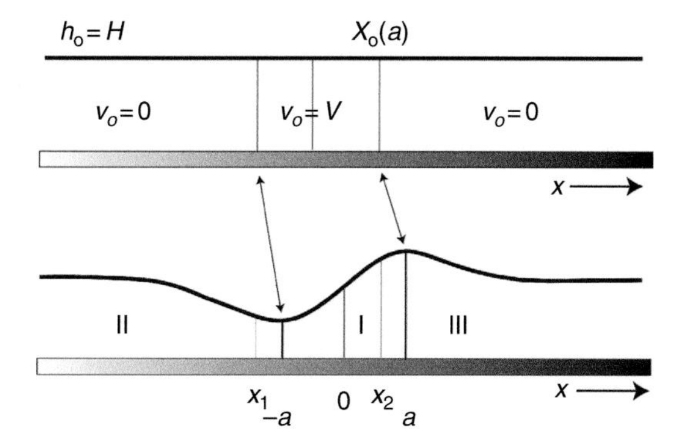

Fig. 7.1 Sketches of the initial state (upper panel) and adjusted state (lower panel) of a shallow-water model. The locations of the boundaries of the adjusted state are labeled as $\pm a$. The moving fluid of the initial state is located in the area $x_1 \leq x \leq x_2$ of the final state. See text for details.

central fluid column in the adjusted state cannot be identified with the central fluid column in the initial state. The fluid columns at the two boundaries of the central strip are $2a$ apart in the adjusted state. Their locations are designated as $x = \pm a$. A particular fluid column would be displaced eastward by a certain unknown distance and become shorter and thereby wider as required by mass conservation. The adjustment problem amounts to determining the zonal displacement of all fluid columns as well as their length and velocity at large time. The lower panel of Fig. 7.1 indicates the general characteristic of the surface elevation in the adjusted state. The domain of the adjusted state has three distinct regions: I, II and III. The symmetry consideration and the incompressibility of the fluid suggest that the width of region I also has a width of $2a$.

Rossby briefly described a procedure for obtaining an exact solution in a footnote of his paper, but he did not carry it out. What he presented is an approximate solution in the two regions outside the central strip with the assumption that the velocity in the central strip remains uniform. He closed the problem by imposing a requirement that the absolute momentum of the fluid in the central strip conserves. Mihaljan (1963) reported the first exact solution of this problem.

In this section, we present a complete analysis of this geostrophic adjustment problem. The equation of motion in the y-direction for a fluid column of unit mass without friction may be used as a starting point of an analysis, viz.

$$\frac{dv}{dt} = -fu. \tag{7.1}$$

Since $u = dx/dt$, we get upon integrating (7.1) from $t = 0$ to a certain large time when a balanced state has been essentially established,

$$v - v_o = -f(X - X_o), \tag{7.2}$$

where $X_o(x)$ is the original position of a fluid column which is located at $X = x$ in the adjusted state; v_o is the initial velocity of the fluid column and should be thought of as a function of X_o. Equation (7.2) is a statement of conservation of absolute momentum in the y-direction of a fluid parcel of unit mass.

By mass conservation, the length (h) and width (Δx) of a fluid column in the adjusted state are related to the counterparts at $t = 0$ by

$$h\Delta x = h_o \Delta X_o \rightarrow h = h_o \frac{dX_o}{dx}. \tag{7.3}$$

Since the flow in the adjusted state is steady and unidirectional, it must be in geostrophic balance. Hence, we have

$$v = \frac{g}{f}\frac{dh}{dx} = \frac{gh_o}{f}\frac{d^2X_o}{dx^2}. \tag{7.4}$$

Writing $\xi \equiv X - X_o$ as the *displacement of a fluid column* in the x-direction from its original location, we treat it as a dependent variable. Then we can identify the location of the left (right) boundary of the originally moving fluid as $x_1 = -a - \xi^*$ ($x_2 = a - \xi^*$) with $\xi^* = X(a) - X_o(a)$. We obtain from (7.2) and (7.4)

$$\frac{gh_o}{f}\frac{d^2\xi}{dx^2} - f\xi = -v_o. \tag{7.5}$$

It is convenient to measure x and ξ in units of a, v in units of V and h in units of H. We thus define non-dimensional variables: $\tilde{x} = x/a$, $\tilde{\xi} = \xi/a$, $\tilde{v} = v/V$ and $\tilde{h} = \frac{h}{H}$. So the boundaries of region I of the adjusted state are $\tilde{x} = \pm 1$. Region II is $-\infty \leq \tilde{x} \leq -1$ and region III is $1 \leq \tilde{x} \leq \infty$. There are effectively four parameters in this system: f, a, gH, V. Therefore, there are two independent non-dimensional parameters: $\lambda = fa/\sqrt{gH}$ and $R = V/fa$. Here R is a Rossby number characterizing the initial disturbance and λ is a ratio of the length scale of the initial disturbance a to the so-called radius of deformation $\sqrt{gH}/f$. The non-dimensional form of (7.5) is (after dropping superscript tilda)

$$\begin{aligned} &\text{in region I,} && \frac{d^2\xi}{dx^2} - \lambda^2\xi = -\lambda^2 R, \\ &\text{in regions II and III,} && \frac{d^2\xi}{dx^2} - \lambda^2\xi = 0. \end{aligned} \tag{7.6}$$

ξ is to be finite as $x \to \pm\infty$. The general solution is then

$$\begin{aligned} &\text{for I,} && \xi = Ae^{\lambda x} + Be^{-\lambda x} + R, \\ &\text{for II,} && \xi = Ce^{\lambda x}, \\ &\text{for III,} && \xi = De^{-\lambda x}. \end{aligned} \tag{7.7}$$

On the basis of (7.4), the corresponding non-dimensional solution of the surface elevation is

$$\begin{aligned} & && h = -\frac{d\xi}{dx} + 1, \\ &\text{for I,} && h = -\lambda\left(Ae^{\lambda x} - Be^{-\lambda x}\right) + 1, \\ &\text{for II,} && h = -\lambda Ce^{\lambda x} + 1, \\ &\text{for III,} && h = \lambda De^{-\lambda x} + 1, \end{aligned} \tag{7.8}$$

where A, B, C, D are integration constants to be determined on the basis of the matching conditions at $x = 0$ and $x = \pm 1$. Since $h = 1$ at $x = 0$, we must have $A = B$. Symmetry considerations also imply that $C = D$. In addition, we have matching conditions that ξ and h must be continuous at $x = \pm 1$ and thereby get

$$\begin{aligned} A = B &= -\frac{R}{2}e^{-\lambda}, \\ C = D &= \frac{R}{2}\left(e^{\lambda} - e^{-\lambda}\right). \end{aligned} \tag{7.9}$$

The non-dimensional form of (7.4) is

$$v = \frac{1}{R\lambda^2}\frac{dh}{dx}. \tag{7.10}$$

So, the explicit form of the complete solution is

$$
\begin{aligned}
&\text{for I}, -1 \le x \le 1 && h = \frac{R\lambda}{2}\left(e^{\lambda(x-1)} - e^{-\lambda(x+1)}\right) + 1, \\
& && \xi = -\frac{R}{2}\left(e^{\lambda(x-1)} + e^{-\lambda(x+1)}\right) + R, \\
& && v = \frac{1}{2}\left(e^{\lambda(x-1)} + e^{-\lambda(x+1)}\right); \\
&\text{for II}, -\infty < x \le -1 && h = -\frac{R\lambda}{2}\left(e^{\lambda} - e^{-\lambda}\right)e^{\lambda x} + 1, \\
& && \xi = \frac{R}{2}\left(e^{\lambda} - e^{-\lambda}\right)e^{\lambda x}, \\
& && v = -\frac{1}{2}\left(e^{\lambda} - e^{-\lambda}\right)e^{\lambda x}; \\
&\text{for III}, 1 < x \le \infty && h = \frac{R\lambda}{2}\left(e^{\lambda} - e^{-\lambda}\right)e^{-\lambda x} + 1, \\
& && \xi = \frac{R}{2}\left(e^{\lambda} - e^{-\lambda}\right)e^{-\lambda x}, \\
& && v = -\frac{1}{2}\left(e^{\lambda} - e^{-\lambda}\right)e^{-\lambda x}.
\end{aligned} \tag{7.11}
$$

The eastward displacement of the central fluid column of the adjusted state and the eastward displacement of the two boundary fluid columns in the central strip of the initial state are given in (7.12a,b) respectively.

$$
\begin{aligned}
\xi(0) &= R\left(1 - e^{-\lambda}\right), \\
\xi(-1) = \xi(1) &= \frac{R}{2}\left(1 - e^{-2\lambda}\right).
\end{aligned} \tag{7.12a,b}
$$

Thus the displacement would be larger for larger R and λ. The dimensional form of (7.12b) is $\xi_{\text{dim}}(a) = \frac{V}{2fa}\left(1 - e^{-2fa/\sqrt{gH}}\right)$. For a disturbance small compared to the radius of deformation, $a \ll \sqrt{gH}/f$, the displacement in units of a becomes $\xi_{\text{dim}}(a) \approx V/\sqrt{gH}$ which is larger for a stronger initial disturbance and a shallower layer of fluid. For a disturbance large compared to the radius of deformation, $a \gg \sqrt{gH}/f$, the displacement would approach the maximum possible value, $\xi_{\text{dim}}(a) \approx V/2fa$. The error in this quantity of Rossby's approximate solution is quite substantial for $\lambda \ge 3$.

7.2.2 Illustrative sample calculations

We use $\lambda = 1.0$ and $R = 1.0$ as a reference case for comparison. The solutions of h and v for this case are shown in Fig. 7.2.

We see that h in the central strip increases from $h(-1) \approx 0.55$ and $h(1) \approx 1.45$, but the gradient of this increase is not uniform. This eastward pressure gradient supports a southerly flow with values between 0.35 and 0.59. Furthermore, h decreases (increases) exponentially at a rate of $\lambda = 1$ away from the eastern (western) boundary of the central strip towards its asymptotic value of $h = 1$. This is the reason why λ^{-1} is called the non-dimensional radius

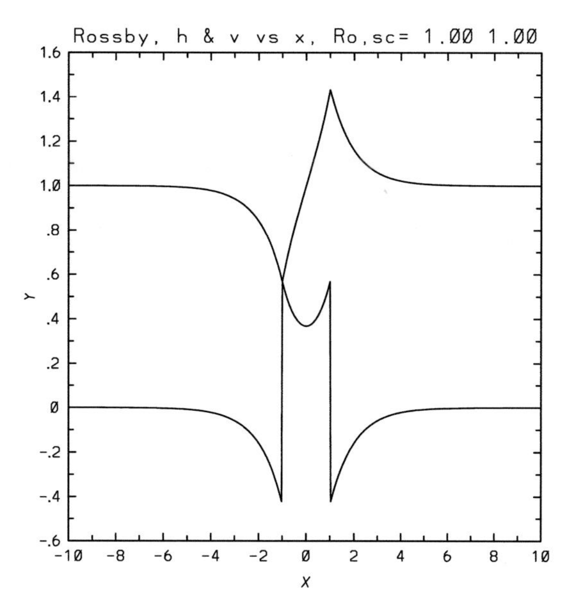

Fig. 7.2 Distribution of the non-dimensional surface elevation (upper curve) and meridional velocity (lower curve) for the case of $R = 1.0$ and $\lambda = 1.0$.

of deformation. The westward pressure gradient outside the central region supports a northerly flow on each side. The maximum magnitude of this northerly velocity is about 0.4. It should be added that a large shear such as those seen in Fig. 7.2 at $x = \pm 1$ in a fluid would be expected to be dynamically unstable. So, we only expect to see a smoothened version of the adjusted state. Furthermore, if we consider friction, it would dissipate the flow after a sufficiently long time. Therefore, we only expect to observe the adjusted state after a time long compared to the transient development but short compared to the dissipative time scale.

To highlight the dependence of the geostrophic adjustment upon the length scale of the initial disturbance, let us increase the length scale a fourfold and keep all other parameters the same as in the reference case. This case is characterized by $R = 0.25$ and $\lambda = 4.0$. The solution is shown over a narrower range of the domain in Fig. 7.3. We see that the adjustment of the pressure field in this case is about the same as that in the reference case. But the wind field has changed much more. The velocity at $x = 0$ is only about $v = 0.02$ and the velocity at $x = \pm 1$ is about $v = 0.5$. In other words, the adjustment is mostly achieved through the velocity field when the length scale of the initial unbalanced disturbance is long compared to the radius of deformation. The solution also indicates that Rossby's approximation about a uniform balanced flow in the central region is less justifiable in a case of longer length scale.

7.2.3 Essence of geostrophic adjustment

The essence of geostrophic adjustment can be most succinctly interpreted from the potential vorticity point of view. In 1938, the concept of potential vorticity was not yet firmly established. What Rossby referred to as "absolute vorticity" in his article is equivalent to what we nowadays call "potential vorticity." The potential vorticity of the

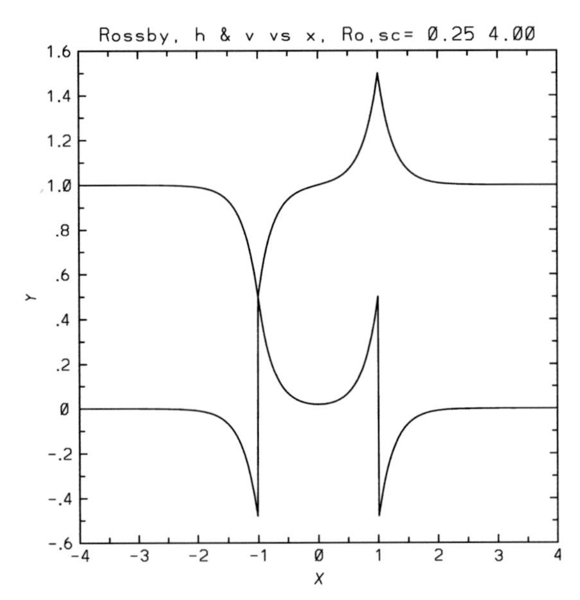

Fig. 7.3 Distribution of the surface elevation (upper curve) and meridional velocity (lower curve) for $R = 0.25$ and $\lambda = 4$.

adjusted state is $\left(1 + R\frac{\partial v}{\partial x}\right)/h$ in units of f/H. According to the solution (7.11), it is equal to 1.0 at all but two points. The exceptions are the two boundaries of the central strip, $x = \pm 1$. The non-dimensional velocity of the initial state and of the adjusted state have a discontinuity equal to 1 at those two boundaries. The PV is therefore infinite at those two points. The distribution of PV can be represented with the Dirac delta function, $\delta(x)$, as

$$\frac{1 + R\frac{\partial v}{\partial x}}{h} = 1 + R(\delta(x+1) - \delta(x-1)). \tag{7.13}$$

It is noteworthy that by taking a derivative of (7.2) with respect to x and making use of (7.3) and (7.4), we would get an equation in dimensional form as

$$\begin{gathered}\frac{\partial v}{\partial x} + f = \left(\frac{\partial v_o}{\partial X_o} + f\right)\frac{dX_o}{dx} = \left(\frac{\partial v_o}{\partial X_o} + f\right)\frac{h}{h_o}, \\ \frac{\frac{g}{f}\frac{d^2h}{dx^2} + f}{h} = \frac{\frac{\partial v_o}{\partial X_o} + f}{h_o}.\end{gathered} \tag{7.14}$$

Equation (7.14) is a statement of conservation of potential vorticity of each fluid column in this shallow-water system. Indeed, one would get the same solution (7.11) by solving (7.14) as a starting point instead of (7.6). In doing so, special care must be taken to impose proper matching conditions at $x = \pm a$ because of the discontinuity in v_o. The matching conditions at $x = \pm a$ are: (i) h is continuous and (ii) dh/dx has a jump related to the discontinuity of the initial velocity.

In light of the discussion above, we may conclude that the mutual adjustment of the velocity and height fields must be such that the PV of every fluid column conserves. This fundamental dynamical constraint of inviscid incompressible fluid makes it possible to deduce the adjusted state directly from an initial state.

7.2.4 Energetics of geostrophic adjustment

The initial amount of kinetic energy of the fluid in a section of length ℓ is $K_o = \rho Ha\ell V^2$. The amount of kinetic energy of the adjusted state in this section is

$$K = \frac{1}{2}\rho Ha\ell V^2 \int_{-\infty}^{\infty} v^2 h\, dx. \tag{7.15}$$

Using the solution (7.11), we obtain

$$K = \frac{1}{2}\rho Ha\ell V^2 \left(\frac{1}{2\lambda} + \left(1 - \frac{1}{2\lambda}\right)e^{-2\lambda}\right). \tag{7.16}$$

The amount of kinetic energy is therefore *reduced* during the adjustment process by

$$\Delta K = K_o - K = \frac{1}{2}\rho Ha\ell V^2 \left(2 - \frac{1}{2\lambda} - \left(1 - \frac{1}{2\lambda}\right)e^{-2\lambda}\right). \tag{7.17}$$

The amount of potential energy in a section of length ℓ in the initial state and adjusted state are

$$\begin{aligned} P_o &= \frac{1}{2} g\rho a\ell H^2 \int_{-\infty}^{\infty} (1)^2 dx, \\ P &= \frac{1}{2} g\rho a\ell H^2 \int_{-\infty}^{\infty} h^2 dx. \end{aligned} \tag{7.18}$$

According to the solution (7.11), the potential energy is *increased* by

$$\begin{aligned} \Delta P = P - P_o &= \frac{1}{2} g\rho a\ell H^2 \int_{-\infty}^{\infty} \left\{h^2 - 1\right\} dx \\ &= \frac{1}{2} H\rho a\ell V^2 \left(\frac{1}{2\lambda} - \left(1 + \frac{1}{2\lambda}\right)e^{-2\lambda}\right). \end{aligned} \tag{7.19}$$

The net loss of energy in the fluid is equal to $(\Delta K - \Delta P)$. This amount of energy must have been transported to a distant location from the initial location by the gravity waves during adjustment. The ratio of K reduction to P increase is

$$\frac{\Delta K}{\Delta P} = \frac{\left(2 - \frac{1}{2\lambda} - \left(1 - \frac{1}{2\lambda}\right)e^{-2\lambda}\right)}{\frac{1}{2\lambda} - \left(1 + \frac{1}{2\lambda}\right)e^{-2\lambda}}. \tag{7.20}$$

This value is equal to ~ 3.8 for $\lambda = 1$. In the limit of $\lambda \ll 1$ (small-scale disturbance), we get $\frac{\Delta K}{\Delta P} \to (-1)$, meaning that the reduction of kinetic energy is about the same as the increase of potential energy in this case. In the limit of $\lambda \gg 1$ (large-scale disturbance), we get $\Delta K/\Delta P \to (4\lambda - 1)$. This amounts to saying $\Delta K \gg \Delta P$ which means that the increase in potential energy is much smaller than the reduction of kinetic energy.

7.3 Complementary Rossby problem

The complementary Rossby problem refers to the problem of adjustment for the case of an initial disturbance that has a non-uniform pressure field without a velocity field. Textbooks typically discuss the adjustment problem for a weak perturbation of this form. This problem is in fact analytically tractable without restricting the magnitude of the initial disturbance to be weak.

7.3.1 An exact analysis

We will analyze this problem for an initial disturbance in the northern hemisphere with $v_o(x) = 0$ and $h_o(x)$ equal to a higher constant value in $|x| \leq a$ than in $|x| \geq a$. The task is to deduce the adjusted state without working out the intermediate states. It is again instructive to anticipate what should happen in the system by rudimentary physical reasoning before undertaking a mathematical analysis. Right after $t = 0$, some fluid must rush outward across $x = \pm a$ because of the pressure differential and undoubtedly would give rise to some turbulent eddies in the immediate neighborhood of $x = \pm a$. Irrespective of the details of such flow, the elevation of the fluid in the central strip must drop due to the outflow. It follows that the two boundaries of the central strip would move outward. The center of the domain remains the same in this case by a symmetry argument. The Coriolis force acting on the eastward moving fluid in $x > a$ would deflect it southward. Likewise, the westward moving fluid in $x < -a$ would be deflected northward. Gravity waves would also be naturally generated and propagate outward from the central strip. Nevertheless, the flow in the modified central strip may be expected to become steady at large time under the strong influence of Earth's rotation.

The initial state in this case is specifically $h_o(x) = A + H$ in $|x| \leq a$ and $h_o(x) = H$ in $|x| \geq a$ with $v_o(x) = 0$ everywhere. Figure 7.4 is a schematic of the initial state and the anticipated adjusted state under consideration.

This system has four parameters: A, H, f, a. We may therefore introduce two independent non-dimensional parameters: $\varepsilon = A/H$ and $\lambda = fa/\sqrt{gH}$; ε is a measure of the intensity of the initial imbalance of the initial state; λ is a ratio of the half-width of the central strip to the radius of deformation. We use the following non-dimensional variables:

$$\tilde{x} = \frac{x}{a}, \quad \tilde{\xi} = \frac{\xi}{a}, \quad \tilde{h} = \frac{h'}{A}, \quad \tilde{h}_{total} = \frac{h}{H}, \quad \tilde{v} = \frac{v}{\varepsilon\sqrt{gH}}, \quad \tilde{Q} = \frac{QH}{f}, \tag{7.21}$$

where Q is potential vorticity. The surface elevation and the velocity in the adjusted state are $\tilde{h}_{total} = \left(1 + \varepsilon\tilde{h}\right)$ and $\tilde{v} = \frac{1}{\lambda}\tilde{h}_{\tilde{x}}$. The potential vorticity is

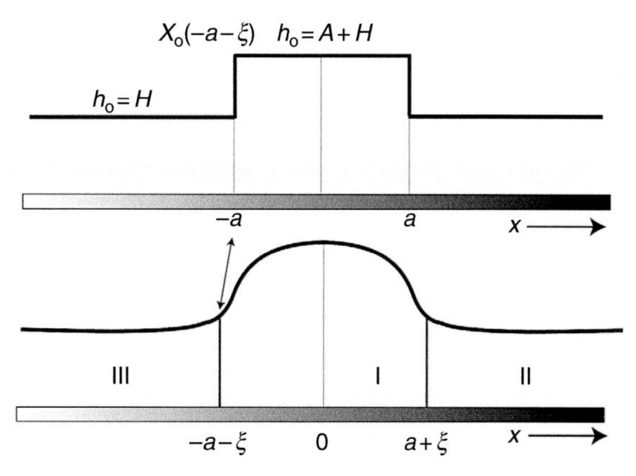

Fig. 7.4 Sketches of the initial state (upper panel) and adjusted state (lower panel) of a shallow-water model. Initial state is characterized by $h_o(x) = A + H$ in $|x| \leq a$ and $h_o(x) = H$ in $|x| \geq a$ and velocity everywhere is $v_o = 0$; ξ is the displacement of the two boundaries.

$$Q = \frac{f + v_x}{h} = \frac{f\left(1 + \frac{g}{f^2} h_{xx}\right)}{h} = \frac{f\left(1 + \frac{\varepsilon}{\lambda^2} \tilde{h}_{\tilde{x}\tilde{x}}\right)}{H\left(1 + \varepsilon \tilde{h}\right)},$$
$$\tilde{h}_{\tilde{x}\tilde{x}} - \tilde{Q}\lambda^2 \tilde{h} = \frac{\lambda^2}{\varepsilon}\left(\tilde{Q} - 1\right). \tag{7.22}$$

When we consider a strong initial disturbance, $\varepsilon = A/H \sim O(1)$, we do not approximate potential vorticity by its quasi-geostrophic form. It is also imperative to determine the displacement of fluid columns from their original positions as part of the unknowns. Because of the innate symmetry in the setup with respect to $x = 0$, it suffices to consider the positive half of an infinite domain. The semi-infinite domain consists of two regions, which are designated as (I) and (II) in a description of the adjusted state; $\tilde{x}_1 \equiv \left(1 + \tilde{\xi}\right)$ is the displaced PV boundary originally at $x = 1$. The non-dimensional potential vorticity (PV) of the initial state is $\tilde{Q} = \dfrac{1}{1 + \varepsilon}$ in region (I) $|x| \leq 1$, and $\tilde{Q} = 1$ in region (II) $|x| \geq 1$. The value of x_1 is itself to be determined as part of the solution. We write the governing equations in the following explicit form (after dropping the tilda notation):

$$\begin{aligned} &\text{in (I)}, \quad 0 \leq x \leq x_1 \quad h_{xx} - b^2 h = -b^2 \text{ with } b^2 = \frac{\lambda^2}{1 + \varepsilon}, \\ &\text{in (II)}, \quad x > x_1, \qquad h_{xx} - \lambda^2 h = 0. \end{aligned} \tag{7.23}$$

The proper boundary conditions are: $h_x = 0$ at $x = 0$ and $h = 0$ as $x \to \infty$. The general solution is then:

$$\begin{aligned} &\text{in (I)} \quad h = 1 + A\left(e^{bx} + e^{-bx}\right), \\ &\text{in (II)} \quad h = Ce^{-\lambda x}. \end{aligned} \tag{7.24}$$

The integration constants A and C can be determined on the basis of the matching conditions at $x = x_1$ which are continuous in both h and h_x. We get $A = -1/2e^{\lambda x_1}$ and $C = e^{\lambda x_1} - \left(e^{\lambda x_1} + e^{-\lambda x_1}\right)/2$. The complete solution is then:

$$
\begin{aligned}
&\text{in } 0 \le x \le x_1, \quad h = 1 - \frac{\lambda\left(e^{bx} + e^{-bx}\right)}{e^{bx_1}(\lambda + b) + e^{-bx_1}(\lambda - b)},\\
&\qquad\qquad\qquad v = -\frac{b\left(e^{bx} - e^{-bx}\right)}{e^{bx_1}(\lambda + b) + e^{-bx_1}(\lambda - b)},\\
&\text{in } x > x_1, \qquad h = \frac{b\left(e^{bx_1} - e^{-bx_1}\right)}{e^{bx_1}(\lambda + b) + e^{-bx_1}(\lambda - b)} e^{\lambda(x_1 - x)},\\
&\qquad\qquad\qquad v = -\frac{b\left(e^{bx_1} - e^{-bx_1}\right)}{e^{bx_1}(\lambda + b) + e^{-bx_1}(\lambda - b)} e^{\lambda(x_1 - x)}.
\end{aligned}
\tag{7.25}
$$

The condition of mass conservation can be used to determine x_1. We thereby obtain

$$
\begin{aligned}
&\int_0^{x_1} (1 + \varepsilon h)dx = \int_0^1 (1 + \varepsilon)dx\\
&\to (1 + \varepsilon)x_1 - \frac{\varepsilon\lambda\left(e^{bx_1} - e^{-bx_1}\right)}{b\left(e^{bx_1}(\lambda + b) + e^{-bx_1}(\lambda - b)\right)} = 1 + \varepsilon.
\end{aligned}
\tag{7.26a,b}
$$

Here x_1 can be readily evaluated as the intersection of two curves that represent the LHS and RHS of (7.26b).

It is instructive to compare this exact solution with the approximate solution for a perturbation. The latter is obtained by using conservation of the *quasi-geostrophic PV* and neglecting the displacement of fluid columns (i.e. setting $x_1 = 1$, a priori). It can be shown that such a solution in the semi-infinite domain is

$$
\begin{aligned}
&\text{in } 0 \le x \le 1, \quad h = 1 - \frac{\left(e^{\lambda x} + e^{-\lambda x}\right)}{2e^{\lambda}}, \quad v = -\frac{\left(e^{\lambda x} - e^{-\lambda x}\right)}{2e^{\lambda}},\\
&\text{in } x > 1, \qquad h = \frac{1}{2}\left(e^{\lambda} - e^{-\lambda}\right)e^{-\lambda x}, \quad v = -\frac{1}{2}\left(e^{\lambda} - e^{-\lambda}\right)e^{-\lambda x}.
\end{aligned}
\tag{7.27}
$$

7.3.2 Illustrative sample calculations

We present some calculations for comparing the approximate with the exact solutions. Those solutions of $h_{total} = 1 + \varepsilon h$ for $\varepsilon = 1$, $\lambda = 0.5$ are plotted in Fig. 7.5. The length scale of disturbance in this case is equal to the radius of deformation. It is found that the PV boundary in the exact solution is displaced to $x_1 \approx 1.5$ meaning that the boundaries have been displaced by 50 percent of the half-width of the central strip. The new boundaries of PV discontinuity are indicated by the two vertical dash lines at $x = \pm x_1$ in Fig. 7.5. The maximum height is $(1 + \varepsilon h(0)) \approx 1.34$. The decrease of mass between $x = 0$ and $x = 1$ from the initial state to the adjusted state is about 0.65. It is indeed equal to the mass in the column between $x = 1$ and $x = x_1$. That is why the linear solution of h_{total} near $x = 1$ is higher than that of the exact solution.

The corresponding distributions of the velocity v are shown in Fig. 7.6.

Both solutions reveal that a shear flow is induced in this problem by the geostrophic adjustment. They confirm the expectation from rudimentary physical reasoning that there

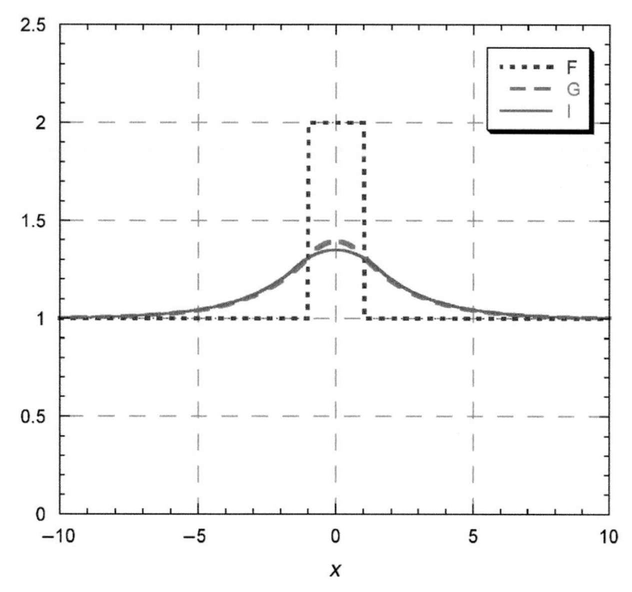

Fig. 7.5 A comparison of the distributions of $h_{total} = 1 + \varepsilon h$, the non-dimensional height of the surface in the initial state (F), the linear solution (G), and the exact solution (I) for the case of $\lambda = 0.5$, $\varepsilon = 1$. Heavy-dot lines are the displaced boundaries.

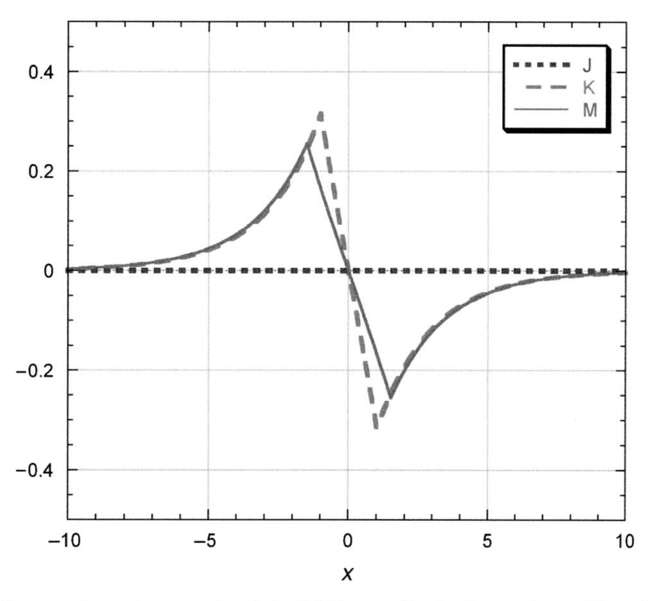

Fig. 7.6 A comparison of the non-dimensional velocity of the initial state (J), the linear solution (K), and the exact solution (M) for the case of $\lambda = 0.5$, $\varepsilon = 1$.

should be a northerly (southerly) jet in the eastern (western) side of the initial disturbance. The displacement of the PV boundaries is more evident in Fig. 7.6. The position of each jet is collocated with the PV boundary in each solution. The strength of the jet is considerably overestimated by the linear solution.

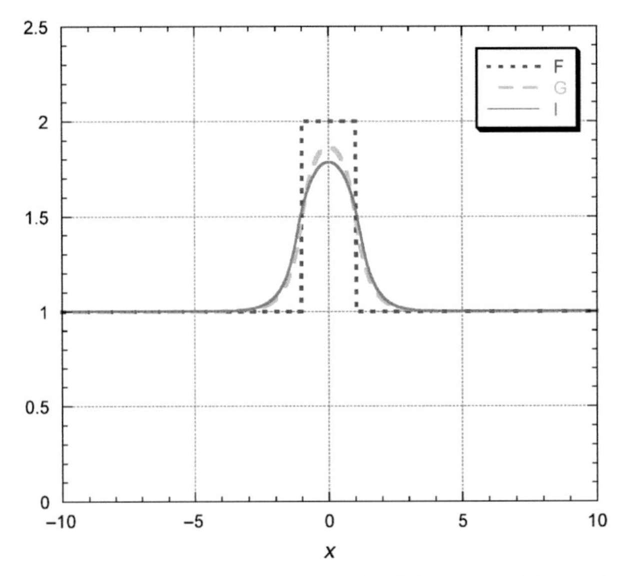

Fig. 7.7 A comparison of the distributions of the non-dimensional height of the surface of the initial state (F), the linear solution (G), and the exact solution (I) for the case of $\lambda = 2.0$, $\varepsilon = 1.0$.

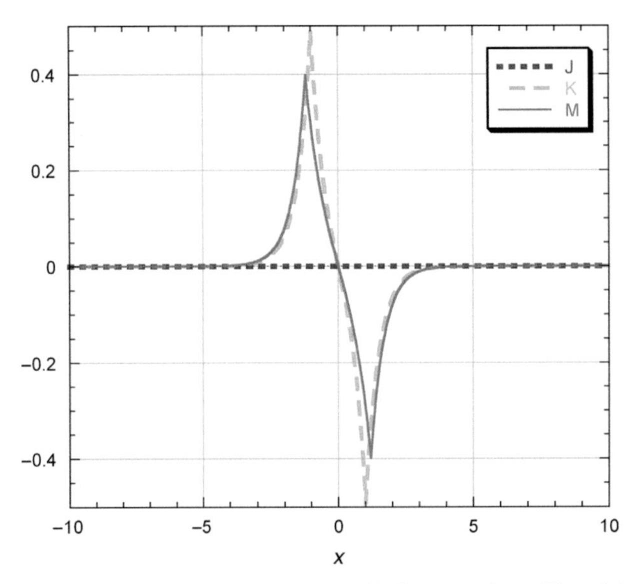

Fig. 7.8 A comparison of the non-dimensional velocity of the initial state (J), the linear solution (K), and the exact solution (M) for the case of $\lambda = 2.0$, $\varepsilon = 1$.

Let us next examine the solutions for the case of a disturbance of larger scale, $\lambda = 2$. This is a case where the length scale of disturbance is equal to four times the radius of deformation. It is found that $x_1 = 1.2$ for this case, meaning that the boundaries have been displaced by only 20 percent of the half-width of the central strip. The counterpart solutions of $h_{total} = 1 + \varepsilon h$ and v are shown in Figs. 7.7 and 7.8 respectively.

Comparing the two adjusted states in Fig. 7.5 and Fig. 7.7, we see that the elevation of the free surface changes by a much smaller extent from the initial configuration in the case of a larger disturbance. The pressure gradient at the PV boundary in the adjusted state is correspondingly much larger. That in turn can support a stronger jet on each side (0.4 vs. 0.25). We may conclude that geostrophic adjustment is achieved mostly by adjusting the velocity field if the length scale of the disturbance is long compared to the radius of deformation. Conversely, geostrophic adjustment is achieved mostly by adjusting the pressure field if the length scale is short compared to the radius of deformation. We have arrived at the same conclusion as in the analysis of Section 7.2. As such, it is a general conclusion of geostrophic adjustment.

7.4 Gradient-wind adjustment problem

To generalize the geostrophic adjustment problem, we consider the adjustment of an initially unbalanced flow that has significant curvature. A prototype atmospheric disturbance of this type is a hurricane. When such a disturbance is significantly disturbed to an unbalanced state by an impulsive external influence such as a sudden change in the precipitation distribution, it should also eventually adjust to a new steady balanced state. The gradient-wind adjustment problem even for the case of an axisymmetric vortex is analytically intractable because of the nonlinear functional relationship between the pressure field and velocity field. It is more general than the previous analyses of adjustment of weak axisymmetric perturbations in the presence of a background vortex.

7.4.1 Model formulation

Let us consider an axisymmetric vortex in a shallow-water model without friction. Suppose gradient-wind balance initially prevails everywhere except in a certain radial band due to an impulsive external forcing of some kind. We again invoke the principle of conservation of potential vorticity of each fluid column. For a cyclonic axisymmetric flow, the conservation of PV can be written as

$$\frac{dv}{dr}+\frac{v}{r}+f=(H+h)Q, \tag{7.28}$$

where r is the radial distance, f the Coriolis parameter, h the departure from a constant reference depth H and v the azimuthal velocity of a fluid column. Using a subscript "o" to indicate the initial state, we then have

$$Q=\frac{\dfrac{dv_o}{dr}+\dfrac{v_o}{r}+f}{H+h_o}$$

as the initial known PV of the same fluid column. The adjusted vortex is assumed to be in gradient-wind balance, namely

$$\frac{dh}{dr}=\frac{f}{g}v+\frac{v^2}{gr}, \tag{7.29}$$

where g is reduced gravity. The values of v and h are expected to be very close to v_o and h_o respectively at a sufficiently large distance $r = r_{\max}$ from the center of the vortex.

Let us consider a Rankin vortex as a balanced flow. The vorticity of this flow is equal to $2V/r_o$ in an inner core, and is zero in the outer region. We measure r in units of r_o, v in units of V, h in units of H and Q in units of f/H; $R = V/fr_o$ and $\lambda = fr_o/\sqrt{gH}$ may be used as the two independent non-dimensional parameters. As elaborated in Section 2.6.1, the non-dimensional properties of a balanced Rankin vortex are

$$h^* = \frac{\lambda^2 R^2}{2}\left(1 + \frac{1}{R}\right)r^2, \quad v^* = r, \quad Q^* = \frac{2R+1}{(1+h^*)} \quad \text{in } r \leq 1;$$

$$h^* = \frac{\lambda^2 R^2}{2}\left(2 - \left(\frac{1}{r}\right)^2\right) + R\lambda^2\left(\frac{1}{2} + \ell n(r)\right), \quad v^* = \frac{1}{r}, \quad Q^* = \frac{1}{(1+h^*)} \quad \text{in } r \geq 1. \tag{7.30}$$

A plot of v^* and $(1 + h^*)$ is shown in Fig. 2.17. The non-dimensional v^* and $(1 + h^*)$ correspond to the v/v and Φ/Φ_{oo}, respectively, labelled in Fig. 2.17.

7.4.2 Analysis

The initial vortex is prescribed as an unbalanced vortex, (a Rankin vortex with a perturbation in the height field) i.e.

$$v_o = v^*, \quad h_o = h^* + \varepsilon \exp\left(-\left(\frac{r-a}{b}\right)^2\right),$$
$$Q = \frac{2R+1}{(1+h_o)} \text{in } r \leq 1, \qquad Q = \frac{1}{(1+h_o)} \text{in } r \geq 1, \tag{7.31}$$

where v^* and h^* are given in (7.30). The second term in h_o represents the initial imbalance. The non-dimensional governing equations take the form of

$$\frac{dv}{dr} = -\frac{v}{r} + \frac{Q}{R}h + \frac{1}{R}(Q-1), \tag{7.32}$$

$$\frac{dh}{dr} = R\lambda^2 v + R^2\lambda^2\frac{v^2}{r}. \tag{7.33}$$

The task is to determine the structure of v and h solely on the basis of the given information about Q which is a function of $v_o(r)$ and $h_o(r)$. A solution must also be compatible with the kinematic constraint that the fluid at the center of the adjusted vortex is at rest.

Equations (7.32) and (7.33) may be combined to one single nonlinear ordinary differential equation (ODE) that governs either v or h. We choose to directly solve (7.32) and (7.33) as a pair of coupled first order ODEs for two unknowns simultaneously. Therefore, we pose the problem of solving (7.32) and (7.33) as an initial value problem in space. The solution is to satisfy the following far field condition and kinematic constraint at the center,

$$\text{at } r = r_{\max}, \quad v = v_* \text{ and } h = h_*,$$
$$\text{at } r \to 0, \quad v \to 0. \tag{7.34}$$

Since we need to consider a domain much larger than the inner core of a vortex, it would be advantageous to introduce a stretch-coordinate so that the inner core can be depicted in high resolution whereas the outer region can be depicted with progressively coarser resolution. Thus, we introduce a new radial coordinate

$$s = \ell n(r). \tag{7.35}$$

The computational domain is $s_{\min} \leq s \leq s_{\max}$ corresponding to a range of radial distance $r_{\min} \leq r \leq r_{\max}$. The computation is done using $s_{\min} = -2$ and $s_{\max} = 2$ corresponding to $r_{\min} = 0.13$ and $r_{\max} = 7.4$. The domain is depicted with $(J+2)$ uniform grids in the stretch-coordinate.

The imbalance in the initial state is strongest at $r = a$ exponentially decreasing away from it. We will consider two distinctly different locations of the initial imbalance; one centering at the edge of the inner core and the other centering at twice the distance further out from the center. It suffices to present the results for one pertinent combination of the parameter values, namely $R = 1$ and $\lambda = 1$. For this condition, the Coriolis force and centrifugal force are equally strong at the edge of the inner core of the initial vortex.

7.4.3 Result

Case (1)

A fairly strong localized imbalance is prescribed with the use of $\varepsilon = 1.0$, $a = 1.0$ and $b = 0.6$. The maximum initial unbalanced depth is then located at the edge of the inner core of the vortex, $r = 1.0$. Rudimentary reasoning suggests that some fluid would move inward leading to an increase in the depth and also possibly an increase in the azimuthal velocity. The PV boundary would be displaced inward as well. We denote the location of the displaced PV boundary by r_1 which is unknown a priori. The properties of the initial state of the vortex under consideration are shown in Fig. 7.9.

The solution is obtained with an iterative algorithm: 225 uniform grids in the s-coordinate are used in this computation. We begin by making a guess of r_1 which allows us to compute a corresponding distribution of PV, $Q(r)$, according to (7.31). We solve (7.32) and (7.33) with a fourth-order Runge–Kutta scheme subject to boundary conditions (7.34) for that specific distribution of Q. The provisional solution of $h(r)$ is then checked to see if it satisfies the mass conservation requirement for that particular value of r_1, namely $\left(\int_0^{r_1} (1 + h)rdr = \int_0^1 (1 + h_o)rdr\right)$. If not, the guess for r_1 is modified and the computation is repeated. This algorithm quickly yields convergence of the result. It is found that $r_1 = 0.81$ is a self-consistent value for this case. The distributions of the total depth of the adjusted vortex, that of the initial vortex and that of a counterpart Rankin vortex are shown in Fig. 7.10.

It is seen that the height field has been considerably modified due to redistribution of mass. The change of potential energy of the vortex in units of $\rho g H^2 r_o^2 \pi$ is

$$\Delta P = \int_{r_{\min}}^{r_{\max}} \left((1 + h)^2 - (1 + h_o)^2\right) rdr = 5.1. \tag{7.36}$$

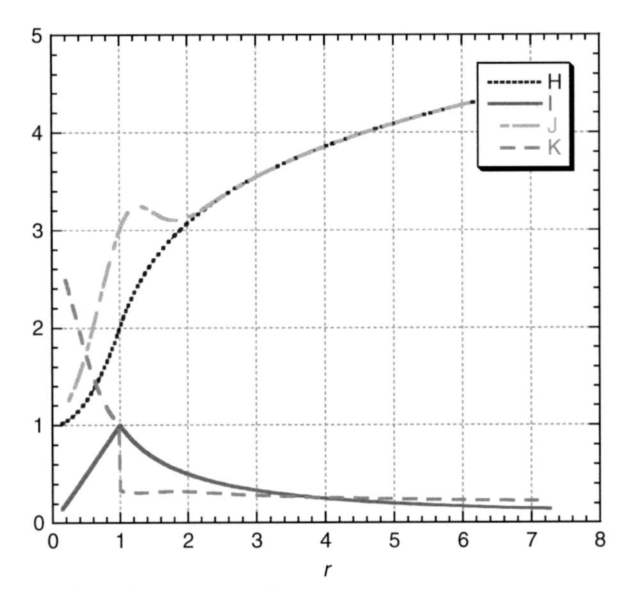

Fig. 7.9 Radial distribution of the non-dimensional depth (J), velocity (I) and potential vorticity (K) of an initial unbalanced vortex and the depth of a corresponding balanced vortex (H) for the case of ($R = 1.0$, $\lambda = 1.0$, $\varepsilon = 1.0$, $a = 1.0$ and $b = 0.6$).

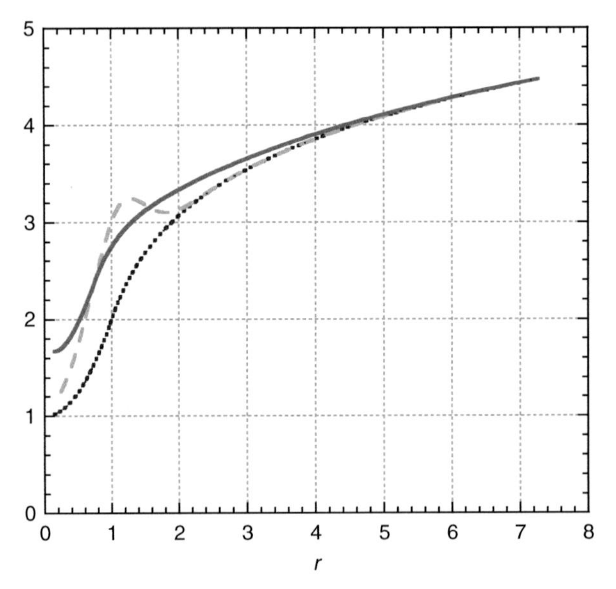

Fig. 7.10 Radial distribution of the non-dimensional total depth of an initial unbalanced vortex (dash), of the adjusted vortex (solid) and of the corresponding Rankin vortex for comparison (dot) for the case of ($R = 1.0$, $\lambda = 1.0$, $\varepsilon = 1.0$, $a = 1.0$ and $b = 0.6$).

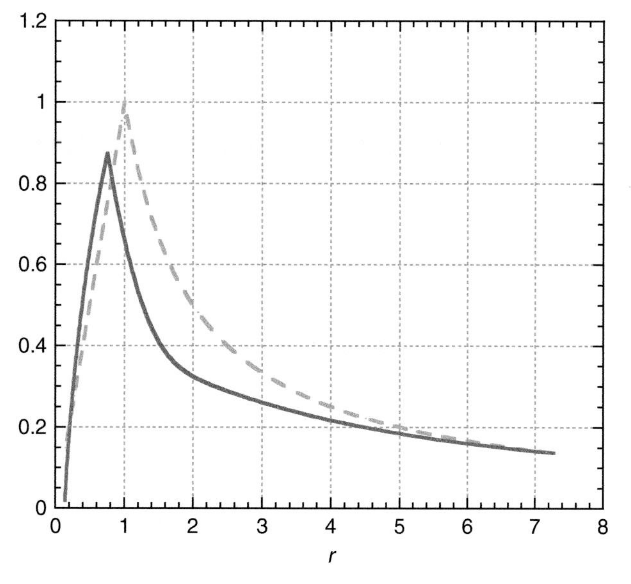

Fig. 7.11 Radial distribution of the non-dimensional velocity field of the initial unbalanced vortex (dash) and of the adjusted vortex (solid) for the case of ($R = 1.0$, $\lambda = 1.0$, $\varepsilon = 1.0$, $a = 1.0$ and $b = 0.6$).

The mass redistribution leads to an increase of pressure gradient inward of $r_1 = 0.81$, but a decrease outward of $r_1 = 0.81$.

The corresponding velocity field is shown in Fig. 7.11. The maximum velocity is now located at $r = r_1 = 0.81$. The adjusted velocity field inward of $r_1 = 0.81$ is enhanced, but the velocity in most of the vortex is substantially weakened.

The change of kinetic energy of the vortex in units of $\rho g H^2 r_o^2 \pi$ is

$$\Delta K = R^2 \lambda^2 \int_{r_{\min}}^{r_{\max}} \left((1+h)v^2 - (1+h_o)v_o^2\right) r dr = -125. \tag{7.37}$$

While the potential energy increases, the kinetic energy decreases as a result of the adjustment process; 120 units of energy must have been transmitted to far distance by gravity waves as a result of the adjustment process. We check to see if the distribution of potential vorticity of the adjusted state,

$$\frac{R\left(\frac{dv}{dr} + \frac{v}{r}\right) + 1}{1+h}$$

agrees with that of the initial state. It is found that except for the displacement of the PV boundary, the two distributions are indeed very close to one another.

Case (2)

It is also of interest to consider a case in which the initial imbalance is located further outside of the inner core. The parameter values are $a = 2$, $b = 0.6$, and $\varepsilon = 1.0$.

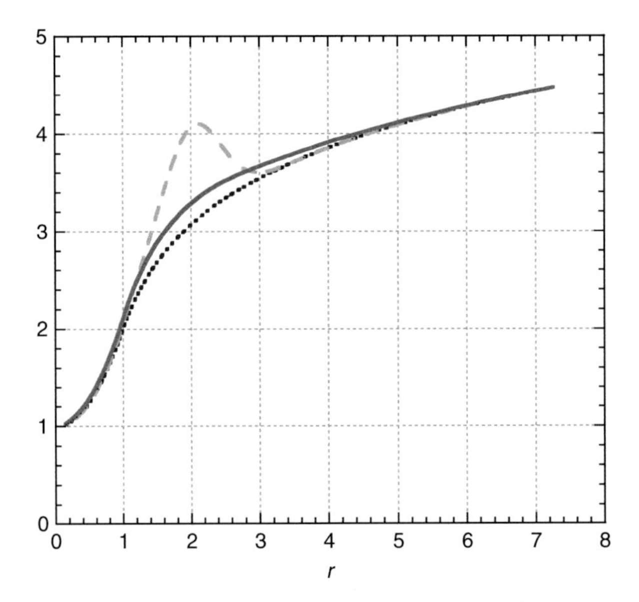

Fig. 7.12 Radial distribution of the non-dimensional total depth of the initial vortex (dash), of the adjusted vortex (solid) and of a corresponding undisturbed vortex (dot) for the case of ($R = 1$, $\lambda = 1.0$, $a = 2$, $b = 0.6$ and $\varepsilon = 1.0$).

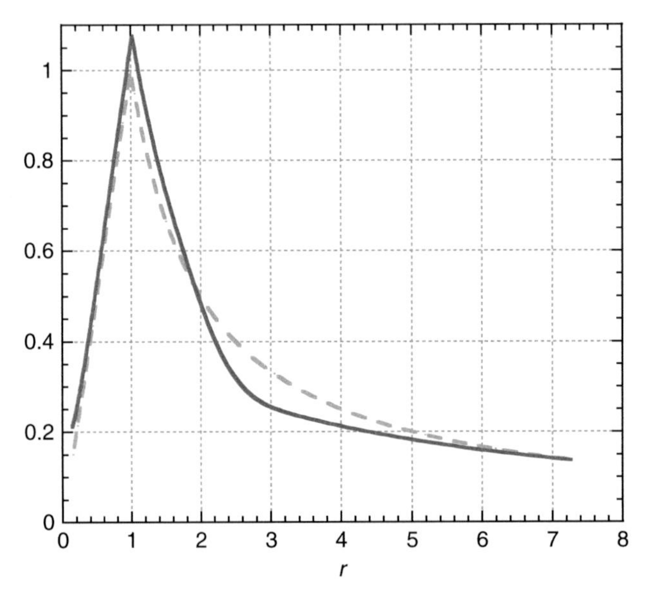

Fig. 7.13 Radial distribution of the velocity field of an initial vortex (dash) and that of the velocity field of the adjusted vortex (solid) for the case of ($R = 1$, $\lambda = 1.0$, $a = 2$, $b = 0.6$ and $\varepsilon = 1.0$).

The center of the initial imbalance is now located at $r = 2$ instead of $r = 1$. The PV boundary is found to shift outward to $r_1 = 1.35$ associated with an outward mass flux. The results for the height and velocity fields of the adjusted vortex are shown in Figs. 7.12 and 7.13.

The change of potential energy of the vortex in units of $\rho g H^2 r_o^2 \pi$ is found to be $\Delta P = -5.3$. A comparison of Fig. 7.13 with Fig. 7.11 reveals the main difference between the adjusted state in case-2 versus that in case-1. It is found that when the initial imbalance is located outside its inner core (i.e. at $r = 2.0$ instead of $r = 1.0$), the wind in the inner core changes very little, but the maximum velocity shifts outward and is enhanced by about 10 percent. The velocity extending out to about $r = 2.0$ is enhanced. The velocity further out is weaker than that in the initial state. Consequently, the zone of relatively strong wind of the adjusted vortex becomes narrower. This is an interesting consequence of the redistribution of mass and related PV. There is a compensating decrease of velocity farther out from the center of the vortex. The change of kinetic energy of the vortex in units of $\rho g H^2 r_o^2 \pi$ is found to be $\Delta K = -57.7$. This change is much smaller than in case-1. It is noteworthy that both the potential energy and kinetic energy of the vortex become smaller after adjustment in this case. Sixty-three units of energy must have been transmitted to far distance by gravity waves.

The characteristics of the response in both cases of initial states suggest one overall conclusion, namely *the azimuthal velocity of the adjusted state on the radially inward (outward) side of the maximum initial unbalanced depth is enhanced (weakened) with respect to that of the initial vortex.* The results for different combinations of the values of R and λ are qualitatively similar to those of the two cases reported above.

Concluding remarks

We have seen that it is feasible to get analytic solutions of the geostrophic adjustment problem for initially unbalanced rectilinear disturbances of any intensity in a shallow-water model. The systematic displacement of fluid columns from their original locations is an integral part of the adjustment process. It follows that the spatial distribution of their potential vorticity in the adjusted state would differ from that in the initial state. The mutual adjustment between the mass and velocity fields is such that the potential vorticity of each fluid column conserves. It is this dynamical constraint that makes it possible to deduce the adjusted state solely on the basis of the information about the initial state without having to determine the intermediate states. The overall conclusion is that the adjustment mostly involves the velocity (mass) field for the case of a large (small) disturbance relative to the radius of deformation.

Symmetry consideration is helpful in an analysis of a physical system that has innate symmetry. For example, the depth of the fluid in the adjusted state has odd symmetry and the velocity field has even symmetry with respect to the center of the domain in the Rossby problem. In the complementary Rossby problem, the elevation of the fluid in the adjusted state has even symmetry, whereas the velocity field has odd symmetry with respect to the center of the domain.

As a generalization of the geostrophic adjustment problem, we have analyzed the adjustment of an initial unbalanced Rankin vortex numerically. It is found that the adjusted state depends very much upon the location of the initial imbalance relative to the inner core of the vortex. There are mass fluxes in both inward and outward directions from the localized initial imbalance. The azimuthal velocity of the adjusted state on the radially inward (outward) side of the initial unbalanced depth is found to be greater (weaker) than that of the initial vortex. Consequently, the belt of strong wind becomes narrower.

8 Instability theories

A general supposition is that the genesis of an atmospheric disturbance is quantitatively understandable in terms of the dynamic instability of a certain background state. There are a number of instability theories in support of this supposition for disturbances of vastly different sizes. To emphasize their distinctiveness as well as interconnectedness, we discuss these theories in four parts (A, B, C and D) of this chapter. Parts A, B and C cover a variety of analyses addressing the instability of different classes of basic shear flows in a dry model setting. In contrast, Part D delineates a type of instability arising from the joint influences of self-induced condensational heating and the basic baroclinicity. A primary focus of most discussions in this chapter is the problem of cyclogenesis in the extratropical atmosphere.

We analyze each type of dynamic instability both as an initial-value problem and as a modal problem. The former would tell us about the characteristics of the transient growth of a disturbance. It would be valuable to be able to establish the analytic necessary condition for modal instability. That however only yields very limited information about the instability. One would still need a modal instability analysis to deduce the number of families of unstable modes and the characteristics of each of those families of modes. The characteristics include the growth rate of each unstable mode, its movement, structure and energetic. It is also possible to deduce a priori which member of a common set of initial disturbances would optimally intensify over a specific time interval. An instability investigation is analytically feasible when the basic state is sufficiently simple. The more complex a basic flow under consideration is, the more we would have to resort to numerical methods. Ultimately, the goal is to understand the dynamical nature of each instability process.

PART 8A SMALL-SCALE AND MESO-SCALE INSTABILITY

This part of the chapter is concerned with instability processes in a dry model setting that are pertinent to small-scale and/or meso-scale disturbances. We begin by considering the simplest possible type of instability in Section 8A.1. It occurs in an atmospheric layer at rest with a basic potential temperature that has a smaller value at greater height. The corresponding unstable disturbance is relevant to thermal plumes. The instability of a layer of rotating fluid with a basic shear flow and neutral stratification is next analyzed in Sections 8A.2.1 and 8A.2.2. We do so from both Lagrangian and Eulerian perspectives.

We focus on a class of disturbances that has the same geometric symmetry as the basic flow itself. Such instability is relevant to the meso-scale rectilinear disturbances common in severe weather. The quantitative results are illustrated in Section 8A.2.3. The impact of a stable stratification in such a system is delineated in Sections 8A.3.1 and 8A.3.2. The corresponding illustrative calculation is presented in Section 8A.3.3.

8A.1 Static instability and the impact of damping

The spontaneous development of disturbances in a layer of compressible fluid in which the basic potential temperature decreases with increasing height is known as *static instability*. Additional complications arise when there is sufficient moisture. These processes are most relevant to small-scale atmospheric disturbances such as plumes and convection in general.

Let us begin by examining what might happen to an air parcel in *a stratified atmosphere without a basic flow* when it is displaced and released at a different elevation from a Lagrangian perspective. Whether the air parcel would tend to move further away from or return to its initial location depends only on the stratification under the influence of gravity. Stratification is measured by $d\bar{\theta}/dz$ where $\bar{\theta}$ is the basic potential temperature. The determining parameter of the system under consideration is $N^2 = \frac{g}{\theta_{oo}}\frac{d\bar{\theta}}{dz}$, the Brunt–Vaisala frequency squared; θ_{oo} is a reference value of potential temperature for the layer and g is gravity. For $N^2 < 0$, the density is *larger* at a higher level. A heavier fluid parcel would naturally tumble down to take the place of a rising lighter fluid parcel elsewhere. The center of mass of the system would be lowered, thereby releasing some gravitational energy to support the intensification of disturbances. Hence, we may expect that a slightly displaced fluid parcel would be further displaced on its own without doing any math. The necessary condition for static instability is that N^2 of the fluid under consideration must have a negative value.

Since an air parcel is infinitesimally small, two simplifying assumptions appear to be plausible. One assumption is that the environment is not disturbed whatsoever by the displacement of an air parcel. The other assumption is that the pressure at a moving air parcel is always equal to the pressure of its environment. The vertical momentum equation for a slightly displaced parcel is then

$$\frac{Dw}{Dt} = g\left(\frac{\bar{\rho} - \rho}{\rho}\right) \approx g\left(\frac{\theta - \bar{\theta}}{\theta_{oo}}\right) \approx -\frac{g}{\theta_{oo}}\frac{\partial \bar{\theta}}{\partial z}\eta$$
$$\rightarrow \quad \frac{d^2\eta}{dt^2} = -N^2\eta. \tag{8A.1}$$

The RHS of (8A.1) is the buoyancy force. A displaced parcel would sinusoidally oscillate about its original level with a period equal to N for $N^2 > 0$. On the other hand, the displacement would be an exponentially increasing function of time for $N^2 < 0$. This means that the lapse rate of the environment, $(-\partial\bar{T}/\partial z)$, exceeds the dry adiabatic lapse rate, $\Gamma_d = g/c_p$. The amplification rate of the displacement would be $\gamma = \sqrt{-N^2}$. Such stratification is called statically unstable stratification.

This line of reasoning can be extended in a number of ways. One extension is to take into consideration moisture in the atmosphere which has a lapse rate, $\Gamma \equiv -\partial T/\partial z$, smaller than $\Gamma_d = g/c_p$. An air parcel would cool at the rate of Γ_d when it is lifted upward. It would eventually become saturated if it is subject to sustained lifting. The accompanying release of latent heat of condensation thereafter would reduce its rate of cooling. The subsequent lapse rate is called moist adiabatic lapse rate, Γ_m which is a function of temperature itself. Thus, so long as we have $\Gamma_m < \Gamma < \Gamma_d$, such a parcel would be able to continue its ascent on its own thereafter. The criterion for this instability can be expressed as $\partial\bar{\theta}_e^*/\partial z < 0$ where $\bar{\theta}_e^*$ is the saturated equivalent potential temperature. This is referred to as *conditional instability.*

Even if the lapse rate of a layer of moist atmosphere is smaller than Γ_m, instability could still occur in it if the whole layer is lifted upward by a sufficient extent. The lower part of the layer has more moisture and would first become saturated. Then, while the upper part of the layer is cooled at the dry adiabatic lapse rate, the lower part is cooled at a smaller moist adiabatic lapse rate. Consequently, the lapse rate of the layer as a whole would become greater than Γ_m and instability would soon follow. The instability in this scenario is referred to as *convective instability.*

We next perform an Eulerian instability analysis for *a two-dimensional non-rotating, stratified, adiabatic, viscous Boussinesq fluid with no basic flow.* For simplicity, we suppose that the viscous coefficient and thermal coefficient have equal value, κ. The domain is an unbounded vertical (x, z) plane. The governing equations for such a perturbation are

$$\begin{aligned} u'_t &= -\frac{1}{\rho_{oo}} p'_x + \kappa\nabla^2 u', \\ w'_t &= -\frac{1}{\rho_{oo}} p'_z + g\frac{\theta'}{\theta_{oo}} + \kappa\nabla^2 w', \\ u'_x + w'_z &= 0, \\ \theta'_t + w'\theta_{oz} &= \kappa\nabla^2\theta', \end{aligned} \tag{8A.2}$$

where $\nabla^2 = \partial^2/\partial x^2 + \partial^2/\partial z^2$ and κ is the damping coefficient. Introducing a streamfunction ψ such that $u' = -\psi_z$ and $w' = \psi_x$, we can readily reduce this system of equations to a single equation

$$L\nabla^2\psi + N^2\psi_{xx} = 0, \tag{8A.3}$$

where $L \equiv \frac{\partial}{\partial t} - \kappa\nabla^2$ and $N^2 = \frac{g}{\theta_{oo}}\frac{d\theta_o}{dz}$ is assumed to be constant. A normal mode solution of (8A.3) is $\psi = A\exp(i(kx + mz - \sigma t))$, provided that the parameters satisfy the following dispersion relation:

$$\sigma^2 + \sigma\left(i2\kappa\left(k^2 + m^2\right)\right) - \left(\frac{N^2k^2}{k^2 + m^2} + \kappa^2\left(k^2 + m^2\right)^2\right) = 0. \tag{8A.4}$$

Suppose the background stratification is such that $N^2 = \frac{g}{\theta_{oo}}\frac{d\theta_o}{dz} = -\gamma^2 < 0$. *If we ignore dissipation for the moment by setting* $\kappa = 0$, the two roots of (8A.4) are simply

$\sigma_{\pm} = \pm i \frac{\gamma k}{\sqrt{k^2 + m^2}}$. The mode associated with σ_- decays and the mode with σ_+ intensifies exponentially in time. This result confirms the necessary condition for instability deduced from the earlier Lagrangian analysis. The growth rate $\gamma k/\sqrt{k^2 + m^2}$ monotonically increases with k. In other words, the smaller a disturbance is, the more unstable it would be. The asymptotic value of the growth rate for $k \to \infty$ is equal to γ in agreement with the result of a Lagrangian analysis for a parcel of infinitesimally small size. But this conclusion is not justifiable for sufficiently small disturbances in a real fluid because the viscous effect would no longer be negligible. In other words, extremely small disturbances are not unstable at all if we take into account the viscous damping however weak it might be. The general solution of the two roots of (8A.4) are

$$\sigma = -i\kappa\left(k^2 + m^2\right) \pm \sqrt{\frac{N^2 k^2}{(k^2 + m^2)}}. \tag{8A.5}$$

The time dependency of one mode is $\exp(-i\sigma_+ t) = \exp\left[\left(-\kappa(k^2+m^2) + \frac{\gamma k}{\sqrt{k^2+m^2}}\right)t\right] \equiv \exp(\varepsilon t)$ where ε is the growth rate. Pancake-like disturbances, $(m \gg k)$, would be stable at large time since $\exp(-i\sigma_+ t) \to \exp\left[\left(-\kappa m^2 + \frac{\gamma k}{m}\right)t\right] \to 0$. Column-like disturbances, $(m \ll k)$, would also be stable because $\exp(-i\sigma_+ t) \to \exp[(-\kappa k^2 + \gamma)t] \to 0$. The modes with certain intermediate values of m relative to k may be unstable for a given small damping coefficient. We therefore anticipate a maximum growth rate within that range of wavenumber space. The scale of the most unstable mode associated with this instability should be dependent on the damping coefficient.

The growth rate is a function of k, m, κ, γ. The functional dependence can be concisely quantified in terms of two non-dimensional parameters. If we measure distance and time in units of m^{-1} and γ^{-1} respectively, we define $\tilde{\varepsilon} = \varepsilon/\gamma$, $\tilde{\kappa} = \kappa m^2/\gamma$ and $\tilde{k} = k/m$. The non-dimensional expression for the growth rate is

$$\tilde{\varepsilon} = \left(-\tilde{\kappa}\left(\tilde{k}^2 + 1\right) + \frac{\tilde{k}}{\sqrt{\tilde{k}^2 + 1}}\right). \tag{8A.6}$$

The variation of $\tilde{\varepsilon}$ with $\tilde{k}$ and $\tilde{\kappa}$ is presented in Fig. 8A.1. What can we learn quantitatively about the instability from this calculation? It tells us that there would be no instability if the non-dimensional damping coefficient $\tilde{\kappa} = \kappa m^2/\gamma$ is greater than about 0.35. That means that for a given unstable stratification (γ), there would be no instability if the damping coefficient is too strong and/or if the vertical scale of the disturbance is too short. When the instability condition is satisfied, the ratio of the wavenumbers of the most unstable mode is $\tilde{k} = k/m \sim 1$ (almost square-shaped disturbance on x–z plane). Such growth rate is of the order of $\varepsilon/\gamma \sim 0.5$. A weaker damping would favor a more column-like disturbance $(k > m)$. The growth rate approaches the asymptotic value of γ.

A similar system consists of a layer of incompressible fluid with density ρ_1 lying over another layer of incompressible fluid with a density ρ_2 in a gravitational field. Let a wavy perturbation of the interface be $\sim e^{ikx}$. The only length scale in this system is k^{-1} which may be regarded as positive without loss of generality. The corresponding Brunt–Vaisala

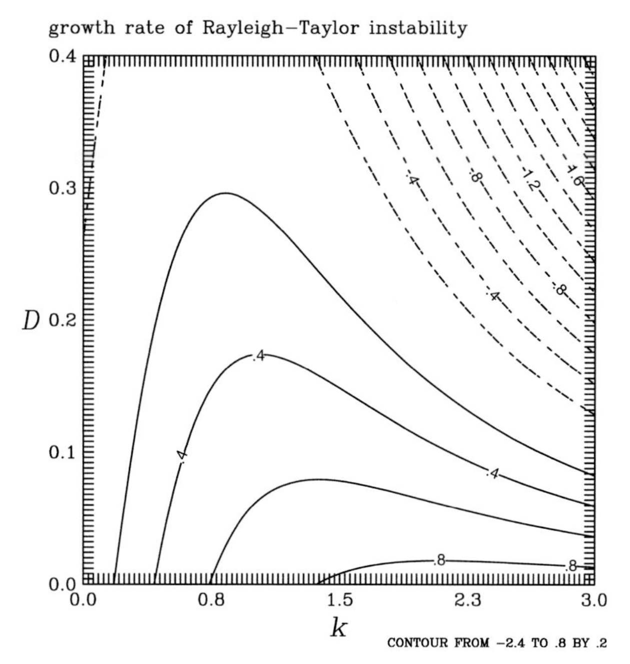

Fig. 8A.1 Variation of the non-dimensional growth rate as a function of the non-dimensional x-wavenumber, $\tilde{k} = k/m$, and viscous coefficient, $D \equiv \tilde{\kappa} = \kappa m^2/|N|$. The zero contour is suppressed.

frequency squared is then $N^2 = \dfrac{gk(\rho_2 - \rho_1)}{(\rho_2 + \rho_1)}$. In light of the discussion above, a disturbance would amplify in time if the upper layer is more dense than the lower layer, $\rho_1 > \rho_2$, so that we have $N^2 < 0$. The instability of this system was first analyzed by Rayleigh (1883). Later, Taylor (1950) pointed out that if a lighter fluid accelerates into a heavier fluid, a similar instability would occur even in the absence of gravity. The instability of this simple two-fluid system is known as *Rayleigh–Taylor instability.*

8A.2 Inertial instability and an application

We next consider a type of instability stemming not from the stratification in conjunction with gravity, but rather from the shear of a basic flow. The instability of a shear flow in a layer of neutrally stratified rotating fluid for a class of disturbances that has the same geometric symmetry as the basic flow itself is known as *inertial instability.* Gravity plays no role in this setting. In the case of a unidirectional shear flow, inertial instability applies to rectilinear disturbances. In a swirling flow, inertial instability is concerned with circularly symmetric disturbances. As different as static instability and inertial instability might appear to be, we will show that they share some intrinsic mathematical similarities. It suffices to discuss inertial instability without dissipation since the influence of dissipation can be inferred later in light of the Rayleigh–Taylor instability.

8A.2.1 Lagrangian analysis

A Lagrangian analysis can also be used to delineate inertial instability. Let us do so in examining axisymmetric disturbances embedded in a balanced vortex in a 2-D rotating homogeneous inviscid fluid, $\hat{v}(r)$ and $\hat{u} = 0$. The momentum equations for axisymmetric flow in this fluid in polar coordinates are

$$\frac{dv}{dt} = -fu - \frac{uv}{r},$$
$$\frac{du}{dt} = -\frac{\partial}{\partial r}\left(\frac{p}{\rho}\right) + fv + \frac{v^2}{r}, \tag{8A.7a,b}$$

where v is the azimuthal velocity and u is the radial velocity. In light of the definition $u = dr/dt$, we obtain an invariance of a fluid parcel by integrating (8A.7a)

$$\left(v + \frac{f}{2}r\right)r = \text{constant}. \tag{8A.8a}$$

Equation (8A.8a) says that the absolute angular momentum of a fluid parcel conserves when it moves either radially inward or outward. Suppose a ring of fluid parcels initially at r_o has azimuthal velocity v_o in equilibrium and is slightly displaced to a location $(r_o + \Delta)$. For a small displacement, $\Delta/r_o \ll 1$, its velocity at the new location according to (8A.8a) would be

$$v \approx v_o - \left(f + \frac{v_o}{r_o}\right)\Delta. \tag{8A.8b}$$

The pressure field associated with a balanced vortex, $\hat{v}(r)$ and $\hat{u} = 0$, is denoted by $\hat{p}$ satisfying

$$0 = -\frac{\partial}{\partial r}\left(\frac{\hat{p}}{\rho}\right) + f\hat{v} + \frac{\hat{v}^2}{r}. \tag{8A.9}$$

Equation (8A.9) indicates that the radial acceleration of a displaced fluid parcel depends on the sum of three forces at its new location: a pressure gradient force, a Coriolis force and a centrifugal force. *We now invoke a key assumption that the pressure gradient force acting on a displaced fluid parcel is the background pressure gradient force at that location.* The pressure gradient force at the new location of the parcel is according to (8A.9) with Taylor's series expansion of $\hat{v}_o$

$$-\frac{\partial}{\partial r}\left(\frac{\hat{p}}{\rho}\right)_{r_o+\Delta} = -f\left(\hat{v}_o + \Delta\frac{\partial \hat{v}}{\partial r}\Big|_o\right) - \frac{\left(\hat{v}_o + \Delta\frac{\partial \hat{v}}{\partial r}\big|_o\right)^2}{r_o + \Delta}. \tag{8A.10}$$

The radial acceleration of the parcel is written as $\dfrac{du}{dt} = \dfrac{d^2r}{dt^2} = \dfrac{d^2\Delta}{dt^2}$. We rewrite (8A.7b) with the use of (8A.8b) and (8A.10) as

$$\frac{d^2\Delta}{dt^2} = -f\left(\hat{v}_o + \Delta\frac{\partial \hat{v}}{\partial r}\Big|_o\right) - \frac{\left(\hat{v}_o + \Delta\frac{\partial \hat{v}}{\partial r}\big|_o\right)^2}{r_o + \Delta} + f\left(v_o - \left(f + \frac{v_o}{r_o}\right)\Delta\right) + \frac{\left(v_o - \left(f + \frac{v_o}{r_o}\right)\Delta\right)^2}{r_o + \Delta}. \tag{8A.11}$$

Since the parcel is initially in equilibrium, we set $v_o = \hat{v}_o$. Equation (8A.11) is the governing equation for one unknown, Δ. For a small displacement, we may approximate (8A.11) by dropping the $O(\Delta^2)$ and higher-order terms and obtain

$$\frac{d^2\Delta}{dt^2} \approx -2\left(\frac{f}{2}+\frac{v_o}{r_o}\right)\left(f+\frac{v_o}{r_o}+\frac{\partial v}{\partial r}|_o\right)\Delta$$
$$= -\frac{1}{r^3}\frac{d}{dr}\left(\left(r^2\eta\right)^2\right)|_o\Delta, \qquad (8A.12)$$

where $\eta = f/2 + v/r$ is the basic absolute angular velocity. The first expression of the RHS of (8A.12) suggests that the radial displacement would increase exponentially in time if the absolute angular velocity, $\left(\frac{f}{2}+\frac{v_o}{r_o}\right)$, and the absolute vorticity of the basic flow, $\left(f+\frac{v_o}{r_o}+\frac{\partial v}{\partial r}|_o\right)$ have opposite signs. This is the *necessary condition for* ***inertial instability*** *of a vortex*. The second expression of the RHS of (8A.12) says that stability requires that the square of the absolute circulation of the basic flow, $\left(r\left(\frac{fr}{2}+v\right)\right)^2$, does not decrease radially outward. This is known as *Rayleigh's circulation criterion* in classical fluid dynamics. The two statements about the necessary condition for inertial instability are of course equivalent.

For a rectilinear basic flow, say in the x-direction denoted by $u(y)$, Δ would stand for a small displacement of a fluid parcel in the y-direction. There is no curvature in this case (formally $r \to \infty$, $\partial/\partial r \to -\partial/\partial y$) and (8A.12) would take on the form

$$\frac{d^2\Delta}{dt^2} \approx -f\left(f-\frac{\partial u}{\partial y}|_o\right)\Delta. \qquad (8A.13)$$

It follows that inertial instability of a parallel shear flow could be expected at a location y_o if the Coriolis parameter f and the absolute vorticity $\left(f-\frac{\partial u}{\partial y}|_o\right)$ have opposite signs. On the basis of (8A.13), we may conclude that the necessary condition for inertial instability is that the absolute vorticity of a basic zonal flow is negative (positive) in the northern (southern) hemisphere. The quantity $\bar{M} \equiv fy - \bar{u}$ is absolute momentum in the x-direction of a parcel in a zonally uniform disturbance in this model. Then $\frac{\partial \bar{M}}{\partial y} = (f - \frac{\partial u}{\partial y}|_o)$ is the absolute vorticity. Thus, it is equivalent to say that the necessary condition for inertial instability, $f\frac{\partial \bar{M}}{\partial y} < 0$, is that the absolute momentum of the basic flow must decrease poleward in both hemispheres.

We can also appreciate the nature of inertial instability from the perspective of balance of forces. Note that the absolute momentum of a displaced parcel of a zonally uniform disturbance conserves, $(dM/dt = du/dt - fv = 0)$. If the momentum of the basic flow is greater further north in the northern hemisphere, a displaced parcel to the north would have smaller absolute momentum than its new environment. Hence, its zonal velocity would be smaller than that of the environment. The northward-oriented pressure gradient force at that location would exceed the Coriolis force associated with the new velocity. Consequently, the displaced parcel would continue to move northward, resulting in instability.

The condition necessary for inertial instability to occur is rare in the atmosphere because the rotation of the Earth is very rapid. Nevertheless, sometimes it appears to prevail at stratospheric levels when there is an easterly jet centering over the equator. The condition of inertial instability could also prevail in the upper troposphere over the Tibetan plateau in summer when surface heating and convective heating are especially strong. The prominent circulation there is known as the "Tibetan high." Some biweekly fluctuations in the weather of southwest China during summer might be attributable to inertial instability of the Tibetan high. They are less restrictively attributable to symmetric instability which will be discussed in Section 8A.3.

8A.2.2 Eulerian analysis

A complete picture of a disturbance undergoing inertial instability can be deduced with an Eulerian analysis. Let us consider *a layer of inviscid homogeneous fluid* of finite thickness, H, instead of a sheet of 2-D fluid. Suppose there is a vertically uniform basic zonal shear flow $(\bar{u}(y), 0, 0)$ in this model setting. This basic flow is assumed to be in geostrophic and hydrostatic balance. The corresponding basic pressure field, $\bar{p}(y,z) = g\rho(H - z) + \hat{p}(y)$, satisfies

$$\begin{aligned} 0 &= -\left(\frac{\hat{p}}{\rho}\right)_y - f\bar{u}, \\ 0 &= -\left(\frac{\bar{p}}{\rho}\right)_z - g, \end{aligned} \tag{8A.14}$$

where $\hat{p}(y)$ is a component of the pressure externally imposed at the top of the fluid layer at $z = H$ to support the basic flow. A weak disturbance that has the same geometric symmetry as the basic flow, $(u'(y,z,t), v'(y,z,t), w'(y,z,t))$, is then governed by

$$\begin{aligned} u'_t + v\bar{u}_y &= fv', \\ v'_t &= -\left(\frac{p'}{\rho}\right)_y - fu', \\ w'_t &= -\left(\frac{p'}{\rho}\right)_z, \\ v'_y + w'_z &= 0. \end{aligned} \tag{8A.15a, b, c, d}$$

Equation (8A.15a,b,c,d) can be reduced to a single equation

$$v'_{yytt} + v'_{zztt} = -f(f - \bar{u}_y)v'_{zz}. \tag{8A.16}$$

Equation (8A.16) is equivalent to $\psi_{yytt} + \psi_{zztt} = -f(f - \bar{u}_y)\psi_{zz}$, where ψ is a streamfunction such that $v' = -\psi_z$, $w' = \psi_y$. It is noteworthy that (8A.16) has the same mathematical form as (8A.3) in the problem of convection without friction. *The quantity $f(f - \bar{u}_y)$ plays the same role as Brunt–Vaisala frequency squared, N^2, in the problem of convection. The y and z axes here are the counterparts of the z and x axes in the convection problem respectively.* On the basis of this mathematical similarity, we may infer that inertial instability would indeed occur when the basic flow is such that $f(f - \bar{u}_y) \equiv -S^2 < 0$ in agreement with the deduction by a Lagrangian analysis. The specially simple case is a basic flow with a sufficiently large uniform shear so that S is a real constant. In an infinite domain, a normal mode solution would then have the form $v' = A\exp(\sigma t)\exp(i(\ell y + mz))$ and the dispersion relation yields $\sigma = \pm Sm/\sqrt{m^2 + \ell^2}$. For $m \gg \ell$, the growth rate would increase with m asymptotically to S. When dissipation is also incorporated, a disturbance of intermediate scale should be the most unstable one as in the case of Rayleigh–Taylor instability. This mathematical similarity between the two physically different forms of instability bridges the physical difference of the two fluid systems.

8A.2.3 Inertial instability of a Gaussian jet

The key parameter $f(f - \bar{u}_y)$ in an atmospheric setting is more typically a function of y rather than a constant. For such a basic flow in this layer of vertically homogeneous fluid, the normal mode solution would have the form of $v'(y, z, t) = V(y)e^{imz}e^{t\sigma}$. The amplitude function is governed by

$$\sigma^2\left(\frac{d^2}{dy^2} - m^2\right)V = f(f - \bar{u}_y)m^2 V. \tag{8A.17}$$

For a finite meridional domain, $-Y \leq y \leq Y$, the appropriate boundary conditions would be $V = 0$ at $y = \pm Y$.

Illustrative analysis

Let us determine the instability properties for an easterly jet of Gaussian shape centering over the equator, $\bar{u} = -u_o \exp(-(y/a)^2)$. To approximate the variation of Coriolis parameter with latitude, we use $f = \beta y$. Then the coefficient on the RHS of (8A.17) is $f(f - \bar{u}_y)m^2 = y^2\beta m^2(\beta - \frac{2u_o}{a^2}\exp(-(y/a)^2))$. It would have negative values to the north of the equator if the jet is sufficiently strong. Measuring distance and velocity in units of L and U, we define the following non-dimensional quantities: $(\tilde{y}, \tilde{a}) = \frac{1}{L}(y, a)$, $\tilde{m} = mL$, $\tilde{\beta} = \beta/UL^{-2}$, $\tilde{\sigma} = \sigma/UL^{-1}$, $(u_o, \tilde{V}) = \frac{1}{U}(u_o, V)$. The non-dimensional form of (8A.17), after dropping the tilde, is

$$\begin{aligned} \sigma^2\left(\frac{d^2}{dy^2} - m^2\right)V &= FV \\ F &= y^2\beta\left(\beta - \frac{2u_o}{a^2}\exp\left(-(y/a)^2\right)\right)m^2, \end{aligned} \tag{8A.18}$$

with boundary conditions: at $y = \pm Y$, $V = 0$. Depicting the domain with J grid points, we recast (8A.18) to a matrix form $\lambda \underline{B}\vec{h} = \underline{A}\vec{h}$ where $\vec{h} = (V_1, V_2, \ldots, V_J)^T$ is the solution vector, $\underline{A}$ and $\underline{B}$ are $J \times J$ matrices. The eigenvalue, $\lambda \equiv \sigma^2$, and the corresponding eigenvector can be readily determined numerically.

The parameter values used in the computation are: $\beta = 2.2, u_o = -1.0$, $a = 0.35, m = 2\pi, Y = 1.0$. The key parameter F has negative values in an equatorial zone (Fig. 8A.2). The largest negative values are located at $y \approx \pm 0.3$.

We present the eigenvalues σ as points on a (σ_r, σ_i) plane (Fig. 8A.3). The unstable modes are stationary ($\sigma_i = 0$). In addition, there are neutral modes propagating in either westward or eastward directions.

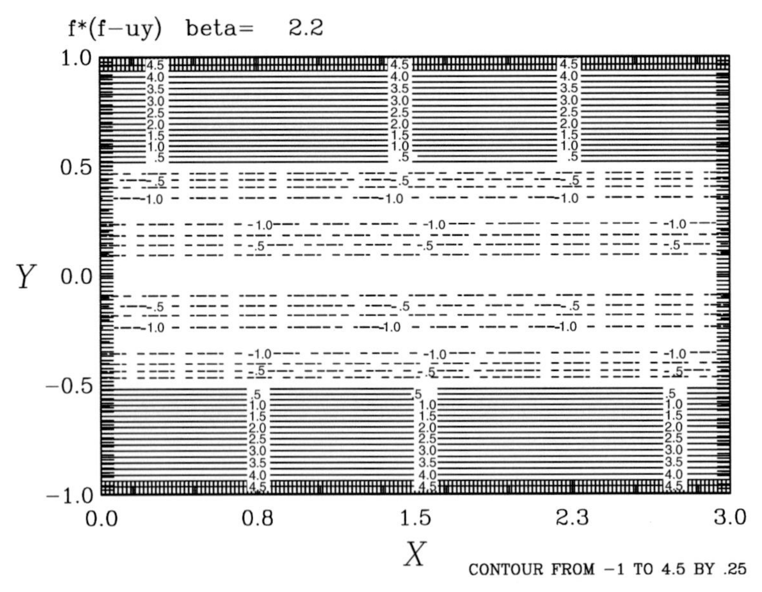

Fig. 8A.2 Distribution of $f\left(f-\bar{u}_y\right)$ of an easterly Gaussian jet over the equator.

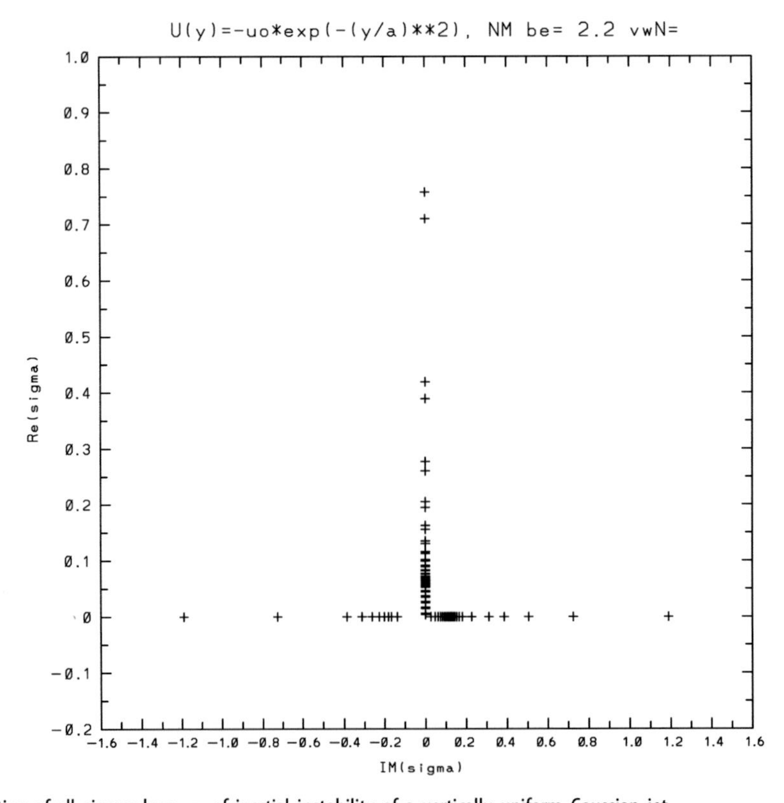

Fig. 8A.3 Distribution of all eigenvalues, σ, of inertial instability of a vertically uniform Gaussian jet.

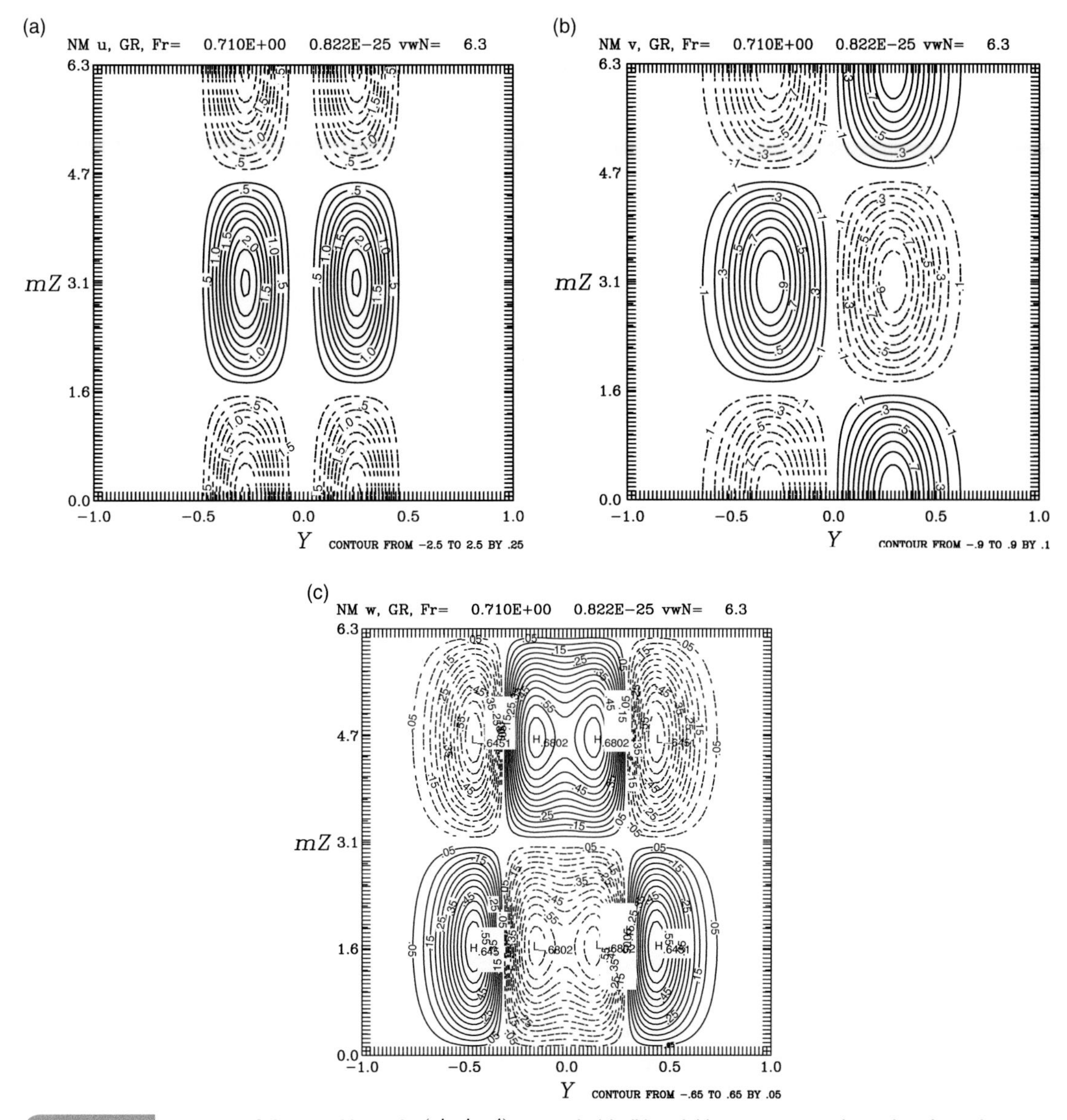

Fig. 8A.4 Structure of the unstable mode (u', v', w') in panels (a), (b) and (c) over one vertical wavelength on the y–z plane associated with inertial instability.

The structure of a normal mode is computed as $(u', v', w') = \mathrm{Re}\{(U, V, W)e^{imz}\}$ where $U = \dfrac{f - \bar{u}_y}{\sigma} V$ and $W = \dfrac{i}{m} V_y$. The results are shown in Fig. 8A.4. The zonal perturbation wind arises from the Coriolis force. The maximum values of u' and v' are located at the two latitudes where $f\left(f - \bar{u}_y\right)$ has largest negative values. The major unstable mode shows that it has a cellular structure on the cross-section across the jet in each tropical region. In view

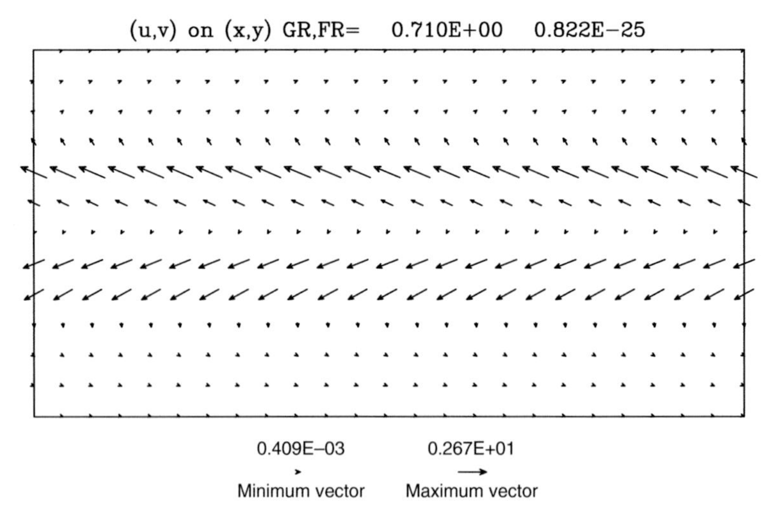

Fig. 8A.5 Distribution of the vector field, (u', v') of an unstable mode at $z = 0$ associated with inertial instability.

of the corresponding vertical motion, w', the disturbance consists of two cells on the y–z plane in each half of the domain.

The product $(u'v')$ at each level is a measure of zonal momentum flux in the y-direction. The unstable disturbance transports zonal momentum away from the jet axis (Fig. 8A.5). Such distribution of Reynold stress in conjunction with the structure of the easterly jet facilitates a conversion of kinetic energy from the basic flow to the disturbance.

8A.3 Symmetric instability and an application

We now discuss a generalized form of inertial instability in a layer of fluid that has a stable stratification as well as a basic flow with horizontal and vertical shear. We focus on the disturbances that have a zonally uniform structure. This is known as *symmetric instability*. When moisture is taken into consideration, the instability is known as *conditional symmetric instability*. Such instability mechanisms are particularly pertinent to meso-scale atmospheric disturbances.

8A.3.1 Lagrangian analysis

The basic state under consideration has a stable stratification $(\theta_z^{(o)} > 0)$ as well as a general zonal shear flow $(\bar{u}, 0, 0)$. We again focus on a class of disturbances that have the same geometric symmetry as the basic flow itself. Suppose the basic flow is in geostrophic balance and has both horizontal and vertical shears. The stratification is uniform. This disturbance would be unstable if the basic shear is strong enough to overcome the stabilizing effect of the stratification. Such instability is referred to as *symmetric instability*.

It is expected to be applicable to the genesis of rectilinear meso-scale disturbances parallel to a background shear flow. To obliquely take into account the influence of stratification in a Lagrangian instability analysis, we deliberately focus on the movement of parcels in this disturbance *on an isentropic surface*. There would be no change in its potential temperature and then the influence of stratification would not be felt in such motions. The analysis for inertial instability in the last section would be directly applicable. Then the necessary condition for symmetric instability is simply $f(f-\bar{u}_y)\big|_{(\theta)} < 0$. In other words, the absolute vorticity on an isentropic surface $(f-\bar{u}_y)\big|_{(\theta)}$ *needs to be negative in the northern hemisphere for symmetric instability to occur*. So, we see that symmetric instability is a close cousin of inertial instability. Recall that potential vorticity is $q = \frac{\vec{\zeta}_a \cdot \nabla\theta}{\rho}$. The PV of the basic state written in isentropic coordinates is $\bar{q} = \frac{1}{\bar{\rho}}(f-\bar{u}_y)\big|_{(\theta)}$ for a basic flow varying only in the y direction. It follows that the necessary condition of symmetric instability is the same as $f\bar{q} < 0$. The existence of negative values of PV in a region would be a necessary condition of symmetric instability.

Note that symmetric disturbances on an isentropic surface are only a subset of possible symmetric disturbances. A general symmetric disturbance may not lie on isentropic surfaces. In other words, a symmetric disturbance may have nonzero potential temperature. The argument used above in deducing the necessary condition of symmetric instability is therefore too restrictive. However, the condition in the form of $f\bar{q} < 0$ is still valid. The basic PV in (y, z) coordinates is

$$\bar{q} = \frac{1}{\rho_o}\left(\bar{\theta}_z(f-\bar{u}_y) + \bar{\theta}_y\bar{u}_z\right), \tag{8A.19}$$

where $\rho_o(z)$ is the background density, $\bar{\theta}(y,z)$ the basic potential temperature and $\bar{u}(y,z)$ the basic zonal flow. Here we see that both the vertical shear and horizontal shear contribute to symmetric instability. The second term is negative because of the thermal wind relation making it easier for symmetric instability to occur.

If the layer of atmosphere under consideration turns out to be saturated with moisture, we may infer that *saturated equivalent potential temperature* should replace potential temperature in the definition of PV in (8A.19). It follows that the necessary condition for symmetric instability under pseudo-adiabatic assumption would be $f\bar{q}_e < 0$, meaning that $\bar{q}_e$ *needs to be negative (positive) in a region of the northern (southern) hemisphere for symmetric instability to occur*. Such instability is called *conditional symmetric instability*. There is supporting evidence for this necessary condition in both observational and numerical simulation studies of precipitation bands.

8A.3.2 Eulerian analysis

An Eulerian instability analysis would give us a complete picture of symmetric instability. Let us consider a rotating Boussinesq fluid with a basic zonal shear flow, $\bar{u}(y,z)$, and a basic potential temperature field, $\left(\theta^{(o)}(z) + \bar{\theta}(y,z)\right)$. The governing equations for a symmetric perturbation in this model are

$$
\begin{aligned}
u'_t + v'\bar{u}_y + w'\bar{u}_z &= fv', \\
v'_t &= -\left(\frac{p'}{\rho}\right)_y - fu', \\
w'_t &= -\left(\frac{p'}{\rho}\right)_z + \frac{g\theta'}{\theta_{oo}}, \\
v'_y + w'_z &= 0, \\
\theta'_t + v'\bar{\theta}_y + w'\theta_z^{(o)} &= 0.
\end{aligned}
\tag{8A.20a, b, c, d, e}
$$

Equation (8A.20d) enables us to introduce a streamfunction on the (y, z) cross-section, $\psi(y, z, t)$, such that $v' = -\psi_z$, $w' = \psi_y$. This system of equations can be reduced to a single equation,

$$\psi_{ttyy} + \psi_{ttzz} = -f(f - \bar{u}_y)\psi_{zz} - N^2\psi_{yy} - 2f\bar{u}_z\psi_{yz} - f\bar{u}_{zz}\psi_y \tag{8A.21}$$

with $N^2 = \frac{g}{\theta_{oo}}\theta_z^{(o)}$. For convenience, we consider a domain $-Y \le y \le Y$ and $0 \le z \le H$ with rigid walls located at $y = \pm Y$. The appropriate boundary conditions are $\psi = 0$ at all boundaries. In the special case of neutral stratification $(N = 0)$ and no vertical shear in the basic flow, (8A.21) would be reduced to (8A.16) in Section 8A.2.2.

Substituting a normal mode solution $\psi(y, z, t) = \varphi(y, z)e^{t\sigma}$ into (8A.21), we get an equation for the amplitude function,

$$\sigma^2(\varphi_{yy} + \varphi_{zz}) = -f(f - \bar{u}_y)\varphi_{zz} - N^2\varphi_{yy} - 2f\bar{u}_z\varphi_{yz} - f\bar{u}_{zz}\varphi_y \tag{8A.22}$$

subject to boundary conditions: $\varphi = 0$ at $y = \pm Y$ and at $z = 0, H$. This is then an eigenvalue–eigenfunction problem with σ being the eigenvalue.

8A.3.3 Symmetric instability of a meso-scale jet

We now apply (8A.22) to a meso-scale jet, $\bar{u}(y, z)$. Suppose the horizontal scale of the jet is L, its vertical scale is D and its maximum speed is V. We define the following non-dimensional quantities: $\tilde{y} = y/L$, $\tilde{z} = z/D$, $(\tilde{f}, \tilde{\sigma}) = \frac{L}{V}(f, \sigma)$, $\tilde{\bar{u}} = \bar{u}/V$, $\tilde{\psi} = \psi/VD$, $\tilde{v} = v/V$, $\tilde{w} = wD/VL$, $\tilde{N}^2 = N^2/V^2D^{-2}$. The vertical shear of the basic flow can be estimated to be $\bar{u}_z = V/D$. Then, the Richardson number is $Ri = N^2/(\bar{u}_z)^2 = N^2D^2/V^2 = \tilde{N}^2$. The Rossby number is $Ro = V/fL = \tilde{f}^{-1}$. The non-dimensional form of (8A.22), after dropping the tilde, is

$$\sigma^2(\varepsilon\varphi_{yy} + \varphi_{zz}) = -f(f - \bar{u}_y)\varphi_{zz} - N^2\varphi_{yy} - 2f\bar{u}_z\varphi_{yz} - f\bar{u}_{zz}\varphi_y, \tag{8A.23}$$

where $\varepsilon = D^2/L^2$. The domain is depicted by 50 grids in both y- and z-directions. The resolution is $\delta z = 0.04$ and $\delta y = 0.08$ corresponding to 80 m and 160 m respectively. We cast (8A.23) with boundary conditions in finite difference form as a matrix equation, which can be readily solved.

We use $D = 2\,\text{km}$, $L = 20\,\text{km}$, $V = 20\,\text{ms}^{-1}$, $f = 10^{-4}\,\text{s}^{-1}$, $N = 10^{-2}\,\text{s}^{-1}$ in the following calculation. The Richardson number is then 1.0 and the Rossby number is 10.

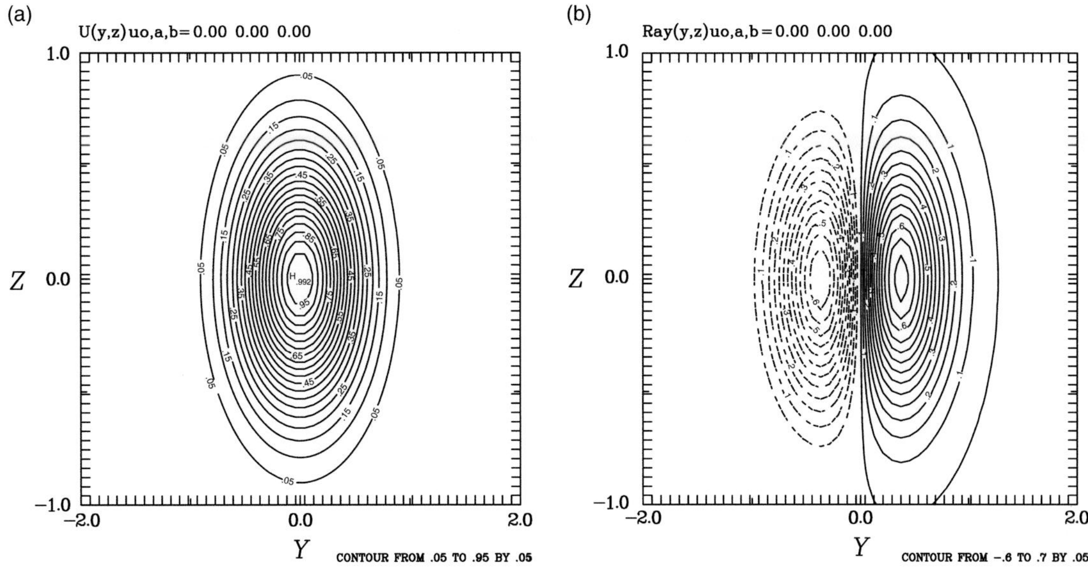

Fig. 8A.6 Structure of (a) the velocity of a meso-scale jet and (b) $\left[\tilde{f}\left(\tilde{f}-\tilde{\bar{u}}_y\right)\right](\delta y/\delta z)^2$.

Therefore, the disturbance is highly non-geostrophic. We consider the following specific basic velocity in non-dimensional form

$$\tilde{\bar{u}} = \exp\left(-\left(\frac{\tilde{y}}{a}\right)^2 - \left(\frac{\tilde{z}}{b}\right)^2\right), \tag{8A.24}$$

where $a = 0.5,\ b = 0.5$ in a domain $-2 \le \tilde{y} \le 2, -1 \le \tilde{z} \le 1$.

The thermal wind equation is $\bar{u}_z \approx -\dfrac{g}{f\theta_{oo}}\bar{\theta}_y$. The corresponding non-dimensional form of the thermal wind equation is $\tilde{\bar{\theta}}_{\tilde{y}} = -\dfrac{V^2\tilde{f}}{gD}\tilde{\bar{u}}_{\tilde{z}}$ with $\tilde{\bar{\theta}} = \bar{\theta}/\theta_{oo}$. For the basic flow (8A.24), we have $\tilde{\bar{\theta}}_{\tilde{y}} = -\dfrac{2V^2\tilde{f}}{gDb^2}\tilde{z}\tilde{\bar{u}}$. Let us also measure $\theta^{(o)}$ used for computing the stratification in units of θ_{oo}. Using a constant Brunt–Vaisala frequency, we have $\theta^{(o)} = \lambda z$. Its non-dimensional value would be $\tilde{\theta}^{(o)} = \theta^{(o)}/\theta_{oo} = \tilde{\lambda}\tilde{z}$ where $\tilde{\lambda} = D\lambda/\theta_{oo}$. Hence, the background potential temperature is

$$\tilde{\theta}_{background} = \tilde{z}\left(\tilde{\lambda} - \frac{2V^2\tilde{f}}{gD}\int_{-Y}^{\tilde{y}} \tilde{\bar{u}}\, d\tilde{y}\right). \tag{8A.25}$$

The structures of this basic flow and the related crucial parameter $\left[\tilde{f}\left(\tilde{f}-\tilde{\bar{u}}_y\right)\right](\delta y/\delta z)^2$ are shown in Fig. 8A.6. We see that the horizontal shear of the jet is strong enough that $\left[\tilde{f}\left(\tilde{f}-\tilde{\bar{u}}_y\right)\right]$ has negative values in a region in the southern half of the jet.

Figure 8A.7 shows that while most symmetric modes are neutral, there are many unstable modes. They are all non-propagating since their σ_i is zero. The most unstable mode has a growth rate ~0.37. Since one unit of time is $L/V = 16.7$ min, the corresponding e-folding time is about 45 minutes. This value is relevant to the development of atmospheric meso-scale disturbances.

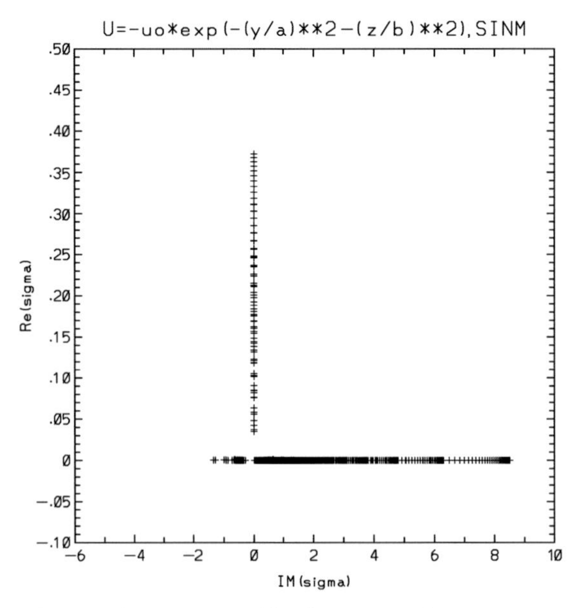

Fig. 8A.7 Distribution of all eigenvalues, σ, of symmetric instability for a meso-scale jet.

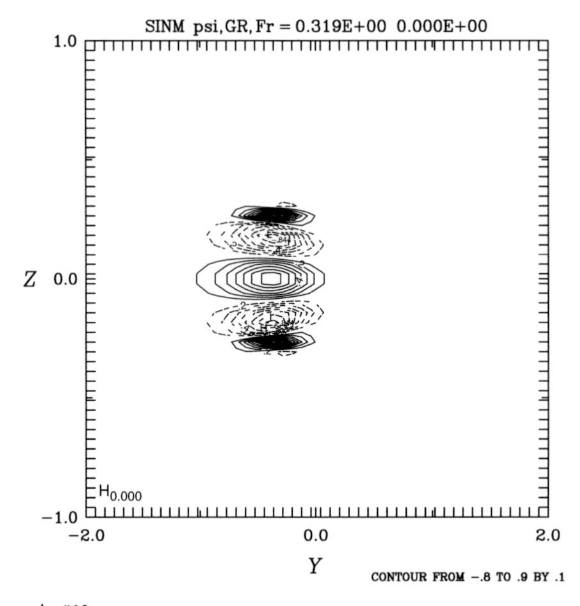

Fig. 8A.8 Structure of unstable mode #10.

The structure of a representative unstable mode is shown in Fig. 8A.8. We see that all unstable modes are located in the region where $\left[f\left(f-\bar{u}_y\right)\right]$ is negative. It confirms that the instability mostly stems from the horizontal shear of the basic flow and is akin to inertial instability. Note that the growth rates of the first 15 unstable modes do not differ by much, ranging from 0.37 to 0.29. The primary family of unstable modes has the structure of a vertical stack of cells. Each cell is relatively thin compared to its width. The cells of the less unstable modes are bigger. A secondary family of unstable modes has the structure of a double-stack of cells of smaller horizontal extent. In light of these results, we may conclude that inertial instability and symmetric instability are fundamentally similar. Furthermore, had we incorporated weak dissipation in the model, the modes with short scales would be attenuated. In other words, dissipation should give rise to a finite scale selection of symmetric instability just as in Rayleigh–Taylor instability.

PART 8B PURELY BAROTROPIC AND PURELY BAROCLINIC INSTABILITY

This part of the chapter discusses many aspects of the instability problem of parallel shear flows in a dry atmospheric model setting. We begin by briefly commenting on the historical highlights of this type of study in Section 8B.1. The instability of a parallel shear flow in a 2-D fluid, known as barotropic instability, is discussed in Section 8B.2. The necessary condition of barotropic instability is first deduced in Section 8B.2.1. It is supplemented with a quantitative modal instability analysis for a Gaussian jet in Section 8B.2.2. This instability is interpreted from two different and complementary perspectives in Sections 8B.2.3 and 8B.2.4. In addition to examining modal growth, it is pertinent to examine the transient growth of a disturbance as an initial value problem. One particularly intriguing aspect of transient growth is the optimal growth for a disturbance over a particular time interval. An analysis of optimal excitation of a barotropic disturbance over a finite time interval in a Gaussian jet is presented in Section 8B.3.

The differences among modal growth, transient growth and optimal growth may be perhaps appreciated using track-and-field as a metaphor. Intensification of an ensemble of disturbances according to their modal growth is like a marathon race in which the most unstable mode would be the eventual winner. Transient growth of a particular disturbance is more akin to sprint. Optimal growth is what one might call an intermediate distance race favoring the runner who has special ability for a specific distance.

The instability of a parallel flow with a vertical shear is known as baroclinic instability. It is most relevant to the problem of cyclogenesis in the extratropics. We discuss all aspects of it in the context of a two-layer quasi-geostrophic model for simplicity. Section 8B.4 elaborates on the physical nature of baroclinic instability for a modal disturbance. A comprehensive modal baroclinic instability analysis is given in Section 8B.5. Section 8B.6 discusses baroclinic instability from the perspective of transient growth with two illustrative analyses. The problem of optimal excitation of a baroclinic disturbance is addressed in Section 8B.7.

This part of the chapter is concluded with a derivation of the necessary condition for instability for a general parallel shear flow in Section 8B.8. This result is reducible to those conditions for the two special cases earlier discussed in Sections 8B.2.1 and 8B.4.1.

8B.1 Historical highlights of past studies

A question of particular relevance to weather forecasting in the first half of the century was "How do transient large-scale atmospheric disturbances such as extratropical cyclones and anticyclones come about?" It was the focus of research for two generations of atmospheric scientists. It was conjectured that the answer is to be found in terms of dynamic instability of an appropriate background shear flow. The early works of the so-called Bergen school in Norway tried to prove that cyclones are unstable modes of the flow associated with an atmospheric front (Bjerknes, 1919). The results did not prove to be encouraging because there is little resemblance between such unstable disturbance and a typical extratropical cyclone. This line of research activities came to a lull in the 1930s. There was renewed interest in the 1940s. One notable publication is concerned with the simplest form of instability, namely that of a parallel flow in a two-dimensional fluid (Kuo, 1949). Such basic flow has only horizontal shear. This is referred to as *purely barotropic instability.* Kuo's result generalizes Rayleigh's result for a non-rotating fluid. This form of instability was shown decades later to be relevant, among other things, to the origin of easterly waves in the tropics.

A break through in the research of cyclogenesis came from two publications concerned with the instability of a parallel flow with only vertical shear in a stratified rotating fluid (Charney, 1947; Eady, 1949). It is referred to as *purely baroclinic instability.* Many researchers in the several decades since then have greatly generalized those investigations. The term *baroclinic instability* is often loosely used in the literature as a referrence to the instability of a general shear flow in which the vertical shear is only one of several but important parameters.

There are perhaps two main reasons as to why it took the research community many decades to reach the current understanding of baroclinic instability. One reason is that the dynamical nature of baroclinic instability per se is rather subtle. The other reason is that it has several distinctly different aspects that could not be delineated in one go. The most subtle aspect of baroclinic instability is that, although internal gravity waves intrinsically exist in a stratified atmosphere, they are superfluous as far as extratropical cyclogenesis is concerned. The coexistence of internal gravity waves and Rossby waves in a rotating stratified fluid greatly complicates a mathematical instability analysis. This complexity was not appreciated in the early attempts when researchers investigated cyclogenesis as shear instability of a flow similar to that around an atmospheric front. Those major unstable modes do not capture the salient characteristics of an extratropical cyclone since they are mostly unstable internal gravity wave modes. It turns out that Rossby waves are the ones primarily involved in baroclinic instability as shown in the seminal publications of Charney and Eady. They formulated models that for the first time filter out the internal gravity wave modes a priori. Those researchers appreciated the importance of the fact that the flow in an extratropical cyclone is a balanced flow. It led them to invoke the constraints of geostrophic balance and hydrostatic balance in the course of simplifying the governing equations. They reduced the original set of governing equations to effectively a quasi-geostrophic potential vorticity equation. Such a derivation is elaborated in Section 6.3. The absence of internal gravity waves in such a model is demonstrated in Section 6.6.

Eady and Charney obtained analytic solutions in their modal instability analyses for a continuous vertical domain. The Eady model has no beta-effect, a uniform stratification, and a basic zonal flow with constant vertical shear between two rigid boundaries. A flow in that model has uniform potential vorticity. The instability analysis is relatively straightforward, but the result is very enlightening. In contrast, the Charney model is more general in that it incorporates the beta-effect, a height-varying stratification, and a constant vertical shear with only a lower rigid boundary. Consequently, a flow in that model has non-uniform potential vorticity. Although the beta-effect alone provides a stabilizing restoring mechanism for neutral Rossby waves, it opens up the possibility of a multitude of unstable modes with different vertical scales. The mathematical analysis is much more involved, as meticulously elaborated in Pedlosky's book (1987, Section 7.8). An unstable normal mode in Charney's model can have any horizontal length scale, with the exception of a set of discrete values. It has a vertical scale that enables its motion to be energy releasing.

Farrell (1989) pointed out that *transient growth* of an arbitrary disturbance can be even more pertinent to forecasting the development of cyclones and anticyclones a few days ahead of time. Furthermore, he showed that it is possible to determine the structure of a baroclinic wave that grows optimally over a specific time interval according to a chosen norm.

Much of the mathematical difficulty in Charney's analysis would be avoided if one limits the vertical degrees of freedom to only two as first shown by Phillips (1954). Although there is only one vertical mode in a two-layer model, it adequately captures the most relevant unstable mode in the Charney model. As such, a two-layer model is a tool of great pedagogical value. We take advantage of the simplicity of a two-layer quasi-geostrophic channel model to analytically investigate all different aspects of purely baroclinic instability with the least effort. All results deduced from this model, of course, have counterparts in a multi-level quasi-geostrophic model. As long as quasi-geostrophic constraint is an acceptable approximation, the results in turn also have counterparts in a general atmospheric model under similar parameter condition.

8B.2 Aspects of barotropic instability

In this section, we analyze the instability of a parallel shear flow in a two-dimensional fluid model in a reentrant channel domain laterally bounded by two rigid walls ($-X \leq x \leq X$, $-Y \leq y \leq Y$). Beta-plane approximation is introduced for the Coriolis parameter, ($f = f_o + \beta y$). The basic flow is denoted by $\bar{u}(y)$, with a corresponding streamfunction $\bar{\psi}(y) = -\int \bar{u} dy$ and relative vorticity $\bar{\zeta}(y) = -\bar{u}_y$. The total streamfunction of the flow is then $\psi_{total} = \bar{\psi} + \psi$. The perturbation velocity is $(u, v) = (-\psi_y, \psi_x)$. A perturbation in this model is governed by the linearized form of the equation for conservation of absolute vorticity, (3.15), in Section 3.3, namely

$$\nabla^2\psi_t + \bar{u}\nabla^2\psi_x + \psi_x(\bar{\zeta}_y + \beta) = 0. \tag{8B.1}$$

It is subject to the following boundary conditions:

$$\begin{aligned} &\text{at } y = \pm Y \quad v = 0 \Rightarrow \psi = 0, \\ &\psi(-X, y) = \psi(X, y). \end{aligned} \tag{8B.2}$$

Note that there is no need to assume quasi-geostrophic balance in this flow.

8B.2.1 Necessary condition for barotropic instability

We first analytically derive the necessary condition for barotropic instability of a given zonally uniform flow. Multiplying (8B.1) by $\frac{\nabla^2\psi}{\bar{\zeta}_y + \beta}$ and integrating over the domain with the use of the boundary conditions, we get

$$\frac{d}{dt}\left\langle \frac{\left(\nabla^2\psi\right)^2}{\bar{\zeta}_y + \beta} \right\rangle = 0. \tag{8B.3}$$

The domain integral is denoted by $\langle\,\rangle$. The derivation of (8B.3) is valid provided that either $\left(\bar{\zeta}_y + \beta\right)$ is not zero anywhere in the domain, or $\nabla^2\psi$ and $\left(\bar{\zeta}_y + \beta\right)$ are both zero at the same point(s). The quantity $\frac{\left(\nabla^2\psi\right)^2}{\bar{\zeta}_y + \beta}$ has the dimension of velocity and is called *pseudo-momentum*. Equation (8B.3) says that the total pseudo-momentum of any perturbation conserves throughout its evolution. This result is rather amazing at first sight, because although $\left(\nabla^2\psi\right)^2$ is positive definite and may change in time everywhere, a disturbance nevertheless has an invariant property, $\left\langle \frac{\left(\nabla^2\psi\right)^2}{\bar{\zeta}_y + \beta} \right\rangle$. This would be possible only if $\left(\bar{\zeta}_y + \beta\right)$ changes sign in the domain so that an increase of $\frac{\left(\nabla^2\psi\right)^2}{\bar{\zeta}_y + \beta}$ in one region would be compensated by a corresponding decrease in another region. In other words, *the existence of* $\left(\bar{\zeta}_y + \beta\right) = 0$ *at one or more latitudes is a necessary condition for instability of a parallel shear flow.* The value of absolute vorticity of the basic flow is an extremum at such a location. This is known as the *Rayleigh–Kuo condition* for barotropic instability of a parallel shear flow, originally deduced in a normal mode analysis. Although this condition enables us to predict whether a given shear flow is unstable or not, it does not tell us much else about the instability, such as the growth rate, movement and structure of the unstable disturbances. The additional information must be determined with a supplemental numerical instability analysis.

8B.2.2 Modal instability properties and structure

For a given steady basic shear flow, the functional dependence of a perturbation in time can be $\exp(i\sigma t)$. This means that it may continually intensify or decay in time when σ has a complex value without changing its structure except possibly in an oscillatory manner. Such a disturbance is called a "*normal mode*" and its time evolution is called modal growth. The amplifying disturbances are called unstable normal modes. We do not need to know about the initial state of a disturbance field in order to determine the most unstable normal mode.

It is convenient to measure distance, velocity, time and streamfunction in units of $L = 1000\,\text{km}$, $U = 10\,\text{m}\,\text{s}^{-1}$, $LU^{-1} = 10^5\,\text{s}$ and $UL = 10^7\,\text{m}^2\,\text{s}^{-1}$ respectively. Let us consider a basic flow in the form of a Gaussian jet that centers at a reference latitude in the extratropics referred to as $y = 0$. It is prescribed in non-dimensional form as

$$\bar{u} = u_o \exp\left(-(y/a)^2\right), \quad \bar{v} = 0 \Rightarrow \bar{\zeta} = 2y\bar{u}/a^2. \tag{8B.4}$$

It is relevant to consider a basic state for a westerly jet defined by $u_o = 3$ and $a = 0.75$ with $f_o = 10$ and $\beta = 1.6$. The minimum value of $\bar{\zeta}$ is about -3.3. It follows that the absolute vorticity is positive everywhere and inertial instability could not occur in this flow according to the discussion in Section 8A.2.1. Furthermore, the maximum value of $\bar{\zeta}_y = \frac{2\bar{u}}{a^2}\left(1 - \frac{2y^2}{a^2}\right)$ is about 10.7 at $y = 0$ and is therefore much larger than $\beta = 1.6$. The impact of the beta-effect is therefore relatively minor in this case. The non-dimensional form of the governing equation (8B.1) is

$$\nabla^2\psi_t = -\bar{u}\nabla^2\psi_x - \psi_x\left(\bar{\zeta}_y + \beta\right). \tag{8B.5}$$

Since the coefficients in (8B.5) are independent of x, a normal mode solution for a given zonal wavenumber k would have the form of $\psi = \phi(y)\exp(\sigma t + ikx)$. There is no restriction to the value of k when we consider an infinitely long channel. For a channel of finite length $(2X)$, we should only consider $k = n2\pi/2X$ with n being an integer so that the cyclical condition would be satisfied. The amplitude of a normal mode is governed by

$$\left(\frac{d^2}{dy^2} - k^2 + \frac{ik\left(\bar{\zeta}_y + \beta\right)}{\sigma + i\bar{u}k}\right)\phi = 0,$$
$$\text{b.c. at } y = \pm Y,\ \phi = 0. \tag{8B.6}$$

Equation (8B.6) can only be solved numerically for the basic flow given by (8B.4). Depicting the meridional domain by J uniform grid points, casting (8B.6) in finite-difference form and applying the boundary conditions, we get a matrix equation

$$A\vec{\phi} = \sigma B\vec{\phi}, \tag{8B.7}$$

where A and B are $J \times J$ matrices. The problem is now reduced to an eigenvalue–eigenvector problem with σ being an eigenvalue and $\vec{\phi}$ the corresponding eigenvector. One can readily solve (8B.7) for all σ and $\vec{\phi}$ with a standard matrix subroutine.

The computation is made for a domain of 6000 km wide, $-3 \le y \le 3$, depicted by $J = 100$ grid points. We first analyze the instability of a wave that has a zonal wavelength of 4000 km, so that the non-dimensional wavenumber is $k = 2\pi/4 = 1.57$. Although there are J numerical eigenvalues, all of them are not necessarily true normal modes. It can be proven that generally there are a number of discrete normal modes and a spectrum of so-called *continuum modes*. The former have a continuous structure, whereas the latter have first-order discontinuity in their spatial structures. It is difficult to analytically establish a priori how many normal modes there are for a given basic flow. By examining the sensitivity of the numerical eigenmodes to the model resolution, we can however identify a subset of them $(M < J)$ as approximate normal modes. They are not sensitive to the resolution used in the computation. The remaining $(J - M)$ solutions are numerical counterparts of the continuum modes, which are sensitive to the resolution. These numerically established modes constitute a complete set of functions for a particular resolution in the sense that any arbitrary disturbance in the model can be represented as a linear combination of them. The evolution of an arbitrary disturbance is equivalent to the evolution of an ensemble of the modes.

We present all eigenvalues as points on a (σ_r, σ_i) plane in Fig. 8B.1. A mode with $\sigma_r > 0$ is an exponentially amplifying normal mode. It is found that there is only one intensifying

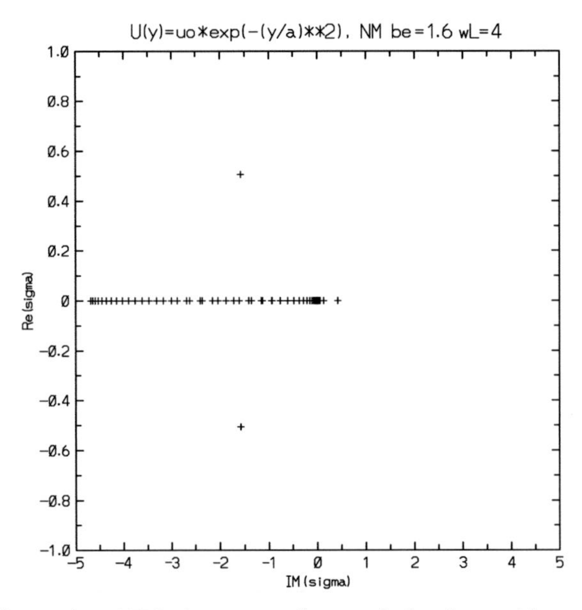

Fig. 8B.1 The eigenvalues for the case $k = 1.57$ in the presence of a westerly Gaussian zonal jet.

mode and one decaying mode in this case. These two numerically obtained eigenvalues are true eigenvalues for we have checked that they are not sensitive to the resolution used in the computation. The remaining 98 numerical eigenvalues are neutral modes ($\sigma_r = 0$). They are the continuum modes.

The growth rate of the unstable mode, σ_r, is about 0.5. The e-folding time of the amplification is therefore about 2 days. The frequency is -1.58 meaning that the unstable mode propagates eastward at a speed equal to $-\sigma_i/k \approx 1.0$ meaning $10\,\mathrm{m\,s^{-1}}$. The period of the wave is ~ 4 days. These characteristics are compatible with some large-scale wave disturbances in the atmosphere.

The eigenfunction of an unstable mode has complex value, $\phi = \phi_r + i\phi_i$. We normalize it so that its total energy is equal to unity. Its structure at a time $t = 0$ is computed as $\psi = \phi_r \cos kx - \phi_i \sin kx$. The shape of the unstable mode is like that of a boomerang (Fig. 8B.2). It tilts in a direction *leaning against* the shear on each side of the westerly jet. This is a signature feature of an unstable mode.

For comparison, let us also examine the structure of the neutral modes. In contrast, a neutral mode has no meridional tilt (Fig. 8B.3). There is a *first-order discontinuity* at two grid points close to $y = \pm y_1 = \pm 0.59$ where $(\sigma_i + \bar{u}k) = 0$. In other words, this mode propagates eastward with a speed $-\sigma_i/k \approx 2.0$ which is equal to the value of the basic flow at $\pm y_1$. Such latitudes are known as critical latitudes. The more grid points we use to depict the domain, the more neutral modes we would get. That is why the precise values of the eigenvalues of particular neutral modes depend on the resolution.

The variation of the growth rate of the unstable mode, σ_r, with its wavelength is an important characteristic of barotropic instability of this Gaussian jet. Its dependence on the beta-effect and the direction of the basic flow turns out to be rather intricate. Additional computations are made to ascertain such functional dependence. Figure 8B.4a

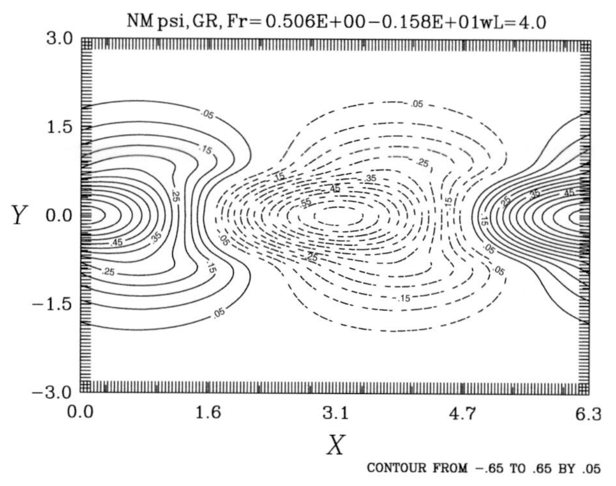

Fig. 8B.2 Structure of the streamfunction of the unstable mode for a westerly Gaussian zonal jet.

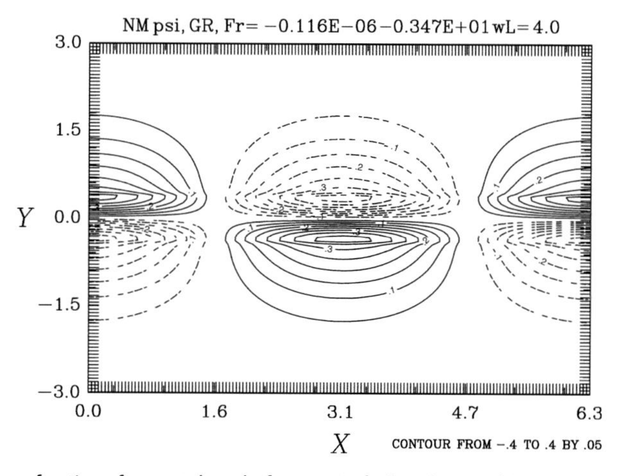

Fig. 8B.3 Structure of the streamfunction of a neutral mode for a westerly Gaussian zonal jet.

shows this variation for the case of a westerly jet with $\beta = 0$. Figure 8B.4b shows the counterpart result with $\beta = 1.6$. There are two branches of unstable modes in each case. In other words, there are two unstable modes in a certain range of wavelength. They differ by their meridional structures. The branch of modes that has larger values of growth rate is a primary branch. In the case of $\beta = 0$, the two curves do not intersect (Fig. 8B.4a). It is seen that the beta-effect is a stabilizing factor since the corresponding growth rates of both branches are larger in the case of $\beta = 0$ than in the case of $\beta = 1.6$. The maximum growth rate of the primary branch for $\beta = 0$ is about 0.75, whereas it is about 0.5 for $\beta = 1.6$. Furthermore, there is a shortwave cutoff in both branches of unstable modes. The minimum wavelength of the primary branch of unstable modes in both cases is about 2.5 (2500 km). The primary branch has a longwave cutoff only in the case of $\beta = 1.6$ at a wavelength of about 7.0 (7000 km). The impact of the beta-effect on the secondary branch is much weaker. The shortwave cutoff is somewhat longer (~5.0 instead of ~4.0) and

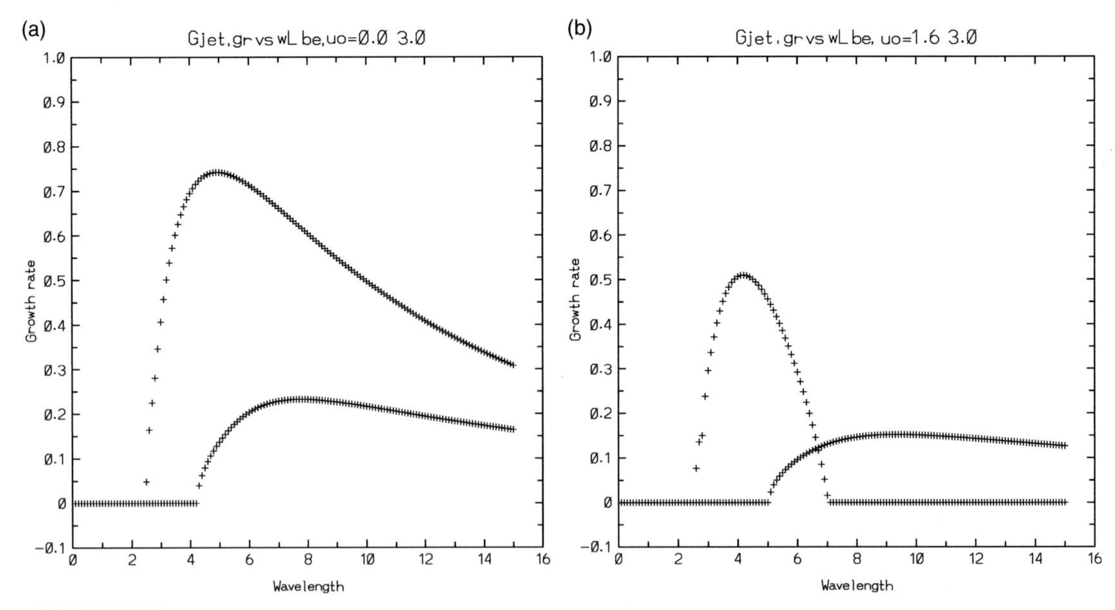

Fig. 8B.4 Variation of the growth rate of unstable modes for *a westerly Gaussian jet* with wavelength in the case of (a) $\beta = 0$ and (b) $\beta = 1.6$.

there is no longwave cutoff for $\beta = 1.6$. Consequently, the secondary branch of modes becomes the sole unstable modes for sufficiently long wavelength when the beta-effect is incorporated. It may be added that all unstable modes propagate eastward as inferred from the results of the imaginary part of the eigenvalue, σ_i. The variation of such frequency with wavelength is not presented in the interest of brevity.

For an *easterly Gaussian jet*, the curves depicting the two branches of unstable modes do not intersect up to wavelength ~15.0 (Fig. 8B.5). The growth rate values of the corresponding modes in each branch are *larger* for $\beta = 1.6$ than for $\beta = 0$. In other words, the beta-effect has a *destabilizing effect* instead (Fig. 8B.5b vs. Fig. 8B.5a). This result is by no means intuitively obvious. These unstable modes propagate westward as indicated by the result of the imaginary part of the eigenvalues largely due to the advective influence of the basic flow.

8B.2.3 Instability from the perspective of energetics

It is instructive to trace the flow of energy in an unstable barotropic wave. We make use of the structure of an unstable disturbance in doing so. Multiplying (8B.5) by ψ, we get

$$\nabla\cdot(\psi\nabla\psi_t) - \nabla\psi\cdot(\nabla\psi_t) = -\bar{u}\left(\psi\nabla^2\psi\right)_x + \bar{u}\psi_x\nabla^2\psi - \psi\psi_x\bar{\zeta}_y. \tag{8B.8}$$

We get from taking the zonal average of (8B.8)

$$-\left[\nabla\psi\cdot(\nabla\psi_t)\right] = \left[\bar{u}\psi_x\psi_y\right]_y - \bar{u}_y\left[\psi_x\psi_y\right], \tag{8B.9}$$

where a zonal average is denoted by a square bracket, [ ]. The boundary conditions have been applied in getting (8B.9). A combination of zonal average and meridional average is

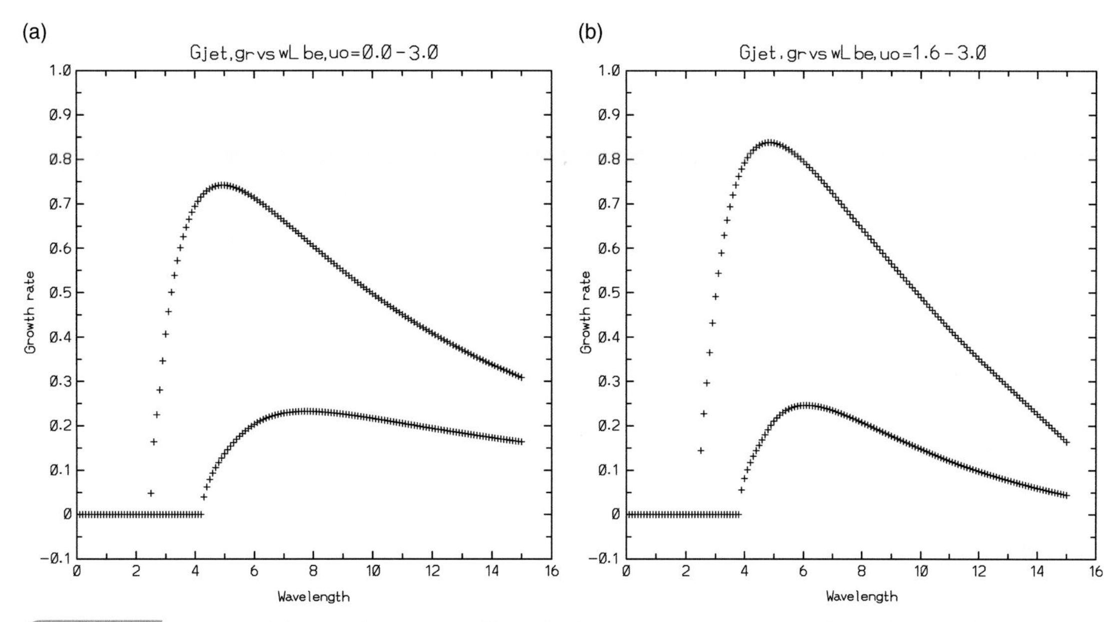

Fig. 8B.5 Variation of the growth rate of unstable modes for *an easterly Gaussian jet* with wavelength in the case of (a) $\beta = 0$ and (b) $\beta = 1.6$.

a domain average. Integrating (8B.9) over the y-domain and applying the boundary conditions at $y = \pm Y$, we then get

$$\frac{dK'}{dt} = -\langle \bar{u}_y u'v' \rangle, \tag{8B.10}$$

where the domain integration is denoted by $\langle\rangle$. $K' = \left\langle \frac{1}{2}(u'^2 + v'^2) \right\rangle$ is the kinetic energy of a wave disturbance with $u' = -\psi_y$ and $v' = \psi_x$. Energetically speaking, a perturbation could intensify in time by extracting energy from the basic flow at a rate represented by the expression on the RHS of (8B.10). The Reynolds stress associated with a perturbation is $-[u'v']$. Its sign depends on the meridional tilt. It follows that a normal mode would intensify in time only if its meridional tilt "*leans against the basic shear*," whereby $-\langle \bar{u}_y u'v' \rangle > 0$. Dynamically speaking, this means that the unstable normal mode must transport zonal momentum in the down-gradient direction of the basic flow. This is why a barotropically unstable wave must have such a signature tilt.

8B.2.4 Instability from the perspective of positive feedback from constituent components (wave resonance mechanism)

The dynamical nature of modal instability of a barotropic flow can be more succinctly interpreted as a process of mutual reinforcement of the constituent components of an unstable mode. Recall that the absolute vorticity of the jet under consideration is $\bar{\zeta}^{(a)} = \beta y + 2u_o y e^{-y^2}$. Its $\bar{\zeta}^{(a)}{}_{\max}$ is located at $y \approx 0.6$ and $\bar{\zeta}^{(a)}{}_{\min}$ at $y \approx -0.6$ (Fig. 8B.6). It follows that we have $\partial\bar{\zeta}^{(a)}/\partial y > 0$ in $|y| < 0.6$ and $\partial\bar{\zeta}^{(a)}/\partial y < 0$ in $|y| > 0.6$.

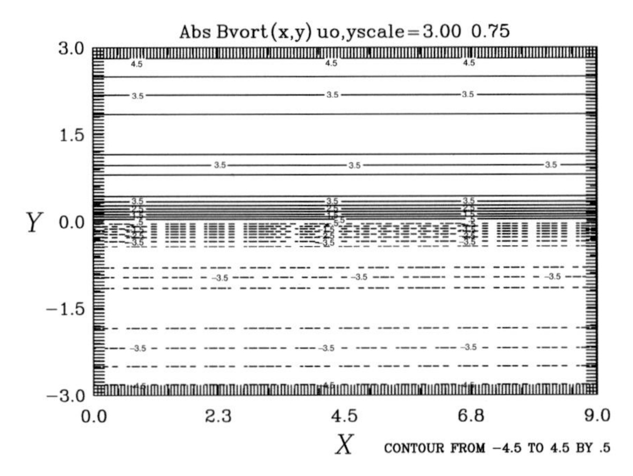

Fig. 8B.6 Distribution of the absolute vorticity of the basic flow.

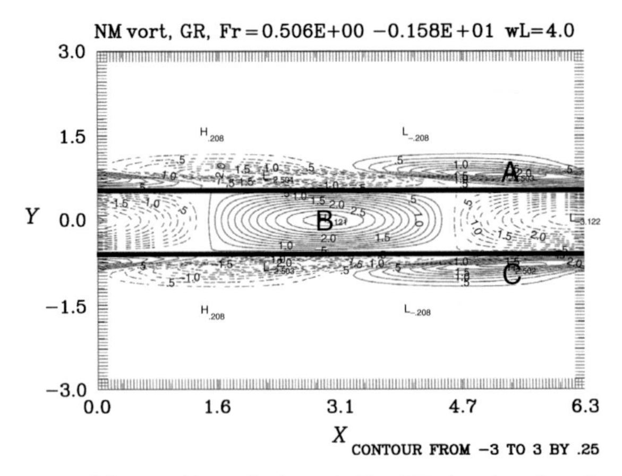

Fig. 8B.7 The structure of the vorticity of the unstable mode shown in Fig. 8B.2. Location of maximum $\bar{\zeta}$ and location of minimum $\bar{\zeta}$ indicated by the two lines.

The corresponding vorticity field of the unstable mode shown in Fig. 8B.2 is presented in Fig. 8B.7.

The meridional tilt of the vorticity contours of the unstable mode has particular implications. The location of maximum $\bar{\zeta}$ is highlighted by a thick solid line and the location of minimum $\bar{\zeta}$ by a thin solid line in Fig. 8B.7. It is instructive to begin by examining the mutual influences among three positive vorticity anomalies exemplified by those labeled as A, B and C. Anomaly B lies on $y = 0$, A above $y \approx 0.6$ and C below $y \approx -0.6$. To every vorticity element $\zeta^{(j)}$ there corresponds a flow component in light of the piece-wise inversion relationship, $\psi^{(j)} = \nabla^{-2}\zeta^{(j)}$. For example, we can visualize a cyclonic flow associated with C. This flow component clearly would bring about positive advection of basic vorticity $\left(-v'\bar{\zeta}_y > 0\right)$ at B. By the same token, the cyclonic flow associated with B also would bring about $\left(-v'\bar{\zeta}_y > 0\right)$ at the center of C. In other words, these two vorticity anomalies mutually reinforce one another through their advection of basic vorticity at each other's location. The same can be said about the mutual influence

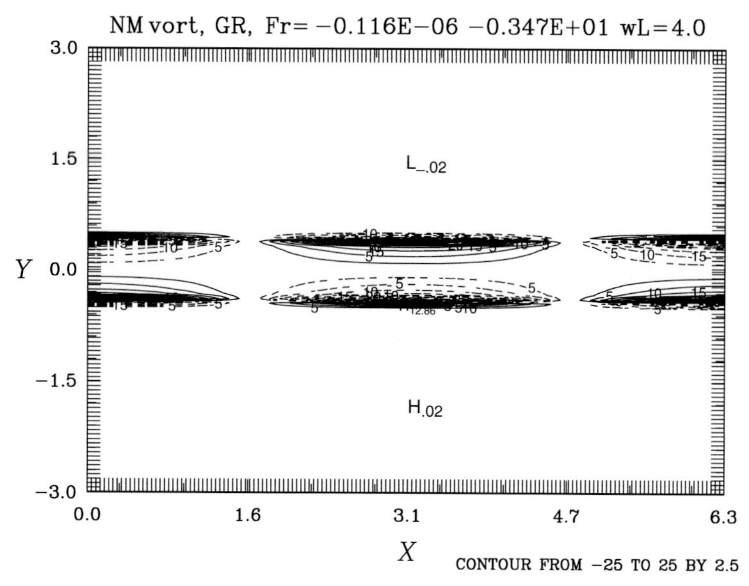

Fig. 8B.8 Structure of the vorticity field of the neutral mode shown in Fig. 8B.3.

between A and B, and also about the mutual influence between A and C although it is considerably weaker because they are farther apart. Thus, all positive vorticity anomalies collectively have positive feedback influence upon themselves. Note that the adjacent negative vorticity anomalies on either side of a positive anomaly have opposite influences and thereby have no net impact. All we have said above about the positive vorticity anomalies applies equally well to the negative vorticity anomalies. Equally important is the fact that all vorticity anomalies of an unstable mode are phase-locked and propagate at a common speed. Their mutual reinforcement would therefore continue indefinitely. That is why the unstable mode would intensify exponentially. This interpretation of shear instability is called "*wave resonance mechanism*." In fact, this interpretation is applicable to any shear instability.

This interpretation also explains why a neutral mode does not intensify or decay. The corresponding vorticity field of the neutral mode in Fig. 8B.3 is shown in Fig. 8B.8. The vorticity changes sign across certain points denoted by $\pm y_1$. The nonzero values of vorticity are highly concentrated near $\pm y_1$. There is effectively a discontinuity in the perturbation vorticity field. This configuration has no meridional tilt and hence does not support mutual reinforcement among the vorticity anomalies. The mode is neutral simply because the wave resonance mechanism does not occur.

8B.3 Optimal growth of barotropic disturbance

It is equally instructive to examine dynamic instability as an initial value problem. An interesting question to address is: "Among all possible members of a well-defined set of initial disturbances, which one would accumulatively intensify the most over a specific time interval τ in the presence of a basic flow according to linear dynamics?" Such growth is referred to as *optimal growth* and such a disturbance is called an *optimal mode*. The task is to determine the structure of the optimal mode and its amplification factor. Taking a time integration approach for this purpose would be cumbersome and is, in fact, not feasible in the case of an infinite set of initial disturbances. Fortunately, there is an elegant way to tackle this problem.

Let us specifically consider a barotropic jet, $U(y)$, in a reentrant channel domain. We focus on a set of initial disturbances with a common zonal wavelength and an arbitrary meridional structure, $\psi(x,y,0) = \xi(y)e^{ikx}$. We wish to determine the particular disturbance which would have intensified more than all others for a specific time interval $t = \tau$. We call it the "optimal mode for $t = \tau$." If we use J interior grid points to depict the domain, the streamfunction of a disturbance at any time would be a J-vector $\vec{\psi}$ and its amplitude would be a J-vector $\vec{\xi}$. The intensity of a disturbance can be measured by the inner product of the streamfunction field (denoted by an angular bracket) with respect to a certain norm. Then the intensity of a disturbance can be evaluated as

$$A(\tau) = \langle \psi, \psi \rangle = \vec{\psi}^H D \vec{\psi}, \tag{8B.11}$$

where $\vec{\psi}^H$ stands for the Hermitian of $\vec{\psi}$. The norm is specified in terms of a matrix operator D. If the Euclidean norm is used, it would be represented by a unit matrix $D = I$ (all diagonal elements are equal to 1.0 and all other elements are 0.0).

We will make use of all normal modes, $\vec{\psi} = \vec{\phi} e^{i(kx+\sigma t)}$, for the basic flow under consideration. The set of normal modes is denoted by $\{\vec{\phi}_j\}, j = 1, 2, \ldots, J$ associated with the J eigenvalues $\{\sigma_j\}$. Using $\{\vec{\phi}_j\}$ as base functions, we can construct any initial disturbance as a linear combination of $\vec{\phi}_j$,

$$\vec{\xi} = \sum_{j=1}^{J} a_j \vec{\phi}_j, \tag{8B.12}$$

where a_j are the projection coefficients. The streamfunction at $t = \tau$ is then

$$\vec{\psi} = \sum_{j=1}^{J} a_j \vec{\phi}_j e^{i\sigma_j \tau} e^{ikx} = P \Lambda \vec{a} e^{ikx} \tag{8B.13}$$

where P is a $J \times J$ matrix which is made up of $\{\vec{\phi}_j\}$ as the columns; Λ is a $J \times J$ matrix of which the diagonal elements are $e^{i\sigma_j \tau}$ and all other elements are zero. The elements of $\vec{a}$ are a_j. The intensity of the disturbance is then,

$$A(\tau) = \vec{\psi}^H D \vec{\psi} = \vec{a}^H \Lambda^H P^H D P \Lambda \vec{a}. \tag{8B.14}$$

The intensity of an initial disturbance is

$$A(0) = \vec{a}^H P^H D P \vec{a}. \tag{8B.15}$$

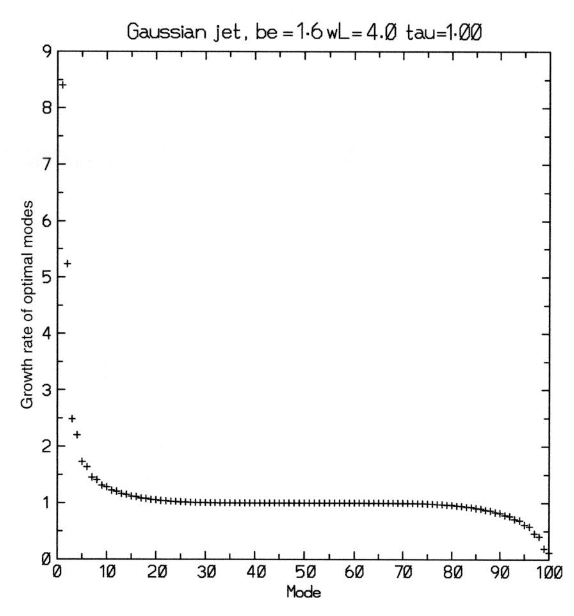

Fig. 8B.9 Amplification factor of all optimal modes for $\tau = 1.0$ for a westerly Gaussian jet.

Then the ratio

$$\lambda \equiv \frac{A(\tau)}{A(0)} = \frac{\vec{a}^H \Lambda^H P^H D P \Lambda \vec{a}}{\vec{a}^H P^H D P \vec{a}} \tag{8B.16}$$

is the amplification factor of a disturbance. It leads to the following eigenvalue–eigenvector problem

$$B(\tau)\vec{a} = \lambda B(0)\vec{a}, \quad \text{where } B(\tau) = \Lambda^H P^H D P \Lambda, B(0) = P^H D P. \tag{8B.17}$$

The largest eigenvalue λ is the maximum amplification factor and the corresponding eigenvector $\vec{a}$ is the optimal mode for $t = \tau$.

Using the projection coefficients of the optimal mode determined on the basis of (8B.17), we can determine the spatial structure of the optimal disturbance in physical space as

$$\psi(x, y, 0) = \sum_{j=1}^{J} \left(\mathrm{Re}\left(a_j \phi_j(y)\right) \cos(kx) - \mathrm{Im}\left(a_j \phi_j(y)\right) \sin(kx)\right). \tag{8B.18}$$

Illustrative calculations

Let us determine the optimal mode for the Gaussian jet represented by (8B.4) among a set of disturbances that have a common zonal wavenumber $k = 2\pi/4$. The domain is depicted by $J = 100$ meridional grids. The optimal-mode eigenvalues, λ, for $\tau = 1$ are computed on the basis of (8B.17). The amplification factor of the mode for optimization time $\tau = 1$ is found to be about 8.4 (Fig. 8B.9). The growth rate of the most unstable normal mode for this basic jet is about 0.5.

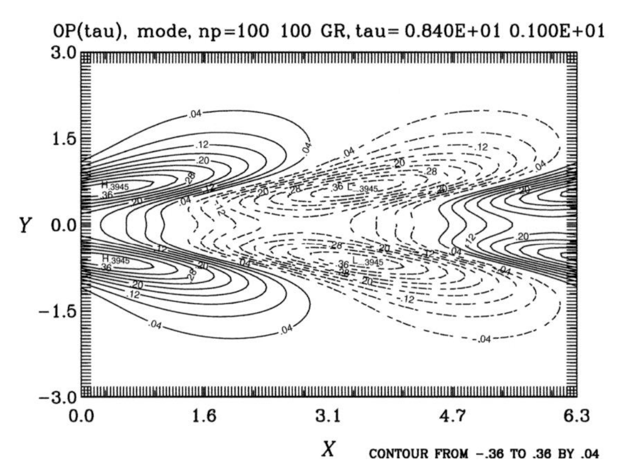

Fig. 8B.10 Structure of the streamfunction of the optimal mode for $\tau = 1.0$ for a westerly Gaussian jet.

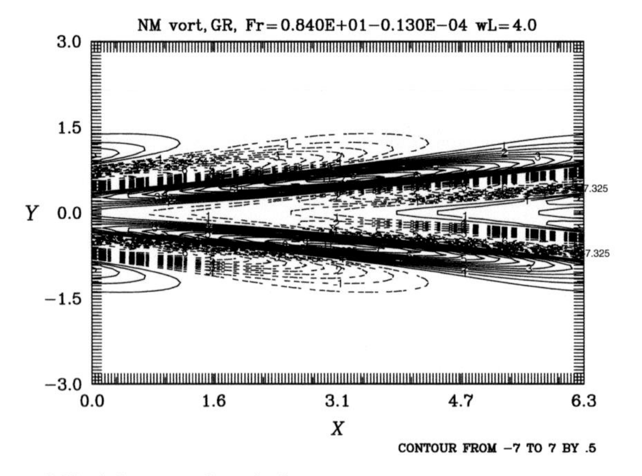

Fig. 8B.11 Structure of the vorticity field of the optimal mode for $\tau = 1$.

That mode would only intensify over a time interval of 1.0 by a factor $\exp(2\tau \cdot (0.5)) = 2.7$. The optimal mode therefore intensifies three times more than the most unstable normal mode. The amplification factor for the remaining optimal modes drops off very quickly.

The structure of the (first) optimal mode is shown in Fig. 8B.10. The degree of similarity/difference between the optimal mode and the most unstable normal mode is noteworthy. The optimal mode has a more pronounced meridional tilt and is strongest at the two latitudes where the absolute vorticity of the basic state has an extremum instead of at $y = 0$ (compare Fig. 8B.10 with Fig. 8B.2). This extensive tilt makes it possible for the disturbance to extract energy from the basic flow rapidly.

The corresponding vorticity field of the optimal mode is shown in Fig. 8B.11. The tilt of the disturbance changes in time due to the straining effect of the basic flow. Consequently, the mutual reinforcement mechanism cannot persist, resulting in a progressive decrease of the instantaneous growth rate in time. The accumulative growth over a time interval of τ is nevertheless larger for this disturbance than any other wave disturbance in this set. Hence, it has a maximum amplification factor at $\tau = 1$.

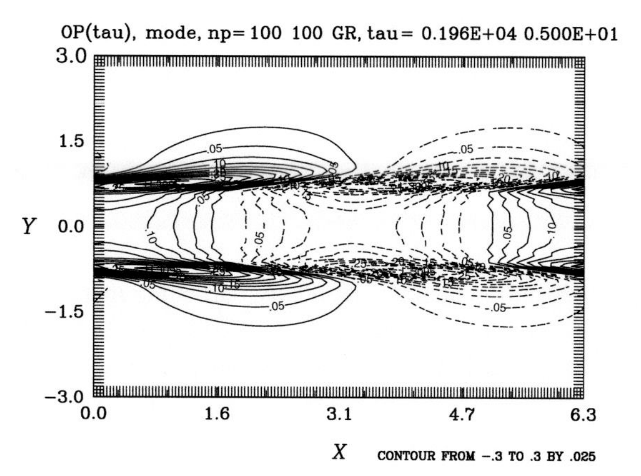

Fig. 8B.12 Structure of the streamfunction of the optimal mode for $\tau = 5$.

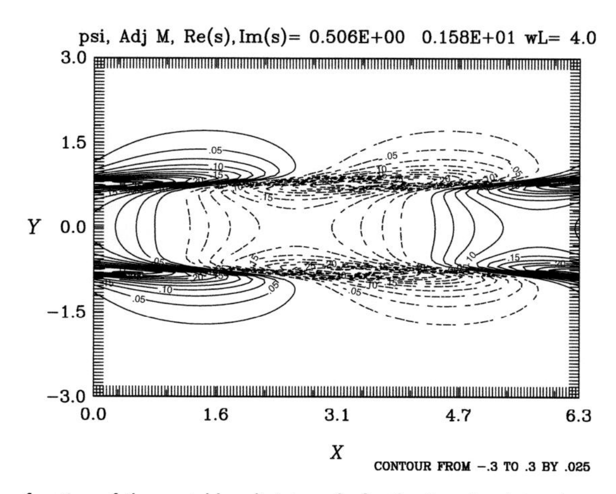

Fig. 8B.13 Structure of the streamfunction of the unstable adjoint mode for the Gaussian jet under consideration.

The amplification factor for a larger value of τ is naturally larger. For example, calculation reveals that the amplification factor of the first optimal mode for $\tau = 5$ is about 1964.0 which is much larger than 8.4 for $\tau = 1$. Figure 8B.12 shows the structure of the first optimal mode for $\tau = 5$. The main features in Fig. 8B.10 are present in Fig. 8B.12, but in a strongly accentuated form.

In the limit of large τ, an optimal mode approaches a corresponding adjoint eigenvector of the system. Since the matrix equation (8B.7) of the modal instability analysis is

$$B^{-1}A\vec{\phi} = \sigma\vec{\phi}, \tag{8B.19}$$

the adjoint eigenvector denoted by $\vec{\eta}$ satisfies

$$\begin{aligned}&\left(B^{-1}A\right)^{T}\vec{\eta} = \varepsilon\vec{\eta},\\ &\text{with } \varepsilon = \sigma^{*},\end{aligned} \tag{8B.20}$$

where the superscript T stands for the Hermitian of the matrix and * stands for the complex conjugate. These analytic relations are confirmed in a computation. Figure 8B.13 shows the corresponding unstable adjoint mode. It is seen that Fig. 8B.13 is indeed very similar to Fig. 8B.12. In the limit of $\tau \to \infty$, the optimal mode would be like Fig. 8B.13.

8B.4 Baroclinic instability in a two-layer QG model

Readers can find analyses of purely baroclinic instability with the Eady model and/or the Charney model in a number of textbooks (e.g. Vallis, 2006, Holton, 1992 and Pedlosky, 1987). Here, we choose to discuss all aspects of baroclinic instability associated with different classes of basic flows in a simplest possible model setting, namely a two-layer quasi-geostrophic model. A representative time-horizontal mean zonal flow in the extra-tropics has a broad vertical structure extending from the surface to the tropopause with a single maximum in between as sketched in Fig. 8B.14a. Therefore, the essential aspect of such a basic state can be represented in a two-layer QG model as shown in Fig. 8B.14b. The task is to ascertain the condition under which such a basic flow would be unstable and to quantify the characteristics of the unstable disturbances in this model setting.

The streamfunction of the baroclinic basic flow is $\bar{\psi}_j = -U_j y$ with $(U_1 - U_3) > 0$ where U_j is the constant basic velocity in the two layers. According to the thermal wind relation, $\frac{\partial \bar{u}}{\partial p} = \frac{R}{f_o p}\frac{\partial \bar{T}}{\partial y}$, the basic temperature at the mid-level must be related to the vertical shear of the basic flow, $\bar{T}_2 = -\frac{(U_1 - U_3) f_o p_2}{R\delta p} y$. Recall that the governing equations for a two-layer quasi-geostrophic model are (6.57a,b) in Section 6.6. The corresponding basic PV is then a linear function of y

$$\bar{q}_1 = (\beta + \lambda^2(U_1 - U_3))y, \quad \bar{q}_3 = (\beta - \lambda^2(U_1 - U_3))y, \tag{8B.21}$$

where $\lambda^2 = \frac{f_o^2}{S(\delta p)^2}$. This basic flow satisfies the adiabatic inviscid QG-PV equation for any value of U_j, because both U_j and $\bar{q}_j$ are linear functions of y satisfying

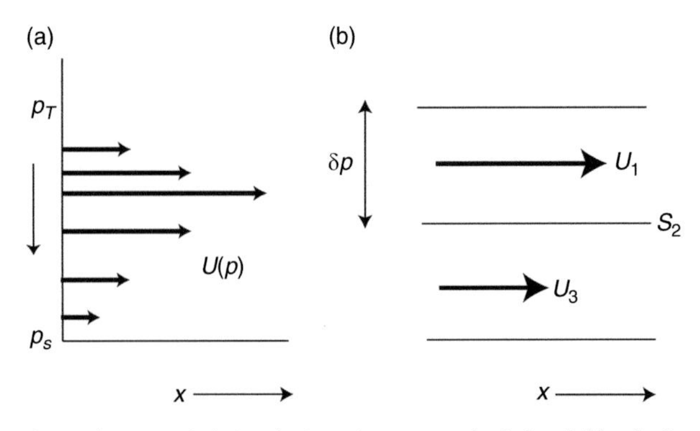

Fig. 8B.14 Schematic of (a) an observed extratropical time-horizontal mean zonal wind and (b) a basic state in a two-layer QG model; tropopause pressure level p_T, surface pressure level p_s, $\delta p = (p_s - p_T)/2$, $f = f_o + \beta y$, stratification at level-2 S_2, basic wind at levels 1 and 3, U_j.

$$J\left(\bar{\psi}_j, \bar{q}_j\right) = 0. \tag{8B.22}$$

It means that the external forcing is implicitly embodied by the existence of the basic flow itself as long as frictional damping is negligible.

A perturbation embedded in this basic flow is governed by the linearized form of the PV equations, viz.

$$\begin{aligned} q'_{1t} + U_1 q'_{1x} + \psi'_{1x}\bar{q}_{1y} &= 0, \\ q'_{3t} + U_3 q'_{3x} + \psi'_{3x}\bar{q}_{3y} &= 0, \end{aligned} \tag{8B.23a, b}$$

where

$$q'_1 = \nabla^2\psi'_1 - \lambda^2\left(\psi'_1 - \psi'_3\right), \quad q'_3 = \nabla^2\psi'_3 - \lambda^2\left(\psi'_3 - \psi'_1\right). \tag{8B.24}$$

It should be pointed out that (8B.23a,b) are valid whether or not U_j and the corresponding $\bar{q}_{jy}$ are constants or functions of y. The domain is a reentrant channel bounded by two rigid walls: $-X \le x \le X$, $-Y \le y \le Y$. The boundary conditions are: (a) all properties of the flow are cyclical at the eastern and western boundaries e.g. $\psi'_j(-X) = \psi'_j(X)$ and (b) no mass flux across the lateral boundaries.

8B.4.1 Necessary condition for baroclinic instability

There is an invariant property of any time-dependent perturbation in this model. It is established by adding the domain average of the product of (8B.23a) and $q'_1/\bar{q}_{1y}$ to the domain average of the product of (8B.23b) and $q'_3/\bar{q}_{3y}$. We obtain

$$\left\langle \frac{q'_1}{\bar{q}_{1y}} q'_{1t} \right\rangle + \left\langle \frac{q'_1}{\bar{q}_{1y}} U_1 q'_{1x} \right\rangle + \left\langle q'_1 \psi'_{1x} \right\rangle + \left\langle \frac{q'_3}{\bar{q}_{3y}} q'_{3t} \right\rangle + \left\langle \frac{q'_3}{\bar{q}_{3y}} U_3 q'_{3x} \right\rangle + \left\langle q'_3 \psi'_{3x} \right\rangle = 0, \tag{8B.25}$$

where the average of a quantity over the horizontal domain is denoted by an angular bracket, $\langle \xi \rangle \equiv \frac{1}{4XY} \int_{-Y}^{Y} \int_{-X}^{X} \xi \, dx \, dy$. By making use of the boundary conditions, we get

$$\left\langle \frac{U_j q'_j q'_{jx}}{\bar{q}_{jy}} \right\rangle = \frac{1}{4XY} \int_{-Y}^{Y} \frac{U_j}{2\bar{q}_{jy}} \int_{-X}^{X} \frac{\partial}{\partial x} \left(q'_j\right)^2 dx \, dy = 0.$$

Also,

$$\begin{aligned} \left\langle q'_1 \psi'_{1x} + q'_3 \psi'_{3x} \right\rangle &= \left\langle \psi'_{1x} \nabla^2 \psi'_1 - \lambda^2 \psi'_{1x}\left(\psi'_1 - \psi'_3\right) + \psi'_{3x} \nabla^2 \psi'_3 - \lambda^2 \psi'_{3x}\left(\psi'_3 - \psi'_1\right) \right\rangle \\ &= \left\langle \psi'_{1x} \nabla^2 \psi'_1 + \psi'_{3x} \nabla^2 \psi'_3 + \lambda^2\left(\psi'_{1x}\psi'_3 + \psi'_{3x}\psi'_1\right) \right\rangle = 0. \end{aligned}$$

Therefore, we get from (8B.25)

$$\frac{d}{dt} \left\langle \frac{q'_1 q'_1}{\bar{q}_{1y}} + \frac{q'_3 q'_3}{\bar{q}_{3y}} \right\rangle = 0. \tag{8B.26}$$

Equation (8B.26) reveals that the quantity $\left\langle \frac{q'_1 q'_1}{\bar{q}_{1y}} + \frac{q'_3 q'_3}{\bar{q}_{3y}} \right\rangle$ is an invariant property of any disturbance whatever the zonally uniform basic flow might be. The quantity $\frac{\rho_o}{2}\left(\frac{q'_1 q'_1}{\bar{q}_{1y}} + \frac{q'_3 q'_3}{\bar{q}_{3y}}\right)$ is called *wave-activity density*.

This invariant property enables us to deduce a necessary condition for purely baroclinic instability. When a basic flow is unstable with respect to a disturbance, the corresponding $\langle q_1' q_1' \rangle$ and $\langle q_3' q_3' \rangle$ would individually increase in time. Then (8B.26) could not be satisfied unless $\bar{q}_{1y}$ and $\bar{q}_{3y}$ have opposite signs. *The necessary condition for baroclinic instability is therefore that the meridional gradient of the basic potential vorticity must change sign in the domain.* It is noteworthy that (8B.3) for a barotropic system is a special case of (8B.26) because the $\bar{q}_j$ would be reduced to the absolute vorticity. Equation (8B.26) highlights the mathematical/dynamical similarity between barotropic instability and baroclinic instability in two distinctly different physical systems.

8B.4.2 Physical nature of baroclinic instability mechanism

When the basic flow under consideration has a westerly shear, $\bar{q}_{1y} = \left(\beta + \lambda^2(U_1 - U_3)\right)$ and $\bar{q}_{3y} = \left(\beta - \lambda^2(U_1 - U_3)\right)$ would have opposite signs only if the shear $(U_1 - U_3)$ is sufficiently large. The baroclinic shear has to be strong enough to overcompensate the stabilizing influences of the stratification and the beta-effect. For example, if we use $\beta = 1.6 \times 10^{-11}\,\mathrm{m}^{-1}\,\mathrm{s}^{-1}$ and $\lambda = 10^{-6}\,\mathrm{m}^{-1}$ (associated with $f_o = 10^{-4}\,\mathrm{s}^{-1}$, $\delta p = 400\,\mathrm{mb}$, $S = -\dfrac{R\bar{T}}{p\bar{\theta}}\dfrac{d\bar{\theta}}{dp} = \dfrac{1}{gT\rho^2}\left(\dfrac{\partial \bar{T}}{\partial z} + \dfrac{g}{c_p}\right) \sim 0.6 \times 10^{-5}\,\mathrm{m}^2\,\mathrm{s}^{-2}\,\mathrm{Pa}^{-2}$ which is equivalent to Brunt–Vaisala frequency $N \sim 1.2 \times 10^{-2}\,\mathrm{s}^{-1}$), the shear must exceed a critical value equal to $(U_1 - U_3)_c = \beta/\lambda^2 \sim 16\,\mathrm{m\,s}^{-1}$ for baroclinic instability to occur. The required shear for baroclinic instability is larger at lower latitudes stemming from the functional dependence of λ upon f_o, $\lambda^2 = \dfrac{f_o^2}{S(\delta p)^2}$. This result is consistent with the observation that synoptic cyclones are seldom observed in the tropics. The explicit form of $\bar{q}_{3y} < 0$ can be rewritten as

$$\beta - \left(\frac{-g f_o \frac{\partial \bar{T}_2}{\partial y}}{R\bar{T}_2 \left(\frac{\partial \bar{T}}{\partial z} + \frac{g}{c_p} \right)_2} \right) < 0.$$

We may then infer that the beta-effect is a stabilizing factor. Furthermore, since $\dfrac{\partial T}{\partial y} \Big/ \left(\dfrac{\partial T}{\partial z} + \dfrac{g}{c_p}\right) = \dfrac{\partial \theta/\partial y}{\partial \theta/\partial z}$, where θ is potential temperature, the inequality above becomes $\dfrac{\beta R\bar{T}_2}{g f_o} - \left(-\dfrac{\partial \bar{\theta}/\partial y}{\partial \bar{\theta}/\partial z}\right)_2 < 0$. Considering two points on a basic isentropic surface dy and dz apart in the y- and z-directions, we have the relation $d\bar{\theta} = 0 = \dfrac{\partial \bar{\theta}}{\partial y} dy + \dfrac{\partial \bar{\theta}}{\partial z} dz$, therefore $\dfrac{-\partial \bar{\theta}/\partial y}{\partial \bar{\theta}/\partial z} = \left(\dfrac{dz}{dy}\right)_{\bar{\theta}}$ is its slope. Thus, the necessary condition of baroclinic instability may be stated in terms of a geometric characteristic of the basic state, viz.

$$\gamma < \left(\frac{dz}{dy}\right)_2 \quad \text{where } \gamma = \frac{\beta R\bar{T}_2}{g f_o}. \tag{8B.27}$$

Typical isentropic surfaces in mid-latitude have $\left(\dfrac{dz}{dy}\right)_2 > 0$ because $\bar{\theta}$ has smaller values at higher latitudes and larger values at greater heights in the troposphere. Thus, a basic $\bar{\theta}$

surface such as the interface of a two-layer QG model slopes northward-upward in a (y, z) cross-section. Its slope must be steeper than a threshold value $\gamma = \frac{\beta R \bar{T}_2}{g f_o} \sim 0.11$ for baroclinic instability to occur. The equivalent angle of this threshold slope relative to the horizon is ~6°. This is the slope of a basic isentropic surface associated with a baroclinic shear for marginal stability, $(U_1 - U_3)_c \sim 16\,\mathrm{m\,s^{-1}}$. For a vertical shear twice as large as the critical value, the slope of the basic isentropic surface would be 0.22 and the corresponding angle would be 12°. The beta-effect increases the threshold slope of the basic isentropic surface to a finite value. If the beta-effect were neglected altogether in the model, the threshold slope is reduced to $\gamma = \frac{\beta R \bar{T}_2}{g f_o} = 0$. It implies that any nonzero basic baroclinic shear would meet the necessary condition for instability in this special case.

Equation (8B.27) by itself is not a sufficient condition for baroclinic instability for any disturbance. For a disturbance to intensify via baroclinic instability, its structure has to have certain characteristics relative to the structure of the basic state. One telltale aspect of its structure is depicted in terms of the trajectories of its air parcels projected on a vertical cross-section normal to the basic flow. There is a wedge area between the threshold slope and the slope of the basic isentropic surface, $\left(\frac{dz}{dy}\right)_2$ on the (y, z) cross-section (Fig. 8B.15a). If the projection of a trajectory of an air parcel moving in a northward-upward direction lies within such a wedge, the air parcel would be less dense than the fluid in its new surrounding. This is the case of a parcel going from A to B in Fig. 8B.15a. This is so because an air parcel of higher potential temperature is less dense than an air parcel of lower potential temperature at the same pressure as readily seen from the definition $\left(\theta = \frac{p}{R\rho}\left(\frac{p_{oo}}{p}\right)^{R/c_p}\right)$. The buoyancy force acting on the air parcel under consideration would be upward and the component of this force in the direction of the trajectory would accelerate the parcel forward. Similarly, the projected trajectory of an air parcel in another part of a disturbance moving in a southward-downward direction within the wedge would find itself denser than its new surrounding, such as the parcel from B to A. The buoyancy force would likewise further accelerate it forward. This positive feedback on the displacement of a parcel by the basic state would lead to continual amplification. In contrast, an air parcel moving outside the wedge area, such as from C to D or vice versa, would experience a negative feedback. Such a parcel would gradually return to its original position. From this perspective of baroclinic instability, the beta-effect reduces the width of the wedge for possible intensification (Fig. 8B.15b) and is therefore indeed a stabilizing factor.

We can mathematically verify the qualitative reasoning above as follows. Let ρ_A be the density of the air parcel at A, θ_A its potential temperature, $(\delta y, \delta z)$ its displacement from A to B in the y- and z-directions, $\delta\rho$ and δp the corresponding changes of its density and pressure. While the potential temperature of a moving air parcel does not change, its pressure does and hence its density upon arriving at B becomes

$$\rho_A + \delta\rho = \rho_A + \left(\frac{\partial \rho}{\partial p}\right)_{\theta_A} \delta p = \rho_A + \left(\frac{\partial \rho}{\partial p}\right)_{\theta_A} \left(\left(\frac{\partial p}{\partial y}\right)_A \delta y + \left(\frac{\partial p}{\partial z}\right)_A \delta z\right), \tag{8B.28}$$

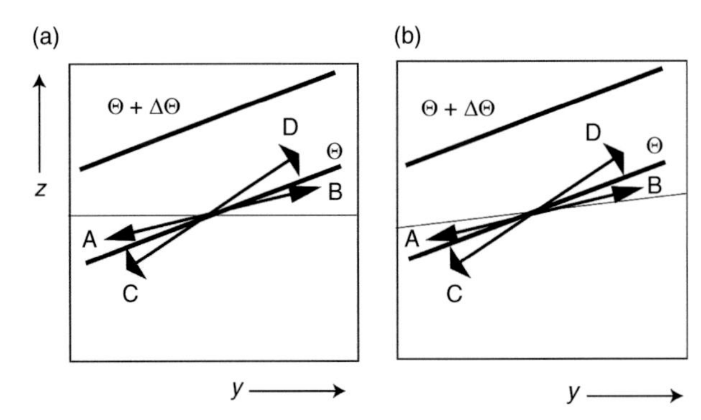

Fig. 8B.15 Schematic of the essence of baroclinic instability: (a) No beta-effect and hence $\gamma = 0$; basic isentropic surfaces $\Theta =$ constant; projected trajectory of air parcel in an unstable disturbance within the "wedge" on a cross-section normal to the basic shear flow (A to B or vice versa) and in a stable disturbance outside the wedge (C to D or vice versa). (b) Counterpart schematic in the case with beta-effect and hence a threshold slope $\gamma \neq 0$.

assuming that its pressure at any instant is never different from its background pressure. The background density at B is

$$\rho_B = \rho_A + \left(\frac{\partial \rho}{\partial y}\right)_A \delta y + \left(\frac{\partial \rho}{\partial z}\right)_A \delta z. \tag{8B.29}$$

Hence the buoyancy force in the z-direction acting on the air parcel is

$$F \equiv \frac{\rho_B - (\rho_A + \delta\rho)}{\rho_A} g = \left[\left(\left(\frac{\partial \rho}{\partial y}\right)_A - \left(\frac{\partial \rho}{\partial p}\right)_{\theta_A}\left(\frac{\partial p}{\partial y}\right)_A\right)\delta y + \left(\left(\frac{\partial \rho}{\partial z}\right)_A - \left(\frac{\partial \rho}{\partial p}\right)_{\theta_A}\left(\frac{\partial p}{\partial z}\right)_A\right)\delta z\right]\frac{g}{\rho_A}. \tag{8B.30}$$

From the definition $\theta = T\left(\frac{p_{oo}}{p}\right)^{R/c_p}$ or equivalently $\rho = \frac{p_{oo}}{R\theta}\left(\frac{p}{p_{oo}}\right)^{c_v/c_p}$, we have $\left(\frac{\partial \rho}{\partial p}\right)_{\theta_A} = \left(\frac{c_v \rho}{c_p p}\right)_{\theta_A}$. Hence, the air parcel experiences *a force in the direction of its displacement* equal to

$$F \sin\phi = \left[\left(\left(\frac{\partial \rho}{\partial y}\right)_A - \left(\frac{c_v \rho}{c_p p}\right)_{\theta_A}\left(\frac{\partial p}{\partial y}\right)_A\right)\delta y + \left(\left(\frac{\partial \rho}{\partial z}\right)_A - \left(\frac{c_v \rho}{c_p p}\right)_{\theta_A}\left(\frac{\partial p}{\partial z}\right)_A\right)\delta z\right]\frac{g \sin\phi}{\rho_A}, \tag{8B.31}$$

where ϕ is the angle between the line of parcel displacement and the y-axis, whereby $\tan\phi = \delta z/\delta y$. Noting that $\frac{\partial \theta}{\partial z} = \frac{\theta}{\rho}\left[\frac{c_v \rho}{c_p p}\frac{\partial p}{\partial z} - \frac{\partial \rho}{\partial z}\right]$, (8B.31) can be written as $F \sin\phi = \left[-\left(\frac{\partial \theta}{\partial y}\right)_A \delta y - \left(\frac{\partial \theta}{\partial z}\right)_A \delta z\right]\frac{g}{\theta_A}\sin\phi$. Since $\left(\frac{-\partial\theta/\partial y}{\partial\theta/\partial z}\right)_{\theta_A} = \left(\frac{dz}{dy}\right)_{\theta_A}$, (8B.31) can be further rewritten as

$$F \sin\phi = \left[\left(\frac{dz}{dy}\right)_A - \frac{\delta z}{\delta y}\right]\delta y \left(\frac{\partial \theta}{\partial z}\right)_A \frac{g}{\theta_A}\sin\phi. \tag{8B.32}$$

It follows that this force would accelerate the air parcel in the direction of its trajectory only if

$$\left(\frac{dz}{dy}\right)_A > \frac{\delta z}{\delta y} = \tan\,\varphi > 0 \text{ for } \beta = 0 \text{ according to (8B.27).} \tag{8B.33}$$

With the beta-effect, (8B.33) is generalized to

$$\left(\frac{dz}{dy}\right)_A > \tan\varphi > \gamma > 0, \qquad \gamma = \frac{\beta \bar{T}_2}{g f_o}. \tag{8B.34}$$

Equation (8B.34) recovers the necessary condition for purely baroclinic instability obtained previously. Whether or not the beta-effect is included, the trajectory of an air parcel in an unstable disturbance must lie within a corresponding wedge region on the cross-section normal to the basic flow as depicted in Fig. 8B.15. This is one way to visualize the essence of baroclinic instability.

8B.5 Modal growth

A modal analysis enables us to address additional questions about the instability: What is the range of parameter values of a basic state in which a disturbance can intensify at large time? How fast would its intensification rate be for a given set of parameters? What are the salient features of an unstable disturbance? How does the structure of an unstable disturbance help us understand the physical nature of baroclinic instability?

8B.5.1 Modal baroclinic instability analysis

One would naturally perform an instability analysis with the use of a channel domain that has a finite width as an analogy to a latitudinal belt in the extratropics, $-Y \leq y \leq Y$. The boundary condition for the streamfunction would be $\psi_j(\pm Y) = 0$. The coefficients of the governing equations (8B.23a,b) are independent of y for the simple basic baroclinic flow under consideration. It follows that the meridional structure of a normal mode would be of the form $\psi_j \propto \cos\left(\frac{m\pi y}{2Y}\right)$, $m = 1, 3, 5\ldots$ One can proceed to determine the instability properties of a normal mode for each value of m separately. However, if we imagine an unbounded domain, i.e. $Y \to \infty$, then even $m = 0$ would be admissible. In doing so, we would be effectively considering 1-D wave disturbances that vary only in the direction of the basic flow. One can readily verify that the larger m of a 2-D normal mode is, the smaller would its growth rate be. So, the growth rate for a disturbance with $m = 0$ is larger than that of any 2-D normal mode with $m = 1, 2, 3\ldots$ All other instability properties of a 1-D normal mode have qualitatively similar counterparts in a 2-D normal mode.

It suffices to analyze the instability properties of 1-D disturbances. Since the coefficients in the governing equations (8B.23a,b) are independent of x, we first write an elementary solution in the form of

$$\begin{aligned} \psi_1'(x,t) &= \xi_1(t)\exp(ikx), \\ \psi_3'(x,t) &= \xi_3(t)\exp(ikx), \end{aligned} \tag{8B.35a, b}$$

where k is an unrestricted wavenumber of the disturbance in the case of an infinitely long channel domain. For a channel of finite length, say $2X$, we should use $k = \frac{2\pi n}{2X}$ with $n = 1, 2, 3 \ldots$ in light of the cyclical condition. Upon substituting (8B.35a,b) into (8B.23a,b), we get

$$\left[-\left(k^2+\lambda^2\right)\frac{d}{dt}+ik\left(\beta_1-U_1\left(k^2+\lambda^2\right)\right)\right]\xi_1+\left[\lambda^2\frac{d}{dt}+ik\lambda^2U_1\right]\xi_3=0,$$
$$\left[\lambda^2\frac{d}{dt}+ik\lambda^2U_3\right]\xi_1+\left[-\left(k^2+\lambda^2\right)\frac{d}{dt}+ik\left(\beta_3-U_3\left(k^2+\lambda^2\right)\right)\right]\xi_3=0, \quad (8B.36a,b)$$

where $\beta_1 = \beta + (U_1 - U_3)\lambda^2$ and $\beta_3 = \beta - (U_1 - U_3)\lambda^2$. By eliminating ξ_1 in (8B.36a, b), we would obtain a single equation for ξ_3, viz.

$$\left(a\frac{d^2}{dt^2}+i2b\frac{d}{dt}-c\right)\xi_3=0, \quad (8B.37)$$

where $a = \lambda^4 - \left(k^2+\lambda^2\right)^2$,
$b=\frac{k}{2}\left(\lambda^4(U_1+U_3)+\left(k^2+\lambda^2\right)\left(\beta_1+\beta_3-(U_1+U_3)\left(k^2+\lambda^2\right)\right)\right)$, and
$c=k^2\left(\lambda^4U_1U_3-\left(\beta_1-U_1\left(k^2+\lambda^2\right)\right)\left(\beta_3-U_3\left(k^2+\lambda^2\right)\right)\right)$. Furthermore, since the coefficients a,b,c are independent of time, ξ_j would have non-trivial solution of the form

$$\xi_1 = Ae^{i\sigma t}, \quad \xi_3 = Be^{i\sigma t}, \quad (8B.38)$$

where A and B are integration constants, provided that σ satisfies a certain condition. Upon substituting (8B.38) into (8B.37), we get the condition

$$a\sigma^2 + 2b\sigma + c = 0. \quad (8B.39)$$

We have reduced the problem to an eigenvalue–eigenvector problem. To each eigenvalue σ there is a corresponding eigenvector, (A,B). The two roots of (8B.39) are

$$\sigma_\pm = \frac{-b \pm \sqrt{b^2 - ac}}{a}. \quad (8B.40)$$

In a certain range of values of k^2 and $(U_1 - U_3)$, $(b^2 - ac)$ would be negative and $\sigma_\pm$ would have complex values. We write the latter as $\sigma_\pm = \sigma_r \pm i\sigma_i$ with $\sigma_i > 0$. The normal mode associated with the σ_- root is $\left(\psi'_1, \psi'_3\right) = (A,B)\exp(\sigma_i t)\exp(i(kx + \sigma_r t))$. It is therefore an exponentially amplifying mode. It propagates in the x-direction at a speed equal to $(-\sigma_r/k)$. The normal mode associated with the σ_+ root, $\left(\psi'_1, \psi'_3\right) = (A,B)\exp(-\sigma_i t)\exp(i(kx + \sigma_r t))$, is an exponentially decaying mode that propagates at the same speed, $\left(-\frac{\sigma_r}{k}\right)$. In the remaining part of the parameter plane, $(b^2 - ac)$ is positive and the corresponding $\sigma_\pm$ has two different real values. Those solutions represent two neutral modes propagating at different speeds.

8B.5.2 Instability properties

The algebraic solution of $\sigma_\pm$ is too complicated to visualize by inspection how σ_r and σ_i would quantitatively vary with the parameters. To develop a feel for the analytic result, we examine the dependence of the roots graphically. The two roots $\sigma_\pm$ are functions of five

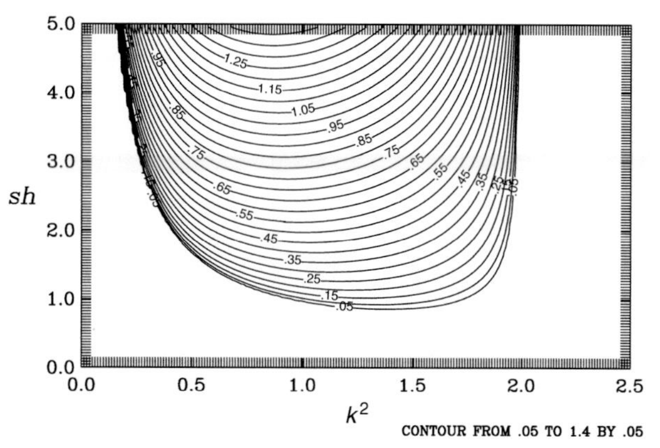

Fig. 8B.16 Variation of the non-dimensional growth rate $\tilde{\sigma}_i$ as a function of $\tilde{k}^2$ and baroclinic shear $sh \equiv (\tilde{U}_1 - \tilde{U}_3)$ for $\tilde{U}_3 = 1.0$ and $\tilde{\beta} = 0.8$.

parameters: $U_1, U_3, \beta, \lambda, k$. These properties then depend on three independent non-dimensional parameters. Measuring distance and velocity in units of $\lambda^{-1} = 10^6\,\text{m}$ and $V = 10\,\text{ms}^{-1}$ respectively, we introduce $k = \lambda\tilde{k}$, $\sigma = V\lambda\tilde{\sigma}$, $U_j = V\tilde{U}_j$, $\beta = V\lambda^2\tilde{\beta}$. For illustration, we present the results of $\tilde{\sigma}_{\pm}$ as a function of $\tilde{k}^2$ and baroclinic shear $(\tilde{U}_1 - \tilde{U}_3)$ for $\tilde{U}_3 = 1$ and $\tilde{\beta} = 0.8$ in Fig. 8B.16.

Figure 8B.16 reveals the following key properties of baroclinic instability in this model:

- There is a curve on the parameter plane along which $\tilde{\sigma}_i$ is equal to zero. This curve separates the parameter conditions for stable modes from those for unstable modes. It establishes the minimum shear required for instability as a function of the wavenumber of disturbances. The modes associated with the parameter conditions slightly above this curve are called *marginally unstable modes*.
- There is no instability for any value of $\tilde{k}$ unless the baroclinic shear exceeds a threshold value, $(\tilde{U}_1 - \tilde{U}_3) = 0.8$. Recall that these results are computed with the use of $\tilde{\beta} = 0.8$. This *minimum baroclinic shear* required for baroclinic instability of waves to occur agrees with the necessary condition for instability deduced in Section 8B.4.1, $\tilde{\beta} - (\tilde{U}_1 - \tilde{U}_3) \leq 0$. Only the wave with a wavenumber $k \approx \lambda\sqrt{0.7}$ (embodying the dependence on β) would be able to intensify when the shear slightly exceeds the minimum required value.
- For a given supercritical shear, $\left(\dfrac{(\tilde{U}_1 - \tilde{U}_3)}{\tilde{\beta}}\right) > 1$, the growth rate is largest for a certain wavenumber, $\tilde{k}_{most}$. That corresponding baroclinic mode would be the most unstable mode. For example, we find $\tilde{k}_{most} = 1.0$ for $(\tilde{U}_1 - \tilde{U}_3) = 3.0$ and the growth rate is about 0.8 in units of $V\lambda$.
- Instability only occurs in a range of wavenumber values $\tilde{k}_{\min} < \tilde{k} < \tilde{k}_{\max} \leq \sqrt{2}$ for a given supercritical shear, where $\tilde{k}_{\min}$ is a longwave cutoff and $\tilde{k}_{\max}$ is a shortwave cutoff. They are both functions of the shear for a given $\tilde{\beta}$.

Next, let us examine the real part of the two eigenvalues which contains information about the phase speed of the unstable normal modes. The latter are computed as $(-\text{Re}\{\tilde{\sigma}_{\pm}\}/\tilde{k})$ (Fig. 8B.17). It is seen that the amplifying and decaying modes for a given parameter

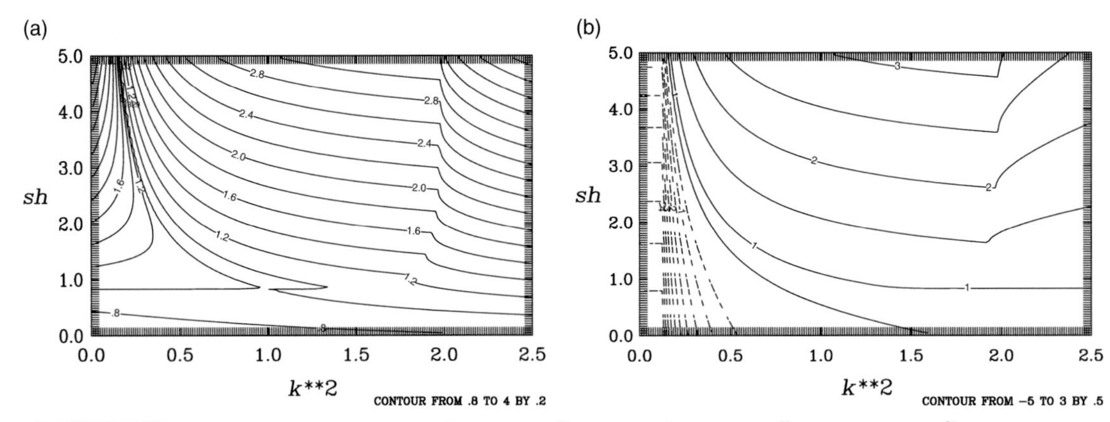

Fig. 8B.17 Phase velocity in x-direction (a) $\left(-\text{Re}\{\tilde{\sigma}_+\}/\tilde{k}\right)$, and (b) $\left(-\text{Re}\{\tilde{\sigma}_-\}/\tilde{k}\right)$ as a function of $\tilde{k}^2$ and baroclinic shear $sh \equiv \left(\tilde{U}_1 - \tilde{U}_3\right)$ for $\tilde{U}_3 = 1.0$ and $\tilde{\beta} = 0.8$.

condition have a common eastward phase velocity. For example, the phase velocity is about 1.9 units of V for $\tilde{k} = 1$ and $\left(\tilde{U}_1 - \tilde{U}_3\right) = 3$. It is noteworthy that the value of this phase velocity is in between $\tilde{U}_1$ and $\tilde{U}_3$. In contrast, the two stable modes have different phase velocities. They coalesce to a common value at the marginal instability condition. For the long stable waves, $k \leq k_{\min}$, the two modes may propagate in opposite directions. For the short stable waves, $k \geq k_{\max}$, both modes propagate eastward.

How relevant are the instability results to cyclogenesis in the atmosphere?

To address this question, we first check whether or not representative values of the parameters associated with a time-horizontal mean state of the extratropical atmosphere satisfy the necessary condition for baroclinic instability. It is indeed the case because

$$\left(\frac{(U_1 - U_3)\lambda^2}{\beta}\right)_{representative} \sim \frac{(30)(1 \times 10^{-12})}{1.6 \times 10^{-11}} \sim 1.9 > 1.$$

The winter mean baroclinic shear in the extratropics is therefore typically supercritical consistent with the frequent formation of cyclones. Let us further check the instability properties. Under this parameter condition, we get from Fig. 8B.16 $k_{most} \sim \lambda\sqrt{4/3} \sim 1.1 \times 10^{-6}\,\text{m}^{-1}$. Therefore, the wavelength of the most unstable wave is about 5.4×10^6 m which is about four times the size of a typical cyclone. The maximum non-dimensional growth rate is about 1.0, which means that $(\sigma_i)_{\max} \sim \beta/\lambda \sim 10^{-5}\text{s}^{-1}$. The corresponding e-folding time, $1/\sigma_i$, is then about 1 day. This value is also compatible with observation. Finally, the expected shortwave cutoff is $k_{\max} = \lambda\sqrt{2} \rightarrow L_{\min} = 2\pi/\lambda\sqrt{2} \sim 4.4 \times 10^6$ m. All disturbances with a wavelength shorter than $L_{\min}$ are stable. The model result is compatible with the observation that cyclones seldom have a scale shorter than 1000 km. The longwave cutoff for a representative shear is $\left(\tilde{k}^2\right)_{\min} \sim 0.5$ and hence $L_{\max} = \dfrac{2\pi}{\lambda k_{\min}} \sim 9 \times 10^6$ m. The model result suggests that a spectrum of waves can spontaneously intensify. This is compatible with the fact that a localized disturbance consisting of a spectrum of constituent wave components can amplify in time as a whole.

The overall agreement between the model instability characteristics and the observed counterparts of baroclinic waves supports the view that the theory of baroclinic instability even in the context of a QG two-layer model is a fundamentally correct theory of cyclogenesis.

8B.5.3 Structure of the unstable baroclinic wave

We can learn a lot more about baroclinic instability from the structure of an unstable mode represented by the eigenvector (*A*,*B*) in (8B.38). While the length of an eigenvector can have an arbitrary value, its orientation is unique. By (8.B36a) we get

$$\frac{B}{A} = \frac{(\tilde{k}^2+1)\tilde{\sigma} - \tilde{k}\left(\tilde{\beta}_1 - \tilde{U}_1\right)(\tilde{k}^2+1)}{\tilde{\sigma} + \tilde{k}\tilde{U}_1}. \tag{8B.41}$$

Without loss of generality, we set $A = 1$ and use (8B.41) to compute $B = |B|e^{i\eta_3}$ associated with each eigenvalue. The streamfunction of a normal mode is $\tilde{\psi}'_1 = \mathrm{Re}\{\exp(i(\tilde{k}\tilde{x} + \tilde{\sigma}\tilde{t}))\}$ and $\tilde{\psi}'_3 = \mathrm{Re}(|B|\exp(i(\tilde{k}\tilde{x} + \tilde{\sigma}\tilde{t} + \eta_3)))$. By the hydrostatic balance, we also have $T'_2 = \dfrac{f_o p_2}{R\Delta p}(\psi'_1 - \psi'_3)$. The non-dimensional temperature perturbation in units of $\dfrac{f_o V\lambda^{-1}p_2}{R\Delta p}$ is

$$\tilde{T}'_2 = \left(\tilde{\psi}'_1 - \tilde{\psi}'_3\right). \tag{8B.42}$$

The structure of the vertical velocity of this unstable baroclinic wave is computed with the linearized omega equation (Eq. (6.58) of Chapter 6) which is

$$\frac{f_o}{\Delta p}\left(\frac{\partial^2}{\partial x^2} - 2\lambda^2\right)\omega'_2 = \lambda^2(U_3 - U_1)\frac{\partial^3(\psi'_1 + \psi'_3)}{\partial x^3} + \lambda^2\beta\frac{\partial(\psi'_3 - \psi'_1)}{\partial x}. \tag{8B.43}$$

The first term on the RHS of (8B.43) represents the effect of thermal advection by a 1-D perturbation. The second term represents the effect of vertical differential advection of planetary vorticity by the perturbation. Using non-dimensional $\tilde{\psi}'_1 = \mathrm{Re}\{\exp(i(\tilde{k}\tilde{x} + \tilde{\sigma}_+\tilde{t}))\}$ and $\tilde{\psi}'_3 = \mathrm{Re}\{\tilde{B}\exp(i(\tilde{k}\tilde{x} + \tilde{\sigma}_+\tilde{t}))\}$ as the solution of the unstable normal mode with eigenvalue $\tilde{\sigma}_+ = \sigma_+/V\lambda$, the solution of non-dimensional omega is $\tilde{\omega}'_2 = \mathrm{Re}\{\tilde{W}\exp(i(\tilde{k}x + \tilde{\sigma}_+ t))\}$ in units of $V^2\lambda^2\Delta p/f_o$, with the non-dimensional amplitude of omega being

$$\tilde{W} = \frac{-i\tilde{k}\left(-\tilde{k}^2(\tilde{U}_3 - \tilde{U}_1)(1+\tilde{B}) + \beta(\tilde{B}-1)\right)}{(\tilde{k}^2+2)}. \tag{8B.44}$$

For illustration, we present the structure of the unstable mode with a wavelength equal to 4.4×10^6 m and hence $k = 1.41 \times 10^{-6}\ \mathrm{m}^{-1}$. The values of the basic state parameters are: $f_o = 1.0 \times 10^{-4}\ \mathrm{s}^{-1}$, $S_2 = 2 \times 10^{-6}\ \mathrm{m}^2\,\mathrm{Pa}^{-2}\,\mathrm{s}^{-2}$, $\Delta p = 5. \times 10^4$ Pa, $\therefore \lambda = 1.41 \times 10^{-6}\ \mathrm{m}^{-1}$, $\beta = 1.6 \times 10^{-11}\ \mathrm{m}^{-1}\,\mathrm{s}^{-1}$, $U_1 = 30.0\ \mathrm{m\,s}^{-1}$, $U_3 = 10.0\ \mathrm{m\,s}^{-1}$.

The domain is $0 \le x \le 2\pi/k$, $0 \le y \le \pi/k$. In using $k = \lambda$, we have $\tilde{k} = 1$. In using $V = 10\ \mathrm{m\,s}^{-1}$, the non-dimensional parameters of the basic state are $\tilde{U}_3 = 1$, $\tilde{U}_1 = 3$,

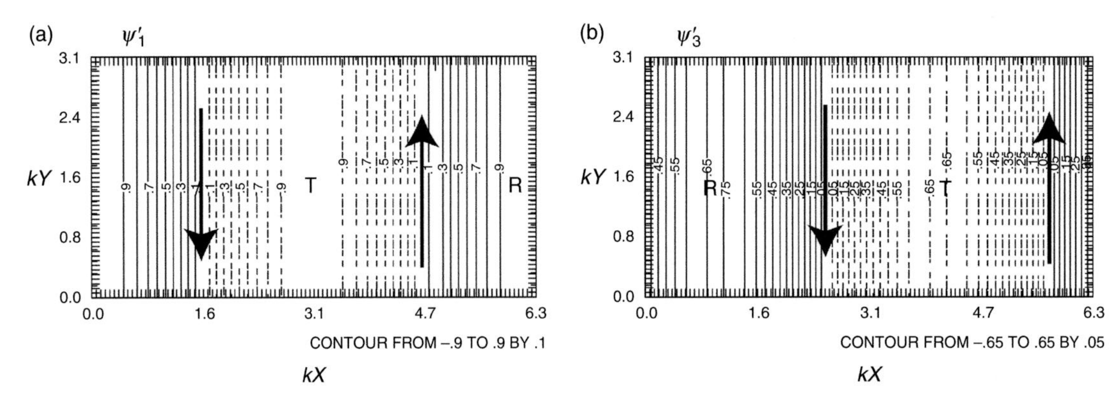

Fig. 8B.18 Structure of the streamfunction of the unstable mode with $\tilde{k} = 1.0$ at (a) upper level and (b) lower level for $\tilde{U}_3 = 1$, $\tilde{U}_1 = 3$, $\tilde{\beta} = 0.8$ in units of $V\lambda^{-1} = 7.1 \times 10^6 \mathrm{m}^2\,\mathrm{s}^{-1}$; trough highlighted by T and ridge by R; arrows indicating maximum velocity.

$\tilde{\beta} = 0.8$. The unstable wave has a complex eigenvalue, $\sigma_+ = -1.47 - i0.51$. The e-folding time of the intensification is about 2 units of $(V\lambda)^{-1}$ and is quite comparable with observed e-folding time during cyclogenesis. This unstable disturbance propagates to the east at a speed of 14.7 m s^{-1} which is slightly smaller than the vertical mean of the basic flow (20.0 m s^{-1}). Without the beta-effect, the phase speed of the unstable wave would be exactly equal to the vertical mean of the basic flow.

The structure of every property in each layer of a 1-D unstable wave mode can be concisely depicted as a sinusoidal curve. But, for clarity, we however choose to plot its structure at the two levels on two planes showing no variation in the meridional direction. It takes up more space in doing so, but that makes it visually easier to appreciate the relative phases of the various properties and their implications. The units of ψ'_j are $V\lambda^{-1} = 7.1 \times 10^6 \mathrm{m}^2\mathrm{s}^{-1}$. The horizontal structure of the $\tilde{\psi}'_1$ and $\tilde{\psi}'_3$ fields of the unstable baroclinic wave are presented in Fig. 8B.18. The unstable mode has a westward vertical tilt in all of its properties, readily identifiable by inspection of Fig. 8B.18. The amplitude of $\tilde{\psi}'_1$ is stronger than that of $\tilde{\psi}'_3$ being 1.0 vs. 0.66 respectively.

The structure of the $\tilde{\omega}'_2$ and $\tilde{T}'_2$ fields are shown in Fig. 8B.19. The units of ω'_2 are $\dfrac{V^2\lambda^2\Delta p}{f_o} = 0.1$ mb s^{-1} and the units of T'_2 are $\dfrac{f_o V\lambda^{-1}}{R} = 2.5$ K. The magnitude of $\tilde{\omega}'_2$ is about 0.85. The maximum vertical velocity at mid-level would be then $w'_2 \approx -\dfrac{\omega'_2}{g\rho} \sim$ 1.7 cm s^{-1}. The magnitude of $\tilde{T}'_2$ is about 0.85, corresponding to 2.1 K. The relative magnitudes of z'_1, z'_3, ω'_2 and T'_2 of the model unstable baroclinic wave are therefore similar to the counterpart properties of an extratropical cyclone. The relative phases among these four properties of the disturbance are particularly informative. A trough in the upper layer is located to the west of a trough in the lower layer. This westward vertical tilt is about one-third of the wavelength. The maximum ascending motion (indicated by $U_{\max}$) is located to the east of the trough in the lower layer by about one-third of a wavelength. The maximum temperature (indicated by $W_{\max}$) is also located to the east of the lower layer trough by

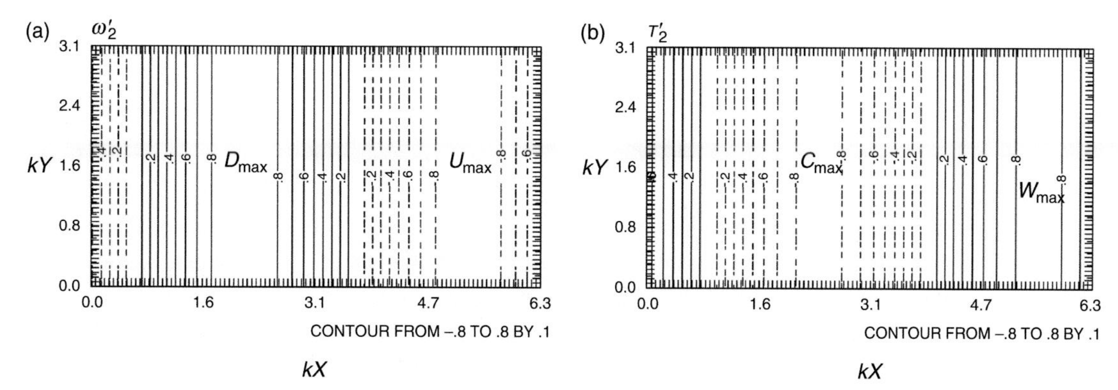

Fig. 8B.19 Structure of the unstable wave with $\tilde{k} = 1.0$ for the basic state $\tilde{U}_3 = 1$, $\tilde{U}_1 = 3$, $\tilde{\beta} = 0.8$, (a) $\tilde{\omega}_2'$ in units of $V^2\lambda^2\Delta p/f_o = 0.1$ mb s^{-1}, U_{max} for maximum ascent and D_{max} for maximum descent and (b) $\tilde{T}_2'$ in units of $\frac{f_o V \lambda^{-1}}{R} = 2.5$ K, W_{max} for maximum temperature and C_{max} for minimum temperature.

about the same amount. Therefore, the ascending motion is to a high degree in phase with the temperature perturbation. We will see that this overall structure is precisely what is essential for a baroclinic wave to intensify. One might add that the decaying normal mode has an opposite vertical tilt.

8B.5.4 Baroclinic instability from the perspective of energetics

The nature of baroclinic instability can be examined from the perspective of energy conversion. The perturbation kinetic energy per unit mass is denoted by K', the perturbation available potential energy by P', the basic kinetic energy by $\bar{K}$, and the basic available potential energy by $\bar{P}$. Note that $\bar{K}$ and $\bar{P}$ are not independent because the basic velocity and temperature are necessarily related by the thermal wind relation. For a 1-D disturbance in a two-layer model, the appropriate definitions for K' and P' are

$$K' = \left\langle \frac{1}{2}\left((v_1')^2 + (v_3')^2 \right) \right\rangle = \left\langle \frac{1}{2}\left((\psi_{1x}')^2 + (\psi_{3x}')^2 \right) \right\rangle,$$

$$P' = \left\langle \frac{R^2}{2S(p_2)^2} (T_2')^2 \right\rangle = \left\langle \frac{\lambda^2}{2} (\psi_1' - \psi_3')^2 \right\rangle, \tag{8B.45}$$

where the angular bracket stands for averaging over a wavelength. Note that the perturbation potential energy is proportional to the variance of the perturbation temperature. The rate of change of K' can be readily derived from the vorticity equations

$$\left(\frac{\partial}{\partial t} + U_1 \frac{\partial}{\partial x} \right) \psi_{1xx}' + \beta \psi_{1x}' = \frac{f_o}{\Delta p} \omega_2', \tag{8B.46}$$

$$\left(\frac{\partial}{\partial t} + U_3 \frac{\partial}{\partial x} \right) \psi_{3xx}' + \beta \psi_{3x}' = \frac{-f_o}{\Delta p} \omega_2'. \tag{8B.47}$$

Multiplying (8B.46) by ψ_1' and (8B.47) by ψ_3', and averaging over a wavelength, we get

$$\frac{dK'}{dt} = C(P', K'),$$
$$\text{where } C(P', K') = \frac{f_o}{\Delta p}\left\langle -\omega_2'(\psi_1' - \psi_3')\right\rangle. \tag{8B.48}$$

In getting (8B.48), we have performed integration by parts and made use of the cyclical condition over a wavelength. The change of perturbation kinetic energy can only come from the perturbation potential energy. We use the notation $C(P', K')$ for the conversion rate from P' to K'. Since $(\psi_1' - \psi_3')$ is proportional to the perturbation temperature at level-2, $C(P', K')$ would have a positive value if there is positive correlation over a wavelength between the vertical velocity and temperature perturbations at mid-level. Hence, K' *of the unstable wave is converted from* P' *by a process of warm air rising and cold air sinking.*

To derive the equation for the rate of change of P', we start with the QG thermodynamic equation

$$\frac{\partial}{\partial t}\left[\lambda^2(\psi_3 - \psi_1)\right] + J\left(\frac{1}{2}(\psi_1 + \psi_3), \lambda^2(\psi_3 - \psi_1)\right) = \frac{-f_o\omega_2}{\Delta p}. \tag{8B.49}$$

The linearized form of this equation for a 1-D disturbance is

$$\frac{\partial}{\partial t}\left[\lambda^2(\psi_3' - \psi_1')\right] + \frac{1}{2}(U_1 + U_3)\lambda^2(\psi_3' - \psi_1')_x + \frac{1}{2}\lambda^2(U_1 - U_3)(\psi_3' + \psi_1')_x = \frac{-f_o\omega_2'}{\Delta p}. \tag{8B.50}$$

Thus, multiplying (8B.50) by $(\psi_1' - \psi_3')$ and averaging over a wavelength would yield

$$\frac{dP'}{dt} = C(\bar{P}, P') - C(P', K'),$$
$$\text{where } \quad C(\bar{P}, P') = 2\lambda^2 U_T\left\langle \psi_{2x}'(\psi_1' - \psi_3')\right\rangle, \tag{8B.51}$$

and $C(\bar{P}, P')$ stands for the conversion rate from $\bar{P}$ to P'. Since it is proportional to $\langle v_2' T_2' \rangle$, it has a positive value if there is a positive correlation over a wavelength between meridional velocity and temperature perturbations at the mid-level. Hence, P' *of an unstable wave is converted from* $\bar{P}$ *by a process of warm air going north and cold air heading south.*

An unstable baroclinic wave has precisely the structure elaborated in the last section required for yielding positive values for both $C(\bar{P}, P')$ and $C(P', K')$. Furthermore $C(\bar{P}, P')$ is necessarily greater than $C(P', K')$, so that $\frac{dP'}{dt}$ and $\frac{dK'}{dt}$ would both have positive values. Finally, adding (8B.48) to (8B.51) yields

$$\frac{d(K' + P')}{dt} = C(\bar{P}, P'). \tag{8B.52}$$

Equation (8B.52) reveals that the ultimate source of energy for baroclinic instability is the potential energy of the basic state in agreement with the qualitative deduction in Section 8B.4.2 concerning the necessary condition of baroclinic instability. Summing up, baroclinic instability occurs because the heat flux by a disturbance from the south to

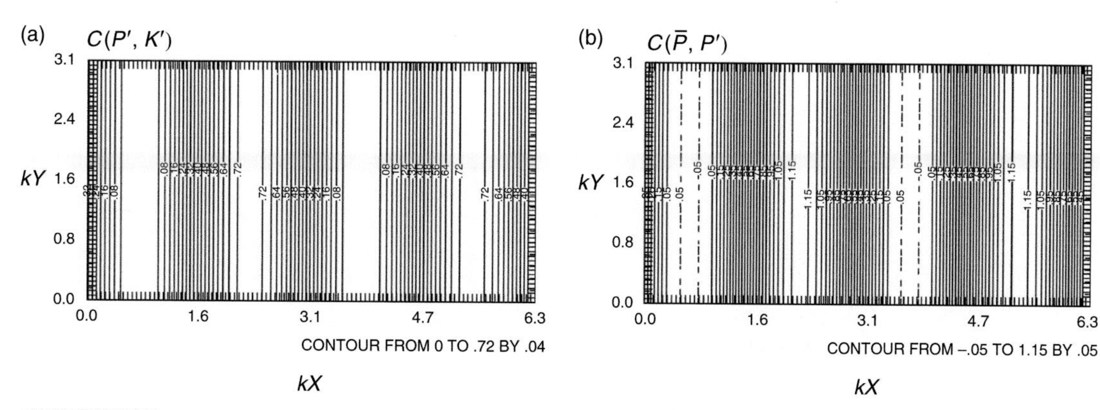

Fig. 8B.20 Distribution of the local values of the conversion rate (a) from perturbation potential energy to perturbation kinetic energy $C\left(\tilde{P}', \tilde{K}'\right) = \left\langle -\tilde{\omega}_2'\left(\tilde{\psi}_1' - \tilde{\psi}_3'\right)\right\rangle$ and (b) from basic to perturbation potential energy $C\left(\tilde{\bar{P}}, \tilde{P}'\right) = \left(\tilde{U}_1 - \tilde{U}_3\right)\left\langle \psi_{2x}'\left(\tilde{\psi}_1' - \tilde{\psi}_3'\right)\right\rangle$ of the unstable mode in units of $V^3\lambda$.

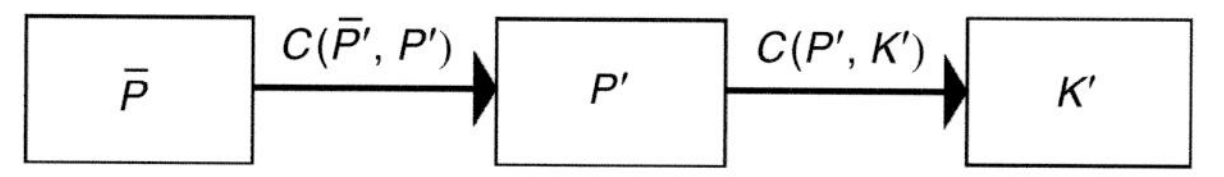

Fig. 8B.21 Schematics of the energetics in baroclinic instability.

the north converts some $\bar{P}$ to P' which is partly converted to K' from the energy perspective.

The spatial distributions of the local kinetic energy and potential energy within a wavelength are quite obvious in light of those of $\tilde{\psi}_j'$ and $\tilde{T}_2'$ (Figs. 8B.19a,b). The local values of the two energy conversion rates in units of $V^3\lambda$ that contribute to $C\left(\tilde{P}', \tilde{K}'\right)$ and $C\left(\tilde{\bar{P}}, \tilde{P}'\right)$ are shown in Fig. 8B.20. The averages of these values in each panel over a wavelength are clearly positive. Furthermore, $C\left(\tilde{\bar{P}}, \tilde{P}'\right)$ is indeed greater than $C\left(\tilde{P}', \tilde{K}'\right)$, so that $\frac{d\tilde{K}'}{dt} > 0$ and $\frac{d\tilde{P}'}{dt} > 0$.

Figure 8B.21 is a schematic for the energetics in baroclinic instability. For baroclinic instability to occur, some basic potential energy is converted to wave potential energy and some of the latter in turn is converted to wave kinetic energy.

8B.5.5 Potential vorticity (PV) transport by an unstable baroclinic wave

Since potential vorticity is a conservative property of every air parcel in the absence of diabatic heating and friction, an unstable wave should give rise to a mixing of PV. An unstable wave must advect basic PV in the down-gradient direction. For the case of an unstable westerly basic shear, we know $\bar{q}_{1y} > 0$ and $\bar{q}_{3y} < 0$. Thus, an unstable baroclinic wave should transport PV southward in the upper layer and northward in the lower layer. The spatial distribution of the local value of PV transport, $v_j'q_j'$, by the unstable mode in

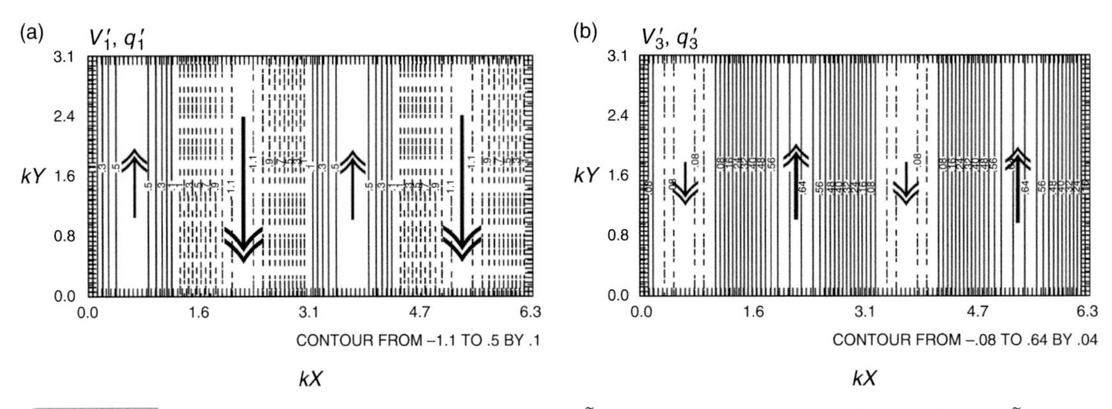

Fig. 8B.22 Potential vorticity flux by the unstable wave with $\tilde{k} = 1.0$ in (a) upper layer and (b) lower layer for $\tilde{U}_3 = 1$, $\tilde{U}_1 = 3$, $\tilde{\beta} = 0.8$. Relative values indicated by length of arrows.

each layer is shown Fig. 8B.22. The result confirms that the zonal average transport of *perturbation PV* by the wave is indeed southward in the upper layer, $[v'_1 q'_1] < 0$, and northward in the lower layer, $[v'_3 q'_3] > 0$.

8B.5.6 Baroclinic instability from the perspective of wave resonance mechanism

Some questions about the instability cannot be satisfactorily addressed from the energetics perspective. For example, why are the very short waves and very long waves dynamically stable in a baroclinic basic flow as we have seen in Section 8B.4? The answers can be deduced from the perspective of potential vorticity dynamics. Let us now see how the positive feedback process works in the context of the unstable wave under consideration. The basic PV and perturbation PV fields of the unstable mode discussed in Sections 8B.4.1 and 8B.4.2 are shown in Fig. 8B.23.

Recall that the QG streamfunction ψ'_j and the QG-PV q'_j are linearly related to one another. This relationship implies that we can identify an element of ψ'_j individually associated with each element of q'_j. Also recall that the perturbation meridional velocity is $v'_j = \psi'_{jx}$. Thus, a cyclonic (anticyclonic) flow element at both levels is associated with each particular positive (negative) PV anomaly. In order to have modal instability, there must be a process of positive feedback among the constituent PV elements. Now let us consider a specific positive q'_1 element in the upper layer, such as the one with maximum value, and the associated cyclonic flow in the lower layer. We see that the southerly flow associated with this q'_1 element would have a positive advection of $\bar{q}_3$ in the region where q'_3 has positive values. Let us also consider a positive q'_3 element in the lower layer and the associated cyclonic flow in the upper layer. Note that $\bar{q}_{1y}$ and $\bar{q}_{3y}$ have opposite signs. The northerly flow associated with this q'_3 element would have a positive advection of $\bar{q}_1$ in the region of the q'_1 element. In other words, those two elements of PV perturbations in the two layers *reinforce* one another via the process of advecting basic PV. These are counter-propagating Rossby

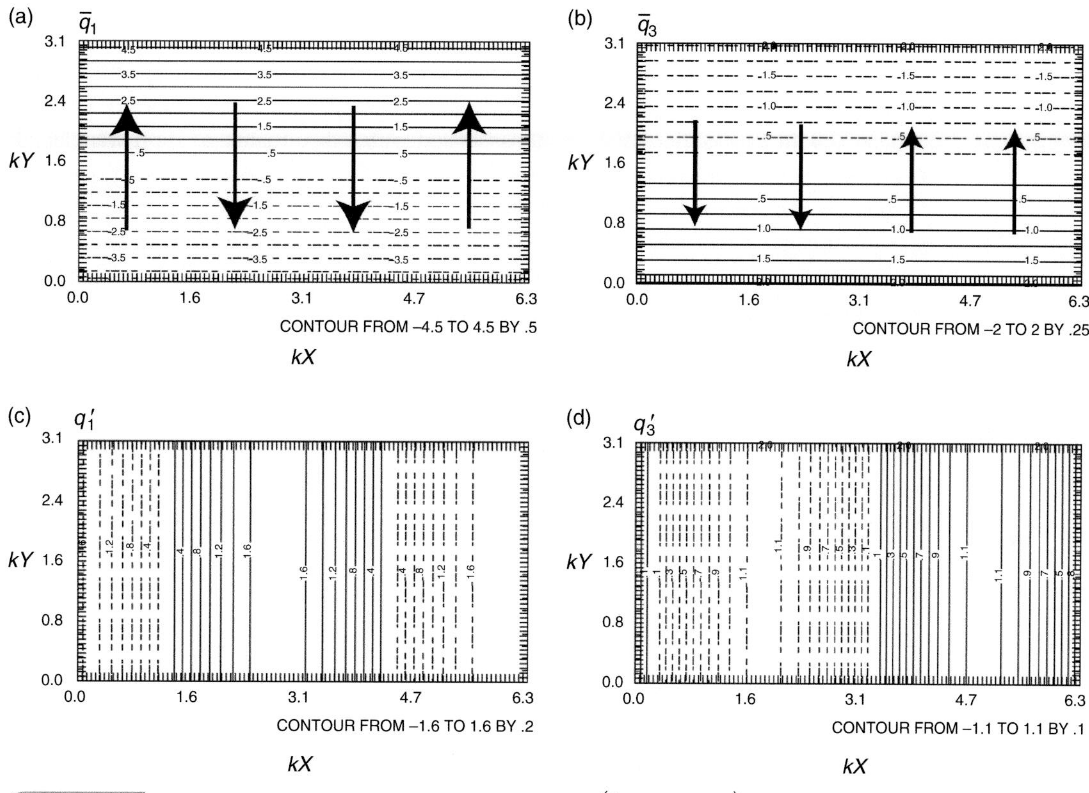

Fig. 8B.23 Basic potential vorticity in the (a) upper layer, $\tilde{\bar{q}}_1 = \left(\tilde{\beta} + (\tilde{U}_1 - \tilde{U}_3)\right)\tilde{y}$, and (b) lower layer $\tilde{\bar{q}}_3 = \left(\tilde{\beta} - (\tilde{U}_1 - \tilde{U}_3)\right)\tilde{y}$, and PV of the unstable mode with $\tilde{k} = 1.0$ in the (c) upper layer $\tilde{q}'_1 = \tilde{\psi}'_{1xx} - \left(\tilde{\psi}'_1 - \tilde{\psi}'_3\right)$, and (d) lower layer $\tilde{q}'_3 = \tilde{\psi}'_{3xx} - \left(\tilde{\psi}'_3 - \tilde{\psi}'_1\right)$ for $\tilde{U}_3 = 1$, $\tilde{U}_1 = 3$, $\tilde{\beta} = 0.8$. Arrows in (a) indicate the flow elements associated with PV elements in (c). Arrows in (b) indicate the flow elements associated with PV elements in (d).

waves (Heifetz *et al.*, 2004). The arrows in panels (a) and (b) highlight the flows associated with PV elements.

Since the two q'_1 and q'_3 elements of an unstable wave move together to the east at the same speed, they are stationary relative to one another. This characteristic results from the combined influence of the intrinsic phase propagation of the constituent wave in each layer and the advective influence of the basic flow. The mutual interaction between them therefore can persist continually. The same can be said about other pairs of PV elements. In other words, the two constituent waves of an unstable mode (one at the upper level and one at the lower level) mutually reinforce one another in a sustained manner making intensification of the entire perturbation possible. This process can be again referred to as *wave resonance mechanism*. The phase velocity and the vertical structure of an unstable wave are precisely what are required for such mutually reinforcing interaction to occur indefinitely.

We are now in a position to address two hitherto unanswered questions.

(i) Why is there a shortwave cutoff ?

The strength of influence at a location by a particular PV element depends on the wavelength and the basic stratification through the parameter λ. The strength of the flow associated with a particular PV element associated with a disturbance of short wavelength would only have significant influence in a relatively short vertical distance away. Hence, a PV element of a short wave in one layer would have weak or no influence upon another PV element of that wave in the other layer. They cannot significantly interact with one another even when they are favorably located with respect to one another. It follows that there exists a shortwave cutoff for baroclinic instability.

(ii) Why is there a longwave cutoff ?

From the perspective of wave resonance mechanism, baroclinic instability would occur only if the constituent disturbance in the upper layer and that in the lower layer are phase-locked, whereby the mutual reinforcement process could persist. Phase-locking would be impossible for very long waves for the following reasoning. The meridional gradient of the upper level PV and lower level PV are $\tilde{\beta}_1 = \tilde{\beta} + \tilde{U}_1 - \tilde{U}_3$ and $\tilde{\beta}_3 = \tilde{\beta} - \tilde{U}_1 + \tilde{U}_3$ respectively. The intrinsic phase velocity of Rossby waves at the two levels would be $-\tilde{\beta}_1/\tilde{k}$ and $-\tilde{\beta}_3/\tilde{k}$ respectively; $\tilde{\beta}_1$ and $\tilde{\beta}_3$ have opposite signs and the corresponding intrinsic phase velocity of the two constituent Rossby waves would be in opposite directions. If the wavelength of the waves is very long (very small k), such intrinsic phase velocities would be much larger than the basic flow. It follows that two constituent disturbances would not be able to phase-lock together. Therefore, there exists a longwave cutoff for baroclinic instability.

8B.6 Transient growth

Making short-range weather forecasts over a specific region requires one to predict how much a particular large-scale flow would change after say 12 or 24 hours. The information about the most unstable normal mode of a background flow would not be relevant if it takes a long time for such a mode to become a strong disturbance. What would be more pertinent is to deduce the transient growth of a disturbance of known structure as an initial value problem. If we integrate for a sufficiently long time, the most unstable normal mode would emerge as the dominant component of the disturbance. We illustrate this with the use of a highly *localized disturbance* embedded in a given baroclinic basic zonal flow.

The results are also presented in non-dimensional form using λ^{-1}, V and Δp as units of distance, velocity and pressure respectively. The non-dimensional streamfunction and p-velocity are $\tilde{\psi}'_j = \dfrac{\psi'_j}{V\lambda^{-1}}$ and $\tilde{\omega}_2 = \dfrac{\omega_2 f_o}{V^2\lambda^2\Delta p}$. The domain is $-X \le \tilde{x} \le X$, and $-Y \le \tilde{y} \le Y$ with $X = 10$ and $Y = 5$. The initial disturbance is introduced in terms of its vorticity field as (without the tilde notation)

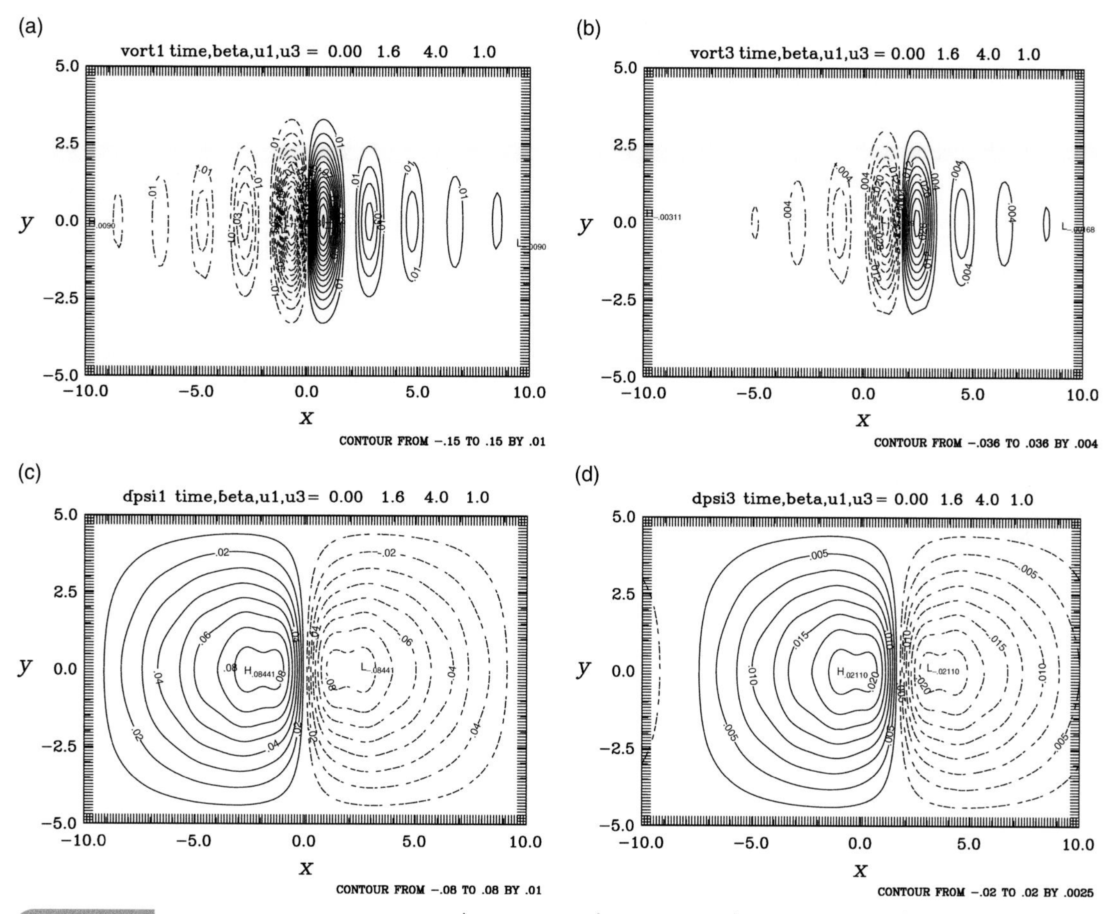

Fig. 8B.24 Initial disturbance in terms of (a) $\zeta_1'(x,y,0)$, (b) $\zeta_3'(x,y,0)$, (c) $\psi_1'(x,y,0)$ and (d) $\psi_3'(x,y,0)$.

$$\begin{aligned} \zeta_1'(x,y,0) &= \sum_{m=1}^{10} \zeta_* \exp\left(-\left(\frac{y}{2}\right)^2\right) \sin\left(m\frac{x2\pi}{2X}\right), \\ \zeta_3'(x,y,0) &= \sum_{m=1}^{10} \zeta_{**} \exp\left(-\left(\frac{y}{2}\right)^2\right) \sin\left(m\left(\frac{x2\pi}{2X}+\theta\right)\right), \end{aligned} \tag{8B.53}$$

with $\zeta_* = 0.02$, $\zeta_{**} = 0.005$, $\theta = -\pi/6$. This initial vorticity field is four times stronger in the upper layer than in the lower layer. It is strongest at $y = 0$ and symmetric about it. Its zonal structure is made up of 10 wave components with a westward vertical tilt of 30° in each component. The structures of such initial vorticity fields in the two layers are shown in Fig. 8B.24a,b. It is localized about the center of the domain. The maximum value of ζ_1' is 0.16. The corresponding streamfunction fields ψ_j' are determined by solving $\nabla^2\psi_j' = \zeta_j'$ numerically as a boundary value problem using boundary condition $\psi_j'(x, \pm Y) = 0$. The results are shown in Fig. 8B.24c,d.

We integrate such a disturbance for 10 days in the presence of an unstable basic flow. An integration is made using $\beta = 1.6$, $U_1 = 4.0$, $U_3 = 1.0$. The constituent waves intensify

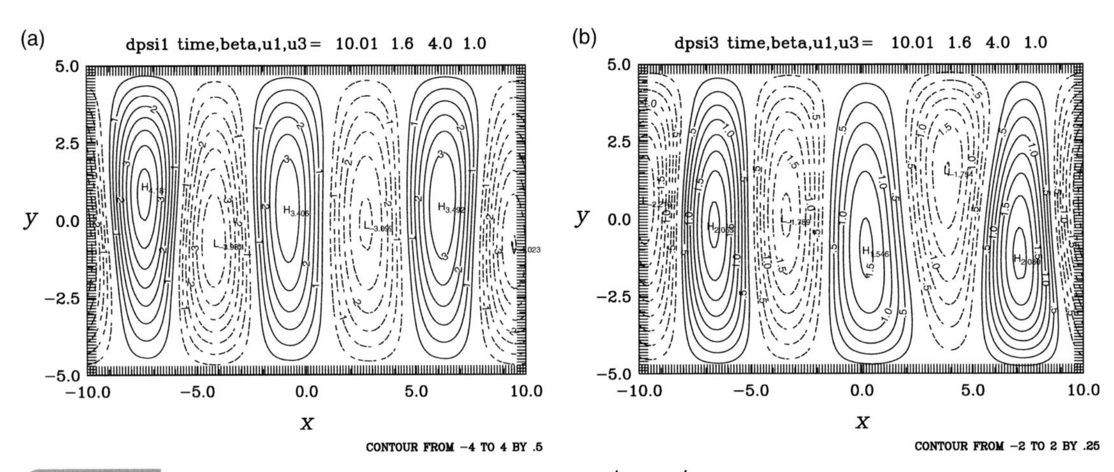

Fig. 8B.25 The structure of the disturbance streamfunction (a) ψ_1', (b) ψ_3' and at $t = 10$ for the case of an unstable basic flow, $\tilde{U}_1 = 4.0$, $\tilde{U}_3 = 1.0$, $\tilde{\beta} = 1.6$.

as in a horse race. By $t = 10.0$, wavenumber-3 emerges as the winner, becoming the dominant component of the disturbance (Fig. 8B.25). The corresponding wavelength is about 7000 km. It has a westward vertical tilt in ψ_j' and its ascending motion ($\omega_2' < 0$) occurs slightly to the east of a lower level trough in ψ_3'.

The intensification of the disturbance as a whole is quantified by the temporal variation of its instantaneous growth rate, $\sigma = \frac{1}{2K}\frac{dK}{dt}$, where K is the total kinetic energy in Fig. 8B.26a. We may infer that the dominant wave component emerges at about $t = 6.5$ (corresponding to 7 days) and it is wavenumber-3. The asymptotic value of this growth rate is about 0.4 since the growth rate begins to decrease at about $t = 9$. The overall evolution of its structure is depicted in terms of $\psi_1'(x, 0, t)$ in Fig. 8B.26b, revealing that the most unstable wave steadily propagates eastward as it intensifies.

One measure of the impact of nonlinear advection of the disturbance itself is the zonal mean part of the disturbance generated during evolution, $[\psi_j'(y, t)]$, as shown in Fig. 8B.27. This initial disturbance is sufficiently weak that the nonlinear dynamics is still negligible throughout this time interval. Nevertheless, the northern half of the domain is being warmed up, $([\psi_1'] - [\psi_3']) > 0$, and the southern half of the domain is being cooled down towards the end of this integration, $([\psi_1'] - [\psi_3']) < 0$, as a consequence of the nonlinear feedback. This negative feedback effect is what one would intuitively anticipate since instability must tend to stabilize the background zonal baroclinic flow.

This example demonstrates that cyclogenesis in a surface baroclinic zone can be triggered by an upper level PV anomaly. Such a phenomenon is often referred to as "type-B cyclogenesis." A positive disturbance initially induced in $\tilde{\psi}_3$ may be viewed as a surface warm temperature anomaly. It is effectively a surface positive PV anomaly. By piece-wise PV inversion, one may expect that the flow at the upper level associated with it would reinforce the upper-level disturbance. This is then in essence the wave resonance mechanism for baroclinic instability.

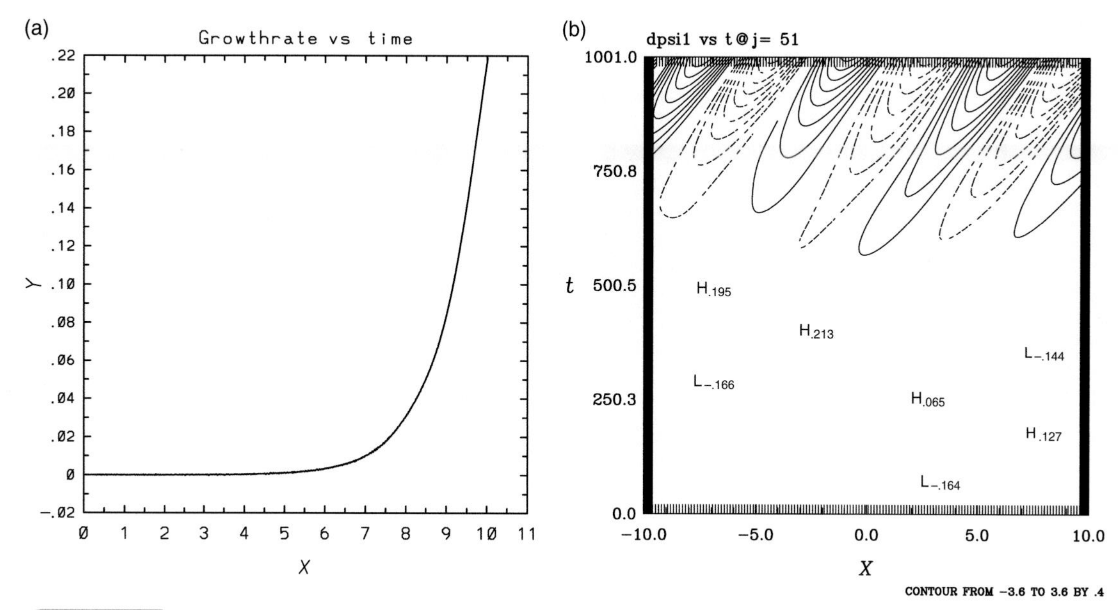

Fig. 8B.26 Variation of (a) the instantaneous growth rate, $\sigma = \frac{1}{2K}\frac{dK}{dt}$, where K is the total kinetic energy of the disturbance and (b) $\psi_1'(x, 0, t)$ for the case of an unstable basic flow, $\tilde{U}_1 = 4.0$, $\tilde{U}_3 = 1.0$, $\tilde{\beta} = 1.6$ (scale of t in panel (b) has been multiplied by a factor of 100).

8B.7 Optimal growth

The formulation of optimal mode analysis in Section 8B.3 is perfectly general. It is neither model specific nor basic flow specific. If we wish to perform an optimal mode analysis for a particular basic flow in a particular model setting, all we need to do is to apply the corresponding eigenvalues and eigenvectors to the equations of the formulation. For a purely baroclinic basic flow in a two-layer QG model considered in Section 8B5.1, there are two eigenvectors $\left\{\vec{\phi}_n\right\}$, $n = 1,2$ for each particular zonal wavenumber k. They are associated with two eigenvalues $\{\sigma_n\}$ which are the two roots given by (8B.40), $\sigma_1 \equiv \sigma_+$ and $\sigma_2 \equiv \sigma_-$. The normal modes of the streamfunction are $\vec{\psi}_n = \vec{\phi}_n \exp(i(kx + \sigma_n t))$ where $\vec{\phi}_n$ is an eigenvector made up of two elements, $\vec{\phi}_n = \begin{pmatrix} \varphi_n \\ \xi_n \end{pmatrix}$. The relation between the two elements of each eigenvector is (8B.41), which is cited below for easy reference

$$\left[-(k^2 + \lambda^2)\sigma_n + k(\beta_1 - U_1(k^2 + \lambda^2))\right]\varphi_n + \left[\lambda^2\sigma_n + k\lambda^2 U_1\right]\xi_n = 0. \tag{8B.54}$$

We may set $\varphi_n = 1.0$ without loss of generality and thereby evaluate the corresponding ξ_n with (8B.54).

The test disturbances under consideration in our optimal mode analysis form an infinite set of disturbances that have a common zonal wavenumber k. They differ by their amplitude

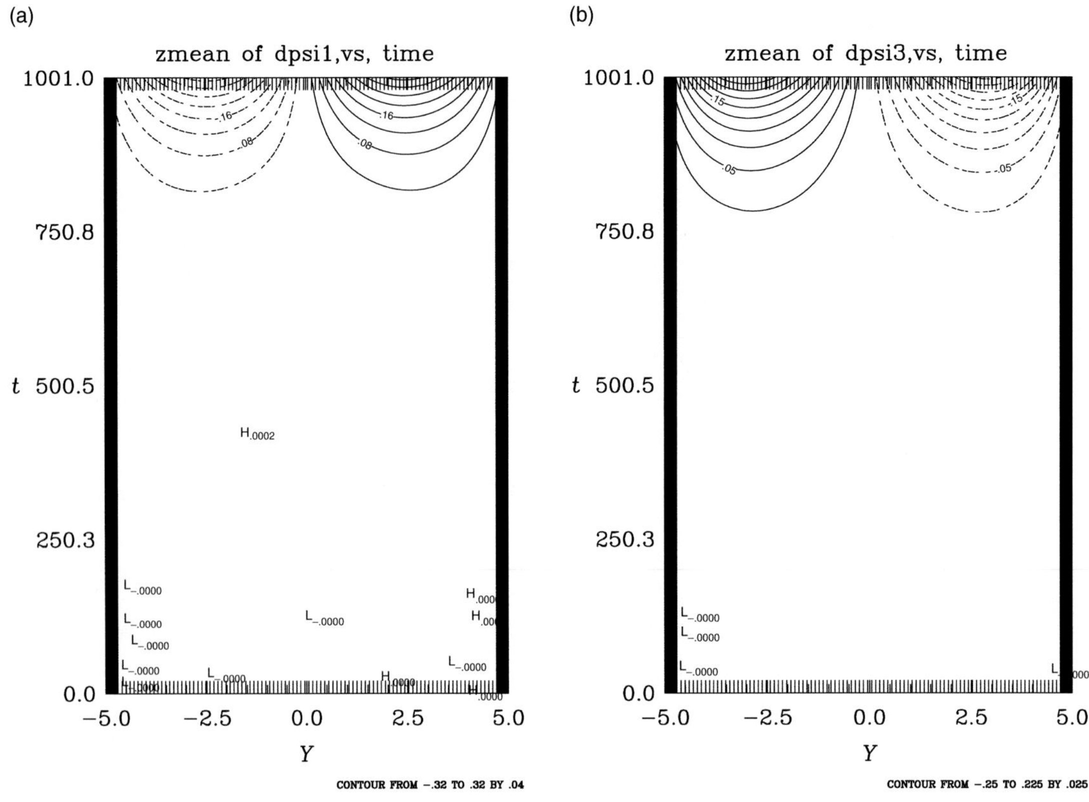

Fig. 8B.27 Temporal variation of the zonal mean part of the disturbance streamfunction, (a) $[\psi_1']$ and (b) $[\psi_3']$ (t has been multiplied by a factor of 100).

and relative phase. According to the formulation discussed in Section 8B.3, an optimal mode analysis amounts to solving an eigenvalue–eigenvector problem. In this case, it is

$$
\begin{aligned}
&B(\tau)\vec{a} = \gamma B(0)\vec{a},\\
&B(\tau) = \Lambda^H P^H D P \Lambda, \qquad B(0) = P^H D P,\\
&P = \begin{pmatrix} \varphi_1 & \varphi_2 \\ \xi_1 & \xi_2 \end{pmatrix}, \qquad \Lambda = \begin{pmatrix} \exp(i\sigma_1\tau) & 0 \\ 0 & \exp(i\sigma_2\tau) \end{pmatrix},
\end{aligned}
\tag{8B.55}
$$

where the superscript H refers to the Hermitian of a vector or matrix; γ is the amplification factor of a disturbance at $t = \tau$; $\vec{a} = \begin{pmatrix} a_1 \\ a_2 \end{pmatrix}$ consists of the projection coefficients of a disturbance with the eigenvectors being used as the base functions. In using the Euclidean norm, we would have $D = \begin{pmatrix} 1 & 0 \\ 0 & 1 \end{pmatrix}$. Then, the matrix $B(\tau)$ has the following analytic form

$$
B(\tau) = \begin{pmatrix} \left(|\varphi_1|^2 + |\xi_1|^2\right)\exp\left(i(\sigma_1 - \sigma_1^*)\tau\right) & \left(\varphi_1^*\varphi_2 + \xi_1^*\xi_2\right)\exp\left(i(\sigma_2 - \sigma_1^*)\tau\right) \\ \left(\varphi_1\varphi_2^* + \xi_1\xi_2^*\right)\exp\left(i(\sigma_1 - \sigma_2^*)\tau\right) & \left(|\varphi_2|^2 + |\xi_2|^2\right)\exp\left(i(\sigma_2 - \sigma_2^*)\tau\right) \end{pmatrix}. \tag{8B.56}
$$

Table 8B.1 Characteristics of the normal modes under consideration

$\sigma_1 \equiv \tilde{\sigma}_+$	φ_1	ξ_1	$\sigma_2 \equiv \tilde{\sigma}_-$	φ_2	ξ_2
−1.0000	1.0	0.0	−0.7333	1.0	0.5

Table 8B.2 Characteristics of the two optimal modes for $\tau = 12$

γ_1	$a_1^{(1)}$	$a_2^{(1)}$	γ_2	$a_1^{(2)}$	$a_2^{(2)}$
17.93	(1.0, 0.0)	(−0.894, 0.0027)	0.0557	(1.0, 0.0)	(0.893, −0.049)

Equation (8B.55) can be readily solved for the eigenvalue γ and the corresponding $\vec{a}$.

A member of this set of arbitrary disturbances is denoted by a vector, $\vec{\psi} = \begin{pmatrix} \psi_1 \\ \psi_3 \end{pmatrix} e^{ikx}$. The index $j = 1{,}3$ refers to the values at the upper and lower layers. The optimal mode at time t can be finally constructed as

$$\vec{\psi} = \sum_{n=1}^{2} a_n \vec{\phi}_n \exp(i(kx + \sigma_n t)), \tag{8B.57}$$

using the $\vec{a}$ associated with the largest value of γ.

When the basic state is unstable, the most unstable normal mode would correspond to the optimal mode at $\tau \to \infty$. In considering the transient growth of a disturbance in the presence of such a basic flow, we could incorporate a sufficiently strong damping that would render the basic flow stable.

Illustrative calculation

We illustrate the optimal mode analysis with an example for the case of a marginally stable basic flow.

The non-dimensional parameters are: $\tilde{U}_1 = 1.8$, $\tilde{U}_3 = 1.0$, $\tilde{\beta} = 0.8$, $\tilde{k} = 1.0$.

The properties of the two normal modes of (8B.56) are summarized in Table 8B.1.

We obtain two modes in an optimal mode analysis for $\tau = 12$. The characteristics of the first optimal mode are shown in the first three columns of Table 8B.2.

The optimal mode is found to intensify by a factor of about 18.0. This result is compatible with the finding obtained by direct integration in Section 8B.4.

In accordance with (8B.57), the structure of the optimal mode at $t = 0$ in physical space can be determined as

$$\begin{pmatrix} \tilde{\psi}'_1 \\ \tilde{\psi}'_3 \end{pmatrix} = \mathrm{Re}\left\{ \left(a_1^{(1)} \begin{pmatrix} \varphi_1 \\ \xi_1 \end{pmatrix} + a_2^{(1)} \begin{pmatrix} \varphi_2 \\ \xi_2 \end{pmatrix} \right) \exp\left(i\tilde{k}\tilde{x}\right) \right\}. \tag{8B.58}$$

The structure is shown in Fig. 8B.28. It is found that $\psi'_3(0)$ is four times stronger than $\psi'_1(0)$ and they are almost 180° out of phase. It is found by numerical integration that by

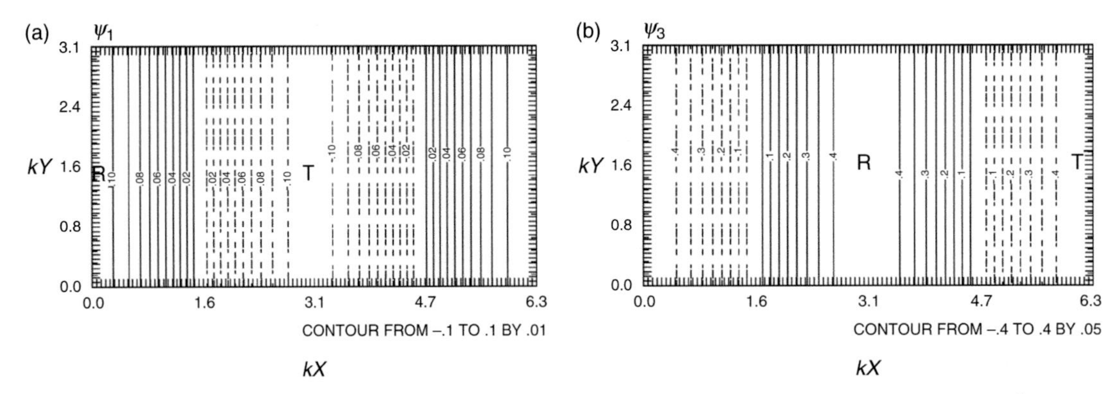

Fig. 8B.28 Structure of (a) ψ_1 and (b) ψ_3 of the optimal mode at $t = 0$ for $\tau = 12.0$ with a stable basic flow $\tilde{U}_1 = 1.8$, $\tilde{U}_3 = 1.0$, $\tilde{\beta} = 0.8$, $\tilde{k} = 1$.

$t = 12$, $\psi_1'(t = 12)$ has become four times stronger than $\psi_3'(t = 12)$ and they have become almost in-phase.

8B.8 Wave-activity density and general necessary condition for instability

The necessary conditions for instability in Sections 8B.2.1 and 8B.4.1 are actually special versions of a necessary condition for instability of a general parallel shear flow in a continuous vertical domain. The existence of such a condition stems from a property of a flow that we have briefly examined in Section 6.6.2. It is the *wave-activity density, A*, defined to be the enstrophy of a disturbance, $\rho_o[q^{*2}]$, divided by the meridional gradient of the potential vorticity of a basic zonal flow, $[q]_y$. To derive this necessary condition, let us consider a disturbance in the presence of a zonal velocity denoted by $U(y,z)$. The density of the basic state is $\rho_o(z) = \rho_{oo}\exp(-z/H)$. The governing equation is the linearized form of Equation (6.57) of Section 6.7. It is cited below for easy reference.

$$q_t^* + Uq_x^* + [q]_y\psi_x^* = 0,$$
$$[q]_y = \beta - U_{yy} - \frac{f^2}{\rho_o}\left(\frac{\rho_o}{N^2}U_z\right)_z, \tag{8B.59}$$
$$q^* = \psi_{xx}^* + \psi_{yy}^* + \frac{f^2}{\rho_o}\left(\frac{\rho_o}{N^2}\psi_z^*\right)_z.$$

The domain is a cyclical channel laterally bounded by two rigid boundaries and vertically bounded by two surfaces, $-X \le x \le X$, $-Y \le y \le Y$, $Z_{bot} \le z \le Z_{top}$. It should be noted that such a general basic flow could be a source of energy both barotropically and baroclinically to an unstable disturbance.

If we multiply (8B.59) by $\rho_o q^*/\left(2[q]_y\right)$ and take the zonal average, we would get

$$\frac{\partial A}{\partial t} + \nabla \cdot \vec{E} = 0,$$

$$A = \frac{\rho_o [q^{*2}]}{2[q]_y} \quad \rho_o[v^* q^*] = \nabla \cdot \vec{E} \quad \nabla = \left(\frac{\partial}{\partial y}, \frac{\partial}{\partial z}\right), \tag{8B.60}$$

$$[v^* q^*] = -[u^* v^*]_y + \frac{f_o g}{\rho_o \theta_{oo}} \left(\frac{\rho_o}{N^2}[v^* \theta^*]\right)_z,$$

where $A(y,z,t)$ is wave-activity density and $\vec{E}$ is the Eliassen–Palm (EP) vector. Wave-activity density has dimension of momentum per unit volume, kg m s^{-1} m^{-3}. Its rate of change is simply the convergence of the EP vector. The domain integral of (8B.60) is

$$\frac{d}{dt} \int_{Zbot}^{Ztop} \int_{-Y}^{Y} A dy dz = -\frac{f_o g}{\theta_{oo}} \int_{-Y}^{Y} \left(\left(\frac{\rho_o [v^* \theta^*]}{N^2} \right)_{Ztop} - \left(\frac{\rho_o [v^* \theta^*]}{N^2} \right)_{Zbot} \right) dy. \tag{8B.61}$$

We have made use of the boundary condition $v^* = 0$ implying $[u^* v^*] = 0$ at $y = \pm Y$. We see from (8B.61) that heat flux at the horizontal boundaries of the domain could give rise to instability of the disturbance. This result was first obtained by Charney and Stern (1962).

It is instructive to consider a special case. The quantity $\rho_o[v^* q^*]/N^2$ at the top boundary of the domain is negligibly small since ρ_o drops off exponentially with height. This quantity at the bottom boundary of the domain however would be zero only if the bottom boundary is parallel to an isothermal surface. In such a case, θ^* at Z_{bot} would be zero as well. The basic flow $U(y,z)$ in this case is known as an *internal jet*. Then (8B.61) would be reduced to

$$\frac{d}{dt} \int_{Zbot}^{Ztop} \int_{-Y}^{Y} A \, dy \, dz = 0. \tag{8B.62}$$

This constraint says that *the total wave-activity density of any perturbation in the presence of an arbitrary internal jet is invariant*. Recall that $A = \rho_o [q^{*2}]/2[q]_y$. Since the enstrophy of a disturbance is a positive definite quantity, it would be increasing in time at an exponential rate everywhere for an unstable wave. This constraint could not be satisfied unless there is *a change of sign in* $[\bar{q}]_y$ *at one or more latitudes within the domain*. When this condition prevails, an increase of the local value of A in one part of the domain where $[\bar{q}]_y$ is positive would be cancelled by a corresponding decrease of the local A in another part of the domain where $[\bar{q}]_y$ is negative. In other words, *a change of sign in* $[\bar{q}]_y$ *is a necessary condition for dynamic instability of an internal jet*. This conclusion was first deduced by Charney and Stern (1962). This is equivalent to the existence of one or more extrema in the distribution of potential vorticity of the basic state. This is the general form of the necessary conditions for instability discussed in Sections 8B.2.1 and 8B.4.1.

Baroclinic instability is still possible even when there is no extremum in the distribution of a basic PV. We also see from (8B.61) that the basic thermal gradient on the lower and upper boundary surfaces of the domain could play a crucial role in supporting instability in that case. A down-gradient eddy heat flux at the lower boundary surface would make the RHS of (8B.61) positive. It follows that the total wave activity would increase in time. The basic state in the models of Eady (1949) and Charney (1947) is of this kind.

Another intrinsic characteristic of instability becomes evident in the following consideration. If we just multiply (8B.59) by q^* and take the domain average of the resulting equation, we would get

$$\frac{1}{2}\frac{d}{dt}\langle [q^{*2}] \rangle = -\left\langle [q]_y [v^* q^*] \right\rangle. \tag{8B.63}$$

The angular bracket stands for the integrals over y and z. We may conclude from (8B.63) that $[q]_y$ and $[v^* q^*]$ must be negatively correlated in the meridional cross-section (i.e. opposite signs on the average) for an unstable disturbance so that the total enstrophy would increase in time. In other words, the eddy flux of PV by an unstable wave must be in the down-gradient direction of the basic PV field. This is compatible with the notions that PV is a conservative property of a fluid and an intensifying disturbance stirs a fluid. A particular example of it has been shown in Section 8B.5.5.

We also see that for a steady wave, the LHS is zero. There would be no eddy PV flux and hence the corresponding Eliassen–Palm vector field must be non-divergent. This conclusion was first derived by Eliassen and Palm (1961) and is known as Eliassen–Palm theorem.

Concluding remarks

In this part of the chapter, we analytically deduce the necessary condition of instability for a parallel shear flow including a purely barotropic or purely baroclinic shear flow. A geometric interpretation of this condition for the case of baroclinic instability is given in terms of the slope of the isentropic surface of the basic state.

The three different aspects of barotropic/baroclinic instability – modal growth, transient growth and optimal growth – have been elaborated, more extensively so for baroclinic instability. We have presented an analytic solution for a 1-D disturbance in the presence of a purely baroclinic basic flow. The properties of the unstable normal mode under relevant parameter conditions are shown to be quite compatible with those of a developing extratropical cyclone. It was also shown that the transient growth of a disturbance depends on its structure and can have a rate faster than that of even the most unstable normal mode. We have further shown how to determine the optimal mode for each optimization time interval without repeatedly doing tiresome explicit numerical integrations. In light of these results, it is justifiable to conclude that the theory of even purely baroclinic instability succeeds in accounting for the most essential characteristics of cyclogenesis in the extratropics.

PART 8C INSTABILITY OF JETS

This part of the chapter is concerned with the instability of two classes of more general basic shear flows: zonally uniform jet and localized jet. The velocity in a uniform jet varies in the vertical and transverse directions, whereas that in a localized jet even varies in all three directions. These features of a basic flow make the equation for a modal instability analysis mathematically inseparable. Although the instability results can still be numerically determined in a straightforward manner, their characteristics are somewhat more subtle. A brief discussion of the general conceptual issues is given in Section 8C.1. We delineate the impact of a weak barotropic shear in an otherwise baroclinic basic flow in Section 8C.2. It is referred to as a barotropic-governor effect. A counterpart problem is concerned with the impact of a weak baroclinic shear in an otherwise barotropic shear flow. We also contrast the instability of a broad baroclinic jet with that of a narrow baroclinic jet in Section 8C.3. The instability of a localized barotropic jet is next examined in Section 8C.4. An instability analysis of a localized baroclinic jet is finally discussed in Section 8C.5 in the context of a specific phenomenon.

8C.1 Nature and scope of the problem

In light of the great success of the classical baroclinic instability theory for cyclogenesis discussed in Part 8B, we can now extend the theory by considering progressively more general classes of basic shear flow. For example, it would be instructive to ascertain the impact of an additional weak barotropic shear on the instability of a basic flow that has a strong baroclinic shear. It is found that such a horizontal shear has a stabilizing influence. It is referred to as a *barotropic-governor effect* (James, 1987). A counterpart problem is concerned with the impact of a weak baroclinic shear on the instability of an otherwise basic flow with a strong barotropic shear. This influence may be similarly referred to as a *baroclinic-governor effect.* The impact of a barotropic shear component on the instability of a baroclinic shear flow should depend not only on its structure but also on its magnitude. There are two scenarios to be considered separately. In one scenario, the barotropic shear component is weak and the barotropic-governor effect would be expected. In the second scenario, the basic flow has a sufficiently strong barotropic shear component. It would be expected to have a destabilizing impact instead. We examine these dynamical issues in Sections 8C.2 and 8C.3.

There are two different aspects of temporal amplification of a disturbance even when we consider a zonally uniform basic shear flow. An initially localized disturbance may intensify in-situ. If the disturbance turns out to have an effective zero group velocity and to be non-dispersive, it would intensify without movement and change of shape. Such temporal amplification is referred to as *absolute instability*. On the other hand, an initially localized disturbance may amplify and simultaneously move in the zonal direction. In an infinitely large domain, the disturbance in a particular longitudinal sector would intensify over a time interval and subsequently vanish. Such instability is referred to as *convective instability*. Merkine (1977) highlighted such differences by introducing these two concepts in an analysis of the asymptotic instability property of a zonally uniform basic flow in a quasi-geostrophic two-layer infinitely long f-plane channel model. In a reentrant channel domain (which is a more appropriate analog of a latitudinal belt), convective instability would dominate at large time because a disturbance can repeatedly go through the eastern and western boundaries. The most unstable normal mode with intrinsic zonal symmetry will eventually emerge as the dominant disturbance regardless of the structure of the initial disturbance.

The large-scale jets in the atmosphere actually have a well-defined localized structure. For example, there are two pronounced localized baroclinic jets in the northern hemisphere on the winter-mean upper tropospheric weather maps. One is located over the Pacific Ocean off the coast of Asia and another over the Atlantic Ocean off the coast of North America as shown in Fig. 9.1(a). The instability of a localized jet is obviously an issue of fundamental dynamical interest. Such instability jointly involves barotropic and baroclinic processes depending upon not only the vorticity field but also the deformation fields of the basic flow. The zonal variation of a jet considerably complicates an instability analysis. The necessary condition for instability of a given localized jet is not known. The distinction between absolute and convective instability is particularly pertinent when we consider the instability of a zonally varying baroclinic basic flow. If its accompanying meridional structure is neglected, the method of WKB (Wentzel–Kramers–Brillouin) approximation can be applied to perform such instability analysis (Pierrehumbert, 1984). This method is justifiable when there is a clear scale separation between the main constituent wave components of an unstable mode and the basic flow itself. In dealing with a general basic flow with shear varying in all three directions, it would be necessary to perform a modal instability analysis as a general boundary value problem as was first done by Fredericksen (1983).

Unstable modes associated with convective instability in a zonally varying basic flow span the whole length of the domain. They are referred to as *global modes*. Those associated with absolute instability are referred to as *local modes*. In Section 8C.4, we examine the instability of a localized barotropic jet without beta-effect as a simplest example of this class of basic flows. In Section 8C.5, we examine the instability of a localized baroclinic jet with a two-layer quasi-geostrophic model. This analysis is made in the context of testing a hypothesis about the dynamical nature of an intriguing atmospheric phenomenon known as mid-winter minimum of the Pacific storm track (Nakamura, 1992).

8C.2 Barotropic-governor effect

Let us now examine the instability of a basic flow that has two components in a two-layer quasi-geostrophic model. One component has purely baroclinic shear and an additional component has purely barotropic shear. We specifically consider the following basic flow

$$U_1 = V\left(u_{1o} + \varepsilon \exp\left(-(y/a)^2\right)\right), \quad U_3 = V\left(u_{3o} + \varepsilon \exp\left(-(y/a)^2\right)\right), \tag{8C.1}$$

where $V, u_{1o}, u_{3o}, \varepsilon, a$ are constants. The corresponding gradient of the basic potential vorticity would be

$$\begin{aligned} \bar{q}_{1y} &= \beta - U_{1yy} + \lambda^2(U_1 - U_3), \\ \bar{q}_{3y} &= \beta - U_{3yy} - \lambda^2(U_1 - U_3). \end{aligned} \tag{8C.2}$$

The governing perturbation equations are

$$\begin{aligned} q'_{1t} + U_1 q'_{1x} + \psi'_{1x}\bar{q}_{1y} &= 0, \\ q'_{3t} + U_3 q'_{3x} + \psi'_{3x}\bar{q}_{3y} &= 0, \end{aligned} \tag{8C.3a, b}$$

and

$$\begin{aligned} q'_1 &= \nabla^2\psi'_1 - \lambda^2(\psi'_1 - \psi'_3), \\ q'_3 &= \nabla^2\psi'_3 - \lambda^2(\psi'_3 - \psi'_1), \end{aligned} \tag{8C.4}$$

with U_j, $\bar{q}_{1y}$ and $\bar{q}_{3y}$ given in (8C.1) and (8C.2). The notations are the same as those introduced in Chapter 6. The domain is a reentrant channel bounded by two rigid walls: $-X \le x \le X$, $-Y \le y \le Y$. The boundary conditions are therefore (i) $\psi'_j(-X) = \psi'_j(X)$ and (ii) $\psi'_j(\pm Y) = 0$. The necessary condition of instability derived in Section 8B.4.1 applies, namely *the meridional gradient of the basic potential vorticity must change sign in the domain*. This condition can be met solely due to the basic vertical shear, $(U_1 - U_3)$, and/or due to the basic horizontal shear, U_{jyy}.

A normal mode solution for this class of basic flows has the form of $\psi'_j = \xi_j(y)e^{i(kx-t\sigma)}$. Let the channel domain be infinitely long for convenience, so that we may use any value for the wavenumber, k. Substituting this solution into (8C.3a,b) and (8C.4), we get the following differential equations that govern the amplitude functions

$$\begin{aligned} \sigma\left(\left(\frac{d^2}{dy^2} - k^2 - \lambda^2\right)\xi_1 + \lambda^2\xi_3\right) &= kU_1\left(\left(\frac{d^2}{dy^2} - k^2 - \lambda^2\right)\xi_1 + \lambda^2\xi_3\right) + k\bar{q}_{1y}\xi_1, \\ \sigma\left(\left(\frac{d^2}{dy^2} - k^2 - \lambda^2\right)\xi_3 + \lambda^2\xi_1\right) &= kU_3\left(\left(\frac{d^2}{dy^2} - k^2 - \lambda^2\right)\xi_3 + \lambda^2\xi_1\right) + k\bar{q}_{1y}\xi_3. \end{aligned} \tag{8C.5}$$

The boundary conditions are $\xi_1(\pm Y) = \xi_j(\pm Y) = 0$. Measuring horizontal distance in units of λ^{-1}, velocity in units of V and time in units of $V^{-1}\lambda$, we introduce the following non-dimensional quantities: $\tilde{k} = k/\lambda$, $\tilde{\sigma} = \sigma/(V\lambda)$, $\tilde{\beta} = \beta/(V\lambda^2)$, $\tilde{U}_j = \tilde{U}_j/V$, $(\tilde{x}, \tilde{y}, \tilde{a}) = \lambda(x, y, a)$. The dependent variables are non-dimensionalized as follows:

$\tilde{\psi}_j = \psi_j / V\lambda^{-1}$, $\tilde{\omega}_2 = \omega_2 f_o / V^2 \lambda^2 \Delta p$, $\tilde{T}_2 = T_2 R / V\lambda^{-1} f_o$. The normal mode solution is $\left(\tilde{\psi}_j, \tilde{\omega}_2, \tilde{T}_2\right) = \left(\tilde{\xi}_j, \tilde{W}_2, \tilde{\theta}_2\right) \exp\left(i\left(\tilde{k}\tilde{x} - \tilde{\sigma}\tilde{t}\right)\right)$.

Using N grid points to depict the domain, we recast the non-dimensional form of (8C.5) with center-difference approximation and make use of (8C.2), (8C.4) and the boundary conditions. The result is a matrix equation $\sigma \underline{B}\vec{F} = \underline{A}\vec{F}$ where $\vec{F}$ is a $2N$-vector containing the unknowns at the grid points; $\underline{A}$ and $\underline{B}$ are $2N \times 2N$ matrices. The $2N$ eigenvalues and the corresponding eigenvectors can be readily evaluated; $N = 99$ grid points are used in the calculations.

Hydrostatic balance implies $\tilde{\theta}_2 = \left(\tilde{\xi}_1 - \tilde{\xi}_3\right)$. The amplitude function $\tilde{W}_2$ can be evaluated in terms of $\tilde{\xi}_j$ on the basis of the linearized omega equation, viz.

$$\left(\frac{d^2}{d\tilde{y}^2} - \tilde{k}^2 - 2\right)\tilde{W} = i\tilde{k}\left(\tilde{U}_3 - \tilde{U}_1\right)\left(\frac{d^2}{d\tilde{y}^2} - \tilde{k}^2\right)\left(\tilde{\xi}_1 + \tilde{\xi}_3\right) + i\tilde{k}\left(\left(\tilde{\beta} - \tilde{U}_{3\tilde{y}\tilde{y}}\right)\tilde{\xi}_3 - \left(\tilde{\beta} - \tilde{U}_{1\tilde{y}\tilde{y}}\right)\tilde{\xi}_1\right). \quad (8C.6)$$

The domain is 8000 km wide and $Y = 4 \times 10^6$ m. It suffices to use $\tilde{W}_2(\pm Y) = 0$ as the boundary condition for (8C.6) since the domain is sufficiently wide. Using $\lambda = 10^{-6}\,\mathrm{m}^{-1}$, $V = 30\,\mathrm{m}^{-1}$ and $\beta = 1.5 \times 10^{-11}\,\mathrm{m}^{-1}\,\mathrm{s}^{-1}$, we have $-4 \leq \tilde{Y} \leq 4$ and $\tilde{\beta} = 0.5$. The tilde superscript is dropped in the rest of this section for convenience.

We use the case of a purely baroclinic basic flow ($u_{1o} = 1.0$, $u_{3o} = 0.2$, $\varepsilon = 0.0$) as a reference for comparison. This basic flow in the upper and lower layers is therefore $30\,\mathrm{m\,s}^{-1}$ and $6\,\mathrm{m\,s}^{-1}$ respectively. All eigenvalues in a calculation are presented in a scatter-plot of its imaginary part versus its real part, $\mathrm{Im}\{\sigma\}$ vs. $\mathrm{Re}\{\sigma\}$. Figure 8C.1a shows that there is one unstable mode for a disturbance of wavelength equal to 5.5 (i.e. 5500 km). This unstable mode has a growth rate equal to 0.1408 corresponding to e-folding time of about 2 days. The zonal phase speed of this unstable mode is $\sigma_r/k = 0.3563$ corresponding to $10.6\,\mathrm{m\,s}^{-1}$. As expected, this unstable mode has a westward vertical tilt and a half-cosine meridional structure that spans the whole domain, centering at the middle of the domain with no horizontal tilt.

The counterpart result for another basic flow characterized by $u_{1o} = 1.0$, $u_{3o} = 0.2$, $\varepsilon = 0.7$ and $a = 1.5$ is shown in Fig. 8C.1b. The basic baroclinic component is the same as before. The barotropic jet component has an intensity equal to 70 percent of the baroclinic flow component in the upper layer and a width equal to 37.5 percent of the domain. We see that there are three unstable modes in this case. The most unstable mode is the counterpart of the one for a purely baroclinic basic flow. Its growth rate is however only equal to 0.0841. In other words, *the barotropic shear reduces the growth rate of this mode from 0.1408 by about 40 percent consistent with the notion of a barotropic governor effect.* Its zonal phase speed on the other hand is considerably increased, $\sigma_r/k = 0.6151$, corresponding to about $18\,\mathrm{m\,s}^{-1}$. The other two unstable modes have substantially smaller growth rates. They have shorter meridional scales and faster zonal phase speeds. These modes are members of two distinct branches of unstable modes destabilized by the barotropic shear of the basic flow. This is one manifestation of a possible destabilizing influence of the additional barotropic shear.

It is instructive to repeat the calculations using a wide range of wavelengths with or without the barotropic jet component in the basic flow. The growth rates of the unstable modes that have the broadest meridional scale for $\varepsilon = 0$ and $\varepsilon = 0.7$ are compared over such a range of wavelengths in Fig. 8C.2a. The results reveal that the additional barotropic

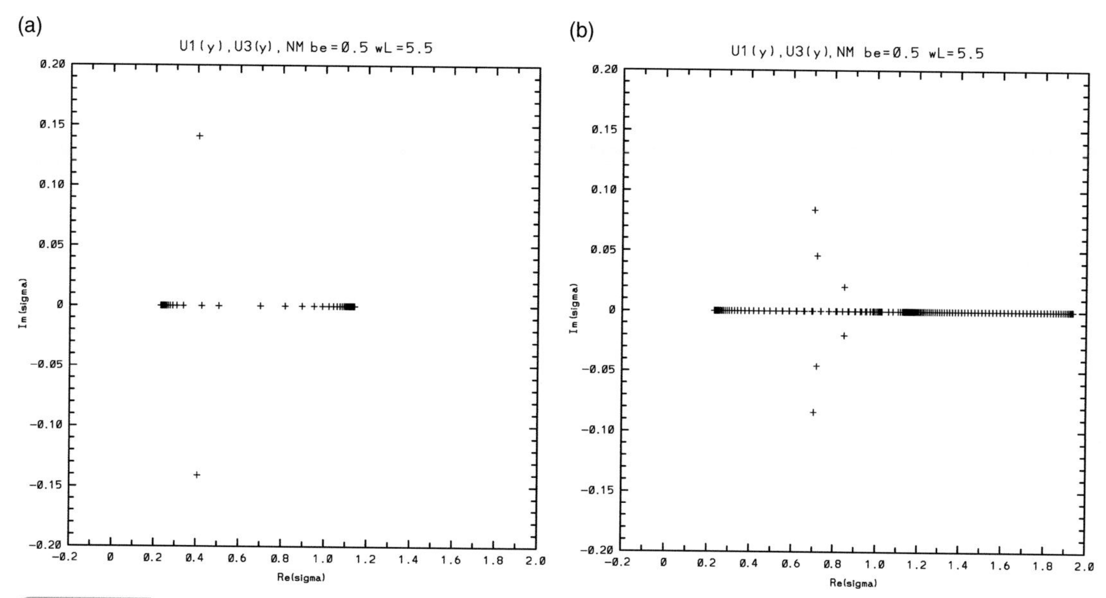

Fig. 8C.1 Distribution of all eigenvalues $\mathrm{Im}\{\sigma\}$ vs. $\mathrm{Re}\{\sigma\}$ in units of $V\lambda$ for wavelength equal to 5.5 in units of λ^{-1}; (a) purely baroclinic basic flow, $u_{1o} = 1.0$, $u_{3o} = 0.2$, $\varepsilon = 0.0$ and (b) the same baroclinic flow with an additional barotropic jet component $\varepsilon = 0.7$ and $a = 1.5$.

shear component of this particular structure and magnitude indeed may be destabilizing or stabilizing dependent on the wavelength of one major branch of normal modes. The shortwave cutoff is slightly reduced from a wavelength of about 4.8 to 4.0 and the longwave cutoff is slightly extended from 8.9 to 9.5. The growth rates for the longwave part of the spectrum for wavelength between 7.5 and 9.5 are also enhanced. But, there is a pronounced stabilizing impact in the range of synoptic-scale wavelengths between 4.9 and 7.5. This result corresponds to James' finding of a barotropic-governor effect. The barotropic jet component also substantially increases the zonal phase speed of the unstable modes of all wavelengths (Fig. 8C.2b). This effect is to be expected since there is a greater advective influence by the basic flow on a disturbance.

In passing, let us also examine how a weak uniform baroclinic shear component would impact upon the instability of an otherwise barotropic jet. For this purpose, we compare the instability properties of two basic flows for a normal mode with a wavelength equal to 4. The new reference basic flow is a purely barotropic shear flow defined by $u_{1o} = u_{3o} = 0$, $\varepsilon = 1.0$, $a = 0.75$. The calculation confirms that the reference basic flow is unstable by virtue of barotropic instability elaborated in Section 8B.2. The structure of ψ_1 is identical to that of ψ_3 so that there is no vertical tilt. Their horizontal tilts on both sides of the jet lean against the basic shear whereby the disturbance extracts kinetic energy from the basic flow. The growth rate is $\sigma_i = 0.1732$ (*e*-folding time of ~2 days) and its frequency is $\sigma_r = 0.5390$ (phase speed of ~$10\,\mathrm{m\,s^{-1}}$). These instability properties are compared with those of another basic flow that additionally has a weak baroclinic component. This flow is characterized by $u_{1o} = 0.2$, $u_{3o} = 0.1$, $\varepsilon = 1.0$, $a = 0.75$. The corresponding unstable mode has similar horizontal tilts but with an additional small *eastward* vertical tilt. It has a

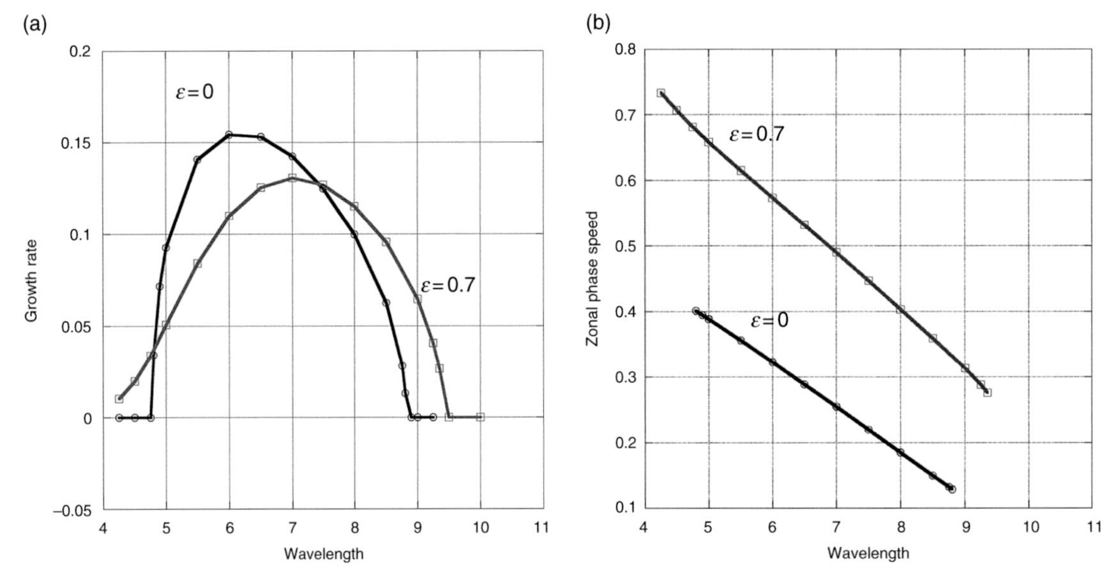

Fig. 8C.2 Comparison of (a) the growth rate, σ_i, in units of $V\lambda = 3 \times 10^{-5}\,\mathrm{s}^{-1}$ and (b) corresponding zonal phase speed, σ_r/k, in units of $V = 30\,\mathrm{m\,s}^{-1}$ of the normal modes that have the broadest meridional scale as a function of wavelength, $2\pi/k$, in units of $\lambda^{-1} = 1 \times 10^6$ m for a purely baroclinic basic flow ($\varepsilon = 0$), or one with a barotropic jet component ($\varepsilon = 0.7$).

slightly smaller growth rate, $\sigma_i = 0.1718$ and a larger frequency, $\sigma_r = 0.7630$. It would be therefore logical to refer to the stabilizing impact of this weak baroclinic shear on the instability of an otherwise barotropic jet as a *baroclinic-governor effect.* Since a purely barotropic jet is a more drastic idealization of the atmospheric zonal flow than a purely baroclinic flow, the notion of a baroclinic-governor effect is likely to be less relevant than a barotropic-governor effect. Nevertheless, it is a bona fide dynamical effect.

8C.3 Instability of baroclinic jets

The observed winter-zonal mean flow has a pronounced baroclinic jet with a distinct core at about 200 mb level over 35° latitude. Both of its horizontal and vertical shears are significant. We may intuitively expect that the instability properties of a baroclinic jet would be dependent on its width for a given vertical structure. For this reason, it would be meaningful to compare the instability properties of a broad baroclinic jet with those of a narrow baroclinic jet. Let us then consider a westerly baroclinic jet that has a Gaussian structure in a quasi-geostrophic two-layer model setting, viz.

$$U_j = V\varepsilon_j \exp\left(-(y/a)^2\right), \tag{8C.7}$$

where $\varepsilon_1 > \varepsilon_3 > 0$. We use $a = 1.5$ to characterize a broad jet and $a = 0.75$ to characterize a narrow jet.

8C.3.1 Energy equations

Following the procedure for deriving the energy equations for the case of a purely baroclinic shear flow (Section 8B.5.4), we obtain the following non-dimensional energy equations of a disturbance embedded in a general baroclinic jet

$$\frac{dK}{dt} = C(P,K) + C(\bar{K},K),$$
$$\frac{dP}{dt} = C(\bar{P},P) - C(P,K), \qquad (8C.8a,b)$$

where$K = \frac{1}{2}\langle (u_1^2 + v_1^2 + u_3^2 + v_3^2) \rangle$, $P = \frac{1}{2}\langle T_2^2 \rangle$, $u_j = -\psi_{jy}$, $v_j = \psi_{jx}$,

$$T_2 = \psi_1 - \psi_3,\ C(P,K) = \langle -\omega_2 T_2 \rangle,\ C(\bar{K},K) = \langle -U_{1y}v_1u_1 - U_{3y}v_3u_3 \rangle,$$
$$C(\bar{P},P) = (U_1 - U_3)\langle v_2 T_2 \rangle.$$

The angular bracket stands for horizontal integration as before. Compared to the energy equations for the classic baroclinic instability, (8C.8a) has one extra term $C(\bar{K},K)$. It represents a conversion rate from the basic kinetic energy to perturbation kinetic energy due to the presence of horizontal shear in the jet. The other two conversion processes are $C(P,K)$ and $C(\bar{P},P)$. The former stands for conversion rate from wave potential energy to wave kinetic energy. The latter stands for conversion rate from the basic potential energy to wave potential energy.

8C.3.2 Instability properties of a broad baroclinic jet

Let us consider a broad baroclinic jet defined by $\varepsilon_1 = 1.5$, $\varepsilon_3 = 0.3$ and $a = 1.5$. The baroclinic shear of this jet is quite strong since its maximum wind at the upper and lower levels are $45\ \mathrm{m\,s^{-1}}$ and $9\ \mathrm{m\,s^{-1}}$ respectively.

Growth rate and phase speed

The calculations in this case reveal that there is one unstable mode over a wide range of wavelengths. The variations of its growth rate and phase speed with its wavelength are shown in Fig. 8C.3a,b. The cutoff wavelengths are about 4.2 and 7.6. The maximum growth rate is about 0.1112 corresponding to an *e*-folding time of about 3 days. The most unstable mode has a wavelength of about 5.5, corresponding to 5500 km. The phase speed of the unstable mode near the shortwave cutoff is about $13\ \mathrm{m\,s^{-1}}$. The longer the wavelength is, the greater the phase speed of an unstable mode would be over most of the range of wavelengths under consideration.

Structure of the most unstable mode

The structure of the streamfunction of the most unstable mode, ψ_j, with a wavelength of $2\pi/k = 5.5$, is shown in Fig. 8C.4a,b. One telltale feature is that the meridional scale of the unstable mode is essentially limited by the width of the basic baroclinic jet. The domain is

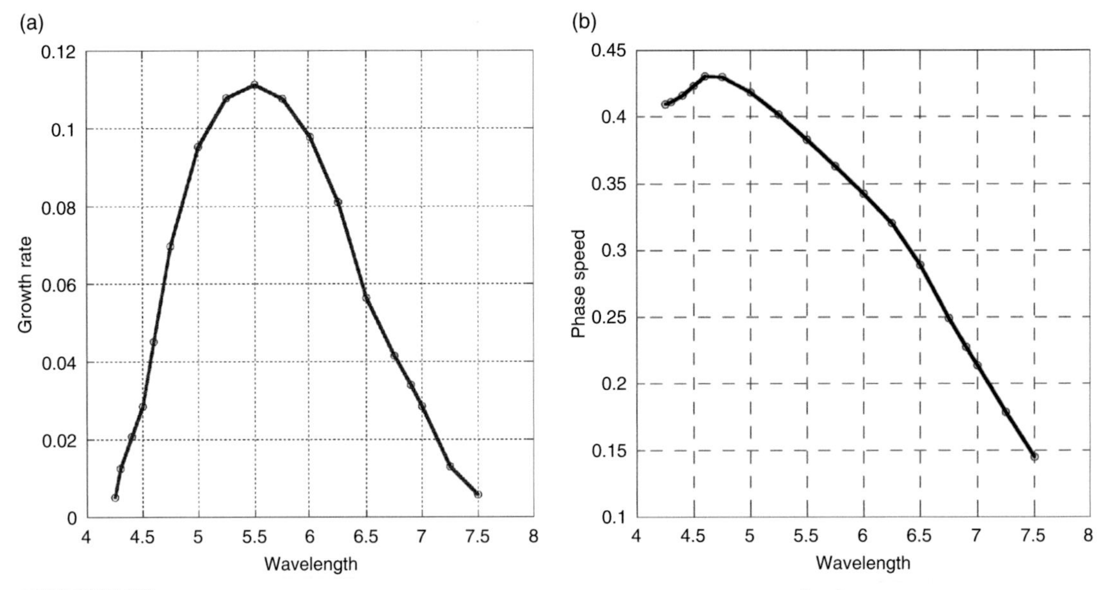

Fig. 8C.3 Variation of (a) growth rate σ_i of the unstable mode in units of $V\lambda = 3 \times 10^{-5}\ \mathrm{s}^{-1}$ (b) phase speed σ_r/k in units of $V = 30\ \mathrm{m\,s}^{-1}$ with wavelength, $L = 2\pi/k$, in units of $\lambda^{-1} = 10^6$ m for a broad baroclinic jet defined by $\varepsilon_1 = 1.5$, $\varepsilon_3 = 0.3$ and $a = 1.5$.

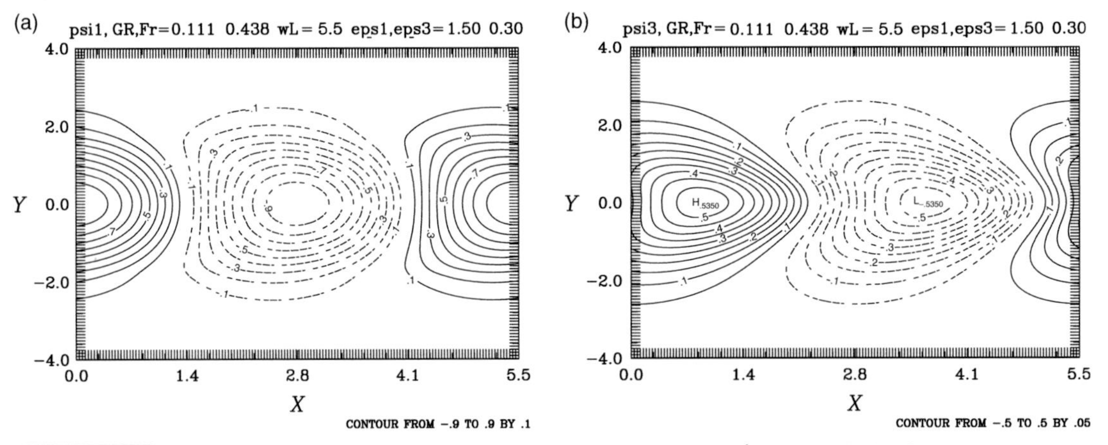

Fig. 8C.4 Structure of (a) ψ_1 and (b) ψ_3 of the unstable mode in units of $V\lambda^{-1} = 3 \times 10^7\ \mathrm{m}^2\,\mathrm{s}^{-1}$ with wavelength $2\pi/k = 5.5$ in units of $\lambda^{-1} = 10^6$ m for a broad baroclinic jet defined by $\varepsilon_1 = 1.5$, $\varepsilon_3 = 0.3$ and $a = 1.5$.

sufficiently wide that the lateral boundaries have little influence on this unstable mode. The horizontal tilts in ψ_j on both sides of the jet align along the direction of the horizontal shear of the jet. Specifically, the tilt is from SW to NE to the south of the jet axis and from NW to SE to the north of the jet axis. Such tilt is more pronounced in $\tilde{\psi}_3$.

We can infer from this horizontal tilt, according to (8C.8a), that u_j and v_j are positively correlated to the south of the jet axis and negatively correlated to the north of the jet axis. A consequence of this is that the wave transports some zonal momentum toward the core of the

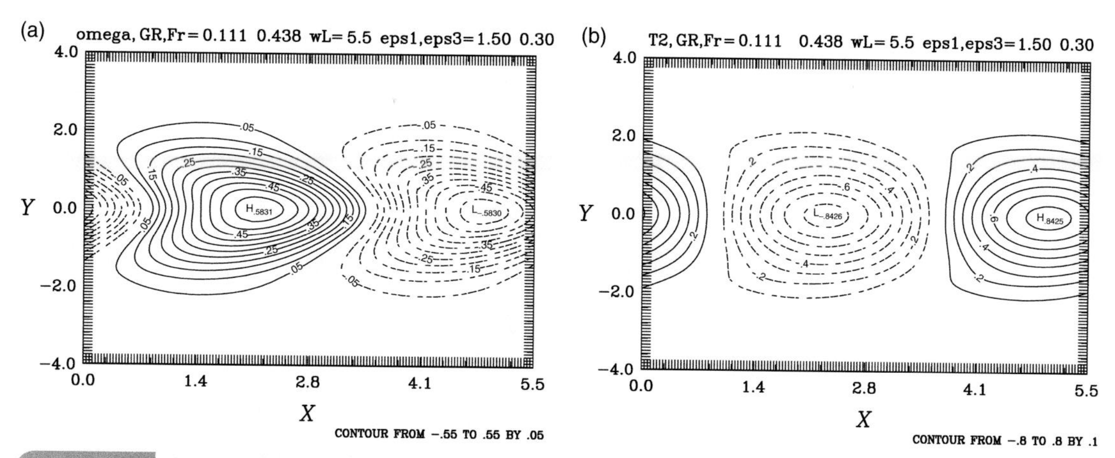

Fig. 8C.5 Structure of the unstable mode with wavelength $2\pi/k = 5.5$ in terms of (a) $\tilde{\omega}_2$ in units of $V^2\lambda^2\Delta p/f_o = 0.36\,\text{Pa s}^{-1}$ and (b) $\tilde{T}_2$ in units of $V\lambda^{-1}f_o/R = 10.4\,\text{K}$ for a broad baroclinic jet defined by $\varepsilon_1 = 1.5$, $\varepsilon_3 = 0.3$ and $a = 1.5$.

jet. In conjunction with the structure of $U_j(y)$, we infer that $C(\bar{K}, K) < 0$ meaning that the unstable mode is transferring some of its kinetic energy to the basic flow.

The vertical tilt of the unstable mode is westward by about one-eighth of its wavelength. The corresponding structures of ω_2 and T_2 are shown in Fig. 8C.5a,b. The ascending motion and warm anomaly are nearly in phase and are located slightly to the east of the lower level trough. We infer that $C(P, K) > 0$ meaning that some wave potential energy is converted to wave kinetic energy. This baroclinic generation of kinetic energy process is strong enough to overcompensate the loss of some of its kinetic energy to the basic state. The structure of ψ_j and T_2 suggests that warm air is going north and cold air is coming south. It follows that $C(\bar{P}, P) > 0$.

The physical nature of the instability of this baroclinic jet from the perspective of energetics of the most unstable mode is summarized in the following schematic diagram (Fig. 8C.6). This unstable wave intensifies ultimately by extracting energy from the potential energy of the basic state, but transfers back some of its kinetic energy to the basic flow. The horizontal shear of this broad baroclinic jet plays a stabilizing role and its width dictates the meridional scale of the unstable mode.

8C.3.3 Instability properties of a narrow baroclinic jet

To demonstrate that a sufficiently strong barotropic shear could play a destabilizing role in the presence of baroclinic shear, we next examine the instability of a narrow baroclinic jet. This jet is only half as wide as that in the last section. All other parameters have the same values as in the last section. The basic jet is then defined by $\varepsilon_1 = 1.5$, $\varepsilon_3 = 0.3$ and $a = 0.75$. The variations of the growth rate and zonal phase speed of a wave disturbance with its wavelength are shown in Fig. 8C.7a,b. The most unstable wave has a shorter wavelength, 4.0 instead of 5.5 for the broad jet. The maximum growth rate is almost twice as large, 0.204 instead of 0.111. The longer the wavelength is, the smaller the zonal phase

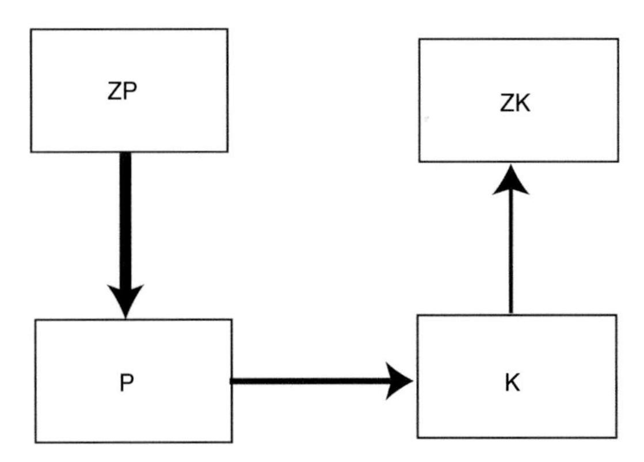

Fig. 8C.6 Schematic of the energetics of the most unstable mode for the basic flow defined by $\varepsilon_1 = 1.5$, $\varepsilon_3 = 0.3$ and $a = 1.5$. (ZP = basic potential energy, ZK = basic kinetic energy.) The width of each arrow qualitatively indicates the relative magnitude of the energy conversion rate under consideration.

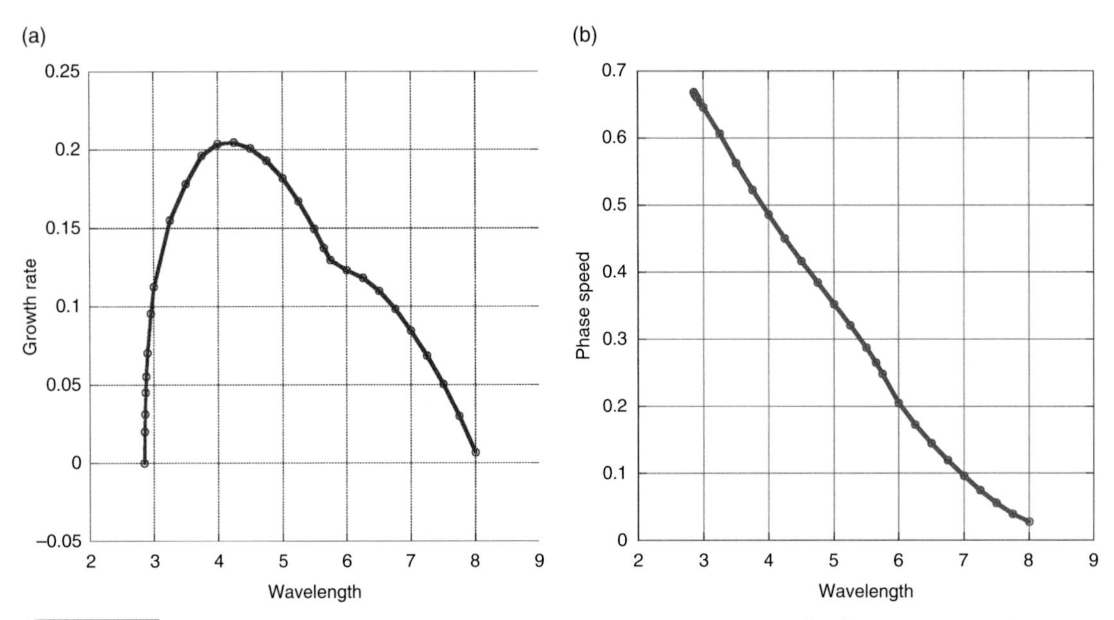

Fig. 8C.7 Variation of (a) growth rate σ_i of the unstable mode in units of $V\lambda = 3 \times 10^{-5}\,\mathrm{s}^{-1}$ (b) phase speed σ_r/k in units of $V = 30\,\mathrm{m\,s}^{-1}$ with wavelength, $L = 2\pi/k$, in units of $\lambda^{-1} = 10^6$ m for a narrow baroclinic jet defined by $\varepsilon_1 = 1.5$, $\varepsilon_3 = 0.3$ and $a = 0.75$.

speed would be. The phase speed for a disturbance of the same wavelength, say 5.5, is smaller, 0.2876 instead of 0.383.

Structure of an unstable normal mode

Let us examine the structure of the unstable mode of wavelength $2\pi/k = 5.5$ for this narrow baroclinic jet. The structure of this unstable mode is totally different than the counterpart unstable mode for a broad baroclinic jet (Fig. 8C.8 vs. Figs. 8.C4 and 8C.5). The

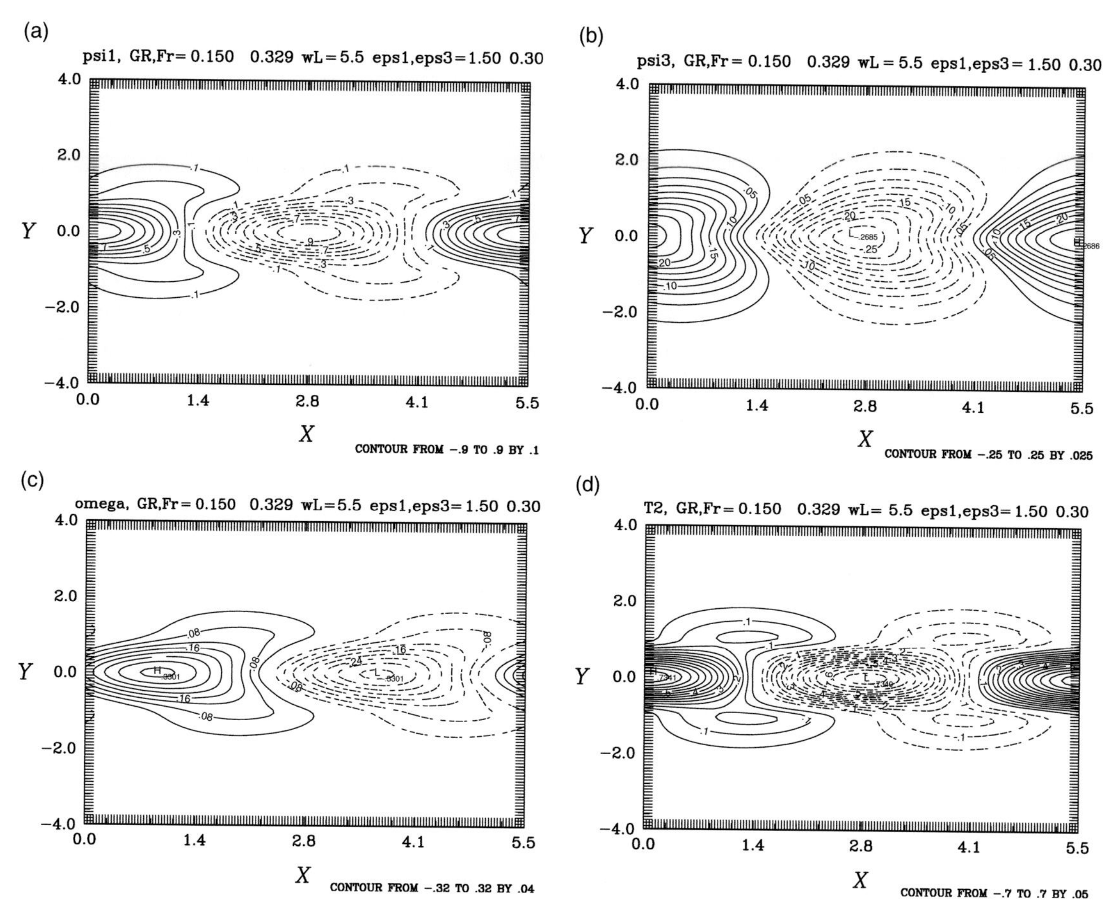

Fig. 8C.8 Structure of the unstable mode with wavelength $2\pi/k = 5.5$ in terms of (a) ψ_1 and (b) ψ_3 of the unstable mode in unit of $V\lambda^{-1} = 3 \times 10^7\ \mathrm{m^2\,s^{-1}}$ (c) ω_2 in unit of $\dfrac{V^2\lambda^2\Delta p}{f_o} = 0.36\ \mathrm{Pa\,s^{-1}}$ and (d) T_2 in unit of $V\lambda^{-1}f_o/R = 10.4\ \mathrm{K}$ for a narrow baroclinic jet defined by $\varepsilon_1 = 1.5$, $\varepsilon_3 = 0.3$ and $a = 0.75$.

streamfunction fields ψ_j now have a SW-NE tilt to the north of the jet axis and a SE-NW tilt to the south of the jet axis. This means that the unstable mode gains kinetic energy by extracting it from the basic flow. The vertical tilt of the unstable mode is very small suggesting that the barotropic process is the primary process of instability. There is actually a positive correlation between ω_2 and T_2 implying that some kinetic energy of the wave is converted to its potential energy. The temperature field in conjunction with the flow field is such that relatively warm air at mid-level flows from the north and relatively cold air from the south on the average. This unstable mode therefore also transfers some of its potential energy to the basic state. This is the energetic manifestation of the baroclinic-governor effect. This transfer of wave potential energy to basic potential energy is however more than compensated by the conversion from kinetic energy to potential energy, so that the disturbance as a whole intensifies. In other words, this unstable mode intensifies barotropically and the baroclinic process plays the role of a stabilizing factor in the case of a narrow baroclinic jet.

8C.4 Instability of a localized barotropic jet

We next consider the dynamics of instability of a basic flow with a higher degree of complexity. It is a barotropic localized jet in a reentrant channel domain of 12 000 km long and 6000 km wide bounded by rigid walls. We measure distance, velocity and time in units of $L = 3000$ km, $U = 25$ m s^{-1} and $LU^{-1} = 1.2 \times 10^5$ s. The non-dimensional variables are defined as $(\tilde{x}, \tilde{y}) = \frac{1}{L}(x, y)$, $\tilde{t} = \frac{U}{L}t$, $\tilde{\psi} = \frac{\psi}{UL}$, $(\tilde{\zeta}, \tilde{f}) = \frac{L}{U}(\zeta, f)$. Hence, the domain is $-2 \le \tilde{x} \le 2$ and $-1 \le \tilde{y} \le 1$. From here on, we drop the tilde superscript on the non-dimensional quantities. An external vorticity forcing, $F(x,y)$, is required to maintain a steady localized jet. Such vorticity forcing would be

$$F = J(\bar{\psi}, \bar{\zeta}), \tag{8C.9}$$

where $J(A, B) = A_x B_y - A_y B_x$ is the Jacobian.

With the total streamfunction and vorticity decomposed as $\psi_{total} = \bar{\psi} + \psi$ and $\zeta_{total} = \bar{\zeta} + \zeta$, the vorticity equation that governs a perturbation is

$$\zeta_t + J(\bar{\psi}, \zeta) + J(\psi, \bar{\zeta}) = 0. \tag{8C.10}$$

According to (8C.10), local change of perturbation vorticity arises from advection of basic vorticity by the perturbation and from advection of perturbation vorticity by the basic flow. Let us consider an especially simple basic flow. Its vorticity, $\bar{\zeta}(x, y)$, is zero in the domain except in two adjacent areas where it has nonzero uniform values. In the absence of beta-effect, the gradient of the absolute vorticity of the basic state would be simply that of $\bar{\zeta}(x, y)$; $J(\psi, \bar{\zeta})$ would be then nonzero only along the boundaries of the two prescribed adjacent areas. Since a localized westerly jet has cyclonic vorticity to its north and anticyclonic vorticity to its south, we specifically prescribe $\bar{\zeta}(x, y)$ as follows

$$\begin{aligned} \bar{\zeta} &= 5, && \text{in area-A,} \\ \bar{\zeta} &= -5, && \text{in area-B,} \\ \bar{\zeta} &= 0, && \text{everywhere else.} \end{aligned} \tag{8C.11}$$

Area-A is a half-elliptical area defined by $(ax)^2 + (by)^2 < r^2$ and $y > 0$ with $a = 0.6$, $b = 1.5$ and $r = 0.35$. Area-B is another half-elliptical area defined by $(ax)^2 + (by)^2 < r^2$ and $y < 0$. The common boundary of area-A and area-B is the line segment defined by $|x| \le r/a$ and $y = 0$. The domain is depicted with 100 grid points in each direction so that the resolution is quite high, $\delta x = 120$ km and $\delta y = 60$ km. The corresponding streamfunction is determined by inverting the vorticity field, $\bar{\psi} = \nabla^{-2}\bar{\zeta}$, subject to a cyclical condition on the zonal boundaries and no mass flux across the meridional boundaries. The vorticity and streamfunction fields of the basic flow under consideration are shown in Fig. 8C.9.

This jet is about 2500 km long and 800 km wide with a maximum velocity of about 21 m s^{-1}. It is perfectly symmetrical about its axis. It is noteworthy that the basic vorticity

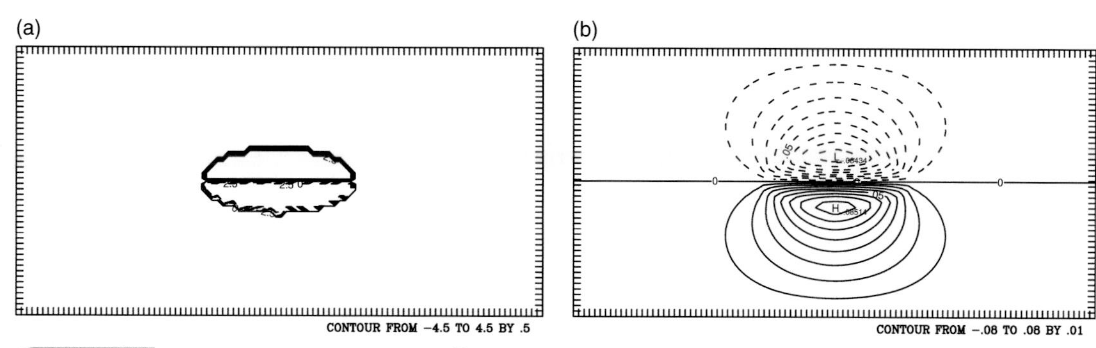

Fig. 8C.9 Distribution of (a) relative vorticity $\bar{\zeta}$ in units of UL^{-1}, and (b) streamfunction of the basic flow, $\bar{\psi}$ in units of UL; domain is $-2 \leq x \leq 2$ and $-1 \leq y \leq 1$.

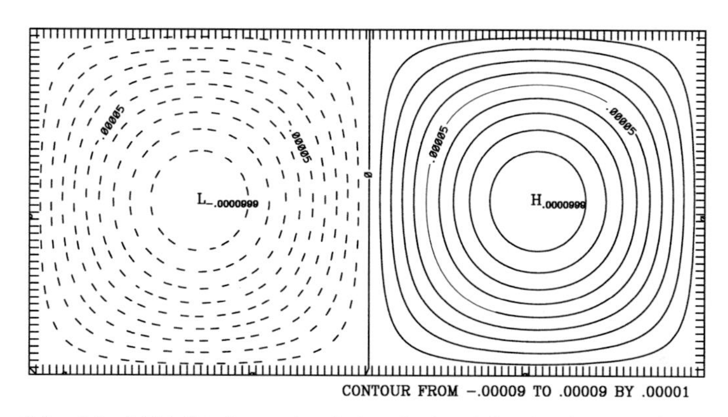

Fig. 8C.10 Structure of vorticity of the initial disturbance; domain is $-2 \leq x \leq 2$ and $-1 \leq y \leq 1$.

has even symmetry in the zonal direction but odd symmetry in the meridional direction with respect to the center point. The latter characteristic is a crucial factor in dictating the structure of unstable disturbances for the following reason. Since the gradient of basic vorticity is nonzero only along the boundaries of the semi-elliptical areas, perturbation vorticity is generated along those boundaries due to advection of the basic vorticity. Rossby waves could intrinsically propagate in a clockwise direction along the boundary of area-A and in an anticlockwise direction along the boundary of area-B. It follows that Rossby waves intrinsically propagate westward along the common boundary of the two semi-ellipses. The westerly jet would however tend to oppose the westward propagation of waves along the common boundary. Hence, the exit point of the localized jet is an accumulation point of wave disturbances. We may therefore anticipate that an unstable mode would be strongest near the exit of a westerly localized jet.

Let us solve (8C.10) as an initial value problem using a single arbitrary weak long wave as an initial disturbance. Its structure is symmetric about the x-axis. The range of values of the non-dimensional vorticity is $-1. \times 10^{-4} \leq \zeta(x, y, 0) \leq 1. \times 10^{-4}$ (Fig. 8C.10).

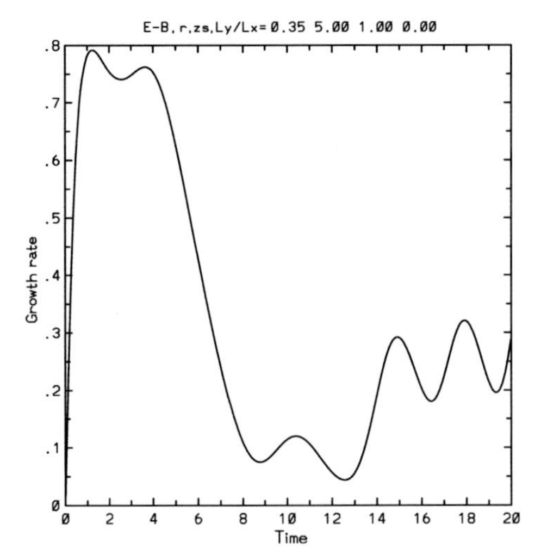

Fig. 8C.11 Variation of the instantaneous growth rate, $\frac{1}{2K}\frac{dK}{dt}$, with time for the basic flow shown in Fig. 8C.1.

8C.4.1 Instability properties

Growth rate

Equation (8C.10) is integrated with a predictor-corrector algorithm using a time step $\delta t = 0.01$, corresponding to about 14 minutes. We use the instantaneous growth rate, $\dfrac{1}{2K}\dfrac{dK}{dt}$, at each time step to portray the evolution where K is the total kinetic energy of the disturbance field. We see that after a rapidly intensifying transient stage, the growth rate evolves toward a slightly oscillating asymptotic value of about 0.25 after about $t = 10$ (Fig. 8C.11). The corresponding e-folding time is about 3.5 days.

Structure of unstable disturbance at large time

The structures of the streamfunction of the unstable disturbance at two time units apart from $t = 14$ to $t = 20$ are shown in Fig. 8C.12. The disturbance at $t = 20$ is close to being the most unstable normal mode. The result confirms that it is a local mode located mostly in the exit region of the jet. The disturbance intensifies about three and half fold during this interval. The corresponding perturbation vorticity is concentrated along the eastern portion of the boundaries of the basic vorticity field.

This intensification can be interpreted from the perspective of wave resonance mechanism in this case of zonally non-uniform basic flow as in the case of zonally uniform basic flow (Mak, 2002).

8C.4.2 Local energetics analysis

We get additional insight into the instability from a diagnosis of the local energetics of the unstable disturbance. The local kinetic energy of a disturbance is $K = \frac{1}{2}u_i'u_i'$ written with the convention that repeat indices refer to summation over $i = 1, 2$ where $u_1' \equiv u'$ and

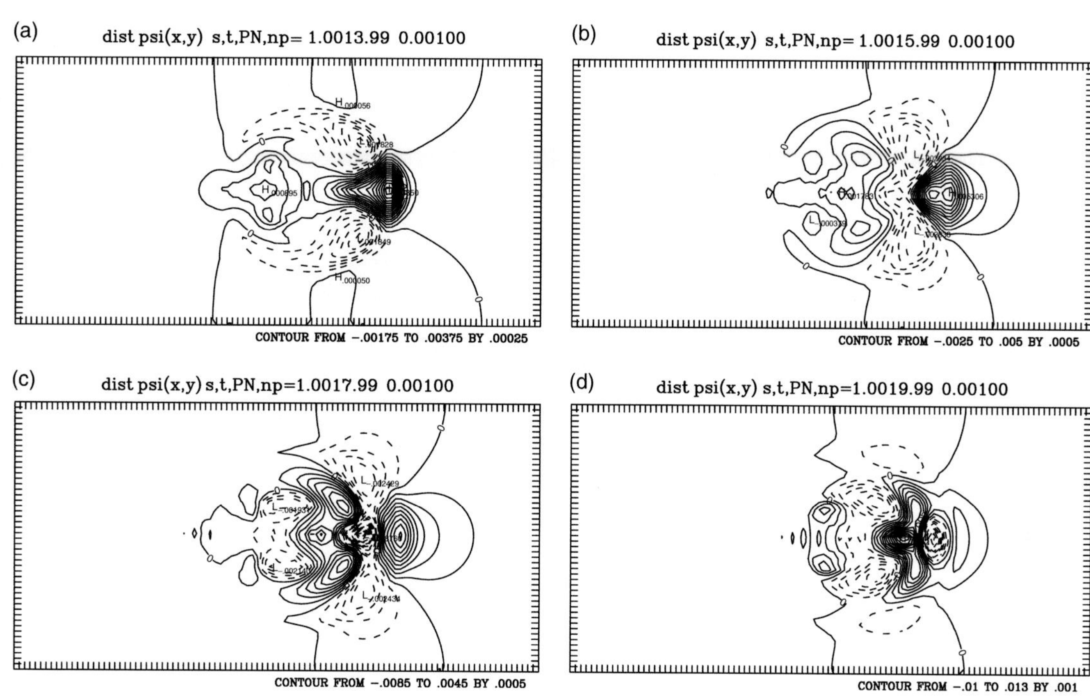

Fig. 8C.12 Structure of the perturbation streamfunction ψ at $t = 14,16,18,20$ in panels (a) to (d); domain is $-2 \le x \le 2$ and $-1 \le y \le 1$.

$u_2' \equiv v'$. The equation for the rate of change of the instantaneous local kinetic energy can be written as (Mak and Cai, 1989)

$$\frac{\partial K}{\partial t} = -u_i' u_j' D_{ij} - \vec{V} \cdot \nabla K - \nabla \cdot \left(\vec{v}' p^{(a)} \right), \tag{8C.12}$$

where $D_{ij} = \left(\partial U_j / \partial x_i + \partial U_i / \partial x_j \right)$ is the rate-of-strain tensor of the basic flow, $\vec{V} = (U_1, U_2) \equiv (U, V)$ is the basic flow, $p^{(a)}$ is ageostrophic pressure. The term $-u_i' u_j' D_{ij}$ represents the local rate of conversion of energy from the basic flow to the perturbation. It can be rewritten as the dot product of two pseudo-vectors, $\vec{E} \cdot \vec{D}$, where $\vec{E} = (E_1, E_2)$, $\vec{D} = (D_1, D_2)$ with $E_1 = \frac{1}{2}(v'^2 - u'^2)$, $E_2 = -u'v'$, $D_1 = U_x - V_y$ and $D_2 = V_x + U_y$. In rewriting this term, we have made use of the non-divergence condition of the basic flow, $\partial U / \partial x + \partial V / \partial y = 0$ and perturbation $\dfrac{\partial u'}{\partial x} + \dfrac{\partial v'}{\partial y} = 0$; E_1 and E_2 embody information of the local shape and orientation of the disturbance velocity field; D_1 and D_2 are the stretching deformation and shearing deformation of the basic flow respectively. Note that the domain integral of the second term and third term are zero. Hence, the net increase of the disturbance energy stems solely from the $\vec{E} \cdot \vec{D}$ term, which can be interpreted as conversion rate from basic kinetic energy to disturbance energy.

The deformation vector field of the basic flow is shown in Fig. 8C.13. The deformation is mostly associated with stretching deformation in the jet entrance and jet exit regions, whereas it is mostly associated with shearing deformation in the central part of a jet.

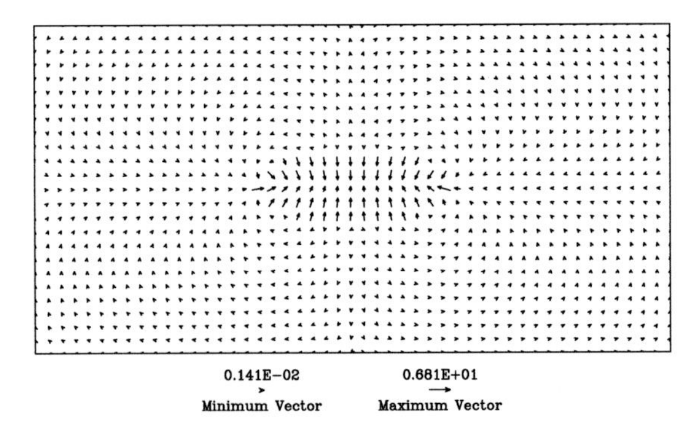

Fig. 8C.13 Deformation vector field of the basic flow, $\vec{D} = (D_1, D_2)$, $D_1 = U_x - V_y$ and $D_2 = V_x + U_y$. The domain is $-2 \leq x \leq 2$ and $-1 \leq y \leq 1$.

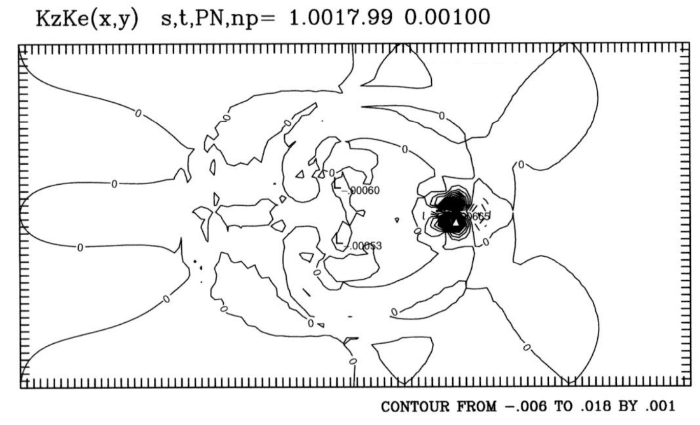

Fig. 8C.14 Distribution of conversion rate from basic kinetic energy to disturbance kinetic energy, $\vec{E}\cdot\vec{D}$ at $t = 18$; domain is $-2 \leq x \leq 2$ and $-1 \leq y \leq 1$.

The distribution of $\vec{E}\cdot\vec{D}$ at $t = 18$ is shown in Fig. 8C.14. It shows that energy is continually converted from the basic flow to the disturbance almost exclusively in a small area near the jet exit. The influence of stretching deformation appears to contribute more to the instability than the shearing deformation. A more detailed interpretation of $\vec{E}\cdot\vec{D}$ is given in Cai (2004).

The ageostrophic pressure is related to the perturbation velocity by

$$-\nabla^2 p^{(a)} = 4u_x U_x + 2U_y v_x + 2V_x u_y - (\beta y v)_x + (\beta y u)_y. \tag{8C.13}$$

The structure of $p^{(a)}$ at $t = 18$ is shown in Fig. 8C.15. This structure shows that it has odd symmetry about the jet axis and consists of a train of dipoles on the eastern part of the localized jet.

The $-\nabla\cdot\left(\vec{v}' p^{(a)}\right)$ term stands for convergence of wave energy flux. The $-\vec{V}\cdot\nabla K$ term represents advection of perturbation energy by the basic flow. The value of wave

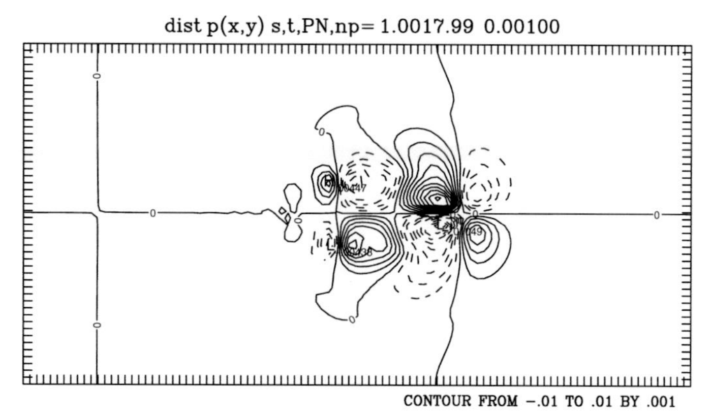

Fig. 8C.15 Distribution of the ageostrophic pressure at $t = 18$ associated with the unstable mode; domain is $-2 \leq x \leq 2$ and $-1 \leq y \leq 1$.

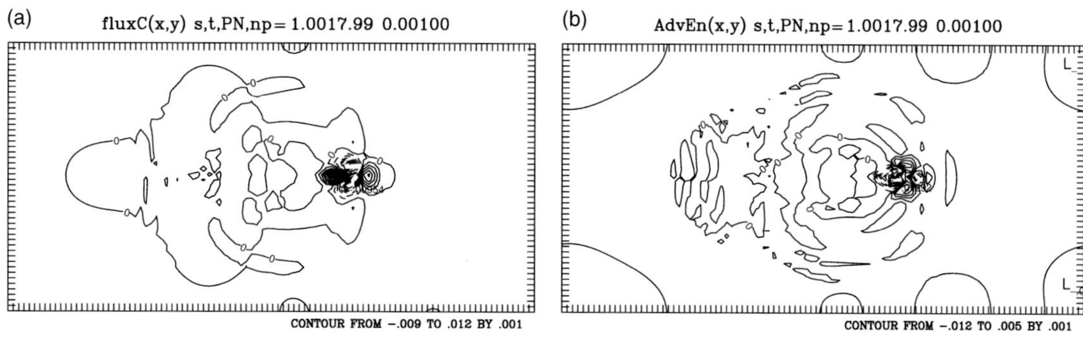

Fig. 8C.16 Distribution of (a) $-\nabla \cdot (\vec{v}' p^{(a)})$ and (b) $-\vec{V} \cdot \nabla K$ at $t = 18$ associated with the unstable disturbance; domain is $-2 \leq x \leq 2$ and $-1 \leq y \leq 1$.

energy flux convergence at $t = 18$ is about two-thirds of the conversion rate, and that of advection is less than one-third of the conversion rate (Fig. 8C.16). These two processes make it possible for the unstable perturbation to spread beyond the areas of energy conversion. In this case, the energy is spread only to a short distance from the jet exit. This aspect of the energetics is the essence of the so-called “downstream development” of disturbances from a localized jet also discussed in Orlanski and Chang (1993).

The distribution of the net rate of change of local kinetic energy at $t = 18$ is shown in Fig. 8C.17, highlighting that the normal mode is highly localized in the exit region of the basic jet.

A counterpart modal instability analysis of a localized jet such as the one under consideration would typically yield a fairly large number of unstable modes. A few of those modes can be stationary while most of them are oscillatory unstable modes (meaning that the eigenvalue has a real part as well as an imaginary part). Depending on the complexity of the basic flow, the unstable normal modes can have rather complex structures.

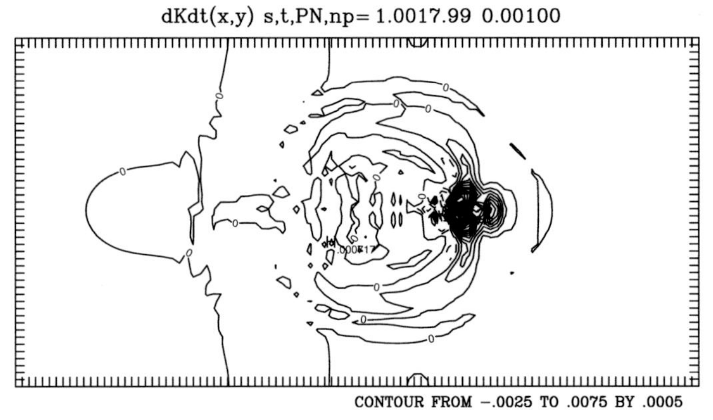

Fig. 8C.17 Structure of the net rate of change of local kinetic energy, $dK/dt = -u_i'u_j'D_{ij} - \vec{V}\cdot\nabla K - \nabla\cdot\left(\vec{v}'p^{(a)}\right)$, of the unstable mode at $t = 18$; domain is $-2 \le x \le 2$ and $-1 \le y \le 1$.

8C.5 Instability of a localized baroclinic jet

We next discuss the dynamics of instability of a localized baroclinic jet in the context of an interesting phenomenon. It was documented that the intensity of the Pacific storm track is weaker in mid-winter than in early or late winter (Nakamura, 1992). Such a feature is somewhat counter-intuitive since the baroclinicity is clearly strongest in mid-winter. It has been suggested that an increase of the barotropic shear in the Pacific jet, in spite of an increase of its baroclinicity from early winter to mid-winter, could naturally give rise to such subseasonal variation. Such an increase in the barotropic shear would give rise to a barotropic-governor effect on the instability process responsible for the storm track itself. This analysis is specifically designed to test this hypothesis for the phenomenon. The discussion is based on an analysis using a two-layer quasi-geostrophic model (Deng and Mak, 2005).

The model domain is a rectangular cyclical channel 24 000 km long and 6000 km wide with two rigid meridional boundaries. Length, velocity and time are measured in units of L, U and L/U respectively. With $U = 10\ \mathrm{m\,s^{-1}}$ and $L = 1000$ km the domain is $|x| \le X = 12$ and $|y| \le Y = 3$. The time unit is $L/U = 10^5\ \mathrm{s} \sim 1$ day. All equations from here on contain non-dimensional quantities. The governing equation with a steady forcing $F(x,y)$ is

$$\frac{\partial q_i}{\partial t} + J(\phi_i, q_i) = F_i, \quad i = 1, 2, \tag{8C.14}$$

where subscripts 1 and 2 refer to the upper and lower levels respectively; q_i and φ_i are the quasi-geostrophic potential vorticity (PV) and streamfunction. The external forcing F_i is required to support a localized jet, expressed in terms of its streamfunction $\bar{\psi}_i$ and potential vorticity $\bar{q}_i$. The forcing F_i is

$$J(\bar{\psi}_i, \bar{q}_i) = F_i, \tag{8C.15}$$

and $\bar{\psi}_i$ and $\bar{q}_i$ in turn are related through

$$\bar{q}_1 = \beta y + \nabla^2 \bar{\psi}_1 + \lambda^2 (\bar{\psi}_2 - \bar{\psi}_1), \qquad \bar{q}_2 = \beta y + \nabla^2 \bar{\psi}_2 - \lambda^2 (\bar{\psi}_2 - \bar{\psi}_1), \quad \text{(8C.16a,b)}$$

where $\beta = \beta_{\text{dim}} L^2 / U = 1.6$ and $\lambda = \lambda_{\text{dim}} L$; $\lambda_{\text{dim}} = f_o / \sqrt{\sigma}(\delta p)$ is the reciprocal of the radius of deformation with f_o, σ, δp being the Coriolis parameter, the static stability parameter, $\sigma = \frac{R}{p}\left(\frac{RT}{pc_p} - \frac{\partial T}{\partial p}\right)$, and δp the thickness of each layer in pressure units. For a static stability parameter value $\sigma = 2.0 \times 10^{-6}\,\text{m}^2\,\text{Pa}^{-2}\,\text{s}^{-2}$, $f_o = 10^{-4}\,\text{s}^{-1}$ and $\delta p = 500$ mb (5×10^4 Pa), we have $\lambda^2 = 2$.

Writing the total field as $\varphi_i^{total} = \bar{\psi}_i + \psi_i$ and $q_i^{total} = \bar{q}_i + q_i$, substituting them into (8C.14) and making use of (8C.15), we obtain the following prognostic equation for perturbation PV q_i,

$$q_{it} + J(\bar{\psi}_i, q_i) + J(\psi_i, \bar{q}_i) = 0, \quad \text{(8C.17)}$$

and ψ_i and q_i are related by

$$q_1 = \nabla^2 \psi_1 + \lambda^2 (\psi_2 - \psi_1), \quad q_2 = \nabla^2 \psi_2 - \lambda^2 (\psi_2 - \psi_1). \quad \text{(8C.18)}$$

The appropriate boundary conditions are

$$\psi_i(-X, y, t) = \psi_i(X, y, t),$$

$$\text{at } y = \pm Y, \quad \frac{\partial \psi_i}{\partial x} = 0, \quad \text{(8C.19a, b, c)}$$

$$\text{at } y = \pm Y, \quad \int_{-X}^{X} \frac{\partial^2 \psi_i}{\partial y \, \partial t} dx = 0.$$

Condition (8C.19c) follows from the requirement that the zonal average zonal velocity at each level conserves. When ψ_i and q_i are known at a given time, the values of q_i at a future time can be obtained by integrating (8C.17). The q_i field can be in turn used to determine ψ_i at that new time with (8C.18) satisfying (8C.19).

8C.5.1 Construction of the basic state

It is necessary to appropriately prescribe the change of a basic flow from early winter to mid-winter. Suppose the basic relative PV, $(\bar{q}_i - \beta y)$, and the corresponding $\bar{\psi}_i$ consist of two components. Thus, we set $\bar{q}_i - \beta y = \hat{q}_i + \tilde{q}_i$ and $\bar{\psi}_i = \hat{\psi}_i + \tilde{\psi}_i$. The zonal mean component of the reference state is denoted by a "hat," and the zonally varying component by a "tilde." We assume that there is a constant zonal mean component in the zonal velocity of the reference state at the two levels, $\hat{U}_i$. We specifically set $\hat{q}_1 = -\hat{q}_2 = \lambda^2 (\hat{U}_1 - \hat{U}_2) y \equiv \lambda^2 \hat{U}_T y$; $\hat{U}_T$ is a parameter for global baroclinicity. It follows that $\hat{\psi}_1 = -\hat{U}_1 y$ and $\hat{\psi}_2 = -\hat{U}_2 y$. We will consider $\hat{U}_1 > 0$ and $\hat{U}_2 = 0$ in our calculations.

Let us consider a simple localized baroclinic jet characterized by a parameter, Δ, a certain distribution of $\tilde{q}_1(x, y)$ and $\tilde{q}_2 = 0$. In particular, we use

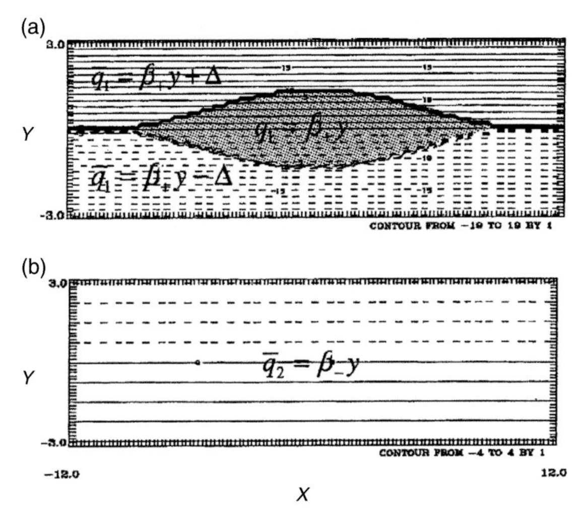

Fig. 8C.18 Distribution of the PV fields $\bar{q}_i$ in the reference state at the (a) upper and (b) lower levels, $\beta_{\pm} = \beta \pm \lambda^2 \hat{U}_T$ (taken from Deng and Mak, 2005).

$$\begin{aligned} \tilde{q}_1(x,y) &= \Delta \quad \text{in a northern region,} \\ \tilde{q}_1(x,y) &= -\Delta \quad \text{in a southern region,} \\ \tilde{q}_1(x,y) &= 0 \ \text{for} \ |y| \leq G, \end{aligned} \tag{8C.20}$$

where
$$\begin{aligned} G(x) &= y_0 - b & &\text{if } x \leq -X + a \text{ or } x \geq X - a, \\ G(x) &= y_0 + b\cos\left(\tfrac{\pi}{X-a}x\right) & &\text{if } -X + a < x < X - a, \end{aligned}$$

with a, b, Δ, y_o being constant parameters. These parameters together with $\hat{U}_T$ control the spatial structure of the basic velocity field. Figure 8C.18 shows the basic PV field for $a = 1.5$, $b = 0.65$ and $y_0 = b + 0.1$.

When Δ is increased as an adjustable parameter, the horizontal shear in the corresponding flow field would become greater, but the shape of the jet remains unchanged. This feature can be verified by direct computation. This is always true when we compare the gradient of a specific PV field with that of another PV field as long as they have the same pattern. However, we cannot make the same inference if two different patterns of PV distribution are compared. This is an important point to keep in mind in order to avoid unnecessary confusion.

It is instructive to focus on the relevant values of two key parameters, $\hat{U}_T$ and Δ. Since $\hat{U}_T$ is a model counterpart of the global measure of baroclinicity, it should increase by about 10 percent from early winter to mid-winter. A meaningful value of Δ should be prescribed on the basis that its corresponding velocity field is compatible with the observed time mean wind field. In particular, we quantify the degree of zonal variation in the basic baroclinicity with the following measure

$$\gamma = \frac{(\bar{U}_1 - \bar{U}_2)_{\max} - (\bar{U}_1 - \bar{U}_2)_{\min}}{(\bar{U}_1 - \bar{U}_2)_{\max}}, \tag{8C.21}$$

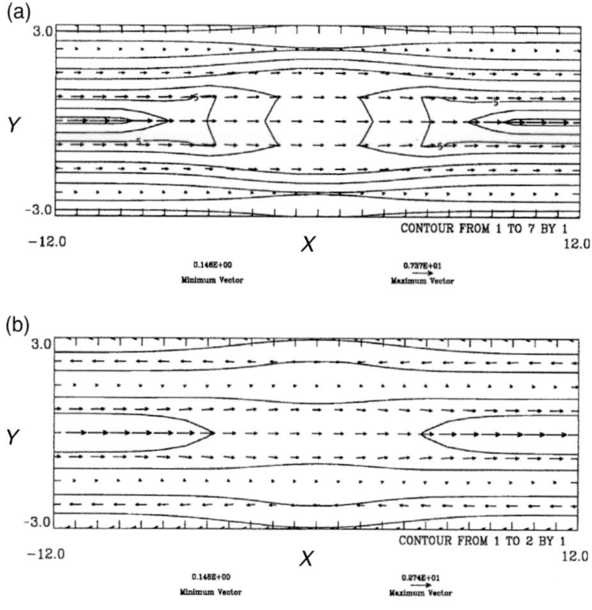

Fig. 8C.19 Non-dimensional speed (contours) and velocity (vectors) of the reference state for the model mid-winter forcing condition in (a) upper level and (b) lower level (taken from Deng and Mak, 2005).

where $\bar{U}_i$ is the zonal velocity of the reference state. If we use $\left(\hat{U}_T, \Delta\right) = (2.2, 5.5)$ as a mid-winter condition, we get $(\bar{U}_1 - \bar{U}_2)_{\max} = 4.5$ and $(\bar{U}_1 - \bar{U}_2)_{\min} = 2.4$. The corresponding value of γ is equal to 0.47. On the other hand, if we reduce $\hat{U}_T$ by 10 percent and double the value of Δ, namely $\left(\hat{U}_T, \Delta\right) = (2.0, 2.5)$, as an early winter condition, we get $\gamma = 0.35$ because the corresponding values for $(\bar{U}_1 - \bar{U}_2)_{\max}$ and $(\bar{U}_1 - \bar{U}_2)_{\min}$ are 3.1 and 2.0 respectively. Therefore, the zonal asymmetry of baroclinicity would only increase by $\dfrac{(0.47 - 0.35)}{0.35} = 34$ percent. Such a change is compatible with the counterpart observed change. It is therefore justifiable to use $\left(\hat{U}_T, \Delta\right) = (2.0, 2.5)$ and $\left(\hat{U}_T, \Delta\right) = (2.2, 5.5)$ in this study as plausible early winter and mid-winter conditions respectively.

Figure 8C.19 confirms that the basic flow is indeed a localized baroclinic jet much stronger in the upper level under the model mid-winter condition, $\hat{U}_T = 2.2$ and $\Delta = 5.5$. The localized jet is about 7000 km long. It has a maximum speed of 73.7 m s^{-1} in the upper level and 27.4 m s^{-1} in the lower level. The jet is centered at $x = \pm 12$ and has a downstream diffluence/confluence region between the two wave-guides at $y = \pm G$. These characteristics are compatible with those of the observed Pacific jet.

8C.5.2 Instability properties

The hypothesis is tested by solving (8C.17) as an initial value problem. We use center-difference to approximate the spatial derivatives and an Euler-backward scheme for time integration. The horizontal domain is depicted with 121×51 grid points. The dimensional grid intervals are $\Delta x = 200$ km, $\Delta y = 120$ km. Such resolution is amply adequate because

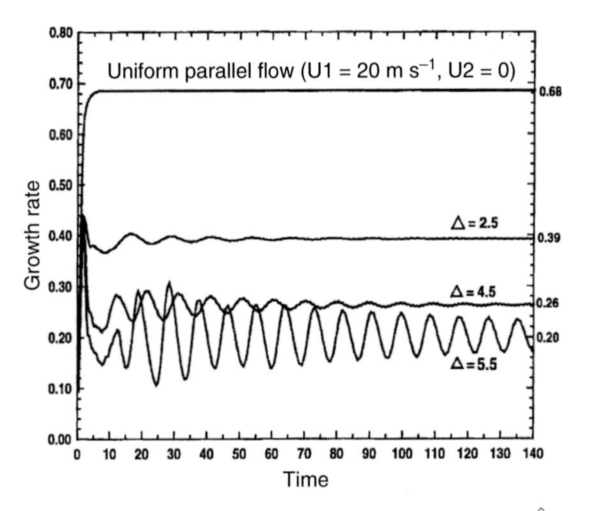

Fig. 8C.20 Variations of the instantaneous growth rate in time for four values of Δ with a fixed $\hat{U}_T = 2$ (taken from Deng and Mak, 2005).

essentially the same instability properties are obtained when only 97×41 grid points are used. The time step is $\delta t = 0.01$, which corresponds to about 17 minutes.

A very weak PV perturbation containing ten wave components of equal amplitude at both levels is used as an unbiased initial disturbance in the analysis, viz.

$$q_i(x, y, 0) = \sum_{k=1}^{10} q_{oi} \sin\left(\frac{k\pi x}{X}\right) \cos\left(\frac{\pi y}{2Y}\right) \tag{8C.22}$$

with $q_{o1} = q_{o2} = 10^{-6}$. In the special case of a horizontally uniform baroclinic flow ($\hat{U}_T \neq 0$, $\Delta = 0$), the condition of marginal instability is $\hat{U}_T = 0.8$ for the values of β and λ under consideration.

Reduction of growth rate

The instantaneous growth rate is calculated as $\sigma = \dfrac{1}{2\langle E\rangle}\dfrac{d\langle E\rangle}{dt}$ where $\langle E\rangle = \iint_{x,y} \left(\sum_{i=1,2} \frac{1}{2}(u_i^2 + v_i^2) + 2\lambda^2\theta^2\right) dx\, dy$ is the domain-integrated energy of the disturbance. Let us begin by examining how the growth rate would vary with Δ for a fixed value of $\hat{U}_T$. The evolution of the instantaneous growth rate for $\Delta = 0.0$, 2.5, 4.5 and 5.5 with a fixed value of $\hat{U}_T = 2.$ are shown in Fig. 8C.20. The asymptotic growth rate is largest for the case of a horizontally uniform baroclinic shear flow ($\hat{U}_T = 2.0$, $\Delta = 0$). It is ~0.68, corresponding to an *e*-folding time of ~1.5 days. For a localized baroclinic jet with $\Delta = 2.5$, the asymptotic growth rate lowers to ~0.39, corresponding to an *e*-folding time of ~2.5 days. For a more localized basic baroclinic jet characterized by $\Delta = 4.5$ and 5.5, the growth rate decreases to 0.26 and 0.20 respectively. As an unstable disturbance propagates through a more localized baroclinic jet, its structure varies more substantially. The rate of extracting energy from the basic flow correspondingly fluctuates more in time. This is the reason why the asymptotic growth rate for the case of $\Delta = 5.5$ oscillates in time.

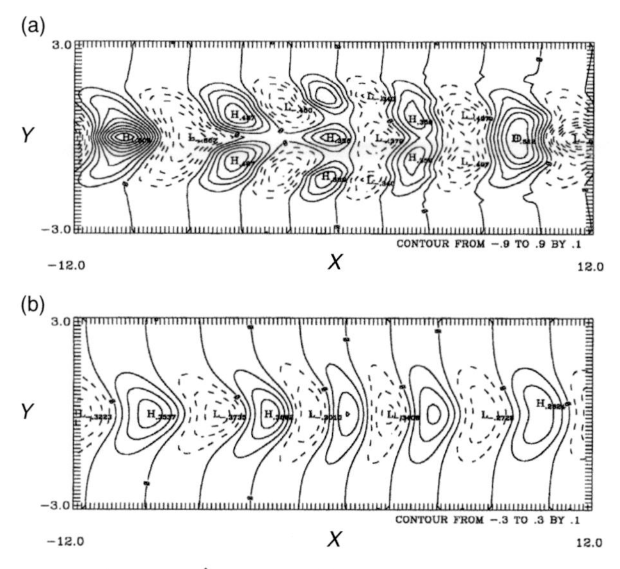

Fig. 8C.21 Normalized perturbation streamfunction for $\hat{U}_T = 2.2$ and $\Delta = 5.5$ at (a) upper level ψ_1 and (b) lower level ψ_2 (taken from Deng and Mak, 2005).

For the mid-winter (January) condition, $\hat{U}_T = 2.2$ and $\Delta = 5.5$, the growth rate is found to be 0.28. Hence, the growth rate is reduced by $\dfrac{0.39 - 0.28}{0.39} = 28$ percent from the model early winter condition to the model mid-winter condition. Such a result is consistent with a reduction of the intensity of storm track. In summary, the linear instability analysis confirms that the growth rate under a plausible model mid-winter condition is indeed significantly smaller than that under a plausible early winter condition.

Structure of unstable disturbance

Additional insight into the nature of the instability for a mid-winter condition may be gained from an examination of its structure and local energetics (generation and redistribution processes). A snap shot of the unstable perturbation streamfunctions ψ_1 and ψ_2 at large time (effectively the most unstable normal mode) is shown in Fig. 8C.21. Here ψ_i is normalized to maximum unit magnitude at the upper level. The dominant wavelength is ~5000 km. The most intense wave-packets are found slightly downstream of the jet core, in the region $-12 \leq x \leq -6$. The disturbance in the upper level is about three times stronger than that in the lower level. There is a distinct westward vertical tilt. The horizontal tilt at both levels is generally in the direction of the basic horizontal shear in the jet exit region (i.e. not leaning against the shear). These tilts have clear implications on the local energetics of the disturbance.

The spatial distributions of the root-mean-square (RMS) of ψ_1 and ψ_2 of the most unstable normal mode can be used as an estimate of a model storm track according to linear dynamics. Since only the pattern of ψ_i is unique, we normalize the RMS values to a maximum of unity at the upper level. Figure 8C.22 shows such a result for ψ_i as calculated from the streamfunction field from $t = 180$ to 240 in our model integration. The result shows a highly

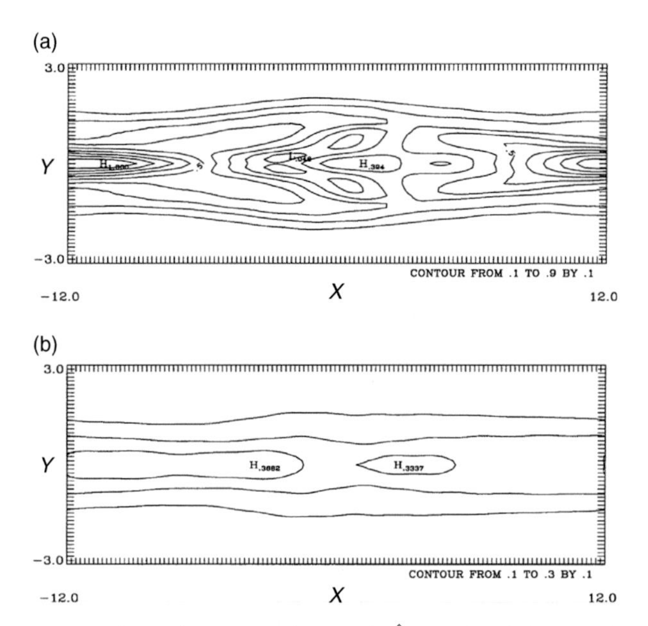

Fig. 8C.22 Normalized root-mean-square of perturbation streamfunction for $\hat{U}_T = 2.2$ and $\Delta = 5.5$ from $t = 180$ to $t = 240$ at the (a) upper level and (b) lower level (taken from Deng and Mak, 2005).

localized model storm track at the upper level with its center being slightly downstream of the background jet. The synoptic variability is much weaker in the diffluent/confluent region of the background flow. The model storm track at the lower level is also located downstream of the jet. The intensity of the lower level storm track is about 35 percent of that of the upper level.

8C.5.3 Local energetics analysis

The vertically integrated local perturbation energy equations for this model can be written as follows:

$$\begin{aligned} \frac{\partial\{K\}}{\partial t} &= \{-\vec{V}\cdot\nabla K\} + \{-\nabla\cdot(\phi_a\vec{v} + \phi\vec{v}_d)\} + \{\vec{E}\cdot\vec{D}\} - \{F_{(3)}T_{(3)}\}, \\ \frac{\partial\{P\}}{\partial t} &= \{-\vec{V}\cdot\nabla P\} + \{\vec{F}\cdot\vec{T}\}, \end{aligned} \tag{8C.23a, b}$$

where $\{K\} = \sum_{i=1,2} K_i = \sum_{i=1,2} \frac{1}{2}(u_i^2 + v_i^2)$ is the total local perturbation kinetic energy with $\vec{v}_i = (u_i, v_i)$ being the perturbation velocity; $\{P\} = 2\lambda^2\theta^2$ is the local perturbation potential energy with $\theta = \frac{\psi_1 - \psi_2}{2}$; $\{-\vec{V}\cdot\nabla K\} = -\sum_{i=1,2} \vec{V}_i\cdot\nabla K_i$ and $\{-\vec{V}\cdot\nabla P\} = -\frac{\vec{V}_1 + \vec{V}_2}{2}\cdot$ $\nabla\{P\}$ are the advection of perturbation kinetic energy and potential energy by the basic flow, respectively; $\vec{V}_i$ is the basic velocity; $\vec{v}_d$ is the irrotational part of the ageostrophic wind and ϕ_a is the ageostrophic part of the geopotential. Then $\{-\nabla\cdot(\phi_a\vec{v} + \phi\vec{v}_d)\} = -\sum_{i=1,2} \nabla\cdot(\phi_{ai}\vec{v}_i + \phi_i\vec{v}_{di})$ is the convergence of energy flux associated with the ageostrophic component of the flow; $\{\vec{E}\cdot\vec{D}\} = \sum_{i=1,2} \vec{E}_i\cdot\vec{D}_i$ is the local

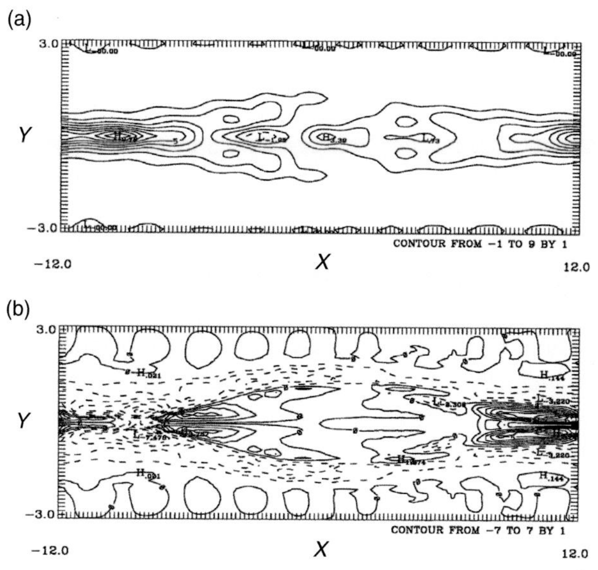

Fig. 8C.23 (a) Baroclinic energy conversion rate $C(P', K')$ and (b) barotropic energy conversion rate $\left\{\vec{E} \cdot \vec{D}\right\}$ for $\hat{U}_T = 2.2$ and $\Delta = 5.5$ (taken from Deng and Mak, 2005).

generation rate of kinetic energy to be calculated as a scalar product of two pseudo-vectors. Vector $\vec{E}_i = \left(\frac{1}{2}(v_i^2 - u_i^2), -u_i v_i\right)$ is a measure of the local structure of the disturbance at each level and vector $\vec{D}_i = \left(\frac{\partial \bar{U}_i}{\partial x} - \frac{\partial \bar{V}_i}{\partial y}, \frac{\partial \bar{V}_i}{\partial x} + \frac{\partial \bar{U}_i}{\partial y}\right)$ is a measure of the stretching and shearing deformation of the basic flow at each level. The vertical velocity at the mid-level is $w = -\omega$. The term $\{\vec{F} \cdot \vec{T}\}$ represents the net generation rate of perturbation potential energy. It has two parts, $\{\vec{F} \cdot \vec{T}\} = \{\vec{F}_H \cdot \vec{T}_H\} + \{F_{(3)} T_{(3)}\}$, where $\{\vec{F}_H \cdot \vec{T}_H\} = -4\lambda^2 \theta J(\psi, \Theta)$ and $\{F_{(3)} T_{(3)}\} = -4\varepsilon^{-1} w\theta$ with $\psi = \frac{\psi_1 + \psi_2}{2}, \Theta = \frac{\psi_1 - \psi_2}{2}$. The former represents a conversion rate from basic potential energy to perturbation potential energy. The latter represents a conversion rate from perturbation kinetic energy to perturbation potential energy. The local conversion from perturbation potential energy to kinetic energy is therefore $C(P', K') \equiv -\{F_{(3)} T_{(3)}\}$. A normalized perturbation with unit domain-averaged perturbation energy will be used in the computations of its local energetics. Figure 8C.23a shows that $C(P', K')$ is positive almost everywhere and has a narrow band of particularly large values distinctly downstream of the jet core. The domain-integrated value of $\langle C(P', K') \rangle$ is 119.6. The baroclinic energy conversion is associated with the westward tilt of the wave-packets at the upper level relative to those at the lower level. A wave-packet also tends to be meridionally strained as it approaches the diffluent part of the basic flow. This process weakens a wave-packet and even splits it into three separate wave-packets as seen in Fig. 8C.21a. Figure 8C.23b shows that there is nevertheless some local barotropic conversion from the basic flow to the disturbance in a narrow zone near $y = 0$ particularly in the region upstream of the jet. The related positive values of $\vec{E} \cdot \vec{D}$ contribute to local elongation of the unstable wave-packets. The domain-integrated value of $\left\langle \vec{E} \cdot \vec{D} \right\rangle$ is however negative, -49.6. This value is significant because

$|\langle \vec{E} \cdot \vec{D} \rangle| / \langle C(P', K') \rangle = 41$ percent. As discussed in the introduction, the nature of instability of a localized jet with such characteristics may be interpreted as a generalized barotropic-governor effect. This stabilization influence is the consequence of an increase in the local deformation of the basic flow from early winter to mid-winter in spite of a simultaneous enhancement of the local baroclinicity.

Concluding remarks

We have closely examined the instability of two idealized zonally varying shear flows in the atmosphere. One is a localized barotropic jet to demonstrate that the notion of wave resonant mechanism is applicable no matter whether the basic flow is zonally uniform. We have shown why an unstable normal mode is naturally located near the exit region of a jet. The other basic flow is a localized baroclinic jet to illustrate the impact of a barotropic-governor mechanism. The results support a hypothesis for the dynamical nature of the observed phenomenon of mid-winter minimum of the Pacific storm track.

PART 8D MOIST BAROCLINIC INSTABILITY

This part of the chapter addresses the problem of instability of large-scale disturbances in a moist atmosphere with the incorporation of condensational heating induced by the disturbance itself. The two major simple schemes of moist parameterization for self-induced condensational heating are first discussed in Section 8D.1. Section 8D.2 discusses an instability analysis of a simple baroclinic shear flow in a moist conditionally unstable quasi-geostrophic model. The model formulation is described in Sections 8D.2.1 and 8D.2.2. The analytic solution of this model for any heating profile is presented in Section 8D.2.3. The solution for the special case of a uniform heating profile is given in Section 8D.2.4. We also present the instability results of several sample unstable disturbances and verify the dynamical nature of the instability in Sections 8D.2.5 and 8D.2.6. This analysis illustrates the fundamental similarity and difference between dry and moist baroclinic instability.

8D.1 Introductory remarks

An intense extratropical cyclone typically brings with it copious rain and/or snow. The large condensational heating associated with such precipitation can be expected to strongly modify the development of a disturbance. Observational studies abound that suggest that the impacts of condensational heating on the rate of intensification, the size, the structure and the movement of extratropical cyclones can be significant (e.g. Gyakum 1983). Nevertheless, the vast majority of the dynamical studies of cyclogenesis in the literature have focused on dry dynamics. One reason is of course that it would be logical to first tackle the problem of cyclogenesis in a dry setting. As we have seen in Parts B and C of this chapter, even the dry dynamics of cyclogenesis has taken much effort of many researchers to unravel. Perhaps a deeper reason is that cyclogenesis in a moist model setting is a virtually intractable problem if it were to be dealt with from first principles. Moist processes give rise to a set of rather messy issues. The scale of a synoptic cyclone differs greatly from that of the clouds embedded in it. The ensemble of clouds comes in very different sizes. They are too small to be individually represented in a model. Each cloud in turn consists of a spectrum of water droplets and/or snow crystals. How the raindrops and snow crystals eventually precipitate out of the clouds in a storm depends on their interaction under the influence of the turbulent flow inside the clouds. Even the physics of phase transition of water on a solid particle (condensation nuclei) is not precisely understood. All those elemental factors have led researchers to experiment with a number of moist parameterization schemes of different complexity. To be sure, any parameterization scheme is intrinsically crude and the corresponding uncertainty can only be assessed a posteriori.

A complex parameterization scheme can be manageable in a numerical simulation of a disturbance. However, only a scheme with a direct link between the diabatic heating and a certain dependent variable of the flow field would be tractable in a theoretical study. The primary objective of a theoretical study of moist baroclinic instability is not so much as to evaluate the consequence of condensational heating on the instability properties of a developing disturbance, but rather to delineate the fundamental dynamical nature of the impacts. For this reason, it would be most preferable to adopt a simple moist parameterization scheme in conjunction with a simplest possible dynamical model.

Theoretical studies of moist baroclinic instability have tried two main approaches. One approach is to parameterize the impact of deep convection from the perspective of so-called *conditional instability of the second kind* (CISK); CISK is a notion that a large-scale disturbance and the embedded ensemble of clouds could intensify together by virtue of a mutually beneficial process. The release of latent heat by the clouds strengthens the flow field of their parent disturbance and the circulation of the latter in turn enhances the moisture supply to the clouds. This general concept was originally invoked in theoretical studies of hurricane formation with an emphasis on the role of frictional convergence of moisture in a boundary layer next to the surface (e.g. Charney and Eliassen, 1964). But frictional convergence is not an intrinsic part of CISK; CISK is meaningful as long as the large-scale disturbance under consideration has an inherent field of convergence and divergence significantly strong in a moist layer near the surface. This would be the case of a baroclinic wave in a moist model setting. This approach is appropriately called *wave-CISK* parameterization. It invokes a release of latent heat proportional to the ascending motion at one specified level identifiable with the top of a moist layer. An integral part of this scheme is to specify a heating profile. A separate parameter is introduced for specifying the heating intensity. This scheme is to be incorporated in a dynamical framework that contains a baroclinic basic flow (Mak, 1982, 1994; Craig and Cho, 1988; Snyder and Lindzen, 1991; Parker and Thorpe, 1995).

In another approach, the diabatic heating is assumed to be proportional to the large-scale vertical motion at every level implicitly requiring that ascending air parcels are saturated. An integral part of such a scheme is to specify a vertical moisture profile. A variation of this approach is to simply use a stratification in an ascending region different from that in a descending region. This approach may be referred to as *large-scale rain* parameterization (Emanuel *et al.*, 1987; Montgomery and Farrel, 1991; Whitaker and Davis, 1994). The impacts of condensational heating on the instability of a baroclinic flow according to these two parameterization schemes turn out to be qualitatively similar. With either scheme, condensational heating gives rise to an increase of the growth rate, a reduction of the length scale of an unstable wave and noticeable modification of its structure and propagation speed.

8D.2 A particular instability analysis

This section discusses a particular moist baroclinic instability analysis. The model framework is the so-called Eady model mentioned in Section 8B.1. It is a quasi-geostrophic model with a continuous vertical domain between two rigid surfaces. The basic state has a baroclinic flow of linear shear with a uniform stratification. There is no beta-effect. This analysis adopts an approximate form of the wave-CISK parameterization. No restriction is imposed on the shape of the heating profile. We first obtain an analytic solution for a perturbation in this model. We later illustrate the quantitative aspect of the analytic solution with the instability properties of unstable modes for a particularly simple heating profile.

8D.2.1 Model formulation

We consider a 2-D disturbance in an Eady model that has a uniform structure in the meridional direction without loss of generality. The zonal domain is $0 \leq x \leq \infty$. Isobaric coordinates are used to depict the vertical domain $p_1 \leq p \leq p_{oo}$. The basic static stability parameter is $S = -\frac{R}{p}\left(\frac{\partial T_o}{\partial p} - \frac{RT_o}{c_p p}\right)$ and the basic zonal baroclinic flow is $\bar{u} = \lambda(p_{oo} - p)$. The latitude of the model is indicated by a Coriolis parameter f. We measure the horizontal distance, velocity, time and pressure in units of L, U, LU^{-1} and p_{oo} respectively. The basic velocity $\bar{u}$, basic shear λ, stratification S, heating rate Q, p-velocity ω and streamfunction ψ are then measured in units of U, U/p_{oo}, $(fL/p_{oo})^2$, fU^2/R, $(U^2 p_{oo}/fL^2)$ and UL respectively. The governing equations are the QG omega equation and the QG-PV equation in the following *non-dimensional form*

$$\frac{\partial^2\omega}{\partial p^2} + S\frac{\partial^2\omega}{\partial x^2} = -2\lambda\frac{\partial^3\psi}{\partial x^3} - \frac{\partial^2}{\partial x^2}\left(\frac{Q}{p}\right),$$
$$\left(\frac{\partial}{\partial t} + \bar{u}\frac{\partial}{\partial x}\right)\left(S\frac{\partial^2\psi}{\partial x^2} + \frac{\partial^2\psi}{\partial p^2}\right) = -\frac{\partial}{\partial p}\left(\frac{Q}{p}\right). \tag{8D.1a, b}$$

The vertical boundary conditions are $\omega = 0$ at $p = p_1,1$ which amount to requiring

$$\left(\frac{\partial}{\partial t} + \bar{u}\frac{\partial}{\partial x}\right)\frac{\partial\psi}{\partial p} + \lambda\frac{\partial\psi}{\partial x} = 0 \tag{8D.2}$$

at the boundaries. We may infer from (8D.1a) that a heating ($Q > 0$) would have a positive feedback influence on the ascending motion ($\omega < 0$). According to (8D.1b), the perturbation PV of an air parcel would be increased where the latent heating increases with height and vice versa. The shape of the heating profile therefore is an important parameter.

8D.2.2 Treatment of the condensational heating

According to the wave-CISK parameterization, the heating rate in the model is written as

$$Q = \begin{cases} -\varepsilon \hat{h} \omega_B & \text{for } \omega_B < 0 \\ 0 & \text{otherwise} \end{cases} \tag{8D.3}$$

where $\hat{h}(p) \neq 0$ is a heating profile in a certain heating layer and ω_B is ω at the top of a surface moist layer $p = p_B$. The positive definite nature of the heating leads to coupling among the different wave components in a disturbance. Writing the p-velocity at $p = p_B$ in terms of Fourier components

$$\omega_B = \int_{-\infty}^{\infty} W_B(k) e^{ik} dk, \tag{8D.4}$$

the heating rate would be

$$Q = \int_{-\infty}^{\infty} \tilde{Q}(k,p) e^{ikx} dk, \tag{8D.5a,b}$$

$$\tilde{Q} = -\frac{\varepsilon}{2} \hat{h} W_B + \text{coupling term.}$$

The coupling term is an integral involving other spectral components $W_B(m)$, $m \neq k$. It would require a fully numerical analysis to quantify the influence of the coupling term. We introduce a simplifying approximation that the feedback effect of the heating induced by a wave disturbance is strongest upon itself. It amounts to focusing on the linear dynamics of the problem. Specifically, we approximate (8D.5b) as

$$\tilde{Q}(k,p) = -\frac{\varepsilon \hat{h}(p)}{2} W_B(k) \tag{8D.6}$$

and proceed to analyze the instability of each spectral component separately.

8D.2.3 General solution

A normal mode solution has the form of

$$(\omega, \psi) = \text{Re}\{(W(p),\ \Psi(p)) \exp(i(kx - \sigma t))\}, \tag{8D.7}$$

where $k = 2\pi/L_{(x)}$ with $L_{(x)}$ being the wavelength of the normal mode. It follows from (8D.1), (8D.2) and (8D.6) that the amplitude functions of a normal mode are governed by

$$\frac{d^2W}{dp^2} - Sk^2W = i2\lambda k^3\Psi - \frac{\varepsilon k^2}{2}hW_B, \tag{8D.8}$$

$$i(\bar{u}k - \sigma)\left(\frac{d^2\Psi}{dp^2} - Sk^2\Psi\right) = \frac{\varepsilon}{2}W_B\frac{dh}{dp}, \tag{8D.9}$$

$$W = 0 \quad \text{at} \quad p = p_1, 1, \tag{8D.10}$$

$$(\bar{u}k - \sigma)\frac{d\Psi}{dp} + \lambda k\Psi = 0 \quad \text{at} \quad p = p_1, 1, \tag{8D.11}$$

where $h = \hat{h}(p)/p$ is simply referred to as the heating profile for convenience. Equations (8D.8) and (8D.9) subject to the boundary conditions (8D.10) and (8D.11) define an eigenvalue–eignfunction problem with σ being the eigenvalue.

We first obtain an integral relation between W and Ψ by applying the method of Green's function to (8D.8), viz.

$$W = \int_{p_1}^{1} G(p, \tilde{p})\left[i2\lambda k^3\Psi - \frac{\varepsilon k^2}{2}W_B h(\tilde{p})\right]d\tilde{p}, \tag{8D.12}$$

where the Green's function satisfying the boundary condition (8D.10) is

$$G(p,\ \tilde{p}) = \begin{cases} G_1(p, \tilde{p}) = \frac{1}{a}\eta(\tilde{p})\xi(p) & \text{for } p < \tilde{p}, \\ G_2(p, \tilde{p}) = \frac{1}{a}\eta(p)\xi(\tilde{p}) & \text{for } p > \tilde{p} \end{cases}$$

$$\xi(p) = e^{m(p-p_1)} - e^{-m(p-p_1)}, \tag{8D.13}$$

$$\eta(p) = e^{m(p-1)} - e^{-m(p-1)},$$

$$a = -2m\eta(p_1), \quad m = k\sqrt{S}.$$

At the level $p = p_B$, we get from (8D.12)

$$W_B = \frac{i2\lambda k^3 \int_{p_1}^{1} G(p_B, \tilde{p})\Psi(\tilde{p})d\tilde{p}}{1 + \frac{\varepsilon k^2}{2}\int_{p_1}^{1} G(p_B, \tilde{p})h(\tilde{p})d\tilde{p}}. \tag{8D.14}$$

Substituting (8D.14) into (8D.9) would then lead to a differential–integral equation for Ψ alone

$$\frac{d^2\Psi}{dp^2} - Sk^2\Psi = \frac{i\lambda k^3\frac{dh}{dp}\int_{p_1}^{1} G(p_B, \tilde{p})\Psi(\tilde{p})d\tilde{p}}{(\bar{u}k - \sigma)\left(1 + \frac{\varepsilon k^2}{2}\right)\int_{p_1}^{1} G(p_B, \tilde{p})h(\tilde{p})d\tilde{p}}. \tag{8D.15}$$

We apply the method of Green's function one more time to (8D.15) so that it would be transformed to an integral equation, viz.

$$\Psi(p) = \frac{\varepsilon\lambda k^3 \int_{p_1}^{1} \frac{K(p,\tilde{p})}{\bar{u}(\tilde{p})k-\sigma}\frac{dh(\tilde{p})}{d\tilde{p}}d\tilde{p}}{1+\frac{\varepsilon k^2}{2}\int_{p_1}^{1} G(p_B,\tilde{p})h(\tilde{p})d\tilde{p}} \times \int_{p_1}^{1} G(p_B\tilde{p})\Psi(\tilde{p})d\tilde{p}, \tag{8D.16}$$

where the second Green's function satisfying the boundary condition (8D.11) is

$$\begin{aligned}
K(p,\tilde{p}) &= \begin{cases} K_1(p,\tilde{p}) = \frac{1}{b}\theta(\tilde{p})\gamma(p) & \text{for } p<\tilde{p}, \\ K_2(p,\tilde{p}) = \frac{1}{b}\theta(p)\gamma(\tilde{p}) & \text{for } p>\tilde{p} \end{cases} \\
\gamma(p) &= e^{m(p-p_1)} + \left(\frac{\sigma-\bar{u}_1 k-\tilde{\lambda}}{\sigma-\bar{u}_1 k+\tilde{\lambda}}\right)e^{-m(p-p_1)}, \\
\theta(p) &= e^{m(p-1)} + \left(\frac{\sigma-\tilde{\lambda}}{\sigma+\tilde{\lambda}}\right)e^{-m(p-1)}, \\
b &= 2m\left[\left(\frac{\sigma-\bar{u}_1 k-\tilde{\lambda}}{\sigma-\bar{u}_1 k+\tilde{\lambda}}\right)e^{m(p_1-1)} - \left(\frac{\sigma-\tilde{\lambda}}{\sigma+\tilde{\lambda}}\right)e^{-m(p_1-1)}\right], \\
\tilde{\lambda} &= \frac{\lambda}{\sqrt{S}}\bar{u}_1 = \bar{u}(p_1), \quad m = k\sqrt{S}.
\end{aligned} \tag{8D.17}$$

Equation (8D.16) is a Fredholm equation.

Analytic solution

The important point to note is that (8D.16) has a degenerate kernel. Multiplying (8D.16) by $G(p_B,\tilde{p})$, integrating the resulting equation over the domain and after cancelling the common factor, $\int_{p_1}^{1} G(p_B,\tilde{p})\Psi(\tilde{p})d\tilde{p}$, from both sides, we finally obtain,

$$1+\frac{\varepsilon k^2}{2}\int_{p_1}^{1} G(p_B\tilde{p})h(\tilde{p})d\tilde{p} = \varepsilon\lambda k^3 \int_{p_1}^{1}\left(\int_{p_1}^{1}\frac{K(p',\tilde{p})}{\bar{u}(\tilde{p})k-\sigma}\frac{dh(\tilde{p})}{d\tilde{p}}d\tilde{p}\right)G(p_B,p')dp'. \tag{8D.18}$$

This is the necessary condition for the existence of a non-trivial normal mode solution with the use of any heating profile h and heating intensity ε. We use it to determine the eigenvalue, σ. Note that $K(p,\tilde{p})$ also depends on σ arising from the boundary condition (8D.11). Equation (8D.18) is valid as long as there is heating, $\varepsilon \neq 0$. In the absence of heating $\varepsilon = 0$, (8D.8) and (8D.9) would become decoupled and the deduction leading up to (8D.18) would be irrelevant.

We next deduce the analytic solution of Ψ. It is useful to note that the integral equation (8D.16) has the following form

$$M(p)\Psi(p) = \int_{p_1}^{1} G(p_B, \tilde{p})\Psi(\tilde{p})d\tilde{p}$$

$$M(p) = \frac{1 + \frac{\varepsilon k^2}{2}\int_{p_1}^{1} G(p_B, \tilde{p})\, h(\tilde{p})d\tilde{p}}{\varepsilon\lambda k^3 \int_{p_1}^{1} \frac{K(p,\tilde{p})}{\bar{u}(\tilde{p})k-\sigma}\frac{dh}{d\tilde{p}}d\tilde{p}}. \tag{8D.19a, b}$$

The RHS of (8D.19a) is a quantity independent of p whatever the solution Ψ might be. It follows that while M and Ψ individually vary with p, their product does not. Hence, we may write

$$M(p)\Psi(p) = M(1)\Psi(1). \tag{8D.20}$$

Furthermore, we may set $\Psi(1) = 1$ as a normalization constant without loss of generality. Then the analytic solution of Ψ is simply

$$\Psi(p) = \frac{\int_{p_1}^{1} \frac{K(p,\tilde{p})}{\bar{u}(\tilde{p})k-\sigma}\frac{dh(\tilde{p})}{d\tilde{p}}d\tilde{p}}{\int_{p_1}^{1} \frac{K(1,\tilde{p})}{\bar{u}(\tilde{p})k-\sigma}\frac{dh(\tilde{p})}{d\tilde{p}}d\tilde{p}}. \tag{8D.21}$$

Thus, once an eigenvalue is known by solving (8D.18), we can use it in conjunction with (8D.21) to determine the structure of $\Psi(p)$.

Using (8D.19a,b) with $\Psi(1) = 1$, we also get from (8D.14) a self-contained expression for W_B as

$$W_B = \frac{i2}{\varepsilon \int_{p_1}^{1} \frac{K(1,\tilde{p})}{\bar{u}(\tilde{p})k-\sigma}\frac{dh(\tilde{p})}{d\tilde{p}}d\tilde{p}}. \tag{8D.22}$$

By substituting (8D.21) and (8D.22) into (8D.12), we then obtain an analytic solution of $W(p)$ as

$$\begin{aligned} W(p) &= \left(\frac{i2\lambda k^3 \int_{p_1}^{1} G(p,\tilde{p}) \int_{p_1}^{1} \frac{K(\tilde{p},p')}{\bar{u}(p')k-\sigma}\frac{dh(p')}{dp'}d\tilde{p}dp'}{\int_{p_1}^{1} \frac{K(1,\tilde{p})}{\bar{u}(\tilde{p})k-\sigma}\frac{dh(\tilde{p})}{d\tilde{p}}d\tilde{p}}\right) \\ &\quad + \left(\frac{-ik^2 \int_{p_1}^{1} G(p,\, \tilde{p})h(\tilde{p})d\tilde{p}}{\int_{p_1}^{1} \frac{K(1,\tilde{p})}{\bar{u}(\tilde{p})k-\sigma}\frac{dh(\tilde{p})}{d\tilde{p}}d\tilde{p}}\right) \\ &\equiv W^{(d)} + W^{(h)}. \end{aligned} \tag{8D.23}$$

Equations (8D.18), (8D.21) and (8D.23) constitute a complete analytic solution of this eigenvalue–eigenfunction problem for any heating profile and heating intensity. Note that setting $p = p_B$ in (8D.23) would recover the necessary condition (8D.18) for the existence of a non-trivial normal mode solution by virtue of (8D.22).

The component $W^{(d)}$ in (8D.23) is directly associated with the vorticity advection process, whereas the component $W^{(h)}$ is directly associated with the diabatic heating. It is tempting to refer to $W^{(d)}$ as a dynamically induced vertical velocity, and $W^{(h)}$ as a diabatically induced vertical velocity for convenience. However, designating them as such does not literally mean that the $W^{(d)}$ component is not influenced by the heating at all, or that $W^{(h)}$ component is not influenced by the dynamical process at all. In reality, they are jointly influenced by the heating and dynamical process as mathematically manifested through their dependence on the eigenvalue σ. The weaker the heating intensity is, the more meaningful their separation in the interpretation above would be. For a relatively weak heating, σ would primarily depend on the dry dynamical process. By the definition of $W^{(h)}$ and with (8D.22), we get $W_B^{(h)} = -W_B \frac{\varepsilon k^2}{2} \int_{p_1}^{1} G(p_B, \tilde{p}) h(\tilde{p}) d\tilde{p}$, the expression (8D.23) at $p = p_B$ can be written as

$$W_B = \frac{W_B^{(d)}}{1 + \frac{\varepsilon k^2}{2} \int_{p_1}^{1} G(p_B, \tilde{p})\, h(\tilde{p}) d\tilde{p}}. \tag{8D.24}$$

8D.2.4 Analysis for the case of a top-hat heating profile

We now examine the analytic solution for a heating profile $h(p)$ that has a simple shape of a top-hat. The cloud base level associated with this profile is indicated by p_* and the cloud top level by p_{**}. Then h is set to unity between p_{**} and p_*, and zero elsewhere. The expression for this heating profile is

$$\begin{aligned} h(p) &= H(p - p_{**}) - H(p - p_*) \\ &= \begin{cases} 1 & \text{for } p_{**} < p < p_* \\ 0 & \text{for } p < p_{**}, \quad p > p_* \end{cases} \end{aligned} \tag{8D.25}$$

where H is the Heaviside unit step function.

CISK-threshold and implication

It is instructive to pause and consider the special case of large-scale moist instability without a basic baroclinic shear, $\lambda = 0$. We distinguish this case from the general case by calling it *pure large-scale wave-CISK* stemming solely from the self-induced condensational heating. It follows from $\lambda = 0$ that we have $W_B^{(d)} = 0$. According to (8D.24), there can be a finite W_B only if the following condition is met:

$$\left(1 + \frac{\varepsilon k^2}{2} \int_{p_1}^{1} G(p_B, \tilde{p}) h(\tilde{p}) d\tilde{p}\right) = 0. \tag{8D.26}$$

Equation (8D.26) is identical to the general necessary condition for moist baroclinic instability (8D.18) for the special case of no baroclinic shear. It can also be viewed as a compatibility condition for a disturbance with a given heating intensity ε and heating profile h.

Let us work out the explicit form of (8D.26). It suffices to focus on a situation that the top of the moist layer coincides with the bottom of the heating layer, $p_B = p_*$. Using the expression for the Green's function $G(p, \tilde{p})$ given in (8D.13) and the heating profile given in (8D.25), we get from (8D.26)

$$\varepsilon = \frac{2\sinh\left(k(1-p_1)\sqrt{S}\right)}{k^2\sinh\left(k(1-p_B)\sqrt{S}\right)\left[\cosh\left(k(1-p_B)\sqrt{S}\right)-1\right]}. \tag{8D.27}$$

Equation (8D.27) may be considered as an equation for the wavenumber of a disturbance under consideration as an implicit function of the other parameters, $k(\varepsilon, p_1, p_B, S)$. That would be the disturbance that can be maintained by self-induced heating alone. But, there may or may not be any value of k that could satisfy (8D.27) for a given heating intensity ε and heating profile h. It was pointed out by Craig and Cho (1988) that there would be no wavenumber that would satisfy (8D.27) unless the heating intensity ε exceeds a certain value. They referred to such a minimum value as the *CISK-threshold*. We will further comment on the implication of this condition when a basic shear is presented below.

Solution for the usual case with a basic baroclinic shear

We now deduce the solution for the case of a moist baroclinic basic state for this heating profile. It follows from (8D.25) that

$$\frac{dh}{dp} = \delta(p - p_{**}) - \delta(p - p_*), \tag{8D.28}$$

where δ is the Dirac delta function. With the use of (8D.25) and (8D.28), the condition for the existence of a non-trivial solution (8D.18) takes on the following simple form

$$1 + \frac{\varepsilon k^2}{2}\int_{p_{**}}^{p_*} G(p_B, p)dp = r_{**}\int_{p_1}^{1} G(p_B, p)K(p, p_{**})dp - r_*\int_{p_1}^{1} G(p_B, p)K(p, p_*)dp, \tag{8D.29}$$

where $r_{**} = \dfrac{\varepsilon\lambda k^3}{\bar{u}_{**}k - \sigma}$ and $r_* = \dfrac{\varepsilon\lambda k^3}{\bar{u}_* k - \sigma}$. The eigenvalue σ is also deeply embedded in the Green's function K. It is a straightforward matter to work out the individual integrals in (8D.29). Special care needs to be taken when we write out the explicit expressions for the two integrals on the RHS of (8D.29) dependent on whether we are dealing with a deep moist layer, $p_B < p_*$ or a shallow moist layer, $p_B > p_*$. The details are given in Mak (1994). In both cases, (8D.29) takes on the form of a fourth-order polynomial in σ symbolically written as

$$A_4\sigma^4 + A_3\sigma^3 + A_2\sigma^2 + A_1\sigma + A_o = 0, \tag{8D.30}$$

where each of the coefficients A_j is a very lengthy algebraic function of the model parameters. There is a dynamical reason for this dispersion relation to be a quartic equation because PV anomalies can only exist at four levels in this model setup. Two of those levels are the bottom and top surfaces of the domain when there is a basic baroclinic shear as elaborated in Section 3.9. Such PV anomalies are generalized PV as discussed in Section 3.9.2. The other two levels are the bottom and top of the heating layer because dh/dp is nonzero only at those levels. Moist baroclinic instability could arise from the mutual reinforcements among those four PV anomalies, some of which may play a more important role than others in a particular unstable mode. Corresponding to the interaction between different pairs of PV anomalies as the dominant interaction, the unstable modes would have different and unique characteristics.

The analytic solutions of Ψ and W for each subinterval separated sequentially by p_1, p_{**}, p_B, p_* and 1 for the case of a deep moist layer or by p_1, p_{**}, p_*, p_B and 1 for the case of a shallow moist layer can be similarly worked out in terms of elementary functions. Some of the details are given in Mak (1994). Thus, all the instability properties of a normal mode can be computed precisely with no difficulty.

8D.2.5 Instability properties

We now examine the instability properties in an illustrative analysis. The following meaningful set of dimensional values of the parameters may be used as a reference case: $p_{oo} = 1000\,\text{mb}$, $p_B = 900\,\text{mb}$, $p_* = 900\,\text{mb}$, $p_{**} = 400\,\text{mb}$, $p_1 = 150\,\text{mb}$, $\lambda = 30/(p_{oo} - p_1)\ \text{m}\,\text{s}^{-1}\,\text{mb}^{-1}$, $S = 0.04\,\text{m}^2\,\text{s}^{-2}\,\text{mb}^{-2}$, $f = 10^{-4}\,\text{s}^{-1}$, $\varepsilon = 1\,\text{K}\,\text{mb}^{-1}$. Using $U = 10\,\text{m}\,\text{s}^{-1}$ and $L = 10^6\,\text{m}$ as the velocity and length scales, the corresponding non-dimensional values of the parameters would be: $p_B = 0.9$, $p_* = 0.9$, $p_{**} = 0.4$, $p_1 = 0.15$, $\lambda = 3.5$, $S = 4.0$, $\varepsilon = 30$. We will compare the basic instability properties as a function of wavelength $L_{(x)} = 2\pi/k$ for different heating intensity ε to ascertain their dependence upon these two key parameters. It is found that there is only one unstable mode for heating intensity less than about $\varepsilon \sim 20$ (corresponding to $Q \sim 54\ \text{K}\,\text{day}^{-1}$). We will shortly verify that this is a moist Eady unstable mode. Figure 8D.1 shows the imaginary part of the four eigenvalues σ_i over a range of weak to moderate heating intensity $0.01 \leq \varepsilon \leq 18$ and a wide range of wavelength. Positive value of $\sigma_i > 0$ is a measure of the growth rate. In this range of heating intensity, there is one amplifying mode, one decaying mode and two neutral modes. For the case of very small heating intensity $\varepsilon = 0.01$, the maximum growth rate is $0.52 \times 10^{-5}\text{s}^{-1}$. For a stronger heating, the growth rate of the unstable mode is larger until it reaches a maximum value of about $2.4 \times 10^{-5}\text{s}^{-1}$ for $\varepsilon = 18$. In other words, there is a 5-fold increase in the growth rate from $Q \sim 0$ to $Q = 48\,\text{K day}^{-1}$. This branch of unstable modes may be called the *moist Eady mode*. It stems from the baroclinic shear of the basic flow but can be significantly enhanced by the self-induced condensational heating. The wavelength of its most unstable mode for $\varepsilon = 0.01$ is 6700 km. This mode becomes shorter for a stronger heating until it reaches a minimum value of about 1400 km at $\varepsilon \approx 15$. In other words, it is reduced by a factor of 4.7. This trend of growth rate enhancement and scale reduction of the moist Eady mode is *reversed* when we further increase the heating intensity. The potential vorticity point of view

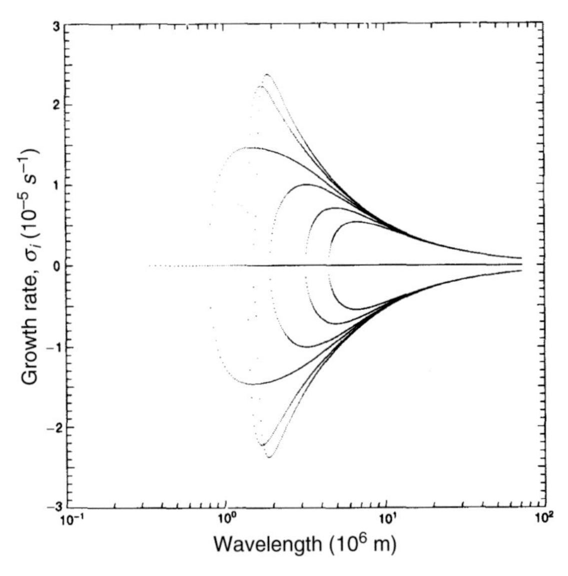

Fig. 8D.1 Variations of the growth rate σ_i of the four eigenmodes with wavelength for six non-dimensional values of the heating intensity, $\varepsilon = 0.01, 8, 13, 15, 17, 18$ ($\varepsilon = 30 \leftrightarrow Q = 80\,\text{K day}^{-1}$). Innermost curve is for the smallest value of ε. Other parameters have values of the reference case. Two modes are neutral and have $\sigma_i = 0$. Each dot is a data point (taken from Mak, 1994).

provides a relatively simple interpretation of the nonlinear dependence of the growth rate on the heating intensity. The interactions among the four PV anomalies give rise to a constructive interference among them when the heating intensity is weak. But such constructive interference does not continue indefinitely as we turn up the heating. Beyond a certain threshold of heating intensity, the strong interference distorts the structure of the moist Eady mode so much that its overall efficiency in extracting energy from the basic state would be progressively smaller resulting in a reversal of the trend.

There are two unstable modes in this system for a heating intensity greater than about $\varepsilon \sim 20$. The impacts of condensational heating are examined in Fig. 8D.2 for (a) $\varepsilon = 20$ and (b) a strong heating $\varepsilon = 60$. Figure 8D.2a shows that in addition to the moist Eady mode, a second branch of unstable modes barely emerges for $\varepsilon = 20$ at a wavelength slightly shorter than the cutoff wavelength of the moist Eady mode. For a stronger heating $\varepsilon = 60$, the existence of the second branch in the subsynoptic range of scales is well established and there is one more branch of unstable modes at even shorter wavelengths. There is a spectral gap between them. These two branches have comparable growth rates but are considerably smaller than that of the moist Eady mode. Their maximum value of growth rate is ~$0.5 \times 10^{-5}\text{s}^{-1}$ which is about four times smaller than the counterpart of the moist Eady mode. It is further found that the spectral gap between the two branches of unstable modes is narrower and may even overlap for a thick moist layer.

Figure 8D.3 shows the corresponding results of the real part of the eigenvalues σ_r. The phase velocity $c = \sigma_r/k$ of the moist Eady mode corresponding to its frequency shown in

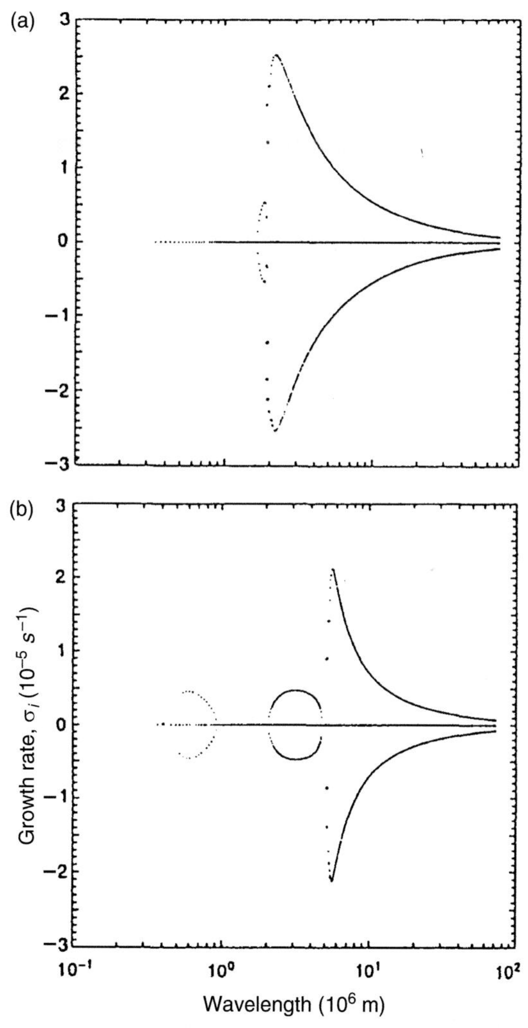

Fig. 8D.2 Variations of the growth rate of the four eigenmodes with wavelength for the heating intensity (a) $\varepsilon = 20$ and (b) $\varepsilon = 60$. $\left(\varepsilon = 30 \leftrightarrow Q = 80\,\text{K day}^{-1}\right)$. Other parameters have values of the reference case (taken from Mak, 1994).

Fig. 8D.3a is slightly higher than that of the dry Eady mode. Condensational heating therefore slightly raises the steering level. It signifies a stronger influence of the PV anomaly at the top of the heating layer. For the case of a strong heating, Fig. 8D.3b reveals that the subsynoptic branch of unstable modes has a frequency higher than that of the moist Eady mode and the other branch of even shorter unstable modes has a frequency lower than that of the moist Eady mode. An example of the former branch of unstable modes with a wavelength of 2×10^6 m has a frequency of $7 \times 10^{-5}\,\text{s}^{-1}$. Its phase velocity is about $c = 2.3$. The corresponding steering flow is equal to the basic flow close to the mid-level between the model tropopause and the top of the heating layer $(\bar{u}(p_1) + \bar{u}(p_{**}))/2 = 2.2$. In other words, we may interpret this unstable mode as

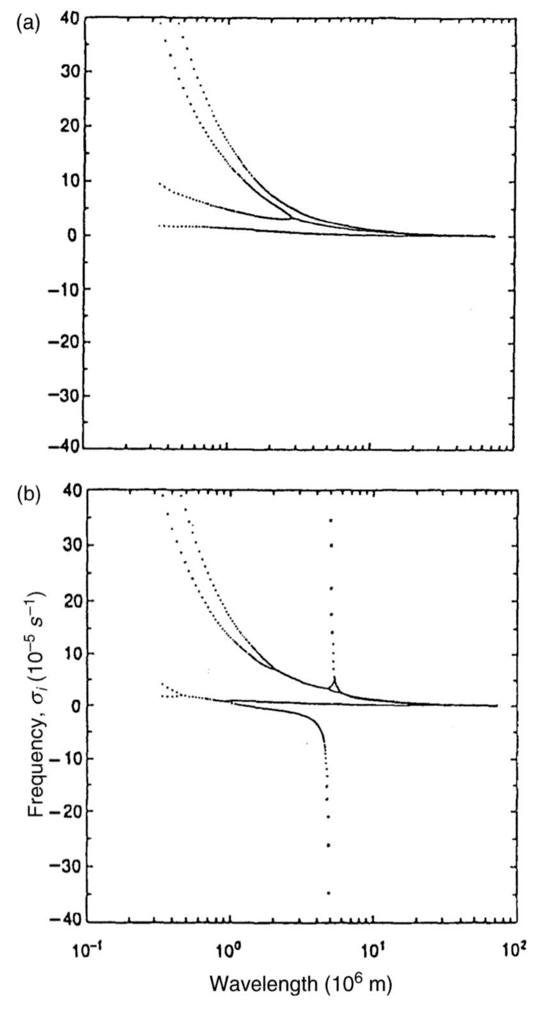

Fig. 8D.3 Variations of the frequency of the four eigenmodes with wavelength for the heating intensity (a) $\varepsilon = 20$ and (b) $\varepsilon = 60$. Other parameters have values of the reference case (taken from Mak, 1994).

primarily arising from positive reinforcement between the generalized PV at the tropopause and the PV anomaly excited at the top of the heating layer. It is referred to as the *diabatically induced upper mode.* An example of the branch of even shorter unstable modes with a wavelength of 5×10^5 m has a frequency of $2 \times 10^{-5}\,\mathrm{s}^{-1}$. The phase velocity of this mode is only about 0.18, which is equal to a steering level close to the mid-level between the model surface and the bottom of the heating layer $(\bar{u}(1) + \bar{u}(p_*))/2 = 0.175$. We may similarly interpret this unstable mode as primarily arising from positive reinforcement between the generalized PV at the surface and the PV anomaly at the bottom of the heating layer. We logically refer to this branch as the *diabatically induced lower mode.* These two branches may be referred to as boundary-CISK modes for short.

Note that the eigenvalue σ only appears on the RHS of the necessary condition for moist instability, (8D.29). The RHS of that equation is proportional to the baroclinic shear

λ apart from the additional dependence of the Green's function $K(p,\tilde{p})$ on it. Therefore, the RHS of that equation is zero when there is no baroclinic shear. The LHS of (8D.29) is the same as the LHS of (8D.26). In other words, the eigenvalue is indeterminate under the condition of CISK-threshold when there is no baroclinic shear. The eigenvalue is expunged from the dispersion relation. This feature appears to be an indication of degeneracy in the limiting case of the moist baroclinic instability problem.

Structure of unstable modes

We next examine the structure of the unstable modes in terms of their streamfunction. The structures of the three sample modes of each branch are shown in Fig. 8D.4. Recall that each eigenfunction is normalized to unity at the bottom boundary, $\Psi(1) = 1$. That is why the values in panel 8D.4b for the diabatically induced upper mode are so much larger. These structures confirm the interpretation that the existence of each mode stems from positive reinforcement between different pairs of PV anomalies of the mode in each case. There is a westward tilt with height in each of these modes suggesting that each mode draws some of its energy from the basic baroclinic shear even when the heating is strong. The westward tilt of the moist Eady mode is greater in the heating layer than below and above in panel (a). This is an indication of the enhancement of the baroclinic instability by the heating. Panel (b) confirms that a *diabatically induced upper mode* is mostly confined between the upper boundary and the top of the heating layer. It has a westward tilt revealing an essential role of the baroclinic shear. Panel (c) confirms that a *diabatically induced lower mode* is mostly confined between the lower boundary and the bottom of the heating layer. It also has a westward tilt revealing an essential role of the baroclinic shear.

The vertical structures of the amplitude and phase angle of the p-vertical velocity $W = W^{(d)} + W^{(h)}$ as well as its two components $W^{(d)}$and $W^{(h)}$of the moist Eady mode are shown in Fig. 8D.5; $|W^{(h)}|$ is stronger than $|W^{(d)}|$ at all levels by more than a factor of two; $W^{(h)}$ has no vertical tilt; $W^{(d)}$ has an eastward tilt of about 100°, leading to an eastward tilt of only about 45° in the total p-velocity W.

The corresponding results for the diabatically induced upper mode are shown in Fig. 8D.6. We see that the phase angle of $W^{(d)}$ is slightly more than 180°out of phase with respect to $W^{(h)}$ at all levels; $|W^{(d)}|$ and $|W^{(h)}|$are comparable at the lower levels; $|W^{(d)}|$ is substantially larger than $|W^{(h)}|$ above the top of the heating layer. Hence, there is considerable cancellation between $W^{(d)}$and $W^{(h)}$ at the lower levels leading to a relatively weak total p-velocity. For the same reason, the total p-velocity is considerably stronger at the upper levels.

The corresponding results for the diabatically induced lower mode are shown in Fig. 8D.7; $|W^{(d)}|$ has substantial value only below about 800 mb; $|W^{(d)}|$ and $|W^{(h)}|$ are comparable at those levels; $W^{(d)}$ and $W^{(h)}$ are about 180° out of phase below about 600 mb. There is then high degree of cancellation between $W^{(d)}$ and $W^{(h)}$, leading to small values of the total p-velocity there. It is interesting to note that although this unstable mode is concentrated between the surface and the bottom of the heating layer, the total p-velocity is strong at the upper levels.

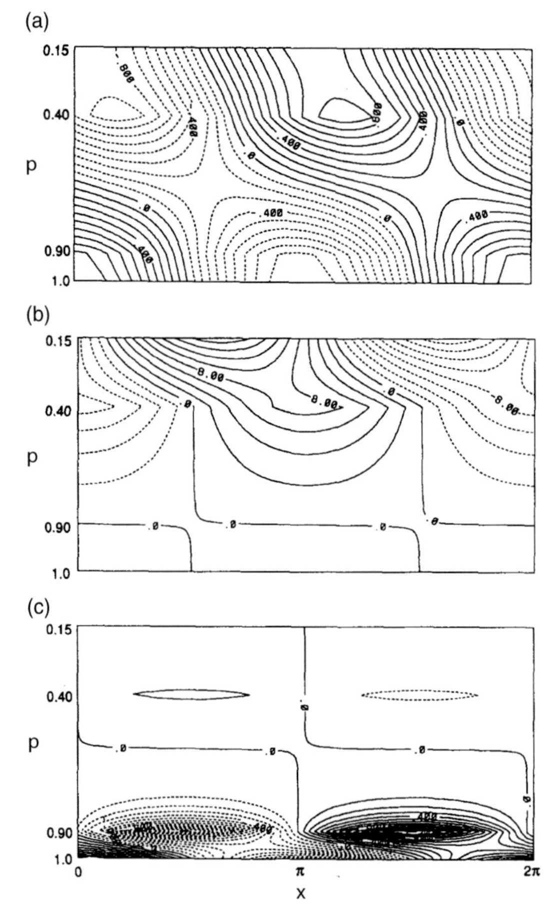

Fig. 8D.4 Structure of the non-dimensional streamfunction of a representative mode from each distinct group of unstable modes for the reference case parametric condition and $\varepsilon = 60$. (a) A moist Eady mode with a wavelength of 6000 km. (b) A diabatically induced upper mode with a wavelength of 3000 km. (c) A diabatically induced lower mode with a wavelength of 600 km.

We have not made an exhaustive diagnosis of the dependence of the instability properties on all parameters for brevity. It could be of interest to delineate the dependence on the thickness of the heating layer $(p_* - p_{**})$.

It is pertinent to point out that W_B of a disturbance would be unbounded according to (8D.14) if its wavenumber satisfies the condition of CISK-threshold, (8D.26). There is an indirect manifestation of this unsettling characteristic in Figs. 8D.2b and 8D.3b. The calculation reveals that the frequency σ_r of the normal modes with a non-dimensional wavelength close to 5.0 for the case of a strong heating $\varepsilon = 60$ is very large. The corresponding imaginary part of the eigenvalue seen in Fig. 8D.2b is very small, $\sigma_i \sim 0$. From (8D.22) we see that the denominator would approach zero for $\sigma \to \pm\infty$ and W_B would be unbounded as well. We do not generally come across this unsettling feature in a calculation unless the wavenumber of the disturbance under consideration happens to

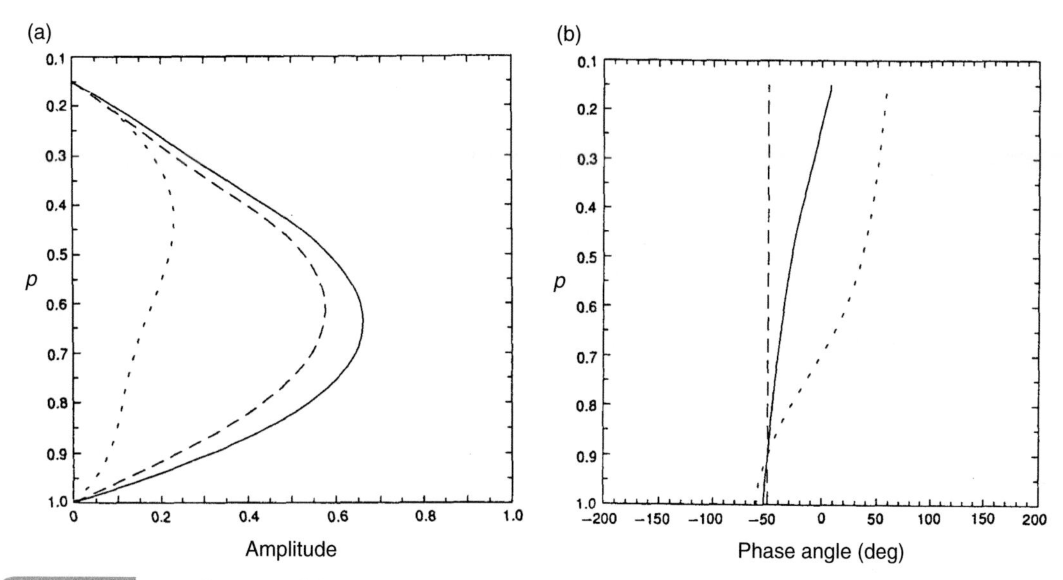

Fig. 8D.5 Vertical structure of the corresponding non-dimensional vertical velocity of the moist Eady mode with a wavelength of 6000 km. (a) Amplitude of W (solid), $W^{(d)}$ (dot) and $W^{(h)}$ (dash). (b) Phase angle of the corresponding three quantities (taken from Mak, 1994).

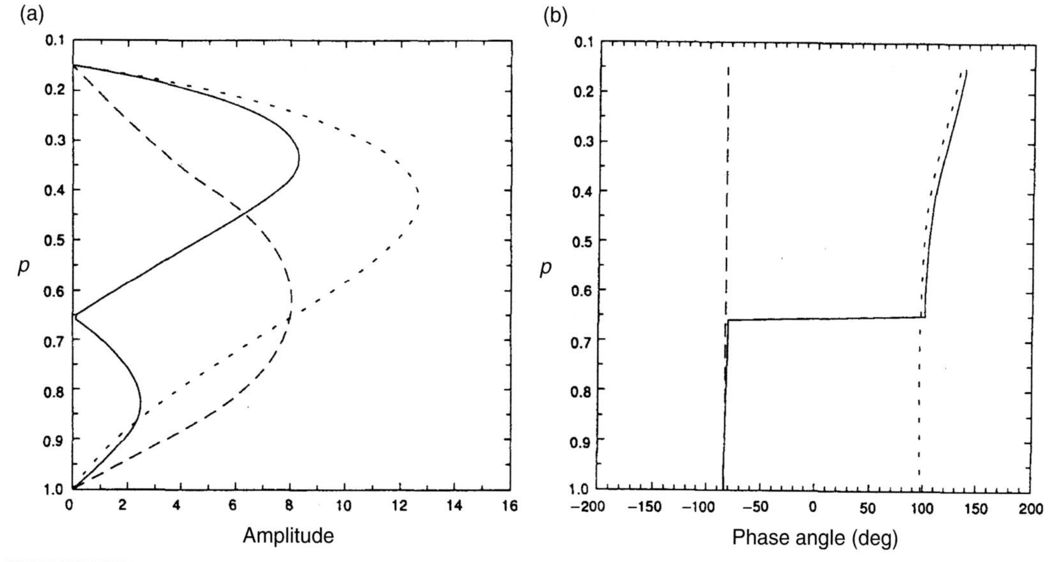

Fig. 8D.6 Vertical structure of the non-dimensional vertical velocity of the diabatically induced upper mode with a wavelength of 3000 km. (a) Amplitude of W (solid), $W^{(d)}$ (dot) and $W^{(h)}$ (dash). (b) Phase angle of the corresponding three quantities; $W = W^{(d)} + W^{(h)}$.

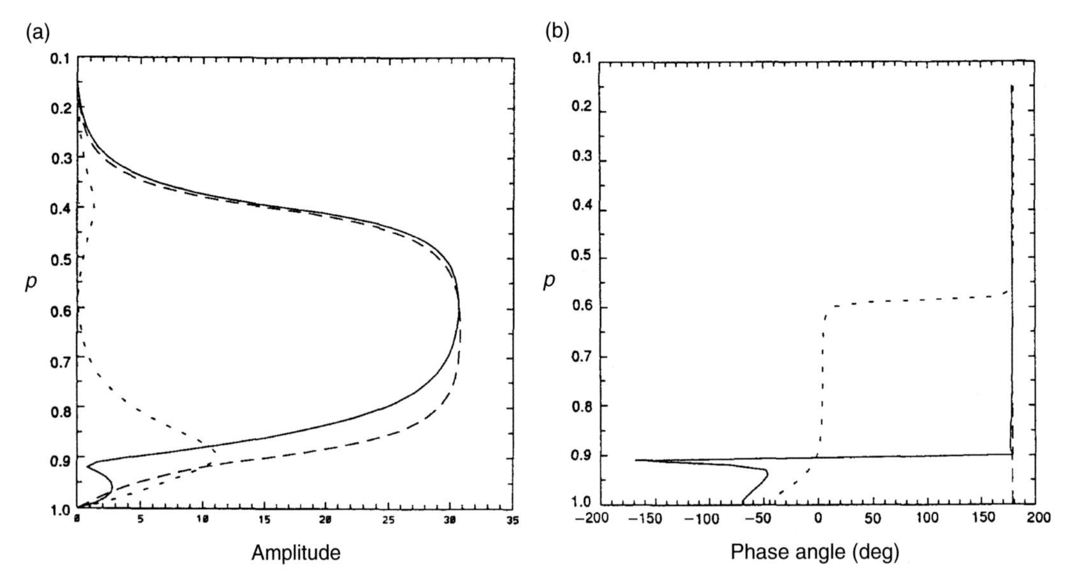

Fig. 8D.7 Vertical structure of the corresponding non-dimensional vertical velocity of the diabatically induced lower mode with a wavelength of 600 km. (a) Amplitude of W (solid), $W^{(d)}$ (dot) and $W^{(h)}$ (dash). (b) Phase angle of the corresponding three quantities; $W = W^{(d)} + W^{(h)}$.

have a value that satisfies the condition of $1 + \frac{\varepsilon k^2}{2} \int_{p_1}^{1} G(p_B, \tilde{p}) h(\tilde{p}) d\tilde{p} = 0$. That is why the solution under almost all conditions is well-behaved. Nevertheless, this unsettling feature exposes an inherent shortcoming of the CISK formalism in the context of the large-scale moist baroclinic instability problem.

8D.2.6 Energetics

Let us finally examine the moist baroclinic unstable modes from the perspective of their energetics. We only examine the conversion rate of kinetic energy K from the available potential energy A. The latter can be partly increased through a conversion from the basic baroclinic state and partly by generation attributable to the diabatic heating. We compute the conversion rate from A to K at each level with the following formula

$$c(A,K) = \int_0^{2\pi} \omega \frac{\partial \psi}{\partial p} dx = \frac{1}{2} \left(\mathrm{Re}\left\{ W^{(d)} \frac{d\Psi^*}{dp} \right\} + \mathrm{Re}\left\{ W^{(h)} \frac{d\Psi^*}{dp} \right\} \right), \tag{8D.31}$$

where the asterisk superscript stands for the complex conjugate of the amplitude function of Ψ. The conversion rate for the disturbance as a whole is $C(A,K) = \int_{p_1}^{1} c(A,K) dp$. The results of $c(A,K)$ for the three modes discussed above are presented in Fig. 8D.8. The upper panel reveals that most of the total conversion takes place within the heating layer in this moist Eady mode. The contribution from the baroclinic process is positive at

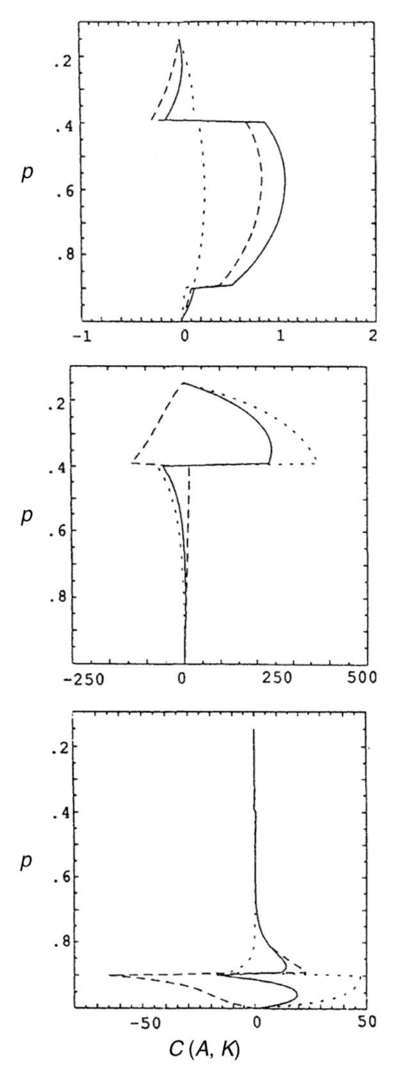

Fig. 8D.8 Vertical distribution of the corresponding non-dimensional conversion rate from the available potential energy to kinetic energy $C(A, K)$ for the three normal modes with a wavelength 6000 km (upper), 3000 km (middle) and 600 km (lower). The solid curve is for the total conversion rate; the dot curve is for the conversion rate associated with $W^{(d)}$; the dash curve is for the conversion rate associated with $W^{(h)}$.

all levels, but is relatively small. The contribution from the heating process is particularly strong in the heating layer, but is negative above the heating layer. The middle panel confirms that the conversion in the diabatically induced upper mode takes place between the top of the heating layer and the upper boundary. The conversion due to the component $W^{(h)}$ associated with the heating is actually negative above the heating layer. It is overcompensated by the large positive conversion due to the component $W^{(d)}$ associated with the dynamical process. But recall that such interpretation of $W^{(d)}$ and $W^{(h)}$ has limited validity because each of them is actually jointly influenced by the dynamical and heating processes. The lower panel confirms that the conversion in this diabatically induced lower mode takes places between the bottom of the heating layer and the lower boundary. It is interesting to see that the conversion below the heating layer is associated with the dynamical component $W^{(d)}$, whereas the conversion within the heating layer is associated with the heating component $W^{(h)}$.

Concluding remarks

We have established an analytic solution for the instability in a moist Eady model for any heating profile. For the special case of a heating profile of a top-hat shape, there are four discrete eigenmodes arising from the existence of potential vorticity anomalies at only four discrete vertical levels. The quantitative results of the instability properties reveal that there is only one unstable mode for a weak to moderate heating. This unstable mode is the moist Eady mode. Under plausible parametric conditions, the growth rate of the most unstable mode is significantly enhanced and its wavelength is significantly reduced when the heating intensity is increased to a moderate value. These findings are compatible with observation. These trends are reversed when the heating exceeds a certain value due to a destructive interference of the interaction among the four PV anomalies. Furthermore, two additional unstable modes in the subsynoptic and shorter wavelength ranges are found for a strong heating intensity. One branch of unstable modes is largely confined between the model tropopause and the top of the heating layer. The branch of even shorter unstable modes is largely confined between the surface and the bottom of the heating layer. These two branches may coexist in some part of the parameter space. The energetics diagnosis confirms the validity of the interpretations summarized above. It also constitutes a separate check of self-consistence of the solution and the illustrative calculations.

This type of analysis can be applied to more general baroclinic model settings in principle. The functional form of the Green's functions would be more complex and would probably have to be evaluated numerically in practice.

9 Stationary planetary wave dynamics

This chapter is concerned with the linear dynamics of stationary planetary waves. We begin by briefly showing some observed key characteristics of such a wave field in Section 9.1. Section 9.2 discusses the conceptual issues. We delineate the dynamical impacts of a zonal mean flow on the propagation of stationary planetary waves in Section 9.3. The excitation of a stationary wave field by a localized thermal forcing or by a localized orography is analyzed separately in Sections 9.4 and 9.5 with a simplest possible model. In each case, we delineate the impacts of the individual dynamical factors such as the beta-effect, the types of shear and the direction of the basic zonal flow on the forced wave field before we examine the combined influence of all of them in the various subsections. The relevance of this model is finally illustrated with a simulation study of the summer mean Asian monsoonal circulation in Section 9.6.

9.1 Observed characteristics of stationary planetary waves

The annual, seasonal or even monthly mean weather maps at different levels have a very broad wavy structure. Such a flow is stationary by definition, $\partial/\partial t = 0$, with respect to an annual, seasonal or monthly time scale respectively. Stationary planetary waves are considerably stronger in winter than in summer simply because the forcings that give rise to them are distinctly stronger. We illustrate the winter planetary wave field with the January mean flow over the northern hemisphere in 2003. The plots are constructed using the NCEP/NCAR Reanalysis Data set. The horizontal structure of the 250 mb geopotential height field, $\bar{Z}_{250}$, and the corresponding departure field from the zonal mean, $\bar{Z}_{250}*$, are shown in Fig. 9.1a,b. The total flow is particularly strong in middle latitudes as evidenced by the two pronounced troughs in Fig. 9.1a. Associated with these troughs are the two major localized winter mean jets. We see in Fig. 9.1b that the January $\bar{Z}_{250}*$ values are greatest around 50N, varying between 350 m and -350 m. The values south of ~20N are quite small.

It is instructive to decompose the $\bar{Z}_{250}*$ field at each latitude into Fourier components, $\bar{Z}^*(m, \phi)$, where m is the zonal wavenumber and ϕ the latitude. The distribution of the amplitude of these components, $|\bar{Z}(m, \phi)|$, as a function of m and ϕ is shown in Fig. 9.2. We see that the quasi-stationary waves are much stronger in the northern hemisphere than in the southern hemisphere. The wavenumber-1 component is particularly strong around 40N, 75N and also around 50S. These waves have significant values only for wavenumbers 1 to 4. Such stationary flow therefore mostly consists of waves of planetary scale and

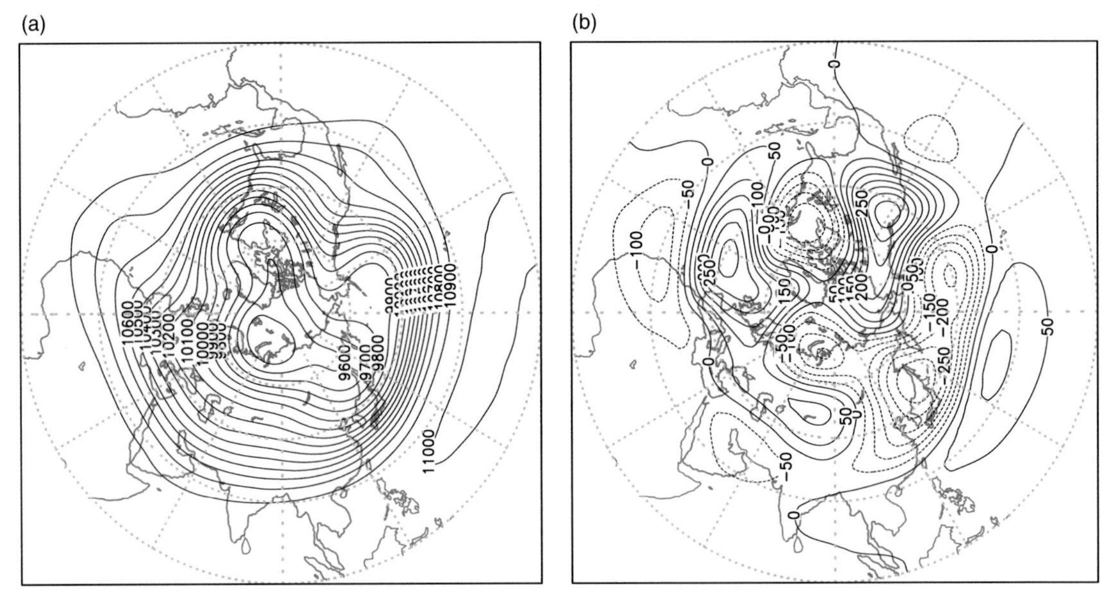

Fig. 9.1 January mean (a) $\bar{Z}_{250}$ and (b) $\bar{Z}_{250}*$ of 2003 of the northern hemisphere in m.

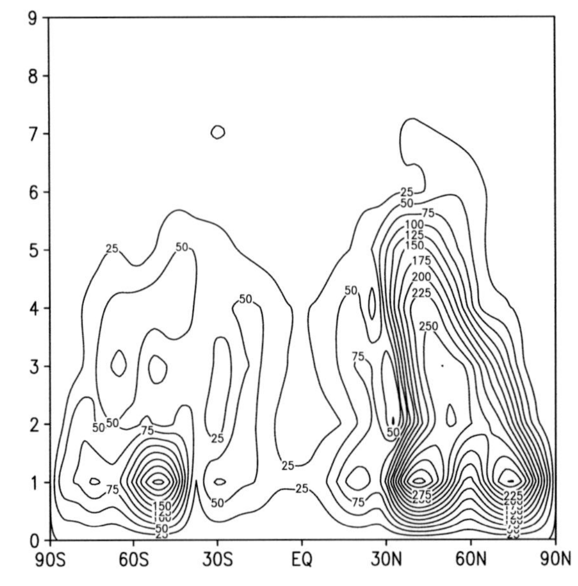

Fig. 9.2 Variation of $|\bar{Z}(m,\phi)|$ of the 250 mb geopotential height in January of 2003 as a function of zonal wavenumber m and latitude ϕ.

a zonal mean flow. Those waves are much stronger in the northern hemisphere than in the southern hemisphere.

The vertical structure of the waves in winter in terms of the geopotential height is shown in Fig. 9.3 with a longitude–pressure cross-section along 50N. The result reveals that while these waves span over the whole troposphere, the ultra-long wave components

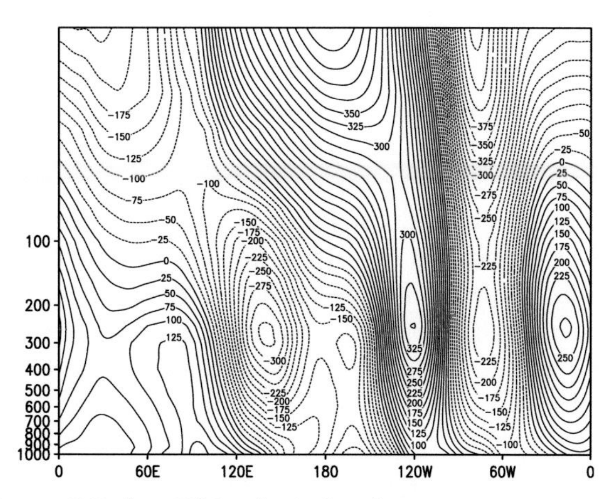

Fig. 9.3 Distribution of $\bar{Z}^*$ in January 2003 along 50N in a longitude and pressure cross-section.

(wavenumber 1 and 2) also intrude into the stratosphere. In contrast, these waves do not penetrate into the stratosphere in the summer season.

In passing, it should be noted that the planetary wave field in the summer season is interesting in its own right even though it is distinctly weaker. It is relevant to the dynamics of the circulation associated with persistent summer drought over North America and Asia. Some model simulation studies have focused on this wave field (e.g. Ting, 1994).

9.2 Introductory remarks about the dynamics of stationary waves

We have made a scale analysis of the vorticity equation for the class of synoptic-scale disturbances in Section 6.2. Let us now contrast that with a counterpart scale analysis for planetary-scale disturbances. Observation suggests that the fundamental scales of planetary-scale disturbances and synoptic-scale disturbances differ on two counts. One is the time scale, which is often considerably longer than the advective time scale, so that planetary-scale disturbances may be regarded as quasi-stationary. The other is of course the horizontal scale, which is $L = 10^7$ m. The scales of horizontal divergence and vorticity in this class are comparable, both being $U/L = 10^{-6}\,\mathrm{s}^{-1}$. Then it follows that $-\vec{V}_2 \cdot \nabla\zeta \sim (U/L)^2 \sim 10^{-12}$, $-w\zeta_z \sim (WU/LH) \sim 10^{-12}$, $v\beta \sim 10^{-10}$, $f\nabla \cdot \vec{V}_2 \sim f_o U/L \sim 10^{-10}$, $\zeta\nabla \cdot \vec{V}_2 \sim (U/L)^2 \sim 10^{-12}$, $(w_y u_z - w_x v_z) \sim (WU/LH) \sim 10^{-12}$. The baroclinic effect is even smaller, $(p_x\alpha_y - p_y\alpha_x) \sim (\delta\rho/\rho^2)(\delta p/L^2) \sim 1 \times 10^{-13}$. Therefore, the balance appears to be $\beta v \sim f(u_x + v_y)$, which is even simpler than that normally incorporated in a quasi-geostrophic model. Nevertheless, we will examine the dynamics of planetary-scale waves with a steady quasi-geostrophic model partly for simplicity and partly for allowing the process of horizontal advection to play a role of first-order importance not implicated by the scale analysis above. Applications using such a model to atmospheric planetary waves have been found meaningful.

While stationary waves have no phase propagation by definition, they are capable of transmitting energy from their source of origin in the meridional and/or vertical directions. It is in this sense that we talk about propagation of stationary waves. The influence of a time-zonal mean flow on the horizontal/vertical propagation of stationary waves is an issue of intrinsic dynamics of the waves. We will first examine the dynamics of their horizontal propagation in a 2-D barotropic model setting in Section 9.3.1. The dynamics of their vertical propagation in a quasi-geostrophic model setting with a continuous vertical domain will be similarly examined in Section 9.3.2. Vertically propagating waves can have a dramatic impact on the stratospheric circulation as elaborated in Section 10.5.

We will next address the question concerning what determines the structure of a stationary wave field. This is one of the classic problems of atmospheric dynamics. The forcing of planetary waves can be thermal and/or mechanical in nature. The thermal forcing partly stems from differential net radiative heating associated with the global distribution of continents and oceans. Another important component of thermal forcing stems from the time mean global distribution of precipitation. The mechanical forcing arises from the time mean surface zonal flow going over the major mountain ranges. Transient eddies also contribute to the thermal and dynamical forcing by virtue of their time mean convergence of heat and vorticity fluxes.

Naturally, there have been many model studies in the past of the global response to an idealized/observed net diabatic heating field and idealized/actual global orography in conjunction with the observed zonal mean flow. Those investigations were performed using models with different approximations, resolution and complexity (an extensive bibliography is given in Wang, 2000). Most past studies focus on the winter planetary waves suggesting that they can be simulated with a reasonable degree of realism even in a linear model. The findings also indicate that the response to the transient eddy forcing is relatively weak in mid-latitudes, but can be more significant in the tropics (e.g. Valdes and Hoskins, 1989).

We will employ a quasi-geostrophic model for the reason elaborated previously. Our objective here is not so much as to discuss the degree of success in simulating a planetary wave field, but rather to shed light on the dynamical nature of different aspects of a response as clearly as possible. The nature of those impacts is not easy to identify when we incorporate all the pertinent physical factors simultaneously. It would suit us better if we break the problem down into its components. A helpful strategy would be to delineate the causal relationships between a response and individual factors in simplest possible but physically recognizable model settings. That would put us in a good position to interpret the key features in a simulation of a stationary forced wave field later on under a general albeit still idealized condition.

We will focus on the responses to the thermal and orographic forcing separately in Sections 9.4 and 9.5. The model has a channel domain, so no attempt is made to capture the features associated with the curvature of the Earth such as the tendency of some wave components to propagate towards the polar region, while others propagate towards the tropics. Another fascinating issue is concerned with how planetary waves inherently interact with the synoptic-scale waves and what the consequences might be. It is a more difficult nonlinear dynamical problem, which will be discussed in Section 11.3.

9.3 Impact of basic zonal flow on wave propagation

In this section, we discuss the dynamics of horizontal and/or vertical propagation of stationary waves through a background shear flow in simple model settings. It is understood that the background shear is sufficiently weak.

9.3.1 Horizontal propagation of stationary waves

The simplest form of propagation of stationary waves through a basic zonal shear flow is examined with a barotropic beta-plane channel model in an effectively unbounded domain. The governing equation for such a disturbance is then the steady vorticity equation linearized about a basic flow $U(y)$. It is a steady version of Eq. (8B.1) in Chapter Part 8B, viz.

$$U\left(\psi_{xx} + \psi_{yy}\right)_x + (\beta - U_{yy})\psi_x = 0, \tag{9.1}$$

where ψ is the perturbation streamfunction. Since the two coefficients in (9.1) do not depend on x, an elementary solution of (9.1) has the form of $\psi(x, y) = \varphi(y)e^{ikx}$. Let us examine what happens to a wave of zonal wavenumber k originating at a meridional location, $y = Y_o$. Substituting this solution into (9.1) would yield an equation that governs its amplitude function, viz.

$$U\frac{d^2\varphi}{dy^2} + \left(\beta - U_{yy} - Uk^2\right)\varphi = 0. \tag{9.2}$$

For a semi-infinite domain either $Y_o \leq y \leq \infty$ or $-\infty \leq y \leq Y_o$. The solution would have an undulating (wave-like) structure in the meridional direction in a region characterized by $\infty > n^2 > 0$ where $n^2(y) = \left(\beta - U_{yy} - Uk^2\right)/U$ is called *refractive-index-squared*. Two limiting values of n are particularly noteworthy. One is $n = \infty$ at a latitude where the basic flow is zero. The other is $n = 0$ at a latitude where the coefficient of the second term in (9.2) is zero. The latitude where $U(y^*) = 0$ is called *critical latitude* and where $\left(\beta - U_{yy} - Uk^2\right)\big|_{y^{**}} = 0$ the latitude is called *turning latitude* for reasons to be elaborated shortly.

The analysis is simpler in the case of a basic flow $U(y)$ that varies smoothly in y. A wave at each location would behave *as if it were in an environment of locally uniform basic flow.* We may then think of this as an initial value problem of determining the trajectory of a wave-packet initially known at $y = Y_o$. The approximate form of the solution is $\varphi = \Phi \exp(i\ell y)$ of which the length scale is distinctly shorter than that of $U(y)$. This is known as a WKBJ solution. The local meridional wavenumber $\ell(y)$ is to be compatible with the dispersion relation of a modal stationary wave wherever the wave-packet might be. The dispersion relation compatible with (9.2) is

$$0 = U - \frac{\beta - U_{yy}}{\ell^2 + k^2} \tag{9.3}$$

implying $\ell^2 = \frac{\beta - U_{yy}}{U} - k^2$. The movement of a wave-packet is its group velocity $(c_{gx}, c_{gy}) = \left(\frac{\partial \sigma}{\partial k}, \frac{\partial \sigma}{\partial \ell}\right)$ with $\sigma = 0 = Uk - \frac{(\beta - U_{yy})k}{\ell^2 + k^2}$. Denoting the position of such a wave-packet as $X(t)$ and $Y(t)$, its movement is governed by

$$\frac{dX}{dt} = \frac{2(\beta - U_{yy})k^2}{(\ell^2 + k^2)^2} = c_{gx}, \qquad \frac{dY}{dt} = \frac{2(\beta - U_{yy})k\ell}{(\ell^2 + k^2)^2} = c_{gy}. \tag{9.4a,b}$$

Since the system is zonally uniform, the zonal wavenumber k of a wave would not change as it propagates. The meridional wavenumber ℓ of a wave-packet would however change as it moves across latitude. Suppose a wave-packet is initially located at $X(0) = X_o$ and $Y(0) = Y_o$. The initial value of ℓ is simply $\ell(0) \equiv \ell^{(o)} = \pm\sqrt{\left(\frac{\beta - U_{yy}}{U}\right)\Big|_{y = Y_o} - k^2}$ according to (9.3). The positive (negative) value of $\ell^{(o)}$ is used if we wish to consider a wave-packet that has its wave-fronts oriented in the NW–SE (SW–NE) direction; $\ell(t)$ should change continuously in time. We can determine the trajectory by numerically integrating (9.4) forward in time.

An integral part of this analysis is to determine how the amplitude of a stationary wave would vary as its meridional wavenumber changes. The variation of this amplitude must be compatible with an intrinsic constraint, namely $\nabla \cdot \vec{E} = 0$ where $\vec{E} = (-\rho_o[u^* v^*], 0)$ is known as the Eliassen–Palm vector for barotropic disturbance and $u^* = -\psi_y$, $v^* = \psi_x$. That is the steady state form of Eq. (10.7a) in Chapter 10 for describing the variation of wave activity. This means that the momentum flux of this wave-packet $[u^* v^*]$ is invariant during its propagation, whereby

$$\frac{\rho_o k}{2} |\Phi|^2 \ell = \left(\frac{\rho_o k}{2} |\Phi|^2 \ell\right)_{t=0}. \tag{9.5}$$

It follows that

$$|\Phi| = |\Phi^{(o)}| \sqrt{\frac{\ell^{(o)}}{\ell}} = |\Phi^{(o)}| \left(\frac{\left(\beta - U^{(o)}{}_{yy} - U^{(o)} k^2\right) U}{\left(\beta - U_{yy} - U k^2\right) U^{(o)}}\right)^{1/4}. \tag{9.6}$$

The local amplitude of this stationary Rossby wave would be then inversely proportional to the square root of its local meridional wave number. Its energy $\left((k^2 + \ell^2)|\Phi|^2\right)$ also varies as a wave-packet propagates, indicating an exchange of energy between a wave-packet and its background zonal flow.

As a wave-packet approaches a latitude y^{**} where $\beta - U_{yy} - Uk^2 = 0$, we have $\ell \to 0$ and hence $dY/dt \to 0$. The movement of the wave-packet in the y-direction would come to a complete stop if it could get to that latitude. However, the amplitude of the wave-packet would increase without bound as it approaches this latitude according to (9.6). Then nonlinear dynamics should become important before a wave-packet could actually reach y^{**}. One conjecture of what might happen is that such a wave-packet would be reflected before it could get there. A computer code meant for tracking the wave propagation should have a logical provision that wave reflection occurs when it gets to within a pre-assigned short distance from y^{**}. The sign of ℓ would be changed and the orientation of the wave-front would be switched like a ray of light bouncing off a mirror. The latitude where $(\beta - U_{yy} - Uk^2)|_{y^{**}} = 0$ is therefore called a "turning latitude."

As a wave-packet approaches a latitude y^* where $U = 0$, ℓ would become progressively larger. This means that the meridional scale of a wave-packet would become progressively shorter. According to (9.4), the wave-packet would cease to move altogether since $dX/dt \to 0$ and $dY/dt \to 0$. The amplitude of the wave would reduce towards zero according to (9.6). One conjecture of what might happen is that the wave-packet would be absorbed by the zonal flow in the neighborhood of a critical latitude. Absorption of the wave could be physically attributable to frictional dissipation in a real fluid however weakly viscous it might be because frictional force becomes important when the length scale of a disturbance is sufficiently short.

In summary, we can compute the trajectory of a stationary wave-packet $X(t)$ and $Y(t)$, and therefore also its wavenumber $\ell(t)$ and amplitude $\Phi(t)$ on a (x, y) plane for given $U(y)$, k, $X(0)$ and $Y(0)$. This method of determining the propagation of stationary waves is known as *ray tracing analysis*.

Illustrative computation

Let us now consider propagation of a stationary Rossby wave through a basic westerly jet embedded in a uniform weak easterly flow, $U = U_o\left(\exp\left(-\left(\frac{y}{L}\right)^2\right) - \varepsilon\right)$ with $0 < \varepsilon < 1$. It would be convenient to measure distance and velocity in units of L and V respectively. Thus, we introduce the following non-dimensional quantities: $(\tilde{k}, \tilde{\ell}) = L(k, \ell)$, $(\tilde{X}, \tilde{Y}) = \frac{1}{L}(X, Y)$, $\tilde{U} = U/V$, $\tilde{\beta} = \beta L^2/V$, $\tilde{t} = \frac{V}{L}t$, $\tilde{y} \equiv \tilde{Y}$. The non-dimensional basic flow is then $\tilde{U} = \tilde{U}_o\left(\exp(-\tilde{y}^2) - \varepsilon\right)$. It follows that we have $\tilde{U}_{\tilde{y}\tilde{y}} = 2\tilde{U}_o(2\tilde{y}^{2-1})\exp(-\tilde{y}^2)$. There are two values of $\tilde{y}^{**}$ such that $\beta + \tilde{U}_o\varepsilon k^2 - \tilde{U}_o\left(4(\tilde{y}^{**})^2 - 2 + k^2\right)\exp\left(-(\tilde{y}^{**})^2\right) = 0$. These are turning latitudes. There are also two critical latitudes at $\tilde{y}^* = \pm\sqrt{-\ln(\varepsilon)}$.

The first calculation is made for the following parameter condition of the basic state: $L = 10^6\,\mathrm{m}$, $U_o = 30\,\mathrm{m\,s^{-1}}$, $U_o\varepsilon = 3\,\mathrm{m\,s^{-1}}$, $\beta = 1.6 \times 10^{-11}\,\mathrm{m^{-1}\,s^{-1}}$. We determine the trajectory of two wave-packets. One has a zonal wavelength equal to $4\pi \times 10^6\,\mathrm{m}$ so that $k = 0.5 \times 10^{-6}\,\mathrm{m^{-1}}$. Another wave-packet has a shorter wavelength with a zonal wave-number $k = 0.75 \times 10^{-6}\,\mathrm{m^{-1}}$. If we use $V = U_o$, the corresponding non-dimensional parameters are $\tilde{U}_o = 1.0$, $\varepsilon = 0.1$, $\tilde{\beta} = 0.53$ and $\tilde{k} = 0.5$ or $\tilde{k} = 0.75$. Then the critical latitudes are at $\tilde{y}^* \approx \pm 1.5$. The turning latitudes for the wave with $\tilde{k} = 0.5$ are at $\tilde{y}^{**} \approx \pm 0.88$ and $\tilde{y}^{**} \approx \pm 1.7$ (Fig. 9.4). The turning latitudes for the wave with $\tilde{k} = 0.75$ are at $\tilde{y}^{**} \approx \pm 0.8$ and $\tilde{y}^{**} \approx \pm 1.6$.

We numerically integrate the non-dimensional governing equations (9.4a,b) with a predictor-corrector scheme (described in Section A.13) using a time step $\delta t = 0.022$. The trajectories of the two wave-packets initially located at $X_o = 0$ and $Y_o = -0.5$ up to $\tilde{t} = 4.5$ (2.8 days) are shown in Fig. 9.5. Each wave-packet propagates northeastward until it approaches its corresponding turning latitude. The longitudinal position of the $\tilde{k} = 0.5$ wave at that time is $\tilde{X} \approx 0.58$. The longitudinal position of the $\tilde{k} = 0.75$ wave is $\tilde{X} \approx 0.9$

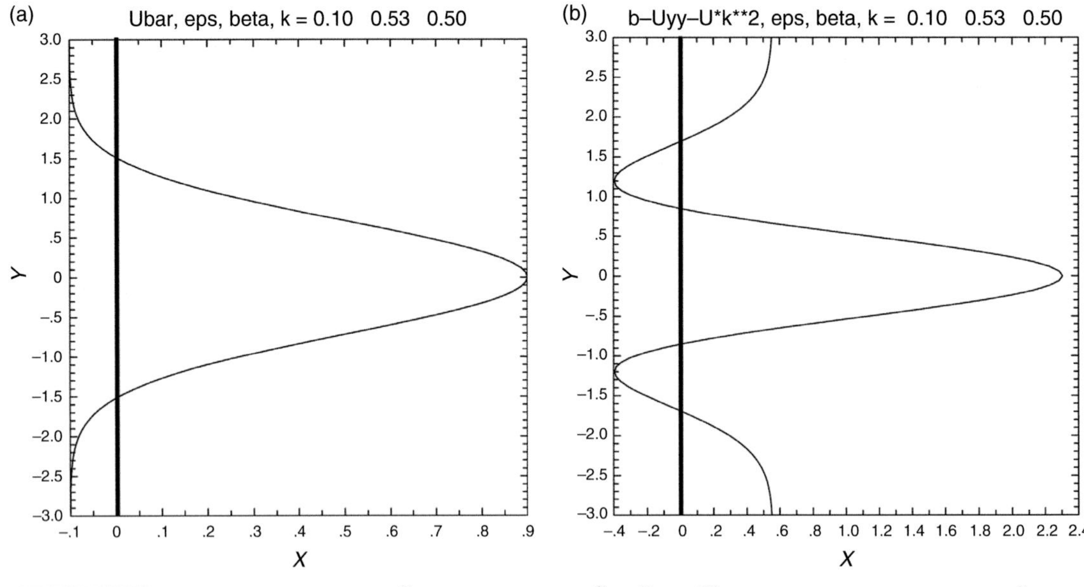

Fig. 9.4 Meridional distribution of (a) $\tilde{U}(\tilde{y})$, and (b) $F(y) = \tilde{\beta} - \tilde{U}_{yy} - \tilde{U}\tilde{k}^2$ for the case of $V = U_o$, $\varepsilon = 0.1$, $\tilde{k} = 0.5$ and $\tilde{\beta} = 0.53$. Thick line highlights $\tilde{U} = 0$ or $F = 0$.

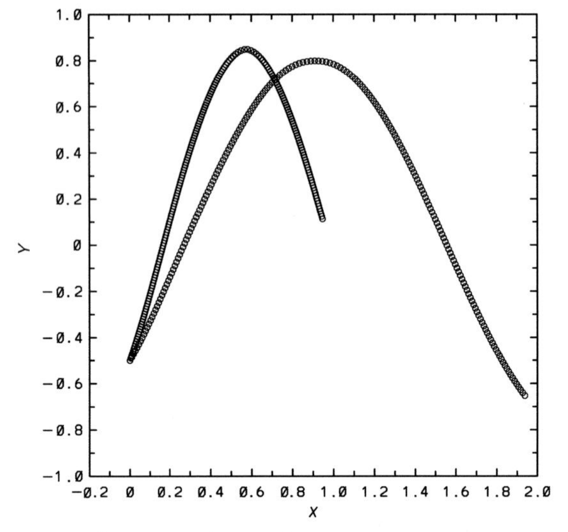

Fig. 9.5 Trajectories of two wave-packets $\left(\tilde{k} = -0.5, \tilde{k} = 0.75\right)$ with $\ell_o > 0$ on the (x,y) plane initially located at $(X, Y) = (0.0, -0.5)$.

when it reaches its turning latitude. The computer code has a provision to switch the sign of the meridional wavenumber when each wave-packet reaches a value of $\tilde{y}$ within a short distance from its $\tilde{y}^{**}$, namely $|\tilde{y}^{**} - \tilde{y}| \leq 0.2$. The result shows that the meridional movement of each wave-packet first slows down. It is subsequently reflected southward. The wave-packet with $\tilde{k} = 0.75$ propagates almost twice the distance to the east than the

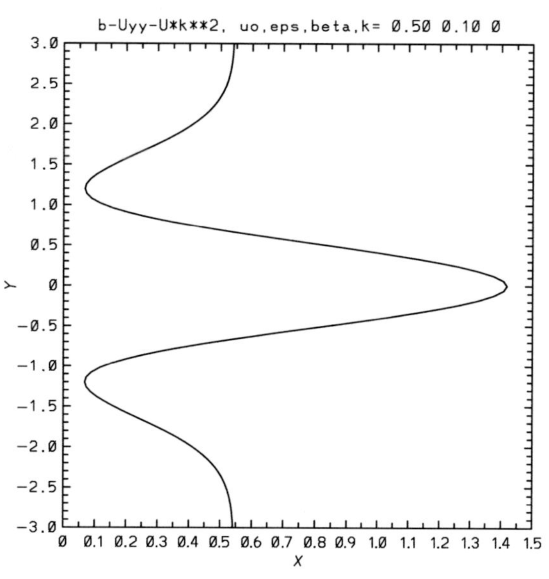

Fig. 9.6 Distribution of $(\tilde{\beta} - \tilde{U}_{yy} - \tilde{U}\tilde{k}^2)$ for $V = U_o/2$, $\varepsilon = 0.1$, $\tilde{k} = 0.5$ and $\tilde{\beta} = 0.53$.

wave-packet with $\tilde{k} = 0.5$ after $\tilde{t} = 30$. Each wave-packet will be reflected again when it reaches the other turning latitude $\tilde{y}^{**}$. It will propagate northward again afterward. Note that the critical latitudes, $\tilde{y}^* \approx \pm 1.5$, lie outside of the turning latitudes and therefore cannot be reached by this wave-packet.

There would be no turning latitudes for this wave-packet if the basic jet is substantially weaker, e.g. $V = U_o/2$, because $(\beta - U_{yy} - Uk^2)$ would be positive everywhere (Fig. 9.6). Then there would be no reflection of the wave-packet in such a case. On the other hand, there are still two critical latitudes since they do not depend on the intensity of the zonal flow. A wave-packet initially located at Y_o between -1.5 and 1.5 would be able to reach one of the critical latitudes, $\tilde{y}^* \approx \pm 1.5$.

We calculate the trajectories of three wave-packets $(\tilde{k} = 0.25, 0.5, 0.75)$ with $\ell_o < 0$ initially located at $(X, Y) = (0.0, 0.5)$ for comparison. The results are plotted in Fig. 9.7a for a time interval up to $\tilde{t} = 30$. A shorter wave propagates faster. Each wave-packet propagates in the southeast direction. As it approaches the critical latitude, its velocity decreases monotonically. Figure 9.7b shows that the normalized amplitude of each wave-packet first increases and eventually decreases towards zero as the wave-packet approaches a critical latitude. The maximum amplitudes of the $(\tilde{k} = 0.25, 0.5, 0.75)$ waves are 1.34, 1.36 and 1.42 respectively.

9.3.2 Vertical propagation of stationary wave

Vertical propagation of a planetary Rossby wave in a continuous vertical domain of a quasi-geostrophic model can be similarly analyzed. The fluid in this model is assumed to have a background density decreasing exponentially with height, $\rho_o = \rho_{oo} \exp(-z/2H)$, where H is the scale height. It suffices to consider a 2-D baroclinic stationary wave $(\partial/\partial t = 0,\ \partial/\partial y = 0)$ of small amplitude in the presence of a basic shear flow, $U(z)$.

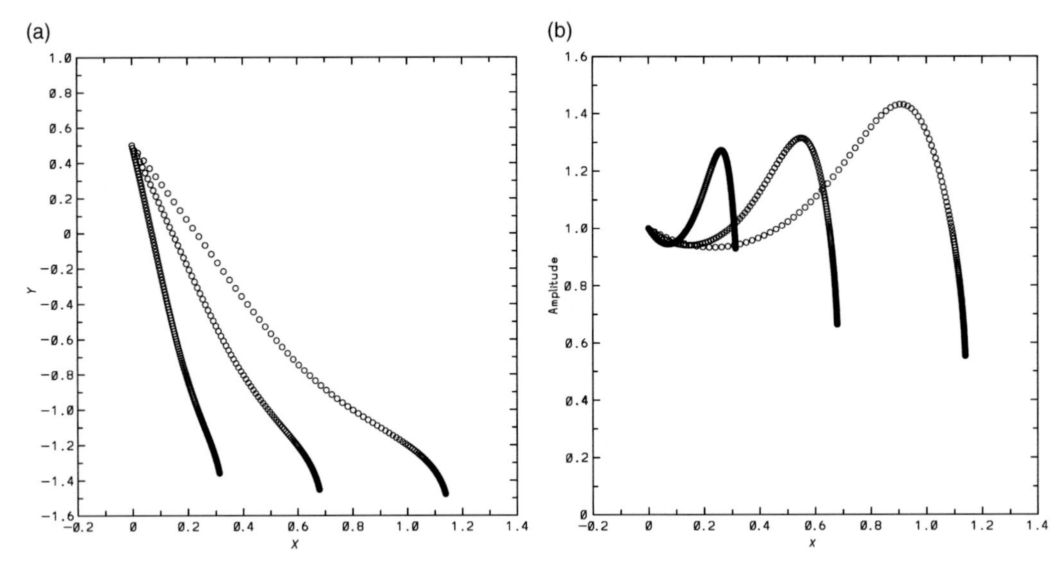

Fig. 9.7 (a) Trajectory of three wave-packets $(\tilde{k} = 0.25, 0.5, 0.75)$ with $\ell_o < 0$ for a time interval of $\tilde{t} = 30$. They are all initially located at $(X, Y) = (0.0, 0.5)$ and (b) the corresponding variations of the normalized amplitude of the wave-packets, $\Phi(t)/\Phi(o)$ according to (9.6). The values are plotted against the x-position of the wave-packets.

The governing equation is the steady linearized form of (6.28) in Section 6.5.3, which can be written in height coordinates as

$$Uq_x + \psi_x\left(\beta - \frac{f^2}{N^2}\left(\frac{\rho_o U_z}{N^2}\right)_z\right) = 0, \tag{9.7}$$

where $q = \psi_{xx} + \dfrac{f^2}{N^2}\left(\dfrac{\partial^2}{\partial z^2} - \dfrac{1}{4H^2}\right)\psi$ is the perturbation PV and N is the Brunt–Vaisala frequency. An elementary solution of (9.7) has the form of $\psi = \varphi(z)e^{ikx}$ with the amplitude function governed by

$$U\frac{d^2\varphi}{dz^2} + \frac{N^2}{f^2}\left(\beta - \frac{f^2}{\rho_o}\left(\frac{\rho_o U_z}{N^2}\right)_z - Uk^2 - \frac{Uf^2}{4H^2N^2}\right)\varphi = 0. \tag{9.8}$$

The solution of (9.8) is undulating (wave-like) in the vertical direction only in a region where $\infty > n^2 > 0$ for

$$n^2(z) = \frac{\left(\beta - \dfrac{f^2}{\rho_o}\left(\dfrac{\rho_o U_z}{N^2}\right)_z - Uk^2\right)\dfrac{N^2}{f^2} - \dfrac{U}{4H^2}}{U},$$

where n is the *refractive index* of this problem. Again we assume that the basic flow $U(z)$ varies smoothly in the z-direction so that there is a distinct scale separation between a wave and the basic flow. Then a modal wave solution has the form of $\varphi = \Phi(z)e^{imz}$, where m also varies in z. The dispersion relation of the wave according to (9.8) is

$$\sigma = 0 = Uk - \frac{Gk}{K_T^2}, \tag{9.9}$$

where $K_T^2 = k^2 + \frac{f^2}{N^2}\left(m^2 + \frac{1}{4H^2}\right)$, $G = \beta - \frac{f^2}{\rho_o}\left(\frac{\rho_o U_z}{N^2}\right)_z$. Now suppose there is a wave source at $Z = Z_o$ at $t = 0$. It is instructive to first examine the simplest case of a uniform basic flow, for which we have $G = \beta$. Equation (9.9) can be rewritten as

$$\frac{f^2}{N^2}m^2 = \frac{\beta}{U} - k^2 - \frac{f^2}{4N^2H^2}. \tag{9.10}$$

If the basic flow is from the east (easterly), the RHS of (9.10) would have a negative value. It follows that m would have an imaginary value, implying that the wave under consideration would not be vertically propagating. If the basic flow is a strong westerly (large positive value), the RHS would be dominated by the last two terms and hence also would have a negative value. There is again no vertical propagation of the Rossby wave. However, if the basic flow is a weak to moderate westerly, the first term on the RHS may be larger than the sum of the last two terms. Then the RHS would have a net positive value, implying that m would have a real value. Rossby waves would be able to propagate vertically in this case. Furthermore, for a given moderate westerly basic flow, the criterion for having a negative value of m^2 is easier to satisfy by smaller values of k^2. In other words, planetary-scale Rossby waves are easier to propagate from the troposphere into the stratosphere when the basic zonal flow is a moderate westerly as in early and late winter. This result was first deduced by Charney and Drazin (1961).

Now let us consider the problem of wave propagation through a vertically varying basic flow $U(z)$. The energy of this wave-packet propagates away from Z_o according to its group velocity, viz.

$$\begin{aligned} \frac{dX}{dt} &= \frac{2Gk^2}{K_T^4}, \\ \frac{dZ}{dt} &= \frac{2Gf^2km}{N^2K_T^4}. \end{aligned} \tag{9.11}$$

The initial conditions are: $X(0) = X_o$ and $Z(0) = Z_o$. The initial value of m must be $m(0) = \pm\sqrt{\frac{N^2}{f^2}\left(\left(\frac{G}{U}\right)|_{z=Z_o} - k^2\right) - \frac{1}{4H^2}}$. The variation of m in time is constrained by the dispersion relation, (9.9). The task is to determine how this wave would propagate away from Z_o. The trajectory of a particular wave-packet $X(t)$ and $Z(t)$ can be similarly evaluated for a given basic flow with a predictor-corrector scheme as we have done in the problem of horizontal propagation in the last section.

This analysis is also applicable to the case of a general basic flow with both horizontal and vertical shears, $U(y, z)$, provided the assumption of scale separation is justifiable. The corresponding refractive index would be a function of both latitude and elevation, $n(y, z)$. The propagation of a stationary planetary wave known at a particular latitude, Y_o, and height, Z_o, in the troposphere can be similarly determined as an initial value problem. We may expect it to penetrate into a stratospheric region bounded by a curve along which $n^2 = 0$ and another curve along which $n^2 \to \infty$. They may be referred to as *critical surface* and *turning surface* respectively.

It should be mentioned in passing that the refractive-index-squared has negative values in the stratosphere in summer because the basic zonal wind is easterly. Consequently, the stationary waves cannot propagate into the stratosphere in summer in contrast to what happens in winter.

9.4 Thermally forced stationary waves

In this section, we analyze excitation of disturbances to a steady thermal forcing under different conditions. The seasonal average thermal forcing is not spatially uniform. The part associated with precipitation is usually strongest in mid-troposphere. Therefore, it would be relevant to delineate the dynamics of a stationary planetary wave field excited by a generic localized forcing with such vertical distribution in a simplest possible model setting. We will only consider a weak to moderate forcing so that the response would be controlled by linear dynamics. Several physical factors are relevant. Apart from the heating and damping, those factors are (i) beta-effect, (ii) uniform component of a westerly basic zonal flow, (iii) basic baroclinic shear and (iv) basic horizontal shear. We will see that the impacts of these factors on a forced response differ not only quantitatively but also qualitatively. The objective is to develop a feel for the nature of the response. This will prepare us to better interpret the various features in a response under a general parametric condition.

9.4.1 Model formulation

A two-layer QG channel model is a simple useful tool for delineating the dynamics of planetary waves. The upper and lower levels of the model are indicated by indices $j = 1$ and 3 respectively. The interface and surface levels are indicated by $j = 2$ and 4 respectively. The domain is bounded by two rigid walls: $-X \leq x \leq X$, $-Y \leq y \leq Y$. A relatively weak and steady thermal forcing Q is introduced at $j = 2$. Dissipation is represented by Rayleigh friction, $\vec{F}_j = -\varepsilon_j \vec{V}_j$, at $j = 1, 3$ with ε_j being a damping coefficient in units of s^{-1}. We consider a basic zonal flow with the form $U_j = u_{jo} + \gamma_j y$. The linear horizontal shear would make it barotropically stable. A sufficiently weak zonal mean vertical shear would make it baroclinically stable, $(u_{1o} - u_{3o})\lambda^2 \beta^{-1} < 1$ where λ^{-1} is the radius of deformation defined by $\lambda^2 = \dfrac{f_o^2}{S_2(\delta p)^2}$, $S = \dfrac{R}{p}\left(\dfrac{RT_o}{c_p p} - \dfrac{dT_o}{dp}\right)$, $T_o(p)$ is the horizontally averaged background temperature and R the gas constant. The steady response is governed by the steady and linearized form of the QG potential vorticity equations with heating and damping, (Eq. (6.57a,b) in Section 6.7)

$$\begin{aligned}
U_1 q'_{1x} + \left(\beta + \lambda^2 (U_1 - U_3)\right)\psi'_{1x} &= -\frac{\lambda^2 RQ}{f_o} - \varepsilon_1 \nabla^2 \psi'_1, \\
U_3 q'_{3x} + \left(\beta - \lambda^2 (U_1 - U_3)\right)\psi'_{3x} &= \frac{\lambda^2 RQ}{f_o} - \varepsilon_3 \nabla^2 \psi'_3, \\
q'_1 &= \nabla^2 \psi'_1 - \lambda^2 (\psi'_1 - \psi'_3), \\
q'_3 &= \nabla^2 \psi'_3 + \lambda^2 (\psi'_1 - \psi'_3),
\end{aligned} \tag{9.12a,b}$$

where ψ'_j and q'_j are the perturbation QG streamfunction and PV respectively, $\nabla^2 = \partial^2/\partial x^2 + \partial^2/\partial y^2$. The subscript x refers to $\partial/\partial x$. Furthermore, the corresponding vertical motion field at the mid-level can be determined with the linearized form of (6.58) in Section 6.7 with heating and damping included, viz.

$$\begin{aligned}\frac{f_o}{\lambda^2 \Delta p}(\nabla^2 - 2\lambda^2)\omega_2 = &\{\beta(\psi'_{3x} - \psi'_{1x}) - U_1\nabla^2\psi'_{1x} + U_3\nabla^2\psi'_{3x}\} \\ &+ \left\{U_3\nabla^2\psi'_{1x} - U_1\nabla^2\psi'_{3x} + 2U_{3y}\psi'_{1xy} - 2U_{1y}\psi'_{3xy}\right\} \\ &+ \{\varepsilon_3\nabla^2\psi'_3 - \varepsilon_1\nabla^2\psi'_1\} - \frac{1}{\lambda^2}\nabla^2 G.\end{aligned} \tag{9.13}$$

We consider a large-scale localized thermal forcing with a two-dimensional Gaussian shape for simplicity,

$$G \equiv \frac{\lambda^2 R Q_2}{f_o} = G^* \exp\left(-\left(\frac{x - x_c}{a}\right)^2 - \left(\frac{y - y_c}{b}\right)^2\right). \tag{9.14}$$

Its central location is (x_c, y_c). Its size and shape are prescribed by a and b as the semi-major and semi-minor axes. Our calculations are made for a heating of circular configuration, $a = b$.

For easy reference, we briefly state the particular choice of non-dimensionalization. The thermal forcing is non-dimensionalized as $\tilde{G} = \frac{G}{G^*}$. All other quantities are non-dimensionalized in terms of a velocity scale V and a length scale λ^{-1}, namely $\tilde{\psi}'_j = \psi'_j/V\lambda^{-1}$, $\tilde{\omega}_2 = \omega_2/V\lambda\Delta p$, $\tilde{q}_j = q_j/V\lambda$, $\tilde{\beta} = \beta/V\lambda^2$, $\tilde{\varepsilon}_j = \varepsilon_j/V\lambda$, $\tilde{U}_j = U_j/V$, $\tilde{u}_{jo} = u_{jo}/V$, $\tilde{\gamma}_j = \gamma_j/V\lambda$, $(\tilde{k}, \tilde{\ell}) = \lambda^{-1}(k, \ell)$, $(\tilde{x}, \tilde{y}) = \lambda(x, y)$. The magnitude of the thermal forcing is prescribed by a non-dimensional parameter $N = G^*/(V\lambda)^2$. The non-dimensional governing equations for the unknowns ψ'_1 and ψ'_3 after dropping the "tilde superscript," are then

$$\begin{aligned}U_1 q'_{1x} + (\beta + (U_1 - U_3))\psi'_{1x} &= -NG - \varepsilon_1\nabla^2\psi'_1, \\ U_3 q'_{3x} + (\beta - (U_1 - U_3))\psi'_{3x} &= NG - \varepsilon_3\nabla^2\psi'_3, \\ \text{with } q'_1 = \nabla^2\psi'_1 - (\psi'_1 - \psi'_3)\ q'_3, &= \nabla^2\psi'_3 + (\psi'_1 - \psi'_3).\end{aligned} \tag{9.15a,b}$$

The boundary conditions are $\psi'_j(X) = \psi'_j(-X)$ and $\psi'_j(\pm Y) = 0$. The omega equation is

$$\begin{aligned}(\nabla^2 - 2\lambda^2)\omega_2 = Ro[&\{\beta(\psi'_{3x} - \psi'_{1x}) - U_1\nabla^2\psi'_{1x} + U_3\nabla^2\psi'_{3x}\} \\ &+ \left\{U_3\nabla^2\psi'_{1x} - U_1\nabla^2\psi'_{3x} + 2U_{3y}\psi'_{1xy} - 2U_{1y}\psi'_{3xy}\right\} \\ &+ \{\varepsilon_3\nabla^2\psi'_3 - \varepsilon_1\nabla^2\psi'_1\} - \nabla^2 G].\end{aligned} \tag{9.16}$$

Method of analysis

A semi-spectral method is used to solve (9.15a,b) for ψ'_j. By decomposing the x-variation of ψ'_j and G at each y into Fourier components, we would get from (9.15a,b) a set of spectral equations for the streamfunction fields of each zonal wavenumber k. The boundary conditions are $\hat{\psi}_j(\pm Y;k)=0$. The differential equations that govern each spectral component are next cast as a matrix equation with the use of center-difference representation of the y-derivatives. Each matrix equation is then solved separately. The inverse Fourier transform of $\hat{\psi}_j(y;k)$ would finally reconstitute the streamfunction fields in physical space, $\psi'_j(x,y)$.

If we fast forward to Section 9.4.7, we would find a fairly general thermally forced steady disturbance under a simple but meaningful parametric condition (Fig. 9.21). It still has a rather complicated structure. It would not be obvious at all as to how one might unambiguously interpret the dynamical nature of such a solution, even though there is unique causality in linear dynamics. The reason is that the solution is made up of an ensemble of wave components associated with a localized forcing. Each of those components is influenced differently by the several physical factors mentioned earlier. To prepare ourselves for making a clear interpretation, we first learn about how each physical factor or simple combinations of them would affect a corresponding forced response. This we will do in the next few subsections. By the time we reach Section 9.4.7, we will be in a good position to interpret without ambiguity whatever numerical solution we might happen to obtain.

9.4.2 Simplest case

The simplest case is one in which there is a thermal forcing and damping with neither a basic flow nor the beta-effect ($U_j=0, \beta=0$). We are effectively considering a hurricane-like disturbance arising from a given thermal forcing. Strictly speaking, a QG model should not be used for a relatively small disturbance. The Equations (9.15a,b) are reduced in this case to

$$\begin{aligned} 0 &= -NG - \varepsilon_1 \nabla^2 \psi'_1, \\ 0 &= NG - \varepsilon_3 \nabla^2 \psi'_3. \end{aligned} \tag{9.17a,b}$$

The solution is symbolically $\psi_3 = \dfrac{N}{\varepsilon_3}\nabla^{-2}G$ and $\psi_1 = \dfrac{-N}{\varepsilon_1}\nabla^{-2}G$. It follows that a circular structure of G would excite an almost circular structure as long as the boundaries have minimal influence. It is obvious that if we were to use an equal value for the damping coefficients at both levels, then we would have $\psi'_1 = -\psi'_3$, meaning that the response would have a strictly baroclinic structure. It follows that there would be a relatively small frictionally induced barotropic part as defined by $\psi_T = (\psi'_1 + \psi'_3)$ when the damping coefficient at the lower level is greater than that at the upper level. The magnitude of ψ'_1 should be stronger than that of ψ'_3. Hence, the response must be *mostly a baroclinic vortex*. From a physical point of view, the flow at the lower level should be cyclonic and the flow at the upper level should be anticyclonic because such heating distribution would generate PV in the lower layer $(q_1 = \nabla^2\psi_1 - \lambda^2(\psi_1 - \psi_3))$ and reduce PV in the upper layer $(q_3 = \nabla^2\psi_3 + \lambda^2(\psi_1 - \psi_3))$. The presence of lateral boundaries does complicate the matter a little bit. One could employ the method of images to obtain an analytic solution that satisfies such lateral boundary conditions. It would be however most convenient to numerically determine such an impact in an illustrative calculation.

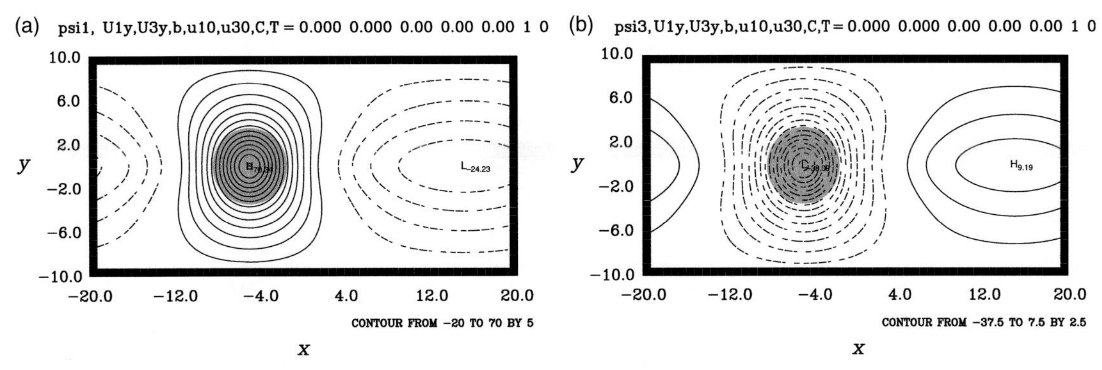

Fig. 9.8 Thermally excited (a) ψ_1 and (b) ψ_3 in units of $V\lambda^{-1}$ for the case of $Ro = 1.0$, $N = 1.2$, $\beta = 0$, $U_j = 0$, $\varepsilon_1 = 0.05$ and $\varepsilon_3 = 0.1$. The area of forcing is indicated by shading.

We consider the following set of parameters: $\lambda^{-1} = 200\,\text{km}$, $a = b = 2\lambda^{-1}$(radius of forcing ~400 km), $x_c = -5\lambda^{-1}$, $y_c = 0$, $Q = 2\times10^{-5}\,\text{K s}^{-1}$ (heating rate ~2K/day), $\Delta p = 4\times10^4\,\text{N m}^{-2}$, $V = 10\,\text{m s}^{-1}$, $(\varepsilon_1)_{\text{dim}} = 2.5\times10^{-6}\,\text{s}^{-1}$ (damping time ~4 days), $(\varepsilon_3)_{\text{dim}} = 5\times10^{-6}\,\text{s}^{-1}$ (damping time ~2 days), $f_o = 5\times10^{-5}\,\text{s}^{-1}$. Then the non-dimensional parameters are $Ro = \dfrac{V}{f_o\lambda^{-1}} = 1.0$, $N = \dfrac{RQ}{f_oV^2} = 1.2$, $\varepsilon_1 = 0.05$ and $\varepsilon_3 = 0.1$. A channel domain of 8000 km long and 4000 km wide is designated as $-20 \le x \le 20$ and $-10 \le y \le 10$; 401 grid points are used in the x-direction for performing Fourier transforms; 201 grid points are used to depict the structure in the y-direction.

The solution of (9.17a,b), ψ_j, indeed verifies that the response is a baroclinic vortical circulation with a cyclonic vortex at the lower level and a stronger anticyclonic vortex at the upper level (Fig. 9.8). The forced vortex is quite strong with extreme vorticity of $\sim5\times10^{-4}\,\text{s}^{-1}$ at the lower level and $-11\times10^{-4}\,\text{s}^{-1}$ at the upper level. The condition of no flow across the lateral boundaries gives rise to a weak but noticeable impact. It gives rise to a small departure from circular symmetry in the main response and a weaker secondary response with opposite rotation away from the region of forcing.

The omega equation (9.16) is reduced to

$$\left(\nabla^2 - 2\lambda^2\right)\omega_2 = Ro\left[\left\{\varepsilon_3\nabla^2\psi_3' - \varepsilon_1\nabla^2\psi_1'\right\} - \nabla^2 G\right]. \tag{9.18}$$

The solution of (9.18) reveals that this disturbance has ascending vertical motion over the area of heating as one would intuitively expect (Fig. 9.9). There is compensating weak descending motion in the rest of the domain. It is found that $(-\omega_2)_{\max}V\lambda\Delta p \sim 2.4\,\text{N m}^{-2}\,\text{s}^{-1}$ (corresponding to $w_2 \sim 0.48\,\text{m s}^{-1}$).

According to (9.18), the vertical motion field is made up of two components; one associated with differential damping and the other with the thermal forcing itself. Specifically, the heating induces ascending motion in the inner region of the heating and descending motion in the outer region. In contrast, there is frictionally induced ascending motion over the whole region of heating. It is noteworthy that the frictionally induced component is two and a half times stronger than the thermally induced component (0.85 vs. 0.33, Fig. 9.10).

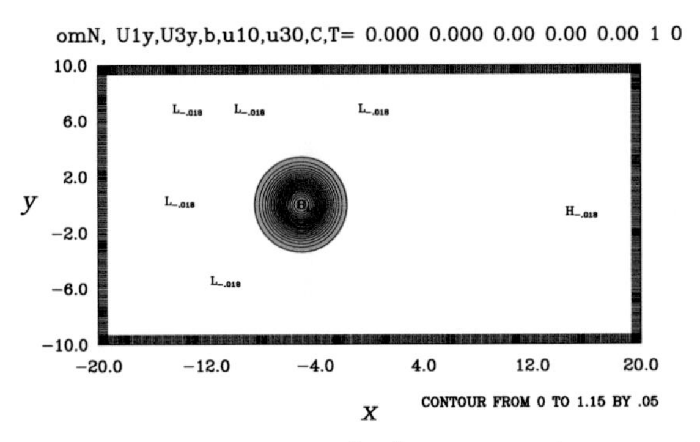

Fig. 9.9 Thermally excited $(-\omega_2)$ in units of $V\lambda\Delta p = 2\,\mathrm{N\,m^{-2}\,s^{-1}}$ for the case of $Ro = 1.0$, $N = 1.2$, $\beta = 0$, $U_j = 0.0$, $\varepsilon_1 = 0.05$ and $\varepsilon_3 = 0.1$. The area of forcing is indicated by shading.

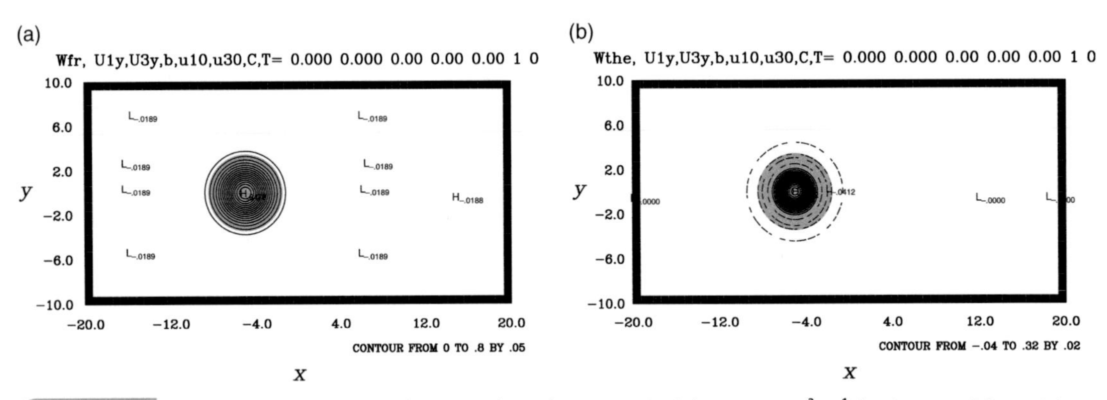

Fig. 9.10 Thermally excited (a) $(-\omega_{fr})$ and (b) $(-\omega_Q)$ in units of $V\lambda\Delta p = 2\,\mathrm{N\,m^{-2}\,s^{-1}}$ for the case of $Ro = 1.0$, $N = 1.2$, $\beta = 0$, $U_j = 0.0$, $\varepsilon_1 = 0.05$ and $\varepsilon_3 = 0.1$. The area of forcing is indicated by shading.

9.4.3 Influence of the beta-effect alone

When the beta-effect is additionally included, the governing equations would be

$$\begin{aligned} \beta\psi'_{1x} &= -NG - \varepsilon_1\nabla^2\psi'_1, \\ \beta\psi'_{3x} &= NG - \varepsilon_3\nabla^2\psi'_3. \end{aligned} \tag{9.19a,b}$$

From these equations we see that the beta-effect would be negligible for a small disturbance. Let us consider a large-scale thermal forcing with a radius equal to $2\lambda^{-1}$ where $\lambda^{-1} = 0.8 \times 10^6$ m. All other parameters have the same values as in Section 9.4.2. Then the values of the non-dimensional parameters would be $Ro = 0.25$, $N = 1.2$, $\beta = 1.0$, $\varepsilon_1 = 0.2$, $\varepsilon_3 = 0.4$. The response still has a strong baroclinic component. The beta-effect strongly influences the response in two ways. It limits the meridional scale of the response and effects a selection of zonal scales in favor of the longer scales (compare Fig. 9.11 with Fig. 9.8). The baroclinic vortical circulation is also shifted westward so that there is a southerly (northerly) flow at the lower (upper) level over the area of thermal forcing.

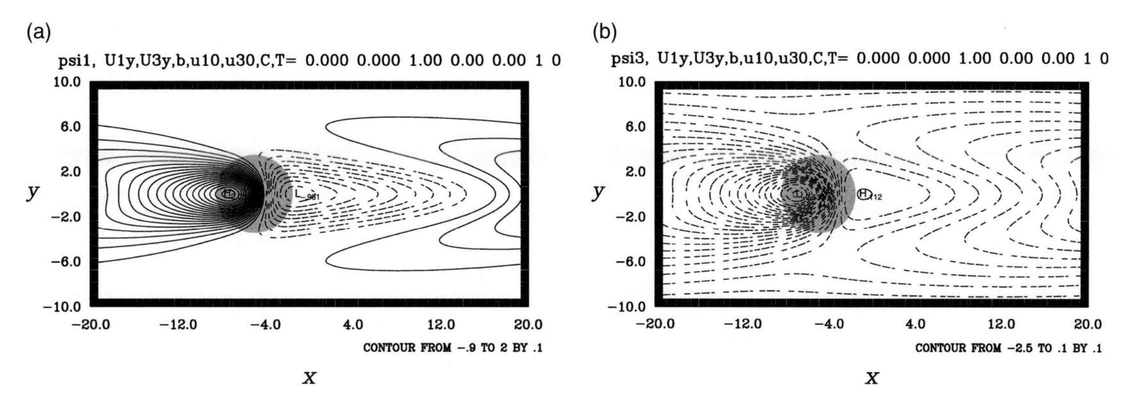

Fig. 9.11 Thermally excited (a) ψ_1 and (b) ψ_3 in units of $V\lambda^{-1}$ for the case of $N = 1.2$, $\beta = 1.0$, $U_j = 0$, $\varepsilon_1 = 0.2$ and $\varepsilon_3 = 0.4$. The area of forcing is indicated by shading.

The dynamical nature of this zonal asymmetry in the response is easy to understand. The tendency of decreasing PV at the upper level over the area of thermal forcing $(-NG < 0)$ must be largely balanced by a positive advection of planetary vorticity due to a northerly flow $(-\beta\psi_{1x} > 0)$. The opposite takes place in the lower layer. In the upstream of the thermal forcing, there is a northerly at the lower level so that the tendency of decreasing PV $(-\beta\psi_{3x} > 0)$ would be balanced by damping $\left(-\varepsilon_3\nabla^2\psi_3 < 0\right)$. The opposite occurs in the upper layer. The response to the east of the thermal forcing can be dynamically interpreted in the same way. Note that the non-dimensional intensity of the response is one order of magnitude weaker than that without the beta-effect. The maximum vorticity of the lower level cyclonic flow is $\sim 1.5 \times 10^{-5}\,\mathrm{s}^{-1}$ and the minimum vorticity of the upper level anticyclonic flow is $\sim -1.7 \times 10^{-5}\,\mathrm{s}^{-1}$.

The omega equation (9.16) in this case becomes

$$\left(\nabla^2 - 2\right)\omega_2 = Ro\left[\left\{\beta\left(\psi'_{3x} - \psi'_{1x}\right)\right\} + \left\{\varepsilon_3\nabla^2\psi'_3 - \varepsilon_1\nabla^2\psi'_1\right\} - \nabla^2 G\right]. \tag{9.20}$$

The total vertical motion is $(-\omega_2)$. The part induced by $\left\{\beta\left(\psi'_{3x} - \psi'_{1x}\right)\right\}$ arises from differential vorticity advection and is referred to as $(-\omega_{dva})$. The part induced by $\left\{\left(\varepsilon_3\nabla^2\psi'_3 - \varepsilon_1\nabla^2\psi'_1\right)\right\}$ stems from differential friction and is referred to as $\left(-\omega_{fr}\right)$. Their results are shown in Fig. 9.12. The maximum ascent is $(-\omega_2)_{\max} V\lambda\Delta p \sim 0.15\,\mathrm{N\,m^{-2}\,s^{-1}}$ (corresponding to $(w_2)_{\max} \sim 0.03\,\mathrm{m\,s^{-1}}$). The differential vorticity advection induces ascending motion over most of the area of heating because $\left\{\beta\left(\psi'_{3x} - \psi'_{1x}\right)\right\} > 0$, whereas the differential friction induces ascending motion over the western part of the area of the heating because $\left\{\left(\varepsilon_3\nabla^2\psi'_3 - \varepsilon_1\nabla^2\psi'_1\right)\right\} > 0$. In addition, weak descent is induced to the northwest and southwest of the heating. It is also interesting to note that the two processes cancel one another outside the area of thermal forcing.

9.4.4 Effect of a constant basic westerly flow alone

We next isolate the impact of a uniform moderate basic westerly flow by using $U_1 = U_3 \equiv U = 0.4$ without the beta-effect. All other parameters have the same

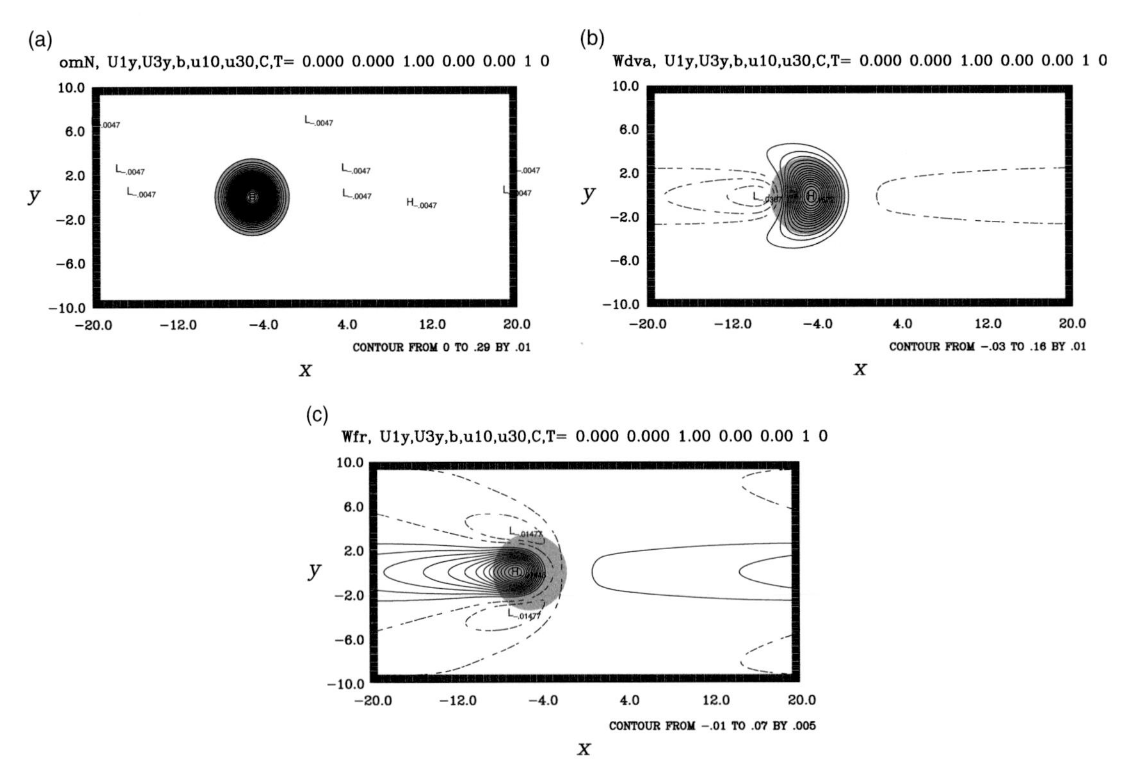

Fig. 9.12 Thermally excited (a) $-\omega_2$, (b) $-\omega_{dva}$ and (c) $-\omega_{fr}$ in units of $V\lambda\Delta p = 0.5\,\mathrm{N\,m^{-2}\,s^{-1}}$ for the case of $Ro = 0.25$, $N = 1.2$, $\beta = 1.0$, $U_j = 0$, $\varepsilon_1 = 0.2$ and $\varepsilon_3 = 0.4$. The area of forcing is indicated by shading.

values as in Section 9.4.3. The governing equations (9.15a,b) would be reduced in this case to

$$\begin{aligned} Uq'_{1x} &= -NG - \varepsilon_1 \nabla^2 \psi'_1, \\ Uq'_{3x} &= NG - \varepsilon_3 \nabla^2 \psi'_3. \end{aligned} \tag{9.21a,b}$$

Such a basic flow merely advects the forced disturbance downstream. One might expect that the heating term NG at level-3 is primarily balanced by the advection term $\left(-Uq'_{3x}\right)$ over the area of thermal forcing. It would follow that large positive value of q'_3 should be to the east of the forcing. Likewise, the large negative value of q'_1 should be also to the east of the forcing. Thus, the baroclinical vortical circulation should be displaced to the east of the thermal forcing in this case. This expectation is confirmed by the results of ψ_j shown in Fig. 9.13. The vortical circulation is again limited to the latitude belt of the heating as in under the influence of the beta-effect alone. It is greatly elongated to the east so that the damping term, $-\varepsilon_j \nabla^2 \psi'_j$, would balance against the advection of PV, $(-U_j q'_{jx})$ outside the area of forcing. The intensity of the disturbance is about the same as in the case with the beta-effect alone. *What we have learned from this illustrative analysis is that while the beta-effect displaces the vortical circulation to the* ***west****, a uniform westerly basic flow displaces the vortical circulation to the* ***east*** *of the thermal forcing.*

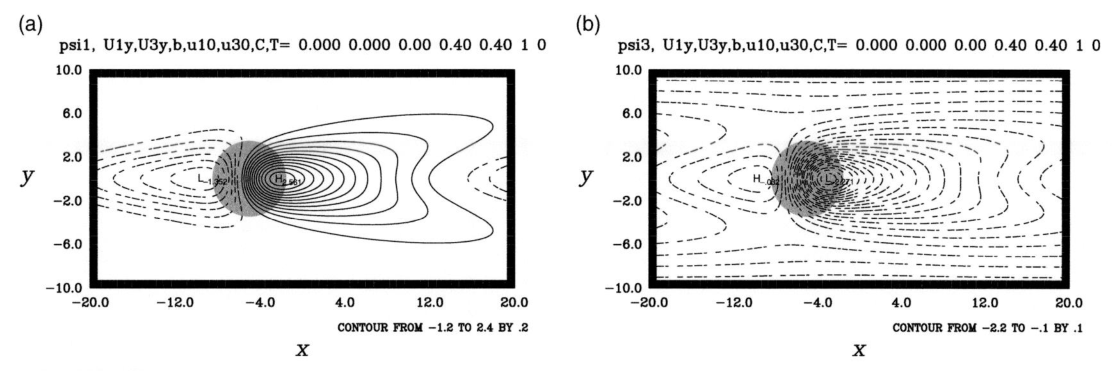

Fig. 9.13 Thermally excited (a) ψ_1 and (b) ψ_3 in units of $V\lambda^{-1}$ for the case of $Ro = 0.25$, $N = 1.2$, $\beta = 0$, $U_j = 0.4$, $\varepsilon_1 = 0.2$ and $\varepsilon_3 = 0.4$. The basic flow is a uniform westerly. The area of forcing is indicated by shading.

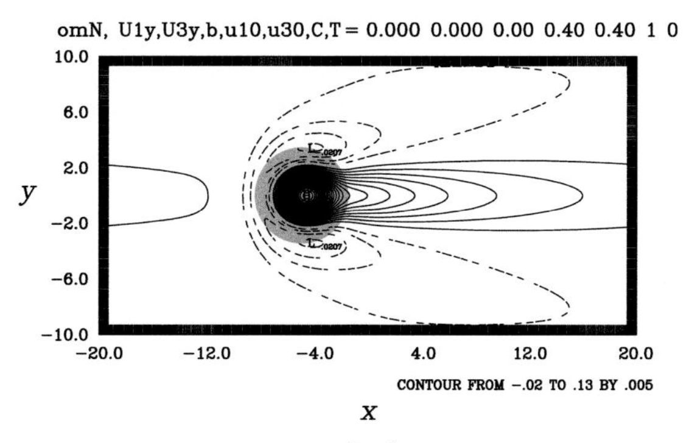

Fig. 9.14 Thermally excited $-\omega_2$ in units of $V\lambda\Delta p = 0.5\,\mathrm{N\,m^{-2}\,s^{-1}}$ for the case of $Ro = 0.25$, $N = 1.2$, $\beta = 0$, $U_j = 0.4$, $\varepsilon_1 = 0.2$ and $\varepsilon_3 = 0.4$. The basic flow is a uniform westerly. The area of forcing is indicated by shading.

It should be emphasized that the response with either the beta-effect alone or a uniform basic flow alone has a vortical structure but not a wavy structure.

It is harder to anticipate the vertical motion field a priori. It turns out to be slightly displaced to the east with a broad distribution in the downstream region. There are in addition two branches of weak descending motion to the north and south of the heating area (Fig. 9.14). The maximum ascent is $\sim 0.066\,\mathrm{N\,m^{-2}\,s^{-1}}$, which is weaker by about a factor of two compared to the case with the beta-effect alone.

9.4.5 Influence of baroclinic shear in a basic flow with the beta-effect

Atmospheric zonal flows typically have a vertical shear. Hence, we next focus on the role of a basic baroclinic flow in the excitation of stationary planetary waves in conjunction with the beta-effect. We prescribe a weak vertical shear, $U_1 = 0.4$, $U_3 = 0.1$. We also reduce the damping coefficients by a factor of four to $\varepsilon_1 = 0.05$ and $\varepsilon_3 = 0.1$ in order to allow more pronounced wavy characteristics. The corresponding damping time scale is 8 days at the lower level and 16 days at the upper level. All other parameters are the same

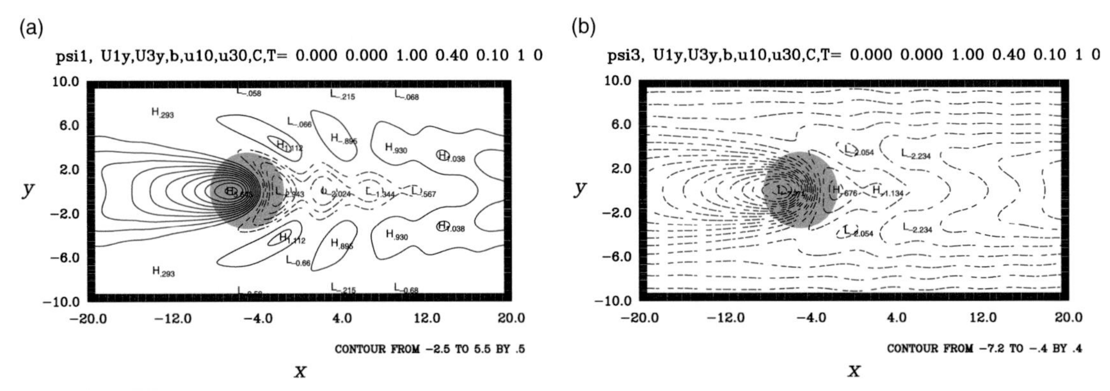

Fig. 9.15 Thermally excited (a) ψ_1 and (b) ψ_3 in units of $V\lambda^{-1}$ for the case of $Ro = 0.25$, $\beta = 1$, $U_1 = 0.4$, $U_3 = 0.1$, $\varepsilon_1 = 0.05$, $\varepsilon_3 = 0.1$. The basic flow only has vertical shear. The area of forcing is indicated by shading.

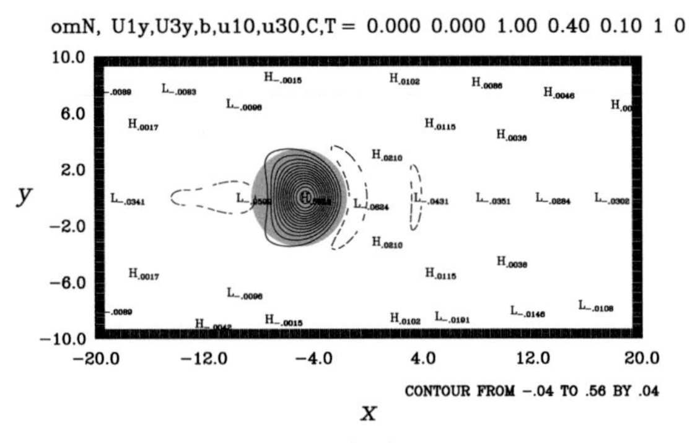

Fig. 9.16 Thermally excited $-\omega_2$ in units of $V\lambda\Delta p = 0.5\,\text{N}\,\text{m}^{-2}\text{s}^{-1}$ for the case of $Ro = 0.25$, $N = 1.2$, $\beta = 1$, $U_1 = 0.4$, $U_3 = 0.1$, $\varepsilon_1 = 0.05$, $\varepsilon_3 = 0.1$. The basic flow only has vertical shear. The area of forcing is indicated by shading.

as in Section 9.4.3. The response in ψ_j (Fig. 9.15) should be compared with Fig. 9.11. The strongest feature is a vortical circulation slightly shifted to the west of the area of thermal forcing. It suggests that the beta-effect dominates over the influence of advection by this zonal flow. This is attributable to the fact that the advection term depends inversely upon the third power of the scale of the flow, whereas the beta-effect depends only inversely on the scale itself. The maximum vorticity is $\sim 4.3 \times 10^{-5}\,\text{s}^{-1}$ at the lower level and $-6 \times 10^{-5}\,\text{s}^{-1}$at the upper level. There are in addition two symmetric wavy packets in ψ_1' about the center of the thermal forcing in the downstream region. This additional wavy feature clearly is a manifestation of forced stationary Rossby waves under the joint influences of the beta-effect and a basic westerly flow. The meridional symmetry of the response stemming from the intrinsic symmetry of the parameters remains intact.

The vertical motion field is shown in Fig. 9.16. Ascent occurs over the area of heating as one would expect. The maximum ascent in this case is $(-\omega_2)_{\max} V\lambda\Delta p \sim 0.29\,\text{m}\,\text{s}^{-1}$. It is interesting to find an area of weak descending motion induced upstream of the heating area. It is associated with the process of differential vorticity advection in a close

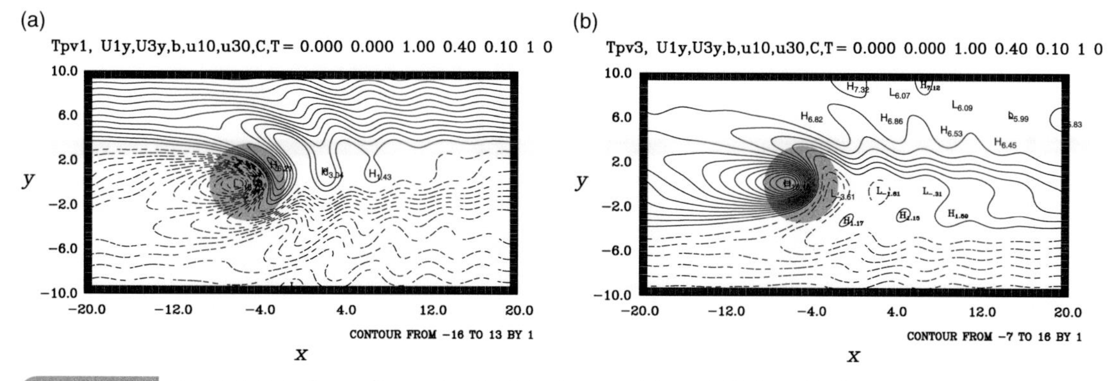

Fig. 9.17 Total PV fields, $(\bar{q}_j + q_j')$, in units of $V\lambda$ for the case of $Ro = 0.25$, $N = 1.2$, $\beta = 1$, $U_1 = 0.4$, $U_3 = 0.1$, $\varepsilon_1 = 0.05$, and $\varepsilon_3 = 0.1$. The basic flow only has vertical shear. The area of forcing is indicated by shading.

examination of the terms. The baroclinic Rossby wave-packet downstream also has a corresponding vertical motion field.

The total PV fields (Fig. 9.17) further highlight the presence of two major components in the forced flow field. The western component is a highly elongated and skewed baroclinic vortical circulation, with positive PV anomaly at the lower level and negative PV anomaly at the upper level. The eastern component of the disturbance field is made up of baroclinic Rossby wave-packets. There is strong advection of negative PV anomaly from the southeast of the thermal forcing at the lower level, and strong advection of positive PV anomaly from the northwest of the thermal forcing at the upper level. The center of the vortical circulation and the root of the wave-packet are collocated since they are excited by a common thermal forcing.

9.4.6 Influence of barotropic shear in a basic flow with the beta-effect

Next, we focus on the influence of barotropic shear in a basic flow on the forced response. The related parameters are $u_{jo} = 0.4$ and $\gamma_j = 0.04$. All other parameters are the same as in the last section. A barotropic shear disrupts the meridional symmetry in the system. The basic zonal flow is zero at the southern boundary and is 0.8 at the northern boundary. The basic vorticity is therefore anticyclonic. The response in ψ_j is shown in Fig. 9.18. We again find two components in the forced response: a baroclinic vortical circulation tilted in the SW-NE direction and a baroclinic wave-packet. They are almost imperceptibly connected, but the discussion in the last two sections reminds us that they are two distinct components. The maximum cyclonic vorticity at the lower level is $\sim 8 \times 10^{-5}\,\text{s}^{-1}$ and the maximum anticyclonic vorticity at the upper level is $\sim -8.7 \times 10^{-5}\,\text{s}^{-1}$. The anticyclonic basic shear is a favorable condition for the baroclinic Rossby waves to propagate to the southeast. The wave-packet is progressively attenuated as it propagates towards the southern boundary, which is a critical latitude. This is consistent with the notion discussed in Section 9.3.1 that stationary waves tend to be absorbed by the basic flow before they get to a critical latitude.

The barotropic shear has a pronounced impact on the vertical motion as well (Fig. 9.19). The descending motion upstream of the thermal forcing is even more pronounced relative

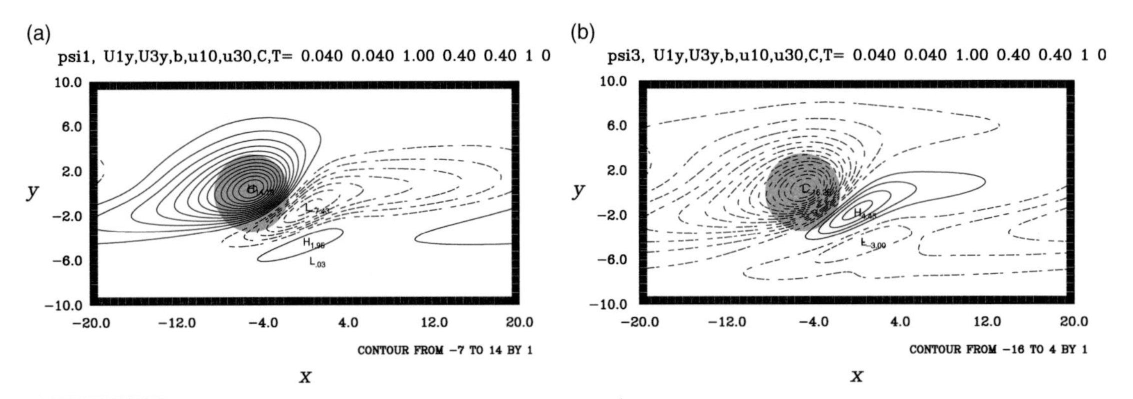

Fig. 9.18 Thermally excited (a) ψ_1 and (b) ψ_3 in units of $V\lambda^{-1}$ for the case of $Ro = 0.25$, $N = 1.2$, $\beta = 1$, $u_{jo} = 0.4$, $\gamma_j = 0.04$, $\varepsilon_1 = 0.05$, $\varepsilon_3 = 0.1$. The basic flow only has horizontal shear. The area of forcing is indicated by shading.

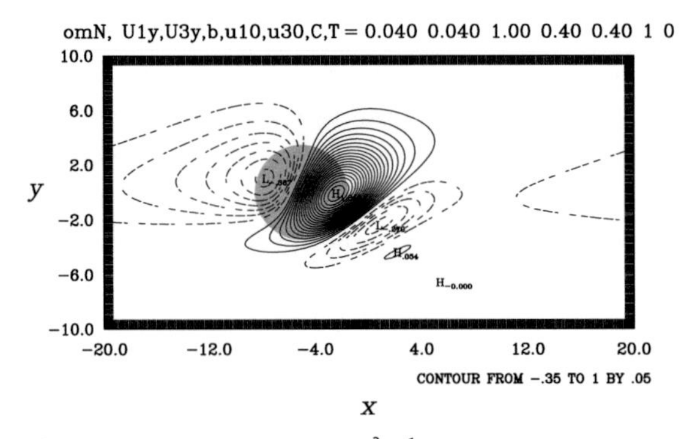

Fig. 9.19 Thermally excited $(-\omega_2)$ in units of $V\lambda\Delta p = 0.5\,\mathrm{N\,m^{-2}s^{-1}}$ for the case of $Ro = 0.25$, $\beta = 1$, $u_{jo} = 0.4$, $\gamma_j = 0.04$, $\varepsilon_1 = 0.05$, $\varepsilon_3 = 0.1$. The basic flow only has horizontal shear. The area of forcing is indicated by shading.

to the ascending motion. Ascent occurs over roughly the eastern half of the area of thermal forcing. The baroclinic Rossby wave-packet has a distinct wavy distribution of ascent/descent to the southeast of the area of thermal forcing. The maximum ascent in this case is $(-\omega_2)_{\max} V\lambda\Delta p \sim 0.25\,\mathrm{N\,m^{-2}s^{-1}}$.

The forced response leaves a strong signature in the total PV fields (Fig. 9.20).

9.4.7 Dynamical nature of a general thermally forced steady disturbance

We are now in a position to interpret a steady planetary wave field under the joint influence of the beta-effect, differential damping and a basic zonal flow with both vertical and horizontal shears. The parameters have the following values: $Ro = 0.25$, $N = 1.2$, $\beta = 1.0$, $u_{1o} = 0.4$, $u_{3o} = 0.1$, $\gamma_j = 0.04$, $\varepsilon_1 = 0.05$ and $\varepsilon_3 = 0.1$. This simply amounts to adding a baroclinic shear to the condition considered in Section 9.4.6.

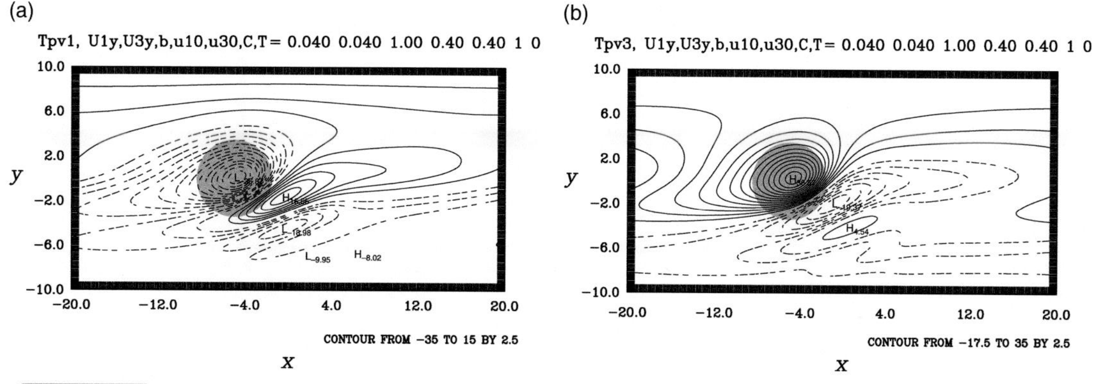

Fig. 9.20 Total PV fields, $(\bar{q}_j + q_j')$, in units of $V\lambda$ for the case of $Ro = 0.25$, $\beta = 1$, $u_{jo} = 0.4$, $\gamma_j = 0.04$, $\varepsilon_1 = 0.05$, $\varepsilon_3 = 0.1$. The basic flow only has horizontal shear. The area of forcing is indicated by shading.

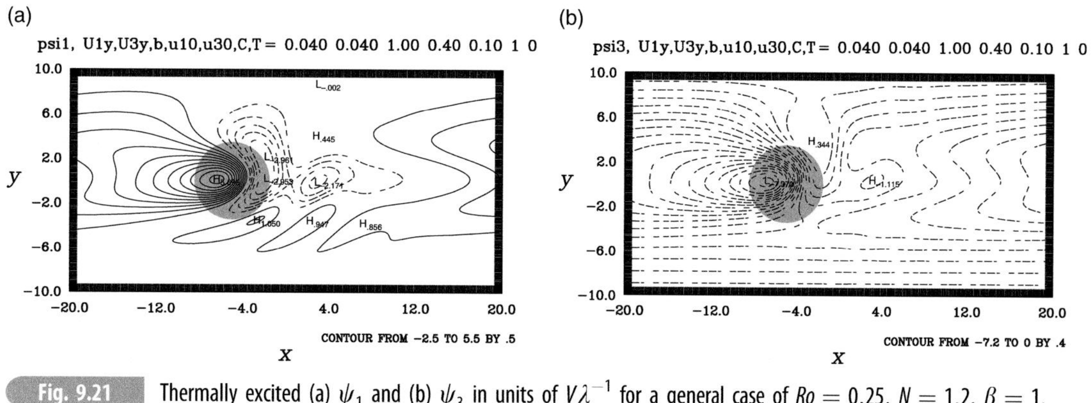

Fig. 9.21 Thermally excited (a) ψ_1 and (b) ψ_3 in units of $V\lambda^{-1}$ for a general case of $Ro = 0.25$, $N = 1.2$, $\beta = 1$, $u_{1o} = 0.4$, $u_{3o} = 0.1$, $\gamma_j = 0.04$, $\varepsilon_1 = 0.05$, $\varepsilon_3 = 0.1$. The area of forcing is indicated by shading.

The results of ψ_j in Fig. 9.21 show a greatly elongated and shifted baroclinic vortical circulation extending to far upstream of the heating area. The maximum cyclonic vorticity at the lower level is $4.5 \times 10^{-5}\ \mathrm{s}^{-1}$, whereas the maximum anticyclonic vorticity at the upper level is $-6.2 \times 10^{-5}\ \mathrm{s}^{-1}$. There are two distinct wave-packets at the upper level propagating to the east of the thermal forcing. One wave-packet is located to the north of the other. The northern wave-packet has a baroclinic structure, whereas the southern wave-packet has a barotropic structure confined at the upper level.

The baroclinic vortical circulation is primarily a response to the thermal forcing under the influence of the beta-effect as fully discussed in Section 9.4.3. The characteristics of the wavy part of the response can be interpreted as follows. First of all, the thermal forcing in conjunction with a basic westerly flow and the beta-effect jointly excite Rossby wave disturbances. It is noteworthy that the southern wave-packet is virtually absent at the lower level. This feature is attributable to the presence of a critical latitude at the lower level, where the zonal basic flow vanishes $U_3(y^*) = 0$ where $y^* = -u_{3o}/\gamma_3 = -2.5$ In contrast, the

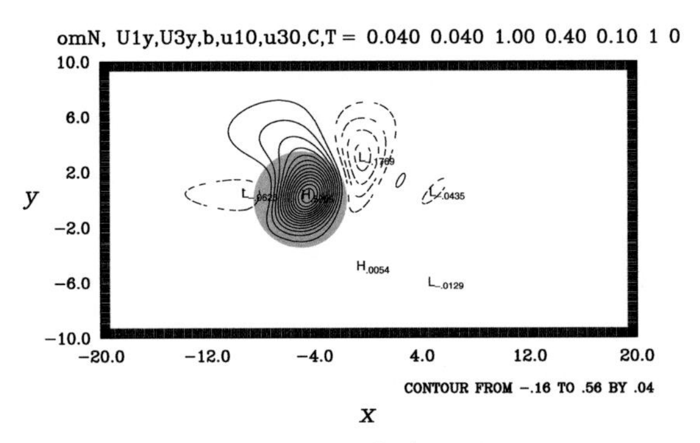

Fig. 9.22 Thermally excited $(-\omega_2)$ in units of $V\lambda\Delta p = 0.5\,\mathrm{N\,m^{-2}s^{-1}}$ for a general case of $Ro = 0.25$, $N = 1.2$, $\beta = 1$, $u_{1o} = 0.4$, $u_{3o} = 0.1$, $\gamma_j = 0.04$, $\varepsilon_1 = 0.05$, $\varepsilon_3 = 0.1$. The area of forcing is indicated by shading.

southern wavy structure of ψ_1' can penetrate far into the southern domain because the critical latitude at the upper level is still at $y = -10.0$.

The northern baroclinic wave-packet first propagates northeastward to about $y = 7.0$ in the downstream region. It then appears to turn to southeastward. We may interpret this feature as a reflection of this wave-packet at a *turning latitude*, y^{**}, where $\beta - U_{yy} - U\hat{k}^2 = 0$. The basic flow under consideration has $U_{yy} = 0$ everywhere. For a barotropic wave, we would use $\hat{k} = k$. For a baroclinic wave, we would use $\hat{k}^2 = k^2 + m^2 + f_o^2/N^2 4H^2$ where m is the vertical wavenumber, N the Brunt–Vaisala frequency and H the scale height of the background density. The dominant wavenumber of the wave-packet is estimated to be $k \approx (2\pi)/(2a)$, then the turning latitude would be $y^{**} \approx 7.5$ without contribution from the second and third terms. This estimate is compatible with the feature of the response under consideration.

The corresponding vertical motion field is shown in Fig. 9.22. The main feature is the ascent over the region of thermal forcing. The maximum ascent is $\sim 0.29\,\mathrm{N\,m^{-2}\,s^{-1}}$. There is an identifiable region of weak descent to the west of the forcing. These two features are associated with the baroclinic vortical circulation discussed in Sections 9.4.5 and 9.4.6. The third pronounced feature is a wavy distribution of alternating ascent and descent downstream in the northern half of the domain. This is clearly an intrinsic property of the baroclinic Rossby wave-packet discussed earlier. As expected, there is no vertical motion associated with the southern wave-packet immediately next to the thermal forcing in the southeast. Therefore, the main features are now well understandable.

The corresponding total PV anomaly fields (Fig. 9.23) are also quite intelligible in light of the discussions above. The positive PV anomaly at the lower level and negative PV anomaly at the upper level over a large part of the area of thermal forcing are associated with the vortical circulation mostly arising from the heating and the beta-effect. The southern wave-packet at the upper level is the signature of a forced barotropic wave-packet. There is no corresponding signal at the lower level. The northern wavy structure is the signature of a forced baroclinic wave-packet.

This concludes our discussion of the dynamical nature of thermally forced stationary planetary waves in the context of a generic model.

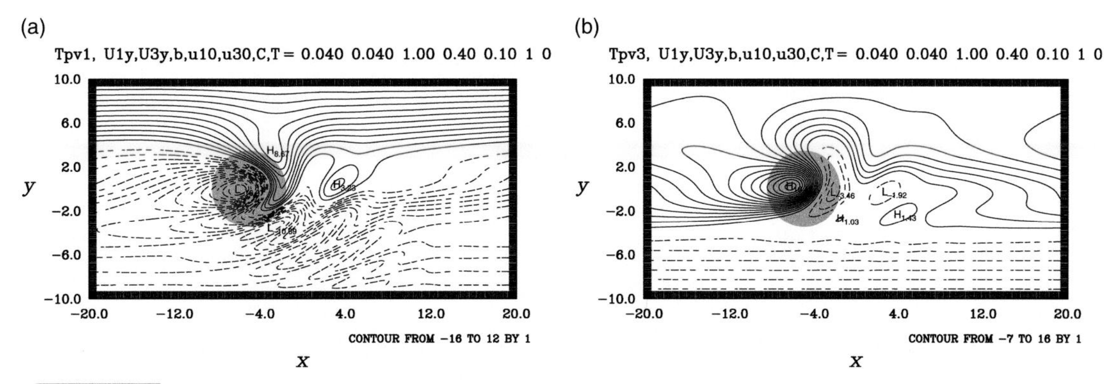

Fig. 9.23 Total PV anomaly fields of the two layers, $(\bar{q}_j + q_j')$, in units of $V\lambda$ for a general case of $Ro = 0.25$, $N = 1.2$, $\beta = 1$, $u_{1o} = 0.4$, $u_{3o} = 0.1$, $\gamma_j = 0.04$, $\varepsilon_1 = 0.05$, $\varepsilon_3 = 0.1$. The area of forcing is indicated by shading.

9.5 Orographically forced stationary waves

We now turn our attention to stationary planetary waves that are excited by a steady zonal flow going over a large-scale localized orography, $h(x, y)$ in a simple baroclinic model. This is an extension of the discussion about the dynamics of orographic waves in a shallow-water model in Section 5.8. A surface flow naturally induces a certain vertical motion field over the orography. It is through such vertical motion that the interior fluid feels the impact of the orography below. The horizontal flow would be partly affected by the mechanism of vortex stretching.

9.5.1 Model formulation

We use *a two-layer quasi-geostrophic model* in an illustrative study of the dynamics of orographically forced stationary planetary waves. The influence of orography is incorporated through the surface vertical velocity w_4, which is related to the horizontal velocity and the orography $h(x, y)$ by the kinematic condition, $w_4 = \left(u_4 h_x + v_4 h_y\right)$; ω_4 is related to w_4 through the hydrostatic balance, $\omega_4 = -g\rho_s w_4$. It appears in the QG vorticity equation at the lower level, $(\nabla^2\psi_3)_t + J(\psi_3, \nabla^2\psi_3 + \beta y) = f_o(\omega_4 - \omega_2)/\Delta p$ and in the thermodynamic equation at the surface, $(\psi_4 - \psi_3)_t + J(\psi_4, \psi_3) = -(S_4 \Delta p/2f_o)\omega_4$. Furthermore, we may use the linearized form, $\omega_4 = -g\rho_s U_4 h_x$, for a sufficiently low topography with a zonal surface flow U_4.

The stable stratification of the basic state S is again introduced as a parameter, the radius of deformation, $\lambda^{-1} = \sqrt{S}\Delta p/f_o$. All other quantities are again non-dimensionalized in terms of a certain velocity scale V and λ^{-1}, namely $\tilde{\psi}_j' = \psi_j'/V\lambda^{-1}$, $\tilde{\omega}_2 = \omega_2/V\lambda\Delta p$, $\tilde{q}_j = q_j/V\lambda$, $\tilde{\beta} = \beta/V\lambda^2$, $\tilde{\varepsilon}_j = \varepsilon_j/V\lambda$, $\tilde{\tau} = V\lambda\tau$, $\tilde{U}_j = U_j/V$, $\tilde{u}_{jo} = u_{jo}/V$, $\tilde{\gamma}_j = \gamma_j/V\lambda$, $(\tilde{k}, \tilde{\ell}) = \lambda^{-1}(k, \ell)$, $(\tilde{x}, \tilde{y}) = \lambda(x, y)$. The height of orography is prescribed in terms of a non-dimensional parameter, $M = \dfrac{g f_o \rho_s}{(V\lambda)\Delta p} h_*$, where h_* is the maximum height. The orography is non-dimensionalized as $\tilde{h} = h/h_*$. The governing equations are then the PV equations for the two layers and the thermodynamic equation at the surface. The non-dimensional linearized governing equations for the three unknowns ψ'_1, ψ'_3 and ψ'_4, after dropping the "tilde superscript," are then

$$\begin{aligned} U_1 q'_{1x} + (\beta + (U_1 - U_3))\psi'_{1x} &= -\varepsilon_1 \nabla^2 \psi'_1, \\ U_3 q'_{3x} + (\beta - (U_1 - U_3))\psi'_{3x} &= -MU_4 h_x - \varepsilon_3 \nabla^2 \psi'_3, \\ U_3 \psi'_{4x} - U_4 \psi'_{3x} &= \frac{1}{2} MU_4 h_x. \end{aligned} \tag{9.22a,b,c}$$

Equation (9.22a,b) can be solved for ψ'_1 and ψ'_3 with the same algorithm applied in Section 9.4; ψ'_4 in turn can be determined with (9.22c). In other words, ψ'_1 and ψ'_3 can be determined without knowing ψ'_4, but ψ'_4 cannot be determined without first knowing ψ'_3.

We specifically consider an orography with a Gaussian shape in the illustrative calculations, namely

$$h(x,y) = \exp\left(-\left(\frac{x - x_c}{a}\right)^2 - \left(\frac{y - y_c}{b}\right)^2\right). \tag{9.23}$$

For $h_* = 1.0$ km, the forcing parameter is then $M = 5.0$. The location, size and shape of the orography are prescribed by $(x_c, y_c) = (-5, 0)$, $a = b = 2$. All other parameters have the same values here as in the analysis of thermally driven stationary planetary waves except of course no heating is included here. We will again first delineate the influences of five physical factors separately: (i) a uniform basic flow alone, (ii) the beta-effect with a westerly basic flow, (iii) the beta-effect with an easterly basic flow, (iv) a basic flow with a baroclinic shear in a westerly basic flow and (v) a basic flow with horizontal shear in a westerly basic flow.

9.5.2 Effect of a uniform westerly basic flow alone

The simplest generic orographic disturbance is one that is excited by a uniform westerly flow in the absence of the beta-effect. The parameters used in the calculation are $M = 5.0$, $U_j = 0.4$, $\beta = 0$, $Ro = 0.25$, $\varepsilon_1 = 0.05$, $\varepsilon_3 = 0.1$. The solutions of ψ_1 and ψ_4 are shown in Fig. 9.24. The structure of ψ_3 is not shown since it is in between those of ψ_1 and ψ_4.

The main feature of this orographic disturbance is an anticyclonic circulation which is slightly skewed and elongated to the west of the orography. Unlike the counterpart thermally excited disturbance, it has an *equivalent barotropic structure*, meaning that it has a small vertical tilt (compare Fig. 9.11 and Fig. 9.24). Such a structure may be mathematically understood as follows. In the event that the damping is weak, (9.22a,b) can be rewritten in this case with $U_j \equiv U$ as

$$\begin{aligned} U\nabla^2(\psi'_1 + \psi'_3)_x &\approx -MUh_x, \\ U(\nabla^2 - 2)(\psi'_1 - \psi'_3)_x &\approx MUh_x. \end{aligned} \tag{9.24a,b}$$

Then the solution would be symbolically $(\psi'_1 + \psi'_3) \approx -M\nabla^{-2}h$ and $(\psi'_1 - \psi'_3) \approx M(\nabla^2 - 2)^{-1}h$. For a large orography, we have $\nabla^{-2} \gg (\nabla^2 - 2)^{-1}$. It

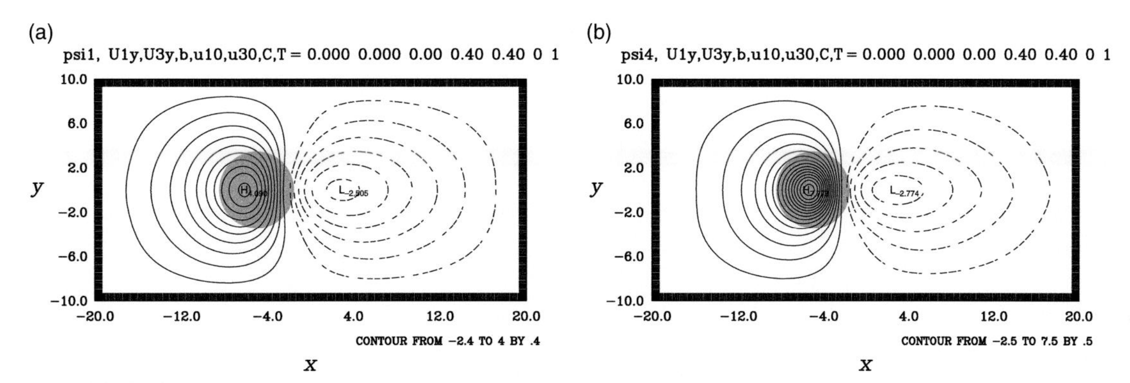

Fig. 9.24 Orographically excited (a) ψ'_1 and (b) ψ'_4 in units of $V\lambda^{-1}$ for the case of a uniform westerly flow with no beta-effect: $Ro = 0.25$, $\varepsilon_1 = 0.05$, $\varepsilon_3 = 0.1$, $\beta = 0$, $U_j = 0.4$, $M = 5$. The orography is indicated by shading.

follows that the barotropic part of the response, $(\psi'_1 + \psi'_3)$, should be considerably stronger than the baroclinic part, $(\psi'_1 - \psi'_3)$.

Since the basic flow is a westerly, the basic streamfunction decreases northward. The total streamfunction contours would bend northward near the orography from the west forming a ridge at the northern tip of the orography and then go southward forming a trough in the immediate downstream region. Such an overall pattern is therefore qualitatively compatible with a typical flow over a large-scale mountain range. In this limited sense, we can say that even this elemental version of the model captures an essential aspect of orographically forced disturbance. Furthermore, since ψ'_4 is proportional to the surface pressure anomaly, it means that the surface pressure is somewhat higher on the west side than on the east side of an orography. There must then be a transfer of zonal momentum from the fluid into the surface. This characteristic is what we have called a *form drag*.

The dynamical nature of the forced disturbance can be further understood in terms of its vorticity and PV fields, $\zeta'_j = \nabla^2\psi'_j$ and q'_j, shown in Fig. 9.25. To begin with, the ζ'_j fields can be inferred from the ψ'_j fields even by inspection. The fact that ψ'_3 is considerably more intense than ψ'_1 would suggest that $q'_3 = \zeta'_3 - \psi'_3 + \psi'_1$ and ζ'_3 should have a similar structure and q'_1 and ζ'_1 should have qualitatively different configurations. This is indeed what we see in Fig. 9.25. We find that the main PV balance at the lower level over the orography is between the advection of the perturbation PV, $-Uq'_{3x}$, and the orographic forcing, $-MU_4h_x$. On the other hand, the balance is between the advection of PV and the damping, $-\varepsilon_3\nabla^2\psi'_3$, downstream of the orography. In order to have a balance between the advection of PV, $-Uq'_{1x}$ and damping $-\varepsilon_1\nabla^2\psi'_1$ everywhere at the upper level, relatively large values of q'_1 have to be located on the eastern part of the orography extending into the near downstream region. The configuration of forced disturbance that we have seen in ψ'_j simply reflects such necessary configurations of q'_j in Fig. 9.25a and b. It is noted in passing that the extreme values of ζ'_1 and ζ'_3 are -1.24 and -2.44 in units of $V\lambda = 1.25 \times 10^{-5}\,\mathrm{s}^{-1}$ respectively; those of q'_1 and q'_3 are 0.42 and -3.90 respectively. The disturbance is meridionally symmetric about $y = 0$ since all parameters have such symmetry in the system.

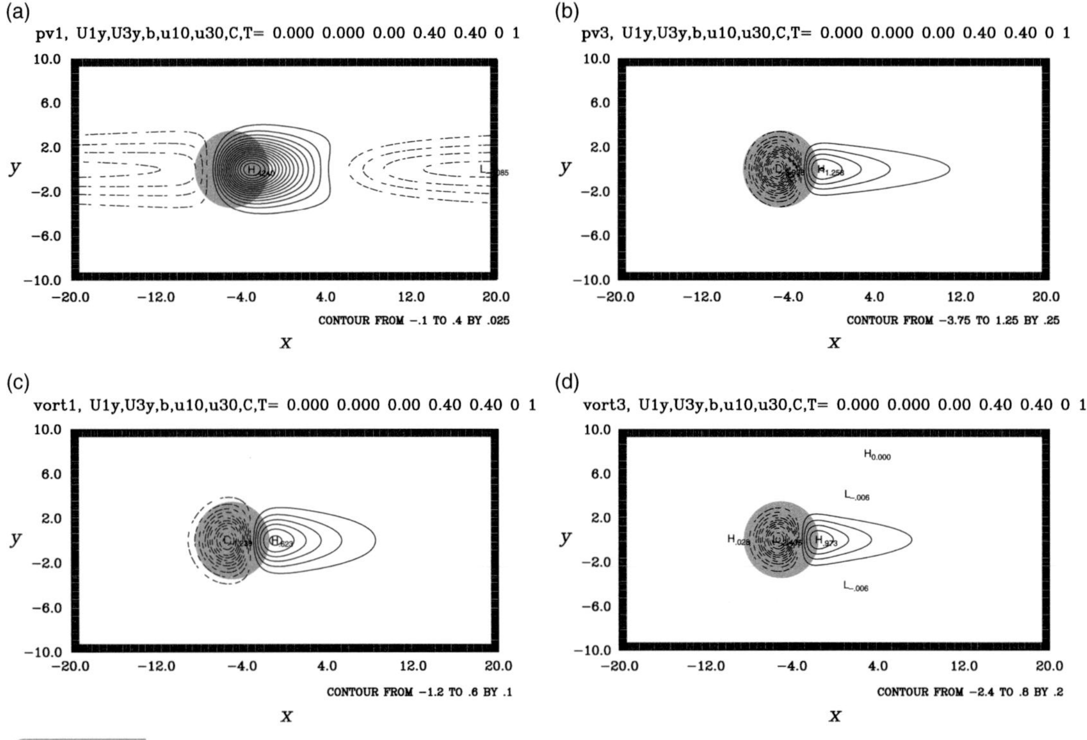

Fig. 9.25 Distributions of (a) q_1', (b) q_3', (c) ζ_1' and (d) ζ_3' in units of $V\lambda$ for the case of a uniform westerly flow with no beta-effect: $Ro = 0.25$, $\varepsilon_1 = 0.05$, $\varepsilon_3 = 0.1$, $\beta = 0$, $U_j = 0.4$, $M = 5$. The orography is indicated by shading.

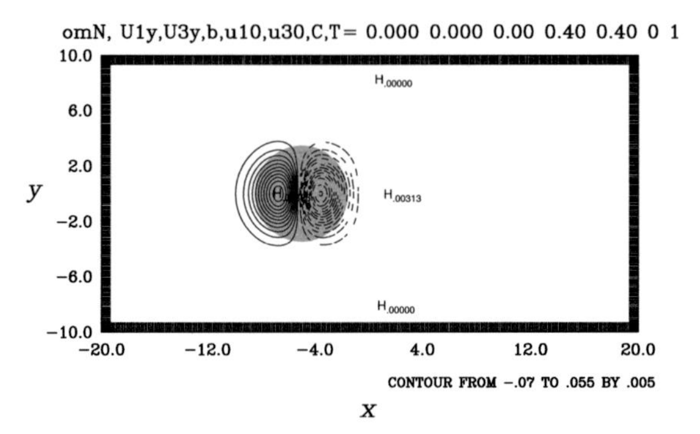

Fig. 9.26 Orographically excited $(-\omega_2)$ in units of $V\lambda\Delta p = 0.5\,\mathrm{N\,m^{-2}s^{-1}}$ for the case of a uniform westerly flow with no beta-effect: $Ro = 0.25$, $\varepsilon_1 = 0.05$, $\varepsilon_3 = 0.1$, $\beta = 0$, $U_j = 0.4$, $M = 5$. The orography is indicated by shading.

The corresponding vertical motion field $(-\omega_2)$ has a dipole structure with ascending (descending) motion on the western (eastern) half of the mountain (Fig. 9.26). This is to be expected since the surface flow must ascend on the upslope side and vice versa. The maximum upward motion is only about $\sim 0.029\,\mathrm{N\,m^{-2}\,s^{-1}}$. It is also easy to see that the vertical motion and the temperature anomaly as measured by $(\psi_1' - \psi_4')$ are negatively correlated. This implies cold air rising and warm air sinking in the disturbance, which is

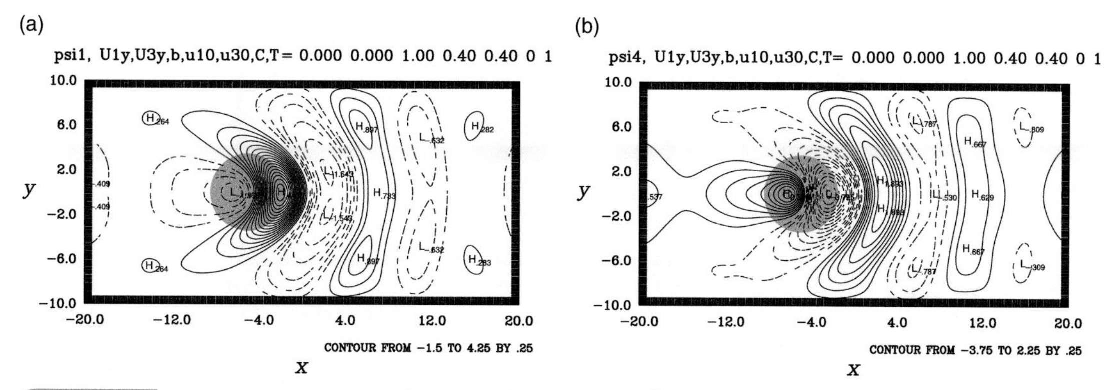

Fig. 9.27 Orographically excited (a) ψ_1' and (b) ψ_4' in units of $V\lambda^{-1}$ for the case of a uniform westerly flow with the beta-effect: $\beta = 1$, $Ro = 0.25$, $\varepsilon_1 = 0.05$, $\varepsilon_3 = 0.1$, $U_j = 0.4$, $M = 5$. The orography is indicated by shading.

therefore baroclinically transferring energy to the basic state. This characteristic highlights the fact that the disturbance is mechanically driven.

9.5.3 Effect of a uniform basic flow with the beta-effect

We next delineate the impact of the beta-effect on orographically forced stationary waves. This analysis is the counterpart of the analysis for a barotropic model setting discussed in Section 5.5. We now use $\beta = 1.0$ for this calculation and leave all other parameters unchanged. The results of ψ_1' and ψ_4' are shown in Fig. 9.27. The intensity of the disturbance at the two levels is comparable. The beta-effect induces two distinct Rossby wave-packets emanating from the orographic forcing. They are symmetric about the center line of the orography; one propagating downstream in the northern half of the domain and the other in the southern half. This is to be expected on the basis of what we have learned about the role of the beta-effect in Rossby wave dynamics in Chapter 5. It is also noteworthy that the dominant length scale of the excited Rossby waves is comparable to the size of the orography itself. Thus, the beta-effect has a unique influence on the scale selection. Furthermore, this orographic disturbance has a baroclinic structure because ψ_1' and ψ_4' are virtually 180° out of phase. This structural characteristic in the forced response is qualitatively different from that in the absence of the beta-effect. The damping progressively attenuates the two wave-packets in the downstream direction.

The counterpart of (9.24a,b) for this case is

$$\begin{aligned}\left(U\nabla^2 + \beta\right)\left(\psi_1' + \psi_3'\right)_x &\approx -MUh_x,\\ \left(U\left(\nabla^2 - 2\right) + \beta\right)\left(\psi_1' - \psi_3'\right)_x &\approx MUh_x.\end{aligned} \qquad (9.25\text{a,b})$$

We see that the beta-effect gives rise to scale selection in the response. The wave component associated with the smallest value of $(U(\nabla^2 - 2) + \beta)$ would be most intense in the response. This is the underlying reason for the mostly baroclinic structure and for a dominant scale corresponding to $(\nabla^2 - 2) \sim -\beta/U$ seen in Fig. 9.27.

The total streamfunction field at the surface $\left(\bar{\psi}_4 + \psi_4'\right)$ shows a deep trough on the eastern side of the orography extending to the downstream region (Fig. 9.28). This flow is

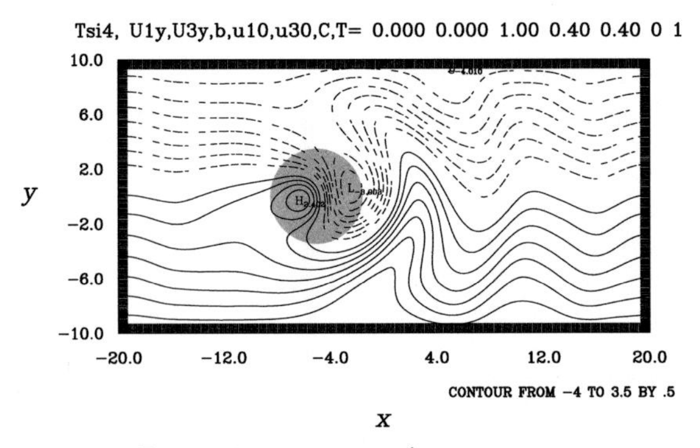

Fig. 9.28 Total streamfunction at the surface $(\bar{\psi}_4 + \psi_4')$ in units of $V\lambda^{-1}$ for the case of a uniform westerly flow with the beta-effect: $\beta = 1$, $Ro = 0.25$, $\varepsilon_1 = 0.05$, $\varepsilon_3 = 0.1$, $U_j = 0.4$, $M = 5$. The orography is indicated by shading.

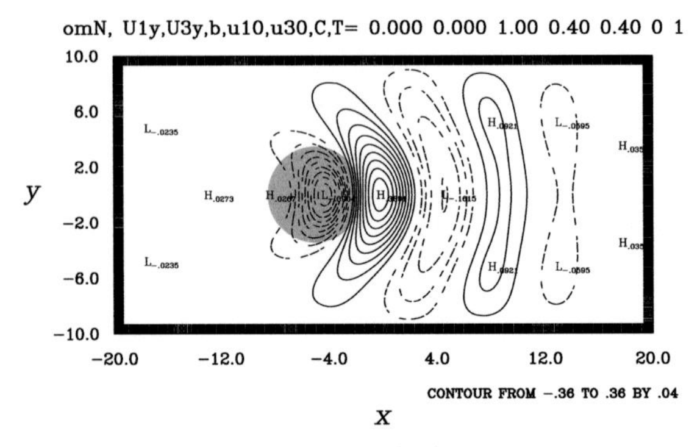

Fig. 9.29 Orographically excited $(-\omega_2)$ in units of $V\lambda\Delta p = 0.5\,\mathrm{N\,m^{-2}s^{-1}}$ for the case of a uniform westerly flow with the beta-effect: $\beta = 1$, $Ro = 0.25$, $\varepsilon_1 = 0.05$, $\varepsilon_3 = 0.1$, $U_j = 0.4$, $M = 5$. The orography is indicated by shading.

quite reminiscent of observed flows in the atmosphere, suggesting that the beta-effect is an important factor in shaping the orographic stationary planetary waves in the atmosphere.

A baroclinic Rossby wave-packet has a corresponding wavy structure in the vertical motion field. Compared to Fig. 9.24, we see that the beta-effect suppresses the ascending motion over the western half of the mountain and induces significant ascending motion downstream of the mountain (Fig. 9.29). The maximum ascent and descent is about $0.18\,\mathrm{N\,m^{-2}s^{-1}}$ which is about four times stronger.

It has been verified with this model that there would be no wavy response if the surface basic flow were from the east as we have found in the analysis discussed in Section 9.3.2. The dynamical reason is the same. In the interest of brevity, the counterpart results for this case are not presented.

9.5.4 Influence of baroclinic shear in a basic westerly flow with the beta-effect

We proceed to delineate the influence of baroclinic shear in a westerly basic flow on orographically forced planetary waves in conjunction with the beta-effect. We change the parameters for the basic flow to $U_1 = 0.8, U_3 = 0.6, U_4 = 0.4$ and leave all other

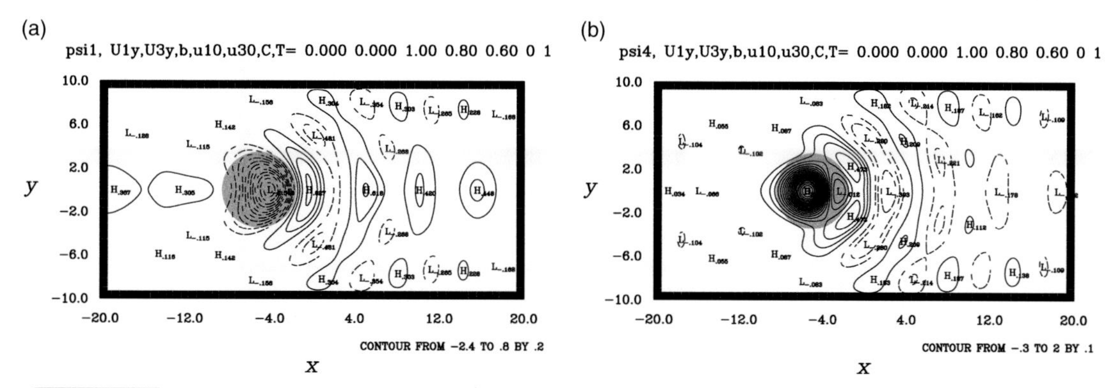

Fig. 9.30 Orographically excited (a) ψ_1' and (b) ψ_4' in units of $V\lambda^{-1}$ for the case of a baroclinic shear in a westerly flow with the beta-effect: $\beta = 1$, $U_1 = 0.8$, $U_3 = 0.6$, $U_4 = 0.4$, $Ro = 0.25$, $\varepsilon_1 = 0.05$, $\varepsilon_3 = 0.1$, $M = 5$. The orography is indicated by shading.

parameters unchanged. The results of ψ_1' and ψ_4' are shown in Fig. 9.30. We see from (9.22a,b) that a westerly vertical shear has the effect of enhancing the beta-effect at the upper level and reducing the beta-effect at the lower level. It follows that Fig. 9.30b should be similar to Fig. 9.24b (the case of no beta-effect), whereas Fig. 9.30a should be more similar to Fig. 9.27a (the case with the beta-effect). Compared with Fig. 9.27, we conclude that this baroclinic shear has three main impacts:

(a) The intensity of the wave field in the downstream region is much weaker (even though the surface basic flow is the same as in the case of no vertical shear).
(b) The length scale of the waves in the downstream region is significantly shorter.
(c) There is virtually no westward displacement of the forced wave pattern at the surface relative to the mountain. We find mixed characteristics in the forced disturbance due to the presence of a baroclinic shear, e.g. a 180° vertical phase difference in the disturbance over the mountain but small phase difference in the wave disturbance downstream. In other words, the barotropic component of the response is stronger in this case.

The total streamfunction field $\left(\bar{\psi}_4 + \psi_4'\right)$ is correspondingly quite different from that without baroclinic shear (Fig. 9.31). The wave field is so much weaker that it is barely visible in the total surface flow field. A baroclinic shear is therefore not a favorable factor for orographic Rossby waves.

The baroclinic shear also has a great impact on the vertical motion field (Fig. 9.32). In contrast to Fig. 9.29, we find ascending motion over the western half of the mountain extending to a short distance upstream, and descending motion over the eastern half of the mountain. The intensity is only about one-third of that without baroclinic shear.

9.5.5 Influence of barotropic shear in a basic westerly flow with the beta-effect

Finally, we examine the influence of barotropic shear in a basic flow in conjunction with the beta-effect on orographically excited stationary waves. The basic flow is the same as the one used in Section 9.4.6, prescribed by $u_{jo} = 0.4, \gamma_j = 0.04$. It has an anticyclonic

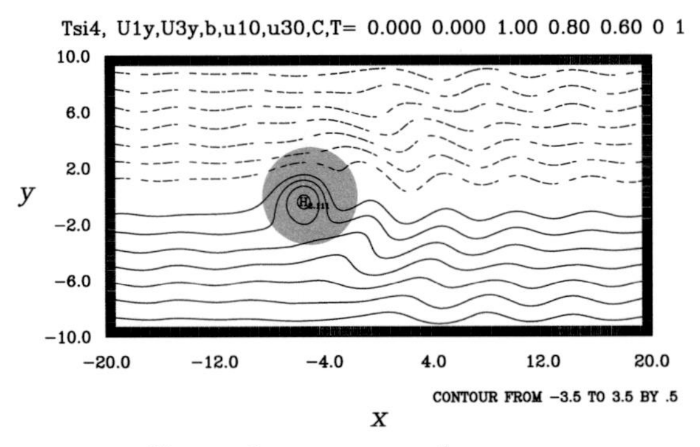

Fig. 9.31 Total streamfunction at the surface $(\bar{\psi}_4 + \psi'_4)$ in units of $V\lambda^{-1}$ for the case of a baroclinic shear in a westerly flow with the beta-effect: $\beta = 1$, $U_1 = 0.8$, $U_3 = 0.6$, $U_4 = 0.4$, $Ro = 0.25$, $\varepsilon_1 = 0.05$, $\varepsilon_3 = 0.1$, $M = 5$. The orography is indicated by shading.

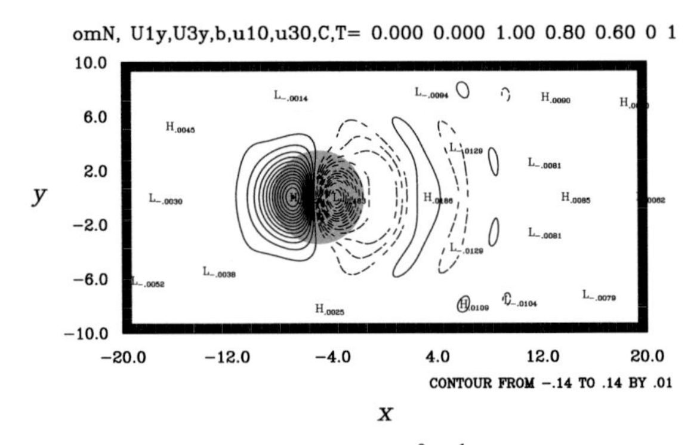

Fig. 9.32 Orographically excited $(-\omega_2)$ in units of $V\lambda\Delta p = 0.5\,\text{N}\,\text{m}^{-2}\,\text{s}^{-1}$ for the case of a baroclinic shear in a westerly flow with the beta-effect: $\beta = 1$, $U_1 = 0.8$, $U_3 = 0.6$, $U_4 = 0.4$, $Ro = 0.25$, $\varepsilon_1 = 0.05$, $\varepsilon_3 = 0.1$, $M = 5$. The orography is indicated by shading.

shear and hence breaks the meridional symmetry of the system. All other parameters have the same values as those used in Section 9.5.4. The critical latitude has no impact on the response because it is located at the southern boundary at the lower level in this parameter setting. The results of ψ'_1 and ψ'_4 show the structure of a baroclinic wave-packet propagating towards the southeast (Fig. 9.33). Therefore, southward propagation of Rossby waves is favored by an anticyclonic shear in the basic flow. In particular, the wave-packet begins with an anticyclonic disturbance in ψ'_4 and a cyclonic disturbance in ψ'_1 over the western half of the orography. This baroclinic structure is opposite of that in the counterpart thermally forced disturbance examined in Section 9.4.6. It highlights the difference between a thermally forced disturbance and an orographically forced disturbance. Since the basic flow increases northward, the forced disturbance tilts in the direction of the basic shear. Its Reynolds stress therefore transports zonal momentum up the gradient of the basic flow implying that this forced disturbance transfers some of its kinetic energy to the basic flow.

The total streamfunction field $(\bar{\psi}_4 + \psi'_4)$ has an intense trough over the eastern part of the orography, much more pronounced than that in the case of baroclinic shear

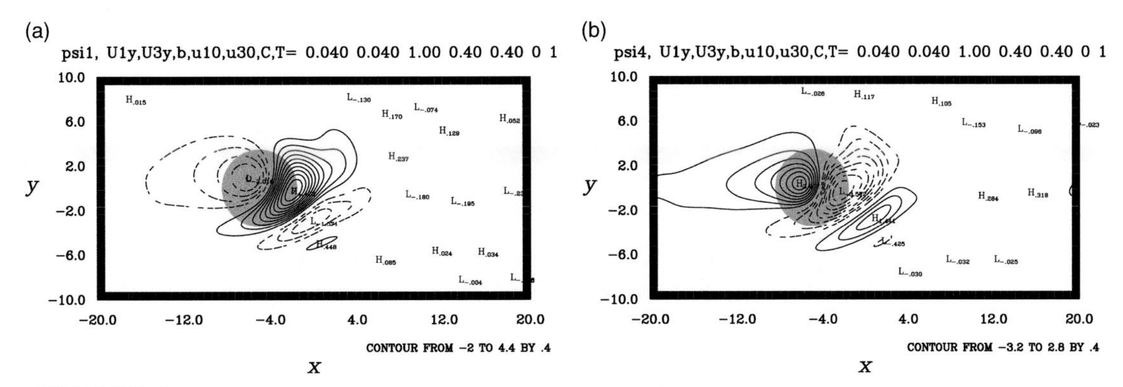

Fig. 9.33 Orographically excited (a) ψ_1' and (b) ψ_4' in units of $V\lambda^{-1}$ for the case of barotropic shear, $\beta = 1$, $u_{jo} = 0.4$, $\gamma_j = 0.04$, $Ro = 0.25$, $\varepsilon_1 = 0.05$, $\varepsilon_3 = 0.1$, $M = 5$. The orography is indicated by shading.

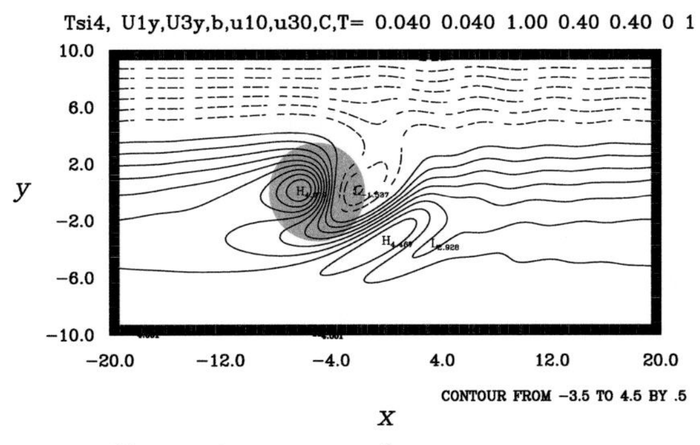

Fig. 9.34 Total streamfunction at the surface $(\bar{\psi}_4 + \psi_4')$ in units of $V\lambda^{-1}$ for the case of a westerly basic flow with barotropic shear, $\beta = 1$, $u_{jo} = 0.4$, $\gamma_j = 0.04$, $Ro = 0.25$, $\varepsilon_1 = 0.05$, $\varepsilon_3 = 0.1$, $M = 5$. The orography is indicated by shading.

(Fig. 9.34 vs. Fig. 9.31). There is an almost northerly flow through the central part of the mountain. The trough wraps around the southern half of the mountain in the southwest direction, giving rise to a strong south-southwesterly surface flow to the southeast side of the mountain. To the east of it, we see a ridge as a part of a progressively weaker wave-packet spreading to the southeast region.

The corresponding vertical motion field is quite different from that in the case of baroclinic shear (Fig. 9.35). We find descent over the eastern half of the orography, about $-0.21\ \mathrm{N\,m^{-2}\,s^{-1}}$ and significant ascent of about $0.14\ \mathrm{N\,m^{-2}\,s^{-1}}$ in the immediate downstream region to the east.

9.6 Illustrative application: mean Asian monsoonal circulation

The *mean Asian summer monsoonal circulation* (*ASM*) has a pronounced stationary planetary wave field. It would be instructive to see if we can simulate the gross features of such an important circulation with the generic model under consideration. Having

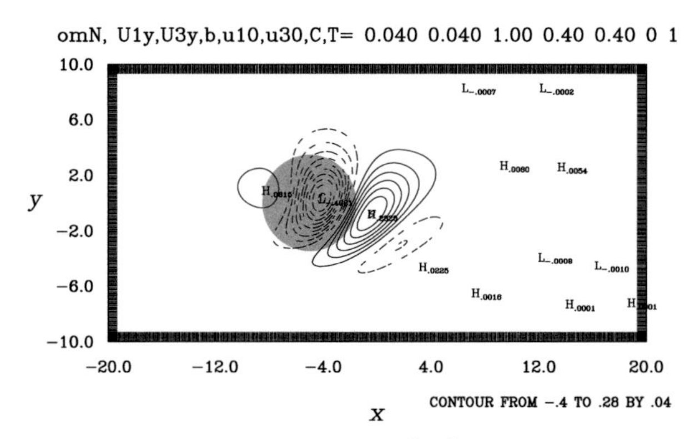

Fig. 9.35 Orographically excited $(-\omega_2)$ in units of $V\lambda\Delta p = 0.5\,\mathrm{N\,m^{-2}\,s^{-1}}$ for the case of a westerly basic flow with barotropic shear: $\beta = 1$, $u_{jo} = 0.4$, $\gamma_j = 0.04$, $Ro = 0.25$, $\varepsilon_1 = 0.05$, $\varepsilon_3 = 0.1$, $M = 5$. The orography is indicated by shading.

examined the responses to the individual physical factors or simple combinations of them in Sections 9.4 and 9.5, we are in a position to interpret the dynamics of a forced circulation under a general parametric condition.

There are several main surface features of ASM. They include a mean surface low-pressure center over southwest China, a southwesterly flow across the Indian subcontinent extending northeastward to the Korean peninsula and a subtropical high-pressure center offshore over the Pacific. Upper-level observation indicates that there is a high-pressure center over the Tibetan plateau and a pronounced trough off the coast of Asia. The southwesterly surface mean flow over the Arabian Sea and the Sea of Bengal and the anticyclonic flow of the adjacent subtropical high over the Pacific produce strong moisture flux into Asia. Such flux of latent heat would be subsequently released over Asia by convection in a favorable environment of large-scale ascent initiated by surface radiative heating. The descent over the subtropical high also would help sustain the anticyclonic flow. This is in effect a self-reinforcing mechanism to help sustain the mean ASM.

It is reasonable to anticipate that the thermal forcing associated with the summer precipitation over the Indian subcontinent, the Bay of Bengal, Southeast Asia and the coastal region of China including the Korean peninsula is essential to ASM. Also important is the surface heating and convective heating over the Tibetan plateau in the summer months. The massive Tibetan plateau should be important and must be incorporated in a model. Other elements of thermal heating and orography are assumed to be of secondary importance and are deliberately left out. The discussion in this section is largely based on a recent investigation (Mak, 2008).

9.6.1 Model formulation

Let us consider a two-layer quasi-geostrophic model which contains a general basic zonal flow $U_j(y)$, a thermal forcing and an orography $h(x, y)$ of relatively simple geometry. The total heating rate at mid-level is $Q_2 = Q_c + \frac{\xi_2}{\tau}(T_* - T_4)$, and surface heating is

$Q_4 = \frac{1}{\tau}(T_* - T_4)$, where τ is a relaxation time, $\xi_2 < 1$. They consist of condensational heating, Q_c, and a heating arising from a surface heating in K s^{-1}. The basic flow has the structure of $U_j = u_{jo} + \gamma_j y$. It is both barotropically and baroclinically stable provided that the prescribed zonal mean vertical shear is sufficiently weak, $(u_{1o} - u_{3o})\lambda^2\beta^{-1} < 1$, where λ^{-1} is the radius of deformation defined before.

The governing equations are the generalized forms of (9.15a,b) and (9.22a,b,c), viz.

$$\begin{aligned} U_1 q'_{1x} + \left(\beta + \lambda^2(U_1 - U_3)\right)\psi'_{1x} &= -\frac{\lambda^2 R Q_2}{f_o} - \varepsilon_1 \nabla^2 \psi'_1, \\ U_3 q'_{3x} + \left(\beta - \lambda^2(U_1 - U_3)\right)\psi'_{3x} &= \frac{\lambda^2 R Q_2}{f_o} - \frac{U_4 f_o g \rho_s}{\delta p} h_x - \varepsilon_3 \nabla^2 \psi'_3, \\ U_4\left(\psi'_{4x} - \psi'_{3x}\right) - \psi'_{4x}(U_4 - U_3) &= \frac{U_4 g \rho_s S_s \delta p}{2 f_o} h_x - \frac{R}{2 f_o} Q_4. \end{aligned} \tag{9.26a,b,c}$$

The vertical motion field at the mid-level can be determined with

$$\begin{aligned} &\frac{f_o}{\lambda^2 \Delta p}(\nabla^2 - 2\lambda^2)\omega_2 \\ &= \{\beta(\psi'_{3x} - \psi'_{1x}) - U_1\nabla^2\psi'_{1x} + U_3\nabla^2\psi'_{3x}\} + \{U_3\nabla^2\psi'_{1x} - U_1\nabla^2\psi'_{3x} + 2U_{3y}\psi'_{1xy} - 2U_{1y}\psi'_{3xy}\} \\ &\quad + \lambda^2 U_4 \psi_{Tx} + \{\varepsilon_3\nabla^2\psi'_3 - \varepsilon_1\nabla^2\psi'_1\} - \frac{1}{\lambda^2}\nabla^2 G_2. \end{aligned} \tag{9.27}$$

The thermal heating, orography and surface temperature are non-dimensionalized as $\tilde{G} = G/G_*$, $\left(G \equiv \lambda^2 R Q_2/f_o\right)$, $\tilde{h} = h/h_*$ and $\tilde{T}_* = T_*/\Theta_*$ where G_*, h_* and Θ_* are the maximum value of each field. All other quantities are again non-dimensionalized in terms of a velocity scale V and a length scale λ^{-1}, namely $\tilde{\psi}'_j = \psi'_j/V\lambda^{-1}$, $\tilde{\omega}_2 = \omega_2/V\lambda\Delta p$, $\tilde{q}_j = q_j/V\lambda$, $\tilde{\beta} = \beta/V\lambda^2$, $\tilde{\varepsilon}_j = \varepsilon_j/V\lambda$, $\tilde{\tau} = V\lambda\tau$, $\tilde{U}_j = U_j/V$, $\tilde{u}_{jo} = u_{jo}/V$, $\tilde{\gamma}_j = \gamma_j/V\lambda$, $\left(\tilde{k}, \tilde{\ell}\right) = \lambda^{-1}(k, \ell)$, $(\tilde{x}, \tilde{y}) = \lambda(x, y)$. The magnitude of the thermal forcing and the height of orography are prescribed by three non-dimensional parameters: $L = \frac{R\lambda}{Vf_o}\Theta_*$, $M = \frac{g f_o \rho_s}{(V\lambda)\Delta p} h_*$ and $N = \frac{G_*}{(V\lambda)^2}$. The non-dimensional governing equations for the three unknowns ψ'_1, ψ'_3 and ψ'_4, after dropping the "tilde superscript," are then

$$\begin{aligned} U_1 q'_{1x} + (\beta + (U_1 - U_3))\psi'_{1x} &= -NG_c - \frac{\xi_2}{\tau} L T_* - 4\frac{\xi_2}{\tau}(\psi'_4 - \psi'_3) - \varepsilon_1 \nabla^2 \psi'_1, \\ U_3 q'_{3x} + (\beta - (U_1 - U_3))\psi'_{3x} &= NG_c + \frac{\xi_2}{\tau} L T_* + 4\frac{\xi_2}{\tau}(\psi'_4 - \psi'_3) - MU_4 h_x - \varepsilon_3 \nabla^2 \psi'_3, \\ U_3 \psi'_{4x} - U_4 \psi'_{3x} &= \frac{1}{2} MU_4 h_x - \frac{1}{4\tau} L T_* - \frac{1}{\tau}(\psi'_4 - \psi'_3), \end{aligned} \tag{9.28}$$

where G_c refers to the condensational heating. Equation (9.28) will be solved with the semi-spectral algorithm as in the previous two sections.

The maximum mean JJA precipitation rate is about $P = 15\,\text{mm day}^{-1}$ over East Asia and $9\,\text{mm day}^{-1}$ over the Indian subcontinent. Therefore, the strength of the heat source is $G_* = 1.7 \times 10^{-10}\,\text{s}^{-2}$ for the former and $G_* = 1.0 \times 10^{-10}\,\text{s}^{-2}$ for the latter. We use $\lambda^{-1} = 10^6\,\text{m}$, $V = 10\,\text{m s}^{-1}$ and hence $Ro = 0.1$, for non-dimensionalization. We use the following values of the other parameters: $\Delta p = 400\,\text{hPa}$, $u_{1o} = 8\,\text{m s}^{-1}$, $u_{3o} = 5\,\text{m s}^{-1}$, $u_{4o} = 3\,\text{m s}^{-1}$, $\gamma_4 = 0.4\frac{u_{4o}}{Y}$, $\gamma_3 = 0.8\frac{u_{3o}}{Y}$, $\gamma_1 = 1.2\frac{u_{1o}}{Y}$ (so that, $0.18 \leq \tilde{U}_4 \leq 0.42$, $0.1 \leq \tilde{U}_3 \leq 0.9$, $-0.16 \leq \tilde{U}_1 \leq 1.76$), $\tilde{\beta} = 2.2$ (for 25° N), $\tilde{\varepsilon}_3 = 3.3$ (3 days damping time), $\varepsilon_1 = 0.\tilde{5}$ (9.20 days damping time), $\Theta = 6\,\text{K}$ and $\tilde{\tau} = 3.0$ (3 days relaxation time). The domain is 40 non-dimensional units long and 20 units wide and is depicted by 401 by 201 grid points. The corresponding resolution is quite adequate, $\delta x = \delta y = 100\,\text{km}$. In order to incorporate an additional zonal mean thermal damping, we use a damping coefficient for the zonal mean component ten times larger than for the other spectral components in the upper layer and five times in the lower layer.

9.6.2 Representation of the model forcing

One heat source is over an area analogous to the Indian subcontinent including the western Bay of Bengal. Another heat source is an elongated region extending from the Korean peninsula, through the eastern coastal plain of China and the Indochina peninsula to the eastern part of the Bay of Bengal. The only orography in the model is a Tibet-like plateau. A surface heating is introduced only on the model Tibetan plateau. The size and shape of each of those elements is represented by a two-dimensional Gaussian function, e.g.

$$G(x, y) = G_* \exp\left(-\left(\frac{x' - x_c}{a}\right)^2 - \left(\frac{y' - y_c}{b}\right)^2\right). \tag{9.29}$$

Its central location is (x_c, y_c). Its size and shape are prescribed by a and b the semi-major and semi-minor axes. Its orientation is prescribed by η, the angle of the x'-axis with respect to the x-axis, through the rotated coordinates, $x' = x\cos\eta + y\sin\eta$ and $y' = -x\sin\eta + y\cos\eta$. A triangular mask additionally imposed on the distribution of the Indian thermal forcing reflects the general shape of the Indian subcontinent. We only set $h_* = 2.5\,\text{km}$ as the maximum elevation of the orography in this QG model. The parameters for the three thermal forcings and orography are listed in Table 9.1.

The configurations of the normalized thermal forcing as well as the model Tibetan plateau are shown in Fig. 9.36. The forcing parameters have the following values: (surface heating) $L = 2.0$, (orography) $M = 12.5$, (condensational heating) $N = 2.2$, (relaxation time) $\tau = 3.0$ and (heat exchange coefficient) $\xi_2 = 0.2$.

9.6.3 Structure of the prototype mean ASM

The results of $\tilde{\psi}_1'$ and $\tilde{\psi}_4'$ fields are shown in Fig. 9.37. The $\tilde{\psi}_3'$ field is similar to the $\tilde{\psi}_4'$ field and is not shown for brevity. One feature of the $\tilde{\psi}_4'$ field is a strong dipole in the region of the model Tibetan plateau largely trapped in the model lower layer. The intensity is $-2.77 \times 10^7\,\text{m}^2\,\text{s}^{-1}$ on the eastern half and $1.4 \times 10^7\,\text{m}^2\,\text{s}^{-1}$ on the western half of the model Tibetan plateau. In addition, there is a weaker elongated low-center over northern India ($-1.93 \times 10^7\,\text{m}^2\,\text{s}^{-1}$) interpreted as a *prototype monsoon low.* There is an

Table 9.1 Parameters for the elements of the forcings; distance is in units of $\lambda^{-1}=10^6$ m. The maximum values of G, h and T_* are normalized to unity

	East Asian interior heating	Indian interior heating	Tibetan plateau elevation and surface heating
Coordinates of the center, (x_c, y_c)	(0.0,0.0)	(−4.0,−1.5)	(−4.0,1.6)
Major semi-axis, a	2.5	2.0	2.2
Minor semi-axis, b	0.6	1.0	0.75
Tilting angle, η	40°	90°	−10°
Coordinates of the corners of a triangular mask	–	(−6.0,0.0), (−2.0,0.0), (−4.0,−4.0)	–

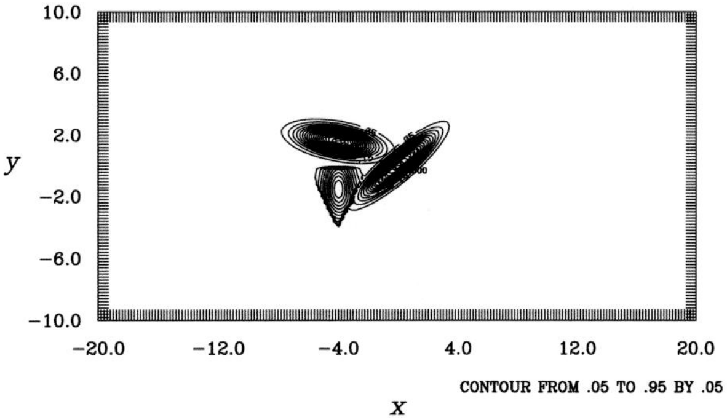

Fig. 9.36 Configurations of the normalized forcing elements in the ASM model. Distance is in units of $\lambda^{-1} = 10^6$ m.

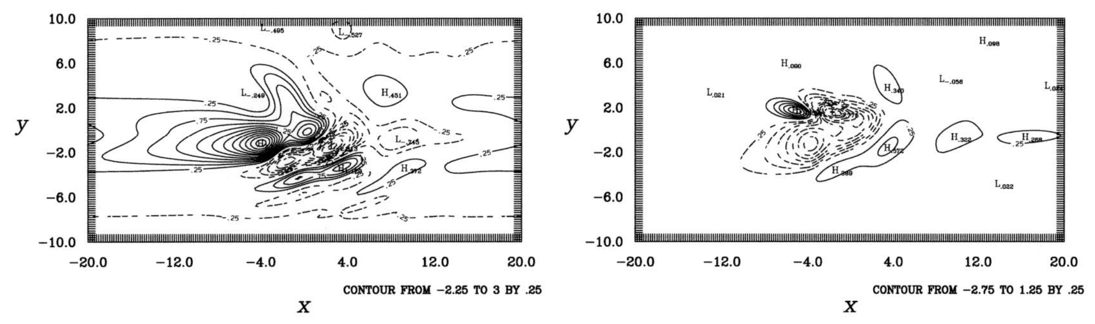

Fig. 9.37 Disturbance streamfunction field in the ASM model (a) at the upper level $\tilde{\psi}'_1$ and (b) at the surface $\tilde{\psi}'_4$ in units of $V\lambda^{-1} = 10^7$ m^2 s^{-1}. Zero contours are omitted. Shaded area indicates the configuration of the combined forcing.

anticyclonic circulation aloft greatly elongated to the west. The high-center in $\tilde{\psi}'_1$ is over northern India with an intensity of 3.1×10^7 m^2 s^{-1}. It is a *prototype Tibetan high*. The monsoon low and the Tibetan high jointly constitute a skewed baroclinic vortical circulation. It should not be viewed in terms of Rossby waves as pointed out in Section 9.4.5. The existence of a baroclinic vortex results from the interior localized heating, which tends to

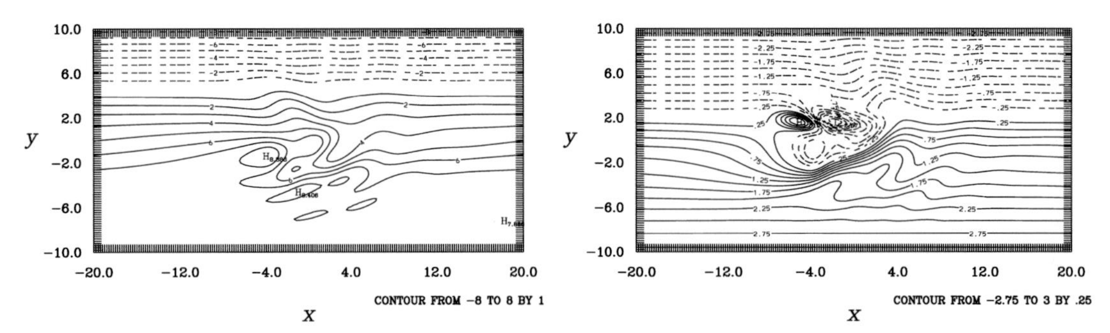

Fig. 9.38 Total streamfunction field in the ASM model (a) at the upper level $(\bar{\psi}_1 + \psi_1')$ and (b) at the surface $(\bar{\psi}_4 + \psi_4')$ in units of $V\lambda^{-1} = 10^7\ \text{m}^2\ \text{s}^{-1}$. Zero contours are omitted. Shaded area indicates the configuration of the combined forcing.

increase the potential vorticity in the lower layer and decrease the potential vorticity in the upper layer. Its extensive elongation to the west stems jointly from the additional influences of the beta-effect and basic baroclinic shear flow.

The disturbance in $\tilde{\psi}_1'$ over the Pacific to the east of the heat source clearly has the signature of a composite wave-packet made up of an ensemble of orographically and thermally excited Rossby waves. The corresponding feature in $\tilde{\psi}_4'$ is a high-center in the subtropical model Pacific region with an intensity of about $0.58 \times 10^7\ \text{m}^2\ \text{s}^{-1}$. This is a *prototype subtropical high*. Above it is a low-center in $\tilde{\psi}_1'$ slightly tilted to the west with an intensity of $-2.3 \times 10^7\ \text{m}^2\ \text{s}^{-1}$. The intensities of these simulated features are compatible with observation.

The total streamfunction field is presented in Fig. 9.38. There is a broad surface trough in $(\bar{\psi}_4 + \psi_4')$ oriented in the WSW–ENE direction stretching from India to central China. The related flow is a westerly over the southern half of the Indian subcontinent. It turns into a southwesterly flow over the Bay of Bengal and the South China Sea. This flow configuration qualitatively agrees with the observed 850 mb wind field and with the simulated 850 mb wind field in some general circulation models (GCM). The absence of a cross-equatorial flow off the model region of eastern Africa is unavoidable in a QG model. But the consequence of the moisture flux by such flow is already embodied in the prescribed heating field. It is noteworthy that there is also a distinct anticyclonic gyre over the western Tibetan plateau. The gyre in this simulation is stronger than in the GCM. There is a distinct high-center in $(\bar{\psi}_1 + \psi_1')$ over India as in the observed 200 mb streamfunction. But ours is a bit weak. The model upper-level monsoon trough further off the coast of Asia seems to be more pronounced than observation.

The corresponding non-dimensional vertical motion field $(-\tilde{\omega}_2)$ is largely compatible with the heat sources (Fig. 9.39). Strong ascending motion occurs along the coastal plain of East Asia, the Bay of Bengal as well as over the eastern two-third of the Indian subcontinent. Some ascending motion spreads to some distance off the coast of Asia. The maximum ascent is about 0.19 over the model Indochina peninsula, equivalent to $2.7\ \text{hPa}\ \text{h}^{-1}$. This is compatible with the observed and GCM-simulated maximum value at 674 mb level over the Indochina peninsula ($\sim 3.0\ \text{hPa}\ \text{h}^{-1}$). However, since the model does not incorporate a heating associated with the maritime continent, there is naturally no counterpart ascending motion in the very low

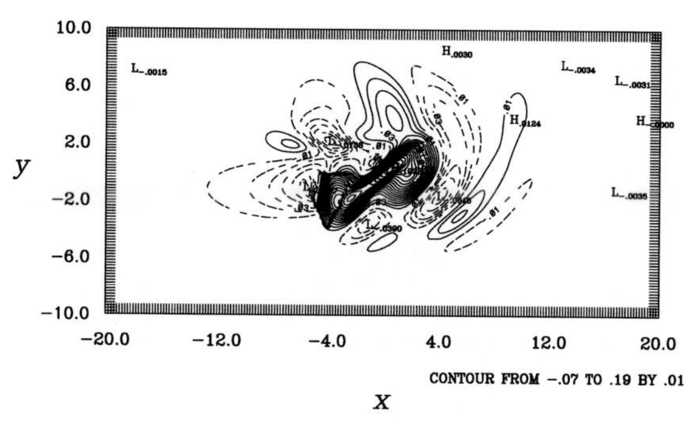

Fig. 9.39 Vertical motion field $(-\tilde{\omega}_2)$ in the ASM model in units of $V\lambda^{-1}\Delta p = 0.4\,\text{Pa s}^{-1}$ equivalent to 14 hPa h^{-1}. Zero contours are omitted. Shaded area indicates the combined forcing.

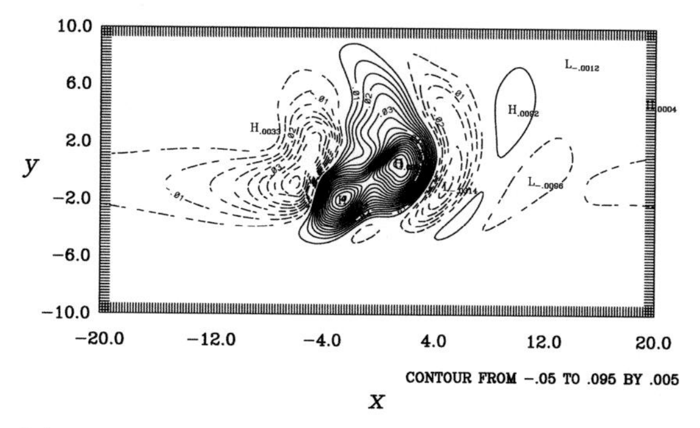

Fig. 9.40 The component of $(-\tilde{\omega}_2)$ in the ASM model associated with the process of vertical differential vorticity advection in units of $V\lambda^{-1}\Delta p = 14\,\text{hPa h}^{-1}$. Zero contours are omitted. The shaded area indicates the configuration of the combined forcing.

latitude region of the model. Orographically induced weak ascending motion occurs on the western half of the Tibetan plateau.

Significant descent occurs over the prototype subtropical high as expected. It is noteworthy to find descending motion over an elongated region to the west of the model Pakistan similar to observation in that arid region. The maximum descent is about -0.07, corresponding to $-1.1\,\text{hPa h}^{-1}$. This result supports the notion that the desert climate of the Middle East could be an integral part of the ASM. Dynamically speaking, this feature is an intrinsic characteristic of the elongated baroclinic vortex. The flow to the west of the vortex gives rise to positive advection of basic absolute vorticity in the lower layer and negative advection in the upper layer. The descending motion at the mid-level is what is needed to satisfy the Sverdrup vorticity balance in each layer, $v\dfrac{\partial \bar{\zeta}_a}{\partial y} \sim f_o\dfrac{\partial \omega}{\partial p}$. In contrast, the dynamics of the flow in the downstream region is primarily characterized by Rossby wave vorticity balance $\left(v\dfrac{\partial \bar{\zeta}_a}{\partial y}\right)_j \sim \left(U\dfrac{\partial \zeta'}{\partial x}\right)_j$. The significant descending motion over the eastern half of the model Tibetan plateau is also a natural consequence of PV conservation.

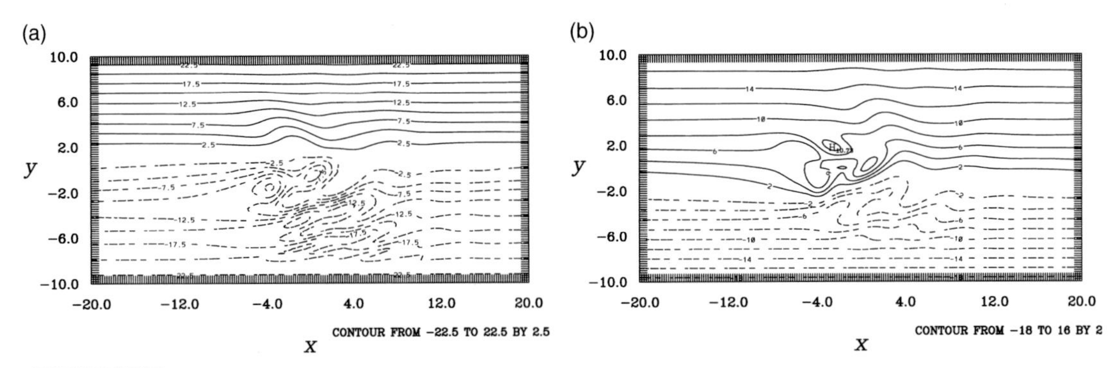

Fig. 9.41 Total potential vorticity field $(\bar{q}_j + q_j')$ in the ASM model at the (a) upper level and (b) lower level in units of $V\lambda = 10^{-5}\,\mathrm{s}^{-1}$. Zero contours are omitted. Shaded area indicates the configuration of the forcing.

It is valid to separately compute the solution of $(-\omega_2)$ associated with each of the five groups of terms on the RHS of (9.2). The component of $(-\tilde{\omega}_2)$ associated with the non-dimensional form of the first group of terms, $\{\beta(\psi'_{3x} - \psi'_{1x}) - U_1\nabla^2\psi'_{1x} + U_3\nabla^2\psi'_{3x}\}$, stems from the process of vertical differential vorticity advection. It turns out to be the largest component and accounts for about 50 percent of the total value (Fig. 9.40). The elongated zone of descending motion in the region over the model Middle East mostly stems from this component. Although the five components are mathematically well defined, they are not fully independent from a physical point of view since the streamfunction depends on the heating itself. The other four components are not presented for brevity.

The total PV fields $(\bar{q}_j + q_j')$ are shown in Fig. 9.41. The PV values in the lower layer over Asia are high, arising from the PV forcing by the mid-tropospheric maximum condensational heating as well as by the Tibetan plateau. There is a tongue of relatively low PV values off the coast of Asia due to PV advection from the south. An opposite distribution of PV exists in the upper layer in that there is a center of low PV over India and a tongue of relatively high PV values off the coast of Asia extending from the northeast. The former directly results from the heating and the latter from the advective process. The pronounced wavy pattern in the southeast sector is the PV signature of the Rossby wave-packet.

This model also enables us to examine the contributions from the individual elements of the thermal forcing to the model ASM. For brevity, those results are not presented, but can be found in Mak (2008).

Concluding remarks

We have illustrated the relevance of a two-layer quasi-geostrophic model for investigating the dynamics of stationary planetary waves with bare essential orographic and thermal forcings. We show that the circulation of the mean ASM can be simulated with some degree of realism. It gives us some insights into the dynamics of this circulation. The solution particularly reveals that the prototype ASM circulation consists of two main components. The western part of it is a greatly elongated and skewed baroclinic vortical circulation. The lower and upper features of the baroclinic vortex are the model counterparts of the monsoon low and the Tibetan high respectively. The eastern part of the model ASM is a composite baroclinic Rossby wave-packet, which manifests itself at the lower and upper levels as the model counterparts of the surface subtropical high and upper level monsoon trough. The center of the skewed vortex and the root of the wave-packet are collocated over East Asia arising from a common source of excitation. This circulation is intelligible in terms of potential vorticity dynamics. Furthermore, the ascending motion field is consistent with the notion that the thermal forcing is largely self-reinforcing. The descending motion over the model Middle East supports the notion that the desert climate is an integral part of the ASM.

10 Wave-mean flow interaction

This chapter is concerned with the dynamics of wave-mean flow interaction in the context of two phenomena, *mean meridional overturning circulation* and *stratospheric sudden warming*. We will examine the former in both Eulerian and Lagrangian senses. Those observed time-zonal mean structures in the troposphere are first examined in Sections 10.1 and 10.2 respectively. A linear theory for the two aspects of this phenomenon is discussed in Section 10.3. The concepts of residual circulation, Eliassen–Palm vector and transformed Eulerian mean (TEM) equations are elaborated in Section 10.3.1. In Section 10.3.2, we present the annual mean distributions of eddy forcing, diabatic forcing and frictional forcing deduced from five years of global data. They are used in this formalism to determine the corresponding annual mean residual circulation in Section 10.3.3. The results reveal the characteristics of such overturning circulation and its extent of agreement with observation. The results also bring to light the relative importance of the several mechanisms in different latitudes responsible for the overturning circulation. The counterpart analyses of the seasonal mean residual circulations are discussed in Section 10.3.4. Section 10.4 discusses the Non-Acceleration Theorem that has a direct bearing on our interpretation of stratospheric sudden warming. In Section 10.5, we discuss the dynamics of such a dramatic transient phenomenon with a model formulation and illustrate it with the results of a seminal study.

10.1 Eulerian mean meridional overturning circulation

The time-zonal mean structure of the velocity (u, v, ω) and potential temperature θ fields as well as the related eddy statistics are the elemental components of the general circulation of the atmosphere. These properties are diagnosed with the use of the NCEP/NCAR (National Center for Environmental Prediction/National Center for Atmospheric Research) Reanalysis data. We report robust statistics of these properties deduced from a five-year subset of data for the period from 2001 to 2005. As such, they may be referred to as climatological statistics. A square bracket denotes the zonal averaging of a dependent variable, $[\xi] = \frac{1}{2\pi}\int_0^{2\pi} \xi d\lambda$. The time-zonal mean overturning circulation on the latitude–pressure cross-section in the Eulerian sense consists of $[\bar{v}]$ and $[\bar{\omega}]$. In light of the continuity equation, we define a streamfunction ψ by the following equations in spherical-isobaric coordinates with standard notations

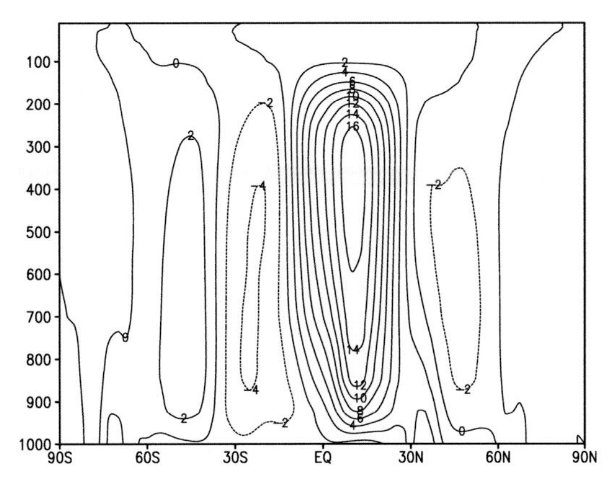

Fig. 10.1 Climatological winter mean streamfunction of the Eulerian meridional circulation in units of 1×10^{10} kg s^{-1}.

$$[\bar{v}(\varphi,p)] = \frac{g}{2\pi a \cos\varphi}\frac{\partial\psi}{\partial p},$$
$$[\bar{\omega}(\varphi,p)] = -\frac{g}{2\pi a^2 \cos\varphi}\frac{\partial\psi}{\partial\varphi}. \tag{10.1a,b}$$

The value of $\psi(\varphi,p)$ may be computed on the basis of the result of $[\bar{v}]$ alone. Using $\psi = 0$ as the boundary condition at the top of a domain and integrating (10.1a) from the top to any isobaric level, we obtain

$$\psi(\varphi,p) = \frac{2\pi a \cos\varphi}{g}\int_0^p [\bar{v}]dp. \tag{10.2}$$

We first examine the climatological winter mean (December, January, February) streamfunction in Fig. 10.1. The dominant feature is the cell in the tropical region of the winter hemisphere extending from about 30° N to about 10° S. The maximum value of ψ is about 17×10^{10} kg s^{-1} implying a clockwise overturning on the meridional plane. It means that this is a thermally direct cell since ascending motion takes place in the warmer equatorial latitudes and descending motion in the cooler subtropics. It vertically extends from the surface to about 100 mb level. There is a similar but weaker cell in the tropical region of the southern hemisphere. These two cells are known as *Hadley cells*, named in honor of George Hadley (1735), who correctly interpreted such cells as thermally forced circulation under a strong influence of the Earth's rotation. The Hadley cell in the winter hemisphere is about twice as wide and four times as intense as the one in the summer hemisphere. The latter is also shallower. It is intriguing to ponder over the dynamics underlying the dramatic difference between the two Hadley cells. We will return to this matter in Section 12.2. Figure 10.1 also clearly reveals the existence of an additional cell in the extratropics of each hemisphere. In contrast, these cells are thermally indirect, known as *Ferrel cells*, named in honor of William Ferrel (1856), who inferred that such cellular circulation must exist. Each Ferrel cell has a width of about 30° latitude and an intensity only about one-eighth of that of the winter Hadley cell.

We would expect that the time-zonal mean overturning circulation is dynamically related to the counterpart distributions of the zonal wind and temperature, $[\bar{u}]$ and $[\bar{\theta}]$. The results

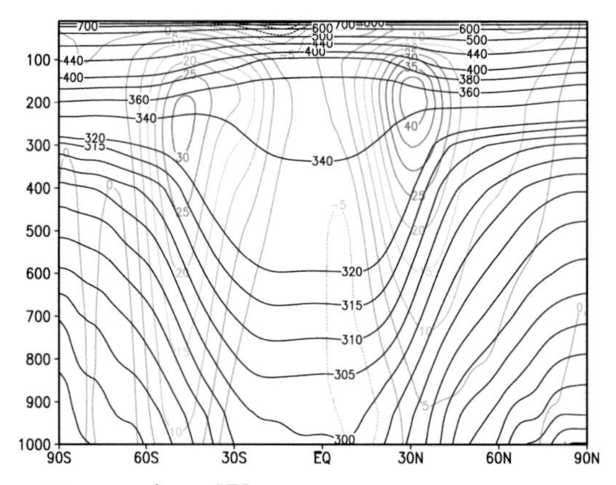

Fig. 10.2 Climatological winter mean $[\bar{u}]$ in m s^{-1} and $[\bar{\theta}]$ in K. See color plates section.

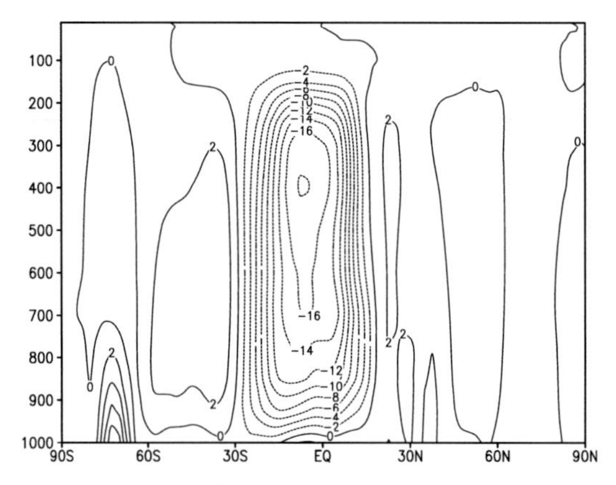

Fig. 10.3 Climatological summer mean streamfunction of the Eulerian meridional circulation in units of 1×10^{10} kg s^{-1}.

of those properties are presented in Fig. 10.2. There is a pronounced westerly jet in the upper troposphere over about 30° N. The maximum velocity of this jet is about 45 m s^{-1}. This localized vertical gradient of $[\bar{u}]$ is located in a strong baroclinic zone as indicated by the pronounced meridional gradient of $[\bar{\theta}]$. Apart from its characteristic poleward decrease, $[\bar{\theta}]$ also distinctly increases with height indicating a strong stable stratification. It is noteworthy that the westerly jet resides at the northern edge of the Hadley cell. This is no coincidence as we will discuss its dynamical nature in Section 12.2. The jet in the summer hemisphere of this season is over about 45° S with an intensity of about 30 m s^{-1}. The corresponding baroclinicity is naturally weaker, but the stratification is about the same in both hemispheres.

Summer mean (JJA) is defined as the average of June, July and August of each year. This period corresponds to the winter season of the southern hemisphere. The climatological summer mean overturning circulation is shown in Fig. 10.3. Again, we find one dominant cell in the tropical region of the winter hemisphere. This Hadley cell spans from about 30° S across the equator to about 15° N. Its intensity is quite similar to that of the winter mean Hadley cell.

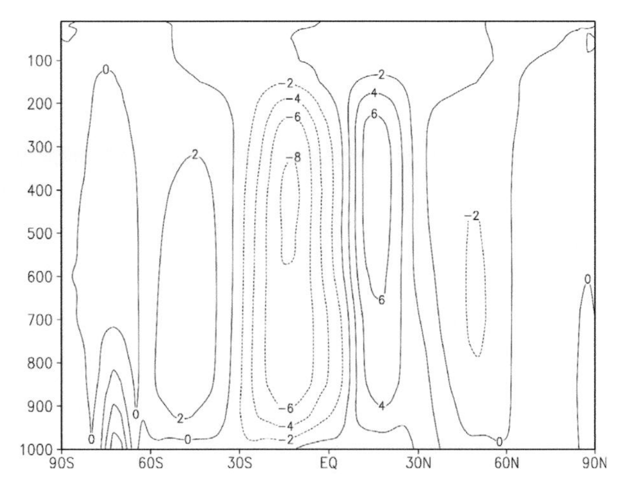

Fig. 10.4 Climatological annual mean streamfunction of the Eulerian meridional circulation in units of 1×10^{10} kg s^{-1}.

The annual mean overturning circulation is shown in Fig. 10.4. As expected, the annual mean structure is essentially the average of Fig. 10.1 and Fig. 10.3. It is much more symmetrical about the equator, although the boundary between the two Hadley cells is located near about 5° N. This pronounced hemispheric asymmetry of the two Hadley cells is ultimately attributable to the fact that the northern hemisphere has much more land mass than the southern hemisphere.

A bigger picture of the global time-zonal mean structure of the atmosphere is shown in Fig. 10.5. It depicts such structure of the zonal velocity and temperature in January and July extending up to about 90 km. This result was obtained in a comprehensive diagnosis by a large group of researchers using a combination of several recent datasets (Randel *et al.*, 2004). The structure of zonal velocity and temperature in the troposphere shown in Fig. 10.2 is compatible with the climatological January counterpart in Fig. 10.5.

Some aspects of the seasonal variations in these structures of the time-zonal mean zonal wind and temperature fields are relatively small, while other aspects are huge. For instance, the gross global structure of the *zonal wind is qualitatively similar in both winter and summer hemispheres in the troposphere, but is qualitatively different in the stratosphere.* While the winter stratospheric zonal wind is westerly, the summer counterpart is easterly. There are corresponding similarities and differences in the zonal mean temperature field. Specifically, while the overall structures of the temperature field in the troposphere and in the equatorial lower stratosphere are qualitatively similar in both seasons, the *meridional gradient of the temperature in the upper stratosphere is qualitatively different in the two seasons*. These differences simply reflect the fact that the zonal wind and temperature fields are in thermal wind balance in all seasons.

The quantitative differences are noteworthy. In January, the tropospheric westerly jet in the northern hemisphere is slightly stronger than the one in the southern hemisphere, about 38 m s^{-1} vs. 30 m s^{-1}. The westerly flow in the northern troposphere extends deep into the stratosphere, where we find another westerly jet of about 47 m s^{-1} at 65 km over 35° N. In contrast, the tropospheric westerly jet at 45° S is below an easterly jet in the stratosphere. The strength of that easterly jet is quite strong ~65 m s^{-1} and is located at about 70 km height over 50° S. The temperature field in January is warmest at the Earth's surface in the

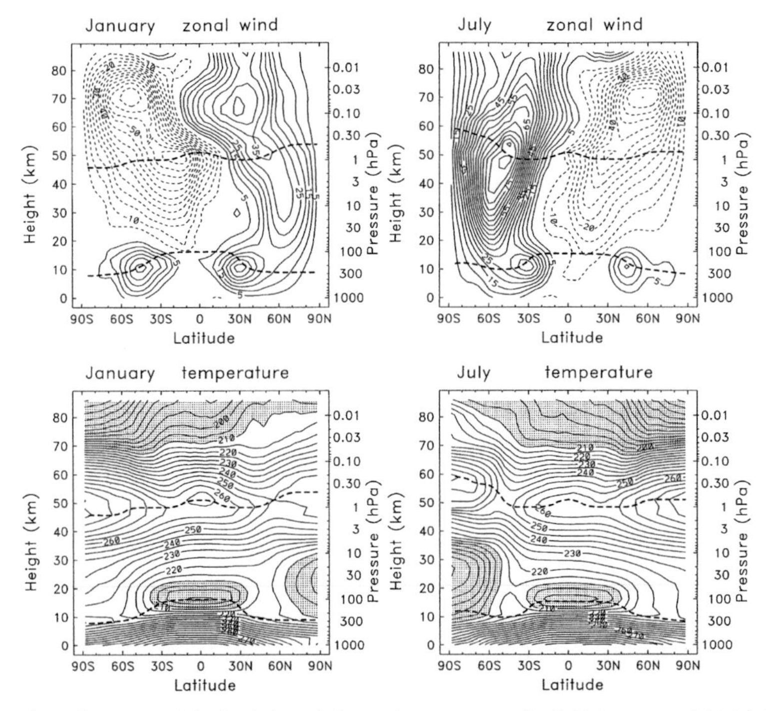

Fig. 10.5 Climatological zonal mean zonal (top) winds and (bottom) temperatures for (left) January and (right) July. Zonal winds are from the URAP dataset (contour interval 5 m s^{-1}, with zero contours omitted). Temperatures are from METO analyses (1000–1.5 hPa), and a combination of HALOE plus MLS data above 1.5 hPa (see SPARC, 2002). The heavy dashed lines denote the tropopause (taken from NCEP) and stratopause (defined by the local temperature maximum near 50 km) (taken from Randel *et al.*, 2004).

equatorial region and decreases poleward. The lowest temperature at the surface is over the winter pole with a value of about 240 K. Temperature decreases with height reaching a minimum at a level indicated by a dashed line, known as the tropopause. The tropopause height varies significantly with latitude. It is at about 16 km over the equator and about 9 km at the poles. The coldest region turns out to be in the lower stratosphere over the equator with a temperature only about 200 K. The temperature increases with height above the tropopause, reaching a maximum value at a level known as the stratopause near about 50 km. The latter is indicated by another dashed line in Fig. 10.5. The highest temperature is about 285 K over the summer pole and is lowest over the winter pole at the stratopause, ~250 K. The stratopause is located at about 50 km high, somewhat higher towards the summer pole. The temperature again decreases with increasing height above the stratopause.

In July, the zonal wind in the troposphere of the northern hemisphere is quite weak. There is still a discernible westerly jet in the upper troposphere with a maximum speed of only about 20 m s^{-1} centering over 45° N. The jet in the southern hemisphere is about 37 m s^{-1} over 30° S. The zonal wind in the stratosphere changes drastically from winter to summer in the same hemisphere. There is an easterly jet in July of about 60 m s^{-1} at a height of 70 km over 55° N and a very strong westerly jet reaching 90 m s^{-1} at about 50 km centering over 50° S. The reversal of wind direction has a far-reaching implication for the vertical propagation of waves in the two seasons. In light of the observed structure of the zonal wind, we would expect that some planetary waves generated in the troposphere would be able to propagate into the stratosphere in winter, but not in summer according to the discussion in Section 9.3.2. There is plenty of observational evidence in support of this expectation.

10.2 Lagrangian mean meridional overturning circulation

The mass transport of air on the meridional plane is measured by the Lagrangian mean, not the Eulerian mean overturning circulation. Their difference stems from the fact that Lagrangian mean overturning circulation includes the mass flux associated with transient motions. For example, wave motions in the oceans and atmosphere could give rise to a mass transport known as Stokes' drift. Although it is not feasible to perform a strictly Lagrangian analysis because the identity of individual fluid parcels cannot be tracked for a long time, the meridional mass flux can be approximately estimated as follows. Since the mass in an elemental "volume" in isentropic coordinates (x, y, θ) is $\left(-\frac{1}{g}\frac{\partial p}{\partial \theta}dx\,dy\,d\theta\right)$, the instantaneous zonal-average mass transport in a layer of thickness $\Delta\theta$ centered at the level θ_1 across a latitude circle is equal to $\left[\int_{\theta_1-\Delta\theta/2}^{\theta_1+\Delta\theta/2} -\frac{1}{g}\frac{\partial p}{\partial \theta}v\,d\theta\right] = \left[-\frac{1}{g}\int_{p_1}^{p_2} v|_{\theta_1}\,dp\right] = \frac{1}{g}[v\Delta p]_{\theta_1}$, where p_1 and $p_2 = p_1 - \Delta p$ are the corresponding pressure at the two isentropic surfaces under consideration. Δp is the thickness of this layer, which in general fluctuates in time with the velocity field. The square bracket stands for zonal averaging. Therefore, what we need to do is to estimate the thickness-weighted meridional velocity at each isentropic level as first reported by Townsend and Johnson (1985). Figure 10.6 shows the isentropic mass transport streamfunction under solstice conditions. It was deduced from the output of a general circulation model (GCM) (Held and Schneider, 1999). It reveals one single cell spanning much of the troposphere in each hemisphere. The winter cell is much stronger and wider. The air in a

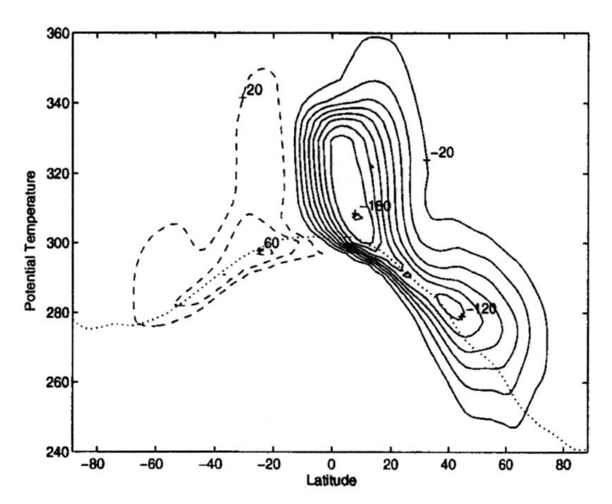

Fig. 10.6 Isentropic mass transport streamfunction for 10^9 kg s^{-1} Jan (50-yr mean of GCM integration); (solid lines) clockwise and (dashed lines) counterclockwise circulation. The dotted line represents the median surface potential temperature (taken from Held and Schneider, 1999).

narrow equatorial belt rises to higher θ-levels primarily due to strong condensational heating. Air in the subtropics sinks to lower θ-levels due to net radiative cooling. This part of the cell very much coincides with the Hadley cell. Hence, the direction of the mean meridional circulation (MMC) in the tropics is by and large the same in both Eulerian and Lagrangian senses. The poleward mass flux in the extratropics continues to move towards the lower θ-levels due to radiative cooling and eventually sinks toward the surface at higher latitudes. The surface branch of this single great big cell must be equatorward. It is confined below the level of mean surface potential temperature, indicated by a dotted line in Fig. 10.6. Such surface mass transport suggests that the equatorward mass flux in the extratropics from high latitudes in the cold sectors of a flow is statistically greater than the occasional poleward mass flux from low latitudes in the warm sectors. The colder air in contact with the Earth's surface is gradually warmed up as it moves southward. Hence it eventually rises from lower θ-levels to somewhat higher θ-levels. The most striking overall feature is that the direction of the Lagrangian MMC in the extratropics is opposite to that of the Eulerian MMC (Ferrel cell).

10.3 Linear theory for the overturning circulation

A significant part of the MMC in both Eulerian and Lagrangian senses can be accounted for in a linear dynamical framework. They are expected to be forced circulations, partly due to the differential net diabatic heating/cooling and partly the collective action of the large-scale waves. The latter is referred to as wave-mean flow interaction. We will invoke a notion of *residual circulation* as an approximate measure of the zonal mean mass transport. We will discuss a formulation for evaluating the residual circulation. We will report sample calculations for the annual and seasonal mean residual circulations in the troposphere. Those results reveal the relative roles of the different mechanisms in the framework of a linear theory. We will defer until Section 12.2 the nonlinear dynamics of the Hadley circulation.

10.3.1 Residual circulation, Eliassen–Palm vector and transformed Eulerian mean equations

We now present a formulation to examine the dynamics of zonal mean circulation on the meridional plane. The primitive equations written in spherical-isobaric coordinates are our starting point.

$$\begin{aligned}
\frac{Du}{Dt} - \frac{uv\tan\varphi}{a} &= -\frac{1}{a\cos\varphi}\frac{\partial\Phi}{\partial\lambda} + fv + F, \\
\frac{Dv}{Dt} + \frac{u^2\tan\varphi}{a} &= -\frac{1}{a}\frac{\partial\Phi}{\partial\varphi} - fu + G, \\
0 &= -\frac{\partial\Phi}{\partial p} - R_{(p)}\theta, \\
\frac{1}{a\cos\varphi}\frac{\partial u}{\partial\lambda} + \frac{1}{a\cos\varphi}\frac{\partial(v\cos\varphi)}{\partial\varphi} + \frac{\partial\omega}{\partial p} &= 0, \\
\frac{D\theta}{Dt} &= \frac{\theta\hat{Q}}{c_pT} \equiv Q,
\end{aligned} \qquad (10.3\text{a,b,c,d,e})$$

where $\frac{D}{Dt} = \frac{\partial}{\partial t} + \frac{u}{a\cos\varphi}\frac{\partial}{\partial\lambda} + \frac{v}{a}\frac{\partial}{\partial\varphi} + \omega\frac{\partial}{\partial p}$, a is the mean radius of the Earth, $\Phi = gz$, z being the height of pressure surfaces, $\theta = T\left(\frac{p_{oo}}{p}\right)^{R/c_p}$, $R_{(p)} = \frac{R}{p_{oo}}\left(\frac{p_{oo}}{p}\right)^{c_v/c_p}$, $f = 2\Omega\sin\varphi$, F and G are the components of the horizontal frictional force, $\hat{Q}$ is the diabatic heating rate in $\mathrm{J\,kg^{-1}s^{-1}}$, Q is the diabatic heating rate in $\mathrm{K\,s^{-1}}$. The model domain is $p_1 \leq p \leq p_o$, $0 \leq \lambda \leq 2\pi$, $-\pi/2 \leq \varphi \leq \pi/2$. The total derivative may be alternatively rewritten in a flux form on the basis of (10.3d), $\frac{D\xi}{Dt} = \frac{\partial\xi}{\partial t} + \frac{1}{a\cos\varphi}\frac{\partial(u\xi)}{\partial\lambda} + \frac{1}{a\cos\varphi}\frac{\partial(\xi v\cos\varphi)}{\partial\varphi} + \frac{\partial(\omega\xi)}{\partial p}$.

A square bracket denotes a zonal mean property and an asterisk denotes a departure from the zonal mean, $\xi^* = \xi - [\xi]$. Each dependent variable is decomposed into a zonal mean part and a departure from it, namely $u = [u] + u^*$, $v = [v] + v^*$ and $\theta = [\theta] + \theta^*$. The departure (wave) part may be regarded as being in quasi-geostrophic balance. Local geostrophic balance and hydrostatic balance are also applicable to the zonal mean zonal flow component. We first perform the zonal-averaging operation on (10.3). Five approximations are introduced:

(i) replacing (10.3b) by local geostrophic balance and retaining the variation of the Coriolis parameter with latitude;

(ii) using $\frac{D\theta}{Dt} \approx \frac{\partial\theta}{\partial t} + \frac{1}{a\cos\varphi}\frac{\partial(u\theta)}{\partial\lambda} + \frac{1}{a\cos\varphi}\frac{\partial(\theta v\cos\varphi)}{\partial\varphi} + \omega\frac{\partial\Theta}{\partial p}$ in (10.3e), where Θ is the background part of the potential temperature field unrelated to the circulation; Θ is set to be the time-horizontal mean background potential temperature. Such an observed quantity is characterized by $\Theta_p < 0$ since the stratification is stable;

(iii) using $[uv] = [u][v] + [u^*v^*] \approx [u^*v^*]$;

(iv) using $[u\omega] \approx 0$, $[v\omega] \approx 0$; and

(v) using $[v\theta] = [v][\theta] + [v^*\theta^*] \approx [v^*\theta^*]$.

Approximations (iii) to (v) amount to neglecting the advective influence of the MMC itself and *reducing the problem to one of linear dynamics*. They are justifiable because the terms involving $[v]$ and $[\omega]$ are small. We would then get

$$[u]_t + \frac{1}{a\cos\varphi}([u^*v^*]\cos\varphi)_\phi - \frac{[u^*v^*]\tan\varphi}{a} = f[v] + [F],$$

$$0 = -\frac{1}{a}[\Phi]_\varphi - f[u],$$

$$0 = -[\Phi]_p - R_{(p)}[\theta], \qquad (10.4\text{a,b,c,d,e})$$

$$\frac{1}{a\cos\varphi}([v]\cos\varphi)_\varphi + [\omega]_p = 0,$$

$$[\theta]_t + \frac{1}{a\cos\varphi}([v^*\theta^*]\cos\varphi)_\varphi + [\omega]\Theta_p = [Q].$$

The zonal mean state of the atmosphere in this model ($[u]$, $[v]$, $[\omega]$, $[\Phi]$, $[\theta]$) is governed by a set of linear partial differential equations. If the domain of our interest is the troposphere, the eddy fluxes, $[u^*v^*]$ and $[v^*\theta^*]$, stem from baroclinic instability occurring in situ. If the domain of our interest is the stratosphere, the eddy fluxes are mostly attributable to those waves that can propagate from below. In trying to understand the dynamics of the zonal mean state per se in either case, we treat the diabatic heating $[Q]$ as well as the eddy fluxes as given and seek to represent the frictional force $[F]$ in terms of the zonal wind $[u]$. The frictional force is expected to be important primarily in a surface boundary layer.

We now define one part of the meridional circulation solely associated with the eddy heat flux denoted by (v_c, ω_c) as

$$\omega_c = -\frac{1}{a\cos\varphi\Theta_p}([v^*\theta^*]\cos\varphi)_\varphi, \qquad v_c = \left(\frac{1}{\Theta_p}[v^*\theta^*]\right)_p, \tag{10.5a,b}$$

so that these quantities satisfy the continuity equation, namely $\frac{1}{a\cos\varphi}(v_c\cos\varphi)_\varphi + \omega_{cp} = 0$. It may be represented in terms of a streamfunction equal to $\xi = \frac{\cos\varphi}{\Theta_p}[v^*\theta^*]$. This flow component would adiabatically compensate for the cooling/heating associated with a divergence/convergence of eddy heat flux. For instance, if there is a divergent heat flux, there would be a descent $(\omega_c > 0)$ to compensate for the cooling tendency because $\Theta_p < 0$. The remaining component of the mean meridional circulation is denoted by $(\tilde{v}, \tilde{\omega})$ in accordance with

$$[v] = v_c + \tilde{v}, \quad [\omega] = \omega_c + \tilde{\omega}. \tag{10.6a,b}$$

The term $(\tilde{v}, \tilde{\omega})$ is known as the *residual circulation*. It can be shown that the residual circulation is a good approximation of the zonal average mass flux as elaborated in the book by Vallis (2006, Section 7.3.3). Since we define ω_c to be a part of the vertical motion that cancels the eddy heat flux divergence, $\tilde{\omega}$ is the remaining part of the vertical motion in response to the diabatic heating and can be also associated with the temporal variation of the zonal mean temperature field; $\tilde{v}$ is the corresponding part of the zonal mean meridional velocity required by mass conservation. We should refrain from jumping to the conclusion that the residual circulation is just a diabatic circulation. It is through the temporal variation of the zonal mean temperature field that the eddy flux of momentum and other dynamical factors would have an impact on the residual circulation. It is imperative to analyze the dynamics of residual circulation on the basis of the equation that governs its structure.

In light of the definitions of (10.5) and (10.6), (10.4a) can be rewritten as

$$\frac{\partial}{\partial t}[u] = f\tilde{v} + [F] + \frac{1}{a\cos\varphi}\frac{\partial(-[u^*v^*]\cos\varphi)}{\partial\varphi} + \frac{[u^*v^*]\tan\varphi}{a} + \frac{\partial}{\partial p}\left(\frac{f}{\Theta_p}[v^*\theta^*]\right).$$

If we define $\hat{\vec{E}} \equiv (\hat{E}_1, \hat{E}_2) = \left(-a[u^*v^*]\cos\varphi, \frac{fa\cos\varphi}{\Theta_p}[v^*\theta^*]\right)$, then the divergence of this vector is

$$\begin{aligned}
\nabla\cdot\hat{\vec{E}} &= \frac{1}{a\cos\varphi}\frac{\partial(\hat{E}_1\cos\varphi)}{\partial\varphi} + \frac{\partial\hat{E}_2}{\partial p} \\
&= \frac{1}{a\cos\varphi}\frac{\partial}{\partial\varphi}\left(-a[u^*v^*]\cos^2\varphi\right) + \frac{\partial}{\partial p}\left(\frac{fa\cos\varphi}{\Theta_p}[v^*\theta^*]\right) \\
&= a\cos\varphi\left\{\left(\frac{1}{a\cos\varphi}\frac{\partial}{\partial\varphi}(-[u^*v^*]\cos\varphi) + \frac{[u^*v^*]\tan\varphi}{a}\right) + \frac{\partial}{\partial p}\left(\frac{f}{\Theta_p}[v^*\theta^*]\right)\right\}.
\end{aligned}$$

Thus, we can rewrite (10.4) as

$$\begin{aligned}
\frac{\partial}{\partial t}[u] &= f\tilde{v} + \left(\frac{1}{a\cos\varphi}\right)\nabla\cdot\hat{\vec{E}} + [F], \\
f[u]_p &= \frac{R_{(p)}}{a}[\theta]_\varphi, \\
\frac{1}{a\cos\varphi}(\tilde{v}\cos\varphi)_\varphi + \tilde{\omega}_p &= 0, \\
\frac{\partial}{\partial t}[\theta] + \tilde{\omega}\Theta_p &= [Q].
\end{aligned} \qquad (10.7\text{a,b,c,d})$$

Here $\hat{\vec{E}}$ is called the *Eliassen–Palm* (EP) *vector* in the (φ, p) plane. Up to this point, we retain the curvature term $\frac{[u^*v^*]\tan\varphi}{a}$. The equations in (10.7) are known as the quasi-geostrophic form of the *transformed Eulerian mean (TEM) equations* in isobaric coordinates. The eddy momentum and heat fluxes appear only in (10.7a) in a single quantity $\left((a\cos\varphi)^{-1}\nabla\cdot\hat{\vec{E}}\right)$, which embodies the influence of the waves on the residual circulation. Had we also neglected the curvature term $\left(-\frac{uv\tan\varphi}{a}\right)$ in (10.4a), the eddy forcing term would be simplified to $\frac{1}{a\cos\varphi}\nabla\cdot\hat{\vec{E}} = \frac{1}{a\cos\varphi}\frac{\partial}{\partial\varphi}(-[u^*v^*]\cos\varphi) + \frac{\partial}{\partial p}\left(\frac{f}{\Theta_p}[v^*\theta^*]\right) \equiv \nabla\cdot\vec{E}$ with $\vec{E} \equiv (E_1, E_2) = \left(-[u^*v^*], \frac{f}{\Theta_p}[v^*\theta^*]\right)$. The simplified form of QG-EP vector is sufficiently accurate in practice since the curvature term is negligibly small for the eddies of synoptic scale. It is, in a way, more consistent to do so since we have not taken into consideration the curvature term $\frac{u^2\tan\varphi}{a}$ in the v-momentum equation. We have already elaborated on the different relationships between the QG-EP vector and other quadratic properties of the wave field in Section 6.6.2.

We should not consider the case of steady state, or else $\tilde{v}$ and $\tilde{\omega}$ can be computed separately with (10.7a) and (10.7d). They would not be compatible except under very special combinations of diabatic heating, EP-vector field and frictional force. For finite temporal variations of the zonal mean field, we may eliminate the time derivative terms between (10.7a) and (10.7d) by virtue of the thermal wind relation (10.7b). This would further narrow our focus on the dynamics of the residual circulation per se on the meridional plane. The resulting equation is then

$$f^2\tilde{v}_p + \frac{R_{(p)}\Theta_p}{a}\tilde{\omega}_\varphi = \frac{R_{(p)}}{a}[Q]_\varphi - f\left(\nabla\cdot\vec{E}\right)_p - f[F]_p; \qquad (10.8)$$

$[F]$, $[Q]$ and $\vec{E}$ are now treated as given. With the introduction of a mass streamfunction of the residual circulation, $\eta(\varphi, p, t)$, satisfying

$$\tilde{v}\cos\varphi = \eta_p, \quad \tilde{\omega}\cos\varphi = -\frac{1}{a}\eta_\varphi, \qquad (10.9)$$

we finally rewrite (10.8) as

$$\begin{aligned}
&f^2\frac{\partial^2\eta}{\partial p^2} + \frac{-R_{(p)}\Theta_p\cos\varphi}{a^2}\frac{\partial}{\partial\varphi}\left(\frac{1}{\cos\varphi}\frac{\partial\eta}{\partial\varphi}\right) \\
&= \frac{R_{(p)}\cos\varphi}{a}\frac{\partial[Q]}{\partial\varphi} - f\cos\varphi\frac{\partial\left(\nabla\cdot\vec{E}\right)}{\partial p} - f\cos\varphi\frac{\partial[F]}{\partial p}
\end{aligned} \qquad (10.10)$$

This is a second-order partial differential equation for one unknown, $\eta(\varphi, p, t)$, which may be interpreted as a response to various types of forcing.

The inhomogeneous terms on the RHS of (10.10) represent three types of forcing. The first type of forcing is thermal in nature. It is proportional to the meridional gradient of the net zonal mean diabatic heating. The second type of forcing is mechanical in nature associated with the large-scale eddies. It is proportional to the vertical gradient of the divergence of the $\vec{E}$ vector field. We refer to it as eddy forcing for short. The third type of forcing is frictional in nature associated with the small turbulent eddies. It is proportional to the vertical gradient of the zonal mean frictional force. We use a Newtonian drag law to parameterize the frictional force for simplicity, namely $F = -\kappa u$. Since frictional force is much larger in a boundary layer near the surface, it would be reasonable to use a vertically varying frictional coefficient. We assume it to be $\kappa = \kappa_o$ for $p_1 \leq p \leq p_s$ and $\kappa(p) = \kappa_o(p/p_1)^2$ for $p \leq p_1$ with $\kappa_o = 2.5 \times 10^{-6}\,\mathrm{s}^{-1}$, $p_s = 1000\,\mathrm{mb}$ and $p_1 = 900\,\mathrm{mb}$. The corresponding damping time is about 4 days. The frictional forcing is evaluated as $f\kappa \cos\varphi [u]_p$. A time-mean component of the forcing would give rise to a time-mean residual circulation. Furthermore, the total streamfunction for the Eulerian MMC is simply $\chi = \xi + \eta$ such that $[\omega] = -\chi_\varphi / a\cos\varphi$ and $[v] = \chi_\varphi / \cos\varphi$. It should be added that there would be temporal fluctuations in η corresponding to particular temporal variations of $[Q]$, $\vec{E}$ and $[u]$.

A set of meaningful boundary conditions for η are:

(a) no circulation across the poles would require $\chi = \xi = \eta = 0$ at $\varphi = \pm\frac{\pi}{2}$,
(b) no vertical motion and negligible heat flux at the top of the domain would require $[\omega] = 0$ and $\xi = 0$ implying $\chi = 0$ and hence $\eta = 0$ at $p = p_1$, and
(c) no vertical motion at the lower boundary would require $[\omega] = 0$ and $\xi = 0$ implying $\chi = 0$ and hence $\eta = -\xi = \dfrac{\cos\varphi}{-\Theta_p}[v^*\theta^*]$ at $p = p_o$.

It follows from (c) that the meridional heat flux at the surface would induce vertical motion in the residual circulation across $p = p_o$. The contours of η would then intersect the $p = p_o$ surface. From a mathematical point of view, the inhomogeneous boundary condition (c) influences the solution as if there were a forcing on an infinitely thin layer next to the boundary.

The dynamics of the residual circulation and the Ferrel cell in the extratropics can be adequately investigated with this model because the eddies are associated with a quasi-geostrophic flow. That is why it is justifiable to have neglected the effect of advection by the weak zonal mean meridional flow itself. This is however not the case for the Hadley cell. Therefore, when we apply the model to a global domain, we should keep in mind the limitation of the result for the tropical part of the domain.

The physical significance of the Eliassen–Palm vector is easier to appreciate in the context of quasi-geostrophic dynamics. We have shown in Chapter 6 that the QG-PV with beta-plane approximation in height coordinates is

$$q = f_o + \beta y + \psi_{xx} + \psi_{yy} + \frac{f_o^2}{\rho_o}\left(\frac{\rho_o}{N^2}\psi_z\right)_z, \qquad u = -\psi_y, \qquad v = \psi_x, \qquad \theta = \frac{f_o \theta_{oo} \psi_z}{g}, \tag{10.11}$$

where $\psi = gp'/f_o$ is the geostrophic streamfunction with p' being the pressure departure from its horizontal mean; ρ_o and N are the background density and Brunt–Vaisala frequency, which are assumed to be at most functions of z. It follows that the eddy flux of QG-PV is simply

$$[v^*q^*] = -[u^*v^*]_y + \frac{f_o g}{\rho_o \theta_{oo}} \left(\frac{\rho_o}{N^2} [v^*\theta^*] \right)_z . \tag{10.12}$$

The RHS of (10.12) has the form of the divergence of a vector,

$$[v^*q^*] = \frac{1}{\rho_o} \nabla \cdot \vec{F},$$
$$\text{where} \qquad \vec{F} = \left(-\rho_o [u^*v^*], \frac{f_o g \rho_o}{\theta_{oo} N^2} [v^*\theta^*] \right), \tag{10.13}$$

with $\nabla = (\partial/\partial y, \partial/\partial z)$. The vector $\vec{F}$ in height coordinates for a beta-plane domain is the counterpart of the EP-vector in isobaric coordinates, $\vec{E}$, introduced in this section. Its y-component is the Reynolds stress of large-scale eddies with a dimension of $\mathrm{kg\,m\,s^{-2}\,m^{-2}}$. Its z-component is proportional to the Coriolis parameter and eddy heat flux and is inversely proportional to stratification. Thus, the divergence of the EP-vector is the eddy flux of PV.

10.3.2 Climatological distributions of eddy, diabatic and frictional forcing

Our goal is to determine the residual circulation satisfying (10.10). It is necessary to first empirically establish the three types of forcing. The distribution of climatological annual mean eddy momentum flux $\left[\overline{u^*v^*}\right]$ has been shown in Fig. 5.19. We see that there is strong poleward flux of zonal momentum in the upper troposphere centering at the jet-stream level (about 200 mb) in the mid-latitudes (from about 20° to 55° latitude) of both hemispheres. The maximum poleward momentum flux is about 55 $\mathrm{m^2\,s^{-2}}$. In the northern hemisphere, the momentum flux is directed up-gradient of the zonal velocity (towards the jet core) to the south of about 30° N and down-gradient (away from the jet core) to the north of ~30° N. For this reason alone, the eddy momentum flux cannot be simply parameterized as a diffusive process. The dynamical nature of such an intriguing feature has been discussed in Section 3.8.5. There is a considerably weaker equatorward eddy momentum flux to the north of 55° N. It is also worthy of note that there is significant eddy momentum flux at relatively high latitudes in the lower stratosphere. For the reason elaborated in Section 9.3.2, the eddy flux at the stratospheric levels is only associated with the planetary-scale waves, whereas the eddy fluxes by the synoptic-scale waves and planetary waves are both large in the troposphere.

The corresponding distribution of the annual-zonal mean eddy heat flux $\left[\overline{v^*\theta^*}\right]$ has been shown in Fig. 5.20 (heat flux is actually $c_p\left[\overline{v^*\theta^*}\right]$ where c_p is specific heat). It is poleward at all latitudes. The largest tropospheric value is at about 850 mb near 55° latitude. Recall that since the time-zonal mean temperature in the troposphere decreases poleward at all latitudes, the eddy heat flux is akin to a diffusive process in sharp contrast to the eddy

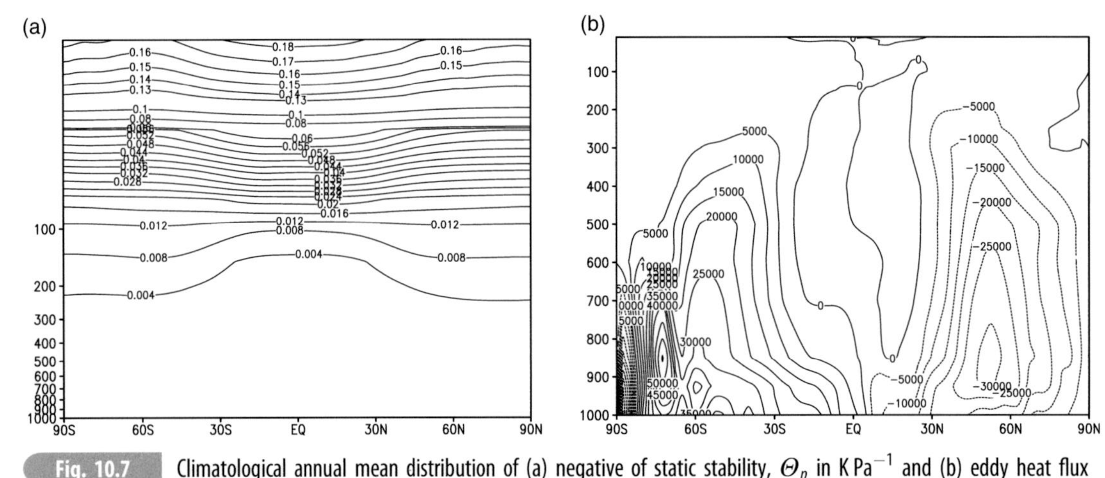

Fig. 10.7 Climatological annual mean distribution of (a) negative of static stability, Θ_p in K Pa^{-1} and (b) eddy heat flux weighted by static stability $\left[\overline{v^*\theta^*}\right]/\Theta_p$ in m s^{-1} Pa.

momentum flux. The values of $\left[\overline{v^*\theta^*}\right]$ are also large at the lower stratospheric levels. Maximum values are found at about 60° latitude indicating a large flux of sensible heat into the two polar regions. We should remind ourselves that when the mass of each atmospheric layer of a given geometric thickness is taken into account, the amount of heat transport in it is actually much larger in the troposphere than that in the stratosphere.

Since the vertical component of the EP-vector actually depends on the eddy heat flux weighted by the static stability, it is instructive to examine $\left[\overline{v^*\theta^*}\right]/\Theta_p$ in isobaric coordinates. The distributions of the annual mean Θ and $\left[\overline{v^*\theta^*}\right]/\Theta_p$ are shown in Fig. 10.7. Since Θ increases monotonically with height and has much larger values in the stratosphere, the largest values of $\left[\overline{v^*\theta^*}\right]/\Theta_p$ are in the lower troposphere (Fig. 10.7b). This statistic has particularly large vertical gradient in the tropical latitudes.

Using the results shown in Figs. 10.6 and 10.7b, we compute the distribution of $\vec{\hat{E}}$ and its divergence in isobaric-spherical coordinates. The domain is the latitude–pressure cross-section, $-\pi/2 \leq \varphi \leq \pi/2$ and $p_1 \leq p \leq p_o$. It should be emphasized that the two components of this E-vector field have different dimensions, namely m^{2}s^{-2} and m s^{-2}Pa respectively. It would be most useful to plot the EP-vectors so that the orientation of the vectors would be independent of the particular units being used. We therefore multiply the vertical component with the aspect ratio of the domain, $\tilde{E}_2 \equiv \dfrac{a}{p_o}E_2$, so that $\tilde{E}_2$ and E_1 have the same dimensions. This would give us an undistorted view of the orientation of the E-vectors in the atmosphere. A negative value of the vertical component is plotted as a component pointing towards lower pressure in both hemispheres. The vector field $\vec{\tilde{E}}_2 = \left(E_1, \tilde{E}_2\right)$ on a cross-section $-\pi/2 \leq \varphi \leq \pi/2$ and $\tilde{p}_1 \leq \tilde{p} \leq 1$ where $\tilde{p}_1 = p_1/p_o$ is shown in Fig. 10.8. $\varphi = \pm\pi/2$ on the horizontal axis are labeled as 90N and 90S respectively, and $\tilde{p} = 1$ on the vertical axis is labeled as 1000 mb in Fig. 10.8.

The annual mean $\vec{\tilde{E}}$ vectors are mostly vertical at the tropospheric levels indicating that the eddy heat flux term is the dominant component. There is a slight indication of the vectors bending towards the equator in the upper troposphere where the eddy momentum flux is significant as seen in Fig. 5.19. The distribution of the divergence of the annual

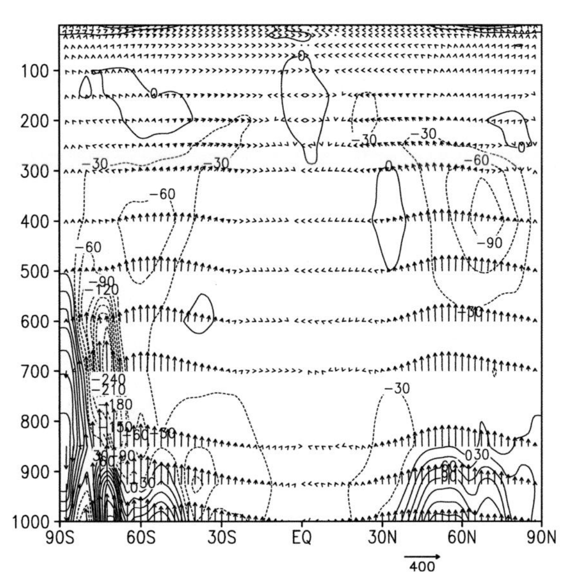

Fig. 10.8 Climatological annual mean distribution of the Eliassen–Palm vector field $\vec{\tilde{E}} = \left(E_1, \tilde{E}_2\right)$ where $\tilde{E}_2 = \dfrac{a}{p_o} E_2$ and its divergence, $\nabla \cdot \vec{E} = \dfrac{1}{a\cos\varphi}\dfrac{\partial(E_1\cos\varphi)}{\partial\varphi} + \dfrac{\partial E_2}{\partial p}$ (contours) in 10^{-6} m s^{-2}. Vertical coordinate is pressure in mb. Horizontal coordinate is latitude.

mean EP-vector field is also shown by the contour lines in Fig. 10.8. We find a maximum positive value at 1000 mb level centering at 60° N and one at 60° S. The large values are in the zone between 30° and 75° in both hemispheres. The value decreases sharply with increasing height up to about 850 mb level. There is convergence of the EP-vectors aloft, particularly at the upper tropospheric levels in mid-latitudes. Again, the large values near the south pole appear to be spurious and are not trustworthy. We examine the implications of this EP-vector field in the context of the dynamical relationship between the eddies and the zonal mean flow.

The net diabatic heating $[Q]$ can be estimated by a *residual method*. The sum of all terms in $\left[\dfrac{D\theta}{Dt}\right]$ at each point may be regarded as an indirect measure of the net diabatic heating rate. Such an annual mean result computed with the same dataset is shown in Fig. 10.9. The major feature is an equatorial zone of significant net diabatic heating in the troposphere between about 15° N and 15° S. There is a slight asymmetry about the equator with a maximum value of about 2 K day^{-1} located at about 450 mb level over 5° N. The strong heating is clearly associated with the two Hadley cells. It is flanked by a weaker net diabatic cooling rate on both sides. The large noisy values at the lower tropospheric levels in the south polar region are not trustworthy. The significant values in the lower stratosphere are expected to have a relatively weak impact because the stratospheric stratification is very large. Also noteworthy is the nearly zero values of net diabatic heating in the extratropics of both hemisphere where the Ferrel cells are located. This heating field is used in the evaluation of the first term of the forcing in (10.10).

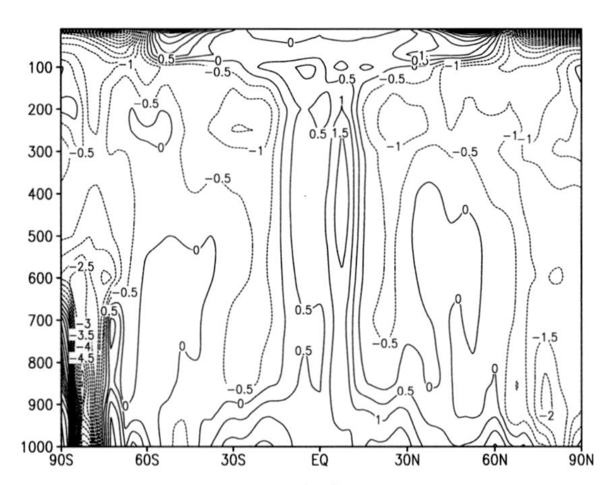

Fig. 10.9 Climatological annual-zonal mean diabatic heating rate $\left[\overline{Q}\right]$ in $10^{-5}\mathrm{K\,s^{-1}}$. Vertical coordinate is pressure in mb. Horizontal coordinate is latitude in degrees.

It is instructive to make a comparison of the three forms of forcings. The climatological annual mean distributions of the diabatic forcing, $\dfrac{R_{(p)}\cos\varphi}{a}\left[\overline{Q}\right]_{\varphi}$, the eddy forcing, $-f\cos\varphi\overline{(\nabla\cdot\vec{E})}_p$, and the frictional forcing, $f\kappa\cos\varphi[\overline{u}]_p$, are shown in Fig. 10.10a,b,c respectively. The diabatic forcing has a characteristic columnar distribution extending from the surface to the tropopause with pronounced latitudinal variations. It has positive values reaching about $7\times10^{-14}\mathrm{m\,s^{-3}Pa^{-1}}$ in a relatively broad belt between 5° N and 25° S and comparable negative values between 5° N and 20° N. The eddy forcing is evaluated using the empirical annual mean $\nabla\cdot\vec{E}$ shown in Fig. 10.8. The extreme values of the eddy forcing are almost twice as large as the diabatic forcing, but they are concentrated mostly in a thin lower tropospheric layer in mid-latitude of both hemispheres. There is still a broad pattern of significant values in the upper troposphere. Recall that the frictional forcing is $-f\cos\varphi\kappa[u]_p$. It is broadly asymmetric about the equator because of its dependence on the Coriolis parameter. It has its relatively small but significant values in the lower troposphere since κ is largest there. Its significant values at the extratropical upper tropospheric levels stem from the large vertical shear. Its values are smaller than the diabatic forcing and eddy forcing.

The distribution of the total forcing is presented in Fig. 10.10d. Its appearance contains features of the three forcings. The response in the residual circulation to a particular forcing depends very much on its location because of the strong dependence on latitude and stratification in the coefficients in the governing equation (10.10). It does not warrant immediately drawing a conclusion that the residual circulation is largely attributable to the eddy forcing in light of the differences in the magnitude of the three forcings.

It is important to pay close attention to the lower boundary condition, $\eta=\dfrac{1}{-\Theta_p}[v^*\theta^*]\cos\varphi$ at $p=p_o$. The annual mean distribution of this quantity is shown by the solid curve in Fig. 10.11. It has significant positive (negative) values in the extratropics of the northern (southern) hemisphere. The maximum magnitude of the annual mean distribution is about $2\times10^4\,\mathrm{m\,s^{-1}\,Pa}$. We will see that an interesting feature of the residual circulation stems from this boundary condition.

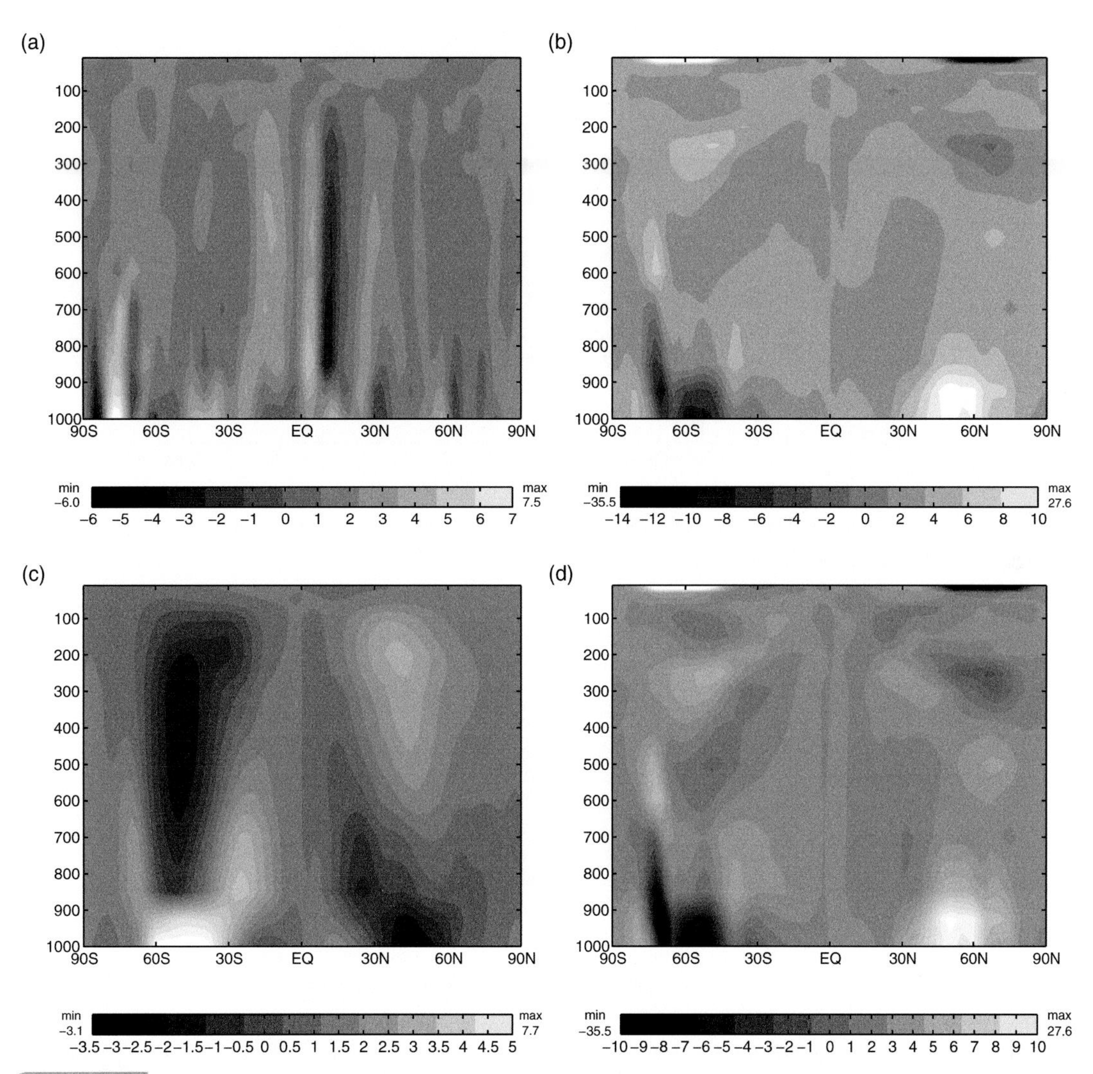

Fig. 10.10 Climatological annual mean (a) diabatic forcing, $\frac{R_{(p)}\cos\varphi}{a}\left[\overline{Q}\right]_{\varphi}$, (b) eddy forcing, $-f\cos\varphi(\overline{\nabla\cdot\vec{E}})_p$, (c) frictional forcing, $f\kappa\cos\varphi[\overline{u}]_p$, and (d) total forcing of the residual circulation. Values in (a) are in units of $10^{-14}\,\mathrm{m\,s^{-3}(Pa)^{-1}}$ and those of (b), (c) and (d) are in units of $10^{-13}\,\mathrm{m\,s^{-3}(Pa)^{-1}}$. Vertical coordinate is pressure in mb. Horizontal coordinate is latitude.

10.3.3 Structure of the annual mean residual circulation

Now we are in a position to solve (10.10) for the annual residual circulation. The annual mean residual circulation in terms of the streamfunction in response to the total forcing is shown in Fig. 10.12. There is essentially one huge cell in each hemisphere. The cell in the southern hemisphere is slightly stronger. There is an ascending branch in the low latitudes. The mass flux is supported by an ascent through p_o in the low latitudes since η increases

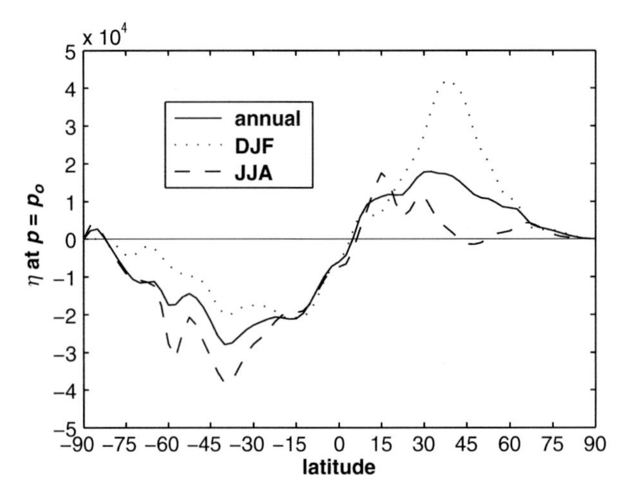

Fig. 10.11 Variation of the climatological annual, summer and winter mean $\eta = \frac{1}{-\Theta_p}[v^*\theta^*]\cos\varphi$ at $p = p_o$ with latitude, in $\mathrm{m\,s^{-1}Pa}$.

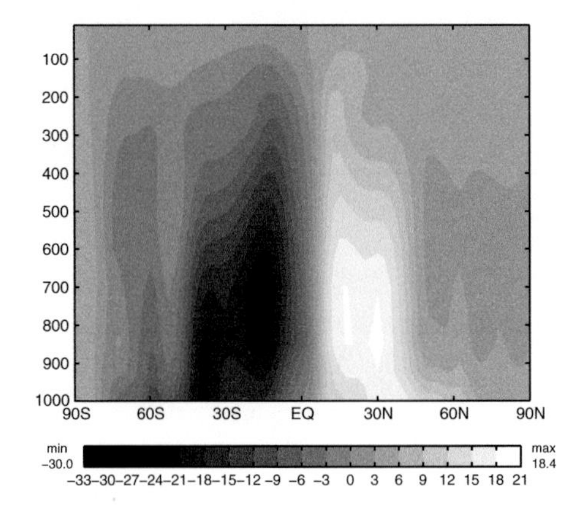

Fig. 10.12 Climatological annual mean streamfunction of the residual circulation in units of $10^3\ \mathrm{m\,s^{-1}\,Pa}$. Vertical coordinate is pressure in mb. Horizontal coordinate is latitude.

poleward. There is a descending branch in the subtropics, which changes to a poleward flux extending to beyond 60° latitude. The value of η at p_o between 35° and 65° decreases with latitude indicating a descent through p_o. By inference, there is an equatorward mass flux in a thin surface layer. It would supply the mass flux for the ascending motion at p_o in low latitudes. The maximum value of the streamfunction in the northern hemisphere is about $24 \times 10^3\ \mathrm{m\,s^{-1}Pa}$. From this we can deduce $\tilde{v}_{\max} \sim 1.2\,\mathrm{m\,s^{-1}}$ and $\tilde{\omega}_{\max} \sim 1.4 \times 10^{-3}\ \mathrm{Pa\,s^{-1}}$ corresponding to $\tilde{w}_{\max} \sim 0.3 \times 10^{-2}\ \mathrm{m\,s^{-1}}$. Its overall structure is reminiscent of Fig. 10.6 deduced by Held and Schneider (1999) from the output of a high resolution GCM under a solstice condition. It is a thickness-weighted mean flow on the vertical cross-section approximately measuring the Lagrangian mean motion on the meridional plane.

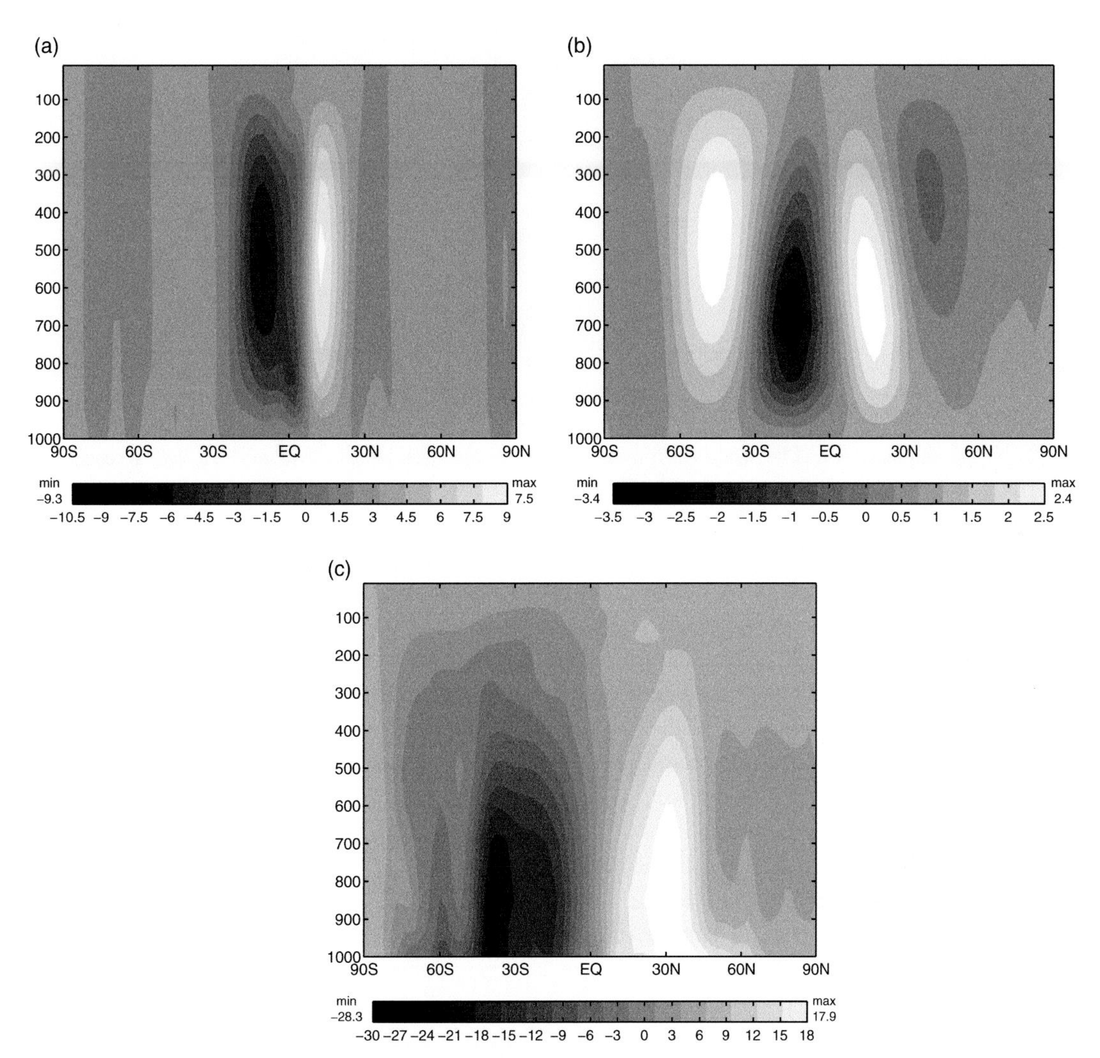

Fig. 10.13 Streamfunction of the climatological annual mean residual circulation associated with (a) diabatic heating forcing, (b) frictional forcing and (c) eddy forcing in units of 10^3 m s^{-1}Pa. Vertical coordinate is pressure in mb. Horizontal coordinate is latitude.

We can determine the streamfunction of a residual circulation in response to each forcing separately because the governing equation (10.10) is a linear partial differential equation. Since the three parts of the forcing are associated with distinctly different physical processes, we must partition the lower boundary condition in a physically consistent manner. As the lower boundary condition stems from the eddy heat flux at p_o, it would be logical to use it exclusively in conjunction with the eddy forcing. We therefore use $\eta = 0$ as the lower boundary condition at $p = p_o$ when we solve for the response to the diabatic forcing or frictional forcing alone.

The streamfunction in response to the diabatic forcing alone is shown in Fig. 10.13a. Because of the dependence on f^2 in the coefficient of $\partial^2/\partial p^2$ and on Θ_p in the coefficient

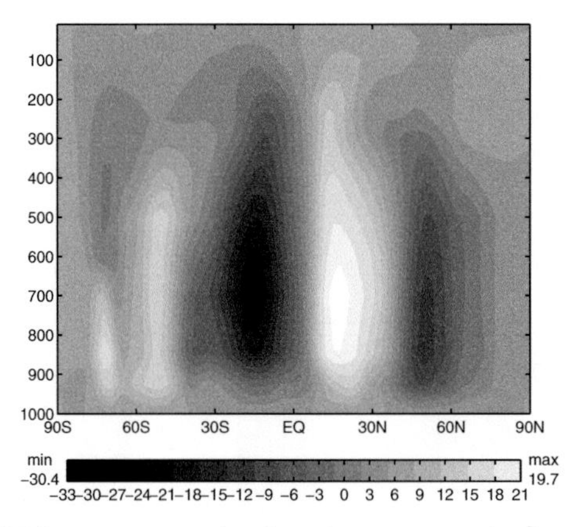

Fig. 10.14 Streamfunction of the model Eulerian mean meridional circulation in units of 10^3 m s^{-1}Pa. Vertical coordinate is pressure in mb. Horizontal coordinate is latitude.

of the other term of the LHS of (10.10), the response $\eta(\varphi, p)$ to the diabatic heating is substantial only in the tropical tropospheric region. We find a cell in each hemisphere rising at the equatorial latitudes, reaching a peak value of 9×10^3 m s^{-1} Pa. It is about 20° latitude wide. The boundary between the two cells is located off the equator over ~5° N. That coincides with the latitude of the maximum annual mean diabatic heating.

The streamfunction of the residual circulation in response to the frictional forcing shows a multi-cell structure, roughly asymmetric about the equator (Fig. 10.13b). A pair of cells exist in the tropical region. The strength of these cells is four times weaker than the diabatically forced cells, about 2×10^3 m s^{-1} Pa. The additional pair of cells extend to fairly high latitude in each hemisphere.

The streamfunction of the residual circulation in response to the eddy forcing in conjunction with the surface eddy heat flux has a totally different structure (Fig. 10.13c). There is one cell in each hemisphere, centering at the subtropical latitude. The strength of this component of the residual circulation is twice as strong as that associated with the diabatic heating, $\sim 20 \times 10^3$ m s^{-1}Pa. There is ascending mass flux in the tropical latitudes and descending mass flux between 30° and 45°, connecting to a poleward flow in the lower troposphere. The latter penetrates to high latitudes. The value of η at p_o decreases poleward implying a downward mass flux across the p_o level. By inference, there must be a surface equatorward return mass flux associated with this component. We may conclude that the poleward mass flux in the lower troposphere and the compensating return equatorward mass flux next to the surface arise largely from the eddy forcing in conjunction with its surface heat flux. The sum of these three panels is of course equal to Fig. 10.12.

We can immediately obtain the streamfunction of the Eulerian mean meridional circulation as $\chi = \eta + \xi$ where $\xi = \dfrac{1}{\Theta_p}[v^*\theta^*]\cos\varphi$. The result shows a thermally direct cell in the tropical region and a considerably weaker thermally indirect cell in mid-latitude of each hemisphere (Fig. 10.14). They are the counterparts of the Hadley cell and the Ferrel cell respectively in this linear dynamical model. In other words, this model result captures the essential aspects of the observed overturning circulation. We may conclude that the dynamics of the Ferrel cell is almost exclusively an eddy driven circulation. Without the eddies, we cannot account for its thermally indirect cellular structure at all. In contrast, the model still can simulate the key features of the Hadley cell without considering the eddies. The results testify the crucial role of the tropical diabatic heating with noticeable contribution from the frictional forcing.

10.3.4 Structure of the seasonal meridional overturning circulation

In response to the strong seasonal variation in the solar insolation, the atmospheric circulation in each hemisphere has pronounced seasonal fluctuation. The land cover in the northern hemisphere is much larger than that in the southern hemisphere. The stationary waves are naturally much stronger in the northern hemisphere than in the southern hemisphere particularly in winter. The zonal mean fluxes by the transient eddies turn out to be not too different in both hemispheres because the zonal mean states of the atmosphere in both hemispheres are comparably unstable baroclinically speaking. It follows that the total eddy forcing should be stronger in the northern hemisphere in winter. Such asymmetry would accordingly manifest in the residual circulation. Only by means of making model computations for both seasons separately would we be able to quantify the seasonal differences in the residual circulation. It suffices to present only the key counterpart figures for each season and discuss the key aspects of the dynamics.

Winter characteristics

The lower boundary condition $\eta = \frac{1}{-\Theta_p}[v^*\theta^*]\cos\varphi$ at $p = p_o$ in winter is shown by the short dashed curve in Fig. 10.11. Its maximum magnitude is twice as large in the northern hemisphere as in the southern hemisphere reaching $4 \times 10^4\,\mathrm{m\,s^{-1}Pa}$ at about 45° N. The location of the maximum magnitude in the southern hemisphere shifts equatorward to about 15° S.

The climatological distributions of the winter mean (DJF) diabatic heating, the $\vec{E}$ vector field and $\nabla \cdot \vec{E}$ are shown in Fig. 10.15a and 10.15b. There is substantial net diabatic heating in the troposphere between 10° N and 20° S and cooling between 10° N and 25° N as well as between 20° S and 35° S. We see in Fig. 10.15b that the eddy fluxes in the winter hemisphere are indeed much stronger than in the summer hemisphere. The $\vec{E}$ vectors at the lower tropospheric levels in the extratropics of the winter hemisphere are longest, point upward and then progressively turn southward in the upper troposphere indicating the direction of the propagation of Rossby waves as explained in Section 6.6.2. The maximum value of the winter (annual) mean $\nabla \cdot \vec{E}$ in the northern hemisphere is about $210 \times 10^{-6}\,\mathrm{m\,s^{-2}}$ $(120 \times 10^{-6}\,\mathrm{m\,s^{-2}})$.

It suffices to just show the climatological winter mean residual circulation in Fig. 10.16. The hemispheric asymmetry is very pronounced in each component of the residual circulation and in the total streamfunction. It is indeed considerably stronger in the northern hemisphere than in the southern hemisphere as expected (45 vs. 27 units of $10^3\,\mathrm{m\,s^{-1}Pa}$). The overall structure of this residual circulation is comparable to the residual mean meridional streamfunction obtained by Held and Schneider (1999) corresponding to

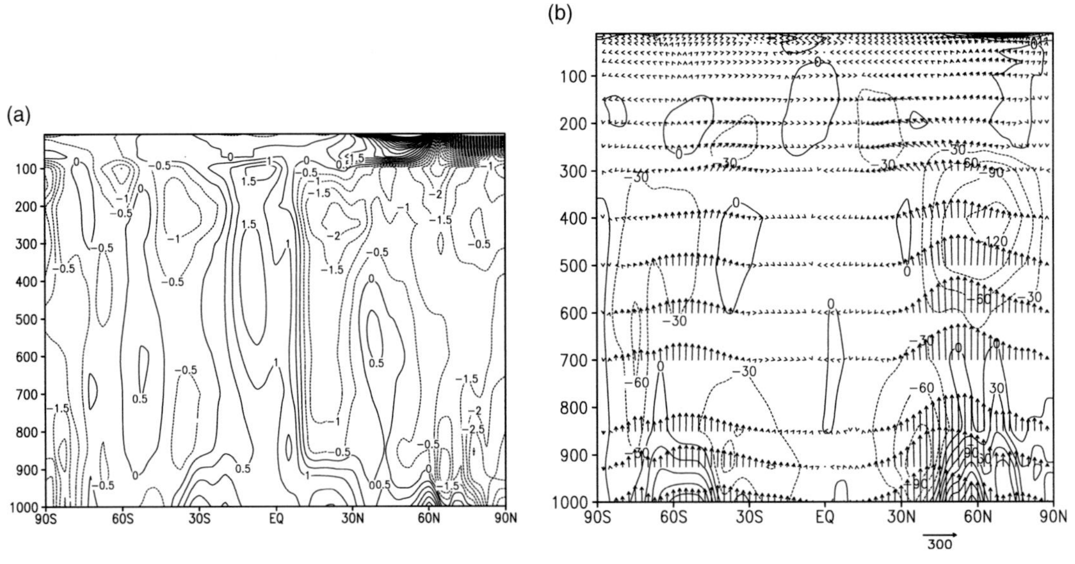

Fig. 10.15 Climatological winter mean (a) diabatic heating rate, $[\overline{Q}]$ in $10^{-5}\,\mathrm{K\,s^{-1}}$, and (b) Eliassen–Palm vector field $\hat{\vec{E}} = (E_1, \hat{E}_2)$ where $\hat{E}_2 = \frac{a}{p_o}E_2$ and its divergence, $\nabla \cdot \vec{E} = \frac{1}{a\cos\varphi}\frac{\partial(E_1\cos\varphi)}{\partial\varphi} + \frac{\partial E_2}{\partial p}$ (contours) in $10^{-6}\,\mathrm{m\,s^{-2}}$. Vertical coordinate is pressure in mb. Horizontal coordinate is latitude.

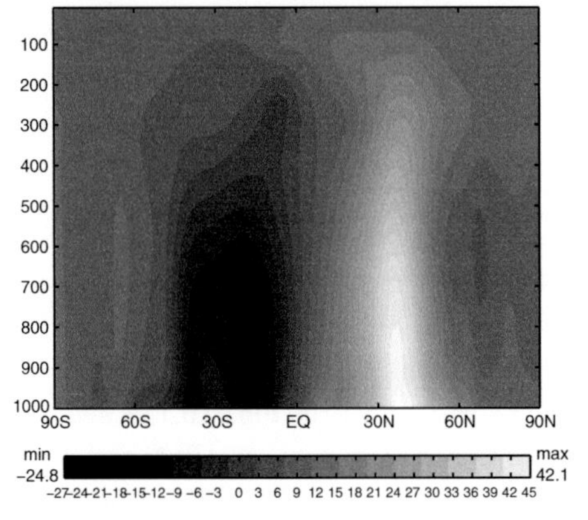

Fig. 10.16 Climatological winter mean residual circulation in units of $10^3\,\mathrm{m\,s^{-1}\,Pa}$. Vertical coordinate is pressure in mb. Horizontal coordinate is latitude.

their isentropic mass circulation (Fig. 10.8). It is also comparable with the result shown in the text by Holton (1992, Fig. 10.10.9) based on the data of Schubert *et al.* (1990).

The streamfunction of the model Eulerian winter mean meridional circulation, $\chi = \eta + \xi$, is shown in Fig. 10.17. The overall structure is quite realistic. The model Hadley cell in the northern hemisphere is much stronger than that in the southern hemisphere. The Hadley cell is much stronger and wider than the Ferrel cell in each hemisphere as we have seen in Fig. 10.1. The Ferrel cell in the winter hemisphere is twice as strong as that in the summer hemisphere as one would expect since the large-scale eddies are much more vigorous in winter.

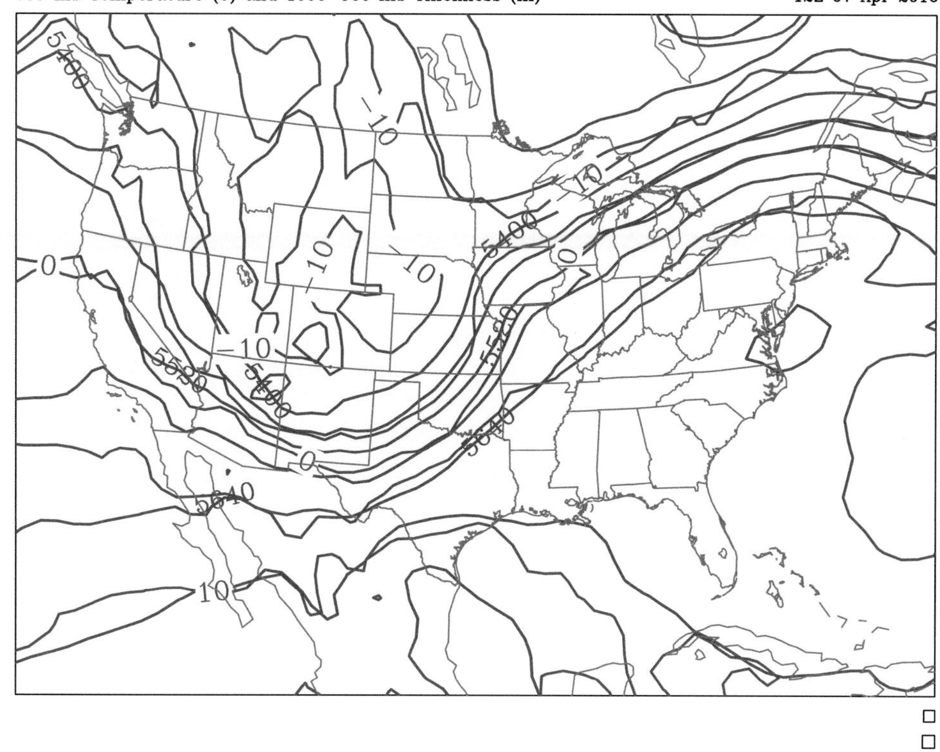

Fig. 2.2 Distributions of temperature at 700 mb (red) and thickness of the layer between 500 mb and 1000 mb (black) on April 7, 2010.

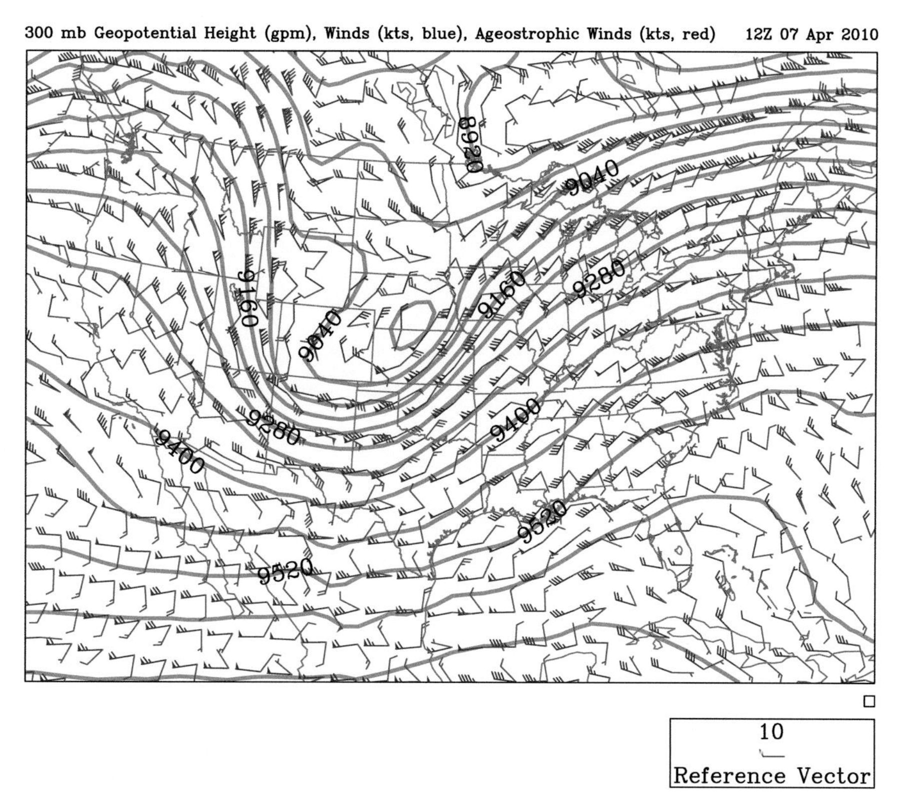

Fig. 2.4 Distributions of the actual wind (blue arrows), the ageostrophic wind (red arrows) and the contours of the geopotential height at 300 mb on April 7, 2010.

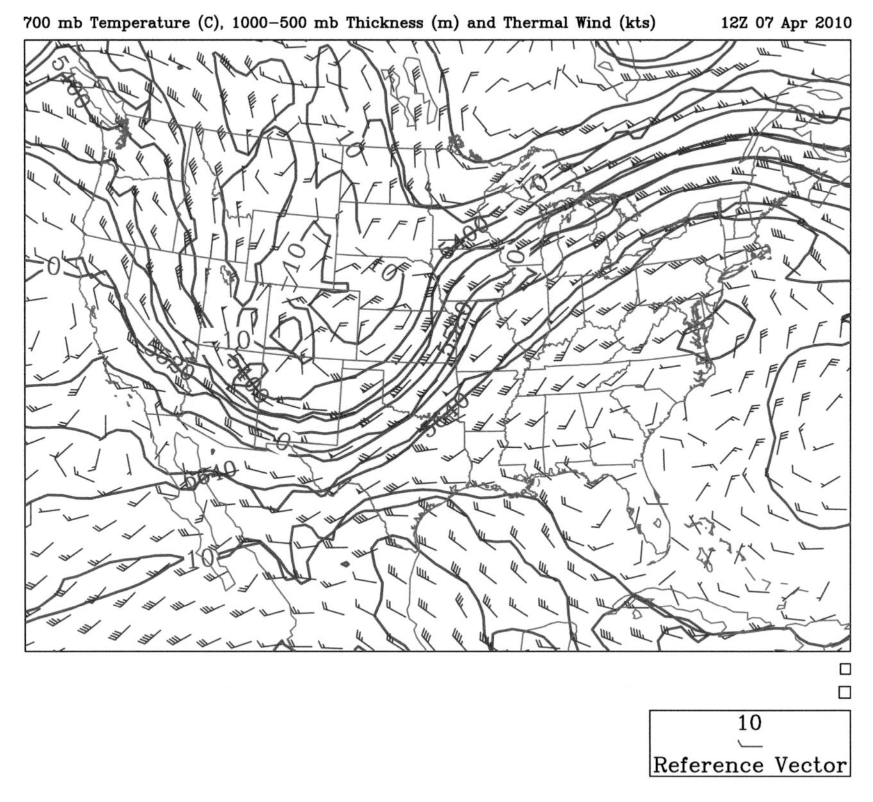

Fig. 2.7 Distributions of 700 mb temperature (red contour), thickness of the layer between 500 mb and 1000 mb (black), thermal wind vector field of this layer (arrows) on April 7, 2010.

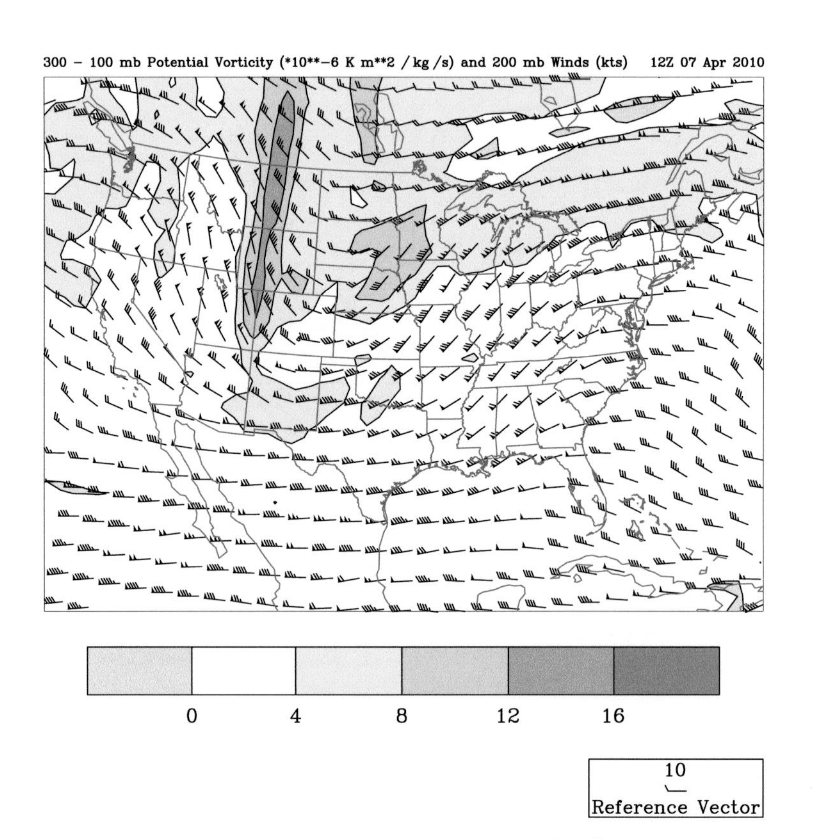

Fig. 3.12 Distributions of potential vorticity of the 300–100 mb layer in 10^{-6} m^2 s^{-1} kg^{-1} K and the 200 mb wind field in knots on April 7, 2010.

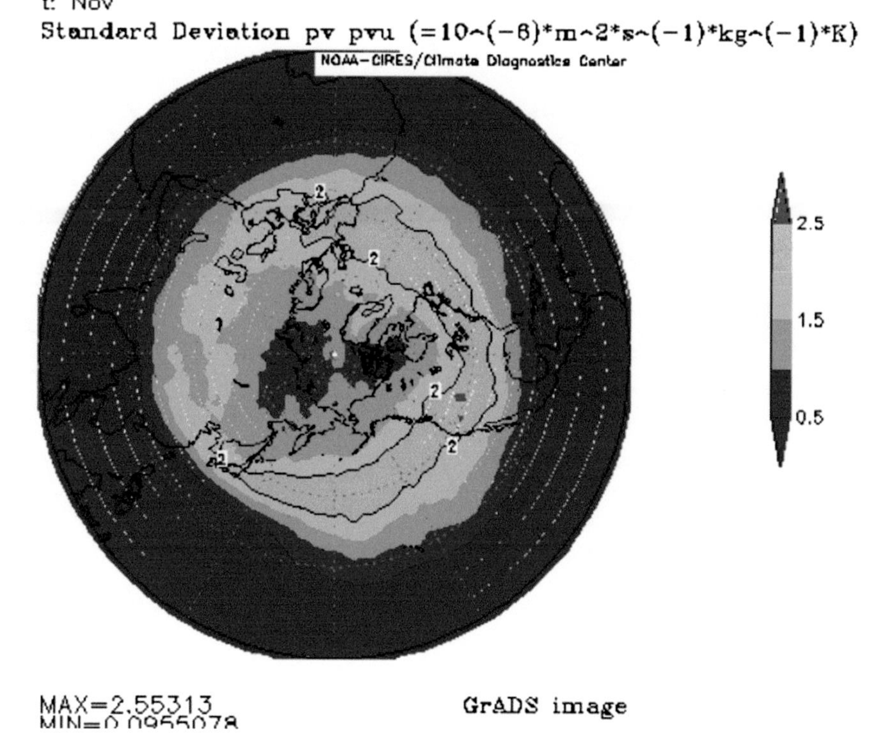

Fig. 3.14 Distribution of the standard deviation of PV on 345 K surface in January.

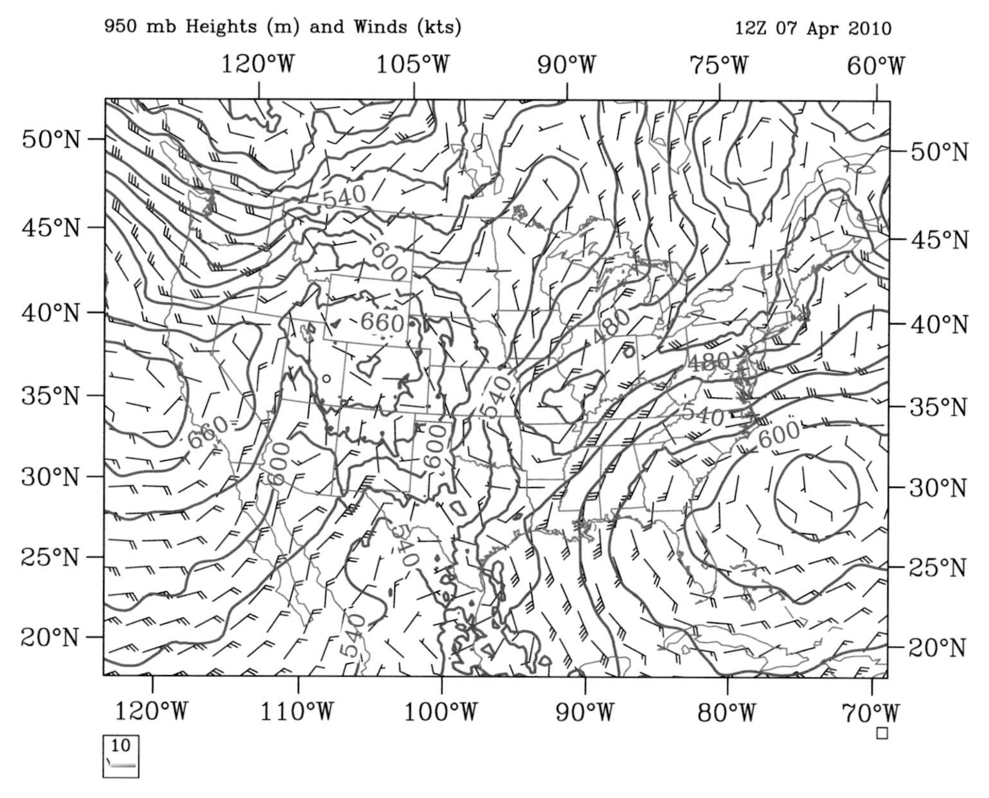

Fig. 4.5 Distributions of the 950 mb height and wind fields on April 7, 2010.

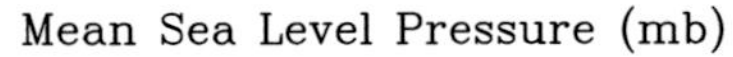

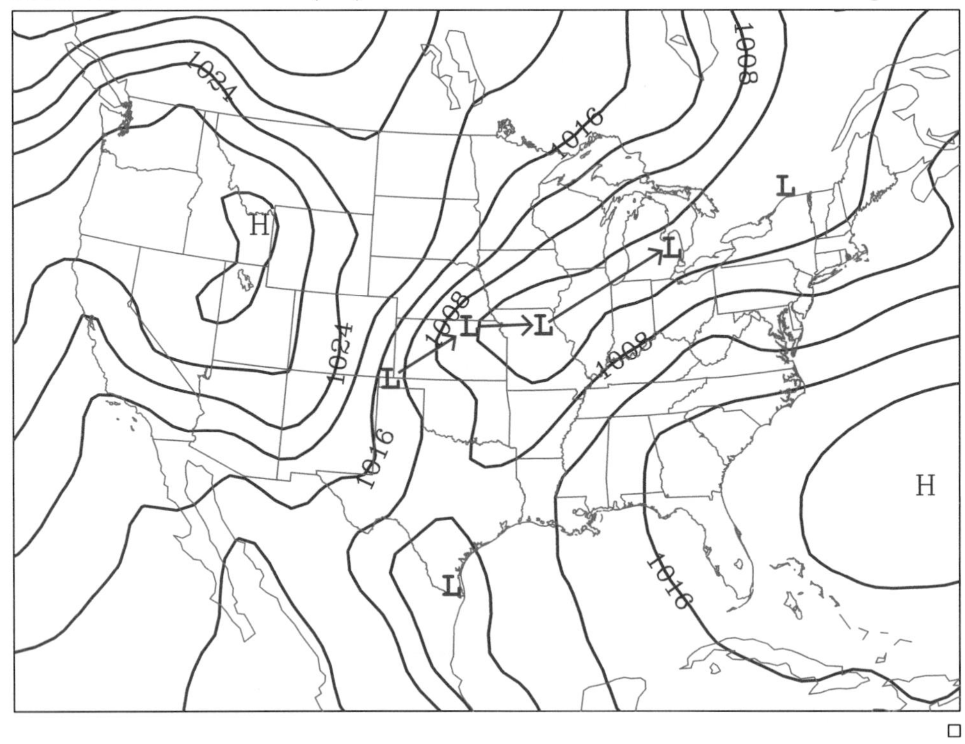

Fig. 6.1 Distribution of Mean Sea Level Pressure over N. America at 12Z on April 7, 2010 and the track of the major surface-low center from April 5 to April 8.

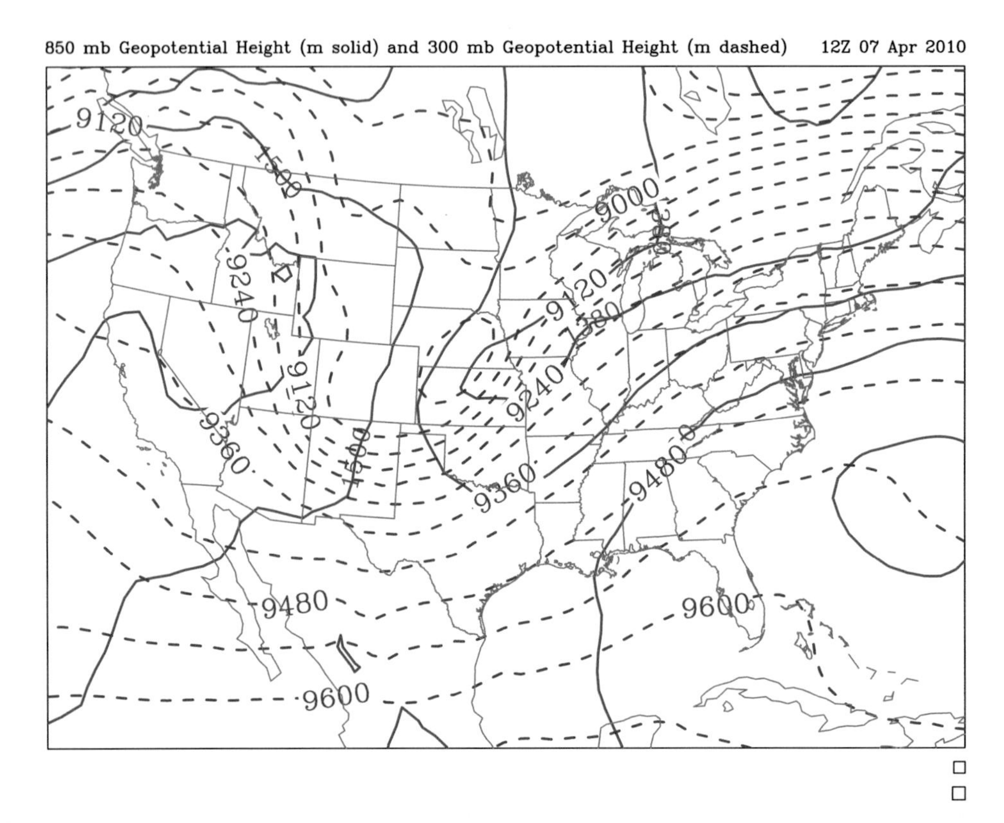

Fig. 6.2 Distributions of the geopotential height field of 850 mb and 300 mb surfaces, Z_{850} and Z_{300}, in meters on April 7, 2010.

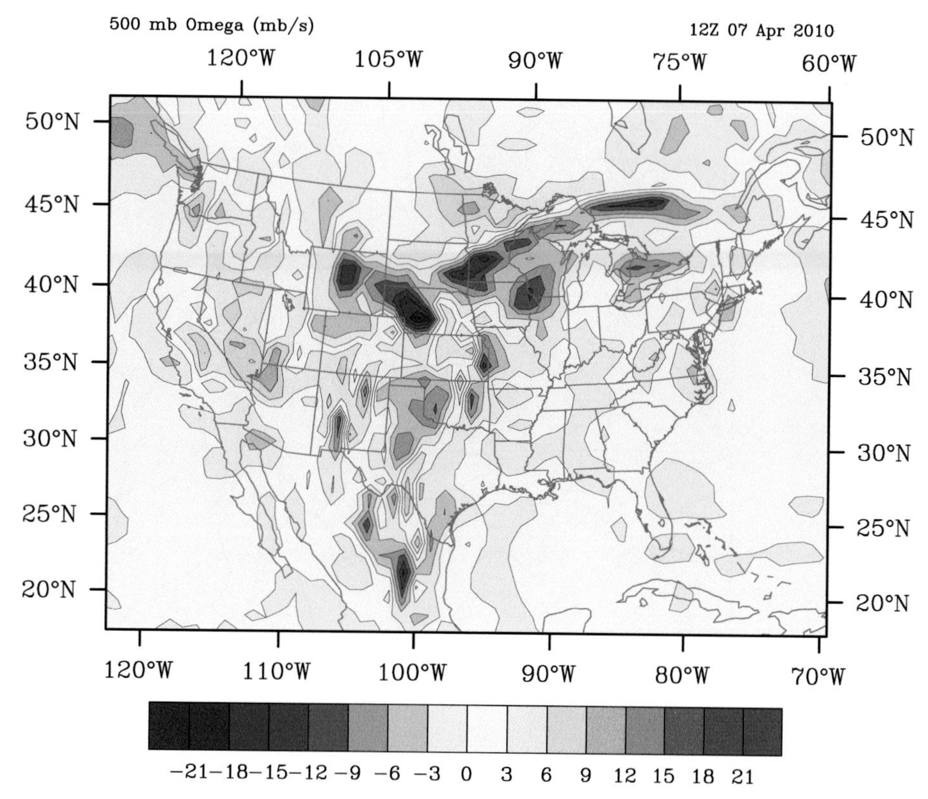

Fig. 6.3 The omega field, ω in mb s^{-1} at 500 mb on April 7, 2010.

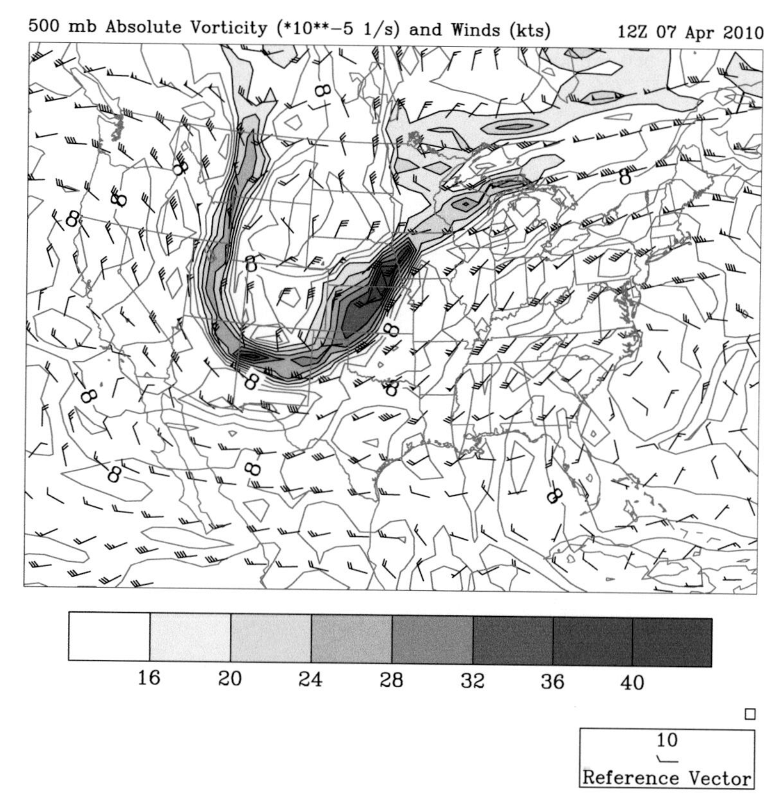

Fig. 6.4 Distributions of 500 mb $\vec{V}$ in knots and ζ_{abs} in 10^{-5} s^{-1} on April 7, 2010.

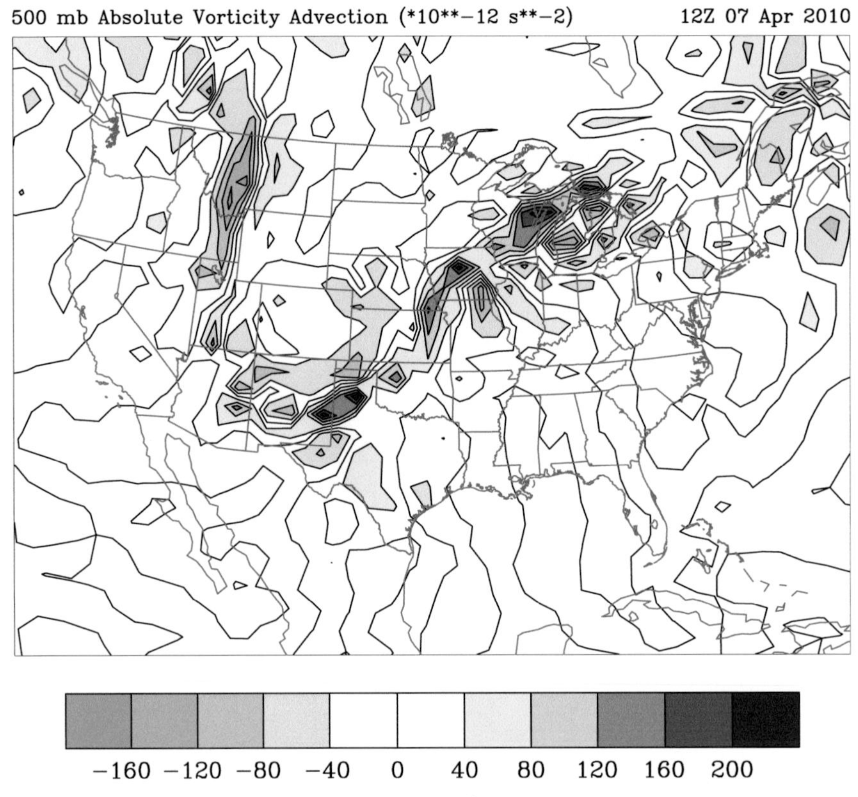

Fig. 6.5 Distribution of advection of absolute vorticity at 500 mb, $-\vec{V}\cdot\nabla\zeta_{abs}(10^{-12}\ s^{-2})$ on April 7, 2010.

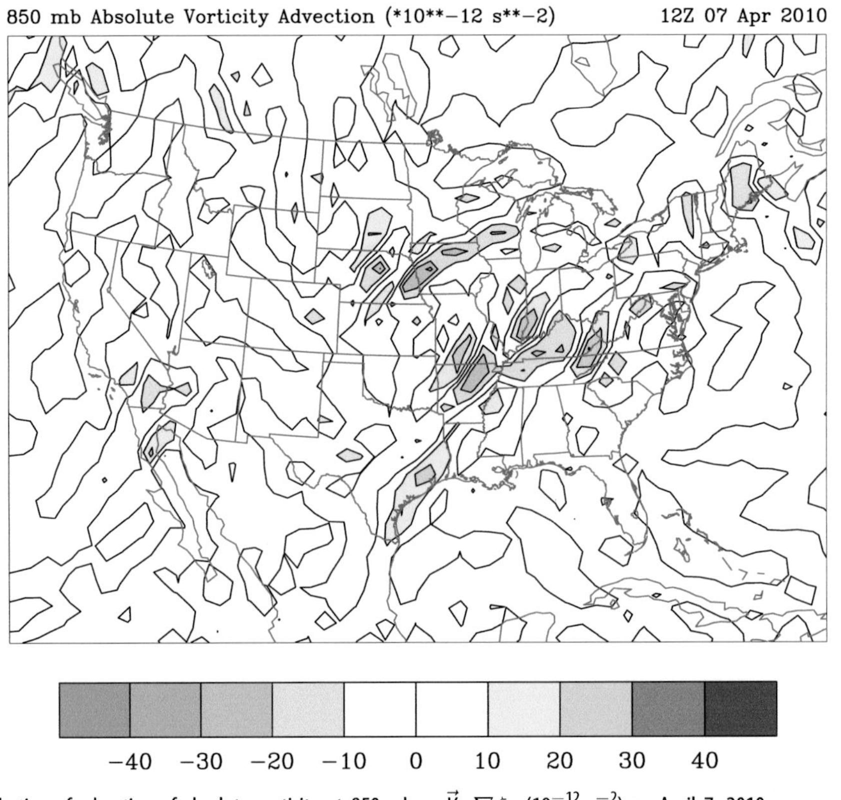

Fig. 6.6 Distribution of advection of absolute vorticity at 850 mb, $-\vec{V}\cdot\nabla\zeta_{abs}(10^{-12}\ s^{-2})$ on April 7, 2010.

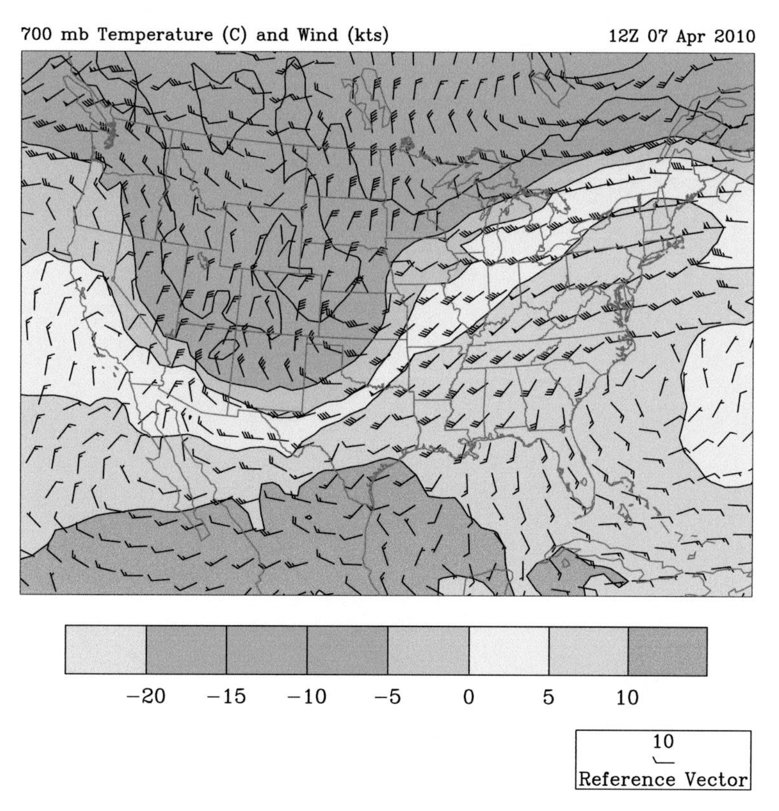

Fig. 6.7 Distributions of wind in knots and temperature in °C at 700 mb on April 7, 2010.

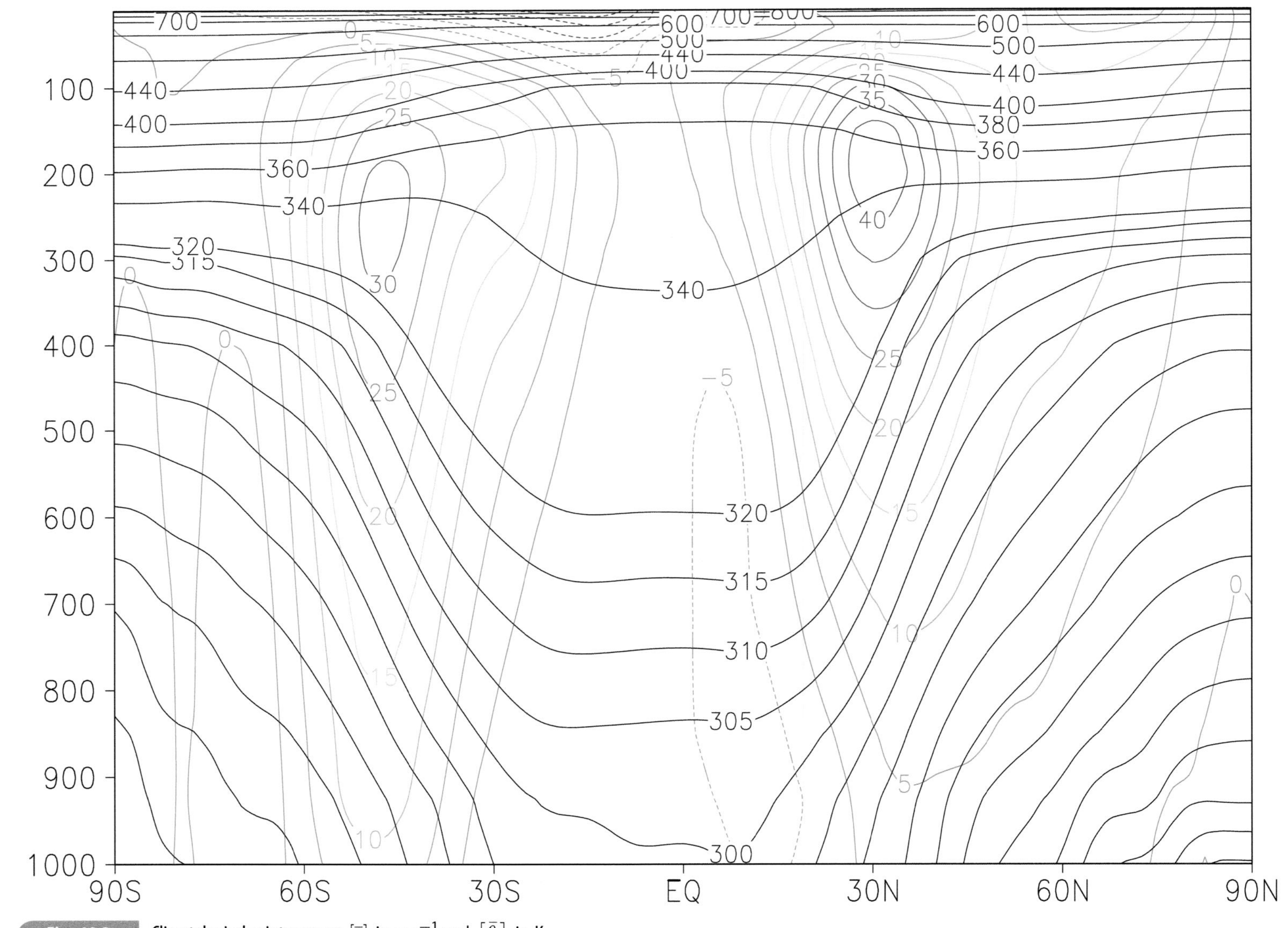

Fig. 10.2 Climatological winter mean $[\bar{u}]$ in m s^{-1} and $[\bar{\theta}]$ in K.

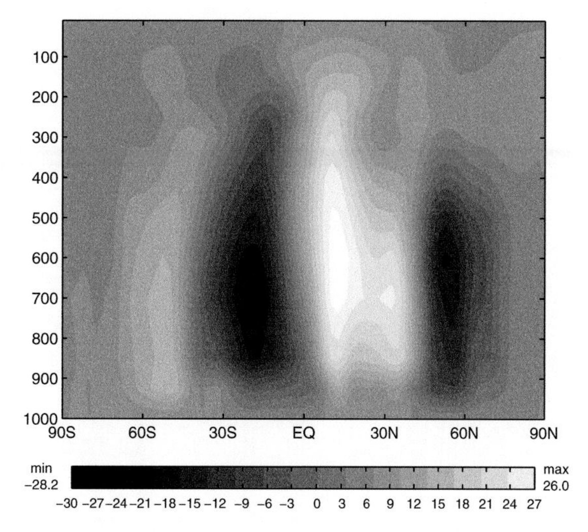

Fig. 10.17 Streamfunction of the model Eulerian climatological winter mean meridional circulation, $\chi = \eta + \xi$, in units of 10^3 m s^{-1} Pa. Vertical coordinate is pressure in mb. Horizontal coordinate is latitude.

Summer characteristics

The lower boundary condition $\eta = \frac{1}{-\Theta_p}[v^*\theta^*]\cos\varphi$ at $p = p_o$ is shown by the long dashed curve in Fig. 10.11. Its maximum magnitude is twice as large in the southern hemisphere as in the northern hemisphere in summer also reaching 4×10^4 m s^{-1} Pa at about 45° S. The location of the maximum magnitude in the winter hemisphere shifts equatorward to about 15° N. The distributions of the climatological summer mean diabatic heating, the $\vec{E}$ vector and its divergence are shown in Fig. 10.18a and 10.18b. Positive values of $[\bar{Q}]$ are found between 5° S and 20° N. A narrow belt of particularly large values centers about 8° N, reaching 3×10^{-5} K s^{-1}. A broader belt of negative values is located between 8° S and 30° S. Fig. 10.18b shows stronger eddy activity in the southern hemisphere as expected. Notice that the eddy activity in the northern hemisphere is still quite substantial, but the total eddy activity in the southern hemisphere in its winter season is weaker than the counterpart in the northern hemisphere. The difference stems from the much stronger stationary waves due to the greater land mass distribution in the northern hemisphere.

It suffices to just show the streamfunction of the climatological summer mean residual circulation in Fig. 10.19. The hemispheric asymmetry is also very pronounced as expected in each component of the residual circulation and therefore in the total streamfunction. For example, there is a two to one ratio in the intensity of the southern hemisphere cell compared to the one in the northern hemisphere. While the mass flux in the northern hemisphere only extends to about 40°N, there is a significant mass transport in the southern hemisphere reaching 60°S. By inference, there is a strong surface branch of equatorward return mass flux in the southern hemisphere in summer.

The streamfunction of the model Eulerian mean meridional circulation in summer also shows a considerably stronger Hadley cell in the winter (i.e. southern) hemisphere

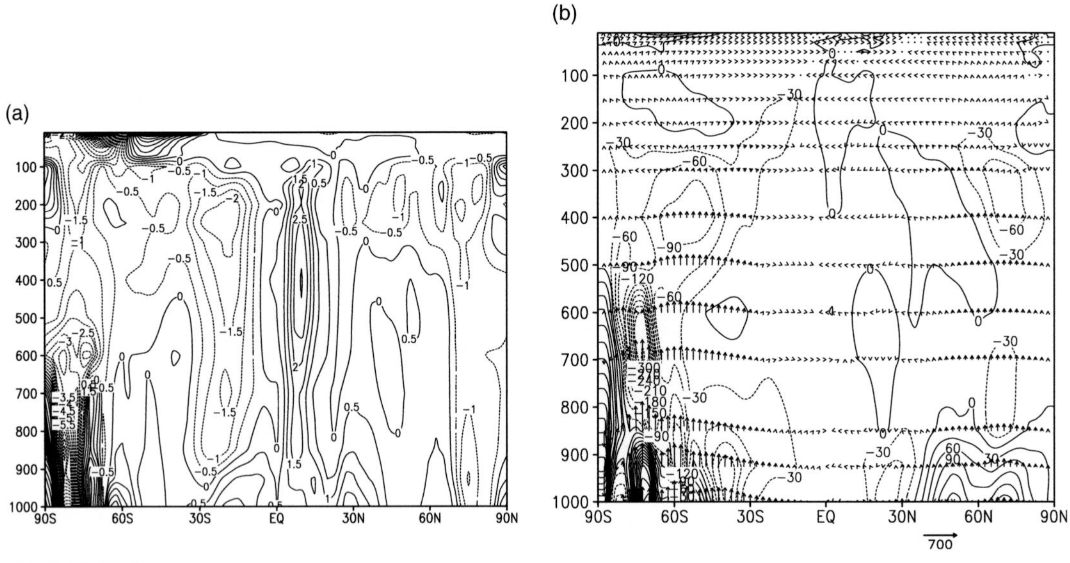

Fig. 10.18 Distribution of the climatological summer mean (a) diabatic heating rate, $[\overline{Q}]$ in 10^{-5} K s^{-1}, and (b) Eliassen–Palm vector field $\vec{E} = (E_1, \hat{E}_2)$ where $\hat{E}_2 = \frac{a}{p_o} E_2$ and its divergence, $\nabla \cdot \hat{\vec{E}} = \frac{1}{a\cos\varphi}\frac{\partial(E_1\cos\varphi)}{\partial\varphi} + \frac{\partial \hat{E}_2}{\partial p}$ (contours) in 10^{-6} m s^{-2}. Vertical coordinate is pressure in mb. Horizontal coordinate is latitude.

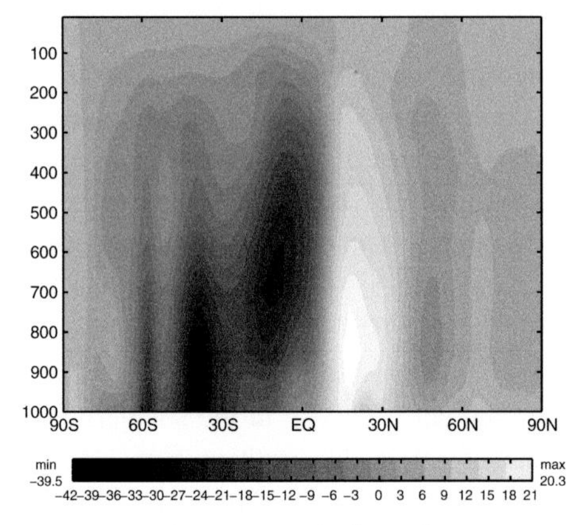

Fig. 10.19 Climatological summer mean residual circulation in units of 10^3 m s^{-1} Pa. Vertical coordinate is pressure in mb. Horizontal coordinate is latitude.

(Fig. 10.20). Again, the model Ferrel cell is much weaker than the Hadley cell in each hemisphere as in the actual atmosphere.

In summary, we have obtained quite convincing results with a relatively simple model for the residual circulation as well as the Ferrel cell for each season and the annual mean. Even the model Hadley cell is very meaningful even though the advective effect of $[\overline{v}]$ and $[\overline{\omega}]$ have not been included in this model. The model results are by and large comparable with the observed counterparts. It tells us that the Hadley cell is primarily forced and sustained by the differential net diabatic heating across the tropics.

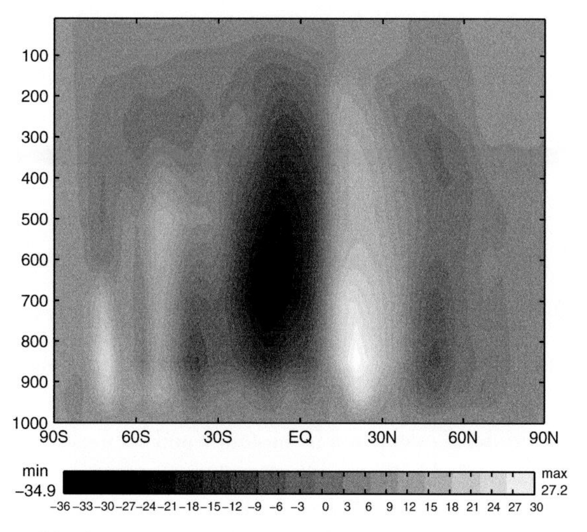

Fig. 10.20 Climatological summer model Eulerian mean meridional circulation $\chi = \eta + \xi$ in units of 10^3 m s^{-1} Pa. Vertical coordinate is pressure in mb. Horizontal coordinate is latitude.

One might add that the empirical distributions of eddy momentum flux and eddy heat flux, exemplified by Figs. 5.19 and 10.7b, are the key ingredients in budget analyses of angular momentum and heat as well as the delineation of the energetics of the atmospheric circulation as a whole (Peixoto and Oort, 1992). One might also add that as far as (10.10) is concerned, there can be contributions from all types of waves including gravity waves in $\nabla \cdot \vec{E}$. High-frequency gravity waves generated in the troposphere can penetrate to mesospheric levels especially in the summer hemisphere. The time-zonal mean flow at those levels is also in thermal wind balance. It follows that this model in principle can also be used to ascertain the role of gravity wave breaking in the maintenance of the mean meridional circulation of the mesospheric region. Such calculation however would not be feasible until there were enough gravity wave data to establish the eddy forcing at those levels. It is now believed to be a process of first-order importance.

10.4 Non-Acceleration Theorem

The impact of large-scale waves on a zonal mean component is succinctly described by the zonal average QG potential vorticity equation (Eq. (6.28) of Chapter 6). Under adiabatic and inviscid conditions, that equation is

$$\frac{\partial [q]}{\partial t} = -\frac{\partial [v^* q^*]}{\partial y}. \qquad (10.14)$$

There is no restriction to the intensity of the large-scale waves as far as (10.14) is concerned; nor is there any restriction to the structure of the zonal mean flow component. Equation (10.14) says that the zonal mean PV would change in time at a rate equal to convergence of eddy flux of PV. In the event that the eddy PV flux is non-divergent, $\left(\frac{\partial[v^*q^*]}{\partial y}=0\right)$, the zonal mean velocity would not change in time, viz.

$$[v^*q^*]\text{independent of } y \Leftrightarrow \frac{\partial[q]}{\partial t}=0. \qquad (10.15)$$

This result is known as the *Non-Acceleration Theorem* (Charney and Drazin, 1961). This mathematical theorem may be interpreted from a physical perspective as follows. Since $[v^*q^*]=-[u^*v^*]_y+\frac{f_o g}{\rho_o\theta_{oo}}\left(\frac{\rho_o}{N^2}[v^*\theta^*]\right)_z$, the condition $[v^*q^*]_y=0$ must be accompanied by a certain distribution of $(-[u^*v^*]_y)_y$ with a compensating distribution of $([v^*\theta^*]_y)_z$. Those eddy fluxes would induce a mean meridional circulation. This theorem would hold only if such meridional circulation in a rotating and stably stratified fluid turns out to counterbalance the impacts of eddy fluxes of heat and momentum upon a zonal mean state. For example, the decelerating tendency due to a divergent eddy momentum flux $(-[u^*v^*]_y<0)$ could be balanced by the accelerating tendency due to the Coriolis force associated with a southerly mean flow $([v]>0)$ in the northern hemisphere. When this occurs, the actual change of zonal wind would be very poorly correlated with the convergence of eddy momentum flux in the data. The cooling tendency due to a divergent eddy heat flux $([v^*\theta^*]_y>0)$ could be balanced by a warming tendency due to the adiabatic warming of a descending motion.

This argument follows directly from the equations for the instantaneous zonal mean state in a quasi-geostrophic framework,

$$\begin{aligned}\frac{\partial[u]}{\partial t}&=f[v]-\frac{\partial[u^*v^*]}{\partial y},\\ \frac{\partial[\theta]}{\partial t}+\frac{N^2\theta_{oo}}{g}[w]&=-\frac{\partial[v^*\theta^*]}{\partial y},\\ \frac{\partial[v]}{\partial y}+\frac{1}{\rho_o}\frac{\partial(\rho_o[w])}{\partial z}&=0.\end{aligned} \qquad (10.16a,b,c)$$

According to (10.16a,b), the zonal mean flow would not change in time, $(\partial[u]/\partial t=\partial[\theta]/\partial t=0)$, only if a mean meridional circulation is induced by the eddy fluxes such that

$$\begin{aligned}[v]&=\frac{1}{f_o}[u^*v^*]_y,\\ [w]&=-\frac{g}{N^2\theta_{oo}}[v^*\theta^*]_y.\end{aligned} \qquad (10.17)$$

It follows that

$$\begin{aligned}\frac{\partial}{\partial y}\nabla\cdot\vec{E}&=\left(\frac{\rho_o}{f_o}[u^*v^*]_y\right)_y-\left(\frac{g\rho_o}{N^2\theta_{oo}}[v^*\theta^*]_y\right)_z\\ &=\rho_o[v]_y+(\rho_o[w])_z\\ &=0\end{aligned}$$

$$[v^*q^*]_y=0\rightarrow[q]_t=0. \qquad (10.18)$$

Ironically, the importance of the Non-Acceleration Theorem is most appreciated in situations when it does not hold. The condition for the Non-Acceleration Theorem does not prevail in the troposphere on the average because the mean eddy PV flux is generally not uniform as evident in the observed EP vector fields (see Fig. 10.7). Recall that there is a maximum positive value of the annual mean $\nabla \cdot \vec{E}$ at the 1000 mb level centering at 60° N as well as at 60° S (Fig. 10.8). The large values span from 30° to 75° latitude in both hemispheres. In other words, $\frac{\partial}{\partial y} \nabla \cdot \vec{E}$ is positive south of about 60° N and negative north of about 60° N in the lower troposphere. Such distributions suggest that the zonal mean PV in mid-latitude would be reduced by the eddies and the zonal mean PV in the polar latitudes would be increased by this process alone. It is noteworthy that $\nabla \cdot \vec{E}$ also has a maximum negative value at around 400 mb at around 60° N and 60° S. So, the waves impact the zonal mean PV at the upper tropospheric levels in the opposite way to the way they do at the lower tropospheric levels. As far as the annual mean state is concerned, these tendencies must be counteracted by diabatic and frictional processes. We have seen that to be the case in the maintenance of the Ferrel cell.

10.5 Stratospheric sudden warming

We next discuss another example of wave-mean flow interaction in the context of a transient phenomenon in the stratosphere. It occurs when the Non-Acceleration Theorem does not hold. In winter, the solar heating over a polar region is very weak leading to a normally poleward decreasing field of zonal mean temperature. Associated with this strong meridional gradient of temperature is a baroclinic shear with a maximum zonal wind at an elevation of about 65 km. This flow is known as polar night jet (see Fig. 10.5). Also, it is observed that the lower part of the stratosphere could drastically change in a matter of days in the midst of winter. The polar temperature in that layer could warm by as much as 40 K in a week and the zonal wind would simultaneously change direction. The normal winter structure would recover shortly afterward. This phenomenon is known as *stratospheric sudden warming*. Such an event is considered *a major warming* mostly observed in the northern hemisphere. Additionally, minor sudden warming may take place a couple of times in both hemispheres in a winter season.

A detailed summary of the observed information about stratospheric sudden warming can be found in Andrews, Holton and Leovy (1987). This dramatic evolution is believed to be attributable to two key processes. One is a forcing in the form of upward propagating planetary waves (mostly azimuthal wavenumber 1 and 2) from the troposphere into the stratosphere. These waves give rise to a poleward flux of heat, which compensates for the prevalent radiative cooling. The second process is that the planetary waves are absorbed by the zonal mean flow at levels where the wave speeds are equal to the zonal mean flow (critical levels). The corresponding $\vec{E}$ vector field would become strongly convergent there and would induce a secondary circulation with descending motion in the polar region and ascending motion further south. The corresponding adiabatic warming associated with the descent would quickly increase the temperature in the polar region. If such warming is sufficiently strong, it would reverse the temperature gradient. As constrained by the thermal wind balance, the westerly zonal flow would change to an easterly. Since large changes in both the zonal mean flow and the eddies occur during stratospheric sudden warming, an analysis of this problem is somewhat more involved than evaluating a residual circulation. The impact of this overturning circulation can be felt at the tropospheric levels. Such dynamical linkage is referred to by some researchers as *downward control*.

10.5.1 Model formulation

In light of the conceptual discussion above, it is imperative to properly simulate the dynamic interaction between a wave field and a zonal mean flow in a model under pertinent conditions. The scale of the circulation under consideration suggests that an adequate model for the stratospheric sudden warming may be constructed in the context of quasi-geostrophic dynamics. We use an asterisk and a square bracket to denote the properties of a wave disturbance and the zonal mean component respectively. The zonal mean flow $[u]$ has a PV field $[q]$ governed by the zonal mean PV equation subject to a diabatic heating $[Q]$ and a zonal frictional force $[F]$. The wave PV field q^* is assumed to evolve according to linear dynamics as a first-order approximation. Hence, we may consider one wave at a time in a rudimentary model. The background density is $\rho_o = \rho_{oo} \exp(-z/H)$ as in a stratospheric layer. The corresponding stratification is $N^2 = \frac{-g}{\rho_o}\frac{d\rho_o}{dz} = \frac{g}{H}$. It follows that pressure exponentially varies with height over a relatively small range of values. It would be convenient to construct a model using height as the vertical coordinate. So a model domain may be taken to be $0 \le \lambda \le 2\pi$, $\varphi_S \le \varphi \le \pi/2$ and $z_{bot} \le z \le z_{top}$. The governing equations for the model written in standard notations are then

$$\frac{\partial [q]}{\partial t} + \frac{1}{a\cos\varphi}\frac{\partial}{\partial\varphi}([v^* q^*]) = f_o \frac{\partial}{\partial z}\left(\frac{[Q]}{\Theta_z}\right) - \frac{1}{a\cos\varphi}\frac{\partial}{\partial\varphi}(\cos\varphi [F]), \tag{10.19}$$

$$\frac{\partial q^*}{\partial t} + [u]\frac{1}{a\cos\varphi}\frac{\partial q^*}{\partial\lambda} + \frac{v^*}{a}\frac{\partial [q]}{\partial\varphi} = 0, \tag{10.20}$$

where $[q] = 2\Omega\sin\varphi + \frac{1}{a^2\cos\varphi}\frac{\partial}{\varphi}\left(\cos\varphi\frac{\partial[\psi]}{\partial\varphi}\right) + \frac{1}{\rho_o}\frac{\partial}{\partial z}\left(\frac{f_o^2\rho_o}{N^2}\frac{\partial[\psi]}{\partial z}\right),$

$$[u] = -\frac{1}{a}\frac{\partial[\psi]}{\partial\varphi}, \quad v^* = \frac{1}{a\cos\varphi}\frac{\partial\psi^*}{\partial\lambda},$$

$$q^* = \nabla^2\psi^* + \frac{1}{\rho_o}\frac{\partial}{\partial z}\left(\frac{f_o^2\rho_o}{N^2}\frac{\partial\psi^*}{\partial z}\right),$$

$$\nabla^2 = \frac{1}{a^2\cos\varphi}\frac{\partial}{\partial\varphi}\left(\cos\varphi\frac{\partial}{\partial\varphi}\right) + \frac{1}{a^2\cos^2\varphi}\frac{\partial^2}{\partial\lambda^2},$$

and ψ is the geostrophic streamfunction. If one derives an equation that governs $\partial[u]/\partial t$ using the transformed Eulerian mean equations as a starting point, the resulting equation would be equivalent to (10.19). Equations (10.19) and (10.20) constitute a system of coupled equations for two unknowns ψ^* and $[\psi]$. What is to be solved is an initial boundary value problem.

The initial field of $[\psi]$ should be patterned after the typical zonal mean stratospheric flow in winter. Let us assume no disturbance initially except at the bottom of the domain. Thus, the initial conditions are

$$[\psi](\varphi, z, 0) = A(\varphi, z), \quad \psi^*(\varphi, z, 0) = 0 \quad \text{for } z > 0. \tag{10.21}$$

The forcing in this problem is introduced through a given stationary wave at the bottom of the domain

$$\psi^*(\varphi, z_{bot}, t) = C(\varphi)\exp(im\lambda) \quad \text{with } m = 1 \text{ or } 2. \tag{10.22}$$

The disturbance is further assumed to vanish at the pole and the southern boundary at all times for simplicity. It follows that the lateral boundary conditions may be prescribed as

$$\begin{gathered} \frac{\partial^2[\psi]}{\partial t \partial \varphi}\left(\frac{\pi}{2}, z, t\right) = 0, \quad \frac{\partial[\psi]}{\partial t \partial \varphi}(\varphi_S, z, t) = 0, \\ \psi^*\left(\frac{\pi}{2}, z, t\right) = 0, \quad \psi^*(\varphi_S, z, t) = 0. \end{gathered} \tag{10.23}$$

The top of the domain is assumed to be sufficiently high that the disturbance does not reach it in the interval under consideration. Hence, the vertical boundary conditions may be prescribed as

$$\begin{gathered} \frac{\partial[\psi]}{\partial t}(\varphi, z_{top}, t) = 0, \qquad \frac{\partial[\psi]}{\partial t}(\varphi, z_{bot}, t) = 0, \\ \psi^*(\varphi, z_{top}, t) = 0. \end{gathered} \tag{10.24}$$

We may seek a solution in the form of

$$\begin{aligned} (\psi^*, v^*, q^*) &= \left(\hat{\psi}, \hat{v}, \hat{q}\right)e^{im\lambda}e^{z/2H}, \\ ([\psi], [u]) &= (\bar{\psi}, \bar{u})e^{z/2H}, \\ [q] &= 2\Omega\sin\varphi + \bar{q}e^{z/2H}, \end{aligned} \tag{10.25}$$

where $\hat{\psi}, \hat{v}, \hat{q}$ have complex values with the understanding $\psi^* = \mathrm{Re}\left\{\hat{\psi}e^{im\lambda}e^{z/2H}\right\}$ in real space. On the other hand, $\bar{\psi}, \bar{u}, \bar{q}$ have real values and are all functions of (φ, z, t). These quantities are interrelated as

$$\begin{aligned} \hat{v} &= \frac{m}{a\cos\varphi}i\hat{\psi}, \\ \hat{q} &= \left(\frac{f_o^2}{N^2}\frac{\partial^2}{\partial z^2} + \frac{1}{a^2\cos\varphi}\frac{\partial}{\partial\varphi}\left(\cos\varphi\frac{\partial}{\partial\varphi}\right) - \left(\frac{m^2}{a^2\cos^2\varphi} + \frac{f_o^2}{4H^2N^2}\right)\right)\hat{\psi}, \\ \bar{u} &= -\frac{1}{a}\frac{\partial\bar{\psi}}{\partial\varphi}, \\ \bar{q} &= \left(\frac{f_o^2}{N^2}\frac{\partial^2}{\partial z^2} + \frac{1}{a^2\cos\varphi}\frac{\partial}{\partial\varphi}\left(\cos\varphi\frac{\partial}{\partial\varphi}\right) - \left(\frac{f_o^2}{4H^2N^2}\right)\right)\bar{\psi}. \end{aligned} \tag{10.26}$$

Writing $\hat{\psi} = \hat{\psi}_r + i\hat{\psi}_i$, $\hat{q} = \hat{q}_r + i\hat{q}_i$, $\hat{v} = \hat{v}_r + i\hat{v}_i$, one gets

$$[v^*q^*] = \frac{m}{2a\cos\varphi}\left(\hat{q}_i\hat{\psi}_r + \hat{q}_r\hat{\psi}_i\right)e^{z/H}$$
$$= \frac{m}{2a\cos\varphi}\mathrm{Im}\left\{\hat{\psi}^{\#}\hat{q}\right\}e^{z/H}, \tag{10.27}$$

where $\hat{\psi}^{\#}$ is the complex conjugate of $\hat{\psi}$. Upon substituting all these quantities into (10.19) and (10.20), we get

$$\frac{\partial\bar{q}}{\partial t} + \frac{m}{a^2\cos\varphi}\frac{\partial}{\partial\varphi}\left(\frac{1}{\cos\varphi}\mathrm{Im}\left\{\hat{\psi}^{\#}\hat{q}\right\}\right)e^{z/2H} = \Sigma,$$
$$\frac{\partial\hat{q}}{\partial t} + \bar{u}\frac{im}{a\cos\varphi}\hat{q}e^{z/2H} + \frac{im\hat{\psi}}{a^2\cos\varphi}\left(2\Omega\cos\varphi + \frac{\partial\bar{q}}{\partial\varphi}e^{z/2H}\right) = 0, \tag{10.28a,b}$$

where Σ stands for the diabatic heating and frictional terms. The problem has been reduced to one of solving (10.28a,b) on the latitude–height section in conjunction with the auxiliary relations for each wavenumber m. This set of equations can be readily solved by a standard numerical scheme. One might add that (10.28a) can be also used to evaluate the residual circulation at any time during the stratospheric sudden warming with the use of the instantaneous eddy forcing.

10.5.2 Matsuno's model results

We use the results of Matsuno's (1971) seminal study for a case to illustrate the dynamical nature of stratospheric sudden warming. He solved a system of equations essentially the same as the one described above in a model domain of $0 \le \varphi \le \pi/2$, $10\,\mathrm{km} \le z \le 110\,\mathrm{km}$. His model did not incorporate the diabatic heating and friction processes. The disturbance is a planetary wave of azimuthal wavenumber 2 with an amplitude of 300 m. The initial zonal mean flow is shown in Fig. 10.21.

The evolution of the wave disturbance for 25 days is shown in Fig. 10.22

We see that the wave propagates upwards into the stratosphere, reaching about 40 km. It strengthens until day 19 when it reaches a critical level. The wave amplitude rapidly decreases thereafter. It is strongest at about 60° N.

The evolution of the zonal mean state is shown in Fig. 10.23.

The zonal mean wind progressively decelerates and rapidly changes over to an easterly on about day 19 coinciding with the time when the wave weakens precipitously. The temperature at the pole warms up rapidly after day 19 and reaches a whopping value of 80° C on day 21. An overview of the temperature change is shown in Fig. 10.24.

The wave field progressively strengthens from day 10 to day 18. Then drastic change takes place from day 18 to day 22. Such evolution is depicted in Fig. 10.25 We see that the height over the pole rises changing from a low to a high mostly in the last four days associated with the rapid decay of the planetary wave.

The results shown in Figs. 10.21 to 10.25 convincingly verify the dynamical interpretation of stratospheric sudden warming discussed earlier. Matsuno did not evaluate the

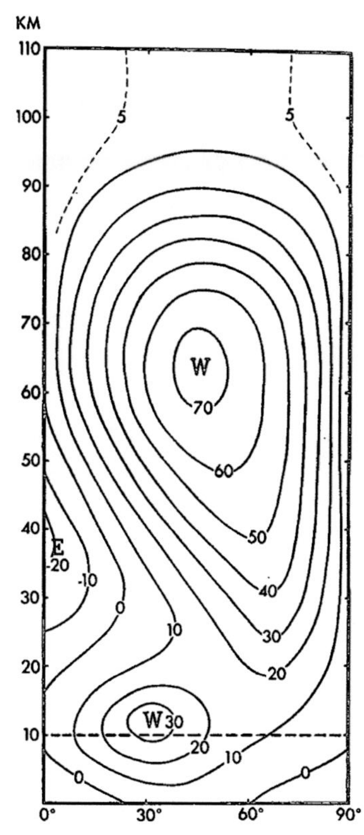

Fig. 10.21 Initial distribution of zonal wind on a latitude–height section in m s^{-1} (taken from Matsuno, 1971).

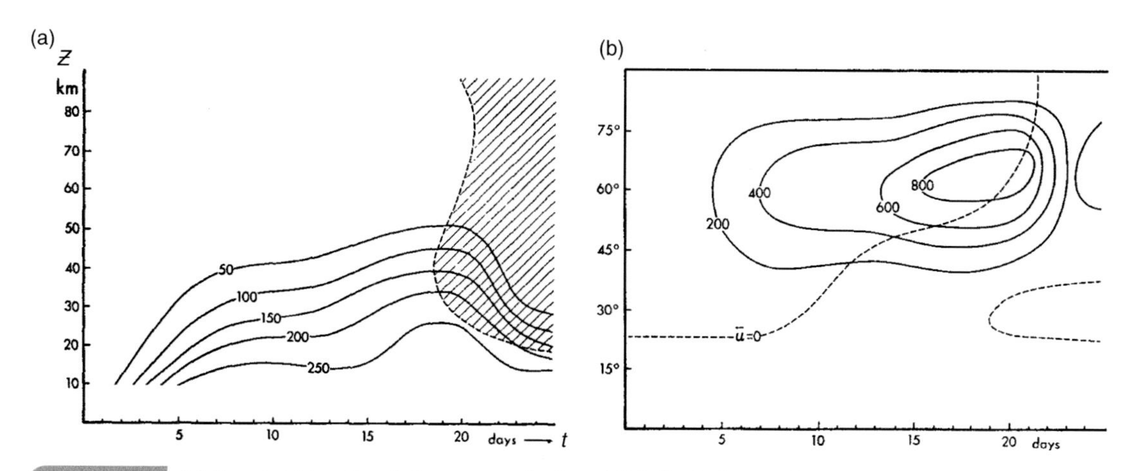

Fig. 10.22 (a) Time–height section of the wave amplitude at 60° N. Hatched area indicates easterly mean flow, (b) Time–latitude section of the disturbed heights (m) of an isobaric surface of about 13 km (taken from Matsuno, 1971).

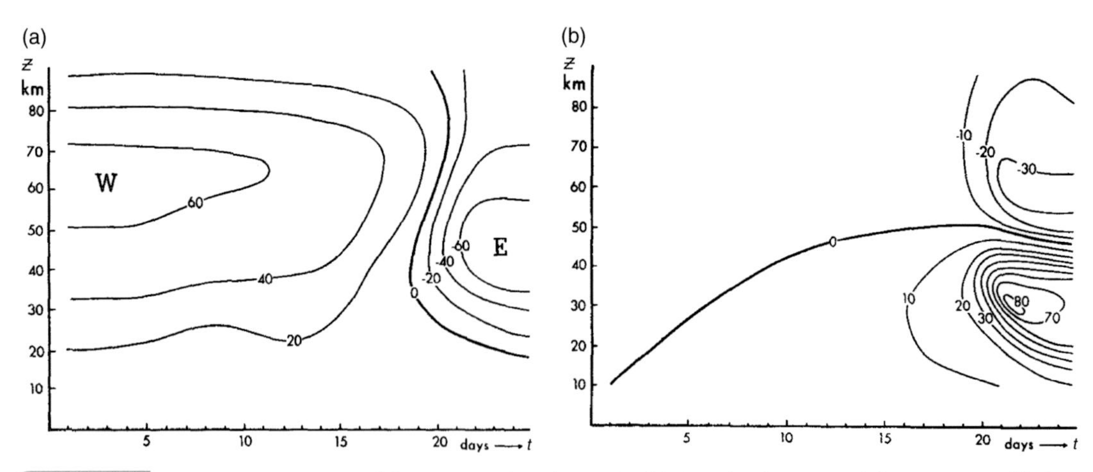

Fig. 10.23 (a) Time–height section of the mean zonal wind at 60° N, (b) Time–height section of the temperature (deviation [°C] from the initial temperature) at the pole (taken from Matsuno, 1971).

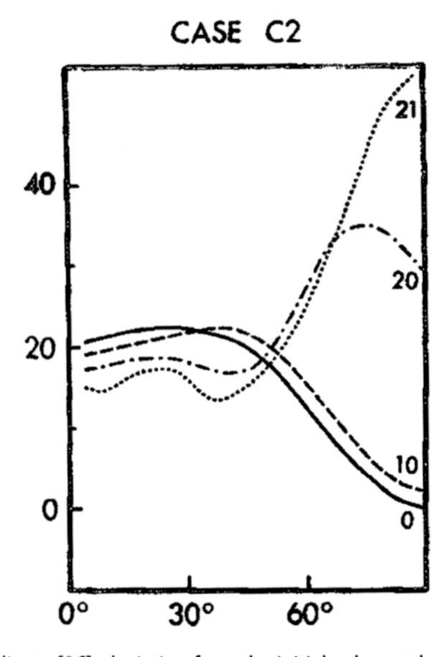

Fig. 10.24 The zonal mean temperature (ordinate [°C]; deviation from the initial value at the pole) at $z = 30$ km, as a function of latitude, on days $t = 0$, 10, 20 and 21 (taken from Matsuno, 1971).

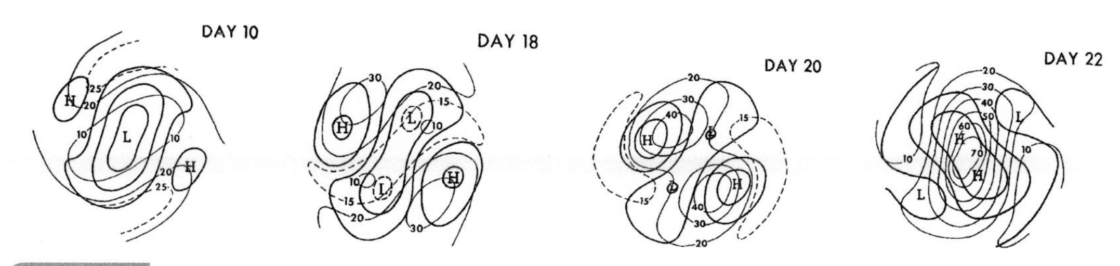

Fig. 10.25 Evolution of isobaric height (550 m contours, thick lines), temperature (°C, thin lines) and the surface of the $p = p_o \exp(-30\text{km}/H) \approx 13$ mb. Temperature is shown as a deviation from its initial value at the pole. The contours cover the area north of 30° N (taken from Matsuno, 1971).

accompanying residual circulation during this model sudden warming. Hsu (1980) presents such information in a primitive-equation model with similar basic approximations. It explicitly confirms the crucial role of adiabatic warming in the polar region. The neglected fully nonlinear dynamics of the wave evolution can be important in a major warming. We will discuss fully nonlinear dynamics of wave-mean flow interaction in the context of three other examples in Chapter 11.

11 Equilibration dynamics of baroclinic waves

This chapter discusses some aspects of nonlinear dynamics of large-scale baroclinic waves. Section 11.1 gives a broad overview of the nature of such nonlinear dynamics and the three analyses selected for illustration. As noted in Chapter 6, the most essential aspects of the dynamics of large-scale baroclinic waves can be adequately investigated with a two-layer quasi-geostrophic model. We begin with a rudimentary discussion of geostrophic turbulence in the context of such a setting in Section 11.2. The first specific analysis presented in Section 11.3 examines the dynamics of the life cycle (LC) of baroclinic waves emerging from an unstable baroclinic jet. The crucial role of an additional barotropic shear leading to the distinct forms of LC1 or LC2 is delineated in Sections 11.3.1 and 11.3.2. We next discuss in Section 11.4 the dynamical symbiotic relation between the synoptic and planetary waves in a forced dissipative flow. The third problem analyzed in Section 11.5 is concerned with the role of nonlinear dynamics in dictating the relative intensity of the two major winter storm tracks of the northern hemisphere.

11.1 Introductory remarks

Short-term fluctuations in large-scale nonlinear atmospheric flows may be regarded as being deterministic. That is after all the raison d'être for the current practice of objective weather forecasting. The dynamical nature of such fluctuations can be interpreted as such. On the other hand, long-term fluctuations of such flows are stochastic in nature. They are referred to as geostrophic turbulence. Such fluctuations should be diagnosed statistically because only their statistical properties are reproducible. The relations among the statistical properties are still intelligible from a dynamical point of view because a physical system is after all constrained by the conservation principles. Nonlinear dynamical problems of both types will be analyzed in this chapter in the context of a simplest possible model setting, namely a two-layer quasi-geostrophic model. To provide some background information about such nonlinear dynamics, we will begin by reviewing the rudiments of geostrophic turbulence.

The first specific problem of analysis is concerned with the short-term evolution of a rapidly intensifying unstable disturbance after it has emerged from the instability of a basic flow. When a weak baroclinic wave in the extratropics intensifies to large amplitude, it equilibrates by nonlinearly interacting with the background flow. It goes through a well-defined life cycle (LC), typically lasting for less than a week. The life cycle of baroclinic waves may be therefore meaningfully examined from a deterministic point of

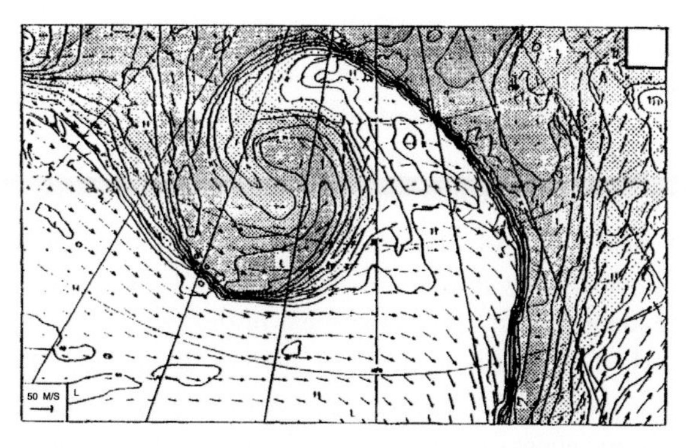

Fig. 11.1 Potential temperature in degrees Celsius on the PV = 2 PVU surface at 12 GMT 28/1/90. Potential temperatures lower than 50 °C are shaded with the lowest values shaded darkest. The contour interval is 5 °C. Potential temperatures greater than 60 °C are not contoured (taken from Thorncroft *et al.*, 1993).

view. The time scales of external forcing and dissipative processes in the context of a dry atmosphere are much longer than a week and therefore are not of primary importance. Observation indicates that there exist two distinct forms of life cycle. One form is characterized by backward-tilted, thinning upper-air troughs advected anticyclonically and equatorward. It is referred to as LC1. Another form is characterized by forward-tilted, broadening troughs that wrap themselves up cyclonically and poleward. It produces major cutoff cyclones in higher latitudes. It is called LC2. One snapshot during the life cycle of a disturbance over the North Atlantic in winter is shown in Fig. 11.1. The disturbance over the central North Atlantic on that day wraps up cyclonically (LC2), whereas the eastern trough over Western Europe has features rather like that in LC1. Such are the key characteristics of developing large-scale disturbances that call for in-depth analyses.

A suitable initial disturbance for consideration is an unstable normal mode of a zonally uniform baroclinic jet. A follow-up question is "how would it evolve subsequently?" Intuition suggests that the disturbance would mature, modify its background flow and decay shortly afterward. When we use Fourier expansion of a flow field, the spectral governing equation for each expansion coefficient contains a series of quadratic terms each being a product of two other expansion coefficients. They represent the horizontal advective process and stand for *triad interactions*, each of which involves three Fourier components (see Section A.15). If the initial disturbance were a single Rossby wave, it would first modify the zonal flow and at the same time generate its first higher zonal harmonic. It would also interact with the spectral components of the zonal mean flow to generate components with the same zonal wavenumber but higher meridional wavenumbers. When it later interacts with its own higher zonal harmonic, additional higher harmonics would be generated. These interactions collectively would generate a particular ensemble of waves consisting of the original wave and its entire set of higher harmonics. The higher the harmonics are, the weaker their amplitudes turn out to be (Cai, 1992). Consequently, the equilibrated wave field of, say, wavenumber 6 in a hemispheric domain would continue to have the gross appearance of wavenumber 6.

The emerging increasingly fine features are due to the presence of the higher harmonics. From a physical point of view, the equilibration process mixes the potential vorticity field. The tendency of homogenizing PV would lead to steadily weakening the wave field as a whole. This sequence of events would therefore give rise to a well-defined life cycle of the original unstable wave.

The problem of life cycle dynamics has been investigated by means of numerical simulation using primitive-equation models (e.g. Thorncroft *et al.*, 1993). Researchers have identified the barotropic shear of the background flow as a major factor in determining whether the life cycle is LC1 or LC2. In Section 11.3, we will present an analysis designed to verify the dependence of the life cycle of an unstable baroclinic wave upon the barotropic shear of its background flow in a two-layer quasi-geostrophic model.

The second specific analysis discussed in this chapter is concerned with the dynamical relation among an ensemble of equilibrated baroclinic waves in a forced dissipative flow. Laboratory studies, such as the rotating annulus experiment, and numerical simulation studies of rotating fluid models have established that there can be different flow regimes. The strength of the forcing and of damping is naturally the key determining factor. Under a progressively stronger forcing, the flow field could change from a zonally symmetric regime, through single steady-wave regime, multiple steady-wave regime, vacillation regime and finally to an effectively chaotic regime (e.g. Lorenz, 1963; Hide, 1966; Mak, 1985; Cai and Mak, 1987). This is also true for a counterpart barotropic fluid system (e.g. Niino and Misawa, 1984; Kwon and Mak, 1988). The atmospheric flow resembles the chaotic regime of a baroclinic system. It has a random appearance and consists of a broad spectrum of components. Similar to the general circulation of the atmosphere, such a spectrum of equilibrated baroclinic wave field would span from the planetary scale to the synoptic scale. Nevertheless, there exists some intrinsic order in such a flow. The relation between the synoptic-scale waves and the planetary-scale waves turns out to be symbiotic in the sense that one would significantly benefit from the other's existence (Cai and Mak, 1990). It has its root in the theory of geostrophic turbulence. We will go over the analysis that establishes such a relation in Section 11.4.

The third specific analysis discussed in this chapter is concerned with the spatial distribution of baroclinic wave activity. Observation indicates that the local variability of all meteorological properties including the wind speed and temperature in the extratropical atmosphere has a distinct regional distribution. As the wind fluctuations at a location are associated with the repeated passage of storms, the regions of local maximum variability are known as *storm tracks*. We observe one storm track over the western part of the N. Pacific and another one over the western part of the N. Atlantic in winter. Such regional distribution reflects the strong influences of the continents and oceans, which are particularly pronounced in the northern hemisphere. The storm tracks are located downstream of two corresponding seasonal mean localized jets. The storms that give rise to a storm track are expected to feed on the localized jet upstream. But surprisingly, the storm track associated with the stronger Pacific jet has been documented to be distinctly less intense than the one associated with the weaker Atlantic jet. To address this apparent paradox, a simple hypothesis has been proposed (Mak and Deng, 2007). We will go over an analysis designed to test this hypothesis in Section 11.5.

11.2 Rudiments of geostrophic turbulence in a two-layer model

Geostrophic turbulence refers to the seemingly random fluctuations of a strongly nonlinear flow in a rapidly rotating, stably stratified fluid. A prime example of that is the general circulation of our atmosphere. The turbulence of such a large-scale three-dimensional geophysical flow field turns out to be more akin to two-dimensional turbulence than the three-dimensional small-scale turbulence. Two-dimensional turbulence is relatively simple because the processes of stretching and tilting of the vortex tubes are absent. In three-dimensional geostrophic turbulence, vortex stretching is still important but not vortex tilting because of the rapid rotation of the planet. The vertical velocity associated with horizontal divergence gives rise to not only vortex stretching but also adiabatic warming/ cooling. Because of the geostrophic and hydrostatic balances, the combined effect of those two processes can be incorporated in terms of the rate of change in one part of the geostrophic potential vorticity due to the advection by geostrophic velocity. Furthermore, the advection of planetary vorticity associated with a varying Coriolis parameter is incorporated in another part of the potential vorticity. These intrinsic logical connections have been elaborated in Chapter 6. Charney (1971) extended the theory of two-dimensional turbulence (Fjortoft, 1953) in a consideration of the turbulence in a quasi-geostrophic flow. He pointed out that each of the two types of turbulent flow has two counterpart conservation constraints. By applying analogous reasoning, he argued that the cascades of energy and potential enstrophy in geostrophic turbulence should be similar to the cascades of energy and enstrophy in two-dimensional turbulence.

We first examine the intrinsic characteristics of two-layer geostrophic turbulence in the absence of forcing, damping and meridional variation in the Coriolis parameter. Those characteristics would be applicable to the range of scales of motions sufficiently far away from the scales at which forcing and damping are important. Such a range of scale is referred to as an *inertial subrange*. The appropriate governing equations are the quasi-geostrophic potential vorticity equations, viz.

$$\frac{\partial q_j}{\partial t} + J\left(\psi_j, q_j\right) = 0, \quad j = 1, 3, \tag{11.1a,b}$$

where $q_1 = \nabla^2\psi_1 - \lambda^2(\psi_1 - \psi_3)$ the PV in the upper layer and $q_3 = \nabla^2\psi_3 + \lambda^2(\psi_1 - \psi_3)$ the PV in the lower layer, ψ_j are the corresponding geostrophic streamfunctions, λ^{-1} is the deformation radius with $\lambda^2 = \dfrac{f^2}{S(\delta p)^2}$ in isobaric coordinates and S is the stratification, $J(\xi, \eta) = \xi_x\eta_y - \xi_y\eta_x$ the Jacobian, $\nabla = (\partial/\partial x, \partial/\partial y)$. For simplicity, the domain is assumed to be doubly periodic, so that cyclic boundary conditions can be used.

It is noteworthy that (11.1) would be greatly simplified under two extreme conditions. When the stratification is very large, we would have $\lambda^2 \to 0$. Physical intuition tells us that the fluid in the two layers would hardly feel the presence of one another; ψ_1 would be the only dependent variable in the simplified (11.1a) and ψ_3 in the simplified (11.1b).

Equations (11.1a) and (11.1b) would be decoupled and become identical in form. They would be reduced to the governing equation for a two-dimensional fluid, namely conservation of vorticity; ψ_1 and ψ_3 would evolve independently and need not be equal in general. When the rotation rate is very fast, the influence of the Coriolis force would be so strong that the fluid in the two layers would be constrained to move as single columns. It follows that we would have $\psi_1 = \psi_3$ everywhere at all times. This is known as the Taylor–Proudman theorem. The model becomes effectively a single-layer fluid model. A turbulent flow under either condition would have the characteristics of two-dimensional turbulence.

11.2.1 Conservation constraints

We first establish the two important invariant properties in this fluid model. A barotropic component ψ and a baroclinic component τ of a flow are introduced to take the place of ψ_1 and ψ_3,

$$\psi = \frac{1}{2}(\psi_1 + \psi_3), \quad \tau = \frac{1}{2}(\psi_1 - \psi_3), \tag{11.2}$$

so that $\psi_1 = \psi + \tau$ and $\psi_3 = \psi - \tau$; ψ is a measure of the vertical mean flow, whereas τ is a measure of the vertical shear. The latter is proportional to temperature according to the thermal wind relation. Then (11.1) can be rewritten as

$$\begin{aligned} &\frac{\partial}{\partial t}\nabla^2\psi + J(\psi, \nabla^2\psi) + J(\tau, (\nabla^2 - 2\lambda^2)\tau) = 0, \\ &\frac{\partial}{\partial t}(\nabla^2 - 2\lambda^2)\tau + J(\psi, (\nabla^2 - 2\lambda^2)\tau) + J(\tau, \nabla^2\psi) = 0. \end{aligned} \tag{11.3a,b}$$

The potential vorticity field of each layer has two components, $\nabla^2\psi$ and $(\nabla^2 - 2\lambda^2)\tau$; q_1 is equal to their sum and q_3 their difference. Their changes in time stem from the advection of them by both ψ and τ according to (11.3). Interchanging $\nabla^2\psi$ with $(\nabla^2 - 2\lambda^2)\tau$, we would convert (11.3a) to (11.3b) and vice versa. Such inherent symmetry facilitates the following deduction.

The total energy of the flow is the sum of the kinetic energy in each layer and the available potential energy, viz.

$$\begin{aligned} E &= \frac{1}{2}\iint dxdy\left(\nabla\psi_1 \cdot \nabla\psi_1 + \nabla\psi_3 \cdot \nabla\psi_3 + \lambda^2(\psi_1 - \psi_3)^2\right) \\ &= \iint dxdy\left(\nabla\psi \cdot \nabla\psi + \nabla\tau \cdot \nabla\tau + 2\lambda^2\tau^2\right). \end{aligned} \tag{11.4}$$

The average total potential enstrophy is defined as

$$\begin{aligned} Z &= \frac{1}{2}\iint dxdy\left(q_1^2 + q_3^2\right) \equiv \frac{1}{2}(Z_1 + Z_3), \\ Z_1 &= \iint dxdy\left(\nabla^2\psi + (\nabla^2 - 2\lambda^2)\tau\right)^2, \\ Z_3 &= \iint dxdy\left(\nabla^2\psi - (\nabla^2 - 2\lambda^2)\tau\right)^2. \end{aligned} \tag{11.5}$$

If we multiply (11.1) by ψ_j, integrate over the domain, make use of the boundary conditions and sum over the two layers, we get

$$\frac{dE}{dt} = 0. \tag{11.6}$$

If we multiply (11.1) by q_j, integrate over the domain and sum over the two layers, we get

$$\frac{dZ}{dt} = 0. \tag{11.7}$$

Equations (11.6) and (11.7) say that this quasi-geostrophic system conserves its total energy and potential enstrophy.

It is convenient to formulate a spectral representation of the flow in this doubly periodic domain with a Fourier transform. The Fourier expansion of ψ and τ is

$$\begin{aligned} \psi &= \sum \psi_k(t)\exp(i\mathbf{k}\cdot\mathbf{x}), \\ \tau &= \sum \tau_k(t)\exp(i\mathbf{k}\cdot\mathbf{x}). \end{aligned} \tag{11.8}$$

Here $\sum$ denotes the summing over all wavenumber vectors $\mathbf{k} = (k_x, k_y) \equiv k\vec{\ell}$, $\vec{\ell}$ is a vector of unit length and $\mathbf{x} = (x, y)$ is the position vector. The expansion coefficients ψ_k and τ_k for a spectral component have complex values.

Let us further assume that the turbulence is isotropic and homogeneous. By that we mean the statistical properties do not depend on the location of the origin and the orientation of the coordinate system. Then a spectrum is only a function of the total wavenumber k. For isotropic and homogeneous turbulence, $\iint \nabla^2 \xi dxdy = \sum k^2|\xi_k|^2$. We would get from (11.4), (11.5) and (11.8)

$$E = \sum \left(k^2|\psi_k|^2 + (k^2 + 2\lambda^2)|\tau_k|^2\right), \tag{11.9}$$

$$Z = \sum \left(k^4|\psi_k|^2 + (k^2 + 2\lambda^2)^2|\tau_k|^2\right). \tag{11.10}$$

Note that the effective wavenumber-squared of a baroclinic spectral component is $(k^2 + 2\lambda^2)$ instead of k^2. Let us introduce a more concise notation for the kinetic energy of a barotropic component at wavenumber $\mathbf{k}$,

$$U_k \equiv k^2|\psi_k|^2, \tag{11.11}$$

and a more concise notation for the energy of a baroclinic component at wavenumber $\mathbf{k}$,

$$T_k \equiv (k^2 + 2\lambda^2)|\tau_k|^2. \tag{11.12}$$

Then (11.9) and (11.10) are simply

$$E = \sum (U_k + T_k), \tag{11.13}$$

$$Z = \sum \left(k^2 U_k + (k^2 + 2\lambda^2) T_k\right). \tag{11.14}$$

In light of (11.3), the spectral representation of a flow immediately suggests that the advective process in physical space is attributable to two types of triad interaction in the spectral space. One type of triad involves three barotropic spectral components,

$$\psi_p, \ \psi_q, \ \psi_k. \tag{11.15}$$

It is called *barotropic triad*. Another type of triad entails one barotropic spectral component and two baroclinic spectral components,

$$\psi_p, \ \tau_q, \ \tau_k. \tag{11.16}$$

It is called *baroclinic triad*. The triads must obey a selection rule that stems from the trigonometric properties of the base functions used in (11.8), viz.

$$\mathbf{p} + \mathbf{q} + \mathbf{k} = 0 \tag{11.17}$$

for the horizontal wave vectors. The evolution of a flow is the consequence of the totality of all possible triad interactions. Each triad conserves energy and potential enstrophy individually whereby (11.13) and (11.14) would be satisfied.

11.2.2 Characteristics of barotropic triad interaction

There is no net change in the total energy or enstrophy of the three barotropic spectral components in each triad. In other words, a barotropic triad is characterized by

$$\begin{aligned} \dot{U}_k + \dot{U}_p + \dot{U}_q &= 0, \\ k^2 \dot{U}_k + p^2 \dot{U}_p + q^2 \dot{U}_q &= 0. \end{aligned} \tag{11.18a,b}$$

The dot refers to time derivative. Equations (11.18a,b) are the same as the counterparts in the case of two-dimensional turbulence. Their implications are most evident from the consideration of a sample triad. For the triad $(k, p = k/2, q = 2k)$, the wave p is twice as long and the wave q is half as long as the wave k. It follows from (11.18)

$$\dot{U}_p = -\frac{4}{5}\dot{U}_k, \qquad \dot{U}_q = -\frac{1}{5}\dot{U}_k. \tag{11.19}$$

We see that if component k loses energy ($\dot{U}_k < 0$), 80% of the loss would go to the longer wave and only 20% of it would go to the shorter wave via this triad interaction.

At the same time, the rates of change of potential enstrophy are

$$p^2 \dot{U}_p = -\frac{k^2 \dot{U}_k}{5}, \qquad q^2 \dot{U}_q = -\frac{4k^2 \dot{U}_k}{5}. \tag{11.20}$$

In other words, the proportion of enstrophy transfers to the other members is the opposite of that of energy transfers. The longer wave receives only 20% of the loss of enstrophy from k, whereas the shorter wave receives 80%. The direction of transfers is also reversible as far as (11.19) and (11.20) are concerned. For example, component k could gain energy ($\dot{U}_k > 0$). In that case, 80% of the gain would come from the longer wave and only 20% of it would come from the shorter wave. The reversibility reflects the fact that the governing equations (11.3a,b) are applicable whether time changes forward or backward. We cannot tell which direction the average cascade by barotropic triad interactions would be until we take into account the additional physical processes that must also take place in a fluid.

11.2.3 Characteristics of baroclinic triad interaction

The barotropic spectral components also interact with the baroclinic spectral components. Conservation of energy and potential enstrophy in a baroclinic triad is described by

$$\begin{aligned} \dot{U_k} + \dot{T_p} + \dot{T_q} &= 0, \\ k^2 \dot{U_k} + (p^2 + 2\lambda^2)\dot{T_p} + (q^2 + 2\lambda^2)\dot{T_q} &= 0. \end{aligned} \tag{11.21a,b}$$

The nature of baroclinic triad interaction strongly depends on the scale of the components involved relative to the radius of deformation. Two wavenumber ranges are of particular interest. For those spectral components much shorter than the deformation radius ($k, p, q \gg \lambda$), (11.21) would be reduced to (11.18) as a good approximation. Therefore, the cascades of barotropic energy, barotropic potential enstrophy, baroclinic energy and baroclinic potential enstrophy in geostrophic turbulence are similar to the energy and enstrophy cascades in two-dimensional turbulence at the high wavenumber part of a spectrum.

More generally, we get from (11.21)

$$\begin{aligned} \dot{U_k} &= \frac{-p^2 + q^2}{2\lambda^2 + p^2 - k^2}\dot{T_q}, \\ \dot{T_p} &= \frac{-2\lambda^2 - q^2 + k^2}{2\lambda^2 + p^2 - k^2}\dot{T_q}. \end{aligned} \tag{11.22}$$

It follows that for those spectral components much longer than the deformation radius ($k, p, q \ll \lambda$), (11.22) would yield

$$\left|\dot{U_k}\right| \ll \left|\dot{T_q}\right|, \qquad \dot{T_p} \approx -\dot{T_q}. \tag{11.23}$$

In other words, the barotropic spectral component in such a baroclinic triad interaction gives or receives only a small amount of energy, but it nevertheless plays a catalytic role. One baroclinic spectral component intensifies at the expense of the other baroclinic spectral component in such an interaction. For example, the interaction among a triad ($k, p = k/2, q = 2k$) in conjunction with $\lambda = 3k$ would yield energy cascades $\dot{U_k} = \frac{-15}{84}\dot{T_p}$ and $\dot{T_q} = \frac{-69}{84}\dot{T_p}$ according to (11.22). If the longest spectral component loses baroclinic energy $\dot{T_p} < 0$ due to this interaction, both shorter components would gain energy; 17.9% of that loss would go to U_k and 82.1% to T_q.

When the longest component p loses energy, it would also lose potential enstrophy. Then both shorter spectral components would gain potential enstrophy from it. The disparity between the change in the enstrophy of the barotropic component and the changes in the potential enstrophy of the two baroclinic components is even more pronounced. For the baroclinic triad under consideration, we get

$$\dot{Z}_k = k^2 \dot{U}_k = \frac{-15}{84} \frac{k^2}{p^2 + 2\lambda^2} \left(p^2 + 2\lambda^2\right) \dot{T}_p = \frac{-15}{1533} \dot{Z}_p,$$

$$\dot{Z}_q = \left(q^2 + 2\lambda^2\right) \dot{T}_q = \frac{-69}{84} \frac{q^2 + 2\lambda^2}{p^2 + 2\lambda^2} \left(p^2 + 2\lambda^2\right) \dot{T}_p = \frac{-1518}{1533} \dot{Z}_p .$$

In other words, only about 1% of the loss of potential enstrophy from component p would cascade to the shorter barotropic component k, whereas 99% of it would be cascaded to the shorter baroclinic component q.

A difference between the cascades in barotropic triad interaction and those in baroclinic triad interaction warrants emphasis. While the intermediate scale component in a barotropic triad would either lose or gain energy/enstrophy, the longest component in a baroclinic triad could be either a beneficiary or a loser. Like barotropic triad interactions, the baroclinic triad interactions are also reversible. It is possible for both shorter spectral components to transfer energy to the longest component. We cannot tell the average direction of the cascades until we take into account the additional physical processes in a real fluid.

11.2.4 Influence of forcing and damping on turbulent cascades

In spite of the intrinsic reversibility of the cascades by triad interactions, there is generally a prevalent direction of the cascades in a real turbulent fluid. It stems from the influence by diabatic and dissipative processes at scales far from each inertial subrange. For that reason, the prevalent direction of a type of cascade is not necessarily the same in different inertial subranges. It calls for a careful consideration of those processes. Dissipation in the atmosphere is attributable to the small-scale eddies which are decidedly nongeostrophic motions. It may be parameterized as a diffusive process, which removes both energy and enstrophy. It is more effective at shorter length scales. It imposes an upper limit of wavenumber in the spectrum of geostrophic turbulence, k_D. Diffusive damping can also remove the energy of a large-scale flow in an Ekman layer next to the surface through the process of Ekman pumping/suction. It is particularly effective in the flow components of very low wavenumber. Another possible form of dissipation is thermal damping of large-scale components. It stems from a greater loss of infrared radiation to space in warm than in cold air and by a similar heat exchange with the Earth's surface. All these processes are irreversible. Irreversibility in nature endows what physicists call an *arrow-of-time*. It imposes statistically preferred directions in the cascades of energy/enstrophy/potential enstrophy in turbulence.

The action of damping in the turbulence of a statistical equilibrium state implies that there must also exist some forcing. For the global atmosphere, an appropriate forcing is zonally uniform so that it would mimic the equator-to-pole differential net radiative heating/cooling. It tends to warm the low latitude region and to cool the high latitude region. That would be able to maintain a certain mean north–south thermal contrast (zonal baroclinicity) across the domain. We may then introduce an external forcing in a model in the form of thermal relaxation of an instantaneous flow towards a suitably prescribed north–south temperature field. This external forcing at the minimum wavenumber k_F would generate baroclinic energy of that scale. Since the same amount of barotropic energy must be removed from the system at k_F in equilibrium, we have

$$\left(\dot{U}_{k_F}\right)_{nc} + \left(\dot{T}_{k_F}\right)_{nc} = 0, \tag{11.24}$$

where nc stands for the net non-conservative processes. The net production of potential enstrophy at k_F is then

$$k_F^2\left(\dot{U}_{k_F}\right)_{nc} + (k_F^2 + 2\lambda^2)\left(\dot{T}_{k_F}\right)_{nc} = 2\lambda^2\left(\dot{T}_{k_F}\right)_{nc} \; 0. \tag{11.25}$$

It has a positive value since potential enstrophy is dissipated at k_D.

Another important internal mechanism would take place when the external heating is sufficiently strong. That mechanism is baroclinic instability. According to the theory of purely baroclinic instability elaborated in Part B of Chapter 8, the significantly unstable waves have a relatively narrow range of scales close to the radius of deformation, λ^{-1}. They are what we call synoptic-scale waves. Notice that this is a linear instability process and is distinct from the multi-step cascade process in the geostrophic turbulence. It directly converts some energy and potential enstrophy from the longest scale (zonal basic state) to the synoptic scale without involving the other scales in between. The instability process results in generating both baroclinic and barotropic energy at the most unstable synoptic scale. Baroclinic instability therefore is an internal mechanism that stirs the atmosphere at synoptic scale. Such stirring in conjunction with diffusive dissipation would jointly dictate the intrinsic cascades of energy and enstrophy by the triad interactions between the synoptic scale and the limiting small scale. Recall from the discussions in Sections 11.2.2 and 11.2.3 that the energy cascade towards the small scale is relatively weak while the enstrophy/potential enstrophy cascade towards the small scale is relatively large by those interactions. Thus, we expect primarily enstrophy (or potential enstrophy) transfer at the high-wavenumber part of the spectrum. On the other hand, the stirring together with Ekman-layer-type dissipation at the low-wavenumber components also would give rise to an intrinsic cascade of barotropic energy from the synoptic scale towards the very long scale. In addition, there should be a cascade of baroclinic energy and potential enstrophy from the scale of external forcing towards the synoptic scale via baroclinic triad interactions. Such a cascade would be able to compensate for the thermal damping at some large scales alluded to earlier and to support the continual cascades of energy and potential enstrophy towards the high-wavenumber components via baroclinic triad interactions.

The overall picture of energy and potential enstrophy cascades in this system of geostrophic turbulence may be succinctly summarized in Fig. 11.2 adapted from Salmon (1980). This figure does not explicitly indicate the removal of baroclinic energy due to thermal damping at some large scales.

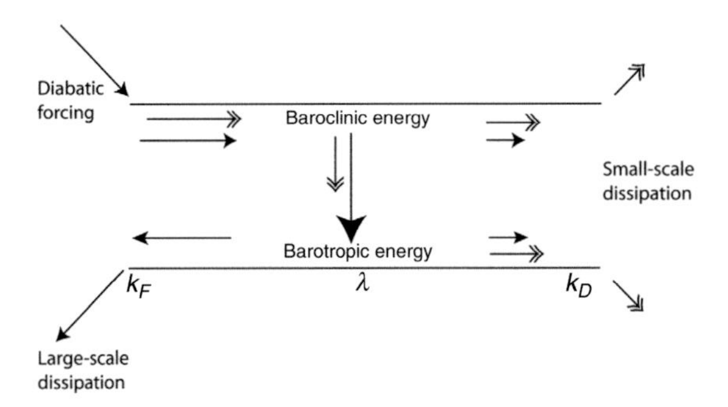

Fig. 11.2 Schematic diagram of energy and potential enstrophy transfer in a two-layer quasi-geostrophic fluid. Horizontal wavenumber $k_F \leq k \leq k_D$, significant damping at $k \geq k_D$, external forcing at k_F, λ^{-1} is radius of deformation. Single arrows indicate energy transfer. Double arrows indicate potential enstrophy transfer (adapted from Salmon, 1980).

11.2.5 Rhines' scale

One might get the impression from Fig. 11.2 that the geostrophic turbulence cascades the energy of the barotropic component of the flow all the way to k_F. But this actually may not take place on a spherical planet because of the influence associated with the spatial variation of the Coriolis parameter, the so-called beta-effect. The reason is that advection of planetary vorticity would be greater than advection of relative vorticity, $\beta\psi_x > \vec{v}\cdot\nabla(\nabla^2\psi)$ at the ultra-long scales. Those scales are $L > \sqrt{u^*/\beta}$ where u^* is a velocity scale. At such length scales, turbulence would progressively degenerate to the orderly Rossby wave motions governed by linear dynamics. The barrier scale is $L \sim \sqrt{u^*/\beta}$ for cascading energy and enstrophy by geostrophic turbulence and is known as Rhines' scale (Rhines, 1975). The more intense the geostrophic turbulence is, the longer would be the length scale energy could be cascaded to.

11.2.6 Spectra of energy and potential enstrophy

We have seen in Sections 11.2.2 and 11.2.3 that the characteristics of the cascades by barotropic triad interactions and baroclinic triad interactions are similar in one inertial subrange, $\lambda < k < k_D$. We may then examine the cascades of barotropic energy and baroclinic energy jointly. There must be cascades of energy and potential enstrophy towards k_D to compensate for the diffusive dissipation. In a state of *statistical equilibrium of isotropic homogeneous geostrophic turbulence*, energy cascade toward higher wavenumbers past each wavenumber k must be the same so that there would be no accumulation of energy at any k. The energy spectrum $E(k)$ depends only on k, ε and η where ε is the rate of energy transfer and η is the rate of potential enstrophy transfer. Since enstrophy is removed at wavenumbers greater than k_D, we have $\eta > k_D^2\varepsilon$. In the limit $k_D \to \infty$, corresponding to vanishingly small viscosity, we must have $\varepsilon \to 0$ otherwise η would become larger and larger without bound. This would contradict the fact that the stirring at λ only supplies energy and potential enstrophy in a finite rate. We conclude that the energy transfer to higher wavenumbers is asymptotically zero and the spectrum $E(k)$ depends only on k and η. The dimension of E, η, ε are $[E] = \mathrm{L}^3\mathrm{T}^{-2}$, $[\eta] = \mathrm{T}^{-3}$, $[\varepsilon] = \mathrm{L}^2\mathrm{T}^{-3}$. By the requirement of dimensional consistency, we deduce that

$$E(k) = C_1 \eta^{2/3} k^{-3}, \tag{11.26}$$

where C_1 is a universal constant. The corresponding enstrophy spectrum is $Z(k) = k^2 E(k) = C_1 \eta^{2/3} k^{-1}$. These spectra are referred to as "−3" and "−1" power laws respectively for this *enstrophy-cascading inertial subrange.*

The situation for the cascades in the low-wavenumber inertial subrange $k_F < k < \lambda$ is less clear. As far as the cascade of barotropic energy is concerned, it is still akin to two-dimensional turbulence. By the same reasoning, the enstrophy transfer vanishes in this subrange for the asymptotic limit $k_F \to 0$. Such a spectrum would depend only on k and ε. Dimensional analysis then yields

$$E(k) = C_2 \varepsilon^{2/3} k^{-5/3}, \tag{11.27}$$

where C_2 is a universal constant. The corresponding enstrophy spectrum is then $Z(k) = k^2 E(k) = C_2 \varepsilon^{2/3} k^{1/3}$. These spectra are referred to as "−5/3" and "1/3" power laws for this *energy-cascading inertial subrange.* The characteristics in the cascades by baroclinic triad interactions in this subrange are different. The energy spectrum may then also depend on η. Dimensional argument would not be sufficient to determine a power law if the spectrum is a function of all three parameters. It is therefore not obvious whether (11.27) would apply to the spectrum of total energy. For the caveats concerning the intrinsic merits/limitation of these theoretical reasonings, readers should consult the more in-depth discussion in the text by Salmon (1998).

A diagnosis of the First GARP Global Experiment (FGGE) data (Boer and Shepherd, 1983) reveals some resemblance of a "−3" power law for the energy spectrum (Fig. 11.3) and a "−1" power law for enstrophy in the high-wavenumber (synoptic-scale) regime in both summer and winter. The results are therefore reminiscent of an enstrophy-cascading inertial subrange. On the other hand, there is no good agreement between the observed spectra and the theoretical spectra in the low-wavenumber (planetary-scale) regime. Energy in this subrange has been confirmed to be cascading in the upscale direction. The seasonal variation of the spectra in this subrange is large. Furthermore, the anisotropy in the turbulence here appears to be substantial.

11.3 Life cycle of baroclinic waves

The characteristics of the life cycle of a baroclinic wave have been found to sensitively depend on the barotropic structure in the basic zonal flow (e.g. Thorncroft *et al.*, 1993). Those results were established with the use of a high-resolution primitive-equation model on a sphere. Since a quasi-geostrophic model adequately represents the main part of the nonlinear advective process, it should be capable of capturing the dynamics of the prototype life cycles of an unstable baroclinic wave. We use it as a learning tool for this

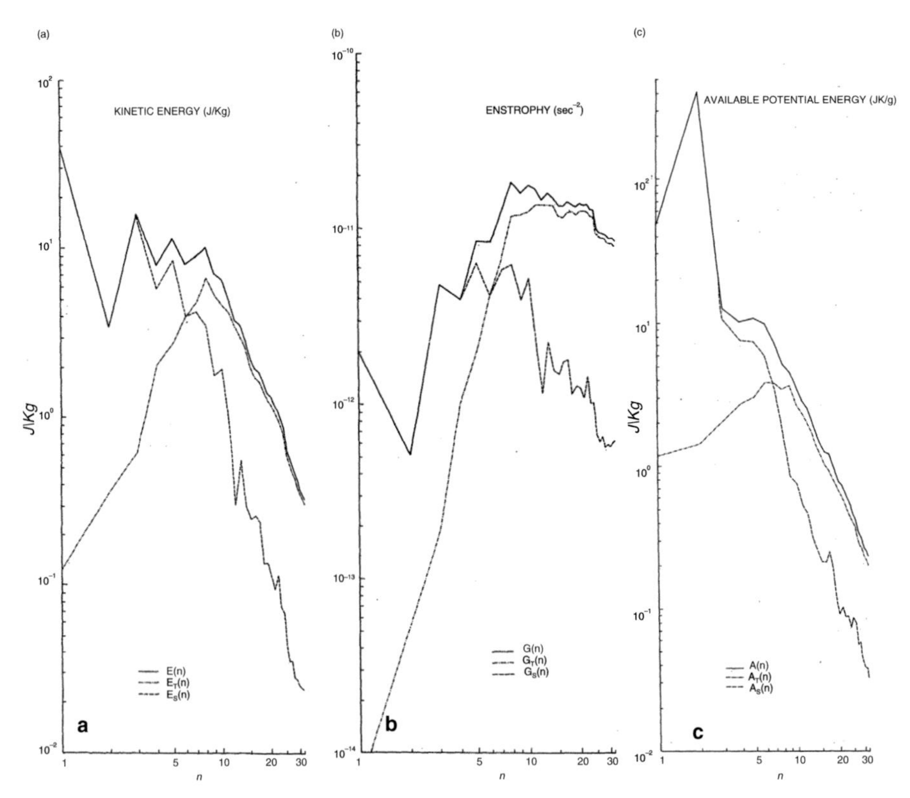

Fig. 11.3 Integrated spectra of (a) kinetic energy $E(n)$, (b) enstrophy $G(n)$ and (c) available potential energy $A(n)$, for January; n is the spherical harmonic index. Stationary and transient components are also shown (Taken from Boer and Shepherd, 1983).

purpose. It is hoped that by quantitatively examining the processes in detail, we will develop a feel for the dynamics of such evolution.

With the notations used in Chapter 6, we define the non-dimensional quantities in a two-layer quasi-geotrophic model as: $\tilde{k} = k/\lambda$, $\tilde{\sigma} = \sigma/(V\lambda)$, $\tilde{\beta} = \beta/(V\lambda^2)$, $\tilde{U}_j = \tilde{U}_j/V$, $(\tilde{x}, \tilde{y}, \tilde{a}) = \lambda(x, y, a)$, $\tilde{\psi}_j = \psi_j/V\lambda^{-1}$, $\tilde{\omega}_2 = \omega_2 f_o/V^2\lambda^2\Delta p$, $\tilde{T}_2 = T_2 R/V\lambda^{-1} f_o$. We consider a background baroclinic jet with a zonal velocity at the two levels as

$$U_j = V\varepsilon_j \exp\left(-(y/a)^2\right), \quad j = 1, 3. \tag{11.28}$$

For $\lambda = 10^{-6}\,\mathrm{m}^{-1}$, $V = 30\,\mathrm{m}^{-1}$, $\beta = 1.5 \times 10^{-11}\,\mathrm{m}^{-1}\,\mathrm{s}^{-1}$, $-Y \le y \le Y$ and $Y = 5 \times 10^6\,\mathrm{m}$, we use $\varepsilon_1 = 1.5$, $\varepsilon_3 = 0.3$, $a = 1.5$, $\tilde{Y} = 5$ and $\tilde{\beta} = 0.5$. We have seen in Section 8C.3.2 that the most unstable baroclinic wave of this flow has a zonal wavelength of 5.5 with a growth rate of about 0.11. This disturbance is symmetric about the center of the jet (Fig. 8C.4 in Section 8C.3.2). This unstable normal mode together with the zonal jet is chosen as the initial state of the model.

The model code previously applied in Section 6.6 is used here to numerically integrate the life cycle of an unstable baroclinic wave. The model resolution is 151 grid points in the

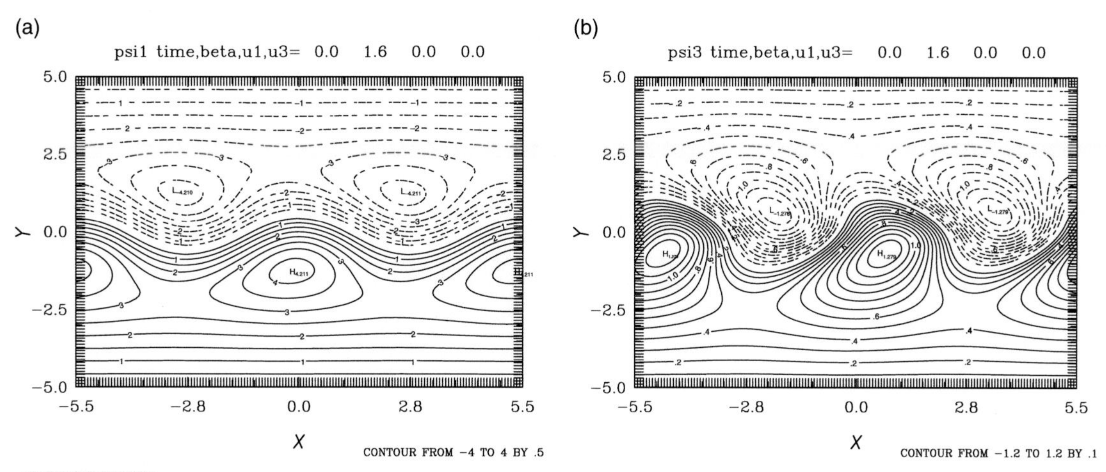

Fig. 11.4 Distribution of the non-dimensional streamfunction at $t = 0$ (a) in the upper layer ψ_1 and (b) in the lower layer ψ_3 in units of $V\lambda^{-1}$.

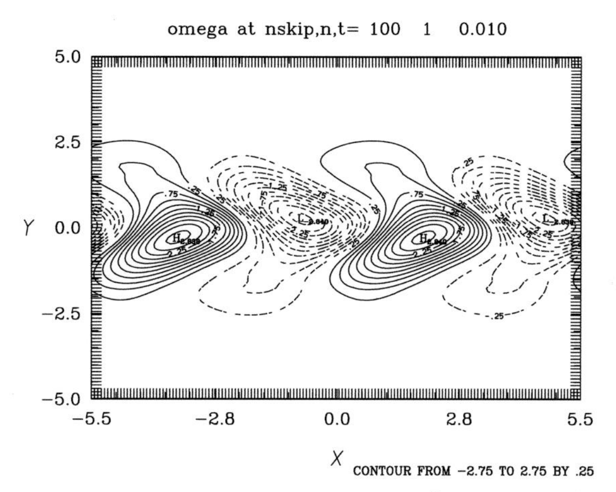

Fig. 11.5 Distribution of the non-dimensional ω_2 field at $t = 0$ in units of $V^2\lambda^2\delta p/f_o$.

x-direction and 101 grid points in the y-direction. Adopting $V = 10\,\text{m}^{-1}$ for non-dimensionalization, we use $\varepsilon_1 = 4.5$ and $\varepsilon_3 = 0.9$ in this calculation. The initial state of the total streamfunction field over a distance of two wavelengths of the unstable wave is shown in Fig. 11.4. The flow is about four times more intense in the upper layer than in the lower layer, since the contour interval is five times larger in panel (a) than in panel (b).

The corresponding ω_2 field is shown in Fig. 11.5. The ascending (descending) motion is located to the east (west) of each trough and somewhat to the north (south) of the jet. The extreme value is about $2\,\text{cm}\,\text{s}^{-1}$.

The variation of the kinetic energy of the flow in time depicts the overall character of the evolution of this flow (Fig. 11.6). It shows that this unstable wave continues to intensify for about two days and then decays subsequently. The main part of the life cycle is pretty much over by about day 6 ($t = 6$).

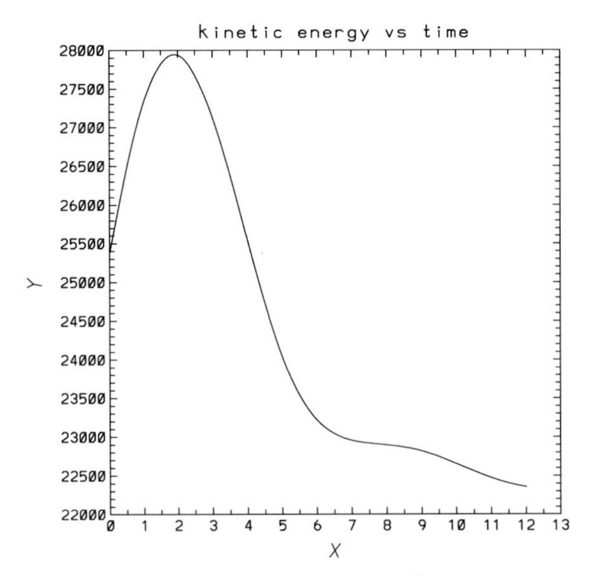

Fig. 11.6 Variation of the kinetic energy of the flow up to $t = 12$ in units of V^2.

The highlights of the life cycle of this baroclinic wave are displayed in terms of the total fields of its potential vorticity and p-velocity q_1, q_3, ω_2 at $t = 4$ in Fig. 11.7. The constant component of PV associated with the Coriolis parameter at a reference latitude is not included in the plots. There are two outstanding features. First, the fundamental harmonic of the wave field remains as the dominant component as dictated by the triad interactions. Second, the initial meridional odd symmetry of the PV distribution about the center line of the domain is preserved throughout the life cycle. The positive PV anomalies in the upper layer roll up cyclonically to the north of the jet axis and the negative PV anomalies roll up anticyclonically to the south of the axis of the jet. They soon assume an appearance of a breaking wave resulting from the excitation of increasingly higher harmonics of the fundamental wave. From a physical point of view, the fine features simply arise from advection of potential vorticity. This process of homogenization of PV is particularly evident in the lower layer. The wave field as a whole progressively becomes less baroclinic and weaker in the course of its life cycle. It should be mentioned that no hyperdiffusion is included in this calculation as is normally done in this type of numerical simulation. Including that would smooth out some of the hyperfine features.

The structure of the p-velocity ω_2 also undergoes great changes. In the first two days or so, each area of ascending or descending motion splits into smaller areas straddling across the axis of the background jet. Subsequently, strong ascent and descent remain mostly along the axis of the jet when the wave has been greatly attenuated. It is interesting to note that while the streamfunction of the wave weakens, the magnitude of its vertical motion field remains largely the same.

The impact of the equilibrated wave on the baroclinic and barotropic components of the zonal mean flow is diagnosed in Fig. 11.8. The evolution of the barotropic zonal mean flow component is depicted in terms of $[\psi_2] = \frac{1}{2}[\psi_1 + \psi_3]$, where the square bracket stands for

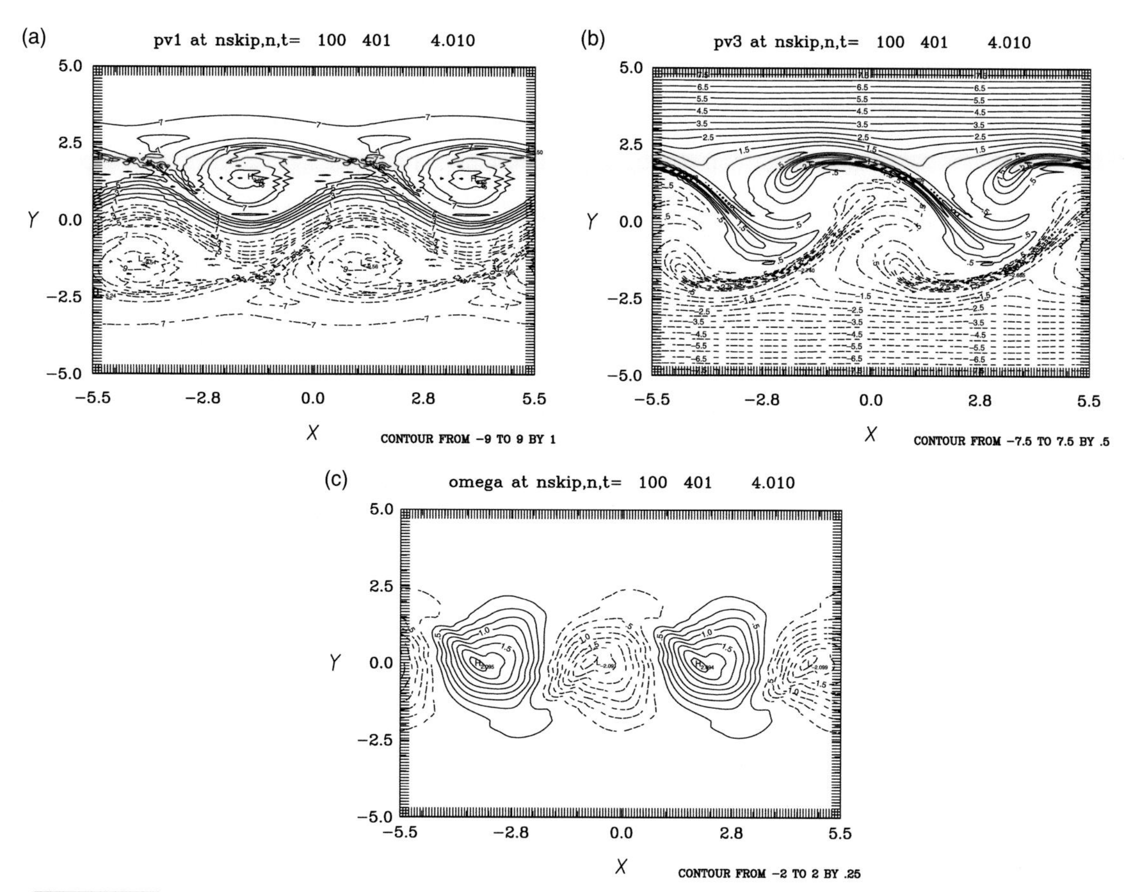

Fig. 11.7 Distribution of the non-dimensional (a) potential vorticity in the upper layer q_1, (b) in the lower layer q_3 and (c) p-velocity ω_2 at $t = 4$ in the reference case.

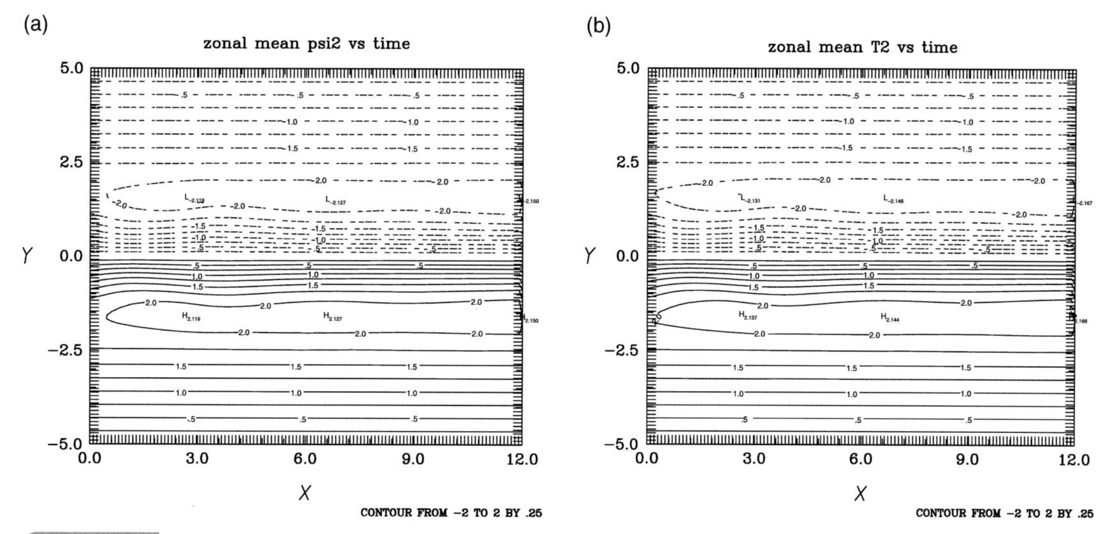

Fig. 11.8 Evolution of the non-dimensional (a) zonal mean barotropic component, $[\psi_2] = \frac{1}{2}[\psi_1 + \psi_3]$, and (b) zonal average baroclinic component, $[\psi_1 - \psi_3]$ during a life cycle.

zonal averaging. The evolution of the baroclinic zonal mean flow component is depicted in terms of $[T_2] = [\psi_1 - \psi_3]$. We see that the barotropic component of the zonal mean jet and the baroclinic shear are both slightly strengthened while the baroclinic wave has virtually attenuated at the end of its life cycle.

For comparison with this control run, additional experiments are made with an *additional* barotropic zonal component added to the initial background flow. The new zonal flow is prescribed as follows

$$U_j = V\left(\varepsilon_j \exp\left(-(y/a)^2\right) + \gamma \tanh(y/b)\right). \tag{11.29}$$

11.3.1 Influence of zonal barotropic cyclonic shear: LC2

We now make another calculation using $\gamma = -1$ in the initial zonal flow prescribed according to (11.29). The other parameters of the zonal flow are $\varepsilon_1 = 4.5$, $\varepsilon_3 = 0.9$, $a = 1.5$ and $b = 1.5$. The initial state consists of the same wave disturbance of the last section and this particular zonal flow. The additional barotropic component is a maximum westerly (easterly) of $10\,\mathrm{m\,s^{-1}}$ ($-10\,\mathrm{m\,s^{-1}}$) in the southern (northern) part of the domain. It has the same meridional scale as the baroclinic jet. The cyclonic shear is strongest at $y = 0$. The additional zonal mean cyclonic vorticity partly negates the anticyclonic vorticity of the baroclinic jet in the southern half of the domain and adds to the cyclonic vorticity of the baroclinic jet in the northern half of the domain. This meridional asymmetry in the zonal mean vorticity is especially pronounced in the lower layer. The meridional asymmetry is evident in the initial streamfunction fields in the two layers (Fig. 11.9).

The corresponding initial ω_2 field is shown in Fig. 11.10. The ascending motion is slightly enhanced and the descending motion is reduced relative to that in the control run (−2.930 vs. 2.784).

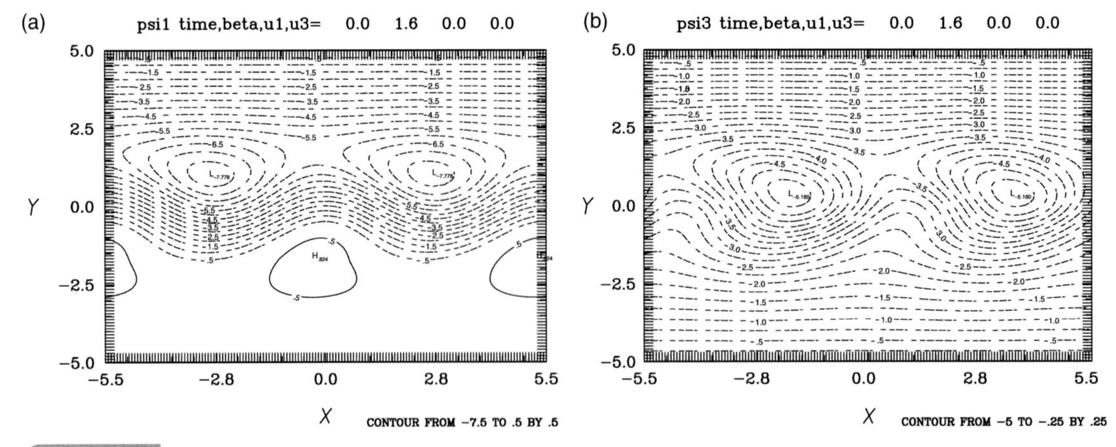

Fig. 11.9 Distribution of the initial non-dimensional streamfunction (a) in the upper layer ψ_1 and (b) in the lower layer ψ_3 in units of $V\lambda^{-1}$ for the case with an additional cyclonic barotropic zonal shear flow component.

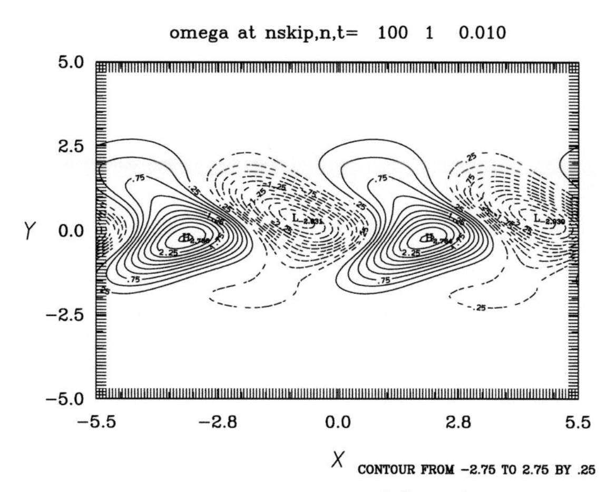

Fig. 11.10 Distribution of the initial non-dimensional ω_2 field in units of $V^2\lambda^2\delta p f_0^{-1}$ for the case with an additional cyclonic barotropic zonal shear flow component.

The initial growth rate is somewhat smaller suggesting that the additional horizontal shear has a stabilizing influence consistent with the barotropic-governor effect discussed in Section 8C.2. The highlights of the life cycle of the baroclinic wave in this case are shown in Fig. 11.11, which is the counterpart of Fig. 11.7 for the control run. The cyclonic disturbance to the north of the baroclinic jet axis is much stronger than the anticyclonic disturbance to the south of the jet axis. The dominant feature is the broadening troughs wrapping themselves up cyclonically and northward analogous to LC2. It is particularly evident in q_3. The corresponding streamfunction field shows the production of a pronounced cutoff cyclone. The asymmetry is also present in the ω_2 field at the mature stage of the wave, say at $t = 2$. At the later stages, the asymmetry in the ω_2 field is much weaker and the extreme values still lie largely on the center-line of the domain. This is the quasi-geostrophic form of the LC2 life cycle. It does capture the main characteristic of LC2. There is considerable resemblance between our quasi-geostrophic LC2 and the plots of potential temperature on the PV= 2 PVU of Thorncroft *et al.* (1993; Fig. 11.11) in their simulation of LC2 with a primitive-equation model.

11.3.2 Influence of zonal barotropic anticyclonic shear: LC1

In contrast to the last experiment, we now make another calculation using $\gamma = 1$ in the initial zonal flow prescribed according to (11.29). This barotropic zonal component has strongest anticyclonic shear at $y = 0$. It has a maximum easterly of $10\,\mathrm{m\,s^{-1}}$ to the south of the domain and a maximum westerly of $10\,\mathrm{m\,s^{-1}}$ to the north. Everything else is the same as in the control run. The initial streamfunction fields in the two layers are shown in Fig. 11.12. We see that Fig. 11.12a is a mirror image of Fig. 11.9a about the center-line of the domain with an eastward shift of half a wavelength and a change of sign; ditto for Fig. 11.12b compared to Fig. 11.9b. In other words, the zonal anticyclonic vorticity is enhanced in the south and the zonal cyclonic vorticity is reduced in the north due to the presence of the additional anticyclonic zonal barotropic flow.

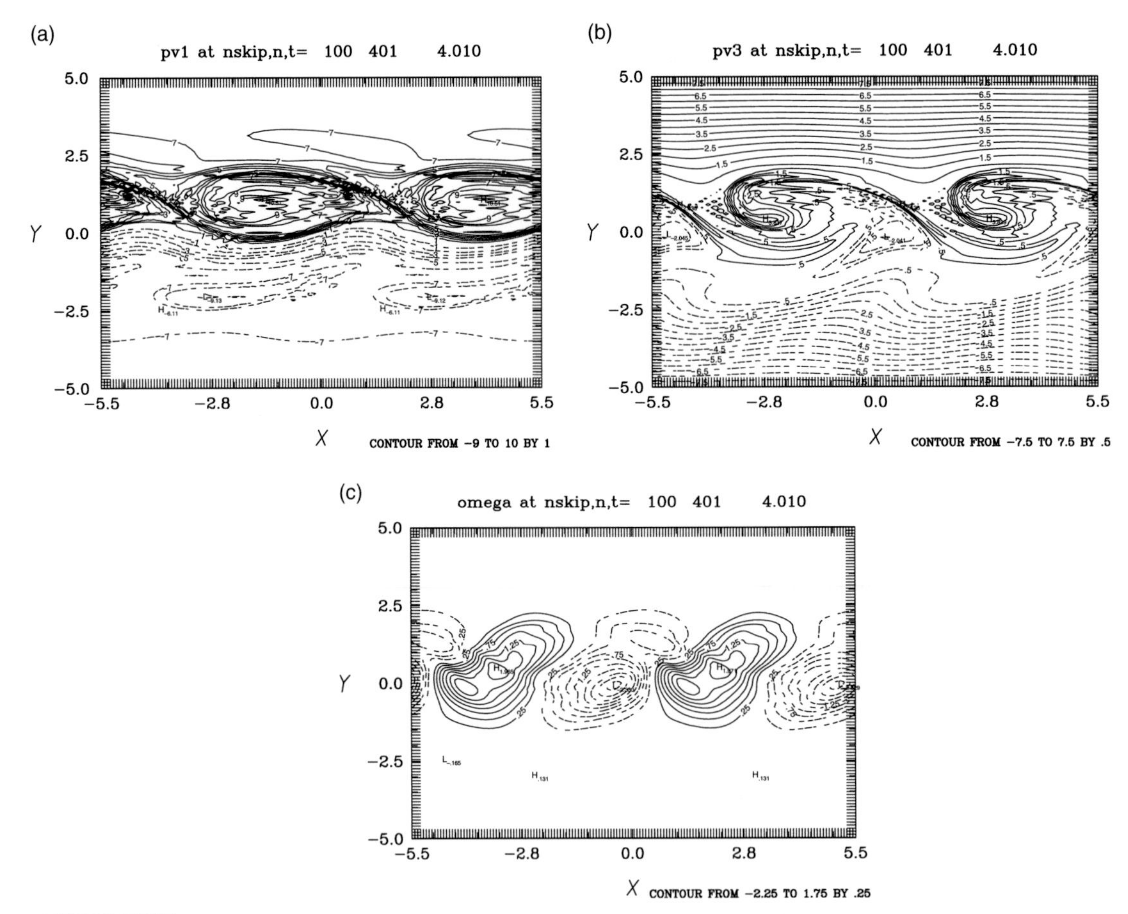

Fig. 11.11 Distribution of the non-dimensional (a) potential vorticity in the upper layer q_1, (b) in the lower layer q_3 and (c) p-velocity ω_2 at $t = 4$ for the case with an additional cyclonic barotropic zonal shear flow component.

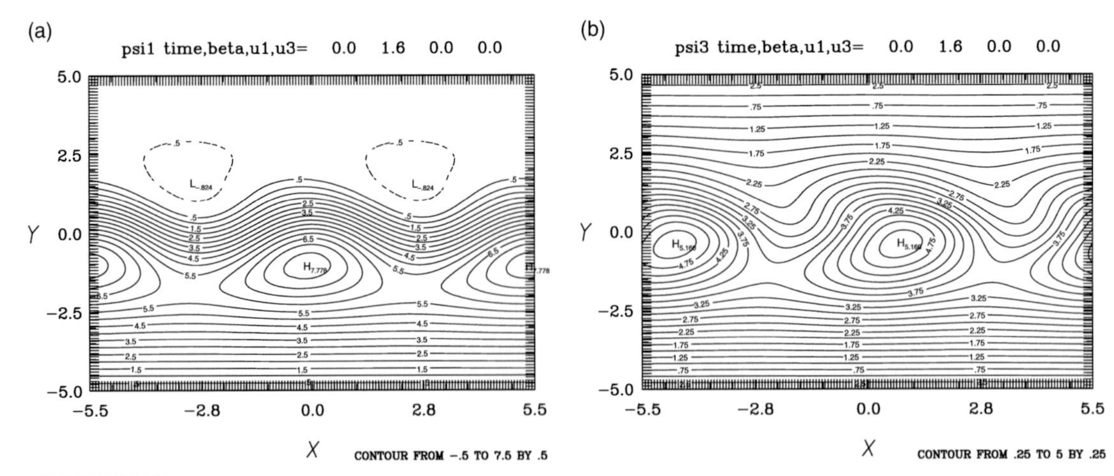

Fig. 11.12 Distribution of the non-dimensional streamfunction at $t = 0$ for the case with an additional anticyclonic barotropic zonal shear flow component, (a) in the upper layer ψ_1 and (b) in the lower layer ψ_3 in units of $V\lambda^{-1}$.

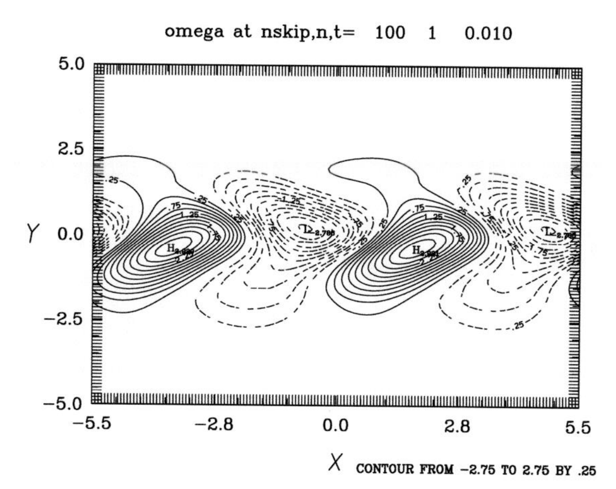

Fig. 11.13 Distribution of the non-dimensional ω_2 field at $t = 0$ in units of $V^2\lambda^2\delta p f_0^{-1}$ for the case with an additional anticyclonic barotropic zonal shear flow component.

The corresponding initial ω_2 field is shown in Fig. 11.13. Again, Fig. 11.13 is a mirror image of Fig. 11.10 with respect to the center-line of the domain with an eastward shift and a change of sign. It means that the configuration of ascending motion here is the same as that of descending motion in the previous case. The extreme descending motion is slightly stronger than the extreme ascending motion (2.930 vs. −2.785) in this case instead.

The life cycle of the baroclinic wave in this case is shown in Fig. 11.14, which is the counterpart of Fig. 11.11. The difference is of course solely a consequence of the difference in the two initial background flows, although the disturbance itself is the same in both experiments. Consequently, every field in Fig. 11.14 is an exact mirror image of the corresponding field in Fig. 11.11 with respect to the center-line of the domain with a change of sign. The anticyclonic disturbance to the south of the baroclinic jet axis is much stronger than the cyclonic disturbance to the north of the jet axis. The dominant thinning upper-air ridge advects anticyclonically and southward. This is most evident in q_3. This is the quasi-geostrophic version of the LC1 life cycle. There is considerable resemblance to the plot of potential temperature on the PV= 2 PVU surface in Thorncroft *et al.*'s (1993; Fig. 11.8) simulation of LC1 with a primitive-equation model.

In retrospect, the difference between LC1 and LC2 could be anticipated a priori. The presence of an additional barotropic zonal flow component with a cyclonic vorticity would naturally enhance the advection of positive PV anomaly and weaken the advection of negative PV anomaly. It is this asymmetry that seems to be the fundamental reason underlying the difference between LC1 and LC2. There is a caveat in this interpretation of the life cycles of baroclinic waves from a quasi-geostrophic perspective. One limitation is that there is inherent similarity between cyclonic and anticyclonic disturbances in QG dynamics when the initial flow has a symmetric structure. Any asymmetric development between them must stem from the asymmetry in the structure of the initial zonal mean flow. Another limitation of a two-layer model is that it cannot depict vertical propagation of baroclinic waves at all. Consequently, the model cannot simulate the observed initial

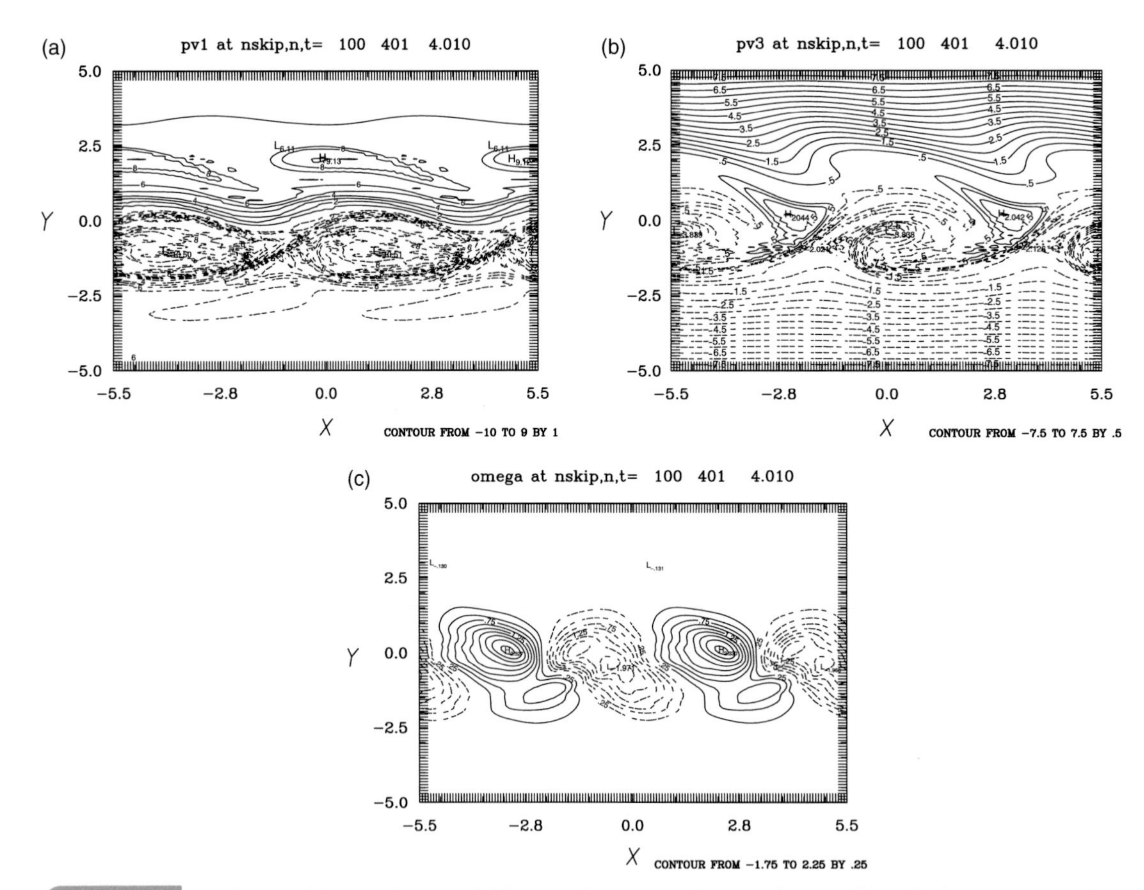

Fig. 11.14 Distribution of the non-dimensional (a) potential vorticity in the upper layer q_1, (b) in the lower layer q_3 and (c) p-velocity ω_2 at $t = 4$, for the case with an additional anticyclonic barotropic zonal shear flow component.

vertical propagation from lower troposphere to upper troposphere as well as the subsequent equatorward propagation of unstable baroclinic waves.

The two factors alluded to above give rise to some important differences between the life cycle of a baroclinic wave in a quasi-geostrophic framework and that in a multi-level primitive-equation framework. The latter simulates a life cycle of a baroclinic wave more faithfully. Nongeostrophic dynamics give rise to sharp frontal development through the positive feedback effect of the momentum and heat advection by the ageostrophic component of the flow itself (to be elaborated in Section 12.1). That would inherently break the symmetry between the cyclonic and anticyclonic structure of a baroclinic wave. It further accentuates the difference between LC1 and LC2. Past studies using the primitive-equation multi-level model have delineated four characteristics of a wave field in each life cycle. They are: exponential normal mode growth, nonlinear saturation in the lower troposphere, upward propagation towards the jet core and a second nonlinear saturation (Thorncroft *et al.*, 1993). Those authors further suggest that the late stages of LC2 are interpretable in terms of a nonlinear reflection scenario of the Rossby-wave critical-layer theory, whereas the late stages of LC1 are more akin to a nonlinear critical-layer absorption scenario.

11.4 Symbiotic relation between synoptic and planetary waves

There is no need to take into consideration the relative phase information among the spectral components in the discussion of isotropic homogeneous geostrophic turbulence. Such information however is relevant to the statistical distribution of actual wave activity because it has pronounced regional variability. It stems from the fact that the continents have significant orography and the oceans have very different thermal properties. Such surface inhomogeneity is particularly important in the northern hemisphere, directly leading to significant variations in the distribution of baroclinicity around each mid-latitude circle. The corresponding background flow is characterized by localized jets, which are made up of quasi-stationary planetary waves and a zonal mean flow. The relative phase angles among the synoptic waves and their phase angles relative to the planetary waves naturally dictate the locations of maximum synoptic wave activity. The resulting dynamical relation between the planetary-scale waves and synoptic-scale waves in a forced dissipative geophysical flow can be quite intricate. The statistical information about the relative phases between these two subsets of waves is therefore more than an academic interest. This is the next problem of nonlinear dynamics we would like to focus on.

The planetary-scale waves in the atmosphere have low frequency, whereas the cyclone-scale waves have high frequency. On the one hand, a localized baroclinic jet would enable an unstable normal mode with a localized structure downstream of the jet to intensify as shown in Section 8C.5. An unstable disturbance has a negative feedback effect on the baroclinic part of the basic flow, but a positive feedback effect on the barotropic part of the basic state. This could be also true for a traveling localized jet in principle. On the other hand, the theory of geostrophic turbulence suggests that there should be also significant transfer of energy from the synoptic-scale waves to planetary-scale waves via triad interactions. The two considerations above jointly suggest *a symbiotic relation between the synoptic-scale and planetary-scale waves in the sense that each benefits from the influence of the other.*

11.4.1 Model formulation

We now go over an analysis that delineates the relation alluded to above in the context of a two-layer quasi-geostrophic model (Cai and Mak, 1990). The forcing of that model is introduced in the form of a zonally uniform meridional temperature gradient associated with a constant baroclinic shear, $2U$. This conceptual framework is described schematically in Fig. 11.15.

The width of the domain is $\pi L = 6000\,\text{km}$ and the length is $2\pi L/\gamma = 30\,000\,\text{km}$. Horizontal distance, velocity and time are measured in units of L, V and $V^{-1}L$ respectively. In particular, V equal to $20\,\text{m}\,\text{s}^{-1}$ is used in the calculation. The spectral form of a quasi-geostrophic two-layer model with forcing and dissipation is used in the analysis. The governing equations for the spectral coefficients of the barotropic and baroclinic components of the departure field from an imposed zonal flow in the model, $\{\psi_{o,k}\}$, $\{\psi_{m,n}\}$,

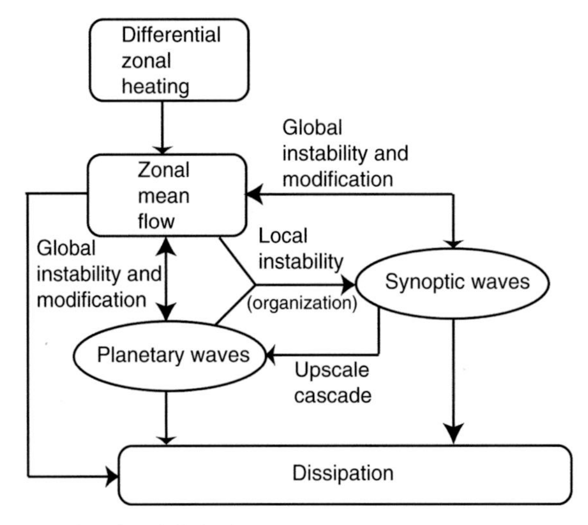

Fig. 11.15 A schematic of the processes in a forced dissipative system.

$\{\theta_{o,k}\}$ and $\{\theta_{m,n}\}$ for $k = 1, 2, \ldots, K$; $m = 1, 2, \ldots, M$; $n = 1, 2, \ldots, N$, are a set of nonlinear ordinary differential equations:

$$\begin{aligned}
\dot{\psi}_{o,k} &= -\left(r + a\lambda_{o,k}^2\right)\psi_{o,k} + F_{o,k}, \\
\dot{\psi}_{m,n} &= -\left(r + a\lambda_{m,n}^2\right)\psi_{m,n} + F_{m,n}, \\
\dot{\theta}_{o,k} &= -\left(r + a\frac{\lambda_{o,k}^3}{2F_o + \lambda_{o,k}}\right)\theta_{o,k} + H_{o,k}, \\
\dot{\theta}_{m,n} &= -\left(r + a\frac{\lambda_{m,n}^3}{2F_o + \lambda_{m,n}}\right)\theta_{m,n} + H_{m,n},
\end{aligned} \tag{11.30}$$

where the subscripts are the zonal and meridional wavenumbers and $\lambda_{m,n} = (m\gamma)^2 + n^2$, r is the damping coefficient of Ekman layer friction, a is the coefficient of hyperdiffusion and F_o is the Froude number. Apart from the forcing term included in $F_{o,k}$ and $H_{o,k}$, $F_{o,k}, F_{m,n}, H_{o,k}, H_{m,n}$ are quadratically nonlinear functions of the unknowns given in Mak (1985). The dot over each unknown on the LHS of each equation above stands for time derivative.

11.4.2 Model results

The resolution we use in making calculations is $M = 16$, and $N = K = 8$. The values of the non-dimensional external model parameters are: $\gamma = 0.4$, $U = 0.9$, $F = 5.0$, $\beta = 2.5$, $r = 0.1$ and $a = 10^{-4}$. The value of the imposed baroclinic shear U is equivalent to a plausible radiative equilibrium temperature contrast of 67 °C across the channel model. The value of the Froude number F_o corresponds approximately to a Brunt–Vaisala frequency of $1.2 \times 10^{-2}\,\mathrm{s}^{-1}$ and that of the beta parameter corresponds to a dimensional

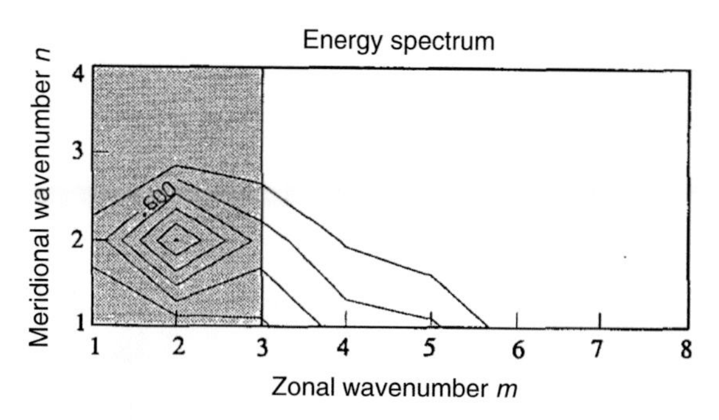

Fig. 11.16 Non-dimensional time-mean energy of the equilibrated flow as a function of the zonal and meridional wavenumbers. The shaded area indicates the planetary-scale waves. Contour interval, $CI = 0.3$ (taken from Cai and Mak, 1990).

value at latitude 45°. The time scale of the Ekman layer friction is about 10 days and that of the hyperdiffusion is about one day for the highest wavenumber component $(m, n) = (16, 8)$.

The model is integrated for 6000 days. The model output is sampled once a day. The circulation from $t = 1000$ to 6000 is diagnosed as the equilibrated flow of this system. The energy of each spectral component is computed as

$$\begin{aligned} E_{o,k} &= \frac{1}{2}\lambda_{o,k}\left(\psi_{o,k}^2 + \theta_{o,k}^2\right) + F_o\theta_{o,k}^2, \\ E_{m,n} &= \lambda_{m,n}\left(\left|\psi_{m,n}\right| + \left|\theta_{m,n}\right|^2\right) + 2F_o\left|\theta_{m,n}\right|^2. \end{aligned} \tag{11.31}$$

Figure 11.16 shows the time-averaged wave energy of the equilibrated state as a function of the zonal and meridional wavenumbers.

In light of Fig. 11.16, all waves with zonal wavenumber $m \leq 3$ may be therefore classified as "planetary-scale waves" and $m \geq 4$ as "synoptic-scale waves." The former are referred to as PW and the latter as SW. The waves in the equilibrated state are intermittently excited as seen in Fig. 11.17.

The time mean spectral energetics of the equilibrated flow among the three subsets of spectral components (net zonal baroclinic shear, synoptic-scale waves and planetary-scale waves) is diagnosed (Fig. 11.18). We see that most of the planetary-scale waves have a net loss of energy to the modified zonal flow except for the (2,1) and (3,2) waves. The planetary-scale waves as a group suffer a net loss of energy to the zonal flow $(C(\overline{U_{\text{mod}}, PW}) = -0.0802)$. The synoptic-scale waves, on the other hand, collectively gain energy from the zonal flow by baroclinic instability $(C(\overline{U_{\text{mod}}, SW}) = 0.0806)$. The synoptic-scale waves supply energy to the planetary-scale waves $(C(\overline{SW, PW}) = 0.0805)$. The standard deviation of these conversion rates are 0.13, 0.14 and 0.14 respectively, whereas their mean values are close to 0.08. The variability of these three energetic terms is therefore large in comparison to the magnitude of their mean values.

It is particularly instructive to statistically measure how often energy transfer occurs in each particular direction among these three subsets of spectral components. The results are shown in Fig. 11.19. The shaded area under each curve is equal to the probability for

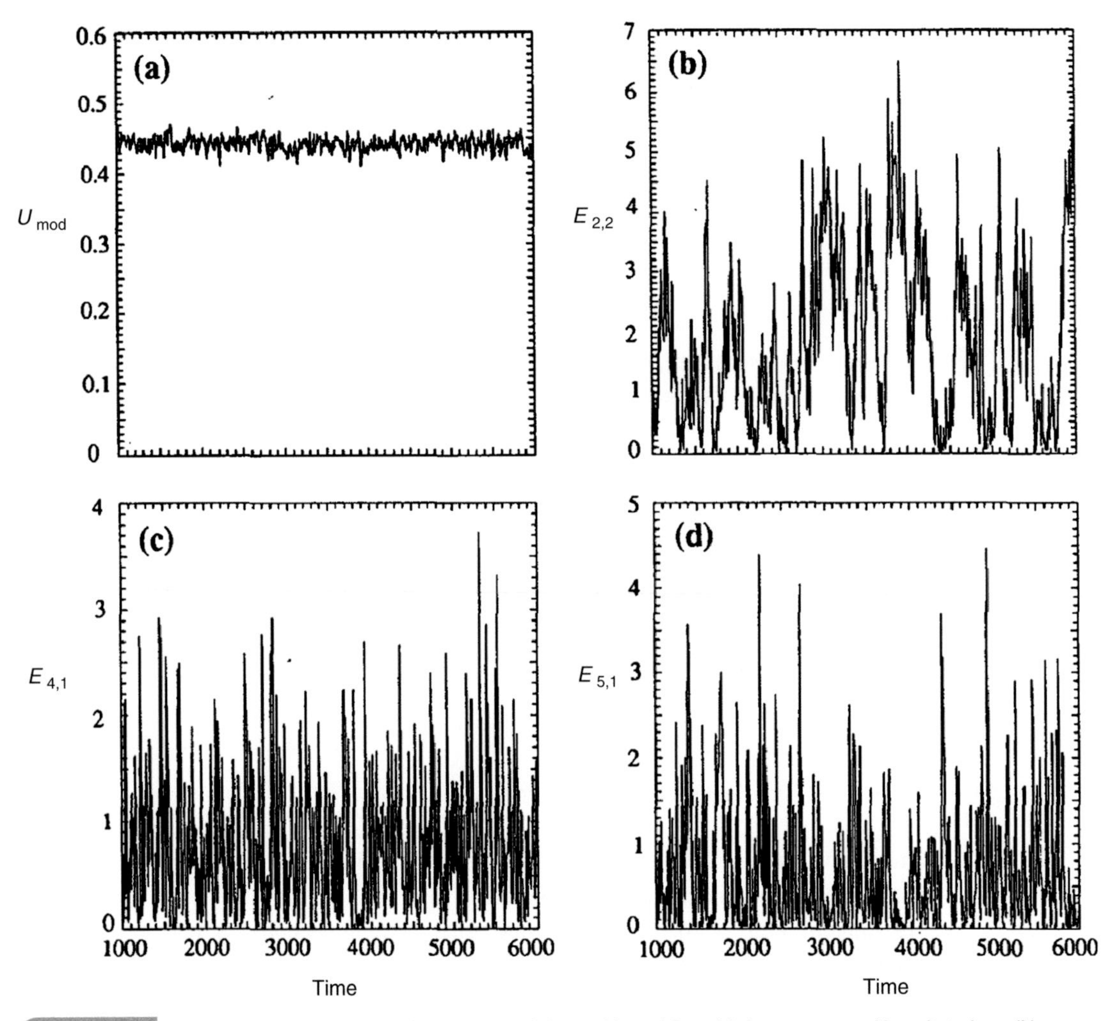

Fig. 11.17 Evolution of the several spectral components of the equilibrated flow, (a) domain-averaged baroclinic shear, (b) energy of the (2,2) wave, (c) energy of the (4,1) wave, (d) energy of the (5,1) wave (taken from Cai and Mak, 1990).

a negative conversion rate and the complementary unshaded area is equal to the probability for a positive conversion rate. The results show that the probability of $C(U_{\text{mod}}, SW) > 0$ is 0.71, that of $C(U_{\text{mod}}, PW) < 0$ is 0.76 and that of $C(SW, PW) > 0$ is 0.72. These three processes are therefore comparatively frequent from a statistical point of view. In other words, SW statistically receives energy from the zonal flow and in turn provides some energy to PW.

To diagnose the relation between the synoptic waves and the planetary waves in this system, we ascertain how the internally forced planetary-scale flow organizes the synoptic waves. It is done by using the dominant member of the planetary waves, the (2,2) wave, as a reference flow component. The objective is to diagnose the *model storm tracks relative to the reference planetary wave*, if they exist. Specifically, we construct a "*phase-shifted flow*" relative to the (2,2) wave at each time t. The time mean circulation of the phase-shifted

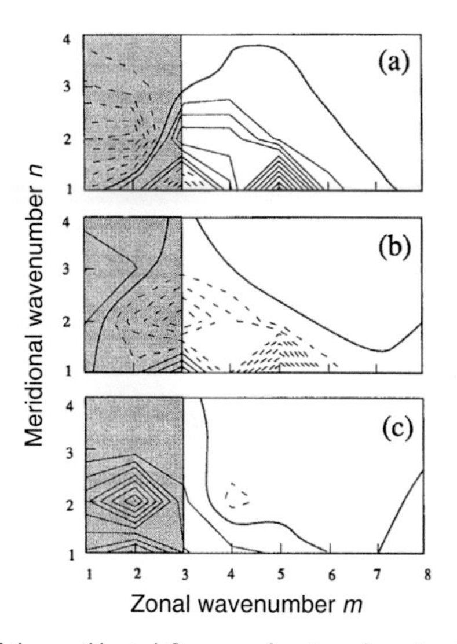

Fig. 11.18 Time mean spectral energetics of the equilibrated flow as a function of zonal and meridional wavenumbers. (a) $\overline{C(U_{\text{mod}}, E_{m,n})}$, net energy conversion from the instantaneous total zonal flow to (m,n) wave. (b) $\overline{C(PW, E_{m,n})}$, energy conversion from the planetary-scale waves to (m,n) wave. (c) $\overline{C(SW, E_{m,n})}$, energy conversion from the synoptic-scale waves to (m,n) wave. The shaded area indicates the planetary-scale waves. $CI = 0.05$ (taken from Cai and Mak, 1990).

flow in the two model layers is a composite flow shown in Fig. 11.20. It naturally consists of a (2,2) wave in conjunction with a zonal flow as expected. Cai and Mak (1990) gives the mathematical details of how this is done.

The anomaly flow is next constructed by subtracting the composite flow from the instantaneous flow. It is used to establish the statistics of the energetics. Of particular interest are the distributions of the time mean local energy of the synoptic-scale anomaly in the equilibrated flow and the standard deviation of such local energy. Figure 11.21a shows two maximum centers downstream of the two jets. Figure 11.21b shows the variability of the local energy measured in terms of the standard deviation. It also has a maximum located downstream of each jet core. This is the sought-after signature of the model storm tracks with a unique location relative to the internally forced planetary-scale flow. In other words, these storm tracks travel with the planetary wave. This is supporting evidence for the hypothesis that the synoptic-scale waves in the equilibrated state of this system intermittently extract a sufficient amount of energy from the modified instantaneous zonal flow to compensate not only for their own dissipative loss of energy but also for providing energy to the planetary-scale waves through the up-scale energy cascade process. As such, the SW and PW are symbiotically related to one another.

A detailed diagnosis of the local energetics of the model confirms that the planetary waves gain energy in the barotropic form. The planetary waves, in turn, create localized strong baroclinic regions whereby the synoptic waves may preferentially intensify downstream of the model planetary jet streams. The cyclone-scale eddies collectively give rise to

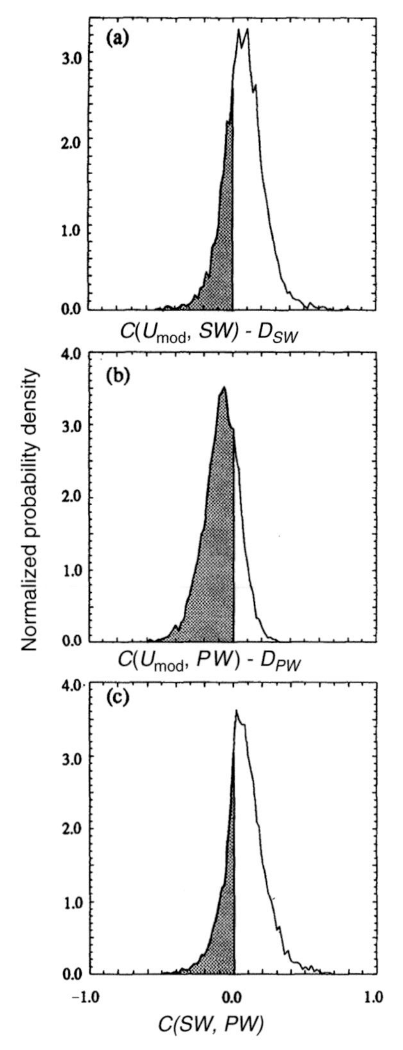

Fig. 11.19 Normalized probability density function of the energetics of the equilibrated flow. (a) Net energy conversion from the instantaneous total zonal flow to the synoptic-scales waves, (b) net energy conversion from the instantaneous total zonal flow to the planetary-scale waves, (c) energy conversion from the synoptic-scale waves to the planetary-scale waves. The shaded area under the curve is equal to 0.29 in panel (a), 0.76 in panel (b) and 0.28 in panel (c) (taken from Cai and Mak, 1990).

two model storm tracks that have a coherent statistical relation with the zonally traveling planetary-scale waves.

Cai and van den Dool (1992) have checked the relevance of the model results above with a diagnosis of 10 years of twice-daily data of the National Meteorological Center (NMC). They introduced a notion of *traveling storm tracks* (the notion of storm tracks itself has been briefly commented on in Section 11.1 and will be further elaborated).

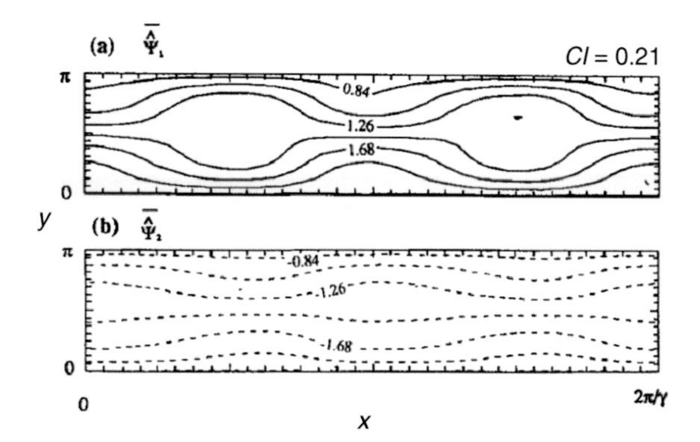

Fig. 11.20 The non-dimensional streamfunction of the phase-shifted composite flow in (a) the upper layer and (b) the lower layer. $CI = 0.21$ (taken from Cai and Mak, 1990).

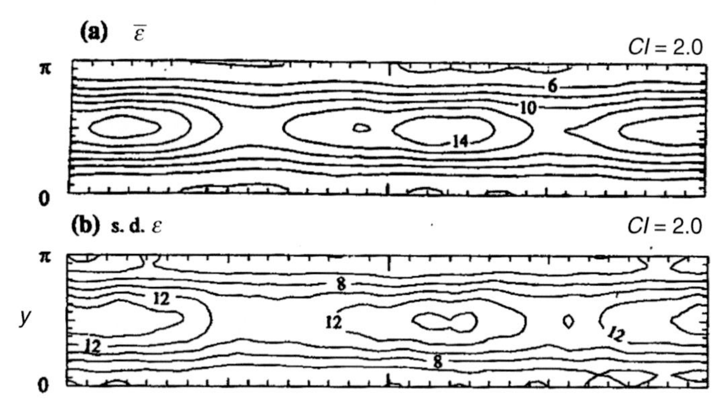

Fig. 11.21 Distribution of (a) non-dimensional time mean local energy of the synoptic-scale anomaly in the equilibrated flow and (b) standard deviation of such local energy (taken from Cai and Mak, 1990).

A special composite technique was devised to identify traveling storm tracks in a framework moving with an individual low-frequency wave of 500-mb geopotential height at 50° N. Their main result is shown in Fig. 11.22. Panel (b) shows that the traveling storm tracks are located in the trough regions of the low-frequency waves. Panel (c) shows that the barotropic feedback effect (i.e. the geopotential tendency due to the vorticity flux) of the traveling storm tracks tends to reinforce the low-frequency waves and to retard their propagation. The baroclinic feedback effect (i.e. the temperature tendency due to the heat flux) of the traveling storms tracks appears to have an out-of-phase relation with the low-frequency waves in temperature from 850 mb to 300 mb. These relationships dynamically resemble that between the climatological stationary waves and the climatological storm tracks. Therefore, this observational study provides evidence in support of the hypothesis that there is a symbiotic relation between the planetary waves and synoptic waves. The local instability theory accounts for the existence of storm tracks as the consequence of the zonal inhomogeneity in either a mean flow or a low-frequency flow.

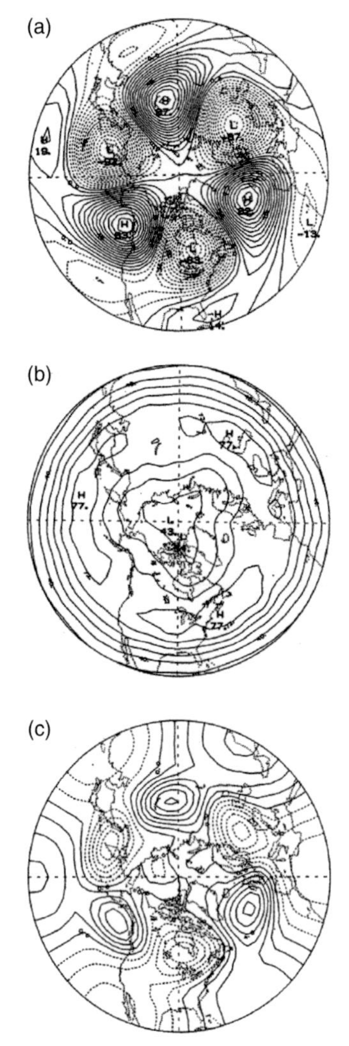

Fig. 11.22 Statistics of the phase-shifted flow of the 200-mb geopotential height with $m_o = 3$. (a) Time mean circulation. (b) Standard deviation of the phase-shifted high-frequency eddies. (c) Tendency field induced by the vorticity flux of the phase-shifted high-frequency eddies. The contour interval for (a) and (b) is 6.0 m and that for (c) is 1.0×10^{-5} m s^{-1}. Only the relative geographical locations in this figure are dynamically meaningful but not the absolute geographical longitudes (taken from Cai and van den Dool, 1992).

11.5 Relative intensity of the winter storm tracks

There are two regions of maximum values of winter mean synoptic-scale root-mean-square 300-mb height, one being over the western Atlantic and the other over the western Pacific (Fig. 11.23). These two preferred geographical locations are known as storm tracks. It also shows that each of the storm tracks is located downstream of a pronounced winter mean jet.

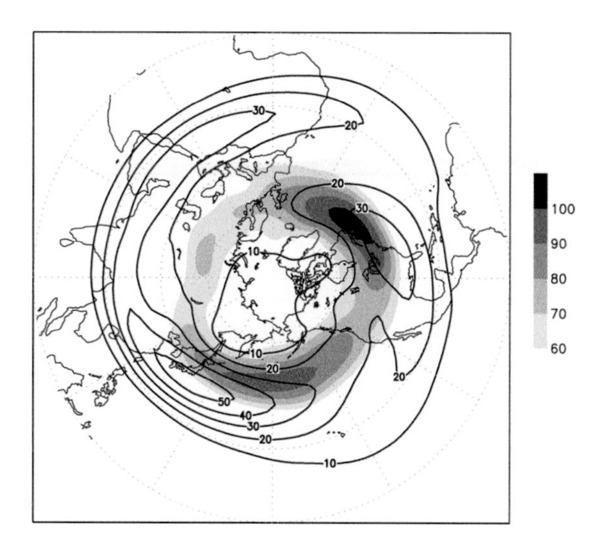

Fig. 11.23 Winter averaged root-mean-square (RMS) of the synoptic 300mb-height in units of m (shading) and the corresponding time mean wind speed in m s^{-1} (contours) (taken from Mak and Deng, 2007).

We now focus on one intriguing aspect of the storm tracks. The local maximum value of RMS of the synoptic 300mb-height in the climatological Atlantic storm track is about 10 percent more intense than that in the Pacific storm track (~105 m vs. 95 m) even though the corresponding Atlantic jet is weaker rather than stronger than the Pacific jet by about 70 percent. Such features are counter-intuitive if one accepted the simple-minded notion that a stronger baroclinic jet would necessarily give rise to a more intense storm track downstream. Evidently, additional physical factors are pertinent. One factor is the role of seeding disturbances from upstream of the storm tracks. The Pacific Ocean is on the downstream side of a huge landmass, Euro-Asia, whereas the Atlantic Ocean is on the downstream side of a much narrower N. America. It follows that the seeding eddies reaching upstream of the Pacific jet after having traversed a huge landmass would be greatly weakened. In contrast, the seeding eddies reaching the upstream of the Atlantic jet after having traversed N. America would be statistically stronger. The difference would be reinforced if we take into consideration the fact that friction is greater over land surfaces than over ocean surfaces. This conceptual consideration suggests a simple hypothesis. The continent–ocean configuration in the NH is a significant contributing factor to the relative intensity of the two storm tracks and the corresponding differential friction would further accentuate their relative intensity. On the other hand, the baroclinicity associated with the Pacific jet is much stronger than that associated with the Atlantic jet. Hence, the intensity of each storm track seems to be dictated by the net effect of two influences. The seeding disturbances over the western Pacific are weaker but they would be re-energized by a stronger Pacific jet, whereas the seeding eddies over the western Atlantic are stronger but they are re-energized by a weaker Atlantic jet. The differences in the horizontal structure of the two jets might be a significant factor as well. The uncertainty concerning the relative intensity of the two storm tracks is a quantitative issue and must be resolved as such.

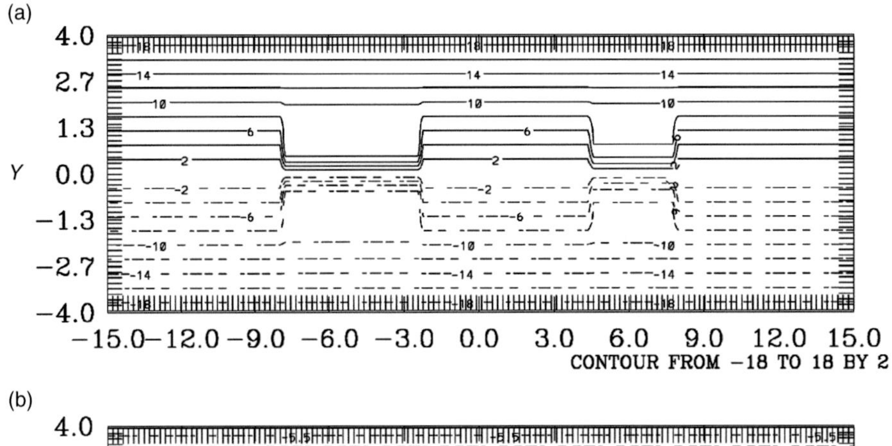

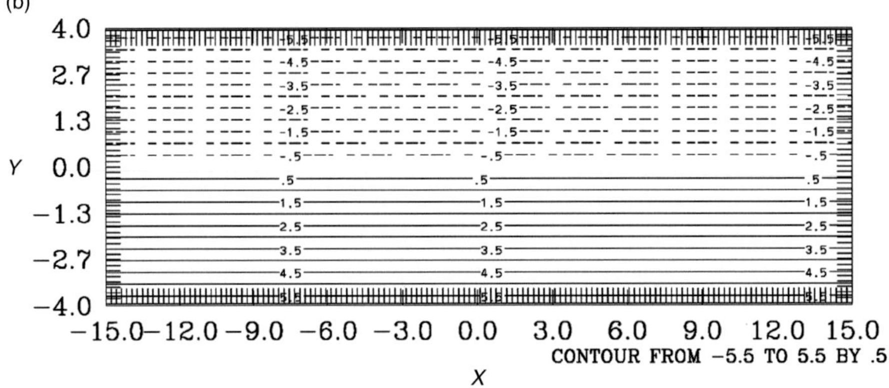

Fig. 11.24 Distribution of the non-dimensional potential vorticity of the reference state in units of UL^{-1} for (a) upper layer and (b) lower layer. Model Euro-Asia spans from $x \approx 11$ through the right boundary ($x = 15$) to $x \approx -8$; model N. America is located between $x \approx 0$ and 5 (taken from Mak and Deng, 2007).

11.5.1 Model formulation

For the purpose of testing the hypothesis above, it suffices to use a two-level quasi-geostrophic beta-plane model driven by a judiciously designed localized forcing. The model domain is a reentrant channel ($-X/2 \leq x \leq X/2$) bounded by two rigid lateral walls ($-Y/2 \leq y \leq Y/2$). It is 30 000 km long and 8000 km wide. We measure velocity, distance and time in units of $U = 10\,\mathrm{m\,s^{-1}}$, $L = 10^6$ m and $U^{-1}L = 10^5$ s respectively. The model North America is located between $x = 0$ and $x = 5$. The model Euro-Asia extends from $x = 10.7$ through the right boundary of the channel ($x = X = 15$) to $x = -7.5$. A fully grid-point representation of the fields is used in the analysis.

An external forcing is introduced by relaxing the instantaneous state towards a reference state, which is prescribed with the climatological winter mean flow as guidance. It is convenient to prescribe an idealized steady zonally non-uniform reference state in terms of its potential vorticity field, $\bar{q}_j$, $j = 1, 2$ (Fig. 11.24). The $\bar{q}_1$ field is made up of a global component, which increases linearly northward, and two localized components with

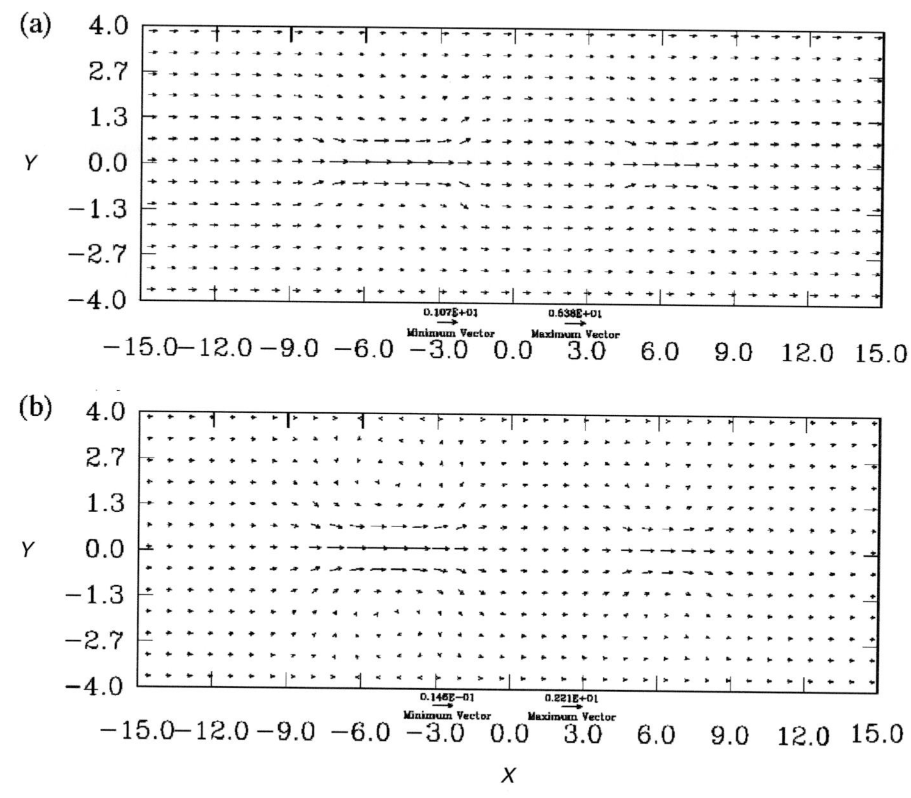

Fig. 11.25 Distribution of the velocity field of the reference state at (a) upper level and (b) lower level in units of U. Max vector in (a) is 0.538 E + 01 and in (b) is 0.221 E + 01 (taken from Mak and Deng, 2007).

a Gaussian structure centered at $y = 0$. The localized component associated with an Atlantic jet is 70 percent as strong as the one associated with a Pacific jet. The $\bar{q}_2$ field is zonally uniform for simplicity. The component associated with the global baroclinicity is strong enough that $\bar{q}_2$ decreases linearly northward. Thus, even in the absence of the localized components, the reference state would be baroclinically unstable in the absence of friction.

The corresponding velocity field of the reference state, $(\vec{V}_j)_{ref}$, obtained by inverting the $\bar{q}_j$ field above, is shown in Fig. 11.25. It has two localized baroclinic jets embedded in a globally uniform baroclinic flow component. The upper level maximum velocity of the model Pacific jet is $54\,\mathrm{m\,s^{-1}}$ and that of the Atlantic jet is $40\,\mathrm{m\,s^{-1}}$. The corresponding lower level values are $22\,\mathrm{m\,s^{-1}}$ and $17\,\mathrm{m\,s^{-1}}$ respectively. The Pacific jet in the reference state is located approximately between $x = -7.0$ and $x = -3.0$, whereas the model Atlantic jet extends approximately from $x = 4.5$ to $x = 7.0$. The influence of the geographical configuration of the continents and oceans is incorporated through such broad structure of the reference flow.

The total flow consists of the reference flow and a departure flow component, $\left(\psi_j\right)_{total} = \bar{\psi}_j + \psi_j$, with potential vorticity $(q_j)_{total} = \bar{q}_j + q_j$ for $j = 1, 2$. The governing equations for the departure fields are the potential vorticity equations at the two levels, viz.

$$\frac{\partial q_j}{\partial t} + J\left(\bar{\psi}_j, q_j\right) + J(\psi_j, \bar{q}_j) + J\left(\psi_j, q_j\right) = -\frac{1}{\tau} q_j - \kappa \nabla^4 q_j,\ j = 1, 2, \tag{11.32}$$

$$\text{with } q_1 = \nabla^2 \psi_1 - \lambda^2(\psi_1 - \psi_2),\quad q_2 = \nabla^2 \psi_2 + \lambda^2(\psi_1 - \psi_2),$$

$$\bar{q}_1 = \beta y + \nabla^2 \bar{\psi}_1 - \lambda^2\left(\bar{\psi}_1 - \bar{\psi}_2\right),\quad \bar{q}_2 = \beta y + \nabla^2 \bar{\psi}_2 + \lambda^2\left(\bar{\psi}_1 - \bar{\psi}_2\right).$$

We use $\lambda^2 = 2.0$ as a parameter for the stratification in units of L^{-2} and $\beta = 1.6$ the beta parameter in units of UL^{-2}. To prevent numerical aliasing, we introduce standard hyper-diffusion with a coefficient $\kappa = 4.1 \times 10^{-4}$ in units of UL^3. The dissipation is associated with the first term on the RHS in the form of linear damping. The frictional coefficient for generic land surfaces may be several times larger than that for water surfaces at moderate wind speeds. Hence, we assume that the damping over continents is four times stronger than over oceans. The damping time scale over the oceanic and land sectors are then set to be $\tau_{land} = 2.5$ and $\tau_{ocean} = 10.0$ in units of $U^{-1}L$.

11.5.2 Model results

The spatial derivatives in the governing equations are cast as a center-difference form. With a weak arbitrary initial disturbance, a backward-Euler scheme is used to do the time integration for 1100 days with a time step of $\Delta t = 0.01$. A record of the evolution of the domain integrated energy of the departure field (not shown) reveals that the model flow essentially reaches an equilibrated state after about 60 days. A record of 1000 days of model output is sufficiently long for determining statistically significant properties of the equilibrated state. An extensive diagnosis of the equilibrated state can be made, including the distribution of various measures of the intensity of eddies, their E-vectors, the local eddy fluxes of heat, momentum and potential vorticity. Their role in shaping the structure of time mean potential vorticity and velocity fields of the equilibrated state can be also ascertained. For brevity, we only discuss three of those results that highlight equilibration dynamics of baroclinic waves.

(i) Model storm tracks

The synoptic eddy component of the flow is first extracted by taking the difference between the total departure field and a 10-day low-pass component. The distributions of the root-mean-square of the filtered meridional velocity at the two levels are shown in Fig. 11.26. The model simulates a storm track in the immediate exit region of the Pacific jet located near $x = -2.0$ with maximum values of $\text{RMS}\{v_1'\} = 1.41$ and $\text{RMS}\{v_2'\} = 0.66$. There is a model Atlantic storm track located near $x = 9.0$ with maximum values of $\text{RMS}\{v_1'\} = 1.33$ and $\text{RMS}\{v_2'\} = 0.67$. These characteristics are reasonable. Given the high degree of idealization in this model, the model storm tracks are probably as realistic as we have a right to hope for. Note that although the Atlantic jet is only 70 percent as strong

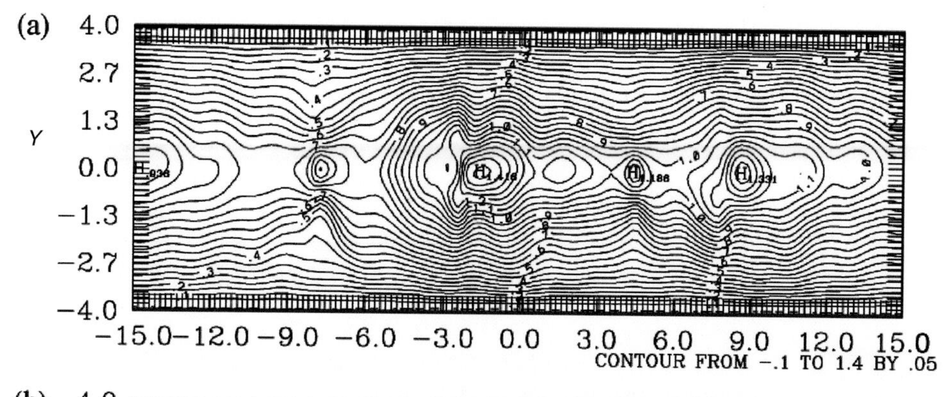

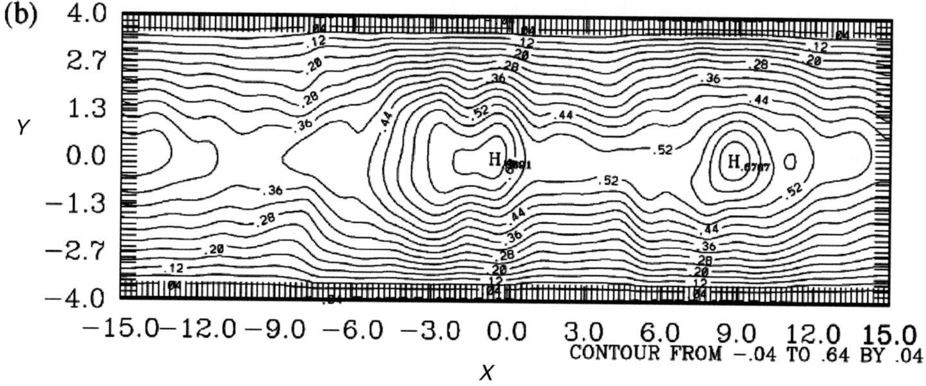

Fig. 11.26 Distribution of the model synoptic eddy activity in terms of (a) upper level RMS$\{v_1'\}$ and (b) lower level RMS$\{v_2'\}$ in units of U (taken from Mak and Deng, 2007).

as the Pacific jet in the reference state, the intensity of the two model storm tracks is virtually the same. This result may be taken as supporting evidence for the hypothesis that the relative intensity of the two storm tracks could be a natural consequence of the geographical configuration and frictional influence of the continents and oceans in NH. It highlights the significant dependence of the Atlantic storm track upon the Pacific storm track.

(ii) Synoptic eddy fluxes of heat and potential vorticity

To further assess the dynamical nature of the model storm tracks, we evaluate the local eddy flux of heat, $\overline{v_m' T_m'}$, where the subscript m refers to the mid-level of the model. Its primary direction is poleward. The strongest heat flux occurs in the two storm track regions in agreement with observation (Fig. 11.27). The maximum value is about 0.27 units of $f_o U^2 L/R$ which amounts to about $10\,\mathrm{ms}^{-1}$ K. It is comparable with the observed winter value at 850 mb.

It is also instructive to examine the synoptic eddy flux of potential vorticity, $\overline{v_1' q_1'}$ and $\overline{v_3' q_3'}$. Figure 11.28 shows that most of the potential vorticity flux takes place in the storm

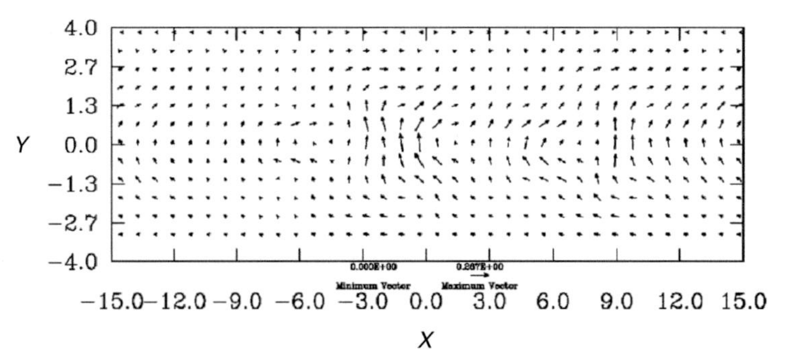

Fig. 11.27 Time mean transport of heat by synoptic eddies in the model as measured by $\overline{v'_m T'_m}$ in units of $f_o U^2 L/R$. Max vector is 0.267E + 00 (taken from Mak and Deng, 2007).

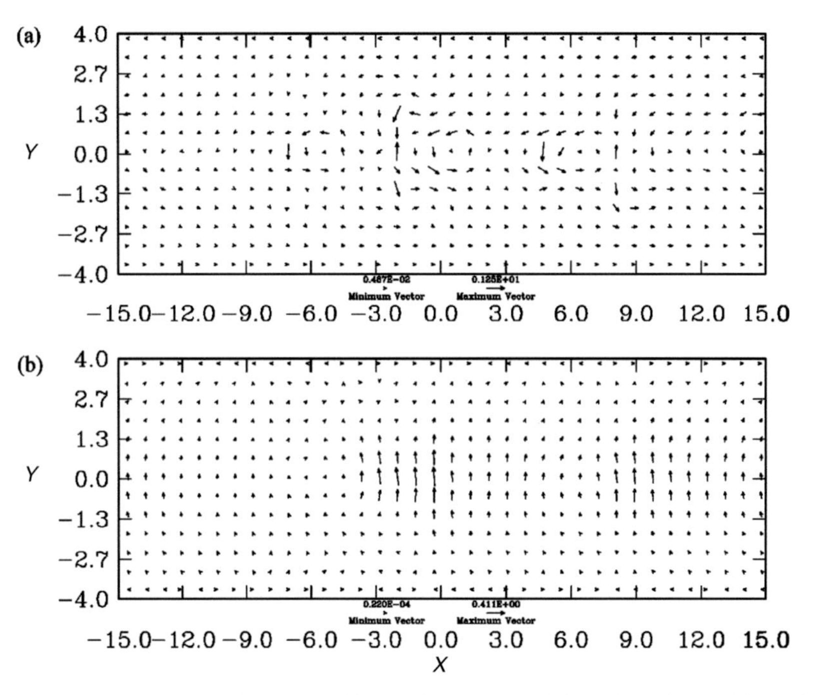

Fig. 11.28 Time mean flux of potential vorticity by synoptic eddies at (a) the upper and (b) the lower levels of the model, $\overline{v'_1 q'_1}$ and $\overline{v'_3 q'_3}$ in unit of $U^2 L^{-1}$. Max vector in (a) is 0.125 E + 01 and in (b) is 0.411 E + 00 (taken from Mak and Deng, 2007).

track regions. There is southward flux at the upper level and northward flux at the lower level. Such directions of flux are opposite of the potential vorticity gradient in the reference state as they should be. The upper level flux is about three times stronger than the lower level flux. All these eddy fluxes have a direct implication on the modification of the statistical average zonal mean state.

(a)

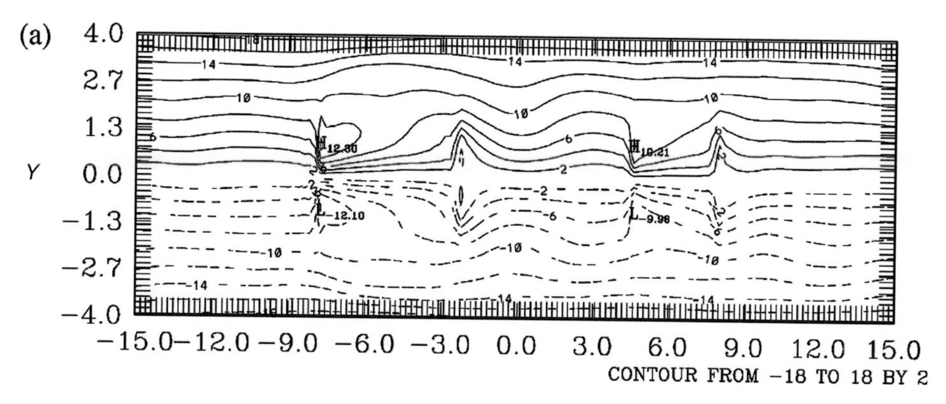

(b)

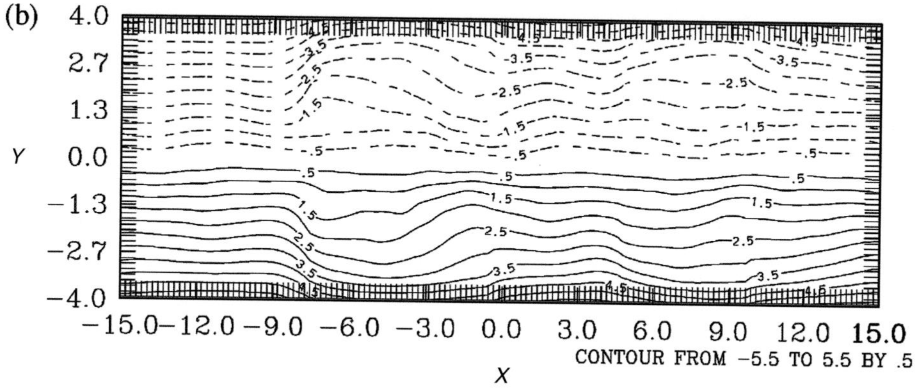

Fig. 11.29 Time mean total potential vorticity of the flow at (a) the upper and (b) the lower levels of the model in units of UL^{-1} (taken from Mak and Deng, 2007).

(iii) Time mean PV and velocity fields of the equilibrated state

Eddies stabilize a flow to the maximum possible extent by reducing the gradient of the background potential vorticity field. The role of eddies in shaping the background flow can be therefore displayed in the time mean potential vorticity field of the equilibrated state at each level (Fig. 11.29). By comparing Fig. 11.29 with Fig. 11.24, we see that the time mean upper level PV field in the eastern half of each jet and its exit region is greatly smoothed by the eddies. The time mean PV at the lower level is also substantially modified relative to the reference state.

To appreciate the impact of eddies on the background flow more directly, we compare the time mean total velocity field of the equilibrated state with that of the reference state (Fig. 11.30 vs. Fig. 11.25). Not surprisingly, the impact is also greatest in the eastern part and exit region of each jet. It is seen that both jets become weaker and shorter compared to those in the reference state. The equilibrated time mean state is expected to be more stable than the reference state. However, it is not completely neutralized since eddies must be continually energized in compensation for the dissipative loss of energy. It should be

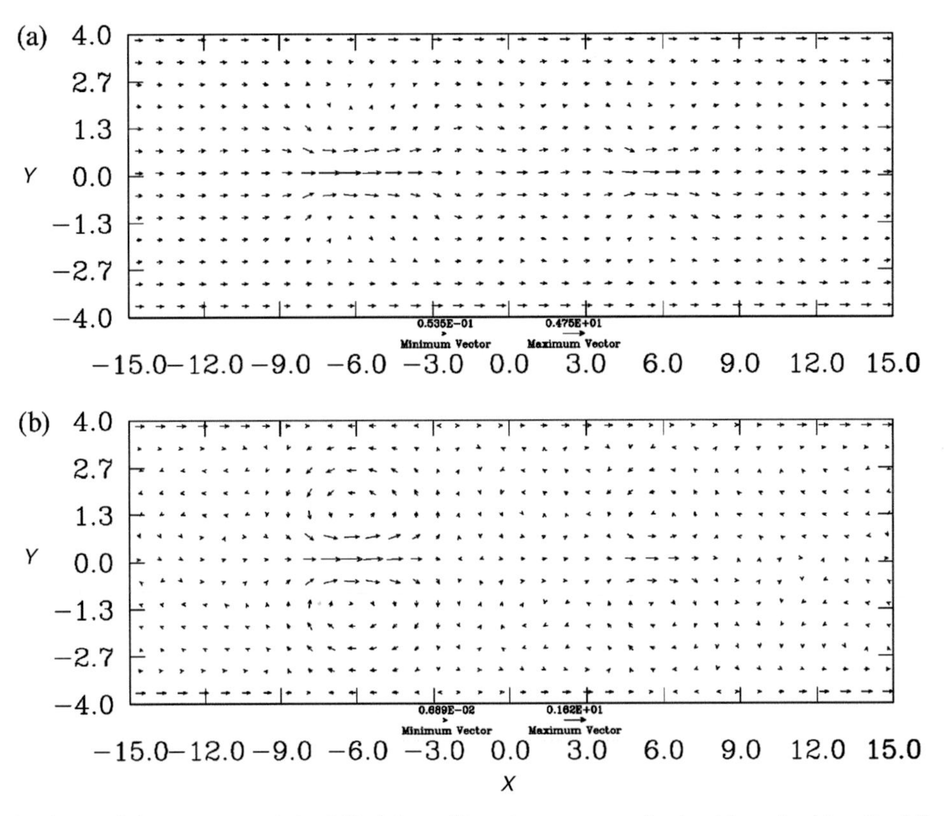

Fig. 11.30 Distribution of the time mean velocity field of the equilibrated state at upper level and lower level in units of U. Max vector is 0.475 E + 01 in (a) and 0.162 E + 01 in (b) (taken from Mak and Deng, 2007).

emphasized that these impacts result not only from the synoptic (high-frequency) eddies but also partly from the low-frequency eddies in the flow. The relationship among the synoptic eddies in the storm tracks, the planetary waves (low-frequency) and the mean flow is symbiotic in character (Cai and Mak, 1990).

12 Nongeostrophic dynamics

This chapter discusses the nature of nongeostrophic dynamics of atmospheric flows in the context of three distinctly different phenomena. The nongeostrophic effect can become important even in a large-scale flow when the momentum advection and heat advection by its ageostrophic velocity component are a significant part of the total advection. This is what we find in surface frontogenesis and in the Hadley circulation. In Section 12.1, we analytically delineate the dynamics of surface frontogenesis arising from the stretching deformation of a background flow in a semi-geostrophic model. In Section 12.2, we deduce analytically the structure of the annual/seasonal mean Hadley circulation in a zonally symmetric two-layer nonlinear primitive-equation model. The nongeostrophic effect is obviously important in small-scale disturbances, particularly when there is not even hydrostatic balance. In Section 12.3, we analyze non-hydrostatic barotropic instability as a possible mechanism for non-supercell tornadogenesis with a Hamiltonian formulation for a layer of barotropic fluid.

12.1 Surface frontogenesis

Surface fronts in the extratropical atmosphere have a local quasi-rectilinear structure across which temperature and wind vary dramatically over a short distance normal to the front. Although the cross-front velocity is considerably weaker than the along-front velocity, it has a strong dynamical impact on frontogenesis because the cross-front thermal contrast is strong. There is a substantial vertical velocity associated with the cross-front velocity especially in the neighborhood of a cold front. It follows that the cross-front advection as well as the vertical advection of momentum and heat are significant. Hence, nongeostrophic dynamics is important for surface atmospheric frontogenesis. In conjunction with the abundance of moisture typically in a frontal region, the cross-front circulation also releases a lot of latent heat and triggers smaller scale disturbances.

Surface frontogenesis can be understood even in the context of dry dynamics. The along-front velocity is largely in geostrophic balance, whereas the cross-front velocity is decidedly not. The latter can however play a central role in frontal development. It is unlikely that a quasi-geostrophic model could adequately capture the essential aspect of frontogenesis because it neglects the advective effect of the ageostrophic component of a flow. There is a model of intermediate complexity, which does incorporate that important process. The approximation incorporated in that model is known as the *geostrophic momentum approximation*. It is based on the following consideration of the exact horizontal vector momentum equation

$$\begin{aligned}
\vec{V} &= \vec{V}^{(g)} + \frac{1}{f}\vec{k} \times \frac{D\vec{V}}{Dt} \\
&= \vec{V}^{(g)} + \frac{1}{f}\vec{k} \times \frac{D}{Dt}\left(\vec{V}^{(g)} + \frac{1}{f}\vec{k} \times \frac{D\vec{V}}{Dt}\right) \\
&= \vec{V}^{(g)} + \frac{1}{f}\vec{k} \times \frac{D\vec{V}^{(g)}}{Dt} - \frac{1}{f^2}\frac{D^2\vec{V}}{Dt^2} \\
&\approx \vec{V}^{(g)} + \frac{1}{f}\vec{k} \times \frac{D\vec{V}^{(g)}}{Dt},
\end{aligned} \tag{12.1}$$

where $D/Dt = \partial/\partial t + \vec{V}_3 \cdot \nabla$, $\vec{V}$ is horizontal velocity and $\vec{V}_3$ is the 3-D velocity. Neglecting the term $\frac{1}{f^2}\frac{D^2\vec{V}}{Dt^2}$ in the last equation is justifiable if the time scale for change in velocity following a fluid particle is considerably longer than $f^{-1} \approx 3$ h. Note that the contribution of the ageostrophic flow to the advection is retained as we would like, but the acceleration is approximated by that of the geostrophic velocity itself. The last step in (12.1) is adopted as an approximate form of the horizontal vector momentum equation. It should be emphasized that no approximation is introduced to the operator *D/Dt* itself. The ageostrosphic thermal advection in the thermodynamic equation is also retained. The vertical momentum equation is approximated by hydrostatic balance. Such a theory is known as *semi-geostrophic theory* as distinct from the quasi-geostrophic theory. The dynamics of frontogenesis in a semi-geostrophic model is nonlinear. It is nevertheless possible to get an analytic solution for it if one judiciously exploits the intrinsic mathematical structure of the system as first demonstrated by Hoskins and Bretherton (1972).

The simplest form of frontal disturbance is one in which all of its properties are uniform in the along-front direction. Let us designate the cross-front, along-front and vertical directions as the x-axis, y-axis and z-axis respectively. The corresponding velocity components are u, v and w. The intensity of a front may be measured in terms of the derivatives of the potential temperature and along-front velocity with respect to x. The rate of frontogenesis can be then measured by the total derivative of such measures of the intensity. In the absence of diabatic and frictional processes, such frontogenetical forcing equations for a two-dimensional front in standard notations are

$$\begin{aligned}
\frac{D}{Dt}\left(\frac{\partial \theta}{\partial x}\right) &= -u_x^{(g)}\theta_x - u_x^{(a)}\theta_x - w_x\theta_z, \\
\frac{D}{Dt}\left(\frac{\partial v}{\partial x}\right) &= -u_x^{(g)}v_x - u_x^{(a)}(f + v_x) - w_x v_z,
\end{aligned} \tag{12.2a,b}$$

where $D/Dt = \partial/\partial t + \left(u^{(g)} + u^{(a)}\right)\partial/\partial x + w\partial/\partial z$ in this case. The superscripts (g) and (a) designate the geostrophic and ageostrophic components of the velocity.

Observation suggests that surface frontogenesis often results from the influence of stretching deformation of a flow on a thermal disturbance. Such a flow component can be represented by a streamfunction of the form $\psi = \alpha xy$ so that $u^{(g)} = -\alpha x$ and $v^{(g)} = \alpha y$. The y-axis is the axis of dilatation and α is the strength of the stretching deformation $\left(u_x^{(g)} - v_y^{(g)} = -2\alpha\right)$. A deformation flow would increase any initial cross-front temperature gradient in a reference frame moving with the basic cross-front velocity due to the impact of the first term of (12.2a), $\left(-u_x^{(g)}\theta_x = \alpha\theta_x\right)$. There would be a similar development in the cross-front shear of the along-front velocity due to the first term of (12.2b), $\left(-u_x^{(g)}v_x = \alpha v_x\right)$. The acceleration of the along-front velocity would in turn increase the cross-front velocity and hence the vertical velocity. Once the ageostrophic velocity increases, it would further enhance the frontogenetic development in accordance with the remaining terms on the RHS of (12.2a,b). The whole process would feed upon itself until friction becomes important and eventually limits how sharp the thermal contrast across a front can be.

12.1.1 Two-dimensional semi-geostrophic model analysis of frontogenesis

We now discuss a specific model analysis of surface frontogenesis in a two-dimensional semi-geostrophic model of an inviscid adiabatic Boussinesq fluid. Mak and Bannon (1984) report both a dry model analysis and a moist model analysis of this problem. The discussion here only dwells upon their dry model analysis. Suppose there is a background large-scale deformation flow, $\bar{u} = -\alpha x$ and $\bar{v} = \alpha y$. The corresponding background pressure field divided by a reference density is $\bar{\phi}(x,y,z) = \phi_o(z) + f\alpha xy - \alpha^2(x^2+y^2)/2$ so that it satisfies the momentum equations for the background flow, $\bar{u}\partial\bar{u}/\partial x = -\partial\bar{\phi}/\partial x + f\bar{v}$, $\bar{v}\partial\bar{v}/\partial y = -\partial\bar{\phi}/\partial y - f\bar{u}$. The component $\phi_o(z)$ is associated with a basic stratification which is statically stable, $\dfrac{g}{\theta_{oo}}\dfrac{d\theta_o}{dz} < 0$, where θ_{oo} is a reference constant value. The basic potential temperature is $(\theta_{oo} + \theta_o(z))$. Embedded in this background flow is a two-dimensional disturbance uniform in the direction of y. The dependent variables of the disturbance are (u, v, w, ϕ, θ), all being functions of x, z and t. We further assume that v is in geostrophic balance and that ϕ and θ are in hydrostatic balance; (u,w) stands for a nongeostrophic component of the flow on the (x,z) cross-section. The governing equations are then

$$
\begin{aligned}
-fv + \frac{\partial\phi}{\partial x} &= 0, \\
\left[\frac{\partial}{\partial t_*} + (-\alpha x + u)\frac{\partial}{\partial x} + w\frac{\partial}{\partial z} + \alpha\right]v + fu &= 0, \\
-\frac{g}{\theta_{oo}}\theta + \frac{\partial\phi}{\partial z} &= 0, \\
\left[\frac{\partial}{\partial t_*} + (-\alpha x + u)\frac{\partial}{\partial x} + w\frac{\partial}{\partial z}\right]\theta + w\frac{d\theta_o}{dz} &= 0, \\
\frac{\partial u}{\partial x} + \frac{\partial w}{\partial z} &= 0.
\end{aligned}
\qquad (12.3a,b,c,d,e)
$$

The problem is to deduce the evolution of a particular initial disturbance to be prescribed later.

Equations (12.3b) and (12.3d) have nonlinear terms associated with the advection by the nongeostrophic component of the flow. We will shortly see that it would be advantageous to transform the equations using the following new independent variables

$$X = x + \frac{v}{f}, \quad Z = z, \quad T = t_* \tag{12.4}$$

where X is called a *geostrophic coordinate*. The transformation relations for the derivatives are $\frac{\partial}{\partial x} = \left(1 + \frac{1}{f}\frac{\partial v}{\partial x}\right)\frac{\partial}{\partial X}$, $\frac{\partial}{\partial z} = \frac{1}{f}\frac{\partial v}{\partial z}\frac{\partial}{\partial X} + \frac{\partial}{\partial Z}$ and $\frac{\partial}{\partial t_*} = \frac{1}{f}\frac{\partial v}{\partial t_*}\frac{\partial}{\partial X} + \frac{\partial}{\partial T}$. The transformed equations would take on a compact form if we adopt the following new dependent variables

$$\begin{aligned} U &= u + w\frac{\partial v}{\partial z}\left(f + \frac{\partial v}{\partial x}\right)^{-1}, \\ V &= v, \\ W &= wf\left(f + \frac{\partial v}{\partial x}\right)^{-1}, \\ \Phi &= \phi + \frac{v^2}{2}, \\ \Theta &= \theta. \end{aligned} \tag{12.5a,b,c,d,e}$$

Then (12.3) would be transformed to

$$\begin{aligned} -fV + \frac{\partial \Phi}{\partial X} &= 0, \\ \left(\frac{\partial}{\partial T} - \alpha X\frac{\partial}{\partial X} + \alpha\right)V + fU &= 0, \\ -\frac{g}{\theta_{oo}}\Theta + \frac{\partial \Phi}{\partial Z} &= 0, \\ \left(\frac{\partial}{\partial T} - \alpha X\frac{\partial}{\partial X}\right)\Theta + WS &= 0, \\ \frac{\partial U}{\partial X} + \frac{\partial W}{\partial Z} &= 0, \end{aligned} \tag{12.6a,b,c,d,e}$$

where $S = f\frac{\partial(\theta_o + \theta)}{\partial Z}\left(f - \frac{\partial V}{\partial X}\right)^{-1}$. The only term that remains to be explicitly nonlinear is WS in (12.6d); S is proportional to the potential vorticity. The case of uniform potential vorticity is of particular theoretical interest since, under this condition, we would be able to treat (WS) as a linear term. Then, the well-developed analytical techniques for solving a linear system would be applicable. Since the initial temperature field yet to be prescribed must have some horizontal structure, S is not strictly uniform. But S would be approximately uniform for a weak initial disturbance. We therefore assume quasi-uniform potential vorticity in this analysis, so that the nonlinear dynamics can be dealt with implicitly in an analytic manner. Under this assumption, the transformation of the independent and dependent variables, as defined in (12.4) and (12.5), enables us to proceed with our task of analytically determining the evolution as an equivalent linear mathematical problem.

It is convenient to work with non-dimensional quantities. Since S has the dimension of stratification, we define an effective Brunt–Vaisala frequency $N = (gS/\theta_{oo})^{1/2}$. The latter is used to define a horizontal length scale NH/f, where H is a vertical length scale of the flow under consideration. It would be logical to use α^{-1} as the time scale. The part of potential temperature associated with the basic stratification is taken to be $\theta_o = Sz$. A set of non-dimensional variables are defined as follows

$$\tilde{X} = \frac{Xf}{NH}, \quad \tilde{Z} = \frac{Z}{H}, \quad \tilde{T} = \alpha T, \quad \tilde{U} = \frac{Uf}{NH\alpha}, \quad \tilde{V} = \frac{V}{NH},$$
$$\tilde{W} = \frac{W}{\alpha H}, \quad \tilde{\Phi} = \frac{\Phi}{N^2H^2}, \quad \tilde{\Theta} = \frac{\Theta}{SH}. \tag{12.7}$$

The non-dimensional form of (12.6) is then (after dropping the tilde)

$$-V + \frac{\partial \Phi}{\partial X} = 0,$$
$$\left(\frac{\partial}{\partial T} - X\frac{\partial}{\partial X} + 1\right)V + U = 0,$$
$$-\theta + \frac{\partial \Phi}{\partial Z} = 0, \tag{12.8a,b,c,d,e}$$
$$\left(\frac{\partial}{\partial T} - X\frac{\partial}{\partial X}\right)\theta + W = 0,$$
$$\frac{\partial U}{\partial X} + \frac{\partial W}{\partial Z} = 0.$$

The operator $(\partial/\partial T - X\partial/\partial X)$ in (12.8b,d) suggests that it would be advantageous if we introduce a further coordinate transformation. We therefore define a length-stretching coordinate in the cross-front direction as

$$\xi = Xe^T, \quad \eta = Z, \quad t = T. \tag{12.9}$$

Then (12.6) can be rewritten as

$$-V + J\frac{\partial \Phi}{\partial \xi} = 0,$$
$$\left(\frac{\partial}{\partial t} + 1\right)V + U = 0,$$
$$-\theta + \frac{\partial \Phi}{\partial \eta} = 0, \tag{12.10a,b,c,d,e}$$
$$\frac{\partial \theta}{\partial t} + W = 0,$$
$$J\frac{\partial U}{\partial \xi} + \frac{\partial W}{\partial \eta} = 0,$$

where $J = e^t$ is the Jacobian of the last coordinate transformation. Note that the only variable coefficient in (12.10) is J. Equations (12.10a,b,c,d,e) can be readily reduced to a single equation

$$\frac{\partial}{\partial t}\left(\frac{\partial^2}{\partial \eta^2}+J^2\frac{\partial^2}{\partial \xi^2}\right)\Phi=0. \tag{12.11}$$

It is a linear partial differential equation.

For simplicity, we consider an initial disturbance that is only a function of ξ prescribed as

$$\theta(\xi,\eta,0)=F, \quad \text{with } F=-\frac{2a}{\pi}\tan^{-1}\xi$$
$$\rightarrow \Phi(\xi,\eta,0)=\eta F. \tag{12.12}$$

This F has asymptotic values equal to $\pm a$ at $\xi=\mp\infty$. We focus on the departure of Φ from its initial field by introducing

$$\chi=\Phi-\eta F, \tag{12.13}$$

which is then zero at $t=0$. The governing equation for χ is then

$$\frac{\partial}{\partial t}\left(\frac{\partial^2\chi}{\partial \eta^2}+J^2\left(\frac{\partial^2\chi}{\partial \xi^2}+\eta\frac{d^2F}{d\xi^2}\right)\right)=0. \tag{12.14}$$

The horizontal boundary conditions are

$$\lim_{\xi\rightarrow\pm\infty}\frac{\partial\chi}{\partial\eta}=0, \tag{12.15}$$

meaning that the potential temperature far away from the frontal zone is unaffected. The lower boundary condition is $w=0$ implying that

$$\frac{\partial^2\chi}{\partial t\partial\eta}=0 \quad \text{at } \eta=0$$
$$\rightarrow \frac{\partial\chi}{\partial\eta}=0 \quad \text{after performing time integration.} \tag{12.16}$$

In other words, the potential temperature at $\eta=0$ in this coordinate is invariant. The upper boundary is formally treated as if it were infinitely far away from the surface. We hence impose the upper boundary condition as

$$\lim_{\eta\rightarrow\infty}\frac{\partial\chi}{\partial\eta} \text{ is finite.} \tag{12.17}$$

The next step is to integrate (12.14) in time and obtain by making use of the initial condition

$$\frac{\partial^2\chi}{\partial\eta^2}+J^2\frac{\partial^2\chi}{\partial\xi^2}=-\eta\left(J^2-1\right)\frac{d^2F}{d\xi^2}. \tag{12.18}$$

The time dependence now only appears parametrically through J and ξ.

We may use a spectral representation of χ and F for their ξ dependence. Let the Fourier transform of χ and F be denoted by $\hat{\chi}(\ell,\eta,t)$ and $\hat{F}(\ell)$. Taking the Fourier transform of (12.18), (12.16) and (12.17) yields for wavenumber ℓ

$$\frac{d^2\hat{\chi}}{d\eta^2} - \tau^2\hat{\chi} = (\tau^2 - \tau_o^2)\eta\hat{F}$$
$$\frac{d\hat{\chi}}{d\eta} = 0 \text{ at } \eta = 0 \qquad (12.19)$$
$$\lim_{\eta\to\infty}\frac{d\hat{\chi}}{d\eta} \text{ is finite,}$$

where $\tau^2 = \ell^2 e^{2t}$, $\tau_o^2 = \ell^2$. For the initial temperature prescribed in (12.12), we have

$$\hat{F} = \frac{ia}{\ell}e^{-|\ell|}, \ i = \sqrt{-1}. \qquad (12.20)$$

The solution of (12.19) is $\hat{\Phi} = \left(\frac{\tau_o^2}{\tau^2} - 1\right)\left(\eta + \frac{1}{\tau}\exp(-\tau\eta)\right)\hat{F}$ which leads to

$$\hat{\theta}(\eta, \tau) = \left[\frac{\tau_o^2}{\tau^2} + \left(1 - \frac{\tau_o^2}{\tau^2}\right)\exp(-\tau\eta)\right]\hat{F}. \qquad (12.21)$$

The inverse Fourier transform of (12.21) is

$$\theta(\xi, \eta, \tau) = -\frac{2a}{\pi}\left[e^{-2t}\tan^{-1}(\xi) + \left(1 - e^{-2t}\right)\tan^{-1}\left(\frac{\xi}{1 + \eta e^t}\right)\right]. \qquad (12.22)$$

The corresponding solution of the vertical velocity is

$$W(\xi, \eta, \tau) = -\frac{\partial\theta}{\partial t}. \qquad (12.23)$$

The corresponding solution of the along-front velocity is obtained from the thermal wind equation. It is found to be

$$V(X, Z, T) = -\frac{2a}{\pi}\left(\frac{Ze^{-T}}{1 + X^2e^{2T}} + \frac{\left(1 - e^{-2T}\right)}{2}\left\{\log\left[\left(1 + Ze^T\right)^2 + X^2e^{2T}\right] - 2T\right\}\right). \qquad (12.24)$$

The corresponding solution for the ageostrophic component of the cross-front velocity is

$$U(X, Z, T) = -V(X, Z, T) + \frac{2a}{\pi}\left(\begin{array}{l}\dfrac{-Ze^{-T}}{1 + X^2e^{2T}} + e^{-2T}\left\{\log\left[\left(1 + Ze^T\right)^2 + X^2e^{2T}\right] - 2T\right\} \\ +\left(1 - e^{-2T}\right)\left[\dfrac{(1 + Ze^T)Ze^T}{(1 + Ze^T)^2 + X^2e^{2T}} - 1\right]\end{array}\right). \qquad (12.25)$$

Equations (12.22) to (12.25) constitute a complete solution of the disturbance in geostrophic coordinates. That is very convenient for numerical evaluation. The solution in physical space can be obtained by graphic transformation using the non-dimensional relation $x = X - V(X, Z, T)$.

12.1.2 Illustrative calculation

To get a feel for the analytic solution above, let us evaluate the solution for a meaningful background state characterized by $N = 10^{-2}\,\mathrm{s}^{-1}$, $f = 10^{-4}\,\mathrm{s}^{-1}$, $\theta_{oo} = 300\,\mathrm{K}$ with a

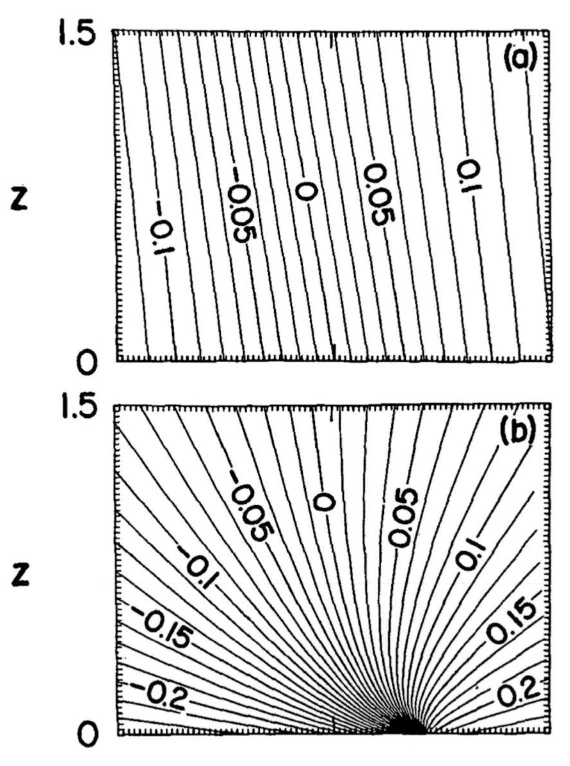

Fig. 12.1 The structure of the potential temperature disturbance at (a) $t=0$ and (b) $t=2.5$. The abscissa is $-1\leq x\leq 1$ (reprinted from Mak and Bannon, 1984).

deformation flow that has a strength equal to $\alpha = 10^{-5}\,\mathrm{s}^{-1}$. The corresponding value of the parameter S is $S = \dfrac{N^2\theta_{oo}}{g} = 3\times 10^{-3}\,\mathrm{m}^{-1}\mathrm{s\,K}$. Suppose the parameter for the initial potential temperature disturbance is $a = -0.25$. The thermal contrast across the domain in the x-direction has a vertical scale of $H = 10\,\mathrm{km}$. The non-dimensional magnitude corresponds to 7.5 K. One unit of distance is $NH/f = 1000\,\mathrm{km}$.

It suffices to compare the various initial fields with their counterparts at $t = 2.5$ shortly before frontal collapse. Figure 12.1 shows such a comparison in the potential temperature departure field. We see a pronounced shallow surface front (about 1 km thick) located at about 0.4 units distance (400 km) to the east of the axis of dilatation of the deformation flow.

To facilitate a direct comparison with observation, we plot the total potential temperature field departure from θ_{oo} at these two time instants in Fig. 12.2.

Figure 12.3 shows a comparison of the corresponding vertical velocity fields. Maximum vertical velocity occurs over the surface front. While the distribution is realistic, its magnitude is rather weak for this particular parameter setting, $\sim 1\,\mathrm{cm\,s}^{-1}$.

A comparison of the structure of the cross-front velocity is shown in Fig. 12.4. One unit is $NH\alpha/f = 10\,\mathrm{m\,s}^{-1}$. So the maximum value at this time is about $5\,\mathrm{m\,s}^{-1}$.

A comparison of the structure of the along-front velocity is shown in Fig. 12.5. One unit is $NH = 100\,\mathrm{m\,s}^{-1}$. So the maximum value at this time is about $30\,\mathrm{m\,s}^{-1}$. This is a rather strong wind.

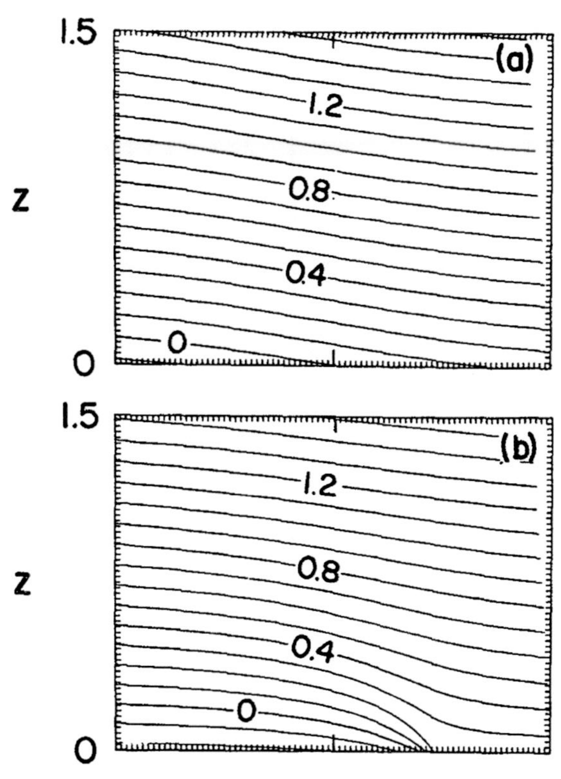

Fig. 12.2 The structure of the total potential temperature departure from the reference value at (a) $t = 0$ and (b) $t = 2.5$, $CI = 0.1$. The abscissa is $-1 \leq x \leq 1$ (reprinted from Mak and Bannon, 1984).

These quantitative results of the model disturbance as a whole are compatible with observation suggesting that this semi-geostrophic model captures the dynamics of surface frontogenesis. The major deficiency is that the vertical velocity is too weak. This is attributable to the fact that condensational heating has not been taken into account in this analysis. Such an interpretation is supported by an extension of this analysis using simple parameterization schemes for condensational heating in Mak and Bannon (1984).

We should be aware that every dynamical process in this model is not necessarily frontogenetic. We are in a position to precisely diagnose the relative importance of the mechanisms throughout the frontal development with the analytic solution. As defined in (12.1), one measure of *frontogenetic forcing* is the rate of change of a fluid parcel's horizontal thermal gradient, $\frac{D}{Dt}\left(\frac{\partial\theta}{\partial x}\right)$. We plot the contribution to this forcing in the intermediate stage of frontal development ($\alpha t_{\text{dim}} = 1.5$) from the confluence and convergence of the cross-front velocity $\left(-\left(\frac{\partial u^{(g)}}{\partial x} + \frac{\partial u^{(a)}}{\partial x}\right)\frac{\partial\theta}{\partial x}\right)$ in Fig. 12.6a. The values are positive and its maximum value is at the surface with a value of about $(1\,\text{k}/100\,\text{km}\,3\,\text{h}^{-1})$. The value decreases with increasing height. The contribution due to the thermal tilting term, $\left(-\frac{\partial w}{\partial x}\frac{\partial\theta}{\partial x}\right)$ is shown in Fig. 12.6b. The values are negative and therefore this

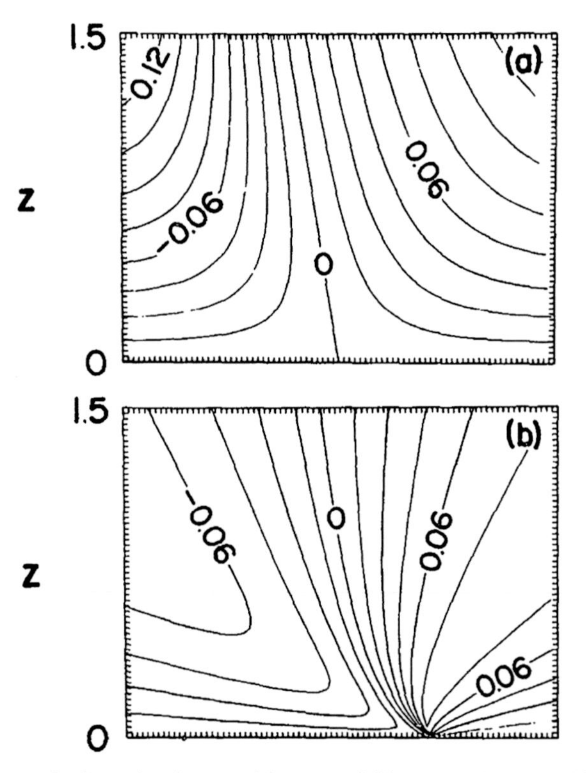

Fig. 12.3 The structure of the vertical velocity disturbance at (a) $t = 0$ and (b) $t = 2.5$, $CI = 0.015$. The abscissa is $-1 \leq x \leq 1$ (reprinted from Mak and Bannon, 1984).

mechanism is *frontolytic*. The overall mechanism leads to frontal development in a shallow layer next to the surface. At this time, the front is located at about 300 km to the east of the axis of dilatation of the geostrophic flow.

Fronts in the atmosphere are not necessarily triggered by a purely deformation flow. The deformation property can be an integral aspect of a 3-D baroclinic wave spontaneously intensifying in a background baroclinic shear flow. Quite realistic frontal development has been numerically simulated in such a setting in a 3-D semi-geostrophic model as well as in a primitive-equation model. The different flows that subsequently give rise to frontal development are reviewed in Hoskins (1982).

12.2 Hadley circulation

We have learned in Chapter 10 that the Hadley cell is the tropical part of the mean meridional overturning circulation. We have examined the observed structure of the annual mean and seasonal mean Hadley cells in Section 10.1. We have shown in Section 10.3 that a linear dynamical model can simulate the gross characteristics of the mean meridional overturning circulation in both Eulerian and Lagrangian

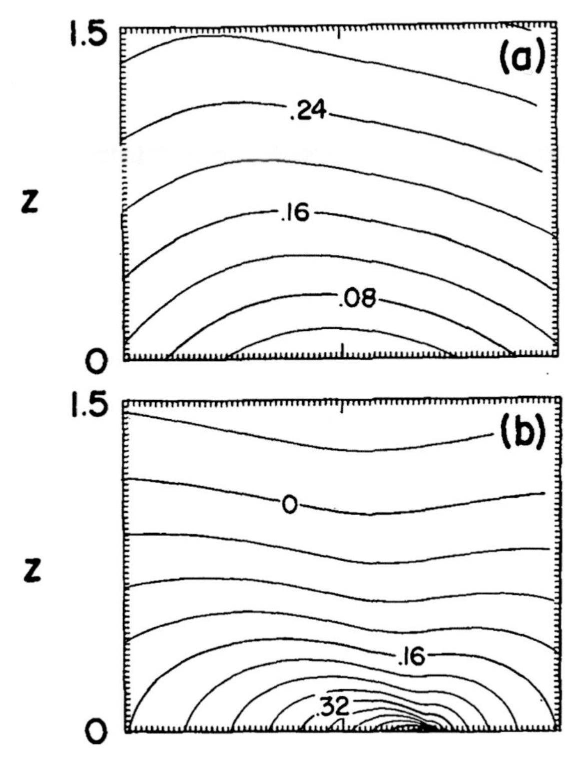

Fig. 12.4 The structure of the cross-front velocity at (a) $t = 0$ and (b) $t = 2.5$, $CI = 0.04$. The abscissa is $-1 \leq x \leq 1$ (reprinted from Mak and Bannon, 1984).

senses. It is however less successful in accounting for the dynamics of the Hadley cell because the nonlinear dynamics associated with the advective effects of $[\bar{v}]$ and $[\bar{\omega}]$ in the tropics is by no means negligible. The dynamics of the Hadley cell is therefore a good example of the importance of nongeostrophic dynamics. A fuller understanding of the Hadley cell rests upon a model that incorporates nongeostrophic dynamics. In such a model, the zonal mean wind and temperature are parts of the unknown and have to be simultaneously determined along with the meridional circulation itself. The whole system with $[\bar{v}]$, $[\bar{\omega}]$, $[\bar{u}]$ and $[\bar{\theta}]$ as unknowns is referred to as the *Hadley circulation.*

12.2.1 Dynamics of the annual mean Hadley circulation

The linear dynamical model of the overturning circulation in Section 10.3 reveals that the diabatic forcing is the most important contributing factor in the tropical region (see Fig. 10.13). This finding confirms that the Hadley cell is a thermally direct circulation which could exist in the tropics even if there were no eddies at all. The observed net diabatic forcing is attributable to the strong convection in a relatively narrow equatorial zone overcompensating the radiative loss to space. Much of the moisture in such precipitation originates from the subtropical ocean surface and is subsequently transported equatorward by the surface trade wind. The latter is an integral part of the Hadley

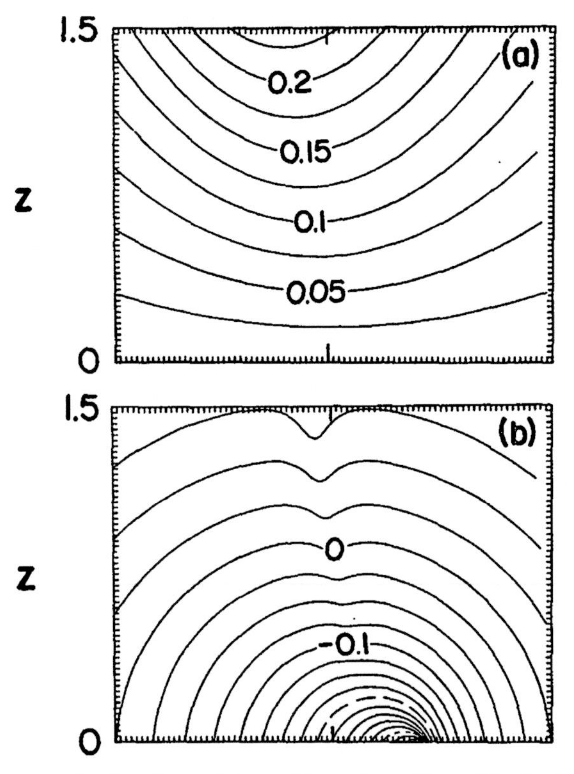

Fig. 12.5 The cross-frontal structure of the along-front velocity at (a) $t = 0$ and (b) $t = 2.5$, $CI = 0.025$. The abscissa is $-1 \leq x \leq 1$ (reprinted from Mak and Bannon, 1984).

circulation itself. The excess heat supports the ascending motion in the equatorial belt, which would in turn support a return flow at upper tropospheric levels towards the subtropics. Conservation of absolute angular momentum of this northward flow would largely dictate the distribution of the upper level zonal wind since friction at the high levels is weak. As this poleward flow progressively weakens, it supports the descending branch of the Hadley cell. Such descent provides adiabatic warming in the subtropics to compensate for the radiative cooling in situ. The overall scheme is that the Hadley cell is ultimately driven by solar radiation, but the actual distribution of the net diabatic heating is partly self-regulated.

We now discuss an approach first developed by Held and Hou (1980) to analytically deduce some key characteristics of the Hadley circulation with a zonally symmetric model driven by an imposed diabatic forcing. By definition, there are no eddies in this model. We partition the time-zonal mean thermodynamic variables as $\theta = \theta_o(z) + \theta'(\varphi, z)$ and $\Phi = \Phi_o(z) + \Phi'(\varphi, z)$ where $\Phi = p/\rho$, φ is the latitude, z is the height; θ_o and Φ_o are the components of a background state. The departure components θ' and Φ' can be assumed to be in hydrostatic balance. The corresponding velocity is (u,v,w). The time-zonal average form of the governing equations with Boussinesq approximation can be written as

$$\frac{v}{a}u_\varphi + wu_z - v\left(2\Omega \sin\varphi + \frac{u}{a}\tan\varphi\right) = F^{(1)}, \tag{12.26}$$

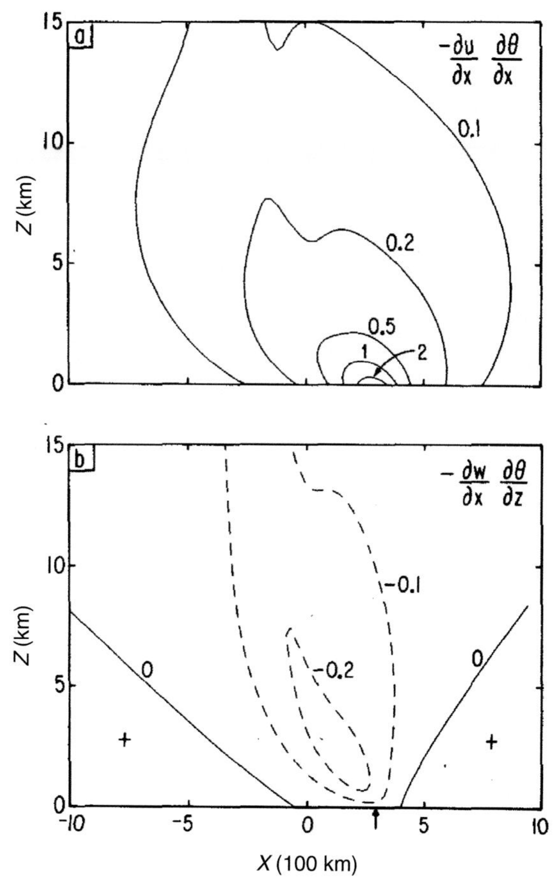

Fig. 12.6 Contour plot of (a) the total deformation term and (b) the thermal tilting term for $\partial\theta/\partial x$ of the frontogenetic processes for the dry case at $\alpha t = 1.5$. The units are $\alpha\theta_{oo}fN/g = 3 \times 10^{-5}\,\mathrm{K}/100\,\mathrm{km\,s}^{-1} \approx 0.33\,\mathrm{K}/100\,\mathrm{km}\,3\mathrm{h}^{-1}$. The contour interval is non-uniform and each contour is labeled. The vertical arrow on the abscissa of the bottom panel locates the surface maximum of $\partial\theta/\partial x$ (reprinted from Bannon and Mak, 1986).

$$\frac{v}{a}v_\varphi + wv_z + u\left(2\Omega\sin\varphi + \frac{u}{a}\tan\varphi\right) = F^{(2)} - \frac{1}{a}\Phi'_\varphi, \tag{12.27}$$

$$\frac{v}{a}\theta'_\varphi + w\theta_{oz} = Q, \tag{12.28}$$

$$\Phi'_z = \frac{g}{\Theta}\theta', \tag{12.29}$$

$$\frac{1}{a\cos\varphi}(v\cos\varphi)_\varphi + w_z = 0; \tag{12.30}$$

Θ is the horizontal average potential temperature, a is the radius of the Earth, Ω is the rotation of the Earth, g is gravity, $F^{(1)}$ and $F^{(2)}$ are frictional forces in the zonal and meridional directions, Q is the net thermal forcing that ultimately gives rise to a Hadley circulation. We represent the net diabatic heating as a thermal relaxation of the potential

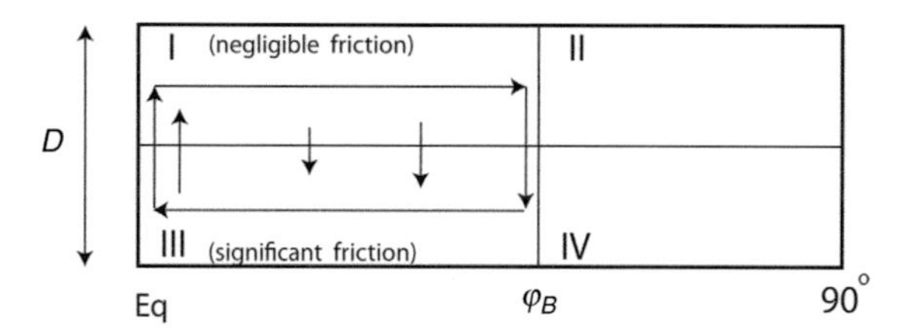

Fig. 12.7 Schematic of a two-layer model for the annual mean Hadley circulation.

temperature $\theta(\varphi, z)$ toward a radiative-convective equilibrium temperature field, $\theta_E(\varphi, z)$, that has hemispheric symmetry,

$$Q = \frac{1}{\tau}(\theta_E - \theta), \tag{12.31}$$

where τ is a relaxation time scale. With the annual mean condition in mind, we prescribe

$$\theta_E = \theta_o(z) + \theta'_E = \theta_o(z) + \Delta\left(\frac{1}{2} - \sin^2\varphi\right). \tag{12.32}$$

The broad vertical structure of the Hadley cell suggests that the essence of its dynamics could be captured even in a two-layer model setting. The hemispheric domain of the model is divided into four sectors (Fig. 12.7). The model Hadley cell exists only in sectors I and III by definition. The sectors II and IV are in thermal equilibrium in this model. Friction is assumed to be significant in the lower layer but not in the upper layer. The problem is to deduce the width of the model Hadley cell, φ_B, the distributions of u, v and w in each sector and the θ' at the interface as functions of latitude.

In sector I where friction is assumed to be negligible, $F^{(1)} = 0$, (12.26) can be rewritten as

$$v\frac{\partial M}{\partial \varphi} = 0, \qquad M = (\Omega a \cos\varphi + u)a\cos\varphi, \tag{12.33}$$

where M is absolute angular momentum per unit mass of air about the rotation axis of the Earth. It follows that air parcels in the upper branch of this model Hadley cell conserve their absolute angular momentum, viz.

$$M(\varphi) = M(0). \tag{12.34}$$

A self-consistent boundary condition of the zonal velocity is $u = 0$ at the equator. Then we get by (12.34)

$$u = \frac{\Omega a \sin^2\varphi}{\cos\varphi}. \tag{12.35}$$

This model u increases strongly with latitude φ, e.g. $u \sim 134$ m s^{-1} at 30°. This value is an upper bound. The observed zonal velocity is much slower partly because the frictional effect by small-scale turbulent eddies is not entirely negligible and more importantly because dynamic instability of the zonal flow would limit how strong it could become.

Frictional force in the zonal direction in sector III (especially the surface friction) is important implying that the surface zonal velocity is weak and the vertical shear $\partial u/\partial z$ is large. Nevertheless, the friction in the meridional direction, $F^{(2)}$, may be still assumed to be much smaller than the pressure gradient force, $-\frac{1}{a}\Phi'_{\varphi}$. Hence, we approximate (12.27) by a geostrophic balance

$$u2\Omega \sin\varphi \approx -\frac{1}{a}\Phi'_{\varphi}. \tag{12.36}$$

The corresponding thermal wind equation is

$$u_z 2\Omega \sin\varphi = -\frac{g}{\theta_{oo}}\theta'_y. \tag{12.37}$$

As long as the model Hadley cell does not extend much beyond 30° latitude, we may further invoke a *small-angle approximation*, namely

$$\sin\varphi \approx \varphi; \;\; \cos\varphi \approx 1. \tag{12.38}$$

Consistent with this approximation, we introduce Cartesian coordinate

$$y = a\varphi, \tag{12.39}$$

with the understanding that we consider a range of latitude such that $y/a \ll 1$. The width of the model Hadley cell is denoted by $Y = a\varphi_B$. For sector I, (12.35) and (12.37) are simplified to

$$u = \Omega a\left(\frac{y}{a}\right)^2, \qquad u_z \sim \frac{u}{D}, \tag{12.40}$$

$$\frac{u}{D}2\Omega\frac{y}{a} = -\frac{g}{\theta_{oo}}\theta'_y, \tag{12.41}$$

$$\therefore \frac{2\Omega^2}{D}\frac{y^3}{a^2} = -\frac{g}{\theta_{oo}}\theta'_y. \tag{12.42}$$

The solution of (12.42) for θ' is

$$\theta'(y) = \theta'(0) - \frac{\Omega^2\theta_{oo}}{2gDa^2}y^4, \tag{12.43}$$

where $\theta'(0)$ is an integration constant yet to be determined. We still need to determine two unknown integration constants: Y and $\theta'(0)$. We can do so by imposing two constraints

(i) Temperature $\theta'(y)$ is continuous at $y = Y$.
(ii) There is no net vertical mass flux in the Hadley circulation.

We satisfy (i) by assuming thermal equilibrium in sector II so that

$$\theta'(Y) = \theta'_E(Y). \tag{12.44}$$

Hence, we have

$$\theta'(0) - \frac{\Omega^2\theta_{oo}}{2gDa^2}Y^4 = \Delta\left(\frac{1}{2} - \frac{Y^2}{a^2}\right). \tag{12.45}$$

Furthermore, we have by (12.28) and (12.31)

$$w\theta_{oz} = \frac{1}{\tau}\left(\theta'_E - \theta'\right). \tag{12.46}$$

Satisfying constraint (ii) requires

$$\int_0^Y w' dy = 0 \Rightarrow \int_0^Y \left(\theta'_E - \theta'\right) dy = 0. \tag{12.47}$$

Thus,

$$\int_0^Y \left(\Delta\left(\frac{1}{2} - \frac{y^2}{a^2}\right) - \theta'(0) + \frac{\theta_{oo}\Omega^2 y^4}{2a^2 gD}\right) dy = 0. \tag{12.48}$$

Upon working (12.48) out, we obtain an expression for the width of the model Hadley cell

$$Y = \sqrt{\frac{5\Delta gD}{3\theta_{oo}\Omega^2}}. \tag{12.49}$$

This result tells us how the width of the model Hadley cell (HC) depends on the parameters. In particular, it is proportional to the square root of the heating intensity and inversely proportional to the rotation rate of the planet. Equations (12.45) and (12.49) also lead to

$$\theta'(0) = \frac{\Delta}{2} - \frac{5gD\Delta^2}{18\theta_{oo}a^2\Omega^2}. \tag{12.50}$$

Finally, we can write the solution for potential temperature in sector I as

$$\theta'(y) = \frac{\Delta}{2} - \frac{5\Delta^2 gD}{18\Omega^2 a^2 \Theta} - \frac{\Delta\Omega^2}{2gDa^2} y^4. \tag{12.51}$$

The main characteristics of the departure of the model potential temperature from the reference potential temperature, $\left(\theta'_E(y) - \theta'(y) = \frac{5gD\Delta^2}{18\Omega^2 a^2\Theta} - \Delta\left(\frac{y}{a}\right)^2 + \frac{\Delta\Omega^2 a^2}{2gD}\left(\frac{y}{a}\right)^4\right)$, are:

- Positive value close to the equator $\Rightarrow$ there is diabatic heating in an equatorial region.
- Decreasing away from the equator.
- Zero at a certain latitude, $\frac{y_*}{Y} = \frac{1}{\sqrt{5}} = 0.45 \Rightarrow$ diabatic heating occurs over slightly less than half of the model HC.
- Negative value further poleward $\Rightarrow$ diabatic cooling occurs in the remaining part of the HC, subtropical region.
- Zero again at the boundary of the model Hadley circulation, $y = Y$.
- Since the area under the curve $\left(\theta'_E(y) - \theta'(y)\right)$ from $y = 0$ to $y = Y$ is zero, there is a balance between the diabatic heating in low latitude regions and the diabatic cooling in the subtropics.

Distribution of the model zonal velocity

Recall that the solution of the zonal velocity in sector I, $y < Y$, is $u = \Omega a \sin^2\varphi / \cos\varphi \approx \Omega a (y/a)^2$. It increases quadratically with distance from the equator. Since we have assumed $\theta' = \theta'_E$ at $y < Y$, there is no diabatic heating or cooling in sector II. The momentum equation in spherical coordinates is approximately

$$\frac{u}{D}\left(2\Omega\sin\varphi + \frac{2u\tan\varphi}{a}\right) = -\frac{g}{a\theta_{oo}}\frac{\partial\theta'}{\partial\varphi} = \frac{2g\Delta}{a\theta_{oo}}\sin\varphi\cos\varphi. \tag{12.52}$$

Then the solution of u in the region $\varphi \geq \varphi_B$ is

$$u = U\cos\varphi,$$
$$U = \frac{\Omega a}{2}\left(-1 + \sqrt{1 + \frac{4gD\Delta}{\Theta\Omega^2 a^2}}\right). \tag{12.53}$$

The absolute zonal velocity is $(\Omega a\cos\varphi + u) = (\Omega a + U)\cos\varphi$. This means that the zonal flow in sector II is in solid rotation. It follows that there is a discontinuity in the model u at $y = Y$

$$u(y)|_{Y-} \neq u(y)|_{Y+}. \tag{12.54}$$

All we can say about the zonal wind in sectors III and IV is that u is weak on the basis of the reasoning above because surface friction has been assumed to be strong.

Distribution of the model vertical velocity

At the mid-level where $v = 0$, the solution for w in sector I, $y < Y$, is

$$w = \frac{1}{\tau\theta_{oz}}(\theta'_E - \theta') = \frac{1}{\tau\theta_{oz}}\left(\frac{5gD\Delta^2}{18\theta_{oo}a^2\Omega^2} - \frac{\Delta y^2}{a^2} + \frac{\Omega^2\theta_{oo}y^4}{2gDa^2}\right). \tag{12.55}$$

There is ascending motion in an equatorial region, $0 < y < y*$ and descending motion in the remaining part of the model Hadley cell, $y* < y < Y$. Hence, $w = 0$ is at $y*/a = \left(\tilde{\Delta}/3\tilde{\Omega}^2\right)^{0.5}$ or $y*/a = \left(5\tilde{\Delta}/3\tilde{\Omega}^2\right)^{0.5} = Y/a$ where $\tilde{\Delta} = \frac{\Delta}{\Theta}$ and $\tilde{\Omega} = \Omega a/\sqrt{gD}$. The ratio of the width of the descending branch to the width of the ascending branch of the model Hadley cell is $\frac{Y - y*}{y*} = \left(\sqrt{5} - 1\right) \approx 1.2$. We can rewrite (12.55) as

$$w = \frac{D}{\tau}\left(\frac{5\tilde{\Delta}^2}{18\tilde{\Omega}^2} - \tilde{\Delta}\tilde{y}^2 + \frac{\tilde{\Omega}^2}{2}\tilde{y}^4\right)\frac{1}{\tilde{\Delta}_v}. \tag{12.56}$$

In writing (12.56), we have used $d\theta_o/dz = \Delta_v/D$ and $\tilde{\Delta}_v = \Delta_v/\theta_{oo}$ is a non-dimensional measure of the vertical thermal contrast associated with the stratification of the background state.

The Hadley circulation can be conveniently depicted in terms of a mass streamfunction $\Psi(y,z)$. The continuity equation (12.30) may be consistently approximated by $v_y + w_z = 0$. Hence, we define

$$v = -\Psi_z, \quad w = \Psi_y. \tag{12.57}$$

Without loss of generality, we use boundary condition, $\Psi(0) = 0$, and get

$$\Psi(y) = \int_0^y w\,dy. \tag{12.58}$$

The intensity of the Hadley circulation is then

$$\Psi_{\max} = \Psi(y*) = \int_0^{y*} w\,dy$$

$$\Psi_{\max} = \frac{8}{45\sqrt{3}}\frac{aD}{\tau}\frac{\tilde{\Delta}^{5/2}}{\tilde{\Omega}^3\tilde{\Delta}_v}. \tag{12.59}$$

Summary of the model results in non-dimensional form

Distance, temperature, streamfunction, zonal velocity and vertical velocity are measured in units of $a, \theta_{oo}, \frac{aD}{\tau}, \Omega a$ and $\frac{\theta_{oo}}{\tau\theta_{oz}}$ respectively; $\theta_{oz} = \frac{\Delta_v}{D}, \tilde{\Omega} = \frac{\Omega a}{\sqrt{gD}}, \tilde{\Delta} = \frac{\Delta}{\theta_{oo}}, \tilde{\Delta}_v = \frac{\Delta_v}{\theta_{oo}}$. The various model results are summarized in Table 12.1.

Illustrative calculation

A reasonable set of parameter values are: $\Theta = 260\,\text{K}$, $\Delta = 80\,\text{K}$, $220 \leq \theta_E \leq 300\,\text{K}$,

$$\theta_{oz} = \frac{\Theta}{g}N^2 = 3 \times 10^{-3}\,\text{Km}^{-1} \sim \frac{\Delta_v}{D} \Rightarrow \Delta_v = 30\,\text{K},$$

$$\tau = 20\,\text{days}, \quad \frac{\Theta}{\tau\theta_{oz}} = 0.05\,\text{m}\,\text{s}^{-1},$$

$$D = 1 \times 10^4\,\text{m}, \quad \Omega = 7.3 \times 10^{-5}\,\text{s}^{-1},$$

$$a = 6.37 \times 10^6\,\text{m}, \quad \frac{aD}{\tau} = 0.37 \times 10^5\,\text{m}^2\,\text{s}^{-1}.$$

Then we get $\tilde{\Delta} = 0.31, \quad \tilde{\Delta}_v = 0.11, \quad \tilde{\Omega} = 1.47$

$$\tilde{Y} = 0.49 \rightarrow Y = 3.1 \times 10^6\,\text{m}, \quad \varphi_B = 29^\circ$$

$$\tilde{u}(\tilde{Y}_-) = 0.24 \rightarrow u(Y_-) = 110\,\text{m}\,\text{s}^{-1},$$

$$\tilde{U} = 0.125 \rightarrow u(Y_+) = 50.7\,\text{m}\,\text{s}^{-1},$$

$$\tilde{\theta}'_E(0) - \tilde{\theta}'(0) = 0.012 \rightarrow \theta'_E(0) - \theta'(0) = 3.2\text{ K}$$

$$\tilde{w}(0) = 0.0123 \rightarrow w(0) \sim 0.06\,\text{cm}\,\text{s}^{-1}$$

$$\tilde{\Psi}_{\max} = 0.0157 \rightarrow \Psi_{\max} = 581\,\text{m}^2\,\text{s}^{-1} \rightarrow v_{\max} \sim \frac{2\Psi_{\max}}{D} \sim 0.15\,\text{m}\,\text{s}^{-1}$$

The distributions of the model θ and θ_E in the region of the model Hadley cell are presented in Fig. 12.8.

Table 12.1 Summary of results in the Hadley circulation model sectors I and II

	Sector I	Sector II
Reference temperature	$\frac{\theta'_E(y)}{\theta_{oo}} = \tilde{\Delta}\left(\frac{1}{2} - \tilde{y}^2\right)$	$\frac{\theta'_E(y)}{\theta_{oo}} = \tilde{\Delta}\left(\frac{1}{2} - \sin^2\varphi\right)$
Width of Hadley cell	$\frac{Y}{a} = \left(\frac{5\tilde{\Delta}}{3\tilde{\Omega}^2}\right)^{0.5}$	–
Width of ascending motion	$\frac{y*}{a} = \left(\frac{\tilde{\Delta}}{3\tilde{\Omega}^2}\right)^{0.5}$	–
Width of descending motion	$\frac{Y-y*}{a} = \left(\frac{\tilde{\Delta}}{3\tilde{\Omega}^2}\right)^{0.5}(\sqrt{5}-1)$	–
Intensity	$\frac{\Psi_{\max}}{aD/\tau} = \frac{8}{45\sqrt{3}}\frac{\tilde{\Delta}^{5/2}}{\tilde{\Omega}^3\tilde{\Delta}_v}$	–
Zonal velocity	$\frac{u}{\Omega a} \equiv \tilde{u} = \tilde{y}^2$	$\tilde{u} = U\cos\varphi$ $\tilde{U} = \frac{1}{2}\left(-1 + \sqrt{1 + \frac{4\tilde{\Delta}}{\tilde{\Omega}^2}}\right)$
Potential temperature	$\frac{\theta'}{\Theta} \equiv \tilde{\theta}'(y) = \frac{\tilde{\Delta}}{2} - \frac{5\tilde{\Delta}^2}{18\tilde{\Omega}^2} - \frac{\tilde{\Omega}^2}{2}\tilde{y}^4$	$\frac{\theta'(y)}{\Theta} = \tilde{\Delta}\left(\frac{1}{2} - \sin^2\varphi\right)$
Vertical velocity	$\frac{w}{\theta_{oo}/\tau\theta_{oz}} = \left(\frac{5\tilde{\Delta}^2}{18\tilde{\Omega}^2} - \tilde{\Delta}\tilde{y}^2 + \frac{\tilde{\Omega}^2\tilde{y}^4}{2}\right)$	–

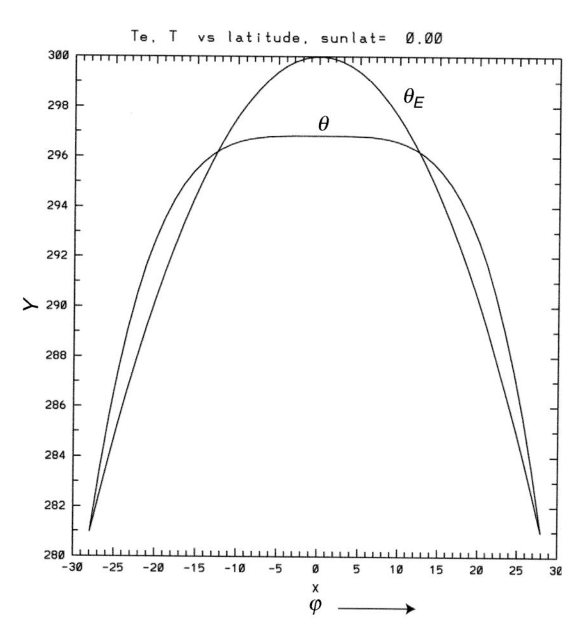

Fig. 12.8 Distribution of the model temperature θ and reference temperature profile θ_E in the region of the model Hadley cell.

The signature feature of Fig. 12.8 is that the area between the θ and θ_E curves where $\theta_E < \theta$ is equal to the area where $\theta_E > \theta$ in the region of the Hadley cell. The distribution of the model u is shown in Fig. 12.9.

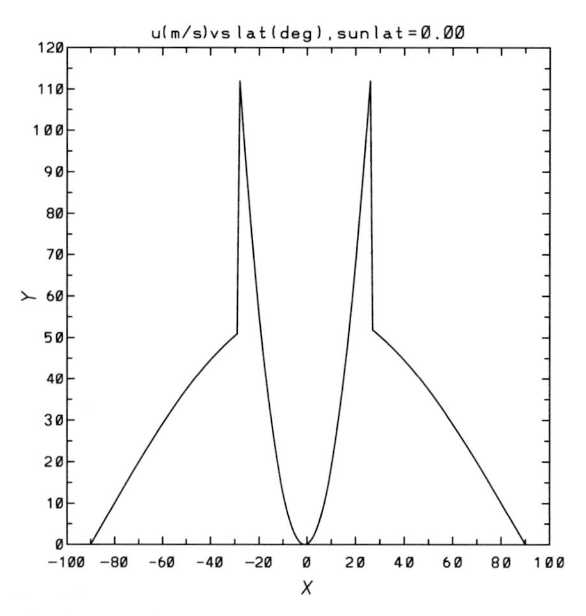

Fig. 12.9 The model zonal wind distribution at the upper level.

The width of the model HC is quite reasonable. The thermal structure in the region of the model HC is compatible with observation. The intensity of the model Hadley circulation is about 10 times too weak. It stems from the use of a smooth distribution of thermal forcing. There is heating from the equator to ~ 12° N and cooling from 12° N to ~30° N. The zonal wind is qualitatively meaningful, but is too strong. The discontinuity is simply a consequence of the assumption for thermal equilibrium in the poleward region. The characteristics of this model Hadley circulation are by and large compatible with observation.

We have not examined the quantitative impact of surface friction in this semi-analytic model analysis. Consequently, we are not in a position to deduce the explicit dependence of the characteristics upon the surface friction. Since the thermal forcing stems from the condensational heating at the ITCZ, a more realistic profile of θ_E would have large values concentrated in a narrow zone. These issues can be quantitatively addressed with the use of a counterpart numerical two-layer model.

12.2.2 Dynamics of the seasonal mean Hadley circulation

We have seen in Figs. 10.1 and 10.3 that the asymmetry of the seasonal mean Hadley circulation with respect to the equator is much greater than that of the diabatic heating itself. This intriguing feature is not what one would intuitively anticipate. We now seek to simulate the structure of a seasonal mean Hadley circulation in the context of a two-layer model. In this analysis, we use the following formula to represent the reference potential temperature $\theta_E(\varphi)$ at any time of the year

$$\theta_E(\varphi) = \theta_{oo} + \Delta\left(2 \sin\varphi^* \sin\varphi - \sin^2\varphi\right); \tag{12.60}$$

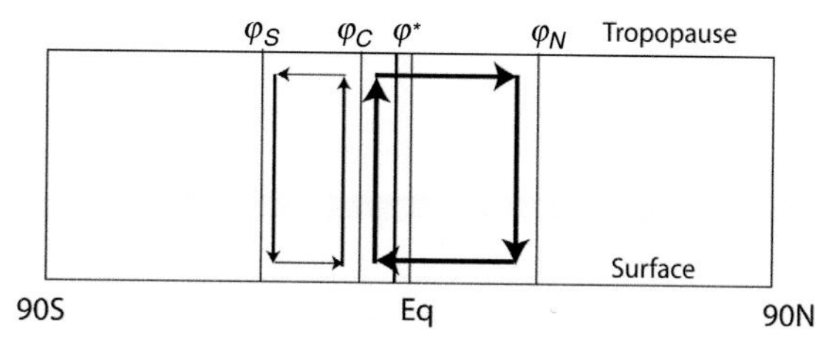

Fig. 12.10 Schematic of the two model Hadley cells in the northern winter. See the text for the meanings of the different quantities.

θ_E has a maximum value at $\varphi = \varphi^*$ at a particular time, with $-\varphi_{\max} \le \varphi^* \le \varphi_{\max}$. We set $\varphi^* = 0$ at the spring and fall equinoxes, $\varphi^* = \varphi_{\max}$ at the summer solstice and $\varphi^* = -\varphi_{\max}$ at the winter solstice. Fig. 12.10 is a schematic of the anticipated seasonal Hadley cell under consideration. The two model Hadley cells in the northern winter span from φ_S to φ_N. The sectors poleward of the two cells are again assumed to be in thermal equilibrium. The four characteristics of this circulation are: φ_N, φ_S, φ_c and $\theta'(\varphi_c)$; φ_c is the latitude separating the winter Hadley cell from the summer Hadley cell and $\theta'(\varphi_c)$ is the potential temperature at φ_c.

This model is used to determine the steady state circulation driven by the thermal forcing of (12.31) with (12.60) at different times of the year. A time-dependent model would be needed to take into account the transient dynamics. But, we assume that the circulation instantaneously adjusts to a sequence of equilibrium states as the distribution of the forcing changes. We specifically use $\varphi^* = \varphi_{\max} \sin(t\pi/365)$ with t being the time in days with $t = 0$ corresponding to the fall equinox. In light of the analysis elaborated in Section 12.2.1, the four unknown constants in the model solution are to be determined on the basis of the following four constraints:

$$\theta'(\varphi_S) = \theta'_E(\varphi_S),\ \theta'(\varphi_N) = \theta'_E(\varphi_N),\ \int_{\varphi_c}^{\varphi_N} (\theta'_E - \theta')d\varphi = 0,\ \int_{\varphi_S}^{\varphi_c} (\theta'_E - \theta')d\varphi = 0. \quad (12.61)$$

With the small angle approximation, (12.60) is approximated by

$$\theta'_E = \Delta\left(2\varphi^*\varphi - \varphi^2\right). \quad (12.62)$$

Then Y_S, y_c, y^* and Y_N correspond to $\varphi_S, \varphi_c, \varphi^*$ and φ_N in Cartesian coordinates. The condition $u = 0$ is now applied at φ_c. Conservation of absolute angular momentum yields for the region $\varphi_S \le \varphi \le \varphi_N$

$$(\Omega a \cos\varphi + u) a \cos\varphi = \Omega a^2 \cos^2\varphi_c$$

$$\begin{aligned} u &= \Omega a \left(\frac{\cos^2\varphi_c - \cos^2\varphi}{\cos\varphi}\right) \\ &\approx \Omega a\left(-\varphi_c^2 + \varphi^2\right) \\ &= \Omega a\left(\frac{y^2}{a^2} - \frac{y_c^2}{a^2}\right). \end{aligned} \quad (12.63)$$

Then the zonal flow is an easterly in $-y_c < y < y_c$. Thermal wind balance is

$$\frac{u}{D} 2\Omega \frac{y}{a} = -\frac{g}{\Theta} \frac{d\theta'}{dy}. \tag{12.64}$$

For $Y_S < y < Y_N$

$$\int_{Y_S}^{y} \frac{-2\Omega^2 \Theta}{gDa^2} y(y^2 - y_c^2) dy = \theta'(y) - \theta'(Y_S). \tag{12.65}$$

It follows that the solution for the potential temperature is

$$\theta'(y) = \theta'(Y_S) - \frac{\Omega^2 \Theta}{gDa^2} \left[\frac{1}{2}\left(y^4 - Y_S^4\right) - y_c^2\left(y^2 - Y_S^2\right) \right]. \tag{12.66}$$

The constraint of continuity in temperature leads to

$$\frac{\Delta}{a^2}\left(2y^* Y_N - Y_N^2\right) = \frac{\Delta}{a^2}\left(2y^* Y_S - Y_S^2\right) - \frac{\Omega^2 \theta_{oo}}{gDa^2} \left[\frac{1}{2}\left(Y_N^4 - Y_S^4\right) - y_c^2\left(Y_N^2 - Y_S^2\right) \right]. \tag{12.67}$$

By defining $x = Y_S/a,\ y = y_c/a,\ z = Y_N/a,\ c = y^*/a$ and $b = \Omega^2 \theta_{oo} a^2 / gD\Delta$, we obtain from (12.67) three constraints written in non-dimensional form as

$$\begin{aligned}
0 &= b\left[\frac{1}{2}\left(z^4 - x^4\right) - y^2\left(z^2 - x^2\right)\right] + 2c(z - x) - z^2 + x^2 \\
0 &= b\left(-\frac{7}{30}y^5 + y^3x^2 - \frac{2}{3}y^2x^3 - \frac{1}{2}yx^4 + \frac{4}{10}x^5\right) \\
&\quad + c\left(y^2 - x^2\right) - \frac{1}{3}\left(y^3 - x^3\right) - (y - x)\left(2cx - x^2\right) \\
0 &= b\left(\frac{1}{10}z^5 + \frac{7}{30}y^5 - \frac{1}{2}x^4z + \frac{1}{2}x^4y - \frac{1}{3}y^2z^3 + y^2x^2z - y^3x^2\right) \\
&\quad + c\left(z^2 - y^2\right) - \frac{1}{3}\left(z^3 - y^3\right) - \left(2cx - x^2\right)(z - y).
\end{aligned} \tag{12.68}$$

In other words, the configuration of the Hadley cells depends on the position of the maximum forcing, $c = y^*/a$ and the intensity of the forcing, $b^{-1} = \dfrac{gD\Delta}{\Omega^2 a^2 \theta_{oo}} = \dfrac{\tilde{\Delta}}{\tilde{\Omega}^2}$.

It is easy to verify that for the special case of $c = 0$, we recover from (12.68) the previous solution for the annual mean width of the model Hadley cell, namely

$$z = -x = \sqrt{\frac{5gD\Delta}{3\Omega^2 \theta_{oo} a^2}}. \tag{12.69}$$

The three nonlinear algebraic equations in (12.68) can be solved for a given value of c by applying Newton's iterative algorithm provided that we start with a good initial guess. We start with a small value of c and repeat the procedure for progressively larger values of c. We can then scan over a wide range of c values.

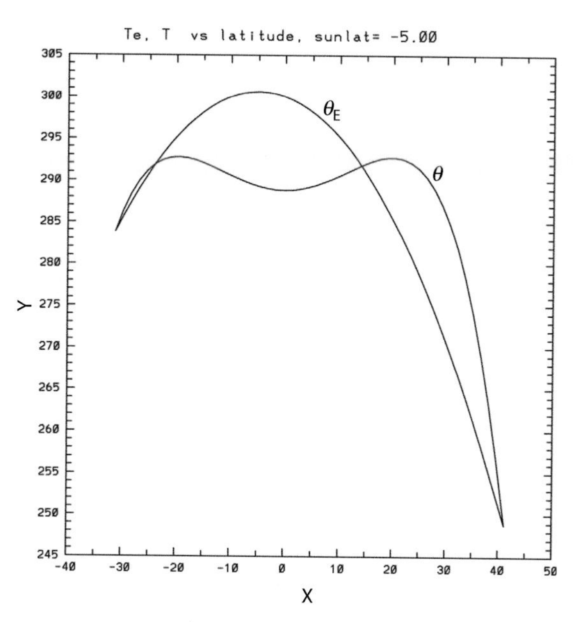

Fig. 12.11 Variations of θ and θ_E with latitude in the northern hemisphere winter season in K.

We present the model solution for the following representative winter condition: $\varphi^* = -5°$, $\Delta = 80\,\text{K}$, $D = 1 \times 10^4\,\text{m}$, $\Theta = 260\,\text{K}$, $\Omega = 7.3 \times 10^{-5}\,\text{s}^{-1}$, $a = 6.3 \times 10^6\,\text{m} \therefore b = \dfrac{\Omega^2 a^2 \theta_{oo}}{gD\Delta} \approx 10$. The solutions of $\theta'(y)$ and $\theta_E(y)$ are shown in Fig. 12.11. There are again equal areas between positive and negative values of $(\theta_E - \theta)$ in each of the Hadley cells. The maximum $|\theta_E - \theta|$ is about 10 K; θ varies very little within the low latitudinal region of the winter Hadley cell. The location of φ_c is such that the positive area and negative area between θ and θ_E of each HC are equal.

The key features of Fig. 12.11 are

(1) φ_N is located at 40° N toward the north pole when φ^* moves to $-5°$
(2) φ_S is located at 30° S toward the south pole when φ^* moves to $-5°$
(3) φ_c is located at 20° S when φ^* moves to $-5°$.

In other words, the model winter Hadley cell is almost six times wider than the summer HC.

The dependence of φ_N, φ_c and φ_S upon the location of maximum thermal forcing φ^* is shown in Fig. 12.12.

Figure 12.13 shows that the maximum westerly at the edge of the winter Hadley cell is about twice as strong as the maximum westerly at the edge of the summer Hadley cell (190 vs. 85 $\text{m}\,\text{s}^{-1}$). There is an easterly flow between 20° N and 20° S. These features are qualitatively compatible with the observed features of the seasonal HC. Although this highly idealized model does not yield quantitatively realistic results, it provides us with great insight into the dynamics. It is amazing to find such a large impact of a rather small asymmetry in the location of maximum θ_E.

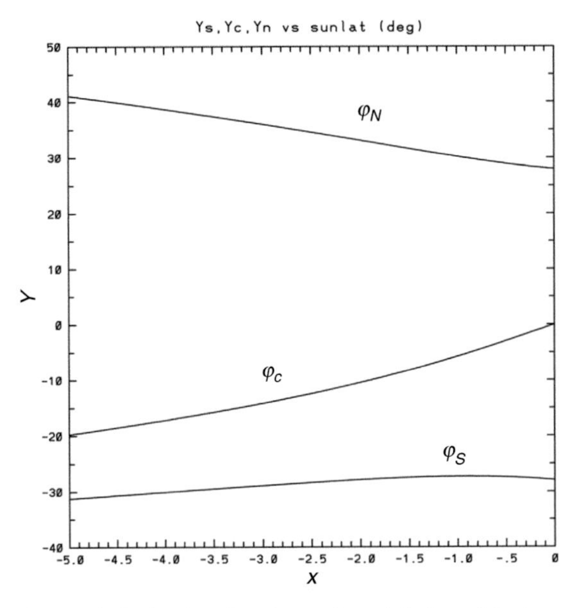

Fig. 12.12 Variation of φ_N, φ_C and φ_S with the location of maximum thermal forcing φ^* in degrees.

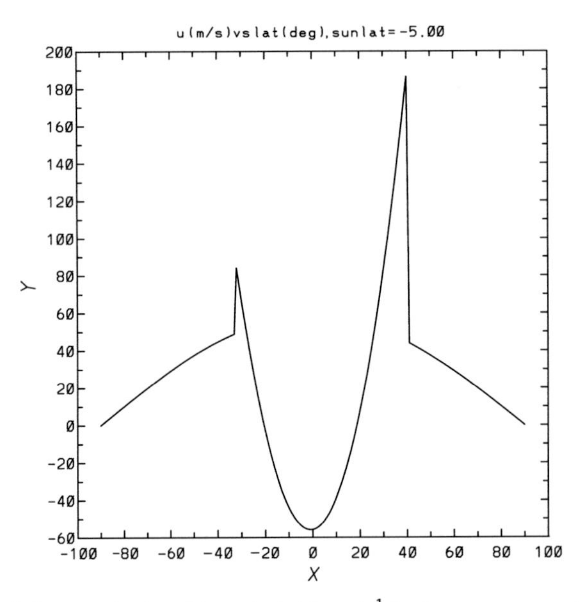

Fig. 12.13 Latitudinal variation of the upper level model zonal wind in m s^{-1} for the case $\varphi* = -5°$.

Furthermore, it is found that the ratio (vertical velocity at equator in winter)/(vertical velocity at equator for annual mean) ~10/2.2 ~ 5≫1. It should be noted that observational as well as modeling studies suggest that the Hadley circulation can be significantly affected by the flux of momentum and heat by large-scale eddies during a part of its annual cycle. It is still a topic of much active research as of 2009.

12.3 Non-supercell tornadogenesis

The importance of nongeostrophic dynamics is next discussed in the context of a type of small-scale disturbances known as a non-supercell tornado (NST). It is a small columnar vortical disturbance. Not only does its Rossby number greatly exceed order unity, but also the non-hydrostatic effect is not negligible. Unlike the more familiar tornadoes spawned in severe weather systems, NSTs are relatively benign. They seem to be quite common in some areas. During a field project near Denver, Colorado, in the summer of 1987, 27 visual vortices were documented in 47 days (Wakimoto and Wilson, 1989). All but one of them were classified as NSTs. They were not associated with accompanying mesocyclones. They all had cyclonic rotation with an average azimuthal shear of $6.2 \times 10^{-2}\ \mathrm{s}^{-1}$. They were typically ≤ 2 km in diameter. Their vertical dimension ranged from 0.8 to 6 km. When detected, a radar vortex was observed an average of 14 min in advance of a visual vortex, suggesting an *e*-folding time of intensification probably between 5 and 10 min. Those NSTs initially formed at low levels (0–2 km) and grew in vertical depth and intensity just prior to vortex sighting. The maximum vertical velocity was comparable to the maximum horizontal wind. The wind of some of the NSTs was so strong that its damage capability was estimated to be as high as F2 (Fujita scale). The nearby clouds were often cumuli or cumulus congestus and the sky was generally not overcast.

The local environment that spawns NSTs has certain signature characteristics. For example, the environment of the NSTs typically had a small convective available potential energy (less than $500\ \mathrm{m}^2\,\mathrm{s}^{-2}$) and a very weak vertical shear. Such conditions are not conducive for supercell thunderstorm formation. An NST environment has two noteworthy common features. One is a significant horizontal shear and the other is a boundary outflow associated with a density contrast in the leading edge surging into the shear flow. Such a surface flow could be associated with a meso-scale cold front or a thunderstorm outflow or an advancing sea-breeze front. Environments with such characteristics, and therefore the occurrence of NSTs, should be present in other parts of the world.

Since strong horizontal shear is typically present in an NST environment, barotropic instability has been suspected as a possible mechanism for its formation. However, the usual two-dimensional barotropic instability is not a satisfactory explanation because an NST has a distinctly columnar structure with an aspect ratio (depth–width) of unity or larger. It follows that the non-hydrostatic effect is likely to be also significant. The non-hydrostatic aspect of the disturbance introduces substantial complication to an instability analysis. Non-supercell tornado is hypothesized to arise from non-hydrostatic barotropic instability. This hypothesis can be succinctly tested with a model developed with a Hamiltonian formulation for a layer of non-hydrostatic barotropic fluid.

12.3.1 Hamiltonian formulation of a model for non-supercell tornadogenesis

The hypothesis can be succinctly tested with a non-hydrostatic barotropic fluid model formulated with the use of the Hamiltonian approach (Mak, 2001). According to *Hamilton's principle of least action*, the evolution of particles in a mechanical system from one instant to another is characterized by a minimum value of its *action*. Hamilton postulated that the action of a system is equal to the total *Lagrangian* (difference between kinetic energy and potential energy) of its constituent components. Hamilton's principle has the advantage of enabling us to derive a set of self-consistent governing equations for a fluid system with the incorporation of any approximation a priori. The summation of the Lagrangian of all fluid parcels in a system takes on the form of an integral. Therefore, applying Hamilton's principle of least action amounts to minimizing an integral, of which the integrand is a function of the dependent variables of the fluid parcels under consideration. Readers are referred to the text by Salmon (1998; Chapters 1 and 8) for a general discussion of Hamiltonian fluid dynamics.

In a Lagrangian description of a fluid, the dependent variables are the locations of each fluid element (x,y,z) with time τ being the independent variable. A fluid system contains an infinite number of fluid elements. Thus, we may describe the fluid elements in a *location space* with continuous *location coordinates* (x, y, z). Implicit to a Lagrangian description is that each fluid element has a distinct identity. Thus, we can also describe the fluid elements in a *label space* with continuous *label coordinates* (a, b, c). The values of (a, b, c) for each fluid element can be arbitrarily assigned, but do not change as it moves about. We therefore may define these coordinates such that the mass of a fluid element is $da \cdot db \cdot dc = d(mass)$. The location of each fluid element at time τ is explicitly indicated as $x(a, b, c, \tau), y(a, b, c, \tau), z(a, b, c, \tau)$. Its mass is also $\rho \cdot dx \cdot dy \cdot dz = d(mass)$ where ρ is the density of the fluid. The location space can be mapped to the label space at any time in accordance with

$$\frac{1}{\rho} = \frac{\partial(x, y, z)}{\partial(a, b, c)}, \tag{12.70}$$

where $\dfrac{\partial(x, y, z)}{\partial(a, b, c)} = \begin{vmatrix} x_a & x_b & x_c \\ y_a & y_b & y_c \\ z_a & z_b & z_c \end{vmatrix}$ is the Jacobian determinant. Taking the derivative of (12.70) with respect to τ would yield a general continuity equation.

Columnar motion in a layer of constant-density fluid is considerably simpler in that the horizontal position of each of its fluid elements is a function of only two of its label coordinates, say a and b: $x(a, b, \tau)$ and $y(a, b, \tau)$. However, its vertical position is necessarily a function of all three label coordinates, $z(a, b, c, \tau)$. The mapping between the two coordinates simplifies (12.70) to

$$\frac{1}{\rho} = \frac{\partial(x, y)}{\partial(a, b)} \frac{\partial z}{\partial c}. \tag{12.71}$$

Since the fluid elements at the bottom surface $z = 0$ remain at the surface, they can be assumed to have a constant value of c, say 0. Likewise, the fluid elements at the free surface have another constant value of c, say 1. It follows that

$$z = ch, \tag{12.72}$$

where h is the depth of a fluid column under consideration and hence

$$h = \frac{1}{\rho}\frac{\partial(a,b)}{\partial(x,y)}. \tag{12.73}$$

Taking the total time derivative of (12.73) would yield

$$\frac{\partial h}{\partial \tau} = \frac{1}{\rho}\frac{\partial}{\partial \tau}\left(\frac{1}{\frac{\partial(x,y)}{\partial(a,b)}}\right) = -\frac{1}{\rho\left(\frac{\partial(x,y)}{\partial(a,b)}\right)^2}\left(\frac{\partial(u,y)}{\partial(a,b)} + \frac{\partial(x,v)}{\partial(a,b)}\right)$$

$$\therefore \frac{\partial h}{\partial \tau} = -\frac{1}{\rho\left(\frac{\partial(x,y)}{\partial(a,b)}\right)^2}\frac{\partial(x,y)}{\partial(a,b)}\left(\frac{\partial(u,y)}{\partial(x,y)} + \frac{\partial(x,v)}{\partial(x,y)}\right) = -h\left(\frac{\partial u}{\partial x} + \frac{\partial v}{\partial y}\right)$$

$$\rightarrow \frac{Dh}{Dt} + h\left(\frac{\partial u}{\partial x} + \frac{\partial v}{\partial y}\right) = 0, \tag{12.74}$$

where $\dfrac{\partial}{\partial \tau} \equiv D/Dt = \partial/\partial t + u(\partial/\partial x) + v(\partial/\partial y)$, $u \equiv Dx/Dt$, $v \equiv Dy/Dt$. This is the continuity equation for the system under consideration.

We next apply Hamilton's principle of least action to derive the momentum equations for a non-hydrostatic columnar flow in this fluid. For a non-hydrostatic flow, the Lagrangian of this layer of fluid is

$$L = \iiint da \cdot db \cdot dc \left\{\frac{1}{2}\left[\left(\frac{\partial x}{\partial \tau}\right)^2 + \left(\frac{\partial y}{\partial \tau}\right)^2 + \left(\frac{\partial z}{\partial \tau}\right)^2\right] - gz\right\}, \tag{12.75}$$

where g is gravity. Upon performing the integration of c with the use of (12.72), we obtain

$$L = \frac{1}{2}\iint da\, db\left[\left(\frac{\partial x}{\partial \tau}\right)^2 + \left(\frac{\partial y}{\partial \tau}\right)^2 + \frac{1}{3}\left(\frac{\partial h}{\partial \tau}\right)^2 - gh\right]. \tag{12.76}$$

The constraint among x, y and h may be incorporated into the Lagrangian with the use of an unknown Lagrange multiplier, $\lambda(a, b, \tau)$,

$$L = \iint da\, db\left\{\frac{1}{2}\left[\left(\frac{\partial x}{\partial \tau}\right)^2 + \left(\frac{\partial y}{\partial \tau}\right)^2\right] + \frac{1}{6}\left(\frac{\partial h}{\partial \tau}\right)^2 - \frac{gh}{2} + \lambda\left[\frac{1}{h} - \rho\frac{\partial(x,y)}{\partial(a,b)}\right]\right\}. \tag{12.77}$$

According to Hamilton's principle, the unknowns $x(t)$, $y(t)$ and $h(t)$ of such a fluid column are such that the definite integral of the action L from $\tau = t_1$ to $\tau = t_2$ has a minimum value. Calculus of variations is the mathematical tool for tackling this problem. We therefore seek

$$\delta\int_{t_1}^{t_2} L\, d\tau = 0 \tag{12.78}$$

for arbitrary independent variations δx, δy, δh, and $\delta\lambda$ subject to the condition that they vanish at the two ends of the time interval, t_1 and t_2.

We apply the extended form of Hamilton's principle to get the governing equations for a system in a canonical form. For this purpose, we introduce three generalized coordinates q_i $(i = 1, 2, 3)$ as

$$q_1 = x, \quad q_2 = y, \quad q_3 = h, \tag{12.79}$$

together with three generalized momenta. The latter are $p_i = \partial\hat{L}/\partial q_i$, where $\hat{L}$ is the corresponding specific Lagrangian. It follows from (12.79) that we have

$$p_1 = \dot{x}, \quad p_2 = \dot{y}, \quad p_3 = \frac{1}{3}\dot{h}. \tag{12.80}$$

Equation (12.78) then takes on the form

$$\delta\int d\tau \iint da\, db\left(\sum_{i=1}^{3} p_i\, \dot{q}_i - H\right) = 0, \tag{12.81}$$

where H is the *Hamiltonian* for each fluid particle,

$$H(q_i, p_i) = \frac{1}{2}\left(p_1^2 + p_2^2 + 3p_3^2\right) + \frac{g}{2}q_3 - \lambda\left[\frac{1}{q_3} - \rho\frac{\partial(q_1, q_2)}{\partial(a, b)}\right], \tag{12.82}$$

subject to arbitrary variations: δq_i, δp_i, and $\delta\lambda$. Upon considering δq_i and δp_i, one readily obtains with integration by parts,

$$\begin{aligned} \delta p_m &: \quad \frac{\partial q_m}{\partial \tau} = p_m, \quad m = 1, 2, \\ \delta p_3 &: \quad \frac{\partial q_3}{\partial \tau} = 3p_3; \end{aligned} \tag{12.83a,b}$$

$$\begin{aligned} \delta q_m &: \quad \frac{\partial p_m}{\partial \tau} = \frac{1}{q_3}\frac{\partial \lambda}{\partial q_m}, \quad m = 1, 2, \\ \delta q_3 &: \quad \frac{\partial p_3}{\partial \tau} = -\left(\frac{g}{2} + \frac{\lambda}{q_3^2}\right). \end{aligned} \tag{12.84a,b}$$

From (12.84b), we also obtain

$$\lambda = -\frac{g}{2}h^2 - \frac{h^2}{3}\frac{\partial \dot{h}}{\partial \tau}. \tag{12.85}$$

Equations (12.83) and (12.84) are the *canonical equations* for this system. The Eulerian form of Equations (12.84a) with the use of (12.85) is

$$\begin{aligned} \frac{Du}{Dt} &= -g\frac{\partial h}{\partial x} - \frac{1}{3h}\frac{\partial}{\partial x}\left(h^2\frac{D^2 h}{Dt}\right), \\ \frac{Dv}{Dt} &= -g\frac{\partial h}{\partial y} - \frac{1}{3h}\frac{\partial}{\partial y}\left(h^2\frac{D^2 h}{Dt}\right). \end{aligned} \tag{12.86a,b}$$

Equations (12.86a,b) are known as Green–Naghdi equations (Green and Naghdi, 1976), which are the momentum equations for a columnar flow. Considering variation $\delta\lambda$ would recover

(12.73) with which we have already obtained the continuity equation. Equations (12.86a,b) together with (12.82) form a closed set of governing equations for this fluid system.

The physical meaning of λ becomes more evident when we combine (12.84a) with (12.83b) and get

$$\frac{\partial}{\partial t}(q_3 p_m) + \nabla \cdot (q_3 p_m \vec{V}) = \frac{\partial \lambda}{\partial q_m}, \quad m = 1, 2, \tag{12.87}$$

where $\vec{V} = (u, v)$. Hence, the Lagrange multiplier λ is interpretable in terms of the total pressure acting on a column of fluid. Its first component $gh^2/2$ is associated with hydrostatic pressure and its second component $\frac{h}{3}\frac{\partial \dot{h}}{\partial \tau}$ is associated with dynamic (non-hydrostatic) pressure. The gradient of this non-hydrostatic pressure is height invariant due to the columnar assumption. Thus, the term $-\frac{1}{3h}\nabla\left[h^2\frac{D^2 h}{Dt^2}\right]$ in (12.86) is in essence the vertical average of the horizontal force associated with the non-hydrostatic pressure.

Since the Hamiltonian (12.82) does not have explicit dependence on time τ or on the label coordinates (a, b), it is said to have time-translation symmetry and particle relabeling symmetry. It follows that this system of governing equations would conserve total mass, total energy and potential vorticity of individual fluid elements. The potential vorticity conservation specifically takes on the form of

$$\begin{gathered} \frac{D}{Dt}\left(\frac{\zeta + \zeta_*}{h}\right) = 0, \\ \zeta_* = \frac{1}{3}\frac{\partial\left(\frac{Dh}{Dt}, h\right)}{\partial(x, y)}, \quad \zeta = \frac{\partial v}{\partial x} - \frac{\partial u}{\partial y}. \end{gathered} \tag{12.88}$$

12.3.2 Model analysis

12.3.2.1 Intrinsic properties

This system of equations has wave solutions in a basic state at rest with a mean depth H_o. It is straightforward to show that the dispersion relation for a perturbation disturbance in the form of $\exp(i(kx + \ell y - \sigma t))$ has three roots, namely,

$$\sigma^2 = \frac{gH_o(k^2 + \ell^2)}{1 + \frac{1}{3}H_o^2(k^2 + \ell^2)}, \tag{12.89}$$

$$\sigma = 0. \tag{12.90}$$

It is instructive to compare (12.89) with the well-known dispersion relation for general surface waves that have a height-varying structure, namely

$$\sigma^2 = g\kappa \tanh(H_o \kappa), \quad \kappa = \sqrt{k^2 + \ell^2}. \tag{12.91}$$

A longwave approximation of (12.91) reduces to that of shallow-water waves

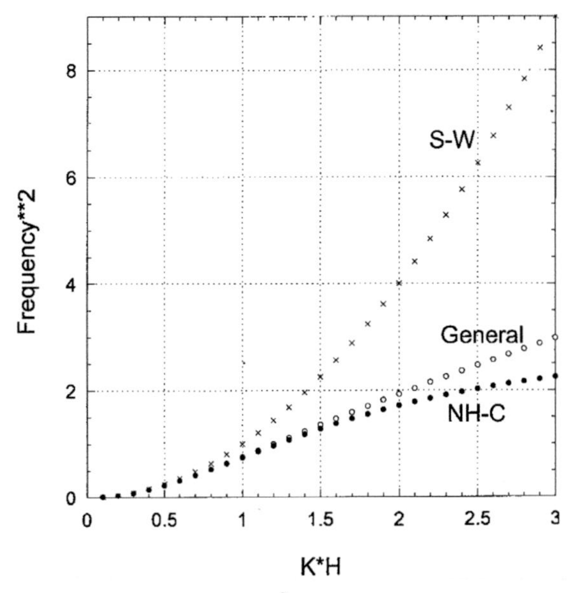

Fig. 12.14 Comparison of non-dimensional frequency-squared $H_o\sigma^2/g$ according to the dispersion relations of this non-hydrostatic columnar model (NH-C, closed circles), of general surface waves (General, open circles), and of a shallow-water model (S-W, crosses) as a function of non-dimensional wavenumber $H_o\kappa$ (taken from Mak, 2001).

$$\sigma^2 = gH_o\kappa^2. \tag{12.92}$$

Figure 12.14 shows the variation of non-dimensional frequency-squared $\sigma^2 H_o/g$ with κH_o in the three different models according to Eqs. (12.89), (12.91) and (12.92) indicated as closed circles, open circles, and crosses, respectively. All three curves approach one another at large wavelength limit, as they should. The shallow-water wave frequency departs rapidly from the general result when $\kappa H_o > 1$, highlighting the fact that the hydrostatic approximation breaks down when the wavelength approaches the layer depth. In contrast, the result obtained in this model is a better approximation of the general result for considerably shorter wavelengths. The range of validity of this model is for waves with κH_o up to about 3. Note that the perturbation pressure and velocity fields of a surface wave are known to decrease exponentially from the surface of the layer with an *e*-folding distance of κ^{-1}. Thus, it would be consistent to approximate such a vertical structure by a column provided that the layer depth is up to two to three units of κ^{-1}. There is a great advantage in doing so because the general three-dimensional problem would be reduced to a much simpler two-dimensional problem. A Hamiltonian formulation enables us to perform such an analysis precisely, self-consistently and economically.

From Fig. 12.14 we see that the non-hydrostatic process has an intrinsic effect of slowing down the propagation of gravity waves. Such an effect is significant when the total horizontal length scale is comparable to the depth of the layer. We will soon see that this effect has an intriguing implication for the instability character of a shear flow in this model. The seemingly spurious steady-state root (12.90) actually represents a degenerate class of vorticity waves in the absence of a basic shear flow.

12.3.2.2 Non-hydrostatic barotropic instability

Let us consider a basic shear flow in this model, $(u, v, h) = (U(y), 0, H(y))$, pertinent to an environment of NST. The basic flow can be an arbitrary non-divergent zonal flow. The non-uniform thickness is intended to be a proxy of a boundary outflow analogous to that introduced in a three-dimensional numerical model (Lee and Wilhelmson, 1997). In this model, the non-uniform depth has to be supported by an external force $F(y)$ in the y-direction equal to

$$F = g\frac{dH}{dy}. \tag{12.93}$$

In the atmosphere, the rapid formation of an NST manifests certain instability in a fairly complicated and relatively slowly evolving background state. The latter may be a much larger disturbance itself and is controlled by some extraneous processes. By ignoring all but one process, we are in a position to test a particular hypothesis for the NST genesis per se. The spatial variation of the basic layer depth is necessarily associated with those extraneous processes. As far as the y-momentum equation in the model is concerned, an "external force" $F(y)$ is to be attributed to those processes so that the pressure gradient force associated with $H(y)$ in the basic state could be balanced. It would be a non-conservative force in the context of a flow in this model. The influence of $F(y)$ on the instability enters entirely through $H(y)$.

Specifically, we consider a monotonic shear zone with a westerly to the south and an easterly to the north, namely,

$$U = -U_o \tanh\left(\frac{y}{L}\right) \tag{12.94}$$

where U_o and L are velocity and length scales, respectively. The layer thickness $H(y)$

$$H = H_o + A\tanh\left(\frac{y}{B}\right) \tag{12.95}$$

varies monotonically across $y = 0$, where B and H_o are two related length scales. Here A is a measure of the total variation of $H(y)$. The linearized forms of (12.74) and (12.86a,b) are then

$$\begin{aligned}
\frac{Du'}{Dt} + v'\frac{dU}{dy} &= -g\frac{\partial h'}{\partial x} - \frac{1}{3H}\frac{\partial}{\partial x}\left[H^2\frac{D}{Dt}\left(\frac{Dh'}{Dt} + v'\frac{dH}{dy}\right)\right],\\
\frac{Dv'}{Dt} &= -g\frac{\partial h'}{\partial y} - \frac{1}{3H}\frac{\partial}{\partial y}\left[H^2\frac{D}{Dt}\left(\frac{Dh'}{Dt} + v'\frac{dH}{dy}\right)\right],\\
\frac{Dh'}{Dt} + v'\frac{dH}{dy} &= -H\left(\frac{\partial u'}{\partial x} + \frac{\partial v'}{\partial y}\right),
\end{aligned} \tag{12.96a,b,c}$$

where $\frac{D}{Dt} = \frac{\partial}{\partial t} + U\frac{\partial}{\partial x}$. For a cyclical channel domain bounded by rigid walls at $y = \pm Y$, we may use the following boundary conditions

$$\text{at } y = \pm Y, \quad v' = 0. \tag{12.97}$$

For the basic state described by (12.94) and (12.95), the system is characterized by seven external parameters: g, Y, H_o, A, B, U_o, L. It follows that we can construct five independent non-dimensional parameters such as

$$\eta = \frac{gH_o}{U_o^2}, \quad \gamma = \frac{H_o}{L}, \quad b = \frac{B}{L}, \quad a = \frac{A}{H_o}, \quad \tilde{Y} = \frac{Y}{L}, \tag{12.98}$$

where η is the inverse of the Froude number. We will only consider a subcritical flow, $\eta > 1$. The dependence on $\tilde{Y}$ is negligible as long as the domain is sufficiently wide. Thus, we effectively need to examine the instability properties in a four-dimensional parameter space. The equations are non-dimensionalized by expressing (x, y), (U, u', v'), t and (h', H) in units of L, U_o, LU_o^{-1} and H_o, respectively. The non-dimensional form of the basic state is

$$\begin{aligned} U &= -\tanh(y), \\ H &= 1 + a\tanh\left(\frac{y}{b}\right). \end{aligned} \tag{12.99a,b}$$

The non-dimensional perturbation equations are simply

$$\begin{aligned} \frac{Du'}{Dt} + v'\frac{dU}{dy} &= -\eta\frac{\partial h'}{\partial x} - \frac{\gamma^2}{3H}\frac{\partial}{\partial x}\left[H^2\frac{D}{Dt}\left(\frac{Dh'}{Dt} + v'\frac{dH}{dy}\right)\right], \\ \frac{Dv'}{Dt} &= -\eta\frac{\partial h'}{\partial y} - \frac{\gamma^2}{3H}\frac{\partial}{\partial y}\left[H^2\frac{D}{Dt}\left(\frac{Dh'}{Dt} + v'\frac{dH}{dy}\right)\right], \\ \frac{Dh'}{Dt} + v'\frac{dH}{dy} &= -H\left(\frac{\partial u'}{\partial x} + \frac{\partial v'}{\partial y}\right). \end{aligned} \tag{12.100a,b,c}$$

Upon making use of (12.100c) in (12.100a,b) and rearranging the terms, one would be able to rewrite the equations systematically as

$$\begin{aligned} L_{11}\left\{\frac{\partial u'}{\partial t}\right\} + L_{12}\left\{\frac{\partial v'}{\partial t}\right\} &= L_{13}(u', v', h'), \\ L_{21}\left\{\frac{\partial u'}{\partial t}\right\} + L_{22}\left\{\frac{\partial v'}{\partial t}\right\} &= L_{23}(u', v', h'), \\ L_{31}\left\{\frac{\partial h'}{\partial t}\right\} &= L_{33}(u', v', h'), \end{aligned} \tag{12.101a,b,c}$$

where L_{ij} ($i = 1, 2, 3$; $j = 1, 2, 3$) are linear differential operators containing spatial derivatives, the parameters and the basic-state variables. The non-hydrostatic effect may be suppressed if we want by setting $\gamma = 0$ in (12.100a,b).

In this analysis, we focus on the modal instability of this system. Hence, we seek a normal mode solution in the form of $(u', v', h') = \left[\hat{u}, \hat{v}, \hat{h}\right]\exp[i(kx - \sigma t)]$. The amplitude functions are then governed by a set of three second-order ordinary differential equations. The y domain is depicted by N uniform grid points at which we define $\hat{v}_j$, $j = 1, 2, \ldots, N$. The boundary conditions (12.97) are then $\hat{v}_1 = 0$ and $\hat{v}_N = 0$. The grid points for $\hat{u}$ and $\hat{h}$ are staggered in between those of $\hat{v}$. It is straightforward to cast (12.100a,b,c) together with the boundary conditions as a matrix equation. The eigenvalues σ and the corresponding

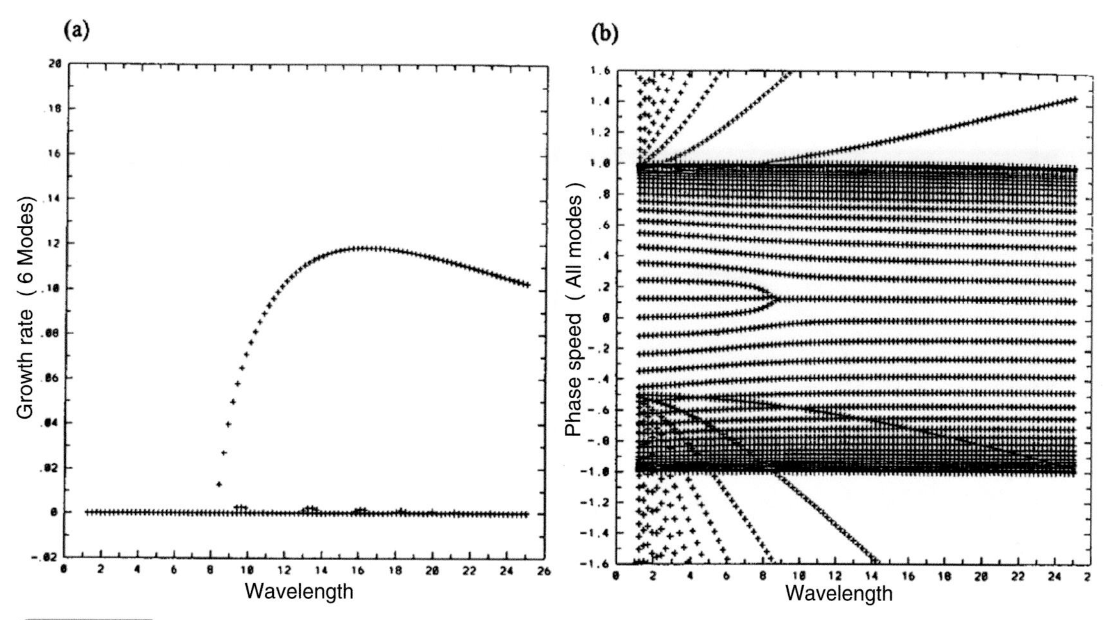

Fig. 12.15 Variation of (a) growth rate of six leading normal modes, and (b) phase speed of all normal modes as a function of zonal wavelength *under hydrostatic conditions* for the reference case. Units in (a) are $L^{-1}U_o = 3 \times 10^{-2}\,s^{-1}$ and in (b) $U_o = 15$ m s^{-1} (taken from Mak, 2001).

eigenvectors can then be readily evaluated. The set of eigenvalues for each parameter setting are sorted in a decreasing order of their imaginary part (i.e., growth rate). We then only need to examine the properties of a few leading eigenmodes.

Illustrative calculation

A relevant set of dimensional values for the parameters are $U_o = 15$ m s^{-1}, $L = 500$ m, $H_o = 2000$ m, $a = A/H_o = 0.25$, and $b = B/L = 1$. A domain width of $\tilde{Y} = 8$ would be sufficiently large. Preliminary testing of numerical convergence establishes that it would be sufficient to use $N = 129$ grid points for depicting an unstable disturbance. Considering the relatively small difference between the density of air in an NST and that of air above, we interpret g as a reduced gravity $g* = g\Delta\rho/\rho_o$ with a value $\Delta\rho/\rho_o \approx 0.01$. It amounts to considering internal gravity waves instead of external gravity waves as intrinsic modes of motion. The reference case is then characterized by $\eta = 3$, $\gamma = 4$, $a = 0.25$, and $b = 1$.

12.3.2.3 Impact of the non-hydrostatic process on the instability

We begin by ascertaining the dynamical impact of the non-hydrostatic process on the instability of a barotropic shear flow. The non-hydrostatic effect is first suppressed by using $\gamma = 0$. Figure 12.15a shows the variation of the imaginary part and hence growth rate of the six leading eigenmodes (σ_i) with wavelength. There is only one dominant branch of unstable modes for this monotonic shear flow. The maximum dimensional growth rate is

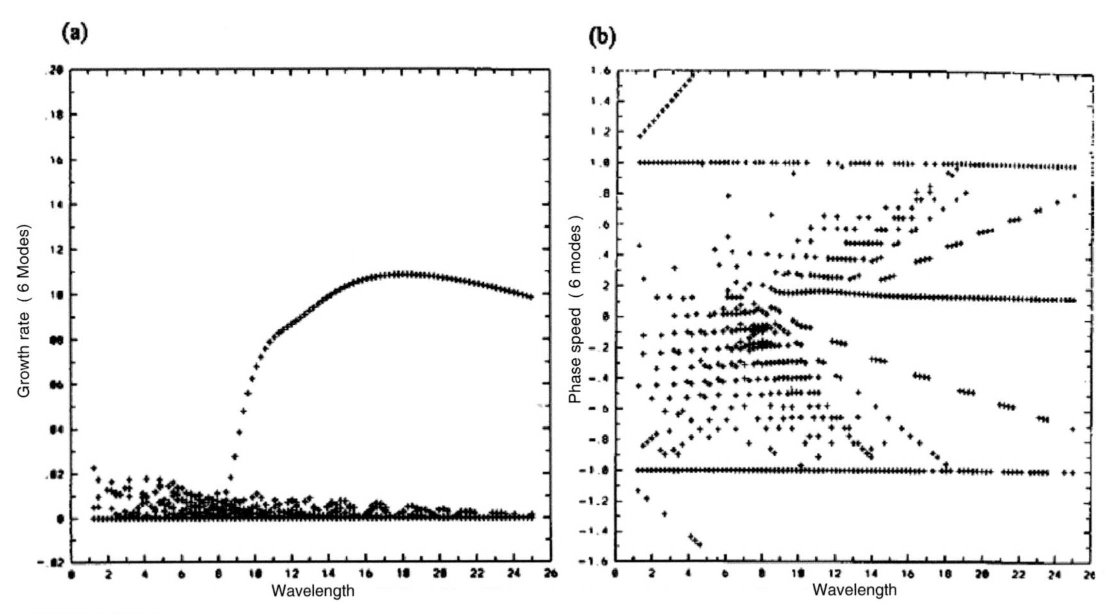

Fig. 12.16 Variation of (a) growth rate and (b) phase speed of six leading normal modes as a function of zonal wavelength *under non-hydrostatic conditions* for the reference case. Units in (a) are $L^{-1}U_o = 3 \times 10^{-2}\ \mathrm{s}^{-1}$ and in (b) $U_o = 15\ \mathrm{m\,s}^{-1}$ (taken from Mak, 2001).

about $0.36 \times 10^{-2}\ \mathrm{s}^{-1}$, corresponding to an *e*-folding time of 280 s. In addition, there are a few negligibly weak unstable modes that are however not a numeric artifact.

We plot in Fig. 12.15b the variation of the phase speed (σ_r/k) of all normal modes in the range of ± 1.6 with wavelength. It reveals three distinct sets of normal modes in the system: vorticity waves, eastward-propagating gravity waves and westward-propagating gravity waves. With only a few exceptions, the phase speeds of the gravity wave modes lie outside the range of those of vorticity waves. The dominant unstable modes have slow phase speeds, ~0.1. We have discussed in Part B Chapter 8 that shear instability may be interpreted as arising from mutual reinforcement of vorticity waves that are counter-propagating and phase-locked. Since the basic flow has a continuous shear, an ensemble of constituent vorticity waves is actually involved. Nevertheless, we may think of them in terms of two equivalent counter-propagating waves. The phase-locking between them enables the structure of the normal mode as a whole to remain unchanged in time while the amplitude intensifies exponentially. Their mutual reinforcement results from the effect of vorticity advection upon one another in the presence of the basic flow. We symbolically refer to such an unstable mode as a $R \leftrightarrow R$ mode.

The counterpart growth rate result under non-hydrostatic conditions is presented in Fig. 12.16a. A comparison between Figs. 12.15a and 12.16a reveals that the growth rate of the dominant branch of unstable modes is slightly reduced, especially for the longer waves (~10 percent). The maximum growth rate still has a corresponding *e*-folding time of 300 s. Such an amplification rate is compatible with that of NST. The reduction of growth rate can be understood from the perspective of energetics, which will be elaborated in

Section 12.3.2.4. Figure 12.16a has a subtle interesting feature. The dominant growth rate curve appears to have a transition from one distinct part to another, merging together at D ~12. To put this in perspective, we note that shear instability arising from mutual reinforcement between two different types of waves is also possible. For example, large-scale shear instability may arise from positive reinforcement between interacting Kelvin and Rossby waves (De la Cruz-Hereida and Moore, 1999). The participating waves in this model are gravity and vorticity waves. They usually hardly interact because gravity waves propagate much faster than vorticity waves. However, the non-hydrostatic effect can make a difference because it greatly slows down the short gravity waves (see Section 12.3.2.1). In the range of wavelengths from about $D = 8$ to about $D = 12$, a sufficiently slow-moving gravity wave and a slow-moving vorticity wave can be phase-locked whereby they can continually extract kinetic energy from the basic flow. Indeed, Fig. 12.16b shows that the short-wavelength part of the dominant branch of unstable modes has phase speed values indistinguishable between those of the vorticity waves and gravity waves. Hence, it would be justifiable to refer to these *gravity–vorticity hybrid modes* symbolically as $G_+ \leftrightarrow R$ modes. In contrast, the long-wavelength part of the dominant branch of unstable modes ($>D = 12$) is made up of entirely slow-moving vorticity waves, namely $R \leftrightarrow R$ modes. They may be referred to as *vorticity modes*.

It is noteworthy that there are many more weakly unstable modes in Fig. 12.16a than in Fig. 12.15a. The reason for this is that under the influence of the non-hydrostatic effect, many more gravity waves are slowed down to phase speeds comparable to those of some vorticity waves. They can therefore resonantly interact. They are located close enough to the strong shear zone that such resonant interaction can tap into the energy of the basic shear for weak amplification.

We next compare the instability properties of a sample *vorticity mode* ($D = 18$) with a sample *gravity–vorticity hybrid mode* ($D = 11$). Their dimensional wavelengths are 9 and 5.5 km, respectively, with corresponding horizontal length scales of 2.2 and 1.4 km. Recall that they have a vertical scale of $H_o = 2$ km. Therefore, this *gravity–vorticity hybrid mode* is within the range of validity of this model, especially considering the fact that its meridional wavelength is shorter than its zonal wavelength. Their non-dimensional growth rates are 0.109 and 0.080 in units of $U_oL^{-1} = 3 \times 10^{-2}$ s^{-1}; whereas their phase speeds are 0.135 and 0.166, respectively, in units of $U_o = 15$ m s^{-1}. Their dimensional e-folding times are then 5.0 and 6.6 min, respectively. Both phase speeds are only about 2 m s^{-1}.

Their structures are however expected to be qualitatively different. Their perturbation zonal velocity u', meridional velocity v', height h' and vertical velocity at the free surface $w' = -H(u'_x + v'_y)$ are shown in Figs. 12.17 and 12.18.

These fields are normalized so that $u'_{max} = 1$. The northern and southern halves of the longer unstable wave (Fig. 12.17, $D = 18$) are essentially similar, confirming that it is indeed a $R \leftrightarrow R$ mode. The northern and southern halves of the shorter unstable mode (Fig. 12.18, $D = 11$) are qualitatively different. While the southern half of the structure is similar to that of the $D = 18$ mode, the northern half of the structure has a distinctly greater SE–NW tilt in each field. The h' field of the gravity–vorticity hybrid mode is almost three times stronger than that of the vorticity mode (0.55 vs. 0.2). The w' field of the hybrid mode is four and a half times stronger than that of the vorticity mode

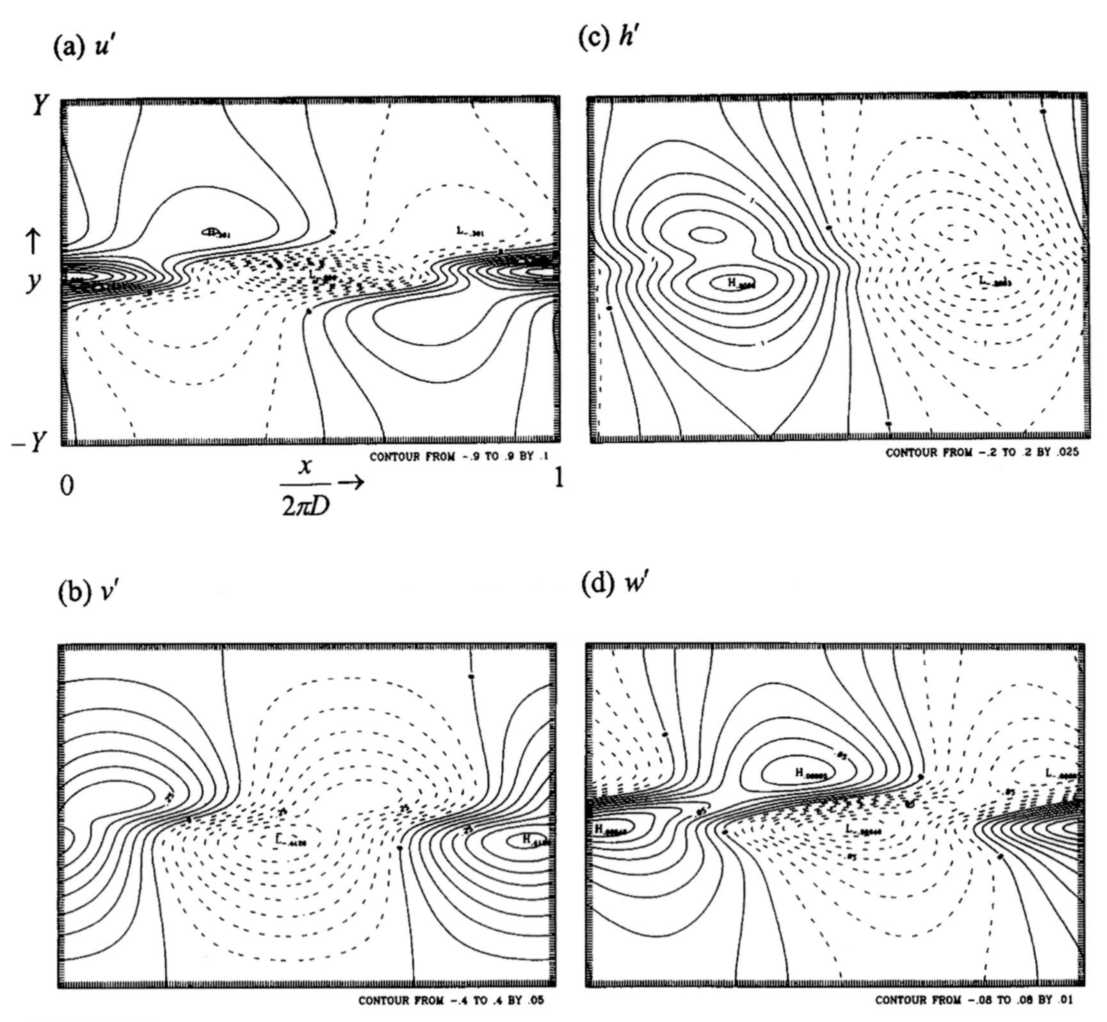

Fig. 12.17 Structure of a sample unstable *vorticity mode* with wavelength $D = 18$ for the reference case: (a) perturbation zonal velocity u', (b) perturbation meridional velocity, (c) perturbation height and (d) perturbation vertical velocity. Normalized to have $u'_{\text{max}} = 1$ (taken from Mak, 2001).

(0.36 vs. 0.08). The ratio of w'_{max} to u'_{max} of the hybrid mode is 0.36 : 1.0, but they are non-dimensionalized differently by a factor of $H_o/L = 4$. The ratio of the dimensional w'_{max} to u'_{max} is therefore 1.4 : 1.0. This feature is quite compatible with observations of NST and is a notable improvement over the result of a counterpart hydrostatic barotropic unstable mode.

The structure has some deficiency in that the maximum vertical velocity is not closely collocated with the maximum vorticity. Such deficiency suggests that additional destabilizing processes such as a weak vertical shear and a weak convective process may bring about a closer collocation between the vorticity and vertical velocity fields. Nevertheless, the overall instability properties of an unstable gravity–vorticity hybrid mode in this simple model setting are reminiscent of those of an NST.

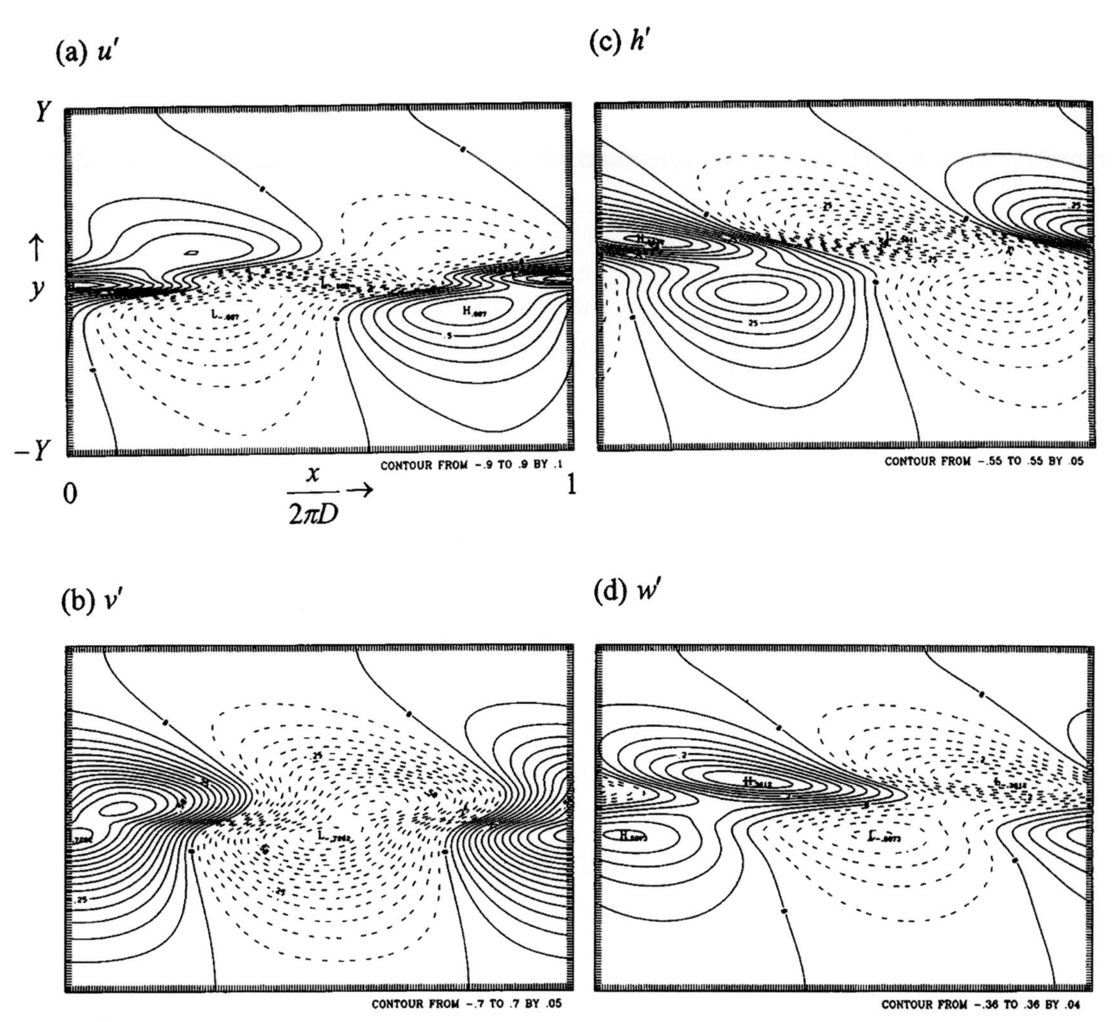

Fig. 12.18 Structure of a sample unstable *gravity–vorticity hybrid mode* with wavelength $D = 11$ for the reference case: (a) perturbation zonal velocity u', (b) perturbation meridional velocity, (c) perturbation height and (d) perturbation vertical velocity. Normalized to have $u'_{\max} = 1$ (taken from Mak, 2001).

12.3.2.4 Energetics and vorticity diagnoses

It is straightforward to derive from (100a,b,c) the following non-dimensional energy equations for any disturbance in this system.

$$\frac{d}{dt}\left\langle \frac{H}{2}\left(u'^2 + v'^2\right) + \frac{\gamma^2 H}{6} w'^2 \right\rangle = \left\langle -H \frac{dU}{dy} u'v' \right\rangle - \eta \langle h'w' \rangle + \eta \left\langle v'h' \frac{dH}{dy} \right\rangle,$$
$$\frac{d}{dt}\left\langle \frac{\eta}{2} h'^2 \right\rangle = +\eta \langle h'w' \rangle - \eta \left\langle v'h' \frac{dH}{dy} \right\rangle. \qquad (12.102\text{a,b})$$

The angular bracket stands for integration over the domain. The wave kinetic energy is $K' = \langle H/2(u'^2 + v'^2) + (\gamma^2 H/6) w'^2 \rangle$ and the wave potential energy is $P' = \langle \eta h'^2/2 \rangle$. The term $\langle -H(dU/dy)u'\upsilon'\rangle \equiv C(K, K')$ stands for the conversion rate from the kinetic energy of the basic flow to the wave kinetic energy. In light of (12.102b), the remaining two terms on the RHS of (12.102a) should be interpreted as one single physical process, $\eta\langle \upsilon'h'(dH/dy)\rangle - \eta\langle h'w'\rangle \equiv C(P', K')$. It stands for the conversion from wave potential energy to wave kinetic energy. The kinetic energy of the basic flow is the sole source of energy that fuels the instability mechanism.

It is instructive to quantitatively examine the energetics of the two sample unstable modes under consideration. For the vorticity mode ($D = 18$, Fig. 12.17) we obtain $K' = 1012$, $P' = 280$, $C(K, K') = 285$, and $C(P', K') = -66$. The positive value of $C(K, K')$ stems from the SW–NE tilt in u' and υ' in the shear zone. The negative value of $C(P', K')$ mostly stems from the positive correlation between the h' and w' fields. Such a disturbance has ascending motion where the fluid layer is deeper. The conversion from K' to P' is an energetic manifestation of the stabilizing effect on the instability by the non-hydrostatic process. For the gravity–vorticity hybrid mode ($D = 11$, Fig. 12.18), we obtain $K' = 2295$, $P' = 1054$, $C(K, K') = 539$, and $C(P', K') = -188$. The ratio of P' to K' of this mode is almost twice as large as that of the vorticity mode. The ratio of $|C(P', K')|$ to $C(K, K')$ is also larger. The non-hydrostatic process and the meridional variation of $H(y)$ have the effect of fractionally converting more of the wave kinetic energy to gravitational potential form and hence reducing the growth rate.

The vorticity dynamics of an unstable mode may be examined with the potential vorticity equation obtained from (12.100), namely

$$\left(\frac{\partial}{\partial t} + U\frac{\partial}{\partial x}\right)\left(\frac{\gamma^2\zeta'_*}{H} + \frac{\zeta'}{H} - \frac{\bar{\zeta}h'}{H^2}\right) + v'\frac{d}{dy}\left(\frac{\bar{\zeta}}{H}\right) = 0, \tag{12.103}$$

where $\zeta' = \upsilon'_x - u'_y$, $\zeta'_* = (1/3)(dH/dy)(\partial w'/\partial x)$, $\bar{\zeta} = -dU/dy$. It can be rewritten

$$\left(\frac{\partial}{\partial t} + U\frac{\partial}{\partial x}\right)\xi' = -v'\bar{\zeta}_y - \left(u'_x + v'_y\right)\bar{\zeta}, \tag{12.104}$$

where $\xi' = \zeta' + (\gamma^2/3)H_y w'_x$, which can be regarded as an extended vorticity incorporating the influence of the non-hydrostatic process. Then the zonal average of the product of (12.103) and ξ' would be

$$\sigma_i|\hat{\xi}|^2 = -\bar{\zeta}_y\mathrm{Re}\left\{\hat{v}\hat{\xi}^*\right\} - \bar{\zeta}\mathrm{Re}\left\{\left(ik\hat{u} + \hat{v}_y\right)\hat{\xi}^*\right\}, \tag{12.105}$$

where σ_i is the growth rate. The hat superscript refers to complex amplitude and the asterisk refers to complex conjugate. Here $|\hat{\xi}|^2$ is then the zonal average extended enstrophy of the perturbation. Thus, *the non-hydrostatic mechanism affects the vorticity dynamics through its impact on the process of extended-vorticity flux and of stretching of the extended vorticity.* The meridional variation of the three terms in (12.105) for the sample unstable gravity–vorticity hybrid mode with $D = 11$ are shown in curves labeled A, B, and C in Fig. 12.19. Their values are confined near the center of the domain, $|y| \le 1$, where the basic shear is maximum. The intensification of the perturbation stems from the vorticity flux term B. The vorticity stretching term C has a substantial negative contribution, consistent with the finding that the non-hydrostatic process has a stabilizing effect.

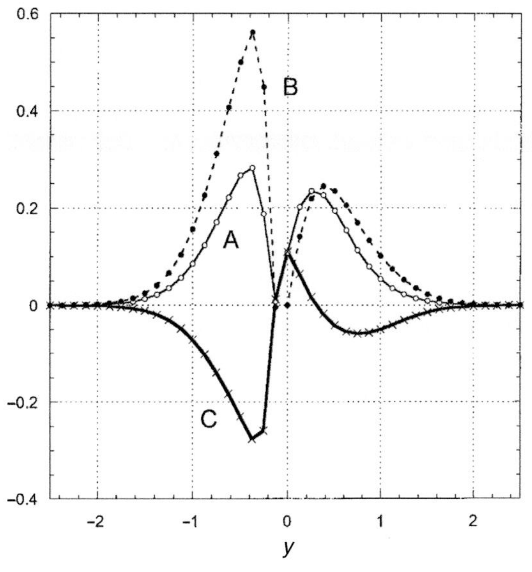

Fig. 12.19 Meridional structure of the three terms in the budget equation of extended enstrophy: (A) $\sigma_i \left|\hat{\xi}\right|^2$, (B) $-\bar{\zeta}_y$ $\text{Re}\left\{\hat{v}\hat{\xi}^*\right\}$ and (C) $-\bar{\zeta}\text{Re}\left\{(ik\hat{u}+\hat{v}_y)\hat{\xi}^*\right\}$ for the unstable *gravity–vorticity hybrid mode* with wavelength $D = 11$ (taken from Mak, 2001).

It appears that the baroclinic effect would be required to generate a positive contribution from the vortex stretching process. This result highlights the limitation of a barotropic model for non-supercell tornadogenesis.

The variations of the instability properties, to the model parameters (a, b, L_o, H_o, U_o) have been examined in Mak (2001). It is found that the values of the parameters in the reference case are close to the optimal condition for instability. Hence, the overall instability properties presented in Section 12.3.2.3 are quite robust. In summary, the analysis above provides supporting evidence of the hypothesis that the non-hydrostatic barotropic instability mechanism is applicable to NST genesis. It captures at least an essential aspect of the genesis. But there are caveats. The a priori columnar assumption excludes us from examining the impact of the vertically varying part of the non-hydrostatic pressure in NST. It however enables us to test a hypothesis and interpret the findings quite simply. The assumption of linear dynamics is obviously only applicable to the initial stage of development. The characteristic of transient growth of NST has yet to be ascertained. Furthermore, this model has left out a host of physical processes, such as surface friction, convergence in the basic flow, spatial variability in density and buoyancy. Convective processes may further favor the gravity–vorticity hybrid modes over the vorticity modes because the former has a stronger vertical velocity field.

Appendix: Mathematical tools

This appendix serves as a quick reference for those readers who might not remember some of the mathematical tools used in this book. This appendix is meant to make the text as self-contained as possible, but not to obviate the need to consult with standard math texts for a deeper appreciation of those tools.

Preamble

An analysis of a light-hearted but non-trivial problem serves to illustrate the essence of calculus and the necessity of using a numerical approach. Suppose there is a postman in a city park walking eastward at a speed U and there is a dog 70° to his left at a distance D away from him at $t = 0$. The dog starts chasing the postman at a speed W always aiming directly at him. (i) Determine and plot the trajectories of the dog and the postman for a time $t = 6$ s with $U = 1.6\,\mathrm{m\,s^{-1}}$, $W = 3.2\,\mathrm{m\,s^{-1}}$, $D = 20$ m. (ii) What is the distance between the dog and the postman at $t = 6$ s?

Let us refer to the positions of the dog and the postman as (x_1, y_1) and (x_2, y_2) in Cartesian coordinates respectively, where the x-axis points eastward. The instantaneous line of sight between the dog and the postman is a line at an angle ϕ measured from the x-axis satisfying $\tan\phi = \dfrac{y_1 - y_2}{x_1 - x_2}$. The movements of the dog and the postman are described by the following differential equations.

$$\frac{dx_2}{dt} = U, \quad \frac{dy_2}{dt} = 0, \quad \frac{dx_1}{dt} = \pm W\cos\phi, \quad \frac{dy_1}{dt} = \pm W\sin\phi.$$

The information about the dog always aiming at the postman is incorporated through the sign on the RHS of the last two equations. From a geometric consideration, the plus (minus) sign should be used whenever the dog is anywhere west (east) of the postman ($x_1 < x_2$ vs. $x_1 > x_2$).

The initial condition of this problem is at $t = 0$, $(x_1, y_1) = (D\cos\beta, D\sin\beta)$ with $\beta = 7\pi/18$ and $(x_2, y_2) = (0, 0)$. To begin with, we integrate the first two equations to get $(x_2, y_2) = (Ut, 0)$ for the trajectory of the postman. The two equations that describe the movement of the dog starting from a general location are coupled through a variable angle ϕ, which is a highly nonlinear function of the dog and postman's coordinates. This nonlinearity makes it difficult, if possible, to get an analytic solution. However, a numerical approach would readily enable us to solve this problem. It suffices to approximate a derivative by the forward-difference form, namely $\frac{df}{dt} \approx \frac{f(t+\delta t)-f(t)}{\delta t}$ for a sufficiently small δt. The time after n time steps is $t = n\delta t$. Denoting the coordinates of the dog and the postman at that time as $x_1^{(n)}, y_1^{(n)}, x_2^{(n)} = Un\delta t, y_2^{(n)} = 0$, we can write the finite-difference form of the remaining three governing equations as:

$$x_1^{(n+1)} = x_1^{(n)} \pm W\delta t \cos\phi^{(n)}, \quad y_1^{(n+1)} = y_1^{(n)} \pm W\delta t \sin\phi^{(n)}, \quad \tan\phi^{(n)} = \frac{y_1^{(n)}}{x_1^{(n)} - Un\delta t}.$$

With the initial condition, we can calculate $(x_1^{(n)}, y_1^{(n)}, \phi^{(n)})$ for $n = 1,2,3,\ldots$ At each time step, it is imperative to ascertain whether the dog is located to the west or east of the postman so that a proper sign can be assigned in the two equations above. This is equivalent to the instantaneous decision made by the dog during the chase. The calculation is done using 50 time steps to represent 6 seconds. The distance between the dog and the postman at the nth time step can be calculated as

$$s^{(n)} = \sqrt{\left(x_2^{(n)} - x_1^{(n)}\right)^2 + \left(y_2^{(n)} - y_1^{(n)}\right)^2}.$$

The answers to the questions are:

(i) The trajectories of the postman and the dog are shown as circles and crosses respectively at equal time intervals for six seconds in Fig. A.1. Note that the dog is east of the postman in the first part of the chase, and is west of the postman in the latter part of the chase. Although the dog's trajectory is a simple smooth curve, I wonder if there exists an analytic expression for it.

(ii) The calculation reveals that the distance between the dog and the postman is not a linear function of time. Figure A.2 indicates that the dog is 1.61 m away from the postman after six seconds. It would take about another second for the dog to catch up with the poor postman.

A.1 Vector analysis

Atmospheric properties are either scalar or vectors, such as temperature and wind respectively. A vector is denoted by $\vec{A} \equiv A\vec{a}$ with A being a measure of its magnitude and $\vec{a}$ an indication of its direction. The projection of a vector $\vec{A}$ onto the three axes of a coordinate system are called its components A_1, A_2 and A_3. In the case of Cartesian coordinates, a vector is $\vec{A} = A_1\vec{i} + A_2\vec{j} + A_3\vec{k}$ where $(\vec{i}, \vec{j}, \vec{k})$ are unit vectors. The following algebraic rules are for vector operations.

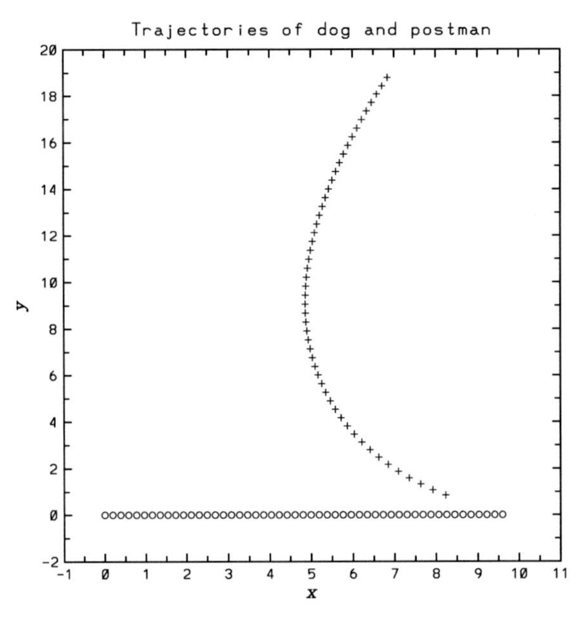

Fig. A.1 Trajectories of the dog (crosses) and postman (circles) in six seconds.

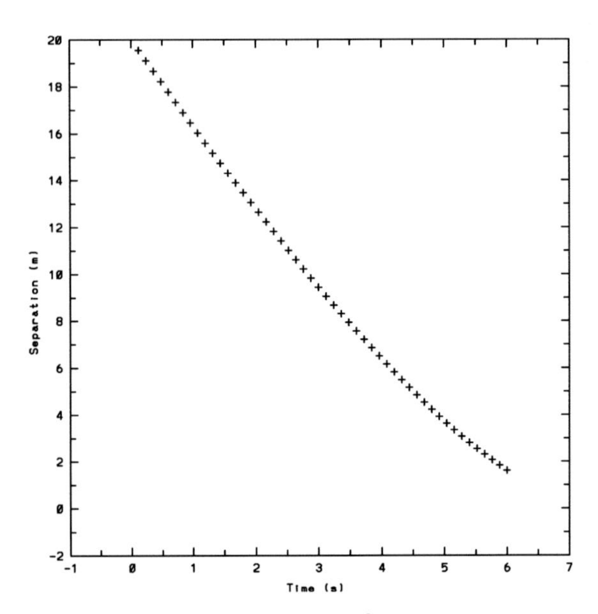

Fig. A.2 The distance between the dog and the postman as a function of time.

Addition/subtraction of two vectors

The sum of two vectors is a vector whose components are equal to the sum of their corresponding components, viz.

$$\vec{C} = \vec{A} + \vec{B} = (A_1 + B_1)\vec{i} + (A_2 + B_2)\vec{j} + (A_3 + B_3)\vec{k}. \tag{A.1}$$

Scalar product of two vectors

Scalar product is one type of multiplication of two vectors that yields a scalar quantity. It is also called "dot product" as defined by

$$\vec{A} \cdot \vec{B} = (A_1 B_1) + (A_2 B_2) + (A_3 B_3). \tag{A.2}$$

It follows that the length of a vector is equal to the square root of the dot product of the vector with itself, $A = \sqrt{\vec{A} \cdot \vec{A}} = \sqrt{\sum_{j=1}^{3} A_j^2}$. The dot product is the same as the product of the length of $\vec{A}$ and the projection of $\vec{B}$ onto $\vec{A}$, viz.

$$\vec{A} \cdot \vec{B} = AB \cos \alpha \tag{A.3}$$

where α is the angle between the two vectors.

Cross product of two vectors

Cross product is another type of multiplication of two vectors that yields a new vector. It is equal to the determinant of a matrix defined as follows

$$\vec{C} = \vec{A} \times \vec{B} = \begin{vmatrix} \vec{i} & \vec{j} & \vec{k} \\ A_1 & A_2 & A_3 \\ B_1 & B_2 & B_3 \end{vmatrix} = (A_2 B_3 - A_3 B_2)\vec{i} + (B_1 A_3 - B_3 A_1)\vec{j} + (A_1 B_2 - A_2 B_1)\vec{k}. \tag{A.4}$$

There is a simple rule for deducing the direction of $\vec{C}$. If we point the fingers of our right hand in the direction of the first vector, $\vec{A}$, and turn them toward the second vector, $\vec{B}$, our thumb would point to the direction of $\vec{C}$. This is called the "right-hand-screw rule." It follows that $\vec{C}$ is perpendicular to both $\vec{A}$ and $\vec{B}$ as highlighted by this literal rule of thumb. Using (A.2) and (A.4), we can verify $\vec{A} \cdot \left(\vec{A} \times \vec{B}\right) = 0$ and $\vec{B} \cdot \left(\vec{A} \times \vec{B}\right) = 0$. The length of $\vec{C}$ is then

$$\left|\vec{C}\right| = AB \sin \alpha \tag{A.5}$$

where α is the angle between $\vec{A}$ and $\vec{B}$. Applying the rule of thumb to the three unit vectors $\left(\vec{i}, \vec{j}, \vec{k}\right)$ of a Cartesian coordinate system, we get

$$\vec{k} = \vec{i} \times \vec{j}, \qquad \vec{i} = \vec{j} \times \vec{k}, \qquad \vec{j} = \vec{k} \times \vec{i}. \tag{A.6}$$

A.2 Complex number and complex function

Although the measurable properties of atmospheric disturbances have real values, it is often convenient to work with functions representing them in terms of complex variables. It is understood that the real part of such a complex function is identifiable with the physically measurable property. A complex number is defined as

$$\begin{aligned} \xi &= a + ib = Ae^{i\theta} = A\cos\theta + iA\sin\theta, \\ i &= \sqrt{-1}, \quad A = \sqrt{a^2 + b^2}, \quad \theta = \tan^{-1}(b/a), \end{aligned} \tag{A.7}$$

where a and b are real numbers. If we visualize a complex number as a 2-D vector, A would be its magnitude and θ its phase angle. If a and b are functions of say x, y and z, $\xi(x, y, z)$ would have a complex value everywhere. It is nevertheless a scalar function, albeit with a complex value at each point. One motivation of using functions of complex variable is that the function $e^{i\theta}$ has simple mathematical properties.

A.3 Derivative and finite-difference

Calculus is an indispensable tool for analyzing the dynamical properties of atmospheric disturbances. The derivative of a function of one independent variable, $f(x)$, with respect to that variable at a point x_1 is defined as

$$\left(\frac{df}{dx}\right)_{x_1} = \lim_{|x_2 - x_1| \to 0} \left(\frac{f(x_2) - f(x_1)}{x_2 - x_1}\right). \tag{A.8}$$

If we deal with a function of two or more independent variables, say $f(x, y)$, then the derivative of f with respect to x at a point (x_1, y_1) is a partial derivative defined as

$$\left(\frac{\partial f}{\partial x}\right)_{x_1, y_1} = \lim_{|x_2 - x_1| \to 0} \left(\frac{f(x_2, y_1) - f(x_1, y_1)}{x_2 - x_1}\right). \tag{A.9}$$

The two quantities in the numerator are evaluated at the same $y = y_1$. Moreover, the derivative of the product of two functions is, for example,

$$\frac{d}{dx}\left(Ae^{i\theta}\right) = e^{i\theta}\frac{dA}{dx} + iAe^{i\theta}\frac{d\theta}{dx}, \tag{A.10}$$

where A and θ are functions of x. For convenience, we often use a subscript for denoting a derivative whether it be an ordinary or partial derivative, $\frac{df(x)}{dx} \equiv f_x$, $\frac{\partial f(x, y)}{\partial x} \equiv f_x$.

An atmospheric property has finite resolution and therefore has meaningful values only at a certain set of grid points in a domain. Suppose two adjacent grid points for a certain value of y are δx apart in the x-direction. A partial derivative of f with respect to x at a point (x_1, y_1) may be evaluated with a *center-difference approximation* of (A.9) as

$$\left(\frac{\partial T}{\partial x}\right)_{x_1 y_1} = \frac{T(x_1 + \delta x, y_1) - T(x_1 - \delta x, y_1)}{2\delta x}. \tag{A.11}$$

A.4 Taylor series expansion

A function $f(x, y)$ in the neighborhood of a point (x_o, y_o) can be expanded into an infinite series as follows

$$\begin{aligned} f(x,y) &= f(x_o, y_o) + (x - x_o) f_x + (y - y_o) f_y \\ &+ \frac{(x - x_o)^2}{2} f_{xx} + \frac{(y - y_o)^2}{2} f_{yy} + \frac{(x - x_o)(y - y_o)}{2} f_{xy} + \cdots \end{aligned} \tag{A.12}$$

All derivatives in (A.12) are to be evaluated at (x_o, y_o). The dots refer to the terms containing higher-order derivatives. The sum of the first three terms would give a good approximate value of $f(x, y)$ in a small neighborhood. Such a truncated series is called a Taylor series expansion. In some problems, we need to incorporate the additional three terms for higher accuracy.

A.5 Del operator: gradient, Laplacian, divergence, curl

The del operator is a vector operator that performs partial derivatives with respect to the independent variables. In Cartesian coordinates, it is defined as

$$\nabla = \vec{i}\frac{\partial}{\partial x} + \vec{j}\frac{\partial}{\partial y} + \vec{k}\frac{\partial}{\partial z}. \tag{A.13}$$

Applying the del operator to a scalar function $f(x, y, z)$ would yield a vector field. It is called the *gradient* of the function

$$\nabla f = \vec{i}\frac{\partial f}{\partial x} + \vec{j}\frac{\partial f}{\partial y} + \vec{k}\frac{\partial f}{\partial z}. \tag{A.14}$$

The del may be applied to a vector field either as a dot product operation or a cross product operation. For example, applying the del as a dot product operator with the gradient of a scalar function would yield

$$\nabla \cdot (\nabla f) = \nabla^2 f = f_{xx} + f_{yy} + f_{zz}. \tag{A.15}$$

This quantity is called the *Laplacian* of the scalar function. Applying the del to a general vector field as a dot product operation would yield

$$\nabla \cdot \vec{A} = A_x + A_y + A_z. \tag{A.16}$$

This quantity is called the *divergence* of a vector field.

When we apply the del to a vector field as a cross product operation, we would get another vector field. The resulting quantity is called the "*curl* of a vector."

$$\nabla \times \vec{A} = \begin{vmatrix} \vec{i} & \vec{j} & \vec{k} \\ \frac{\partial}{\partial x} & \frac{\partial}{\partial y} & \frac{\partial}{\partial z} \\ A_1 & A_2 & A_3 \end{vmatrix} = \vec{i}\left(A_{3y} - A_{2z}\right) + \vec{j}(A_{1z} - A_{3x}) + \vec{k}\left(A_{2x} - A_{1y}\right). \tag{A.17}$$

Note that we have

$$\nabla \cdot \left(\nabla \times \vec{A}\right) = \left(A_{3y} - A_{2z}\right)_x + (A_{1z} - A_{3x})_y + \left(A_{2x} - A_{1y}\right)_z = 0 \text{ and}$$

$$\vec{A} \cdot (\nabla \times \vec{A}) = 0 \text{ for any vector } \vec{A}.$$

A.6 Line, surface and volume integrals

The counterpart of a derivative of a vector function $\vec{B}(x, y, z)$ is an integral of it from one end of an open curve in space Γ to the other end. It is called a line integral defined as

$$\int_{\Gamma} \vec{B} \cdot d\vec{\ell}, \tag{A.18}$$

where $d\vec{\ell}$ at a point under consideration is an infinitesimally short vector element in the direction tangential to the curve Γ. This integral is then the sum of the product of the tangential component of $\vec{B}$ along the curve and $|d\vec{\ell}\,|$. When Γ is a closed loop, (A.18) would be written as $\oint_{\Gamma} \vec{B} \cdot d\vec{\ell}$ with the convention that the integral is taken in the anti-clockwise direction along Γ.

The counterpart of a second derivative of a function $f(x, y)$ is a surface integral of it over an area A

$$\oiint_{A} f(x, y) dx\, dy. \tag{A.19}$$

A volume integral is an integral of a function over a volume V, viz.

$$\iiint_{V} f(x, y, z) dx\, dy\, dz. \tag{A.20}$$

A.7 Stokes' theorem and Gauss' theorem

Here we summarize the interrelationships among the line, surface and volume integrals of certain properties of a vector field. A line integral of any vector field $\vec{B}$ tangential to a closed curve Γ in the anticlockwise direction is equal to the surface integral of the normal component of the curl of $\vec{B}$ over the surface A enclosed by Γ. It is known as *Stokes' theorem* and is written as

$$\oint_{\Gamma} \vec{B} \cdot d\vec{\ell} = \oiint_{A} (\nabla \times \vec{B}) \cdot \vec{n}\, da, \tag{A.21}$$

where $\vec{n}$ is a unit vector outward normal to each elemental area da of the surface.

An integral of the divergence of any vector field $\vec{F}$ over a volume V is equal to the surface integral of the outward normal component of the vector itself $(\vec{F} \cdot \vec{n})$ over the surface S enclosing the volume under consideration; $\vec{n}$ is the outward normal unit vector and ds is the elemental area. It is known as *Gauss' theorem* and is written as

$$\oiiint_{V} (\nabla \cdot \vec{F})dV = \oiint_{S} \vec{F} \cdot \vec{n}\, ds. \tag{A.22}$$

It is also known as the *divergence theorem.*

A.8 Method of separation of variables

The formulation of some problems in atmospheric sciences leads to linear partial differential equations (PDE) with constant coefficients. We can solve them with the method of separation of variables. We illustrate it with a second-order PDE such as

$$\frac{\partial^2 f}{\partial x^2} - 4\frac{\partial^2 f}{\partial y^2} = 0 \tag{A.23}$$

for a certain domain with appropriate boundary conditions.

Case (A.1)

In the case of a closed domain, $0 \le x \le 1$ and $0 \le y \le 1$, the problem would be well posed if we impose boundary conditions for the unknown such as $f = 0$ along all boundaries. We may seek a solution in the form of $f = X(x)Y(y)$. Substituting it into the differential equation, we get

$$\frac{X''}{X} - 4\frac{Y''}{Y} = 0. \tag{A.24}$$

It follows that $X''/X = 4Y''/Y =$ constant, which may be positive or negative. Not knowing any better at the moment, let us try $Y''/Y = c^2$, hence $X''/X = 4c^2$, where c is real. Then the solution is $f = (Ae^{cy} + Be^{-cy})(De^{2cx} + Ee^{-2cx})$. Now we try to impose the boundary conditions.

$$f(0,y) = 0 = (Ae^{cy} + Be^{-cy})(D+E) \to E = -D,$$

$$f(1,y) = 0 = (Ae^{cy} + Be^{-cy})\left(e^{2c} - e^{-2c}\right)D.$$

This condition cannot possibly be satisfied unless $D = -E = 0$ leading to a trivial solution. Therefore we must alternatively set $Y''/Y = -b^2$ and $X''/X = -4b^2$ with b being real. The corresponding solutions are $Y = G_1 \cos by + G_2 \sin by$ and $X = F_1 \cos 2bx + F_2 \sin 2bx$. Then the general solution is $f = F_1G_1 \cos 2bx \cos by + F_1G_2 \cos 2bx \sin by + F_2G_1 \sin 2bx \cos by + F_2G_2 \sin 2bx \sin by$, where F_1, F_2, G_1, G_2 are four constants of integration and b is a constant.
Now we impose the four boundary conditions:

$$\begin{aligned}
0 &= f(0,y) = F_1G_1 \cos by + F_1G_2 \sin by, \\
0 &= f(1,y) = F_1G_1 \cos 2b \cos by + F_1G_2 \cos 2b \sin by + F_2G_1 \sin 2b \cos by \\
&\quad + F_2G_2 \sin 2b \sin by, \\
0 &= f(x,0) = F_1G_1 \cos 2bx + F_2G_1 \sin 2bx, \\
0 &= f(x,1) = F_1G_1 \cos 2bx \cos b + F_1G_2 \cos 2bx \sin b + F_2G_1 \sin 2bx \cos b \\
&\quad + F_2G_2 \sin 2bx \sin b.
\end{aligned}$$

The first and third conditions could be satisfied by seting $F_1 = G_1 = 0$,
The second and fourth conditions would become

$$0 = f(1,y) = F_2G_2 \sin 2b \sin by,$$

$$0 = f(x,1) = F_2G_2 \sin 2bx \sin b.$$

They would be satisfied only if $b = n\pi$, $n = 1, 2, 3$. Then the solution is

$$f(x,y) = A \sin(2n\pi x) \sin(n\pi y), \quad n = 1, 2, 3. \tag{A.25}$$

Here F_2 and G_2 appear as a single product denoted by A, which may have an arbitrary value. We conclude by saying that the solution is a set of periodically varying functions in both x- and y-directions with discrete wavelengths. The members of the set are indicated by different values of n. If y represents time, then (A.25) stands for 1-D standing oscillations as exemplified by a string tied at two ends at $x = 0, 1$.

Case (A.2)

In the case of an unbounded domain, $-\infty \le x \le \infty$ and $-\infty \le y \le \infty$, we only require the solution to be finite everywhere so that it could represent a physical property.

This is a much weaker constraint. We may again seek solutions of the form $f = X(x)Y(y)$ so that $X''/X = 4Y''/Y =$ constant. Suppose we set $Y''/Y = c^2, X''/X = 4c^2$. Then the solution would be $f = (Ae^{cy} + Be^{-cy})(De^{2cx} + Ee^{-2cx})$. We may immediately conclude that such a solution would be incompatible with the requirement that the solution is finite. So we again set $Y''/Y = -b^2$ and $X''/X = -4b^2$. The elementary solution can be written in the following form,

$$f = \mathrm{Re}\left\{Ae^{i(b(2x+y))}\right\} = A\cos(b(2x+y)),$$
$$\text{or } f = \mathrm{Re}\left\{Be^{i(b(-2x+y))}\right\} = B\cos(b(2x-y)), \tag{A.26}$$

where $i = \sqrt{-1}$, and A, B, b are real constants. Re{} stands for taking the real part of a complex quantity inside the bracket. Since the requirement is that the solution has to be finite, any values of A, B and b are acceptable. For $A = B = 1$, the solution is $f = \cos(b(2x+y))$. It is equal to 1 along a set of lines $(b(2x+y)) = n\pi$, $n = 0, 2, 4, \ldots$ These lines are oriented in the SE–NW direction. The solution is equal to -1 along another set of lines $(b(2x+y)) = m\pi$, $m = 1, 3, 5, \ldots$ It therefore has a wavy structure. The x-wavelength is $2\pi/2b$ and the y-wavelength is $2\pi/b$. Similarly, there is another independent wave-like solution, $f = \cos(b(2x-y))$. The contours are oriented in the SW-NE direction. If y represents time, then (A.26) would stand for 1-D waves propagating in the x-direction. We should be aware that $f = \sin(b(2x+y))$ and $f = \sin(b(2x-y))$ are also solutions, but they are not independent of the two discussed above.

A.9 Matrix, eigenvalue, eigenvector, normal modes and non-normal modes

A matrix is an ordered set of numbers arranged in the form of a two-dimensional array with m rows and n columns. It is called a $m \times n$ matrix and is denoted by $\underline{A}$. We only mention the few rules actually applied in some discussions/analyses in this book. A fundamental property of a matrix is its *determinant*. It is a number obtained as the sum of all possible products in each of which there appears one and only one element from each row and each column. Each such product is assigned a plus or minus sign according to the so-called *Cramer's rule*. We have already invoked the notion of determinant of $\underline{A}$ in Section A.1. When a $m \times n$ matrix $\underline{A}$ multiplies a $n \times m$ matrix $\underline{B}$, the result is a new $m \times m$ matrix $\underline{C}$. We say that $\underline{A}$ transforms $\underline{B}$ to $\underline{C}$.

$$\begin{aligned}\underline{A}\cdot\underline{B} &= \begin{pmatrix} A_{11} & A_{12} \\ A_{21} & A_{22} \\ A_{31} & A_{32} \end{pmatrix}\cdot\begin{pmatrix} B_{11} & B_{12} & B_{13} \\ B_{21} & B_{22} & B_{23} \end{pmatrix} \\ &= \begin{pmatrix} A_{11}B_{11}+A_{12}B_{21} & A_{11}B_{12}+A_{12}B_{22} & A_{11}B_{13}+A_{12}B_{23} \\ A_{21}B_{11}+A_{22}B_{21} & A_{21}B_{12}+A_{22}B_{22} & A_{21}B_{13}+A_{22}B_{23} \\ A_{31}B_{11}+A_{32}B_{21} & A_{31}B_{12}+A_{32}B_{22} & A_{31}B_{13}+A_{32}B_{23} \end{pmatrix} \\ &\equiv \underline{C}.\end{aligned} \tag{A.27}$$

When $\underline{B}$ is a n-vector, $\underline{A}$ would simply transform it to a m-vector $\underline{C}$.

In atmospheric analyses, we typically deal with square matrices, $m = n$. For each matrix $\underline{A}$, there exists a special set of vectors $\vec{\xi}$ such that

$$\underline{A}\vec{\xi} = \lambda\vec{\xi}, \tag{A.28}$$

where λ is a constant. Equation (A.28) says that the vector $\vec{\xi}$ is transformed by a matrix operator $\underline{A}$ to another vector, which is parallel to itself. Such a special vector is called an *eigenvector* of the matrix $\underline{A}$. The proportionality constant λ is called an *eigenvalue*. There are at most m distinct values of λ. To every eigenvalue, there is an $\vec{\xi}$ as the corresponding eigenvector. The eigenvalues and the eigenvectors are intrinsic characteristics of a matrix operator. We deal with this class of problems in analyses of dynamical instability of atmospheric flows. Eigenvectors are also known as *normal modes* in the context of a dynamical analysis of a disturbance.

How we solve an eigenvalue–eigenvector problem is illustrated below with a simple 2×2 matrix, $\underline{A} = \begin{pmatrix} a_{11} & a_{12} \\ a_{21} & a_{22} \end{pmatrix}$. The eigenvector is denoted as $\vec{\xi} \equiv \begin{pmatrix} \xi_1 \\ \xi_2 \end{pmatrix}$ associated with an eigenvalue λ satisfying

$$\underline{A}\begin{pmatrix} \xi_1 \\ \xi_2 \end{pmatrix} = \lambda\begin{pmatrix} \xi_1 \\ \xi_2 \end{pmatrix}. \tag{A.29}$$

A non-trivial solution of $\vec{\xi}$ exists only if the determinant of the coefficient matrix, $(\underline{A} - \lambda\underline{I})$, vanishes, where $\underline{I} = \begin{pmatrix} 1 & 0 \\ 0 & 1 \end{pmatrix}$ is the *unit matrix*. According to Cramer's rule, the determinant of $(\underline{A} - \lambda\underline{I})$ for the 2×2 matrix under consideration is

$$(a_{11} - \lambda)(a_{22} - \lambda) - a_{21}a_{12} = 0. \tag{A.30}$$

The two roots of the quadratic equation (A.30) are therefore the eigenvalues. The magnitude of an eigenvector can be arbitrary, but its direction is unique since ξ_2 is related to ξ_1 by $\xi_2 = \xi_1 \frac{\lambda - a_{11}}{a_{12}}$. Therefore, when we determine an eigenvector, we may set $\xi_1 = 1$ without loss of generality. It follows that the eigenvectors corresponding to the two eigenvalues λ_1 and λ_2 are

$$\vec{e}_1 = \left(1, \frac{\lambda_1 - a_{11}}{a_{12}}\right), \quad \vec{e}_2 = \left(1, \frac{\lambda_2 - a_{11}}{a_{12}}\right). \tag{A.31}$$

Representation of a vector in terms of eigenvectors

Any given vector in a 2-D vector space may be represented as a linear combination of the two eigenvectors $\vec{e}_j$ of a certain matrix operator, e.g.

$$\vec{B} = c\vec{e}_1 + d\vec{e}_2; \tag{A.32}$$

$\vec{e}_1$ and $\vec{e}_2$ are called base vectors in this context. The constants c and d are called projection coefficients, which can be easily obtained when $\vec{e}_1$ and $\vec{e}_2$ are orthogonal and normalized, $(\vec{e}_i \cdot \vec{e}_j = 0,\ i \neq j;\ \vec{e}_i \cdot \vec{e}_j = 1,\ i = j)$. They are $c = \vec{e}_1 \cdot \vec{B}$ and $d = \vec{e}_2 \cdot \vec{B}$.

A matrix $\underline{A}$ is said to be a *normal matrix* if

$$\underline{A}\,\underline{A}^T = \underline{A}^T\underline{A}, \tag{A.33}$$

where $\underline{A}^T = \begin{pmatrix} a_{11} & a_{21} \\ a_{12} & a_{22} \end{pmatrix}$ is called the transpose of $\underline{A}$ when its elements are real. The condition (A.33) requires

$$a_{12}^2 = a_{21}^2,\ a_{12}a_{22} + a_{11}a_{21} = a_{11}a_{12} + a_{21}a_{22}. \tag{A.34}$$

That means $\underline{A}$ must be a real symmetric matrix. A significant property of a normal matrix is that its eigenvectors are orthogonal to one another. It is illustrated by the following example. For $\underline{A} = \begin{pmatrix} 1 & 3 \\ 3 & 2 \end{pmatrix}$, then the eigenvalues are $\lambda_1 = \frac{3}{2} + \sqrt{\frac{37}{4}}$ and $\lambda_2 = \frac{3}{2} - \sqrt{\frac{37}{4}}$. The corresponding eigenvectors are $\vec{e}_1 = \left(1, \frac{1}{3}\left(\frac{1}{2} + \sqrt{\frac{37}{4}}\right)\right)$ and $\vec{e}_2 = \left(1, \frac{1}{3}\left(\frac{1}{2} - \sqrt{\frac{37}{4}}\right)\right)$. It follows that $\vec{e}_1 \cdot \vec{e}_2 = 0$. In contrast, the eigenvectors of a *non-normal matrix* are not orthogonal to one another. The orthogonality of the eigenvectors has far-reaching implications in applications. If the elements of a matrix $\underline{A}$ are complex, condition (A.33) means that $\underline{A}$ is a Hermitian matrix. Then $\underline{A}^T = \begin{pmatrix} a_{11} & a_{21}^* \\ a_{12}^* & a_{22} \end{pmatrix}$ where a_{ij}^* is the complex conjugate of a_{ij}.

The properties mentioned above are general in the sense that there exist counterpart results for a square matrix of any dimension, $m > 2$. Indeed, it is even true when the matrix has infinite dimension. In that limit, a matrix operator is effectively a differential operator. The counterpart of a matrix equation above is a differential equation,

$$D\{f\} = \lambda f, \tag{A.35}$$

where D denotes a differential operator. After taking into account the appropriate boundary conditions, we can proceed to solve (A.35). The problem is to determine the eigenvalues and eigenfunctions of the differential operator.

A.10 Fourier transform and spectral representation of a field

A generalization of (A.32) is a representation of a function as a linear combination of a set of base functions. This is called a spectral representation of the function. It is based on the concept of Fourier transform. A Fourier transform exists for any integrable function, $f(x)$, (that is $\int_{-\infty}^{\infty} |f| dx$ has a finite value). For all practical purposes, atmospheric fields are integrable. A Fourier transform pair are

$$\begin{aligned} \hat{f}(k) &= \frac{1}{2\pi}\int_{-\infty}^{\infty} f(x)e^{-ikx}dx, \\ f(x) &= \int_{-\infty}^{\infty} \hat{f}(k)e^{ikx}dk, \end{aligned} \tag{A.36}$$

where k is the x-wavenumber. For example, for $f(x) = \exp(-x^2)$ we have $\hat{f}(k) = \frac{1}{2\sqrt{\pi}}\exp(-k^2/4)$. This method is also widely used in diagnostic studies of atmospheric data. It can be readily extended to a function of several dimensions. When we deal with $f(t)$ as a deterministic function, the Fourier transform is $\hat{f}(\sigma)$ where σ is frequency. When we deal with data containing noise as in a typical time series $f(t)$, it is a random function. In that case, it would be necessary to perform spectral smoothing of $\left|\hat{f}(\sigma)\right|^2$ in order to get a statistically robust distribution of the power in the signal (energy) as a function of frequency. The result is called the power spectrum of $f(t)$.

A.11 Poisson problem

A boundary value problem known as the Poisson problem is commonly encountered in atmospheric analysis. A generic form of it written in Cartesian coordinates for a rectangular domain, $-X \le x \le X$ and $-Y \le y \le Y$, is

$$\psi_{xx} + \psi_{yy} = F, \tag{A.37}$$

where $F(x, y)$ is a given function that represents certain known processes. When the model domain is an analog of a latitudinal belt, we impose a cyclical boundary condition in the zonal direction $\psi(-X) = \psi(X)$. If Ψ stands for the streamfunction of a flow, we would regard the lateral boundaries as rigid so that $\psi(\pm Y) = 0$ would be a reasonable lateral boundary condition. Using $(I+1)$ grids in the x-direction and $(J+2)$ grids in the y-direction to depict the domain including the boundaries points, we recast (A.37) as a set of inhomogeneous algebraic equations by approximating the derivatives in center-difference form. The boundary conditions can be incorporated in this set of equations. The set of resulting equations is a matrix equation

$$\underline{A}\vec{\psi} = \vec{F} \tag{A.38}$$

where $\underline{A}$ is a $N \times N$ matrix where $N = I \times J$ associated with the Laplacian operator; $\vec{F}$ is a N-vector and $\vec{\psi}$ is a N-vector containing all the unknowns at the interior points. The solution is then $\vec{\psi} = \underline{A}^{-1}\vec{F}$ and $\underline{A}^{-1}$ is called the inverse of $\underline{A}$ obtainable by a standard subroutine for matrix inversion. This is an example of forced response in a linear atmospheric system.

When a large number of grid points are required to adequately depict a domain, the matrix may be too large to deal with for evaluating its inverse by an available computer. In that case, we would take an intermediate step first. We would represent the x-dependence of F and Ψ by Fourier decomposition at each value of y as discussed in Section A.10. Each of the Fourier coefficients of Ψ is then separately determined for a corresponding coefficient of F. The dimension of the matrix to be inverted would be much smaller, $J \times J$. The final step would be to evaluate the inverse Fourier transform to get the unknowns $\psi_{i,j}$ in physical space.

A.12 Method of Green's function

The problem in Section A.11 can be solved by a different method. Let us write the Fourier transform with respect to x of (A.37) as

$$-k^2\hat{\Psi} + \frac{d^2\hat{\Psi}}{dy^2} = \hat{F}(y;k) \tag{A.39}$$

where k is a zonal wavenumber. For convenience, suppose the lateral domain is infinitely wide. So the lateral boundary conditions would be $\hat{\Psi}(\pm\infty;k) = 0$. We seek a solution for each wavenumber k in an integral form as

$$\hat{\Psi} = \int_{-\infty}^{\infty} G(y,y')\hat{F}(y')dy', \tag{A.40}$$

where $G(y,y')$ is known as *Green's function* satisfying

$$-k^2G + \frac{d^2G}{dy^2} = \delta(y'). \tag{A.41}$$

The function $\delta(y')$ is known as the *Dirac delta function* at $y = y'$. It is singular at $y = y'$ and has the property of $\int_{-\infty}^{\infty} \delta(y)dy = 1$. The general solution of (A.41) is

$$G(y,y') = Ae^{ky} + Be^{-ky}. \tag{A.42}$$

We impose the boundary conditions $G(\pm\infty) = 0$ so that

$$\begin{aligned} G &= G_1(y,y') = Ae^{ky} \quad \text{for } y \le y', \\ G &= G_2(y,y') = Be^{-ky} \quad \text{for } y \ge y'. \end{aligned} \tag{A.43}$$

The integration constants A and B can be determined on the basis of two matching conditions at $y = y'$. One condition is that the Green's function is continuous at $y = y'$, leading to

$$G_1 = G_2 \text{ at } y = y'. \tag{A.44}$$

The other condition is that it must be compatible with the property of the Dirac delta function. If we integrate (A.41) over an infinitesimal interval across $y = y'$, we would get

$$\frac{dG_2}{dy}\Big|_{y'} - \frac{dG_1}{dy}\Big|_{y'} = 1. \tag{A.45}$$

By applying the conditions (A.44) and (A.45), we then obtain $A = \frac{1}{2ke^{ky'}}$ and $B = \frac{1}{2k}e^{ky'}$. Hence the Green's function is

$$\begin{aligned} G_1(y,y') &= \frac{1}{2k}e^{k(y-y')} \quad \text{for } y \le y', \\ G_2(y,y') &= \frac{1}{2k}e^{-k(y-y')} \quad \text{for } y \ge y'. \end{aligned} \tag{A.46}$$

This method is used in the analysis of moist baroclinic instability discussed in Chapter Part 8D.

A.13 Predictor-corrector algorithm for integrating ordinary differential equations

A simple numerical method for solving an initial value problem is used in a number of applications in this book. It is one of the so-called *predictor-corrector methods*. Suppose the problem is to solve

$$\frac{d\xi}{dt} = f(\xi, t),$$
$$\text{at } t = 0, \qquad \xi(0) = A, \tag{A.47}$$

where f may be any function of ξ. We use the following notation: $\xi(t_n) \equiv \xi_n$ and $t_n = n\delta t$ where δt is the time step, $n = 0, 1, 2, 3 \ldots$ The initial condition is $\xi_o = A$. A forward-difference representation of (A.47) at the nth time step is

$$\xi_{n+1} - \xi_n = f(\xi_n, t_n)\delta t. \tag{A.48}$$

We first compute a provisional value of the unknown at t_{n+1} with the formula

$$\xi_{n+1}^{(1)} - \xi_n = f(\xi_n, t_n)\delta t. \tag{A.49}$$

Next we use the provisional value $\xi_{n+1}^{(1)}$ to get a second provisional value of the unknown at the next time step, viz.

$$\xi_{n+2}^{(2)} - \xi_{n+1}^{(1)} = f\left(\xi_{n+1}^{(1)}, t_{n+1}\right)\delta t. \tag{A.50}$$

The final value of the unknown at the nth time step would be calculated as

$$\xi_{n+1} = \frac{1}{2}\left(\xi_{n+2}^{(2)} + \xi_n\right). \tag{A.51}$$

One of the more accurate numerical methods for solving this type of problem is the fourth-order Runge–Kutta scheme. There is too much detail to review here. We apply it in Chapter 7 to solve the gradient wind adjustment problem.

A.14 Calculus of variations

The method of determining the functions that minimize or maximize a *functional* (a function of functions) in the form of a definite integral is called *calculus of variations*. It is used in an application of the Hamiltonian formulation of fluid mechanics discussed in Chapter 12. We illustrate this methodology for the case of a single function with one independent variable, $u(x)$. Let u' be the derivative of u with respect to x. The functional under consideration is $I(x, u, u')$

$$I = \int_a^b F(x, u, u')dx, \tag{A.52}$$

where F is a given function with $u(a) = A$, $u(b) = B$; a, b, A and B are given constants. Different functions $u(x)$ subject to the same constraints at the two ends of the interval would lead to different values of I. The task is to deduce the particular function that would yield either a maximum or a minimum value of I.

The method of calculus of variations enables us to derive a differential equation that the unknown function u must satisfy. Let us denote the set of test functions by $u(x) + \delta u$, all being equal to A at $x = a$ and B at $x = b$; δu is an arbitrary *variation of* $u(x)$ vanishing at $x = a$ and $x = b$. We write $\delta u \equiv \varepsilon\eta(x)$ without loss of generality; η can be therefore any arbitrary twice-differentiable function with the constraints $\eta(a) = 0$ and $\eta(b) = 0$; ε is a parameter which may differ from one function to another. Thus, the integral parametrically varies with ε. Its derivative with respect to ε is

$$\frac{dI}{d\varepsilon} = \int_a^b \left[\frac{\partial F(x, u, u')}{\partial u}\frac{\partial u}{\partial \varepsilon} + \frac{\partial F(x, u, u')}{\partial u'}\frac{\partial u'}{\partial \varepsilon}\right]dx. \tag{A.53}$$

The function that maximizes or minimizes I has $\varepsilon = 0$ satisfying the condition $dI/d\varepsilon = 0$. Such a variation of I is said to be stationary. Thus, we obtain

$$I'(0) = \int_a^b \left[\frac{\partial F}{\partial u}\eta + \frac{\partial F}{\partial u'}\eta'\right]dx = 0. \tag{A.54}$$

It should be emphasized that the partial derivatives $\partial F/\partial u$ and $\partial F/\partial u'$ are to be worked out treating x, u and u' as independent variables. Furthermore, integrating the second term in (A.54) by parts yields

$$\int_a^b \frac{\partial F}{\partial u'}\eta' dx = \left[\frac{\partial F}{\partial u'}\eta\right]_a^b - \int_a^b \frac{d}{dx}\left(\frac{\partial F}{\partial u'}\right)\eta\, dx = -\int_a^b \frac{d}{dx}\left(\frac{\partial F}{\partial u'}\right)\eta\, dx. \tag{A.55}$$

It follows from (A.54) and (A.55) that

$$\int_a^b \left[\frac{\partial F}{\partial u} - \frac{d}{dx}\left(\frac{\partial F}{\partial u'}\right)\right]\eta\, dx = 0. \tag{A.56}$$

Since $\eta(x)$ is an arbitrary function, (A.56) can be satisfied only if the coefficient of η vanishes, viz.

$$\frac{d}{dx}\left(\frac{\partial F}{\partial u'}\right) - \frac{\partial F}{\partial u} = 0. \tag{A.57}$$

This is the differential equation that u must satisfy. It is known as the *Euler–Lagrange equation*. Because of the dependence of u and u' on x, we may further write

$$\frac{d}{dx}\left(\frac{\partial F}{\partial u'}\right) = \frac{\partial}{\partial x}\left(\frac{\partial F}{\partial u'}\right) + \left[\frac{\partial}{\partial u}\left(\frac{\partial F}{\partial u'}\right)\right]\frac{du}{dx} + \left[\frac{\partial}{\partial u'}\left(\frac{\partial F}{\partial u'}\right)\right]\frac{du'}{dx}. \tag{A.58}$$

Hence, an explicit form of (A.58) is

$$F_{u'u'}\frac{d^2u}{dx^2} + F_{uu'}\frac{du}{dx} + (F_{xu'} - F_u) = 0. \tag{A.59}$$

So the required $u(x)$ is governed by a second-order ordinary differential equation (A.59) subject to boundary conditions $u(a) = A$ and $u(b) = B$.

The algorithm above can be readily extended to cases where the integral is a functional depending on several functions of multiple independent variables, e.g. $I(x, y, u, v, u_x, u_y, v_x, v_y)$. By considering a variation of u and a variation of v separately, one would deduce two coupled governing equations for u and v in this case.

A.15 Triad interactions

A major type of nonlinear dynamics of a flow stems from its advective process. When we use a spectral representation of a flow, the advective process may be interpreted as interactions among the spectral components that satisfy certain algebraic conditions. Each interaction involves three components and is called a triad interaction. We invoke this notion in the discussion in Section 11.2. For illustration, let us derive the spectral governing equation for an inviscid two-dimensional fluid. In physical space, the governing equation is

$$-\zeta_t = \psi_x\zeta_y - \psi_y\zeta_x, \tag{A.60}$$

where the streamfunction Ψ and vorticity ζ are related by

$$\zeta = \nabla^2\psi. \tag{A.61}$$

For simplicity, we consider a doubly periodic domain $0 \le x \le X$ and $0 \le y \le Y$. We now represent a flow in terms of a complete set of orthogonal functions such as

$$\{\cos\phi_j\}, \quad \phi_j = k_jx + \ell_jy, \quad j = 1, 2, 3, \ldots$$

$$k_j = \frac{2\pi j}{X}, \quad \ell_j = \frac{2\pi j}{Y}, \quad \vec{K}_j = (k_j, \ell_j) \text{ wavenumber vector.}$$

The jth member of the set is referred to as spectral component-j. We can spectrally expand Ψ and ζ with these base functions as

$$\psi(x, y, t) = \sum_j a_j(t)\cos\phi_j, \tag{A.62}$$

$$\zeta(x, y, t) = \sum_j b_j(t)\cos\phi_j. \tag{A.63}$$

By (A.61), we have

$$b_j = -K_j^2 a_j. \tag{A.64}$$

Equation (A.60) can be then written as

$$\sum_m \frac{da_m}{dt} K_m^2 \cos\phi_m = \sum_m \sum_n -a_m a_n K_n^2 (k_m \ell_n - k_n \ell_m) \sin\phi_m \sin\phi_n. \tag{A.65}$$

The indices on the RHS may be interchanged. Hence, (A.65) can be rewritten as

$$\begin{aligned} &\sum_m \frac{da_m}{dt} K_m^2 \cos\phi_m \\ &= \sum_m \sum_n \frac{1}{2}(a_m a_n)(K_m^2 - K_n^2)(k_m \ell_n - k_n \ell_m)(\cos\varphi_{m+n} - \cos\varphi_{m-n}) \\ &= \sum_m \sum_n a_m a_n B_{m,n}(\cos\phi_{m+n} - \cos\phi_{m-n}), \end{aligned} \tag{A.66}$$

where

$$\phi_{m+n} = \phi_m + \phi_n, \tag{A.67}$$

$$B_{m,n} = \frac{1}{2}(K_m^2 - K_n^2)(\vec{K}_m \times \vec{K}_n) \cdot \vec{z}, \tag{A.68}$$

with $\vec{z}$ being a vertical unit vector. The base functions are orthogonal, viz.

$$\langle \cos\phi_j \cos\phi_m \rangle = \frac{1}{2}\delta_{j,m}, \tag{A.69}$$

where the angular bracket stands for domain average; $\delta_{j,m} = 0$ if $j \neq m$, otherwise $= 1$. Multiplying (A.66) by $\cos\phi_j$, averaging over the domain and making use of the orthogonal properties, we finally get

$$\frac{da_j}{dt} = \frac{1}{K_j^2} \sum_m \sum_n \frac{1}{2} a_m a_n B_{m,n}(\delta_{j,m+n} - \delta_{j,m-n}), \qquad j = 1, 2, 3 \ldots \tag{A.70}$$

This is the spectral form of (A.60) for component-j. Equation (A.70) says that the amplitude of component-j would change due to the totality of all possible interactions between a component-m and a component-n satisfying the condition either $n = j - m$ or $n = m - j$; $B_{m,n}$ is called the *interaction coefficient* between component-m and component-n. It measures the effectiveness of such an interaction; $B_{m,n} = 0$ for either when $K_m = K_n$ in this particular case or when $\vec{K}_m$ and $\vec{K}_n$ are parallel.

Summing up, a spectral component can be altered by interaction between two components if the sum or difference of their wavenumber vectors is equal to its own wavenumber vector. These interactions are called triad interactions. Their effectiveness is quantified by a corresponding interaction coefficient.

References

Andrews, D. A., J. R. Holton and C. B. Leovy (1987). *Middle Atmospheric Dynamics*. Academic Press.

Bannon, P. R. and M. Mak (1986). A diagnosis of moist frontogenesis with an analytic model. *J. Atmos. Sci.* **43**, 2017–2022.

Bjerknes, J. (1919). On the structure of moving cyclones. *Gesfysiske Publikasioner*, **1**.

Boer, G. and T. Shepherd (1983). Large-scale two-dimensional turbulence in the atmosphere. *J. Atmos. Sci.* **40**, 164–184.

Bretherton, F. P. (1966). Critical layer instability in baroclinic flows. *Quart. J. Roy. Meteorol. Soc.* **92**, 325–334.

Cai, M. (1992). A physical interpretation for the stability property of a localized disturbance in a deformation flow. *J. Atmos. Sci.* **49**, 2177–2182.

Cai, M. (2004). Local instability dynamics of storm tracks. In *Observation, Theory, and Modeling of Atmospheric Variability*, ed. X. Zhu *et al.*, World Scientific Publishing pp. 3–38.

Cai, M. and M. Mak (1987). On the multiplicity of equilibria of baroclinic waves. *Tellus* **39A**, 116–137.

Cai, M. and M. Mak (1990). Symbiotic relation between planetary and synoptic scale waves. *J. Atmos. Sci.* **47**, 2953–2968.

Cai, M. and H. M. van den Dool (1992). Low frequency waves and traveling storm tracks. Part II: Three-dimensional structure. *J. Atmos. Sci.* **49**, 2506–2524.

Chang, E. K. M., S. Lee and K. L. Swanson (2002). Storm track dynamics. *J. Climate* **15**, 2163–2183.

Charney, J. G. (1947). The dynamics of long waves in a baroclinic westerly current. *J. Meteor.* **4**, 135–162.

Charney, J. G. (1971). Geostrophic turbulence. *J. Atmos. Sci.* **28**, 1087–1095.

Charney, J. G. and P. G. Drazin (1961). Propagation of planetary-scale disturbances from the lower into the upper atmosphere. *J. Geophys. Res.* **66**, 83–110.

Charney, J. G. and A. Eliassen (1964). On the growth of the hurricane depression. *J. Atmos. Sci.* **21**, 68–75.

Charney, J. G. and M. Stern (1962). On the stability of internal baroclinic jets in a rotating atmosphere. *J. Atmos. Sci.* **19**, 159–172.

Craig, G. and H. R. Cho (1988). Cumulus convection and CISK in midlatitudes, Part I. Polar lows and comma clouds. *J. Atmos. Sci.* **45**, 2622–2640.

De la Cruz-Heredia, M. and G. W. K. Moore (1999). Barotropic instability due to Kelvin wave-Rossby wave coupling. *J. Atmos. Sci.* **56**, 2376–2383.

Deng, Y. and M. Mak (2005). An idealized model study relevant to the dynamics of the midwinter minimum of the Pacific storm track. *J. Atmos. Sci.* **62**, 1209–1225.

Eady, E. T. (1949). Long waves and cyclone waves. *Tellus* **1**, 33–52.

Edmon, H. J. Jr., B. J. Hoskins and M. E. McIntyre (1980). Eliassen-Palm Cross Sections for the Troposphere. *J. Atmos. Sci.* **37**, 2600–2616.

Ekman, V. W. (1905). On the influence of the Earth's rotation on ocean currents. *Arch. Math. Astron. Phy.* **2**, 1–52.

Eliassen, A. and E. Palm (1961). On the transfer of energy in stationary mountain waves. *Geofys. Publ.* **22** (8B.3), 1–23.

Emanuel, K. A. (1979). Inertial instability and mesoscale convective systems. Part I. Linear theory of inertial instability in rotating viscous fluids. *J. Atmos. Sci.* **12**, 2425–2449.

Emanuel, K. A., M. Fantini and A. J. Thorpe (1987). Baroclinic instability in an environment of small stability to slantwise moist convection, Part I. Two-dimensional models. *J. Atmos. Sci.* **44**, 1559–1573.

Ertel, H. (1942). Ein neuer hydrodynamischer Wirbesatz. *Meteorol. Z.* **59**, 277–281.

Farrell, B. E. (1989). Optimal excitation of baroclinic waves. *J. Atmos. Sci.* **46**, 163–172.

Ferrel, W. (1856). An essay on the winds and currents of the oceans. *Nashville Journal of Medicine and Surgery*.

Fjortoft, R. (1953). On the changes in the spectral distribution of kinetic energy for two dimensional nondivergent flow. *Tellus* **5**, 225–230.

Frederiksen, J. S. (1983). Disturbances and eddy fluxes in Northern Hemisphere flows: Instability of three dimensional January and July flows. *J. Atmos. Sci.* **40**, 836–855.

Gill, A. E. (1982). *Atmosphere-Ocean Dynamics*. International Geophysics Series, vol. 30, Academic Press.

Green, A. E. and P. M. Naghdi (1976). A derivation of equations for wave propagation in water of variable depth. *J. Fluid Mech.* **78**, 237–246.

Gyakum, J. R. (1983). On the evolution of the QEII storm, II. Dynamic and thermodynamic structure. *Mon. Wea. Rev.* **111**, 1156–1173.

Hadley, G. (1735). Concerning the cause of the general trade winds. *Philos. Trans.*, **39**.

Haynes, P. H. and M. E. McIntyre (1990). On the conservation and impermeability theorems for potential vorticity. *J. Atmos. Sci.* **16**, 2021–2031.

Heifetz, E., C. H. Bishop, B. J. Hoskins and J. Methven (2004). The counter-propagatinjg Rossby perspective on baroclinic instability. I: Mathematical basis. *Quart. J. Roy. Meteorol. Soc.* **130**, 211–231.

Held, I. M. and A. Hou (1980). Nonlinear axially symmetric circulations in a nearly inviscid atmosphere. *J. Atmos. Sci.* **37**, 515–533.

Held, I. M. and T. Schneider (1999). The surface branch of the zonally averaged mass transport circulation in the troposphere. *J. Atmos. Sci.* **56**, 1688–1697.

Hide, R. (1966). Review article on the dynamics of rotating fluids and related topics in geophysical fluid mechanics. *Bull. Amer. Meteorol. Soc.* **47**, 873–885.

Holton, J. R. (1983). The influence of gravity waves breaking on the general circulation of the middle atmosphere. *J. Atmos. Sci.* **40**, 2497–2507.

Holton, J. R. (1992). *An Introduction to Dynamic Meteorology*, 3rd edn. Academic Press.

Hoskins, B. J. (1975). The geostrophic momentum approximation and the semi-geostrophic equations. *J. Atmos. Sci.* **32**, 233–242.

Hoskins, B. J. (1982). The mathematical theory of frontogenesis. *Annu. Rev. Fluid Mech.* **14**, 131–151.

Hoskins, B. J. and F. P. Bretherton (1972). Atmospheric frontogenesis models: Mathematical formulation and solution. *J. Atmos. Sci.* **29**, 11–37.

Hoskins, B. J., M. E. McIntyre and A. W. Robertson (1985). On the use and significance of isentropic potential vorticity maps. *Quart. J. Roy. Meteorol. Soc.* **111**, 877–946.

Hsu, C-P. F. (1980). Air parcel motions during a numerically simulated sudden stratospheric warming. *J. Atmos. Sci.* **37**, 2768–2792.

James, I. N. (1987). Suppression of baroclinic instability in horizontally sheared flows. *J. Atmos. Sci.* **44**, 3710–3720.

James, I. N. (1994). *Introduction to Circulating Atmospheres*. Cambridge University Press.

Kuo, H. L. (1949). Dynamic instability of two-dimensional nondivergent flow in a barotropic atmosphere. *J. Meteorol.* **6**, 105–122.

Kwon, H. J. and M. Mak (1988). On the equilibration in nonlinear barotropic instability. *J. Atmos. Sci.* **45**, 294–308.

Lee, B. D. and R. B. Wilhelmson (1997). The numerical simulation of non-supercell tornadgenesis. Part I: Initiation and evolution of pretornadic misocyclone and circulations along a dry outflow boundary. *J. Atmos. Sci.* **54**, 32–60.

Lindzen, R. S. and A. Hou (1988). Hadley circulations for zonally averaged heating centered off the equator. *J. Atmos. Sci.* **45**, 2416–2427.

Lorenz, E. (1963). The mechanics of vacillation. *J. Atmos. Sci.* **20**, 448–464.

Mak, M. (1982). On moist quasi-geostrophic baroclinic instability. *J. Atmos. Sci.* **39**, 2028–2037.

Mak, M. (1985). Equilibration in nonlinear baroclinic instability. *J. Atmos. Sci.* **42**, 1089–1101.

Mak, M. (1994). Cyclogenesis in a conditionally unstable moist baroclinic atmosphere. *Tellus* **46A**, 14–33.

Mak, M. (2001). Nonhydrostatic barotropic instability: Applicability to nonsupercell tornadogenesis. *J. Atmos. Sci.* **58**, 1965–1977.

Mak, M. (2002). Wave-packet resonance: Instability of a localized barotropic jet. *J. Atmos. Sci.* **59**, 823–836.

Mak, M. (2008). Dynamics of the mean Asian summer monsoon in a maximally simplified model. *Q. J. R. Meteorol. Soc.* **134**, 429–437.

Mak, M. and P. R. Bannon (1984). Frontogenesis in a moist semigeostrophic model. *J. Atmos. Sci.* **41**, 3485–3500.

Mak, M. and M. Cai (1989). Local barotropic instability. *J. Atmos. Sci.* **46**, 3289–3311.

Mak, M. and Y. Deng (2007). Diagnostic and dynamical analyses of two outstanding aspects of storm tracks. *Dyn. of Atmos. and Oceans* **43**, 80–99.

Marshall, J. and R. A. Plumb (2008). *Atmospheric, Ocean and Climate Dynamics: An introductory text*. Elsevier Academic Press.

Matsuno, T. (1966). Quasi-geostrophic motions in the equatorial area. *J. Meteorol. Soc. Jpn.* **44**, 25–43.

Matsuno, T. (1971). A dynamical model of the stratospheric sudden warming. *J. Atmos. Sci.* **28**, 1479–1494.

Merkine, L. (1977). Convective and absolute instability of baroclinic eddies. *Geophys. Astrophys. Fluid Dyn.* **9**, 129–157.

Mihaljan, J. M. (1963). The exact solution of the Rossby adjustment problem. *Tellus* **15**, 150–154.

Montmomery, M. T. and B. F. Farrel (1991). Moist surface frontogenesis associated with interior potential vorticity anomalies in a semigeostrophic model. *J. Atmos. Sci.* **48**, 343–367.

Nakamura, H. (1992). Midwinter suppression of baroclinic wave activity in the Pacific. *J. Atmos. Sci.* **49**, 1629–1642.

Niino, H. and N. Misawa (1984). An experimental and theoretical study of barotropic instability. *J. Atmos. Sci.* **41**, 1992–2111.

Orlanski, I. and E. K. M. Chang (1993). Ageostrophic geopotential fluxes in downstream and upstream development of baroclinic waves. *J. Atmos. Sci.* **50**, 212–225.

Parker, D. J. and A. J. Thorpe (1995). Conditional convective heating in a baroclinic atmosphere: A model of convective frontogenesis. *J. Atmos. Sci.* **52**, 1699–1711.

Pedlosky, J. (1987). *Geophysical Fluid Dynamics*, 2nd edn. Springer-Verlag.

Peixoto, J. P. and A. H. Oort (1992). *Physics of Climate*. American Institute of Physics.

Phillips, N. A. (1954). Energy transformations and meridional circulations associated with simple baroclinic waves in a two-level, quasi-geostrophic model. *Tellus* **VI**, 273–286.

Pierrehumbert, R. T. (1984). Local and global baroclinic instability of zonally varying flow. *J. Atmos. Sci.* **41**, 2141–2162.

Randel, W. J. and I. M. Held (1991). Phase speed spectra of transient eddy fluxes and critical layer absorption. *J. Atmos. Sci.* **48**, 688–697.

Randel, W., P. Udelhofen, E. Fleming, *et al.* (2004). The SPARC intercomparison of middle-atmosphere climatologies. *J. Climate* **17**, 986–1003.

Rayleigh, Lord (John William Strutt) (1883). Investigation of the character of the equilibrium of an incompressible heavy fluid of variable density. *Proc. Lond. Math. Soc.* **14**, 170–177.

Rhines, P. B. (1975). Waves and turbulence on a beta-plane. *J. Fluid Mech.* **69**, 417–443.

Rodwell, M. J. and B. J. Hoskins (1996). Monsoons and the dynamics of deserts. *Quart. J. Roy. Meteorol. Soc.* **122**, 1385–1404.

Rossby, C.-G. (1936). Dynamics of steady ocean currents in the light of experimental fluid mechanics. *Mass. Inst. of Technology and Woods Hole Oc. Inst. Papers in Physical Oceanography and Meteorology* **5 (1)**, 1–43.

Rossby, C.-G. (1938). On the mutual adjustment of pressure and velocity distributions in certain simple current systems, *II. J. Marine Res.* **1**, 239–263.

Rossby, C.-G. *et al.* (1939). Relations between variations in the intensity of the zonal circulation of the atmosphere and the displacements of the semi-permanent centers of action. *J. Marine Res.* **2**, 38–55.

Salmon, R. (1980). Baroclinic instability and geostrophic turbulence. *Geophys. Astrophys. Fluid Dyn.* **15**, 167–211.

Salmon, R. (1998). *Lectures on Geophysical Fluid Dynamics*. Oxford University Press.

Schneider, T., I. M. Held and S. T. Garner (2003). Boundary effects in potential vorticity dynamics. *J. Atmos. Sci.* **60**, 1024–1040.

Schubert, S., C.-K. Park, W. Higgins, S. Moorthi and M. Suarez (1990). An Atlas of ECMWF Analysis (1980–87) Part I- First Moment Quantities. *NASA Technical Memorandum* 100–747.

Schubert, W. H., J. J. Hack, P. L. Silva Dias and S. R. Fulton (1980). Geostrophic adjustment in an axisymmetric vortex. *J. Atmos. Sci.* **37**, 1464–1484.

Snyder, C. and R. S. Lindzen (1991). Quasi-geostrophic wave-CISK in an unbounded baroclinic shear. *J. Atmos. Sci.* **48**, 76–86.

Starr, V. P. (1968). *Physics of Negative Viscosity Phenomena*. McGraw-Hill.

Taylor, G. I. (1950). The instability of liquid surfaces when accelerated in a direction perpendicular to their planes. *Proc. Roy. Soc. Lond. A* **201** (1065), 192–196.

Thomson, W. (Lord Kelvin) (1879). On gravitational oscillations of rotating water. *Proc. Roy. Soc. Edinburgh* **10**, 92–100.

Thorncroft, C. D., B. J. Hoskins and M. E. McIntyre (1993). Two paradigms of baroclinic-wave life-cycle behaviour. *Quart. J. Roy. Meteorol. Soc.* **119**, 17–55.

Ting, M. F. (1994). Maintenance of northern summer stationary waves in a GCM. *J. Atmos. Sci.* **51**, 3286–3308.

Townsend, R. D. and D. R. Johnson (1985). A diagnostic study of the isentropic zonally averaged mass circulation during the first GARP global experiment. *J. Atmos. Sci.* **42**, 1565–1579.

Valdes, P. J. and B. J. Hoskins (1989). Linear stationary wave simulations of the time-mean climatological flow. *J. Atmos. Sci.* **46**, 2509–2527.

Vallis, G. K. (2006). *Atmospheric and Oceanic Fluid Dynamics*. Cambridge University Press.

Wakimoto R. and J. W. Wilson (1989). Non-supercell tornadoes. *Mon. Wea. Rev.* **117**, 1113–1140.

Wang, H. (2000). Understanding the maintenance and seasonal transitions of the climatological stationary waves in the atmosphere. Ph.D. thesis, University of Illinois, Urbana-Champaign, IL, USA.

Whitaker, J. S. and C. A. Davis (1994). Cyclogenesis in a saturated environment. *J. Atmos. Sci.* **51**, 889–907.

Index